Pam White

STARCH:
Chemistry and Technology

SECOND EDITION

CONTRIBUTORS

Douglas A. Corbishley
James R. Daniel
William M. Doane
Paul L. Farris
A. C. Fischer, Jr.
Larry E. Fitt
Dexter French
Douglas L. Garwood
C. W. Hastings
Bienvenido O. Juliano
Keiji Kainuma
H. M. Kennedy
Robert E. Klem
J. W. Knight
Norman E. Lloyd
Merle J. Mentzer
William Miller
Eugene L. Mitch
C. O. Moore
William J. Nelson
R. M. Olson
Felix H. Otey
John F. Robyt
Robert G. Rowher
Morton W. Rutenberg
R. V. Schanefelt
Jack C. Shannon
Eileen Maywald Snyder
Daniel Solarek
J. V. Tuschhoff
Stanley A. Watson
Roy L. Whistler
Austin H. Young
Henry F. Zobel

STARCH:
Chemistry and Technology

EDITED BY

ROY L. WHISTLER
DEPARTMENT OF BIOCHEMISTRY
PURDUE UNIVERSITY
WEST LAFAYETTE, INDIANA

JAMES N. BEMILLER
DEPARTMENT OF CHEMISTRY AND BIOCHEMISTRY
SOUTHERN ILLINOIS UNIVERSITY
CARBONDALE, ILLINOIS

EUGENE F. PASCHALL
MOFFETT TECHNICAL CENTER
CORN PRODUCTS
SUMMIT-ARGO, ILLINOIS

SECOND EDITION

1984

ACADEMIC PRESS, INC.
Harcourt Brace Jovanovich, Publishers
Orlando San Diego New York
Austin Boston London Sydney
Tokyo Toronto

ACADEMIC PRESS, INC.
Orlando, Florida 32887

United Kingdom Edition published by
ACADEMIC PRESS, INC. (LONDON) LTD.
24/28 Oval Road, London NW1 7DX

Library of Congress Cataloging in Publication Data
Main entry under title:

Starch.

Includes index.
1. Starch. I. Whistler, Roy Lester. II. BeMiller, James N. III. Paschall, Eugene F.
TP248.S7S7 1984 664'.2 82-25311
ISBN 0-12-746270-8

PRINTED IN THE UNITED STATES OF AMERICA

87 88 9 8 7 6 5 4 3

CONTENTS

Chapter I

History and Future Expectation of Starch Use

Roy L. Whistler

Chapter II

Economics and Future of the Starch Industry

Paul L. Farris

Chapter III

Genetics and Physiology of Starch Development

Jack C. Shannon and Douglas L. Garwood

Chapter IV

Enzymes in the Hydrolysis and Synthesis of Starch

John F. Robyt

Chapter V

Starch Oligosaccharides: Linear, Branched, and Cyclic

Keiji Kainuma

Chapter VI

Molecular Structure of Starch

Roy L. Whistler and James R. Daniel

Chapter VII

Organization of Starch Granules

Dexter French

Chapter VIII

Fractionation of Starch

Austin H. Young

Chapter IX

Gelatinization of Starch and Mechanical Properties of Starch Pastes

Henry F. Zobel

Chapter X

Starch Derivatives: Production and Uses

Morton W. Rutenberg and Daniel Solarek

Chapter XI

Chemicals from Starch

Felix H. Otey and William M. Doane

Chapter XII

Corn and Sorghum Starches: Production

Stanley A. Watson

Chapter XIII

Tapioca, Arrowroot, and Sago Starches: Production

Douglas A. Corbishley and William Miller

Chapter XIV

Potato Starch: Production and Uses

Eugene L. Mitch

Chapter XV

Wheat Starch: Production, Modification, and Uses

J. W. Knight and R. M. Olson

Chapter XVI

Rice Starch: Production, Properties, and Uses

Bienvenido O. Juliano

Chapter XVII

Acid-Modified Starch: Production and Uses

Robert G. Rohwer and Robert E. Klem

Chapter XVIII

Starch in the Paper Industry

Merle J. Mentzer

Chapter XIX

Applications of Starches in Foods

C. O. Moore, J. V. Tuschhoff, C. W. Hastings, and R. V. Schanefelt

Chapter XX

Starch and Dextrins in Prepared Adhesives

H. M. Kennedy and A. C. Fischer, Jr.

Chapter XXI

Glucose- and Fructose-Containing Sweeteners from Starch

Norman E. Lloyd and William J. Nelson

Chapter XXII

Industrial Microscopy of Starches

Eileen Maywald Snyder

Chapter XXIII

Photomicrographs of Starches

Larry E. Fitt and Eileen Maywald Snyder

LIST OF CONTRIBUTORS

Numbers in parentheses indicate the pages on which the authors' contributions begin.

DOUGLAS A. CORBISHLEY (469), Research Department, Industrial Starch and Food Products Division, National Starch and Chemical Corporation, Bridgewater, New Jersey 08876

JAMES R. DANIEL[1] (153), Department of Biochemistry, Purdue University, West Lafayette, Indiana 47907

WILLIAM M. DOANE (389), Northern Regional Research Center, Agricultural Research, Science and Education Administration, U.S. Department of Agriculture, Peoria, Illinois 61604

PAUL L. FARRIS (11), Department of Agricultural Economics, Purdue University, West Lafayette, Indiana 47907

A. C. FISCHER, JR. (593), Technical Sales Service, Corn Products, Summit-Argo, Illinois 60502

LARRY E. FITT (675), Moffett Technical Center, Corn Products, Summit-Argo, Illinois 60502

DEXTER FRENCH[2] (183), Department of Biochemistry and Biophysics, Iowa State University, Ames, Iowa 50011

DOUGLAS L. GARWOOD[3] (25), Department of Horticulture, The Pennsylvania State University, University Park, Pennsylvania 16802

C. W. HASTINGS (575), A. E. Staley Manufacturing Company, Decatur, Illinois 62525

BIENVENIDO O. JULIANO (507), Chemistry Department, International Rice Research Institute, Los Baños, Laguna, Philippines

KEIJI KAINUMA (125), National Food Research Institute, Tsukuba-Gun, Ibaraki-Ken 305, Japan

[1]Present address: Department of Foods and Nutrition, Purdue University, West Lafayette, Indiana 47907.

[2]Deceased.

[3]Present address: Garwood Seed Co., Stonington, Illinois 62567.

H. M. Kennedy[4] (593), Research Department, Acme Resin Co., Division of CPC International, Forest Park, Illinois 60130

Robert E. Klem (529), Grain Processing Corporation, Muscatine, Iowa 52761

J. W. Knight (491), Fielder Gillespie Ltd., Sydney, New South Wales, 2000 Australia

Norman E. Lloyd[5] (611), Clinton Corn Processing Company, A Division of Standard Brands, Inc., Clinton, Iowa 52732

Merle J. Mentzer (543), CPC Latin America, Division of CPC International, 1106 Buenos Aires, Argentina

William Miller (469), Tapioca Associates, Inc., Wilton, Connecticut 06897

Eugene L. Mitch[6] (479), Chemical Operations, Boise Cascade Corp., Portland, Oregon 97201

C. O. Moore (575), A. E. Staley Manufacturing Company, Decatur, Illinois 62525

William J. Nelson[5] (611), Clinton Corn Processing Company, A Division of Standard Brands, Inc., Clinton, Iowa 52732

R. M. Olson (491), Moffett Technical Center, Corn Products, Summit-Argo, Illinois 60502

Felix H. Otey (389), Northern Regional Research Center, Agricultural Research, Science and Education Administration, U.S. Department of Agriculture, Peoria, Illinois 61604

John F. Robyt (87), Department of Biochemistry and Biophysics, Iowa State University, Ames, Iowa 50011

Robert G. Rowher (529), Grain Processing Corporation, Muscatine, Iowa 52761

Morton W. Rutenberg (311), Research Department, Industrial Starch and Food Products Divisions, National Starch and Chemical Corporation, Bridgewater, New Jersey 08807

R. V. Schanefelt (575), A. E. Staley Manufacturing Company, Decatur, Illinois 62525

[4]Present address: Grain Processing Corporation, Muscatine, Iowa 52761.

[5]Present address: Corporate Technology Group, Wilton Technology Center, Nabisco Brands, Inc., Wilton, Connecticut 06897.

[6]Present address: Process Chemical Development, Chemax, Inc., Portland, Oregon 97210.

JACK C. SHANNON (25), Department of Horticulture, The Pennsylvania State University, University Park, Pennsylvania 16802

EILEEN MAYWALD SNYDER (661, 675), Moffett Technical Center, Corn Products, Summit-Argo, Illinois 60502

DANIEL SOLAREK (311), Research Department, Industrial Starch and Food Products Division, National Starch and Chemical Corporation, Bridgewater, New Jersey 08876

J. V. TUSCHHOFF (575), A. E. Staley Manufacturing Company, Decatur, Illinois 62525

STANLEY A. WATSON (417), Ohio Agricultural Research and Development Center, The Ohio State University, Wooster, Ohio 44691

ROY L. WHISTLER (1, 153), Department of Biochemistry, Purdue University, West Lafayette, Indiana 47907

AUSTIN H. YOUNG (249), A. E. Staley Manufacturing Company, Decatur, Illinois 62525

HENRY F. ZOBEL (285), Moffett Technical Center, Corn Products, Summit-Argo, Illinois 60502

PREFACE

A major change has occurred during the past few years in the way starch is processed from corn, in the way starch is used, and in the way it is chemically modified for speciality use. In fact, corn wet milling has changed so dramatically that it might be said an entire new wet-milling industry has been created.

When the first edition of this treatise appeared in 1965–1966, the wet-milling industry was just beginning to move from batch operations, tables, and screens to continuous processing. Innovations have now brought the industry to a highly sophisticated level with up-to-date, state-of-the-art, continuous, economical processes for conversion of corn to starch and on to D-glucose of excellent quality. A significant further development was the introduction of advanced enzyme engineering to convert high-purity D-glucose to a mixture of D-glucose and D-fructose equivalent to invert sugar from sucrose, thereby opening to the corn industry the hitherto unavailable but enormous market in sweeteners. Solid entry into this market is provided by the stability of corn supply and the low cost of producing D-glucose–D-fructose syrup. The result is to provide a sound domestic supply of sweetener.

Marketing of corn starch derivatives has also changed, not only with a limitation of the number of derivatives offered but with some derivatization being transferred from starch producers to starch users. This is especially marked in starches used by the paper industry, but perhaps will occur with other large-scale users of starch. The present trend is for starch producers to sell unmodified pearl starch to paper producers who make their own particular modification in slurry and capture the produced solubles through absorption on the paper fibers and clay particles. Hence, solubles are not lost but are saved to serve functionally in the sheet formation. The capture of solubles results in a cost savings. On-site modification of starch is economical also because the product need not be dried for shipment but can be used before secondary effects bring about property changes.

While the number of starch derivatives has been reduced, the quality of derivatives has been vastly increased due to new reaction procedures that yield products fitting specific user requirements.

The first two chapters deal with history and present economics of the starch industry. The third chapter gives in detail the genetics and development of starch. Chapters IV and V describe the enzyme chemistry of starch. Chapters VI–IX present chemical and physical information on the structure and behavior of the starch granule and on the fractionation, structure, and properties of starch molecules. Chapters X and XI describe reactions of starch, its conversion to specific derivatives, and their applications. Chapters XII–XVI describe the production and use of the commercial starches: corn, tapioca, arrowroot, sago, rice, wheat, and potato. Chapters XVII–XX give specific attention to the use of starch in the paper industry, in foods, and in adhesives, with attention also to acid modification of starch. Chapter XXI describes the process for conversion of starch to D-glucose and to D-glucose–D-fructose sweeteners. The final two chapters give excellent photographs of starches and present techniques for first examination of a starch.

It is hoped that this addition may be useful to research workers, chemical engineers, technical sales personnel, and people associated with the legal profession who wish to learn more about starch.

With emphasis on the fundamental and practical aspects of starch, information on analytical and experimental laboratory procedures is somewhat restricted. A complete treatment of the specialized methods needed in work with starch and starch fractions is given in "Methods in Carbohydrate Chemistry," Volume IV (Roy L. Whistler, ed., Academic Press, New York, 1964), which should be used as companion volume to the present review.

Although documentation of material presented is quite complete, descriptions should not be construed as indicating that the use of the procedures or processes described are free from patent restrictions.

The editors are grateful to the authors for their ready response to the request to contribute and for their kindness and understanding in accepting editorial efforts.

Gratitude is expressed to the individual starch manufacturers. Each member company of the Corn Refiners Association has been enthusiastically interested in this undertaking.

ROY L. WHISTLER
JAMES N. BEMILLER
EUGENE F. PASCHALL

CONTENTS OF PREVIOUS VOLUMES

Volume I
Fundamental Aspects

Volume II
Industrial Aspects

CHAPTER I

HISTORY AND FUTURE EXPECTATION OF STARCH USE

By Roy L. Whistler

Department of Biochemistry, Purdue University, West Lafayette, Indiana

I. Introduction

Although the past is usually prologue to the future, industrial usage of starch will be greater than normal methods of economic projection would indicate. Even though starch usage has shown a steady, rapid rate of growth over an extended period of time, new events are increasing demand for starch in an ever upward direction.

The birth of enzyme engineering made possible low cost conversion of starch to D-glucose and on to an equilibrium mixture of D-glucose and D-fructose equivalent in sweetness to invert sugar from cane or beet. This process alone made possible an immediate seizure of 30% of the sucrose market in the United States and doubled the amount of starch produced by the wet milling industry. With development of even more sophisticated methods of enzyme chemistry, it will become possible to transform starch into novel molecules possessing new properties suitable for entirely new applications.

A second event that makes projections for starch use difficult, but still suggests an upward demand, is the increased price of energy. This places new usage requirements on low cost starch to serve as a source of alcohol, components for plastics, special absorbents, paper extenders, oil well drilling muds, and as an additive of tertiary oil recovery systems, the potential requirement for the latter being enormous.

A third, large area of starch usage, more accurately predictable, is as a basic

STARCH, 2nd ed.

ISBN 0-12-746270-8

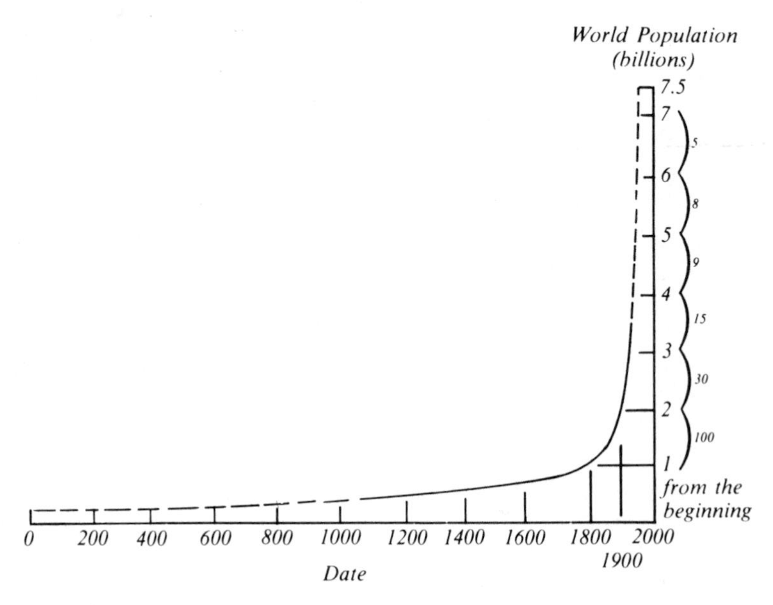

FIG. 1.—World population growth.

food to supply nutritional requirements of the growing world population (Fig. 1). Traditionally, carbohydrates supply 80% of the calories consumed by the human population with two-thirds of these calories coming from starch. Now that starch can be economically converted to sweeteners, it will supply a greater proportion of nutritional calories. Then too, as meat costs rise, the amount of carbohydrates consumed in the diet of populations in developed countries may increase over present levels.

Each of these considerations indicates a firm place for starch as a future industrial raw material.

II. Early History

Starchy foods have always been an item in the diet of man. It is natural, therefore, that the practical use of starch products, and later of starch itself, should have developed in an early period. Some developments are cloaked in the predawn twilight of the unrecorded past. Strips of Egyptian papyrus, cemented together with a starchy adhesive and dated to the predynastic period of 3500–4000 B.C., give evidence of the early use of starch. However, a description of starch and its application is not found until much more recent times. The historian and philosopher, Caius Plinius Secundus (Pliny the Elder, 23–74 A.D.), described documents of 130 B.C. made by sizing papyrus with modified wheat starch to produce a smooth surface. The adhesive was made from fine ground wheat flour, boiled with a dilute solution of vinegar. The paste was spread over strips of papyrus; the strips beaten with a mallet and further strips lapped over the edges to give a broader sheet. Pliny stated that the 200-year-old sheets which he observed were still in good condition. Pliny also described the use of starch to whiten cloth and to powder hair. Chinese paper documents of about 312 A.D. are reported to contain starch size (*1*). Later, Chinese documents were first coated with a high-fluidity starch to provide resistance to ink penetration and then covered with powdered starch to provide weight and thickness. Starches from rice, wheat, and barley were commonly used at that time.

A procedure for starch production was given in some detail in a Roman treatise by Cato in about 184 B.C. (*2*). Grain was steeped in water for 10 days, pressed and mixed with fresh water, then filtered on a linen cloth, after which the starch in the filtrate was allowed to settle, washed with water, and finally dried in the sun.

Modified starches used for adhesives or to provide a sweet molasses developed at an early period. Hydrolysis was a common method of modification, and vinegar as well as amylolytic enzymes were used. Abu Mansur (*3*), an Arabian teacher and pharmacologist, was acquainted with the use of saliva on starch for producing an "artificial honey" used for treating wounds.

Starch was introduced into Northern Europe to stiffen linen, possibly early in the fourteenth century. Colored and uncolored starches were used as cosmetics. Uncolored starch was used principally as a hair powder. Blue starch was employed by the Puritans until its use was banned by Queen Elizabeth in 1596. Yellow starch was quite fashionable until a notorious woman prisoner wearing a bright yellow-starched ruffle was publicly executed. Red starch cosmetics remained in fashion for many years.

Leeuwenhoek (*4*), the inventor of the microscope, made intensive and accurate observations of starch granules in 1719 and a dictionary (*5*) describing starch and its manufacture was published in 1744.

As starch became a more important industrial commodity, work was done on its modification. This included the great discovery of Kirchoff in 1811 that sugar could be produced from potato starch by acid-catalyzed hydrolysis. Then, there was the accidental discovery of the torrefaction method for producing dextrins, now termed British gums. In 1821, a fire occurred in a Dublin textile mill that utilized starch as a size. After the blaze was extinguished, one of the workmen noticed that some of the starch had been turned brown by the heat and dissolved easily in water to produce a thick, adhesive paste. The roasting of new starch was repeated and the product was shown to have the useful properties previously observed. In this way, heat dextrinization became known and subsequently elaborated into wide usage.

In Europe, early use of wheat and barley starch gave way to white potato starch, which was produced in large quantities in The Netherlands and Germany.

III. American Development

The first American wheat starch plant was started by Gilbert (*6*) at Utica, New York in 1807, but was converted to one producing corn starch in 1849. The change from wheat to corn starch began with Thomas Kingsford's development in 1842 of a manufacturing process in which crude corn starch was purified by alkaline treatment. The wheat starch plant of George Fox started in 1824 at Cincinnati was also converted to a corn starch plant in 1854. The William Colgate wheat starch plant built in Jersey City in 1827 was changed to a corn starch plant in 1844. Thomas Kingsford was hired by the Colgate Company in 1832 and became its superintendent in 1842. Another early wheat starch plant was that of T. Barnett built in Philadelphia in 1817 before moving to Knowlton, Pennsylvania, in 1879, and ceasing operation in 1895. In that year, there were five wheat starch and sixteen corn starch plants operating in the United States. The corn starch plants produced 200 million pounds of starch per year.

Potato starch manufacture began in 1820 in Hillsborough County, New Hampshire. Usage of potato starch grew rapidly until in 1895, sixty-four factories were in operation, of which fourty-four were in Maine. They produced 24 million pounds of starch per year during approximately 3 months of operation. Most of the starch was sold to the textile industry.

Rice starch manufacture, using caustic treatment of rice, began in 1815. However, production did not increase significantly and the little rice starch later used was mainly imported.

Following Kirchoff's finding (*7*) in 1811 that sweet dextrose (D-glucose) could be produced by acid-catalyzed hydrolysis of starch, factories were built to produce sweet syrups. Within a year, factories were built in Munich, Dresden, Bochman, and Thorin. By 1876, Germany alone had forty-seven dextrose syrup factories using potato starch to produce 33 million pounds of syrup and 11 million pounds of solid sweetener.

An American syrup plant of 30 gallon per day capacity was started in 1831 at Sacket Harbor, New York, but soon failed. A large plant was built by the Hamlins in 1873 at Buffalo, New York. In 1883, the Chicago Sugar Refinery Company began the manufacture of dextrose but soon added starch and modified starches.

In 1880, there were one-hundred forty starch plants producing corn, wheat, potato, and rice starches. Most were small and located along the Atlantic coast. The largest was the Glen Cove, New York, plant producing 65 million pounds of corn starch per year.

Thomas Kingsford and Son built a starch plant at Oswego, New York, in 1849 that continued to operate until destroyed by fire in 1903. The corn starch plant built in 1855 by Wright Duryea of Glen Cove, Long Island, became the largest starch factory in the United States in 1891, grinding 7000 bushels of corn per day. First designated the Duryea Starch Works, the name was changed in 1881 to the Glen Cove Manufacturing Company.

The corn starch plant built in Indianapolis, Indiana, in 1867 by William F. Piel, Sr., was destroyed by fire in 1872. A second plant was closed by the city because of the environmental problem of fermentation odor from gluten overflow water. This led to the construction of a larger and more efficient plant in 1873.

By 1890 the number of American starch plants had decreased to eighty, producing 240 million pounds of starch per year. Although many small plants were built, they could not compete, and in 1890 a consolidation of several occurred to become the National Starch Manufacturing Company of Kentucky, representing 70% of the corn starch capacity. The consolidation included the Duryea plant at Glen Cove, the Piel plant at Indianapolis, and eighteen plants in Ohio, Illinois, Iowa, and Indiana. The firm built a new modern plant at Des Moines, Iowa, in 1899. Some of the independents then combined in the same year to form the United Starch Company. Among these independents were Thomas Kingsford and Sons, the U.S. Sugar Refining Company at Waukegan, Illinois, founded in 1888, the Argo Starch Company at Nebraska City, Nebraska, founded in 1894, and the Sioux City Starch Company, founded in 1896. In 1900 the United Starch Company and the National Starch Manufacturing Company combined to form the National Starch Company of New Jersey. Kingsford reported (*6*) that in 1895 there were sixteen corn starch factories producing 200 million pounds of starch per year and grinding 29,000 bushels of corn per day.

In 1902, the Glucose Sugar Refining Company joined with the National Starch Company to form the Corn Products Company that then represented 80% of the corn starch industry with a daily grind of 65,000 bushels. The Corn Products Company also included the Illinois Sugar Refining Company at Pekin, Illinois, and 49% of the stock of the New York Glucose Company formed by E. T. Bedford and Associates at Edgewater, New Jersey. A disastrous price war then ensued between Corn Products Company and the New York Glucose Company because of the refusal of the latter to allow Corn Products Company more than 50% stock ownership. The result was a merger in 1906 of the two combatants to form the Corn Products Refining Company, with a grinding capacity of 140,000 bushels of corn per day, which soon was reduced to 110,000 bushels or 74% of the United States total.

Also, in 1906 the Western Glucose Company was incorporated to become later the American Maize Company, the National Candy Company was incorporated to become later the Clinton Sugar Refining Company, and the Staley Company of Decatur, Illinois, began. In 1913 antitrust action caused the Corn Products Refining Company to spin off the Granite City, Illinois, the Davenport, Iowa, and the Oswego plants.

In 1958 Corn Products Refining Company acquired Best Foods Company and changed the parent name once again to Corn Products Company.

Anheuser-Busch Companies, Inc., which built a cornstarch plant in 1923 at St. Louis, Missouri, sold its only starch-producing facility at Lafayette, Indiana, to the A. E. Staley Manufacturing Company in 1981.

1. Present American Companies

Today, CPC International Inc. (Corn Products Company) has plants at Argo, Illinois; Pekin, Illinois; North Kansas City, Missouri; Stockton, California; and Winston-Salem, North Carolina.

The A. E. Staley Manufacturing Company that built its Decatur, Illinois plant in 1912 now has plants at Decatur, Illinois; Morrisville, Pennsylvania; Lafayette, Indiana; and Loudon, Tennessee.

American Maize-Products Company, started at Roby, Indiana, in 1908, now has plants at Hammond, Indiana and Decatur, Alabama.

Penick and Ford Ltd., beginning in 1902 as the Douglas Starch Company at Cedar Rapids, Iowa, is now a subsidiary of Univar Corporation.

Clinton Corn Processing Company, Inc., with plants at Clinton, Iowa, and Montezuma, New York, formerly a subsidiary of Nabisco Brands Incorporated, has been leased to ADM Foods.

National Starch and Chemical Corporation, with a plant at Indianapolis, Indiana, is a subsidiary of Unilever Corporation.

The Hubinger Company, with a plant at Keokuk, Iowa, is a subsidiary of H. J. Heinz Company.

ADM Foods, a division of Archer Daniels Midland Company, with plants at Cedar Rapids, Iowa, and Decatur, Illinois, has leased Clinton Corn Processing Company, Inc.

Amstar Corporation has a plant at Dimmitt, Texas.

Cargill, Incorporated has plants at Cedar Rapids, Iowa; Dayton, Ohio; and Memphis, Tennessee.

IV. Waxy Corn

Starch has many applications in the food and non-food industry, but usual starches are mixtures of linear and branched polysaccharides. These molecules have very different physical properties, and starches composed of only one component have special properties that open new and broader applications and lead to specialized uses for which regular starches are unsuited.

One unusual genetic variety of corn arose in China among the corn plants transferred from the Americas. This was a corn whose starch granules contained no linear molecules, but only branched molecules. It stained red with iodine, not blue as do ordinary starches. When the corn kernel was cut with a knife, the cut surface appeared shiny as though it contained wax. Hence the corn was called waxy corn, though no wax was present. Waxy corn was brought to the United States in the first years of the twentieth century and remained a curiosity at experiment stations until World War II cut off the supply of tapioca from the East Indies. During a search for a replacement, waxy corn starch was found suitable. Iowa state geneticists developed the waxy corn in their possessions to a high-yielding hybrid. It was introduced as a contract crop and has continued to serve as a valuable speciality starch. It has continued in use because of its unique properties and because the corn crop also supplies quality oil and protein. Although other similar starches, such as waxy sorghum and glutinous rice, are composed only of branched molecules, they have not had the industrial acceptance of waxy corn.

V. High-Amylose Corn

The linear polysaccharide of normal starches has the common property of linear molecules in its ability to form junction zones, crystallize in part, and form films and even fibers of high strength and flexibility. Amylose can simulate the behavior of cellulose in many important applications. Thus, in 1946, R. L. Whistler and H. H. Kramer, a geneticist, set about to produce a corn modification that would be the opposite of waxy corn, in which the starch would be composed only of linear (amylose) molecules. Much was learned about the effect of specialized genes on endosperm composition. The two investigators were able to raise the amylose content from the normal of 25% to 65%. Toward the end of their work, and subsequently, others entered into genetic programs; the amylose content was increased to 85%, with 65–70% being common in varieties made commercially.

As world demands for polymers grow and prices of polymers made from petrochemical sources increase, it may be expected that starch amylose, and likely high-amylose corn, will assume a stronger place as an industrial commodity.

VI. Future of Starch

Changing world economics is making it more practical to obtain chemicals from agriculture. Both academic and industrial investigators are giving more attention to developing technologies for converting agricultural products to chemicals and to methods for modifying starch, cellulose, and sucrose.

Starch is the lowest priced and most abundant worldwide commodity. It is produced in most countries and is available at low cost in all countries. Its level price over many years is impressive and makes it especially attractive as an industrial raw material (Table I). Its production by the wet-milling industry has continued to increase (Table II) and may increase at a faster rate as starch takes

Table I

Corn Price

Year	*Dollars per pound*
1972	0.021
1973	0.037
1974	0.053
1975	0.048
1976	0.045
1977	0.037
1978	0.042
1979	0.048
1980	0.051
1981	0.043

Table II

Wet-Milled Corn

Year	*Bushels* $\times 10^6$	*Pounds* $\times 10^9$
1972	265	14.8
1976	350	19.6
1977	365	21.0
1978	400	22.4
1979	430	24.1
1980	480	26.9 (projected)

Table III

U.S. Population Sweetener Consumption (Pounds per Capita)

Year	*Glucose syrup*	*Glucose–fructose syrup*	*Sucrose*
1960	9.4	0	97.4
1965	12.3	0	96.8
1975	16.2	6.8	91.5
1978	17.6	13.9	91.5
1980	18.2	15.5	91.3 (projected)

more of the sweetener market (Table III) and as governments subsidize ethanol production.

VII. References

(*1*) J. Wiesner, *Papier-Fabr.*, **9,** 886 (1911).

(*2*) Marcus Porcius Censorius Cato, "De Agriculture," 184 B.C., "Scriptores rei Rustica."

(*3*) Abu Mansur Muwaffak, of Hirow, North Persia (975 A.D.). Translation in Vol. III of Kobert's "Historische Studien," Halle (1893) as quoted by R. P. Walton, *in* "A Comprehensive Survey of Starch Chemistry," Chemical Catalog Co., New York, 1928, Part 1, p. 236.

(*4*) A. van Leeuwenhoek, "Epistolae Physiologicae super compluribus Naturae Arcanus," Epistolae XXVI, 1719.

(*5*) "Universal Lexikon des Gegenwart vergangenheit," H. A. Pierer, ed., Altenburg, Germany, Vol. 1, 1733; Vol. 15, 1737; Vol. 39, 1744.

(*6*) T. Kingsford, *in* "One Hundred Years of American Commerce," M. DePew, ed., D. O. Haynes, New York, 1895, p. 456.

(*7*) G. S. C. Kerchoff, *Acad. Imp. Sci. St. Petersbourg, Mem.*, **4,** 27 (1811); see Schweigger's *Beitr. Chem. Physik* (Nurenberg), **4,** 27 (1811), *in* R. P. Walton, "A Comprehensive Survey of Starch Chemistry," Chemical Catalog Co., New York, 1928, Part 2, p. 1.

CHAPTER II

ECONOMICS AND FUTURE OF THE STARCH INDUSTRY

BY PAUL L. FARRIS

Department of Agricultural Economics, Purdue University, West Lafayette, Indiana

I. INTRODUCTION

This chapter highlights the economic features of the starch industry in the United States, giving particular emphasis to demand prospects and industry organization. The starch industry, as defined here, refers to producers of commercial starch. Man consumes starch both from foods to which commercial starch, or products made from commercial starch, have been added, as well as from starch-bearing plants. Commercial starch also is an important ingredient in manufacturing a wide range of industrial products, such as paper, textiles, and building materials.

Commercial starch in the United States is made primarily from corn by the wet-milling industry. The term "wet-milling" is used because the corn is wet when it is ground and water is used as the suspension medium during most of the other operations.

Corn and relatively small quantities of sorghum (milo) grains are the basic materials used by the corn wet-milling industry. Annual utilization of corn by this industry fluctuated between 55 and 83 million bushels during the 1930s. Utilization increased rapidly during the 1940s and has continued to rise steadily through the 1970s (Table I).

Of the corn produced for grain in the United States, the percentage used by the

STARCH, 2nd ed.

ISBN 0-12-746270-8

corn wet-milling industry has shown a slightly rising trend, from around 3.5% in the 1930s to over 5.5% in the 1970s. However, because much corn is used on the farms where it is produced, the proportion sold from farms that has been taken by the wet-milling industry has been around 9% in the last decade. The proportion was about 20% in the 1940s. The proportion fluctuated greatly during the 1930s, dropping as low as 12% and rising to over 30% in the major drought years of 1934 and 1936.

A typical bushel of corn weighing 56 pounds yields about 34 pounds of starch, 2 pounds of oil, 11 pounds of animal feed (gluten and hull), and nine pounds of

Table I

Corn Produced for Grain, Corn Sold from Farms, and Wet-Process Grindings for Selected Periods, 1929–1970, and Annual, 1963–1981[a]

Year[b]	Corn produced for grain (million bushels)	Corn sold from farms (million bushels)	Wet-process corn grindings		
			Million bushels	Percent of production	Percent of sales from farms
1929–1938 (av.)	1986.2	415.2	69.6	3.5	16.8
1939–1941 (av.)	2321.0	536.0	103.7	4.5	19.3
1942–1947 (av.)	2645.6	624.0	123.2	4.7	19.7
1948–1952 (av.)	2925.4	913.6	126.3	4.3	13.8
1953–1962 (av.)	3294.1	1356.5	150.0	4.5	11.0
1963–1970 (av.)	4240.4	2184.0	210.3	5.0	9.6
1963	4019.2	1875.1	194.5	4.8	10.4
1964	3484.3	1703.5	200.7	5.8	11.8
1965	4102.9	2013.9	204.3	5.0	10.1
1966	4167.6	2105.3	208.0	5.0	9.9
1967	4860.4	2597.7	210.0	4.3	8.1
1968	4449.5	2354.9	207.0	4.7	8.8
1969	4687.1	2557.5	216.0	4.6	8.4
1970	4152.2	2263.7	242.0	5.8	10.7
1971	5646.3	3199.8	246.0	4.4	7.7
1972	5579.8	3251.8	284.0	5.1	8.7
1973	5670.7	3543.2	295.0	5.2	8.3
1974	4701.4	2946.7	315.0	6.7	10.7
1975	5840.8	3719.7	343.0	5.9	9.2
1976	6289.2	3959.8	362.0	5.8	9.1
1977	6505.0	3956.7	380.0	5.8	9.6
1978	7267.9	4432.1	400.0	5.5	9.0
1979	7938.8	4962.6	430.0	5.4	8.7
1980	6644.8	4145.8	465.0	7.0	11.2
1981	8201.0	[c]	500.0	6.1	—

[a] Source: References *1*, p. 14, and *2*.
[b] Years beginning October 1.
[c] Series discontinued.

Table II

Corn Starch[a] Shipments by the Corn Wet-Milling Industry (Census Industry 2046)[b]

Year	Corn starch[a] shipments: Quantity, million pounds	Corn starch[a] shipments: Value, million dollars	Total value of primary product shipments, million dollars	Value of corn starch[a] shipments as percent of total value of primary product shipments, %
1977	5486.4	408.2	1938.6	21.1
1972	3588.3	208.8	786.7	26.5
1967	3119.0	199.5	646.6	30.9
1963	2563.1	177.3	547.2	32.4

[a] Includes sorghum (milo) starch.
[b] Source: Reference *4*.

water (*3*). Some of the starch is converted to other products, such as sweeteners, by corn refiners. During the 1970s, about 15 pounds of starch per bushel of wet-process corn grindings were shipped as starch, and the remainder was used by corn refining companies to manufacture other products. Starch shipments, as a percent of the total value of products shipped by the industry, declined from around one-third in 1963 to about one-fifth in 1977 (Table II).

II. Statistical Estimation of the Demand for Starch

Because starch use permeates the entire economy, the demand for starch in any particular year depends rather directly on the level of national income and output. The growth of the starch industry relative to the growth of the general economy depends on a number of factors, including changes in the composition of national income and output, changes in the technology of industrial processes, the development of new products, and changes in the availability and prices of substitutes for starch.

The most important general indicator of change in the demand for starch is the gross national product (GNP) of the United States (Table III). From the early 1930s to the mid-1960s, the quantity of corn processed by the corn wet-milling industry rose at about two-thirds the rate of increase in GNP. Growth in utilization of corn by the wet-milling industry began to accelerate in the mid-1960s, reflecting the development and improvement of high-fructose corn syrup and major market expansion for this new product. Corn wet-grindings approximately doubled between 1966 and 1980 while GNP (in constant dollars) rose about 50%. (The price of corn tends to affect inversely the amount of corn taken by the wet-milling industry, but the influence is not great.)

Table III

Price of Corn, U.S. Gross National Product, and Wet-Process Grindings: Calendar Years 1929–1966

Year	Price of corn, No. 3 Chicago, dollars/bushel[a]	BLS wholesale price index 1967 = 100	Price of corn deflated by BLS wholesale price index, dollars/bushel	GNP in 1972 prices billion dollars[c]	Corn wet-process grindings, million bushels[d]
1929	0.934	49.1	1.902	314.6	86.6
1930	0.820	44.6	1.839	285.2	75.7
1931	0.517	37.6	1.375	263.3	66.9
1932	0.305	33.6	0.908	227.1	61.8
1933	0.398	34.0	1.171	222.1	75.1
1934	0.649	38.6	1.681	239.0	65.9
1935	0.813	41.3	1.969	260.5	58.3
1936	0.836	41.7	2.005	295.4	74.5
1937	1.030	44.5	2.315	309.2	68.4
1938	0.545	40.5	1.346	296.4	73.3
1939	0.500	39.8	1.256	318.8	77.2
1940	0.632	40.5	1.560	343.3	81.7
1941	0.705	45.1	1.563	398.5	110.3
1942	0.833	50.9	1.636	460.3	130.3
1943	1.046	53.3	1.962	530.6	128.5
1944	1.146	53.6	2.138	568.6	119.9
1945	1.168	54.6	2.139	560.0	119.1
1946	1.524	62.3	2.446	476.9	120.6
1947	2.034	76.5	2.659	468.3	139.3
1948	2.031	82.8	2.453	487.7	109.9
1949	1.312	78.7	1.667	490.7	116.2
1950	1.478	81.8	1.807	533.5	131.4
1951	1.793	91.1	1.968	576.5	129.0
1952	1.768	88.6	1.995	598.5	126.1
1953	1.562	87.4	1.787	621.8	130.3
1954	1.573	87.6	1.796	613.7	130.9
1955	1.383	87.8	1.575	654.8	137.9
1956	1.415	90.7	1.560	668.8	141.4
1957	1.272	93.3	1.363	680.9	139.4
1958	1.228	94.6	1.298	679.5	144.1
1959	1.201	94.8	1.267	720.4	153.0
1960	1.135	94.9	1.196	736.8	153.5
1961	1.108	94.5	1.172	755.3	157.1
1962	1.107	94.8	1.168	799.1	171.4
1963	1.235	94.5	1.306	830.7	184.9
1964	1.229	94.7	1.298	874.4	193.9
1965	1.278	96.6	1.323	925.9	204.9
1966	1.344	99.8	1.347	981.0	203.6

[a] Source: Reference *5*.
[b] Source: Reference *6*.
[c] Source: Reference *7*.
[d] Source: References *1*, p. 38, and *8*, p. 30.

To quantify the influences of these factors, a multiple regression equation was fitted to data for the 1929–1941 and 1953–1966 years. The data for 1942–1952 were omitted because of economic dislocations associated with the war and early postwar years. Years after 1966 were not included in the time span selected for regression analysis because the growth in corn sweetener production clearly had changed. The variables, all on a calendar year basis, were as follows:

Y = Corn wet-process grindings, in millions of bushels. This variable is used as a measure of the quantity of starch and starch related products produced in the United States.

X_1 = Price of corn, No. 3 at Chicago, in dollars deflated by BLS Wholesale Price Index (1967 = 100). This variable is used to reflect the real cost of the raw product input.

X_2 = GNP in billions of dollars in constant 1972 prices. This variable is used to reflect the composite real demands for starch.

X_3 = Time (29, 30, 31, etc., for appropriate years). This variable is to account for long-time trends in technology, changes in the composition of national industrial activity and income, and other systematic influences associated with the passing of time.

The estimated relationship is given in Eq (1) with the standard errors in parentheses.

$$Y = \underset{(13.787)}{59.505} - \underset{(4.123)}{8.128X_1} + \underset{(0.025)}{0.252X_2} - \underset{(0.486)}{1.288X_3}; \; R^2 = 0.985 \qquad (1)$$

Here we see that for one billion dollars increase in GNP, there was associated with it an average increase in wet process grindings of 252,000 bushels of corn. A change of one cent per bushel in the price of corn was associated with an opposite change of about 81,000 bushels in the utilization of corn by this industry; assuming constant levels of corn prices and GNP, there was a tendency for annual wet process grindings to decline by about 1.29 million bushels each year.

At average corn prices for the 27-year-period, these relationships indicate a very inelastic response to price of quantity of corn taken by the industry. A 1% change in the price of corn was associated with a change in the opposite direction of about 0.10% in wet-process grindings. On the other hand, a 1% change in GNP was associated with a 1.12% change in the same direction of wet-process grindings. (These elasticities were measured at mean levels of all independent variables for the 1929–1941 and 1953–1966 periods.)

III. Projected Future Volumes of Corn Likely to Be Used by the Wet-Milling Industry

In projecting future growth of the corn wet-milling industry, the total market potential was divided into two components, the composite demand excluding high-fructose corn syrup and a separate estimate for high-fructose corn syrup. These two separate estimates were then combined into overall industry projections in 1985 and 1990.

The regression equation estimated above was employed in projecting industry demand exclusive of high-fructose corn syrup. In order to examine the reliability of the regression relationships derived from the historical period ending in 1966, a prediction was made for 1978. In that year, with GNP at $1,438.6 billion (in 1972 dollars) and corn at $1.10 per bushel (in 1967 dollars or $2.25 in current dollars), predicted wet-process grindings were about 313 million bushels.

High-fructose corn syrup (HFCS) sales were 1.25 million tons (dry basis) in 1978 (*9*). Assuming 33.35 pounds per bushel (*10*), 1978 utilization of corn for high-fructose corn syrup (HFCS) required about 75 million bushels of corn. Adding this figure to the utilization estimate given by the regression equation, 313 million, gives a total estimated corn grind for 1978 of 388 million bushels. The actual grind was reported to be 400 million bushels, indicating the regression equation prediction was too low. This might have occurred because either some omitted influence was involved, the basic underlying relationship had changed, or the linear relationships assumed in the regression model were not appropriate. Thus, in employing the equation for projections to 1985 and 1990, the estimates could be more likely to err on the low than on the high side.

Because of the phenomenal growth of HFCS beginning in the mid-1960s, the projection of the separate demand for this new product is added to the regression projections exclusive of HFCS. By 1981, HFCS consumption alone was over 23 pounds per person, which was about 18% of total caloric sweetener consumption (Table IV). Total corn sweeteners, including HFCS, rose from 13.5% of total caloric sweetener consumption in the United States in 1966 to 35% in 1981.

In developing a projection for HFCS consumption to add to the composite projection obtained with the regression equation, there is an implicit assumption that the HFCS market demand is a net addition to demand for corn sweetener products. This assumption may not be entirely accurate; however, the principal growth in demand for HFCS is expected to be as a substitute for sugar. Carman and Thor indicated that per capita consumption of glucose corn syrup, dextrose, and minor caloric sweeteners is likely to continue at present levels. (See reference *11*, p. 50.)

Nevertheless, the growth rate of high-fructose corn syrup consumption is at some point expected to reach an overall market saturation potential. Carman and Thor predicted that the most likely ceiling values would be in the range of 20% to 30% of total sweetener sales, which indicates considerable further growth from the approximately 18% market penetration achieved by HFCS in 1981. (See reference *11*, p. 44.) Taking various factors into account, Carman and Thor estimated that corn used for the manufacture of high-fructose corn syrup would range between 173 and 241 million bushels in 1985 and from 197 to 288 million

Table IV

Corn Sweetener Consumption in the United States, 1963–1981[a]

	Corn sweeteners					
	Corn syrup					
Year	*High-fructose, lb/capita*	*Glucose, lb/capita*	*Crystalline dextrose, lb/capita*	*Total corn sweetners, lb/capita*	*Total caloric sweeteners, lb/capita*	*Corn sweeteners as % of total caloric sweeteners, %*
1963	—	9.9	4.3	14.2	113.3	12.5
1964	—	10.9	4.1	15.0	113.5	13.2
1965	—	11.0	4.1	15.1	113.9	13.3
1966	—	11.2	4.2	15.4	114.4	13.5
1967	0.1	11.9	4.2	16.2	116.1	14.0
1968	0.3	12.6	4.3	17.2	118.0	14.6
1969	0.5	13.2	4.5	18.2	120.8	15.1
1970	0.7	14.0	4.6	19.3	122.5	15.8
1971	0.9	14.9	5.0	20.8	124.3	16.7
1972	1.3	15.4	4.4	21.1	124.9	16.9
1973	2.1	16.5	4.8	23.4	125.6	18.6
1974	3.0	17.2	4.9	25.1	121.8	20.6
1975	5.0	17.5	5.0	27.5	118.0	23.3
1976	7.2	17.5	5.0	29.7	124.4	23.9
1977	9.5	17.6	4.1	31.2	126.8	24.6
1978	12.1	17.8	3.8	33.7	126.6	26.6
1979	14.9	17.9	3.6	36.4	127.1	28.6
1980	19.2	17.6	3.5	40.3	125.3	32.2
1981	23.3	17.8	3.5	44.6	125.5	35.5

[a] Source: Reference 9, p. 27.

bushels in 1990. Midpoints in these ranges were added to the projections for ongoing composite demands for corn to be used in making other products produced by the wet-milling industry (Table V).

The combined estimates indicate a potential growth in demand for corn by the wet-milling industry to about 600 million bushels in 1985 and 700 million bushels or more in 1990. If the projections are realized, the industry will have grown 50% above its 1978 level by 1985 and 75% by 1990.

1. Evaluation of the Projections

A number of considerations are involved in making projections. One of these is the selection of the period from which projections are to be made, another is the determination of relationships within the period. Still another is the necessity

Table V

Projected Wet-Process Corn Grindings in 1985 and 1990 Based on Selected Assumptions

GNP (real) projected from 1976–1978 average at the rate of	Projected wet-process corn grindings, millions of bushels					
	1985			1990		
	Excluding HFCS[c]	HFCS[d]	Total	Excluding HFCS[c]	HFCS[d]	Total
3%/year[a]	378	207	585	441	242	683
4%/year[b]	413	207	620	509	242	751

[a] This was the approximate annual growth rate which prevailed between 1969–1971 and 1976–1978.

[b] This was the approximate annual growth rate which prevailed between 1959–1961 and 1969–1971.

[c] Based on relationships shown in the regression equation, with the price of corn assumed at $1.10/bushel in 1967 dollars.

[d] Midpoints of the ranges, 173 to 241 million tons in 1985 and 197 to 288 million tons in 1990, projected by Carman and Thor (*11*), p. 52.

of projecting the variables that have been found to be related to the variable in which one is interested. Judgment is involved at each step. Nevertheless, projections can be useful in indicating probable trends associated with the selection of certain underlying assumptions.

The regression projections are based on a period spanning several years in which abnormal or unusual years were eliminated. The linear regression model used is a relatively simple one; nevertheless, it is believed to be a logical expression of relationships between the dependent variable and variables systematically related to the growth of the starch industry.[1] The projections of the price of corn in real terms is within current expectations of trends in grain prices in the future. The alternative projections of GNP are the most important and most crucial. It is believed that they realistically bracket the probable future growth trend of the economy based on past experience.

In order to evaluate the projections further, some qualitative appraisal must be

[1]The uses of alternative variables such as the Federal Reserve Board Indexes of Industrial Production, total durable manufactures, and nondurable manufactures were explored. Exploratory analyses were also conducted using alternative time periods and nonlinear formulations of the variables. These explorations did not promise results more useful in prediction than the model selected. Moreover, for purposes of projecting the dependent variable, projections of values of the independent variables in the model selected, the price of corn, GNP, and time, can probably be made and evaluated more readily than projections of independent variables considered to be logical alternatives.

made of influences associated with possibilities of changes in the composition of national output, changes in industrial processes, and changes in the prices and availability of substitutes for starch. Insofar as changes in the composition of gross national output are concerned, the most likely trend is for a continuing rise in the primary starch-using industries.

The growth in edible uses of starch, primarily as sweeteners, seems highly promising. The total industry may be stimulated toward added growth as corn is used as an ingredient in production of alcohol for fuel. The potential of alcohol in the liquid fuel market will depend basically on the relative prices of alternative fuels and on the combination of public policies with respect to subsidies and incentives which prevail relative to liquid energy.

In summary, there appears to be considerable growth potential for the corn wet-milling industry in the foreseeable future. There will certainly continue to be challenges to markets for starch by various substitute products. However, new starch products are also being found to penetrate markets previously closed to starch. While it is impossible to foresee what technological breakthroughs may be forthcoming, there are certainly possibilities for starch to maintain or even to improve its position relative to the substitute products. The expected growth of the economy and the changing composition of its output make the projections seem realistic and portend a substantial growth in the market for starch in the 1980s.

2. Foreign Trade in Starch

Both potato starch and tapioca were imported in small quantities very early in American history. Potato starch came from Germany and tapioca from the Netherlands East Indies. From 1895 to 1905, tapioca came to be used both industrially and for food, and imports increased to 30 to 35 million pounds a

TABLE VI

Starch Exports and Imports, Selected Years[a]

	Exports of corn starch, including milo		*Imports of starch*		
Year	*Millions of pounds*	*Percent of industry shipments*	*Tapioca (cassava), millions of pounds*	*All other, millions of pounds*	*Total, millions of pounds*
1977	81.6	1.5	82.7	56.8	139.5
1972	56.2	1.6	141.8	55.4	197.2
1967	64.7	2.1	304.1	37.0	341.1
1963	53.9	2.1	85.9	10.8	96.7

[a] Source: References *12* and *13*.

year. During this period, the export market for corn starch was also developed, reaching 100 million pounds in 1915. Exports of starch exceeded imports until after 1930, when imports increased markedly and exports fell. Toward the latter part of the 1930s, exports of corn starch began rising again. However, exports account for a relatively small, and declining, share of total U.S. production (Table VI). Currently, imports of starch consist mainly of tapioca, mostly from Thailand. Foreign trade in starch is expected to continue to account for only minor quantities relative to the total U.S. industry.

IV. Organization of the Corn Wet-Milling Industry (see also pages 4–6)

The corn wet-milling industry lends itself to large-scale output. Corn product refining is a highly technical business producing a wide range of food and industrial products from virtually a single raw material. Consequently, a large share of the industry's output is produced by a few firms. The largest four companies accounted for 63% of total industry production in 1977, and the largest eight companies accounted for about 89% (Table VII). The share of industry sales accounted for by the leading firms as a group has declined; however, only a dozen or so companies account for essentially the entire industry output. Most of the companies that are in operation today had their origins during the early years of the twentieth century. Nevertheless, several of the firms have been acquired by other companies, and most of the remaining ones have become

Table VII

Percentage of Total Value of Shipments Accounted for by Largest Companies in the Corn Wet-Milling Industry (SIC 2046), Selected Years[a]

			Share of value of shipments			
Year	*Number of companies*	*Total value of shipments, millions of dollars*	*Four largest companies, %*	*Eight largest companies, %*	*Twenty largest companies, %*	*Primary product specialization, %*
1977	24	2014.8	63	89	99+	94
1972	26	832.3	63	86	99+	93
1967	32	751.3	67	89	99+	84
1963	49	622.4	71	93	99	83
1958	53	522.0	73	92	99	91
1954	54	458.4	75	93	99	91
1947	47	460.0	77	95	99+	88
1935	NA	NA	79	95	NA	NA

[a] Source: Reference *14*, pp. 9–14.

increasingly diversified. As a result, corn refining activities for the most part have become divisions within large conglomerate type firms.

The industry had its origin in the East in the 1840s. The entry of new firms accompanied by severe price cutting set the stage for the formation of the National Starch Manufacturing Company in 1890, which brought together some twenty plants that controlled 70% of the output. Cutthroat competition was not stopped, however, and a period of combination and recombination followed which resulted in the formation of the Corn Products Refining Company in 1906. The new company controlled 64% of starch output and 100% of glucose (corn syrup) output.

Nevertheless, relatively easy entry into the industry and new consolidations gradually eroded the position of Corn Products, and this company's share of corn wet grindings dropped from 95% to 65% by 1914. Corn Products, however, was looked upon as a monopoly, and in 1916, a United States District Court ordered its dissolution. The case was concluded in 1919 when the defendant withdrew an appeal to the United States Supreme Court and consented to dispose of two dextrose mills and of the old National Starch Company.[2] Corn Products Refining Company remained the dominant firm in the industry. Its corn grind represented 60% of the total in 1918, 51% in 1923, and declined gradually to around 45% in 1945, about 45% in the 1950s and probably less in more recent times.

1. Rates of Return in Corn Refining

Because most firms in the corn wet-milling industry are diversified, their reported measures of profitability, which are for the firms' entire operations, may obscure earnings in their corn refining activities. However, because of the leading position of CPC International (formerly Corn Products Refining Company) and the fact that this firm's business remains heavily oriented toward corn refining, the firm's reported data may give a general indication of earnings in corn refining.

During the 1930s, the ratio of net income to net worth for Corn Products Refining Company averaged 9.2%, compared with 7.2% for leading manufacturing corporations (Table VIII). In the World War II and early postwar years, the average earning rate increased, but not as much as for leading manufacturing corporations. After the mid-1950s, the rate of return for Corn Products Refining Company rose above the average of leading manufacturing corporations, and this company continued to outperform these other manufacturers through 1980.

However, rates of return vary from year to year, caused in part by fluctuations in the price of corn and a relatively stable price for starch. According to data from one corn refining company, there have been periods of three to four years without a change in the price of starch, and the amounts of the changes were

[2]This background information is mainly from references *15*, pp. 257–286, and *16*, pp. 201–220.

Table VIII

Ratio of Net Income to Net Worth for CPC International and Leading Manufacturing Corporations: Annual Average for Selected Periods, 1933–1980

	Ratio of net income to net worth,[a] annual average	
Period	CPC International	Leading manufacturing corporations
1971–1980	15.7	14.6
1956–1970	14.0	12.0
1941–1955	11.1	13.0
1933–1940	9.2	7.2

[a] Source: References *17* and *18*.

relatively small during the 1950s and 1960s. On October 23, 1952, this company's price of pearl starch was reduced from 6.47 to 6.32 cents per pound. The next price changes were to 6.42 cents per pound in 1956 and to 6.22 cents per pound in 1960. By 1964, the price was 5.58 cents per pound. The price then went to 6.08 cents per pound in 1965 and to 6.38 cents in 1967, where it remained until 1970. The price then moved periodically higher during the 1970s, reaching 12.6 cents per pound in mid-1979. Selective allowances were made from quoted prices principally during the latter part of the 1960s, but allowances apparently were discontinued in the 1970s.

Because corn is a major cost component, amounting to 76% of the cost of materials consumed by the industry in 1977, and because the price of corn is much more variable than the price of starch, changes in corn prices tend to be related inversely to changes in earnings of corn refining companies. The relationship is illustrated in regression equation (2) which is based on annual data for 1934–41 and 1953–1980. (The World War II and early postwar years, including the Korean Conflict, were omitted because of the substantial economic dislocations which prevailed during that period. Standard errors are given in parentheses.)

$$\Delta Y = \underset{(0.349)}{0.0366} - \underset{(2.635)}{7.029}\,\Delta X_1 + \underset{(14.680)}{6.567}\,\Delta X_2;\; R^2 = 0.18 \tag{2}$$

ΔY = Change from the preceding year in the rate of return on net worth for CPC International

ΔX_1 = Change from the preceding year in the logarithm of the deflated price of corn (for the marketing year beginning the prior October 1, deflated by the B.L.S. Wholesale Price Index)

ΔX_2 = Change from the preceding year in the logarithm of U.S. Gross National Product (in constant 1972 dollars)

The coefficient of the variable reflecting changes in corn prices was inversely related, as hypothesized, in a statistically significant way to changes in rate of return on net worth for the leading corn refining company. The influence of economic activity, as measured by changes in real U.S. Gross National Product, on the rate of return was positive, as expected, but the coefficient was not statistically significant. Several other factors not included in the regression equation, such as variations in labor costs, work stoppages, market conditions, and a variety of influences peculiar to the operations of that particular company undoubtedly affected its earnings as well.

Further evidence which is consistent with the hypothesis that ingredient costs are inversely related to rates of return is seen in Census data. The cost of corn, including sorghum grain, amounted to 42.5% of the value of primary product shipments of the corn wet-milling industry in 1963; 44.7% in 1967; 42.8% in 1972, and 45.9% in 1977. In the years when grain costs were relatively low (1963 and 1972), rates of return for CPC International rose from the preceding year; and in the other two Census years, when grain costs were relatively high, rates of return decreased from the preceding year.

2. Future Industry Organization

Looking ahead, the industrial organization and behavior of the corn wet-milling industry seems unlikely to change drastically in the foreseeable future. Further product diversification may occur as the firms succeed in developing new products for a growing and changing consumer market. However, the opposite trend may occur in manufacturing plants. Most industrialists believe that unmodified pearl starch will increasingly be the principal polysaccharide produced in wet-milling plants. Production of starch derivatives will probably occur to a greater extent in other types of plants or in starch consuming industries, such as the paper industry. Plant automation is expected to increase, and new semi-wet-milling technologies may lead to increased efficiency in pearl starch production.

The conglomerate character of firms in the industry is expected to continue to prevail. The share of industry business accounted for by the largest four firms will probably decline somewhat further as other firms in the industry increase their investments in corn refining capacity. Growth in plant capacity, given the economies of scale and the highly technical nature of the business, is expected to occur primarily within existing firms rather than through new entrants. The total number of firms in the industry is believed not likely to change significantly, although an open question could be whether corn refining operations of existing companies might be acquired by other companies.

Pricing behavior seems unlikely to change greatly from past practices. Product prices might show more frequent upward adjustments than in the 1950s and 1960s if price inflation in the general economy is relatively higher than in the 1950–1970 period. Earning rates in corn refining operations will likely continue to compare favorably with those of most other segments of American industry.

V. References

(*1*) "Grain and Feed Statistics through 1956," U.S. Department of Agriculture, Statistical Bulletin No. 159 (Rev. May 1957).
(*2*) "Feed Situation," U.S. Department of Agriculture, various issues for indicated years.
(*3*) "Tapping the Treasure in Corn," Corn Refiners Association, Washington, D.C., 1976.
(*4*) "Census of Manufacturers," Bureau of the Census, U.S. Department of Commerce, for indicated years.
(*5*) "Agricultural Statistics," U.S. Department of Agriculture, issues for indicated years.
(*6*) "Wholesale Prices," Bureau of Labor Statistics, U.S. Department of Labor, issues for indicated years.
(*7*) "Survey of Current Business," U.S. Department of Commerce, October 1978.
(*8*) "Grain and Feed Statistics through 1966," U.S. Department of Agriculture, Statistical Bulletin No. 410, September 1967.
(*9*) "Sugar and Sweetener Report," U.S. Department of Agriculture, SSR-Vol. 7, No. 2, May 1982.
(*10*) G. J. Cubenas and L. F. Schrader, "Simulation Model of the HFCS Production Process." See reference *9*, p. 68.
(*11*) H. F. Carman and P. K. Thor, "High Fructose Corn Sweeteners: Economic Aspects of a Sugar Substitute," Giannini Foundation of Agricultural Economics, Information Series 79-2, Division of Agricultural Sciences, University of California Bulletin 1894, July 1979.
(*12*) "U.S. Exports," Bureau of the Census, U.S. Department of Commerce, FT-410, issues for various years.
(*13*) "U.S. Imports," Bureau of the Census, U.S. Department of Commerce, FT-246, issues for various years.
(*14*) "Concentration Ratios in Manufacturing," Special Report Series MC77-SR-9, 1977 Census of Manufactures, Bureau of the Census, U.S. Department of Commerce, May 1981.
(*15*) S. N. Whitney, "Antitrust Policies," Vol. II, The Twentieth Century Fund, New York, 1958.
(*16*) M. W. Watkins, "Industrial Combinations and Public Policy," Houghton Mifflin, Cambridge, England, 1927.
(*17*) "Moody's Industrial Manual," Moody's Investors Service, Inc., New York, for indicated years.
(*18*) "Monthly Economics Letter," Citibank, New York, April issues for indicated years.

CHAPTER III

GENETICS AND PHYSIOLOGY OF STARCH DEVELOPMENT

By Jack C. Shannon and Douglas L. Garwood

Department of Horticulture, The Pennsylvania State University, University Park, Pennsylvania

STARCH, 2nd ed.

ISBN 0-12-746270-8

I. Introduction

Starch, a common constituent of higher plants, is the major form in which carbohydrates are stored. Starch in chloroplasts is transitory and accumulates during the light period and is utilized during the dark. Storage starch accumulates in reserve organs during one phase of the plant's life cycle and is utilized at another time. Starches from reserve organs of many plants are important in commerce.

The pathway of starch synthesis is complex and not completely understood. Although gross starch structure is similar in various species, variation in granule structure and in starch fine structure is well documented and described elsewhere in this volume. Variation can be associated with plant species, cultivars of a species, environment in which a cultivar is grown, and genetic mutations.

This chapter first reviews nonmutant starch granule composition and development and then focuses on genetic mutants and how they have been useful in understanding the complexity of polysaccharide biosynthesis and development. Due to space limitation, attention is given only to a few of the plant species which are important sources of commercial starch production; the discussion will focus on maize (*Zea mays* L.) because of the many known endosperm mutants of maize which affect polysaccharide biosynthesis. Although developing maize kernels have been used for many of the investigations of starch biosynthesis, the information gained probably applies to other species, and these effects are illustrated whenever appropriate. As a result of this approach, it has been necessary to be selective in choosing examples to illustrate general trends in the genetics and physiology of starch development. Apology is given to other authors whose papers could also have been used to illustrate similar points.

No chapter can adequately cover all aspects of starch development, biosynthesis, and genetics. Readers wishing more detailed information should consult the books by Badenhuizen (*1*), Radley (*2*), and Banks and Greenwood (*3*); and review papers by Creech (*4*), Nelson (*5*), Preiss and Levi (*6*), Juliano (*7*), Marshall (*8*), Preiss (*285*), Preiss and Levi (*286*), and Banks and Muir (*287*).

II. Occurrence

1. General Distribution

Starch can be found in all organs of most higher plants (*1, 9*). Organs containing starch granules include pollen, leaves, stems, woody tissues, roots, tubers, bulbs, rhizomes, fruits, flowers, and the pericarp, cotyledons, embryo, and endosperm of seeds. These organs range in chromosome number from the haploid pollen grain to the triploid endosperm, the main starch storing tissue of cereal grains.

In addition to higher plants, starch is found in mosses and ferns and in some protozoa, algae, and bacteria (*9*). Some algae, namely the Cyanophycae or blue-green algae (*10, 11*), and many bacteria produce a reserve polysaccharide similar to the glycogen found in animals (*9, 12*). Both starch and a water-soluble polysaccharide, similar to glycogen and termed phytoglycogen, occur in sweet corn and other maize genotypes (*13*). A glycogen-type polysaccharide also has been reported in the higher plant *Cecropia peltata* (*14*).

Badenhuizen (*9*) has classified starch producing species into two groups. In the first group, starch is formed in the cytosol of a cell, while in the second group, starch is formed within plastids.

2. Cytosolic Starch Formation

Starch granules are formed in the protozoa *Polytomella coeca* (*12, 15*), but other species of protozoa produce amylopectin-type polysaccharides, glycogen, or laminaran (*12, 15*).

The red algae, Rhodophyceae, produce a granular polysaccharide called Floridean starch on particles outside the chloroplasts. In many of its properties, this starch resembles the amylopectin of higher plants, but in other properties, it is intermediate between amylopectin and glycogen. Floridean starch contains no amylose (*10, 16*). Free polysaccharide granules are also produced in the Dinophyceae, but the chemical nature remains uncertain (*10*).

Starchlike substances are produced in several species of bacteria (*9, 12*). For example, *Escherichia coli* produces a linear polyglucan (*12, 17*). *Corynebacterium diphtheriae* produces a starchlike material, and *Clostridium butyricum* produces a polyglucan with some branching (*12*). *Neisseria perflava* produces a polyglucan intermediate in structure between amylopectin and glycogen (*17*); however, recent work shows that the structure more closely approaches that of glycogen (*18*).

3. Starch Formed in Plastids

Starch is formed in chloroplasts of moss, fern, and green algae (*9*). Chlorophyceae (green algae) starch is similar to that of higher plants, and several species have been used in studies of starch biosynthesis (*10, 16*). Other classes of algae which produce starch are Prasinophyceae (*10, 19*) and Cryptophyceae (*19, 20*).

In the plastids of higher plants, starch granules can be classified as transitory or reserve (*1*). Transitory starch granules accumulate for only a short period of time before they are degraded. Starch formed in leaf chloroplasts during the day, which is subsequently hydrolyzed and transported to other plant parts at night in the form of simple sugar, is an example of transitory starch. Transitory starch is also formed in lily (*Lilium longiflorum*) pollen during germination of the pollen grains (*21*). Transitory and reserve starch granules can be differentiated by the fact that transitory starch granules lack the species specific shape associated with reserve starch granules. Furthermore, when exogenous sugar is supplied, the number, but not the size, of granules in a chloroplast will increase while the reverse occurs in amyloplasts (*1*).

Reserve starch is usually formed in amyloplasts, although it is occasionally formed in chloroamyloplasts. These are chloroplasts which have lost their lamellar structure and subsequently start producing fairly large reserve starch granules (*1*). Chloroamyloplasts form starch independent of photosynthesis. They have been described in tobacco (*Nicotiana tabacum* L.) leaves, *Aloe* leaves and flowers, central pith of potato (*Solanum tuberosum*) fruit, *Pellionia* and *Dieffenbachia* stems, and other tissues (*1, 9*). Such sources of reserve starch are insignificant, however, when compared to the reserve starch formed in roots, tubers, and seeds.

III. Cellular Developmental Gradients

To properly evaluate data relating to reserve starch development and composition, cellular development of tissues in which this starch is formed must be appreciated. Enlarging potato tubers (*22*) and endosperms of developing maize (*23–27*), rice (*Oryza sativa* L.) (*7*), sorghum (*Sorghum bicolor* (L.) Moench) (*28–30*), wheat (*Triticum aestivum* L.) (*31, 32*), rye (*Secale cereale* L.) (*33*), triticale (X *Triticosecale* Wittmack) (*34*), and barley (*Hordeum vulgare* L.) (*35*) kernels are composed of a population of cells of varying physiological ages.

In maize kernels, the basal endosperm cells begin starch biosynthesis late in development and contain small starch granules (*23, 26, 27, 36*). Peripheral maize endosperm cells, which are the last to develop, also contain small starch granules (*24, 26, 27, 36*). Thus, a major gradient of cell maturity from the basal endosperm to the central endosperm and a minor gradient from the peripheral cells adjacent to the aleurone layer inward exist in *normal* (nonmutant) maize endosperm. A similar cellular developmental gradient occurs in sorghum (*28–30*).

In barley, starch formation begins at the apex of the grain and around the suture across the central region (*35*). Deposition occurs last in the youngest cells near the aleurone layer (*35*). Related gradients occur in rice (*7*), rye (*33*), triticale (*34*), and wheat (*31, 37*).

Since all endosperm cells are not the same age, the physiologically younger cells may undergo the same developmental changes in starch biosynthesis as older cells, but at a later time in grain or kernel development. Shannon (*38*) divided 30-day-old *normal* maize kernels into seven endosperm zones and found that the sugar and starch composition of the lower zone corresponds to that found in whole endosperms 8, 10, and 12 days post-pollination, while the carbohydrate composition of upper zones is similar to that in the kernels 22–28 days post-pollination. When starch granules from 36-day-old *normal* maize kernels were separated into different size classes, a decline in apparent amylose percentage with decreasing granule size was observed, which reflected the characteristics of unfractionated starch isolated from endosperms earlier in kernel development (*39*). Although variations in granule size occur throughout the endosperm, starch granules within a given cell of *normal* maize endosperm are similar in size (*25, 36*).

The existence of cellular developmental gradients has two important ramifications when studying the genetics and physiology of starch development. First, evaluations of developing tissue using whole tissue homogenates are based on polysaccharides and enzymes isolated from cells of differing physiological age. Thus, such whole tissue data represents only an average stage of cellular development at the date of sampling. Second, tissue that does not reach maturity because of environmental or other reasons will differ in composition from mature tissue, and variation in starch composition can occur between samples.

As tissues storing reserve starch develop and the cells fill with starch granules, the starch concentration, expressed as a percentage of tissue weight, increases. For example, the starch content of potatoes increases from 5 to 18% of the fresh weight as tuber size increases from 0–1 cm to 10–11 cm (*40*). In maize, numerous workers have demonstrated a similar increase with data reported by Wolf and co-workers (*41*) and Early (*42*) being typical. At 7–10 days post-pollination, starch comprises less than 10% of kernel weight. This percentage increases to 55–60% by 30–35 days, and then remains fairly constant until maturity. The starch content of barley kernels rises in a sigmoid pattern with time, and 95% is deposited between 11 and 28 days after ear emergence (*43*). Similar increases are observed in the reserve starch concentration in other species (*44–47*).

IV. Nonmutant Starch Granule Polysaccharide Composition

1. Polysaccharide Components

Nonmutant or *normal* reserve and transitory starch granules are composed primarily of amylose and amylopectin. Amylose is essentially a linear polymer consisting of (1→4)-linked α-D-glucopyranosyl units. Amylopectin is a branched polymer of α-D-glucopyranosyl units primarily linked by (1→4) bonds with branches resulting from (1→6) linkages (*48, This Volume,* Chap. VI). Properties of these two major starch components are summarized in Table I.

To determine the relative amounts of amylose and amylopectin in starch and the properties of these components, starch granules must first be isolated and purified from the plant species to be studied (*3, 49*). Fractionation of the starch into its components can be achieved through two basic methods involving either selective leaching of the granules or complete granule dispersion (*3, 49, 50*). Methods based on granule dispersion are more satisfactory (*50*). Fractionation methods have been extensively reviewed (*3, 50–53, This Volume,* Chap. VIII). Thus, only the basic aspects of these methods needed to establish a framework for discussing the starch composition of different species and genotypes will be presented. Methods for dispersing the granule have included autoclaving in water, solubilization in cold alkali, treatment with liquid ammonia, and solubilization in dimethyl sulfoxide, with the latter method preferred (*49*). Details are discussed in Chapter VIII.

Table I

Properties of the Amylose and Amylopectin Components of Starch[a]

Property	*Amylose*	*Amylopectin*
General structure	Essentially linear	Branched
Color with iodine	Dark blue	Purple
λ_{max} of iodine complex	~650 nm	~540 nm
Iodine affinity	19–20%	<1%
Average chain length (glucose residues)	100–10,000	20–30
Degree of polymerization (glucose residues)	100–10,000	10,000–100,000
Solubility in water	Variable	Soluble
Stability in aqueous solution	Retrogrades	Stable
Conversion to maltose by crystalline β-amylase	~70%	~55%

[a] Adapted from Marshall (*8*), Williams (*48*), and Radley (*2*).

Once dispersed, the differential iodine-binding properties of amylose and amylopectin (Table I) can be utilized to estimate the amount of linear polysaccharide present in the starch without fractionating the starch (*54*). Amylose can be determined either by measuring the absorbance of the starch–iodine complex (blue value procedure) and relating this absorbance to that obtained for amylose and amylopectin standards (*55–61*) or by the method of potentiometric iodine titration in which the amount (mg) of iodine bound per 100 mg of polysaccharide is determined and this amount is related to the amount bound by an amylose standard (*3, 54, 62, 63*). For nonmutant starches, these procedures give similar results (*54*); however, absolute results can vary with both procedures, depending on the iodine-binding properties of the amylose and amylopectin standards. Lansky and co-workers (*64*), for example, showed that iodine affinities for purified amyloses could range from 18.5 to 20.0% with some amylose subfractions having iodine affinities of 20.5–20.8%. Furthermore, the amylose content estimated by all of the procedures based on iodine complex formation should be considered "apparent amylose" (*65*); the occurrence of branched chain components with long external chains would result in an overestimation of the amylose content (*54, 66*). Likewise, the presence of short-chain-length amylose would cause the amylose content to be underestimated (*54*) because absorption of the starch–iodine complex is reduced when the average degree of polymerization is less than 100 (*67*). These limitations should be remembered when amylose percentages are presented.

Dispersed starch can be separated into the amylose and amylopectin components by adding a polar organic substance such as thymol or *n*-butanol to produce an insoluble amylose complex (see *This Volume*, Chap. VIII). This initial precipitate is usually purified by solubilizing the complex and recomplexing it. Amylopectin may be recovered from the initial supernatant by freeze-drying or by precipitation with alcohol (*3, 49, 50*). Alternatively, the amylopectin component can be removed first from the dispersion by high-speed centrifugation followed by the addition of a polar organic substance to precipitate the amylose from the supernatant (*68*). Recently, dispersed starch was fractionated using Sepharose column chromatography (*69, 70, 71*). All these procedures will permit quantitative estimation of the amount of amylose in the starch.

Amylose and amylopectin isolated following fractionation consist of a population of molecules that vary in their degree of polymerization (Table I). For example, amylose can be subfractionated into a graded series of molecular sizes (*64, 72, 73*); the amylopectin fraction also has a broad distribution of molecular weights (*74, 75*). In addition to heterogeneity of molecular sizes, amylose also appears to consist of a mixture of both linear and slightly branched chains, the proportions of which may vary with the source of the starch and with the maturity of the source (*3*). In short, both fractions are both polymolecular and polydisperse.

From this work it can be concluded that starches cannot be divided sharply into amylose and amylopectin, but that the two major fractions blend into each other through intermediate fractions. The presence of intermediate polysaccharides in the starch granule is apparent from the Sepharose elution profile of *normal* maize starch when compared to the profile of a mixture of purified amylose and amylopectin (*69, 71*). Based on indirect evidence from iodine affinities, Lansky and co-workers (*64*) suggested that 5–7% of *normal* maize starch consists of material intermediate between the strictly linear and highly branched fractions. Subsequently, several "non-amylopectin" types of branched polysaccharides have been recovered by various modifications of the previously described fractionation procedures. For example, Erlander and co-workers (*76*) recovered a low-molecular-weight component from the supernatant following amylose precipitation with thymol and removal of amylopectin by centrifugation. The polysaccharide remaining in the supernatant had a β-amylolysis limit and degree of branching similar to that of amylopectin. Perlin (*68*) obtained an intermediate component following removal of amylopectin by centrifugation and precipitation of amylose with amyl alcohol. The polysaccharide remaining in the supernatant was more highly branched than amylopectin, based on reduced β-amylolysis limits, and was of lower molecular weight. A related highly branched polysaccharide with viscosity similar to amylopectin was recovered from the supernatant following recomplexing the amylose fraction of starch from potato tuber, rubber (*Havea brasiliensis*) seed, barley kernels, and oat (*Avena sativa* L.) kernels (*77, 78*). A "loosely" branched polysaccharide related to amylopectin, but with greater average chain lengths and higher β-amylolysis limits, was recovered from rye and wheat starches (*78*) and from *normal* maize starch (*79*). 'Hinoat' oat starch was found to contain 26% of an intermediate molecular weight, branched starch component following Sepharose 2B gel filtration chromatography, while wheat starch contained 10% of a similar fraction (*80*).

Another polysaccharide reported in small amounts in starch of nonmutant rye (*78*), wheat (*78*), and maize (*81*) is short-chain-length amylose. In *normal* maize starch, this linear polysaccharide has an average chain length of 58 (*81*).

2. Species and Cultivar Effects on Granule Composition

The amylose concentration in nonmutant reserve starch of higher plants varies, depending upon the species and cultivar from which the starch is isolated. Deatherage and co-workers (*82*) analyzed starch from 51 species and reported a range of 11–37% amylose. A summary of data from the literature for 23 species indicates a range of 11–35% amylose (*48*). Starches of six species of legumes investigated had amylose contents which varied from 29% to 37% (*83*).

Almost as much variation for amylose percentage has been observed among cultivars of a single species, as observed among species. For example, amylose percentage of starch ranges from 20% to 36% for maize (399 cultivars) (*82, 84*), from 18% to 23% for potatoes (493 cultivars) (*85*), from 21% to 35% for sorghum (284 cultivars) (*82, 86, 87*), from 17% to 29% for wheat (167 cultivars) (*82, 88*), from 11% to 26% for barley (61 cultivars including 5 genetic lines) (*89, 90*), from 8% to 37% for rice (74 cultivars) (*91–94*), and from 34% to 37% for eight cultivars of peas (*Pisum sativum* L.) (*46, 82*). Because of the variation in amylose concentration among species and among cultivars within a species, no average amylose percentage will be meaningful for nonmutant starches per se or for nonmutant starches of a given species. However, all nonmutant starches can be characterized as having more amylopectin than amylose.

Species and/or cultivar differences also are observed in other starch properties and in the properties of isolated amylose and amylopectin. To illustrate, purified amylose samples have been shown to differ in β-amylolysis limit and average degree of polymerization (*3, 46, 94*). Purified amylopectin samples also have been shown to differ in β-amylolysis limit, average length of unit chain, and viscosity (*3, 46, 48, 94, 95*).

3. Developmental Changes in Granule Composition

Increased amylose concentrations have been observed with increasing age of the tissue from which the starch was isolated for various plant species. Several authors (*39, 41, 96–98*) have reported increased amylose concentrations in maize endosperm during kernel development. For example, Tsai and co-workers (*98*) reported an amylose increase from 9% to 27% from 8 to 28 days post-pollination. The amylose concentration in potato starch increases from 12% in 0–1 cm tubers to 20% in 15–16 cm tubers (*40*). In starch from cassava (*Manihot utilissima*) roots harvested at various maturities, significant variation in amylose concentration (16–17%) has been observed; however, the net increase from 5 to 9 months of age amounts to only 0.3% (*99*). In starch of developing rice grains, amylose increases from 23% to 27% in cultivar 'IR8' from 4 to 39 days post-pollination (*100*) and from 30% to 37% from 3 days post-pollination to maturity in cultivar 'IR28', with 41% observed 7 days post-pollination (*44*). Amylose concentration has been shown to increase in developing wheat endosperms by various workers (*45, 101–105*); however, the amount of increase varies with the initial sampling date and the cultivar examined. In starch of developing barley kernels, amylose concentration increases from 16% to 28% from 9 to 46 days after anthesis (*106*), from 13% to 25% and from 14% to 26% for two cultivars during a 12-week period (*107*), and from 14% to 22% from 14 to 30 days after ear emergence, with the concentration remaining constant from 30 days until maturity (*43*). The amylose concentration in smooth-seeded pea starch increases from 15% in 2–6 mm peas to 37% in 11–12 mm peas (*46*). Developmental differences also are observed in other starch properties and in the properties of isolated amylose and amylopectin (*40, 46, 99, 100, 103–105*).

Similar increases in amylose concentration are observed as a function of increasing granule size when granules from a developing tissue at a single stage of development are separated into various size classes (Table II). Since the smaller granules have lower amylose percentages similar to those of nonseparated starch granules from younger tissue, the smaller granules presumedly are isolated from the physiologically younger cells present in the developing tissue (see Section III). This effect of granule size on amylose concentration is not applicable to the small starch granules found in mature wheat and barley endosperms, since the small and large populations have similar properties (*45, 108, 109*). In barley and wheat, these small granules are formed late in the growth cycle and represent a second discrete population and not immature granules (*108, 109*).

Because amylose concentration varies with maturity of the tissue, starches from tissues that do not reach maturity will be altered in their physicochemical properties from the corresponding mature starch.

4. Environmental Effects on Granule Composition

Growing conditions associated with different locations, years, planting dates, etc., can also affect the polysaccharide composition of nonmutant granules. Location and year of production and environmental conditions affect amylose concentration in rice (*91, 110*) with milled samples of 'IR8' rice ranging from 27% to 33% amylose (*111*). The amylose content of 'Selkirk' wheat grown at four locations ranged from 23.5% to 24.7% (*88*) and of 'Katahdin' potatoes

Table II

Amylose Content of Starch Granules of Various Size Classes Isolated from Maize Endosperm 36 Days Post-Pollination and from Intermediate Size (5–6 cm) Potato Tubers

Maize[a]		Potato[b]	
Granule size, μm	*Amylose, %*	*Average granule size, μm*	*Amylose, %*
Unfractionated	25.4	28 (Unfractionated)	17.2
10 to 20	26.4	37	19.5
5 to 10	23.0	28	18.0
Less than 5	20.5	16	16.7
		10	16.0
		7	14.4

[a] Data from Boyer and co-workers (*39*).
[b] Data from Geddes and co-workers (*40*).

grown at three locations from 21% to 24% (*82*). The amylose concentration in starch from 30 samples of 'Compana' barley representing different environmental and cultural practices ranged from 19% to 23% (*89*). Limited variation was seen for amylose percentage in maize starch from plants grown for 3 years in each of eight states. Year effects ranged from 26.2% to 26.8% averaged over locations, and location effects ranged from 25.5% to 27.7% averaged over years (*112*). Although present, environmental effects are not as large as those associated with cultivars or cultivar maturity.

V. Nonmutant Starch Granule and Plastid Morphology

1. Description

Reserve starch granules in higher plant tissues develop in organelles called amyloplasts (*1*). An amyloplast may contain one starch granule, or it may contain several granules, depending on the plant species or genetic mutant. When only one granule is produced in an amyloplast, such as in wheat, potato, barley, maize, pea, and others, it is called a simple granule (*9*). When two or more granules occur in one amyloplast, they form the parts (granula) of one compound granule. Such granules are often rounded at first but become angular as they pack together within the plastid. The granula of the compound granule are separated by a narrow layer of stroma (*9*). Examples of species having compound granules include rice, oats, cassava, sweet potato (*Ipomoea batatas* L.), sago (*Metroxylon* sp.), and dasheen (*Colocasia esculenta*). Badenhuizen (*9*) terms semi-compound granules as those that are initially compound, but become united by the deposition of a common surrounding layer of starch. Starch granules from the bulb of *Scilla ovatifolia* are semi-compound (*9*). Goering and co-workers (*113, 114*) described the presence of large "starch chunks" in seeds of *Amarenthus retroflexus* (pigweed). The starch chunks are composed of many small granules cemented together with amorphous starch (*114*) and can be considered as semi-compound granules.

Wheat, rye, and barley produce two types of granules. The first granules produced in the endosperm cells develop into large lenticular granules (*9*). However, about two weeks after initiation of the first granules, additional small granules are produced within evaginations of the original amyloplasts, and then these separate from the original amyloplasts by constriction (*115*). The secondary granules are generally spherical and remain small. Although two basic size classes exist, no abrupt size change occurs, and some intermediate size granules are present (*32, 116*). In mature barley kernels, the large granules constitute about 90% of the total starch volume, but represent only 12% of the total number of granules (*117*). Large starch granules in 17 wheat flours averaged 12.5% of the total granule number while accounting for 93.0% of the starch granule weight (*116*). Many of the large lenticular granules of wheat and barley have an equatorial groove or furrow (*118, 119*). Buttrose (*118*) suggested that starch synthesizing enzymes may be concentrated within the equatorial groove.

2. Species and Cultivar Effects on Granule Morphology

Size and shape of reserve starch granules are extremely diverse and are species specific (*9*). This diversity is illustrated in photographs of starch granules from over 300 species and varieties (*120*). Microscopic characteristics of various starches are also summarized by Moss (*121*) and Kent (*122*). (See also *This Volume*, Chaps. XXIII and XXIV.) The scanning electron microscope (SEM) has been used to show the topography of starch granules. Hall and Sayre published SEM pictures of various root and tuber starches (*123*), cereal starches (*119*), and 16 other miscellaneous starches (*124*). (See also *This Volume*, Chap. XXIV.)

Smaller granules are often found in tissues of species producing compound granules such as rice (*119*), malanga (*Xanthosoma sagittifolium*) (*123*), and cowcockle (*Saponaria vaccaria* L.) (*124*). As noted earlier, the secondary granules in barley, wheat, and rye remain small, with most less than 10 μm in diameter (*122*). On the other extreme, large granules in potato tubers can exceed 120 μm in diameter (*123*). Starch granules from most species are non-uniform in size. The amount of this variation can be seen, for example, by examining granule size distributions for wheat (*125*), rye (*125*), triticale (*125*), potato (*40*), barley (*43*), maize (*39, 41*), and dropwort (*Filipendula vulgaris*) (*126*).

Differences in average starch granule size in cultivars from a single species also have been reported. For example, average starch granule diameter ranged from 8.2 to 17.5 μm in 12 sorghum cultivars (*86*); from 17.8 to 25.6 μm in six triticale cultivars (*125*); and from 3.8 to 5.7 μm in 10 rice cultivars (*94*).

In addition to having an effect on amylose percentage (Section IV), varying environmental conditions also affected average starch granule diameter. Data on average starch granule size for rice cultivars grown in two different seasons (*92*) and dropwort grown at varying fertility levels (*126*) illustrate this effect.

In contrast to the species specific shape and size of reserve starch granules, transitory starch granules in chloroplasts appear similar in all species (*9*). In chloroplasts, the assimilatory starch granules are very small and disc-shaped (*1*).

3. Developmental Changes in Average Starch Granule Size

As tissues storing reserve starch mature, starch (Section III) and amylose (Section IV) concentrations increase. Similarly, average starch granule size increases with increasing age of the storage tissues. Such increases have been documented in maize (*39, 41*) and rice (*100*) endosperm, in potato tubers (*40*), and in pea cotyledons (*46*). This trend does not apply to average starch granule size in barley, wheat, and rye where a second population consisting of a large number of small granules are formed late in development. In kernels of these species, average granule size initially increases; however, as the small granules are formed, average granule size decreases (*43, 45*). In barley the maximum average granule diameter of 10.5 μm was observed 16 days after ear emergence (*43*).

4. Formation and Enlargement of Nonmutant Granules

Plant cells have several types of plastids, such as proplastids, chloroplasts, chloroamyloplasts, chromoplasts, and amyloplasts, depending on the species and tissues. Badenhuizen (*1*) contends that, although certain plastids do not form starch under natural conditions, they all can be induced to form starch by floating tissue pieces on a sugar solution. Although starch can be produced in a variety of plastids when supplied with sugar, chloroplasts and amyloplasts are the primary sites of starch accumulation in nature. Transitory starch is produced in chloroplasts. In this case, small granules are formed between the lamellae of the chloroplasts during periods of excessive photosynthetic assimilate production (*1*). During extended light periods, the number of small granules in a chloroplast increases, but granule size remains relatively small (*1*). Starch granules in chloroplasts are partially degraded at night to supply sugars for translocation. The control of starch synthesis and degradation in chloroplasts will be discussed in Section VI.

Reserve or storage starch accumulates in specialized leucoplasts called amyloplasts and occasionally in chloroamyloplasts. Amyloplasts are organelles bounded by a double membrane which develop from proplastids. Duvick (*127*) studied early plastid and starch development in maize endosperm cells. He described small filaments which developed knobs in maize endosperm cells. Then, according to him, starch granules formed within the filament knobs. Based on more recent electron microscopic studies (*1, 118, 128*), the filaments observed in living cells by Duvick (*127*) apparently are proplastids developing into amyloplasts (knobbed filaments). Proplastids and young amyloplasts in fixed sections have very irregular shapes and likely assume various shapes (amoeboid) in the living cell (*128*).

The inner membrane of young amyloplasts from barley (*118*) and maize (*128, 129*) have been shown to be extensively invaginated to form tubuli, stroma lamellae, or vesicles. Badenhuizen (*129*) observed that starch granules are formed in the "pockets" provided by the lamellar structure. He suggested that these pockets appear to be necessary for the initiation of starch granule formation, perhaps by promoting locally elevated concentrations of enzymes and substrates. The inner membrane of chloroplasts contain the specific translocators necessary for transfer of metabolites between the chloroplast stroma and the cytosol (*130*). It is assumed, based on the similarity between chloroplasts and amyloplasts, that the inner membrane of the amyloplasts also would function in regulation of metabolite transfer. Thus, the invaginations of the inner membrane noted above would effectively increase the area of the membrane and perhaps allow for more effective substrate transfer into amyloplasts (*118*).

The stroma (ground substance) of amyloplasts appears homogeneous by electron microscopic examination (*1, 128*). However, Badenhuizen (*9*) suggests that it contains inorganic and organic substances such as lipids, sugars, proteins, nucleotides, amino acids, nucleic acids, and inorganic ions. Liu and Shannon (*131*) confirmed the presence of many of these compounds plus various phosphorylated intermediates of gluconeogenesis in isolated starch granule preparations. (See also Section VI.) Presence of these various metabolites and proteins supports the suggestion that starch synthesis, not simply accumulation, occurs within the amyloplasts (*1*). Badenhuizen (*1, 9, 129*) has observed granular particles in the amyloplast stroma of tissue fixed in potassium permanganate. Although the granular structure observed may have been an artifact caused by the fixation procedure, it did show the presence of stroma material that accumulated in amyloplasts and then declined with the formation of the starch granule (*9*). The accumulation and decline of these particles also occurs during starch granule growth (*9*). Badenhuizen (*9*) called these particles coacervate droplets and concluded that they become attached to the periphery of the starch granule. Based on this observation, Badenhuizen (*1, 9*) concluded that starch molecules are produced in the amyloplast stroma and then the completed molecules become part of the growing starch granule. Shannon and co-workers (*132*) exposed maize plants to $^{14}CO_2$ and determined the distribution of ^{14}C in the amylose and amylopectin components of starch 1–6 h later. They found that the specific activity (^{14}C/mg of polysaccharide) of amylose and amylopectin increased at a similar rate, and that the radioactivity was distributed throughout the polysaccharide molecules. They concluded that these data supported Badenhuizen's (*1, 9*) suggestion that starch molecules are completely synthesized in the amyloplast stroma and then are deposited on the granule surface. Once the polysaccharides are part of the granule, there was no evidence of subsequent conversion of amylose to amylopectin (*132*). This is in contrast to conclusions drawn from long-term ^{14}C labeling studies of wheat starch (*133, 134*). In these studies, amylose appeared to be synthesized first, and then transformed into amylopectin. These differences may be due to the different species used or to the widely different sampling times.

VI. Polysaccharide Biosynthesis (see also Chapter IV, Section V)

1. Enzymology

Enzymes responsible for the synthesis of transitory starch in leaves and storage starch in seeds, tubers, etc., are generally considered to be the same in both types of tissues (*1*). Chloroplast starch is synthesized and accumulates during the light period when photosynthetic carbon fixation exceeds the demand for the assimilates by the plant, and it is hydrolyzed at night or when the assimilate demand exceeds current carbon fixation. Thus, synthesis and degradation or mobilization of transitory starch in chloroplasts are finely regulated. In storage tissues, starch synthesis is the predominant function of the amyloplast enzymes during tissue development. Thus, it is likely that the activity of amyloplast enzymes may be regulated by a mechanism different from that in chloroplasts.

Several pathways of starch synthesis have been proposed (*135–137*). Most of the evidence for these pathways comes from studies of isolated enzymes. With such *in vitro* studies, cellular compartmentation is destroyed, and one can only speculate as to the specific *in vivo* pathway of starch synthesis and the possible effector compounds which may be regulating synthesis. In spite of this shortcoming, such proposed pathways are important in focusing the investigator's attention on areas of research needed in the future. One such pathway is given in Figure 1. All the enzymes in this pathway have been measured in extracts from maize endosperm and other starch storing tissues.

For purposes of this discussion, sucrose is considered as the primary substrate for starch biosynthesis. In storage tissues, the UDPG formed from the action of sucrose synthase (*135, 138*) (Fig. 1, enzyme 6) can be utilized directly for starch synthesis by granule-bound starch synthase (Fig. 1, enzyme 9), or it can be converted to D-glucose 1-phosphate (G-1-P) by UDPG pyrophosphorylase (Fig. 1, enzyme 7) (*135*). Evidence that sucrose synthase may be significant in the production of substrates for starch synthesis *in vivo* comes from the observation that the maize endosperm mutant *shrunken* (*sh*), which causes a 40% reduction in kernel starch relative to *normal,* also causes a reduction in sucrose synthase activity (*139*). G-1-P can also be produced from sucrose via the combined action of invertase, hexokinase, phosphoglucoisomerase (for the D-fructose moiety), and phosphoglucomutase (Fig. 1, enzymes 1,2,4,5, respectively). The G-1-P produced by either mechanism can function as a substrate for phosphorylase (Fig. 1, enzyme 11) (*138, 140*). G-1-P also can serve as substrate for ADPG pyrophosphorylase (Fig. 1, enzyme 8) which yields ADPG. ADPG can serve as substrate for both granule bound and soluble starch synthase (Fig. 1, enzymes 9 and 10) (*141*). Thus, phosphorylase, starch-granule-bound starch synthase, and

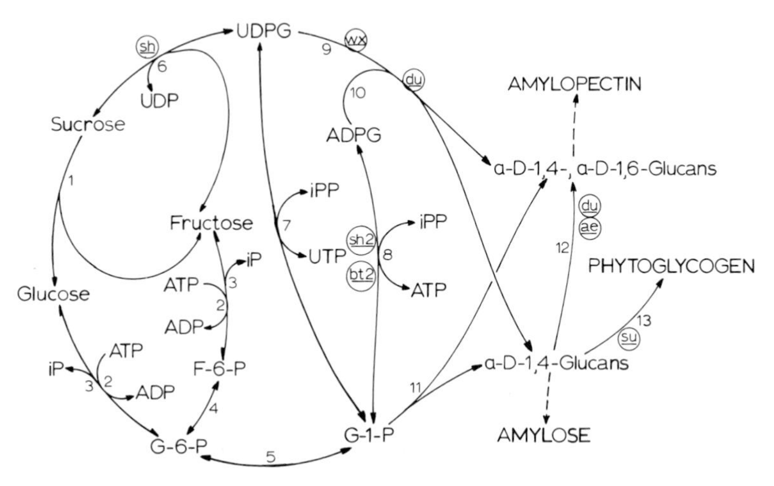

FIG. 1.—Hypothetical scheme for the conversion of sucrose to amylose, amylopectin, and phytoglycogen in *Zea mays* L. kernels [adapted from Boyer and Shannon (*296*)]. The endosperm mutant symbols are positioned at the enzyme sites shown to be affected by the different mutations (see Table X). Enzymes indicated by the small numbers are as follows: 1 = invertase (EC 3.2.1.26, β-D-fructofuranosidase); 2 = hexokinase (EC 2.7.1.1); 3 = hexose-6-phosphatase (EC 3.1.3.9, glucose-6-phosphatase); 4 = glucosephosphate isomerase (EC 5.3.1.9); 5 = phosphoglucomutase (EC 2.7.5.1); 6 = sucrose synthase (EC 2.4.1.13); 7 = UDP-glucose pyrophosphorylase (EC 2.7.7.9, glucose-1-phosphate uridylyltransferase); 8 = ADP-glucose pyrophosphorylase (EC 2.7.7.27, glucose-1-phosphate adenylyltransferase); 9 = starch granule bound starch synthase (EC 2.4.1.21); 10 = soluble starch synthase; 11 = starch phosphorylase (EC 2.4.1.1); 12 = Q-enzyme (EC 2.4.1.18, 1,4-α-D-glucan branching enzyme); 13 = phytoglycogen branching enzyme.

soluble starch synthase are all capable of catalyzing the *in vitro* synthesis of (1→4)-α-D-glucosidic bonds. The question may be asked whether one of these is the predominant enzyme for starch synthesis *in vivo*.

Prior to about 1960, phosphorylase was the only known enzyme capable of producing (1→4)-α-D-glucan polymers (*140, 142*). Multiple forms of phosphorylase have been reported in maize (*143, 144*), potatoes (*145*), *Vicia faba* (*146–148*), *Pisum sativum* (*149*), and *Phaseolus vulgaris* (*146*). Certain of these phosphorylase enzymes increase during the period of active starch synthesis and deposition, and it has been suggested that these phosphorylases may play an important role in starch synthesis (*143–148*). However, others have discounted phosphorylase as a starch synthetic enzyme because of the unfavorable inorganic phosphate (P_i) to G-1-P ratio in plant cells (*6*). Starch synthesis from G-1-P by phosphorylase requires a relatively high concentration of G-1-P. Although it has been shown that the concentration of P_i is many fold higher than that of G-1-P in whole cell homogenates (*131*), some investigators contend that at the site of synthesis, the amyloplast stroma, the P_i to G-1-P ratio may favor synthesis (*1, 3, 143*). In immature maize endosperm, the concentration of P_i is over 100 times higher than that of G-1-P (*131, 150*). Likewise, a recent analysis of components associated with starch granules isolated nonaqueously from maize kernels showed a similar very high P_i to G-1-P ratio (*131, 150*). Thus, in maize kernels, it has been concluded that *in vivo* starch synthesis by phosphorylase is highly unlikely (*131, 150*).

The discovery in 1957 of the role of nucleoside diphosphate sugars in carbohydrate metabolism stimulated greater interest in polysaccharide biosynthesis. Much of the early work came from L. F. Leloir's laboratory, and he summarized many of his contributions to the advancement of carbohydrate metabolism when he accepted the 1970 Nobel Prize in chemistry (*141*). Many studies of the starch synthases have been reported and recently reviewed (*6, 138, 151, 152, 286*).

Starch synthases from photosynthetic tissues and the soluble starch synthases from storage tissues are virtually specific for ADPG (*153*). Starch granule-bound starch synthases from storage tissues may utilize both ADPG and UDPG, but ADPG appears to be the preferred substrate (*154*). Tsai (*155, 156*) reported that the starch granule-bound starch synthase more actively transfers the D-glucose from UDPG and ADPG to amylose, while the soluble starch synthases preferentially transfer glucose to amylopectin *in vitro*.

The *waxy* (*wx*) mutants of rice (*157*) and maize (*156, 158, 159*) have only very low starch granule-bound starch synthase activities. Akatsuka and Nelson (*160*) suggested that the normal allele at the *wx* locus is either the structural gene for the starch granule-bound starch synthase or that it specifies receptor sites on the starch granule to which the enzyme binds. Recently, Nelson and co-workers (*159*) reported the presence of a minor starch synthase activity in *wx* starch granules. They speculated that a similar minor synthase activity is present in nonmutant starch granules also. The *wx* mutants which accumulate starch with approximately 100% amylopectin contain two soluble ADPG starch synthases (*156, 157, 161*).

Although granule bound starch synthases from storage tissues can utilize UDPG *in vitro*, it appears that ADPG is the predominant substrate for starch synthesis *in vivo*. Evidence for this comes from the observation that the maize endosperm mutants *shrunken-2* (*sh2*) and *brittle-2* (*bt2*) are lacking (*162*) or show very low (*163, 288*) ADPG pyrophosphorylase activity. However, *sh2* endosperms have normal levels of UDPG and CDPG pyrophosphorylases (*162, 164*). The fact that *sh2* mutant kernels are very low in starch and high in sucrose relative to *normal* kernels underscores the importance of ADPG as the primary substrate for starch synthesis (*6*).

Both phosphorylases and starch synthases catalyze the addition of D-glucose to the nonreducing ends of primer maltooligosaccharide molecules through $(1\rightarrow4)$-α-D-glucosidic bonds. Thus, branching enzyme activity is necessary to produce amylopectin. Haworth and co-workers (*165*) first reported such a transglycosylase, Q-enzyme (Fig. 1, enzyme 12), in 1944. This enzyme produces

(1→6)-α-D branches by cleaving a fragment from the linear chain and transferring it to the number six position of a glucose residue of the growing polymer (*152, 166, 167*). Multiple branching enzyme activities have been reported in spinach leaf (*168*), *smooth-seeded pea* (*Pisum sativum* L.) (*289*), and maize endosperm (*169*) extracts. When separated by DEAE-cellulose chromatography, one peak of branching activity is coincident with one of the soluble starch synthase activity peaks (*169*). This starch synthase–branching enzyme mixture, in the presence of high salt (0.5 *M* sodium citrate), is capable of catalyzing the synthesis of an amylopectin-like polyglucan in the absence of added malto-oligosaccharide primer. Schiefer and co-workers (*170*) electrophoretically separated extracts from developing maize endosperm and identified several bands of starch synthase activity. Based on the iodine staining properties of the product, they concluded that certain starch synthases are bound to branching enzyme and others are free from branching enzyme. They suggested that, in the cell, starch synthases may exist in a free form or as a branching enzyme–starch synthase complex. The *ae* mutant, which accumulates starch with a higher amylose percentage than *normal,* also has more free starch synthase relative to the branching enzyme-starch synthase complex (*170*). Also, the *su* mutant, which produces the more highly branched polyglucan, phytoglycogen, produces a brown staining band indicative of the presence of a phytoglycogen branching enzyme–starch synthase complex (*170*). Boyer and Preiss (*171*) chromatographically separated branching enzymes from *normal* and *ae* and found that one of the branching enzyme fractions (fraction IIb) is absent in *ae*. In view of the possible presence of a branching enzyme–starch synthase complex *in situ* and the stimulatory effect of citrate on the polyglucan synthetic activity of this complex *in vitro,* Boyer and co-workers (*96*) suggested that citrate or some similar metabolite may stabilize the branching enzyme–starch synthase complex *in situ.*

Amylose and amylopectin appear to be synthesized simultaneously *in vivo.* Shannon and co-workers (*132*) found that the amylose and amylopectin components of starch from developing maize kernels harvested one to 36 h after exposure of the intact plant to $^{14}CO_2$ had similar specific radioactivities at each sampling time. Furthermore, the radioactivity was distributed throughout the amylose and amylopectin molecules. These data fit a model suggested by Badenhuizen (*1*) in which the polysaccharides are synthesized in the matrix of the amyloplast followed by crystallization of the completed molecules onto the starch granule. This study (*132*) supports the suggestion of Schiefer and co-workers (*170*) that a branching enzyme–starch synthase complex may exist *in situ;* if so, free amylose would not be an intermediate in the synthesis of amylopectin.

Other enzymes of possible significance in starch synthesis or degradation have been reported and reviewed by Marshall (*8*). A disproportionating enzyme, D-enzyme (EC 2.4.1.25), was first reported in potatoes (*172*) and has been reported in sweet corn. The significance of this enzyme in starch synthesis is unknown, but it may function in the mobilization of starch to phytoglycogen in the *su* mutant. Plant debranching enzyme (R-enzyme, EC 3.2.1.9) may also be functioning in starch mobilization to phytoglycogen in *su* (*173*). However, in *normal* cereal kernels, it functions during germination (*8*), and it is not known to have a starch synthetic function. Erlander (*174*) proposed a mechanism for the synthesis of starch from glycogen, but Marshall (*8*) considers it an unlikely scheme. Amylosucrase (EC 2.4.1.4) is a bacterial (*Neisseria perflava*) enzyme that brings about the synthesis of a high-molecular-weight glycogen or amylopectin-type α-glucan from sucrose (*17*). This enzyme catalyzes the production of α-glucans from sucrose, with no indication of mediation by UDPG, ADPG, or G-1-P (*18*). The glycogen-like product of amylosucrase indicates either that the enzyme catalyzes the formation of both (1→4)-α-D and (1→6)-α-D linkages, or that it exists as a polymerizing–branching enzyme complex (*18*). Okada and Hehre (*18*) pointed out that the extent of *in vitro* D-glucose polymerization catalyzed by amylosucrase (250 parts D-glucose transferred for each part of preformed polysaccharide) is much greater than the D-glucose transfers reportedly obtained by the combined sucrose synthase–starch synthase system from rice grains (1 part transferred per 100 parts precursor) or sweet corn and beans (1 part transferred per 5000 parts precursor). Although amylosucrase has not been found in higher plants, Okada and Hehre (*18*) wonder whether the sucrose to starch conversion process in intact plants has been adequately defined. Additional studies are needed to establish whether an amylosucrase-type enzyme is functioning in the *in vivo* synthesis of starch in higher plants.

2. Compartmentation and Regulation of Starch Synthesis and Degradation in Chloroplasts

Photosynthesis occurs in organelles called chloroplasts. Chloroplasts develop from proplastids and are bounded by a double membrane (*1*). The outer membrane is freely permeable to small molecules, but the inner membrane is a functional barrier between the chloroplast stroma and the cytosol (*130*). The inner membrane is the site of specific metabolic transport systems. Specific translocators described in chloroplast envelopes are the phosphate translocator, dicarboxylate translocator (*175*), and an ATP translocator (*176*). The presence of two amino acid translocators in chloroplast membranes also has been suggested (*177*). The chloroplast membrane is essentially impermeable to free sugars such as D-glucose, D-fructose, and sucrose. Hexose phosphates and pentose phosphates cross the membrane very slowly (*178, 179*). Dihydroxyacetone phosphate (DHAP), 3-phosphoglycerate (3-PGA), and P_i move in and out freely via the phosphate translocator (*175*). The dicarboxylate translocator facilitates the rapid exchange of dicarboxylates such as malate, oxaloacetate, succinate, α-ketoglutarate, and fumarate, and related amino acids such as asparate and glutamate. ATP can cross the membrane by means of the ATP translocator, but this is not the major system for the transfer of energy from the chloroplast to the cytoplasm (*178–180*).

Chloroplasts which produce starch under natural conditions reduce carbon by the Calvin cycle of photosynthesis. The mesophyll cells of so called C_4 plants, such as maize, fix carbon by the C_4 pathway of photosynthesis (*181*), but they generally produce very little starch (*182, 183*). Rather, the four carbon acids produced in the mesophyll cells are transferred to the bundle sheath cells where they are decarboxylated and the resulting CO_2 is refixed by the Calvin cycle enzymes in the bundle sheath chloroplast (*181*). Bundle sheath chloroplasts of C_4 plants accumulate starch (*183*). In the Calvin cycle, the enzyme ribulose-bisphosphate carboxylase catalyzes the carboxylation and cleavage of ribulose 1,5-bisphosphate to yield two molecules of 3-PGA (*181*). The 3-PGA molecules thus produced by photosynthesis are (*a*) used as a substrate for the Calvin cycle, (*b*) unloaded from the chloroplasts via the phosphate translocator (*175*), and/or (*c*) utilized in the production of chloroplast starch. The triose phosphates transferred from the chloroplasts are used for the synthesis of sucrose in the cytoplasm by the combined activity of enzymes of gluconeogenesis and sucrose synthesis (*179*). Similar gluconeogenetic enzymes, components of the Calvin cycle, produce F-6-P which can then be converted by glucosephosphate isomerase into G-6-P, which in turn is converted to G-1-P by phosphoglucomutase. The G-1-P then can be utilized in the production of starch directly by phosphorylase, or more likely it is converted by ADPG pyrophosphorylase into ADPG, which is the substrate for the starch synthases. Although there is still disagreement among investigators as to the relative importance of the phosphorylases and starch synthases in the *in vivo* synthesis of starch in storage tissues, most researchers agree that in chloroplasts, D-glucose is polymerized by the bound and soluble starch synthases from ADPG and that phosphorylase functions in starch mobilization (*6*).

Chloroplast starch is degraded at night or during periods when assimilate demand exceeds current photosynthetic production (*6*). Mobilization of chloroplast starch involves enzymes which cleave the (1→4)-α-D-glucosidic bonds and the (1→6)-α-D-glucosidic branches. The released sugars or sugar phosphates (G-1-P) must then be converted to triose phosphates which can be transported to the cytosol via the phosphate translocator. Preiss and Levi (*6*) recently reviewed the various hydrolyses possibly involved in starch degradation and concluded that "little definitive work has been done on the localization or regulation of the enzyme of the first steps of starch degradation."

α-Amylase is generally accepted as one of the most important enzymes in the hydrolysis of storage starch granules, with β-amylase and phosphorylase being less important (*6*). Most recent studies indicate that phosphorylase is the primary enzyme involved in hydrolysis of leaf starch (*184*). Levi and Priess (*184*) suggested that the small amounts of maltose found in pea chloroplasts during starch degradation may have been produced from G-1-P and D-glucose by maltose phosphorylase rather than from the action of α-amylase or β-amylase. Although phosphorylase appears to be responsible for cleavage of the (1→4)-α-D-glucosidic bonds of starch, there is no definitive information on enzymes responsible for hydrolysis of the (1→6)-α-D-linkages in amylopectin. It is assumed that one of the debranching enzymes is involved (*185*).

As noted earlier, fine control is necessary to regulate the synthesis and degradation of transitory starch in chloroplasts. Current evidence indicates that starch synthesis is regulated by the production of the substrate ADPG by ADPG pyrophosphorylase. Priess and co-workers have extensively studied the pyrophosphorylases from bacteria, green algae, chloroplasts, and storage tissues (*186–188, 285*). In all photosynthetic tissues tested, ADPG pyrophosphorylase was activated by glycolytic intermediates and inhibited by P_i. An early product of photosynthesis, 3-PGA is the most effective activator of ADPG pyrophosphorylase from green algae and leaves (*186*). Thus, it is generally accepted that ADPG pyrophosphorylase functions as a regulatory enzyme which responds to positive (3-PGA) or negative (P_i) allosteric effectors (*6, 184, 186–188*). During periods of excess carbon fixation, 3-PGA increases in the chloroplasts which stimulates the production of ADPG from ATP and G-1-P resulting in an increased synthesis of starch (*186*). In the dark, the photosynthetic production of 3-PGA ceases, the level of ADPG declines, and starch synthesis ceases (*186*). Also in the dark, P_i in chloroplasts increases 30–50%, and the pH of the chloroplasts' stroma declines. Preiss and Levi (*6*) suggested that the lower pH may enhance activity of certain starch hydrolases, and the increased P_i and lower ADPG concentrations may stimulate starch hydrolysis by phosphorylase. However, they (*6*) add that definitive studies on the regulation of chloroplast starch degradation are lacking.

3. Compartmentation and Regulation of Starch Synthesis in Amyloplasts

Amyloplasts are organelles specialized for the accumulation of starch in storage cells. They develop from proplastids, as do chloroplasts, and are bounded by double membranes (*1*). Amyloplasts develop into chloroplasts under certain conditions (*189*) and vice versa (*1*). Thus, it is assumed that the nature of the amyloplast envelope may be like that of the chloroplast with a similar membrane transport system (*190*). The inner membrane of young amyloplasts from barley (*118*) and maize (*128*) has been shown to be extensively invaginated to form tubuli, stroma lamellae, or vesicles. Buttrose (*118*) suggested that if the inner membrane is the one limiting uptake, the increased area resulting from the invaginations of the inner membrane would allow for more rapid uptake of sugars or sugar phosphates into the amyloplast stroma. However, since we now know that chloroplast membranes are relatively impermeable to neutral sugars and hexose phosphates, it is more likely that carbohydrates may enter the amyloplasts as triose phosphates via the phosphate translocator (*131, 150, 190*).

Amyloplasts containing starch granules are extremely fragile, and attempts to aqueously isolate intact amyloplasts have been disappointing. Therefore, it has not been possible to directly study the membrane translocators in isolated amyloplasts. This problem has also made attempts to study the compartmentation of enzymes in amyloplasts very difficult. However, papers from Viswanathan's laboratory (*191–193*) reported the presence in isolated amyloplasts of all enzymes necessary to convert glucose and fructose to starch. Unfortunately, attempts to repeat Viswanathan's results in my laboratory were completely unsuccessful (Shannon, unpublished data). Williams and Duffus (*194*) isolated "amyloplasts" from barley endosperm using an aqueous media and studied the distribution of enzymes of carbohydrate metabolism and starch synthesis between the amyloplasts and the cytosol. They concluded that the production of G-1-P and ADPG from sucrose occurs in the cytosol. However, they presented no evidence that the isolated amyloplasts were intact and indeed contained the plastid stroma enzymes. Fishwick and Wright (*290*) were able to purify sufficient amyloplasts from potato (*Solarum tuberosum* L.) to characterize the lipids of the amyloplast envelope membrane. However, they pointed out that the yield of intact amyloplasts rarely exceeded 16%.

Because of the difficulty in isolating intact amyloplasts for studying enzyme compartmentation, we (*131, 150*) approached the question of what reactions of starch synthesis are occurring in amyloplasts by determining the metabolite composition of nonaqueously isolated "amyloplasts." For this work, maize endosperm slices were quick frozen and freeze dried, and the starch granules were isolated using glycerol and 3-chloro-1,2-propanediol (*150*). These granule preparations were relatively free of cytoplasmic and nuclear contamination, based on RNA and DNA content, respectively, but they contained the metabolites which are closely associated with the starch granules *in situ* and which become fixed to the granules during freeze drying. These components are thought to represent the *in situ* metabolites of the amyloplast. The starch granule preparation contained neutral sugars, P_i, intermediates of gluconeogenesis, organic acids, and amino acids, as well as adenosine and uridine nucleotides and nucleoside diphosphate sugars (Table III). Some of these constituents, such as the neutral sugars, malate and inorganic phosphate, accumulate to relatively high levels. Over 30% of the cellular malate and P_i and over 15% of the intermediates of gluconeogenesis from DHAP to G-1-P were recovered with the starch granules. The starch granule preparation was also relatively rich in the adenosine and uridine nucleotides. Approximately 10% of the cellular ADPG and UDPG were recovered in the starch granule preparation. The lower percentage of these two

Table III

Quantity and Percent of Various Cellular Constituents Associated with Nonaqueously Isolated Starch Granules[a]

Constituent	nmole/mg starch	%[b]	Constituent	nmole/mg starch	%[b]	Constituent	nmole/mg starch	%[b]
Sucrose	62.50	13	AMP	2.66	29	Lysine	1.21	21
Glucose	21.30	26	ADP	1.09	15	Histidine	0.25	25
Fructose	11.40	16	ATP	2.43	22	Arginine	T[c]	T
G-1-P	0.10	25	UMP	2.79	24	Aspartic acid + unknown	8.96	38
G-6-P	4.44	21	UDP	1.45	15	Threonine + serine	47.50	32
F-6-P	0.93	28	UTP	3.83	19	Glutamic acid	14.91	40
FDP	0.24	16	ADP-Glc	0.68	9	Proline	4.17	39
3-PGA	0.82	7	UDP-Glc	1.91	11	Cysteine	T	T
DHAP	0.20	27	NAD	0.80	10	Glycine	2.63	51
G-3-P	0.03	7	NADP	0.24	17	Alanine	28.11	43
PEP	0.34	19				Valine	2.48	32
P_i	15.15	34				Methionine	1.07	35
Citrate	0.85	6				Isoleucine	0.39	26
Malate	17.39	34				Leucine	0.57	27
Pyruvate	0.11	14				Tyrosine	0.48	29
						Phenylalanine	0.42	29

[a] Data adapted from Liu and Shannon (*131*).
[b] % of the total cellular constituents in the granule preparation.
[c] Trace amount.

primary substrates of starch synthesis in the amyloplasts may mean that their rate of utilization in starch synthesis is higher than their utilization in other pathways outside the amyloplasts. More than 20% of all free cellular amino acids were recovered in the starch granule preparation. Amounts of threonine plus serine, alanine, and glutamic acid were high in both the cellular and starch granule preparation, compared to amounts of the other free amino acids (Table III).

Based on the metabolite compartmentation in the glycerol-isolated starch granules, the mechanism of carbohydrate metabolism in corn endosperm cells given in Figure 2 was proposed (*131, 150*). In this scheme, the hexoses are converted to DHAP in the cytoplasm via glycolysis (Fig. 2, enzymes 2,4,14,16, and 17), the DHAP moves via the phosphate translocator into the amyloplast where it is converted into starch by the combined action of gluconeogenesis (Fig. 2, enzymes 17,16,15,4, and 5) and starch synthesis (Fig. 2, enzymes 7,8,9 and 10). Q-enzyme is also required for amylopectin production.

The high recovery of the cellular malate (34%) in the starch granule preparation may indicate the functioning of a dicarboxylate translocator in the amyloplast membrane. Citrate is excluded from chloroplasts (*175*) and also is

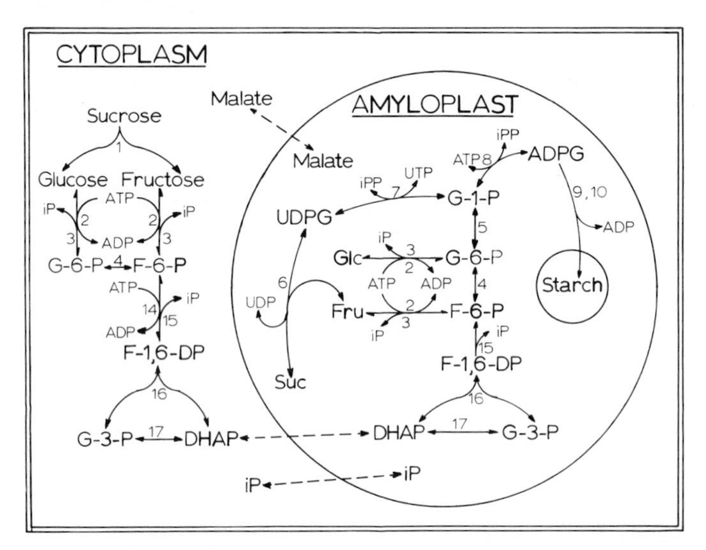

FIG. 2.—Proposed compartmentation of metabolites and enzymes of carbohydrate metabolism and starch synthesis between the cytoplasm and amyloplasts of maize endosperm cells. Adapted from Shannon (*297*). Enzymes indicated by the small numbers are as given in Figure 1 plus the following: 14 = 6-phosphofructokinase (EC 2.7.1.11); 15 = fructose-1,6-diphosphatase (EC 3.1.3.11, fructose-bisphosphatase); 16 = fructose-diphosphate aldolase (EC 4.1.2.13, fructose-bisphosphate aldolase): 17 = triosephosphate isomerase (EC 5.3.1.1).

excluded from amyloplasts, with only 6% of the total being recovered in the granule preparation (Table III).

Several conclusions can be drawn from an examination of the compartmentation of metabolites in the starch granule preparation. (a) The quantity of G-6-P is over 40 times greater than that of G-1-P. Thus, G-1-P is utilized almost as rapidly as it is produced, indicating that phosphoglucomutase (enzyme 5), the enzyme catalyzing the reversible interconversion of G-6-P and G-1-P, may be one of the rate limiting steps in starch synthesis. (b) The quantity of G-1-P is very low (0.1 nmoles/mg starch) compared to inorganic phosphate (15.2 nmoles/mg starch). Because the P_i to G-1-P ratio is over ten times higher than the equilibrium constant for phosphorylase, starch breakdown would be favored (*6*). Thus, these data do not support a starch synthetic function for phosphorylase in maize kernels. (c) The quantity of 3-PGA in the starch granule preparation is relatively low and represents only 7% of that in the whole cell. As noted earlier, 3-PGA is an allosteric effector compound which stimulates *in vitro* production of ADPG by chloroplast derived ADPG pyrophosphorylase (*186, 188*). The ADPG pyrophosphorylase (Fig. 2, enzyme 8) from maize endosperm is also stimulated by 3-PGA, but to a much lesser extent (*195*). From the results of this compartmentation study, it appears that 3-PGA is not functioning *in situ* as an effector metabolite for ADPG pyrophosphorylase in maize endosperm cells. (d) Citrate enhances gluconeogenesis by stimulating hexose diphosphatase (*187, 188*) (Fig. 2, enzyme 15); it has been shown to stimulate the *in vitro* activity of unprimed starch synthase (*161, 196*). Because malate rather than citrate accumulates in the amyloplasts (Table III), we suggest that, in the intact cell, malate may be functioning in the control of gluconeogenesis and in the stabilization of the starch synthase–branching enzyme complex. Boyer and Preiss (*197*) found that malate was 80% as effective as citrate in stimulating the *in vitro* activity of unprimed starch synthase.

VII. Mutant Effects

Maize is unique among higher plants relative to the number of mutants which have been identified and extensively examined. Several mutants effect the quantity and quality of carbohydrates in the triploid endosperm. Furthermore, these mutants often modify kernel development (*26, 27*), mature kernel phenotype (*198*), and starch granule morphology (*26, 71, 199*). The *shrunken* (*sh*), *shrunken-2* (*sh2*), *brittle* (*bt*), and *brittle-2* (*bt2*) mutants condition an accumulation of sugars at the expense of starch. The *shrunken-4* (*sh4*) mutant, which causes a reduction in starch accumulation, was originally thought to be a phosphorylase mutant (*200*), but it was later shown to affect the quantity of pyridoxal phosphate (*201*), thus, reducing the activities of several endosperm enzymes such as phosphorylase, which require pyridoxal phosphate. The *sh* mutant causes a reduction in sucrose synthase activity (*139*), while *sh2* and *bt2* each lack (*162*) or have very low levels of ADPG pyrophosphorylase (*163*). The genetic lesion associated with *bt* is unknown at this time.

Mutants affecting endosperm protein production include *opaque-2* (*o2*), *opaque-6* (*o6*), *opaque-7* (*o7*), *floury* (*fl*), *floury-2* (*fl2*), and *floury-3* (*fl3*). These mutants all cause a reduction in the prolamin (zein) fraction of storage proteins and a compensatory increase in albumin and globulin fractions (*5*). The mutant *soft starch* (*h*) causes a loose packing of starch in the endosperm cells, but has not been related to any major change in storage proteins (*202*), starch composition (*203*), or starch granule structure (*199*).

Because the primary focus of this section is mutant effects on polysaccharide composition, mutants in maize which cause changes from *normal* in amylose percentage and phytoglycogen production will be reviewed. The maize mutants in this group include *waxy* (*wx*), *amylose-extender* (*ae*), *sugary* (*su*), *sugary-2* (*su2*), and *dull* (*du*). These mutants alone and in various multiple mutant combinations have dramatic effects on kernel development, starch granule development and morphology, and polysaccharide composition (*4*). As will be pointed out subsequently, certain mutants cause the production of polysaccharides differing from standard amylose and amylopectin in molecular weight and degree of branching.

From 0 to 12 days post-pollination, little or no detectable differences are observed between *normal* and these mutants (except *su*) with respect to kernel and amyloplast development. The various mutant effects thus become expressed during the major period of starch accumulation. The mutant *su* differs from *normal* and the other mutants by initially producing compound granules (*27, 128*).

Saussy (*26*) made an extensive survey of mutant effects on maize endosperm and starch granule development at 16 and 27 days post-pollination. 'IA5125' versions of *normal, ae, du, su,* and *wx* singly and in double, triple, and quadruple combinations (except *su wx* and *ae su wx*) were studied. *Normal* and all mutant genotypes exhibited the major gradient of starch granule development from the kernel base (least mature) to the central crown region of cells (most mature) described in Section III. Two basic types of minor gradients of starch granule development were observed. The type I minor gradient is similar to that described for *normal,* with an increase in cellular maturity inward from the peripheral cells adjacent to the aleurone layer (*24, 27*). The type II minor gradient is similar to type I along the peripheral endosperm and toward the interior for a few cell layers (variable with the genotype), but then an abrupt decrease in the volume of cellular inclusions occurs. These differences in minor gradients and other specific mutant effects will be noted in discussion of the mutants.

For convenience, the effects of the various mutants and mutant combinations including information on kernel phenotype, starch granule size and physical properties, WSP concentration, amylose percentage, and relative sizes and iodine binding capacity of polysaccharides following separation by Sepharose column chromatography are summarized in Tables IV, V, VI, VII, and VIII. Current information on the specific mutants singly and in combination and information on similar mutants in other species when such mutants are known is presented below.

1. Waxy

Waxy (*wx*) or *glutinous* (*gl*) loci have been identified in maize, sorghum, rice, barley, and Job's tears (*Coix lachryma-jobi*) (*209, 210*). These mutants produce starch granules in the endosperm and the pollen which stain red with iodine and which contain nearly 100% amylopectin; however, starch granules in other plant tissues contain both amylose and amylopectin and stain blue with iodine (*210*). Red-staining starches have been reported in other plant species, but these have not been characterized (*210*). Floridian starch found in the red algae also resembles amylopectin and lacks amylose (*10, 16*).

Phenotypically, *wx* kernels are full and often appear opaque (Table IV) (*198, 210, 211*). Starch and dry weight production in *wx* kernels are equal to that in *normal* kernels and increase at similar rates (*41, 44, 87, 96, 212, 213*). Sugar and WSP (Table VI) levels are also similar to *normal* in immature (*204, 208*) and mature kernels (*205*).

The *wx* mutant is epistatic to all other known mutants relative to the lack of accumulation of amylose (*4, 71*). Multiple mutants containing *wx* and *ae* have been reported to produce amylose (Table VII); but as will be pointed out in discussing the *ae wx* genotype, this apparent amylose, as measured by iodine

Table IV

Mature Kernel Phenotype of Normal and Selected Single, Double, Triple, and Quadruple Recessive Maize Genotypes[a]

Genotype	*Gene expressed*[b]	*Kernel phenotype*[c]
Normal	None	Translucent
wx	*wx*	Opaque
ae	*ae*	Tarnished, translucent, or opaque; sometimes semi-full
su	*su*	Wrinkled, glassy; S.C.[d]: Not as extreme
su 2	*su 2*	Slightly tarnished, often etched at base
du	*du*	Opaque to tarnished; S.C.: Semi-collapsed, translucent with some opaque sectors
ae wx	C	Semi-full to collapsed, translucent or glassy, may have opaque caps; S.C.: Slightly fuller, etched, translucent to glassy
ae su	C	Not quite as full as *ae*, translucent (tarnished in S.C.), may have opaque caps
ae su 2	*ae*	Translucent or opaque, etched base
ae du	C	Translucent, not as full as *ae*; S.C.: Etched, translucent, or tarnished
du su	*su*	Wrinkled, glassy (Duller than *su*); S.C.: Extremely wrinkled, glassy
du su 2	C	Translucent, etched
du wx	C	Semi-collapsed, opaque; S.C.: Shrunken, opaque
su wx	*su*	Wrinkled, glassy to opaque
su 2 wx	*wx*	Opaque, often etched
su su 2	*su*	Wrinkled, glassy
ae du su	*su*	Wrinkled, translucent; S.C.: Slightly wrinkled, translucent
ae du su 2	C	Semi-collapsed, translucent
ae du wx	C	Shrunken, opaque to tarnished; S.C.: Semi-collapsed, tarnished
ae su su 2	C	Partially wrinkled, translucent to tarnished
ae su wx	C	Semi-collapsed, opaque to translucent; S.C.: Etched, semi-full, translucent
ae su 2 wx	C	Etched, semi-full or wrinkled, translucent
du su su 2	*su*	Wrinkled, glassy
du su wx	*su*	Wrinkled, glassy
du su 2 wx	C	Semi-collapsed, opaque, etched
su su 2 wx	*su*	Wrinkled, glassy
ae du su wx	C	S.C.: Etched, semi-full, translucent to tarnished

[a] Adapted from Garwood and Creech (*198*).

[b] If one gene is responsible for the phenotype, that gene is listed. "C" signifies a complementary expression giving a new phenotype differing from the phenotypes of the stocks possessing the individual genes.

[c] Kernels approach full size unless indicated as semi-full, semi-collapsed, shrunken, or wrinkled.

[d] S.C. means the phenotype observed in sweet corn inbreds.

Table V

Mean Starch Granule Size, Birefringence End-Point Temperature (BEPT), and X-Ray Diffraction Pattern of 14 Maize Genotypes 24 Days Postpollination[a]

Genotype	Granule size, μm		BEPT	X-Ray pattern
	Minimum	Maximum		
Normal	7.99 fgh	8.53 jkl	70.3 bcdef	A
wx	8.61 gh	9.41 l	74.3 ef	A
ae	5.56 bcd	6.32 defg	97.7 h	B
su	3.06 a	3.52 a	69.0 abcde	A
su 2	7.68 fg	9.14 kl	63.7 a	A
du	5.19 bcd	5.98 cdef	70.7 bcdef	A
ae wx	5.67 bcd	6.03 cdef	82.3 g	B
ae su	5.34 bcd	8.20 ijk	88.0 g	B
ae su 2	5.57 bcd	6.46 defg	87.7 g	B
ae du	5.42 bcd	6.54 efgh	73.3 def	B
du su	3.11 a	3.85 ab	73.7 def	A[b]
du su 2	5.97 cde	8.79 jkl	63.3 a	A
du wx	6.18 de	6.86 fgh	76.7 f	A
su su 2	2.56 a	2.98 a	67.3 abcd	—

[a] Adapted from Brown and co-workers (*199*). Means followed by the same letter are not significantly different at the 1% level using Duncan's Multiple Range Test.

[b] Weak crystalline A pattern.

binding, is due to a loosely branched polysaccharide having long external chains (*69, 70, 214*). Owing to the lack of amylose, *wx* granules stain reddish-purple with iodine, although some *wx* granules have blue-staining cores (*1, 207, 210*).

Starch granules from maize, sorghum, and rice kernels homozygous for *wx* have been reported to have from 0–6% amylose (*82, 86, 87, 91, 94, 215*). This apparent amylose content may be due to the measurement technique used, to the effect of non-waxy starch granules from maternal tissue, to differences in the degree of branching, or to the presence of some linear material as suggested by blue-staining cores. If present, this linear material is minimal, for no amylose peak is observed in chromatographic profiles of *wx* starch (*69, 70, 71, 216*). These differences in apparent amylose content involve both cultivar and environmental effects as previously described for *normal* starch (*82, 86, 87, 91, 94, 215*). Alleles at the *wx* locus can also vary in amylose percentage with *wx-a* having 2–5% amylose compared to 0% in *wx-Ref* (*217, 218*). *Waxy* amylopectins vary in β-amylolysis limit, average chain length, and molecular size (*94, 219, 220*) as previously described for *normal* starch. The *normal* (*Wx*) allele is

Table VI

Water-Soluble Polysaccharide (WSP) Concentration in Immature and Mature Kernels of 26 Maize Genotypes[a]

	Immature		*Mature*
Genotype	*10% ETOH*[b]	*$HgCl_2$*[c]	*10% ETOH*[d]
Normal	3	0	4
wx	3	0	7
ae	4	0	6
su	28	25	19
su 2	2	—[e]	4
du	2	0	5
ae wx	6	0	5
ae su	4	0	5
ae su 2	4	—	6
ae du	7	0	5
du su	29	18	18
du su 2	3	—	4
du wx	11	2	8
su wx	29	19	17
su 2 wx	3	—	5
su su 2	31	—	19
ae du su	16	7	9
ae du su 2	10	—	8
ae du wx	4	Trace	6
ae su su 2	11	—	9
ae su wx	12	7	10
ae su 2 wx	6	—	7
du su su 2	35	—	32
du su wx	38	28	30
du su 2 wx	14	—	8
su su 2 wx	39	—	18

[a] All mutants were in a genetic background related to the single cross W23 × L317. Data expressed as percentage of kernel dry weight.

[b] Data adapted from Creech (*204*). WSP extracted with 10% ethanol.

[c] Data adapted from Black and co-workers (*13*). WSP extracted with aqueous $HgCl_2$; an aliquot hydrolyzed with H_2SO_4 and the increase in reducing sugar determined.

[d] Data adapted from Creech and McArdle (*205*). WSP extracted with 10% ethanol.

[e] Genotype not included in study.

Table VII

Apparent Amylose Percentages of Various Maize Genotypes Determined Using Iodine Binding Procedures[a]

Genotype	Reference		
	Kramer et al. (206)	Seckinger and Wolf (207)	Holder et al. (208)
Normal	27	27	29
wx	0	—[b]	<1
ae	61	57	60
su	29	—	33
su 2	40	28	38
du	38	35	34
ae wx	15	26	26
ae su	60	—	51
ae su 2	54	45	56
ae du	57	50	45
du su	63	13	40
du su 2	47	—	46
du wx	0	—	2
su wx	0	—	0
su 2 wx	0	—	0
su su 2	55	30	41
ae du su	41	—	28
ae du su 2	48	23	37
ae du wx	—	—	2
ae su su 2	54	—	31
ae su wx	13	4	14
ae su 2 wx	—	28	28
du su su 2	73	—	44
du su wx	0	—	0
du su 2 wx	0	—	0
su su 2 wx	0	—	0

[a] Colorimetric measurement of starch–iodine complex used to estimate apparent amylose percentages. Genotypes were not incorporated into an isogenic background.

[b] Genotype not included in study.

not completely dominant to the *wx* allele, and amylose percentage is reduced by several percent in the *Wx wx wx* endosperm genotype (*96, 221–224*).

The increase in average granule size during kernel development of *wx* maize and the final granule morphology of *wx* granules are similar to that of *normal* (*41, 199*). Also, as reported for *normal,* the average size of *wx* granules varies with the cultivar (*86, 94, 215*) and environmental conditions (*215*). Birefringence of *normal* and *wx* granules is similar; however, iodine staining reduces the

Table VIII

Amylose Percentage of Starch from 16 Maize Genotypes Determined Following Sepharose 2B-CL Column Chromatography and the Peak Fraction's Absorbance Maxima and Extinction Coefficients (E) of the Polysaccharide–Iodine Complex[a]

Genotype	*Amylose, %*	*Peak fraction (tube no.)*	*Maximum absorbance, nm*	*E at 615 nm*
Normal	29	14	510–540	22
		31	640	121
wx	0	14	470–480	24
ae	33	13	540–550	43
		33	600	92
su	65	13	480–530	28
		28	640	97
du	55	14	480–500	35
		28	640	104
ae wx	0	13	530–540	49
		21	530–540	39
ae su	28	13	540–550	48
		21	540–560	51
		29	640	95
ae du	47	14	530–540	40
		31	640	90
du su	70	14	540–570	47
		29	640	92
du wx	0	14	470–480	20
su wx	0	14	495–505	32
ae du su	31	14	530–550	49
		23	540–560	40
		31	640	83
ae du wx	0	14	460–500	33
		24	460–500	22
ae su wx	0	13	540–550	43
		21	530–540	34
du su wx	0	14	460–480	17
		25	450–470	14
ae du su wx	0	13	<400	25
		24	450–470	19
su phytoglycogen	0	13	≤ 400	16
		22	≤ 400	6
Amylose–amylopectin 1:1 mixture	51	14	470–530	31
		31	640	103

[a] Maize genotypes converted to the IA5125 sweet corn inbred background. Date adapted from Yeh and co-workers (*71*).

intensity of birefringence of *normal* granules, but has little effect on that of *wx* granules (*71*). The BEPT of *wx* granules is similar to *normal*, and both have A-type x-ray diffraction patterns (Table V).

Kernels of *wx* have the major and minor (type I) developmental gradients characteristic of *normal* kernels (*26, 27*). Saussy (*26*) observed the presence of occasional starch granules surrounded with phytoglycogen; however, this was due to the sweet corn background used in her study and not to the *wx* gene itself. Simple, spherical starch granules are initially produced in *wx* kernels, and these increase in size and, in many cells, become irregular in shape due to extensive cell packing (*26, 27*). Boyer and co-workers (*27*) reported that all starch granules were initiated at essentially the same time and that there was no evidence of additional granules (secondary granule initiation) being initiated later in development. Saussy (*26*) reported secondary granule initiation in *wx* as well as in *normal* and most other mutant genotypes. Boyer and co-workers (*27*) used *wx* in a dent background, while Saussy (*26*) used a sweet corn background.

As noted in Section VI, *wx* mutants of maize (*158*) and *gl* mutants of rice (*157*) lack the major starch granule-bound starch synthase activity. However, *wx* maize granules do contain a minor granule-bound ADPG–starch synthase (*159*) and two soluble ADPG–starch synthases (*161*).

2. Amylose-Extender

Mutant genes, which cause an increase in apparent amylose percentage in starch of pea cotyledons and of maize and barley pollen and endosperm, have been reported (*225*). High-amylose maize is homozygous for the *ae* gene, and the mature kernels are sometimes reduced in size (Table IV). High-amylose peas are homozygous for the *rugosus* (*r*) gene and have a wrinkled, collapsed phenotype (*226*), while high-amylose barley kernels appear similar to *normal* and are homozygous for the *amylose-1* (*amy-1*) gene (R. F. Eslick, personal communication). Starch and dry weight production are reduced and sugars increased in these high-amylose genotypes compared to nonmutant kernels or seeds. The rate of starch increase is also slightly reduced (*46, 96, 97, 106, 204, 208*). Apparent amylose content increases with increasing maize and barley kernel age and with increasing pea diameter, reaching values of 45–69% (*46, 96, 97, 106*). The *normal* alleles are not completely dominant to the recessive *ae* and *amy-1* alleles, since two doses of the recessive allele (i.e., *Ae ae ae*) result in a 2–8% increase in apparent amylose content compared to the *normal* genotype which lacks the recessive allele (*96, 223, 227, 228*). Extensive variation in apparent amylose concentration occurs compared to the amount observed in *normal* genotypes (see Section III). For example, variation is observed for amylose concentration as a function of the maize inbred crossed with *ae* (*229–233*) with a range of 36.5–64.9% reported (*230*). Minor modifying genes in the various inbreds have been proposed as a possible cause of the variation (*229–233*). Such modifying genes have been utilized to produce a series of hybrids which differ in apparent amylose concentration from 50% to 75% (*3*). Differences also have been associated with *ae* alleles arising as independent mutants, with *ae-i1* and *ae-i2* conditioning lower amylose percentages than five other alleles when compared in two isogenic backgrounds (*234*). Amylose percentage also varies 17% among wrinkled seeded pea cultivars (*46, 82*).

Not only does variation occur between *ae* inbred lines and hybrids (i.e., background or modifier effects) and *ae* alleles, but an 8–14% range also exists within an inbred line homozygous for *ae* and grown at a single location in a single year (*230, 233*). This is likely due to a combination of error in amylose determination, segregation of modifier genes which were not yet homozygous (*235*), and the microenvironment of each plant.

Significant differences in *ae* amylose percentage result from both location and year of production with the effect of location considerably greater than that of years (*112, 236*). Later planting dates are associated with higher *ae* amylose percentages; however, poorer agronomic performance negates the value of the increase (*237*). Minor mechanical damage to the plants has little effect on amylose concentration with only a 1–3% reduction caused by extreme leaf defoliation (*238*).

Variation in amylose concentration is also observed between butt, center, and tip zones within individual *ae* ears with the highest percentage in kernels taken from the butt of the ear and the lowest percentage in kernels from the tip zone (*239*). In addition, when the endosperm tissue is divided into tip, middle, and crown portions, the middle portion is highest in amylose percentage within each ear zone (*239*).

The amylose percentages presented above are all based on "blue-value" tests. Yeh and co-workers (*71*) recently employed column chromatography using Sepharose 2B-CL to fractionate the starch polysaccharides from mature *normal*, *ae*, and 14 other maize endosperm mutants. Column fractions were reacted with iodine, and the absorptions at 560 and 615 nm were determined. Any fraction having a higher absorption at 560 than at 615 nm was classified as amylopectin; and, conversely, fractions with a higher absorption at 615 nm were considered to be amylose. Based on the elution profile, she found considerable carbohydrate intermediate in size (up to fraction 25) between amylopectin and amylose, which absorbed higher at 560 than at 615 nm. These fractions appear similar to the loosely branched amylopectin described for *ae wx* starch (*69, 214*) and suggested for *ae* amylopectin (*240, 241*). Whistler and Doane (*79*) isolated such a polymer from *ae* starch. Low-molecular-weight polymers similar to the short chain amylose described by Banks and Greenwood (*3*) eluted near the end of the profile; these polymers had a higher absorbance at 560 nm than at 615 nm and by definition were not included as amylose. Based on Yeh's calculation, *ae* starch contains 33% amylose (Table VIII) (*71*). If the low-molecular-weight polymers eluting after amylose are included, the amylose percentage increases to 41%, which is still much lower than amylose percentages based on blue-value measurements (Table VII). Similar low amylose percentages are obtained following gel filtration after debranching by isoamylase (*216*). Because the long external chains of loosely branched polysaccharides complex iodine (*69*), they contribute to the estimate of amylose percentage as measured by the blue-value procedure. Although the amylose percentage based on Yeh's (*71*) and Ikawa and co-workers' (*216*) procedures may not be exact, they probably represent a much closer estimate of the true amylose content of *ae* starch than do blue-value estimates.

Starch granule preparations from *ae* kernels generally contain two distinct geometric forms, spherical and irregular (*26, 36, 97, 225, 242*). Irregular granules vary in shape, but often are elongated and nonbirefringent. Sometimes spherical granules also develop elongated extensions of amorphous, nonbirefringent starch (*39*). The proportion of irregular granules in *ae* starch has been reported to vary from 0% (*26, 71, 97*) to 100% (*243*) and was shown to increase during kernel development (*27, 97*), with increasing apparent amylose content (*97, 225*) and with physiological age of the cells (*36*). The proportion of irregular granules depends on the completeness of starch isolation, on the classification criteria used (*242*), and on the inbred background (*26, 71*). Average *ae* starch granule size increases with kernel development; however, *ae* granules are smaller than *normal* at all developmental stages (*39, 199*). Boyer and co-workers (*36*) reported a two-phase growth pattern consisting of spherical granule initiation and growth followed by a secondary initiation of irregular granules. Sandstedt (*244*) also indicated that *ae* irregular granules are surrounded by spherical granules within an endosperm cell. There is considerable cell to cell variation in the presence and proportion of irregular granules (*36, 244*); but in kernels harvested 36 days post-pollination, the proportion of irregular granules is highest in the more mature endosperm cells (*36*).

Inbred background apparently influences the morphology of the irregular granules produced by *ae*. The elongated amorphous granules noted above occur when the *ae* mutation is incorporated into dent backgrounds (*27, 36*). However, when *ae* is incorporated into the sweet corn inbred 'IA5125' and the *su* mutant deleted, no elongated amorphous granules are found at 16 or 27 days after pollination (*26*) or at maturity (*71*). Secondary granule initiation does occur, and the irregular granules are more blocky in appearance (*26*). Kernels of *ae* have the major developmental gradient and type I minor gradient characteristic of *normal* (*26, 27*).

Starch granules from *ae* kernels have a much higher BEPT than *normal* or the other mutants (Table V). Also, based on 14 genotypes studied, the B-type x-ray diffraction pattern appears to be unique to *ae* and the *ae* containing genotypes (Table V).

In high-amylose barley (*228*) and wrinkled-seeded peas (*46*), average granule size is less than in *normal* with high-amylose starch granules smaller at all stages of development. High-amylose barley starch granules are more irregular than *normal* (*228*). High-amylose pea starch granules often develop a very irregular system of fissures, making them superficially resemble compound granules (*1, 46, 124*).

Based on the accumulation of loosely branched amylopectin in *ae* (*240, 241*) and *ae wx* (*69, 214*) genotypes, Boyer and co-workers (*96*) suggested that the *Ae* allele affects the degree of branching of amylopectin by controlling the quantity of an effector, at the site of starch synthesis, which stabilizes a starch synthase–branching enzyme complex. According to this suggestion, the enzyme complex is needed for production of *normal* amylopectin. The *ae* allele-gene product may block effector accumulation resulting in increased free branching enzyme and free starch synthase. This is in agreement with the observations of Schieffer and co-workers (*170*) who showed with zymograms that starch synthases of *ae* have increased activity in the free synthase bands and a decreased activity in the starch synthase–branching enzyme complex bands. Boyer and co-workers (*96*) suggested that citrate may function as the effector compound which stabilizes the enzyme complex *in situ*. However, malate rather than citrate was shown to accumulate in *normal* maize amyloplasts, and Liu and Shannon (*131, 150*) suggested that malate may be the effector compound functioning *in situ*. Although the *ae* mutant is clearly influencing the efficiency of branching, Boyer and co-workers (*96*) point out that their hypothesis for *ae* action is only a suggestion and that the assignment of a positive gene product to *ae* awaits direct evidence.

Recently, Boyer and Priess (*169*) reported the presence of three forms of branching enzyme in extracts from *normal* maize endosperm. One component, fraction IIb, coelutes with the citrate-stimulated, "unprimed" starch synthase. When the branching enzymes from *ae* kernels were similarly separated, the total activity was only 20% of *normal*, and there was a complete absence of branching enzyme fraction IIb (*171*). Based on these results, Nelson (*5*) concluded that the missing branching enzyme activity in *ae* could explain the effects of *ae* on the polysaccharides formed. More recently, Boyer and co-workers (*291*) showed that, with increasing doses of the recessive alleles, *ae* in maize and r_a in peas, there was an increase in the linearity of amylopectin produced. In maize, this *ae* effect on amylopectin was apparently due to the deficiency of branching enzyme IIb (*171*). Hedman and Boyer (*292*) reported a near-linear relationship between increasing dosage of the dominate *Ae* allele and branching enzyme IIb activity and suggested that *ae* is the structural gene coding for branching enzyme IIb. The high-amylose wrinkled pea, Progress #9, has greatly reduced levels of branching enzyme (*289*). Thus, the presence of modified amylopectin in the "high-amylose" mutants in both species owes to the reduced activity of branching enzyme.

3. Sugary

The standard sweet corn of commerce is homozygous recessive for *su*. The main effect associated with *su* mutants in maize and sorghum is the synthesis and accumulation of phytoglycogen to 25% or more of the kernel dry weight (Table VI) (*13, 205, 245, 246*).

Phytoglycogen consists of α-D-glucosyl units linked by (1→4) and (1→6) bonds. Its structure is similar to that of amylopectin, except that phytoglycogen is more highly branched and is extracted as the major component of the water-soluble polysaccharide (WSP) fraction in sweet corn (*13, 247, 248*). Mature *su* sorghum and maize (Table IV) kernels are wrinkled and have reduced amounts of dry matter (*205, 208, 212, 249*). Their sugar content is higher and their starch content much lower than in *normal* maize (*204, 205, 208, 250–252*) or sorghum (*246, 249, 253*). Starch concentration in *su* maize expressed as a percentage of dry weight increases until 15–20 days post-pollination, and then remains constant (*41, 204, 251, 254*). Total polysaccharide concentration, however, increases through 30–40 days post-pollination due to increases in phytoglycogen concentration, with total carbohydrate percentage approaching that in *normal* kernels (*41, 204, 251, 254*). At maturity, the total carbohydrate percentage is equal to (*252, 254*) or less than that in *normal* kernels (*205, 246*), depending on the genetic background. However, absolute amounts are reduced, reflecting the reduced dry matter in *su* kernels. In general, maize kernels from dent lines homozygous for *su* contain more sugar and less phytoglycogen and starch than kernels of a sweet corn line (*204, 205*).

The amylose percentage of starch, as measured by iodine binding, from *su* kernels averages somewhat higher than the percentage from *normal* kernels (Table VII), and the amylose percentage has been reported to increase with advancing kernel age (*41, 255*). Although the data in Table VII represent data from several studies over several years, other investigators have reported widely different amylose percentages in *su* starch (*71, 206, 208, 251, 256–258*). These have varied from 0% amylose (*251*) to 65% amylose (*71*). The 65% amylose, reported by Yeh and co-workers (*71*) (Table VIII), was based on calculations from a Sepharose separation of the starch polysaccharides. Similarly, the amylose percentage of starch from *su* sorghum kernels varied from near *normal* (*86*) to somewhat higher than *normal* (*253*). The widely differing amylose percentages probably relate to kernel age and methods of starch isolation and measurement. Possible reasons for these discrepancies will be discussed in more detail after considering the morphological changes occurring in the developing *su* kernel.

The morphology and development of *su* maize plastids and kernels is well established (*1, 9, 23, 26, 27, 128*). Immediately prior to initiation of starch synthesis in an endosperm cell, the proplastids collect around the nucleus as in *normal* (*26, 27, 128*). From one to several small starch granules then form in each amyloplast (*128*). During development, granules enlarge only slightly (*41, 199, 255*), reaching an average diameter of 3.6 μm at maturity (*41*). However, within the more mature cells of the central crown region, the initially formed starch granules are degraded and are replaced with phytoglycogen (*26, 27, 128*). Thus, within developing kernels, plastid types range from amyloplasts with compound starch granules, to amyloplasts containing phytoglycogen and a few small starch granules, to amyloplasts containing phytoglycogen plus many very small starch granules and/or granule fragments, and to plastids containing only phytoglycogen (*26, 27, 128*). The cells with the different plastid types are located in specific regions of the endosperm and apparently are related to the physiological age of the cells with phytoglycogen plastids being in the most mature cells (*23, 26, 27*). The *su* kernels go through the major and minor developmental sequence characteristic of *normal,* except that, as the cells mature, they fill with phytoglycogen rather than starch (*26, 27*). Later in kernel development, phytoglycogen plastids in some cells appear to rupture (*26, 27, 128*). The released material, thought to be phytoglycogen, was described as a dense-staining "rosette" material (*128*) similar in appearance to animal glycogen (*259*). Thus, phytoglycogen appears to accumulate in both plastids and the cytoplasm, with that in the cytoplasm possibly arising from ruptured plastids.

Owing to the small size of *su* starch granules (Table V) and their partially degraded reminants, difficulties are encountered in isolating a starch sample which is representative of that in the total population of cells found in the endosperm. With procedures involving starch-tabling, up to 90% of the starch can be lost (*260*), and similar losses of the smaller particles would be expected with isolation procedures based upon low-speed centrifugation or gravity sedimentation. Particles staining both red and blue with iodine have been observed *in situ* and in isolated granules (*26, 260, 261*). Thus, the granules differ from each other, and loss of small granules and granule particles probably results in isolated granule preparations which are not representative of the total granule population. Differences in isolation procedures used by different investigators may explain some of the discrepancy in amylose percentages reported for *su* starch. The amylose percentage in the starch also is affected by the completeness of phytoglycogen removal. Polysaccharide particles smaller than starch granules have been observed in *su* kernels (*26*) and also have been isolated from immature kernels (*258, 293*). These intermediate particles, composed of phytoglycogen and amylose (*258*), cause a further difficulty in accurately estimating the amylose percentage in starch and in the characterization of phytoglycogen. If these particles are considered to be starch granules, the amylose content will be underestimated. If they are collected with the phytoglycogen fraction, amylose will be found, a phenomena which has been reported (*262, 293*). Thus, kernels homozygous for the *su* gene cannot be considered to contain only phytoglycogen and starch granules, but also must be considered to have a range of particles with intermediate composition resulting from the partial conversion of starch granules into phytoglycogen.

Several investigators (*13, 171, 263–265*) have reported the presence of a branching enzyme (phytoglycogen branching enzyme) in *su* kernels, in addition to Q-enzyme, which is capable of forming a phytoglycogen-like polysaccharide from amylose *in vitro*. Black and co-workers (*13*) observed the presence of phytoglycogen branching enzyme in all maize genotypes containing phytoglycogen and in two mutants (*du* and *wx*) which do not accumulate phytoglycogen. Of the three branching enzymes present in maize kernels, Boyer and co-workers (*294*) suggest that branching enzyme I plays a major role in phytoglycogen formation. However, there is a specific interaction between branching enzyme I and starch granules from *su* kernels. For example, treatment of *su* starch granules with this enzyme causes the formation and release of phytoglycogen-like glucans, but no soluble glucan was released from enzyme-treated non-mutant starch granules (*294*). Black and co-workers (*13*) concluded that the gene *su* is not the controlling factor, either in the formation of phytoglycogen or of the phytoglycogen branching enzyme. Nelson (*5*) agrees that the *su* locus is not the structural gene for the phytoglycogen branching enzyme.

A complex multiple allelic series exists at the *su* locus in maize, and four phenotypic categories have been established for mature kernels based on examination of 12 independently occurring mutations (*266*). For most alleles, mature kernels resemble the reference allele, *su-Ref,* discussed in the preceding paragraphs (Table IV) (*266*). Kernels of three alleles, including *su-am* (*amylaceous*), are near-normal in appearance and are best observed as double mutants with *du* or *su2* (*261, 266–268*). Kernels of *su-st* (*starchy*) vary from near-normal to slightly wrinkled with *su-st* recessive to *su-Ref* in some backgrounds (*266, 269*). The fourth class, represented by *su-Bn2* (*Brawn-2*), has a kernel phenotype intermediate between *su-Ref* and *su-am* (*266*). This phenotype complexity is

Table IX

Dry Weight and Carbohydrate Composition of Kernels Sampled 20 Days Postpollination for Alleles at the Sugary Locus Converted to the W64A Dent Inbred Background[a]

Sugary allele	*Ears sampled, no.*	*Kernel weight, mg*	*Glucose*[b]	*Fructose*[b]	*Sucrose*[b]	*WSP*[b]	*Starch*[b]
su-Ref	3	27	45	39	245	130	77
su-Bn2	8	33	41	36	177	55	241
su-st	7	27	60	54	124	122	191
su-am	7	36	60	52	78	4	356

[a] D. L. Garwood and S. F. Vanderslice (*295*).
[b] Milligrams per gram of dry weight.

reflected in the carbohydrate composition conditioned by these alleles with composition ranging from that of *normal* for *su-am* to that exhibited by *su-Ref* (Table IX). Based on the existence of multiple alleles which condition unique carbohydrate compositions, Garwood and Vanderslice (*295*) hypothesized that the *su* locus is a regulator locus.

An independent recessive modifier of the *su* locus, named *sugary enhancer* (*se*), has been described in the sweet corn line 'IL677a' (*270, 271*). The resulting *su se* genotype accumulates high sugar levels similar to *sh2* and also high levels of phytoglycogen similar to *su-Ref* (*254, 270, 271*).

4. Sugary-2

Kernels of the maize endosperm mutant *su2* have a slightly tarnished phenotype (Table IV) and are similar to *normal* in soluble sugar, WSP (Table VI), and starch concentrations during development (*204, 208*) and at maturity (*205*). Kernel dry weight is often (*204, 208, 212*), but not always (*205*), reduced. Starch granule size (Table V) (*204, 256*) and rate of size increase during development (*199*) are similar to *normal;* however *su2* granules have extensive internal fractures (*260*). Starch from *su2* endosperms is 10–15% higher in apparent amylose content than is *normal* starch (Table VII), with the *normal* (*Su2*) allele completely dominant to *su2* (*222–224*). As with other genotypes, year of production (*224*), *su2* allele examined (*223*), the background into which *su2* is incorporated (*230*), and different ears within a *su2* inbred (*230*) affect apparent amylose percentage. Although *su2* starch composition is altered, purified *su2* amylose and amylopectin have properties similar to those of *normal* amylose and amylopectin (*256*). Singh (*253*) has described a sorghum mutant similar to *su2*, which also has nonmutant levels of reducing sugars, WSP, and starch, but is higher in sucrose and amylose percentage.

Brown and co-workers (*199*) reported that starch granules from 18- and 24-day-old *su2* kernels are weakly birefringent and have an A-type x-ray pattern (Table V), in contrast to the B-type pattern reported for starch granules from mature *su2* kernels (*256, 272*). The BEPT of *su2* granules is lower than that of *normal* granules (Table V) (*206, 273*). Based on these granule properties, the *su2* gene has been suggested to cause a reduction in the molecular association between the starch molecules of the granule (*199*); however, no genetic lesion has been established for *su2*.

5. Dull

The *du* mature kernel phenotype varies with background, ranging from full size to semi-collapsed (Table IV). The presence of the "normal appearing" form is best detected in combination with *su-am* (*261, 267, 274*). The more extreme expression may be associated with the presence of a dominant *dull-modifier* gene

(*274*). Mature kernel dry weight of *du* also varies, with some weights similar to those of *normal* (*204, 213*) and others significantly less (*212*). The sugar concentration is slightly higher and the starch concentration lower than *normal* in both immature (*204, 208*) and mature (*205*) kernels.

The amylose percentage of *du* starch in a dent background is 5–10% higher than the percentage in *normal* starch (Table VII). Yeh and co-workers (*71*) reported 55% amylose in starch from mature *du* kernels in a sweet corn background (Table VIII). Differences in these values may be due to the Sepharose separation procedure used by Yeh and co-workers (*71*) or to a background effect. The *normal* (*Du*) allele is completely dominant to *du* for amylose percentage (*221, 223, 224*). The amylose percentage is affected by the *du* allele (*223*), by the background (*230*), and by the year of production (*224*). Although the amylose percentage is higher than in *normal,* the polysaccharide components have similar properties (Table VIII) (*256*).

Most *du* granules are similar in shape, size, birefringence, and iodine staining to *normal* granules (*26, 256, 260*); however, some irregularly shaped granules and spherical granules, which have little or no birefringence, have been reported (*26, 260*). Average *du* starch granule size is smaller than *normal* granule size (Table V) (*204*). Cell to cell variation in granule size and morphology has been reported (*260*). BEPT and x-ray diffraction patterns are similar for *du* and *normal* (Table V) (*206*).

Saussy (*26*) studied *du* kernel and plastid development in a sweet corn background. The *du* kernels have a major developmental gradient similar to *normal* except for the presence of slender, thin-walled cells near the developing embryo which appear partially compressed. Although *du* kernels in a dent background do not accumulate phytoglycogen (*13*), those in a sweet corn background do have cells in the central endosperm with plastids containing phytoglycogen and one or two small starch granules (*26*). Secondary initiation of granules has been observed in some cells (*26*). Kernels of *du* have a type II minor developmental gradient from the outside toward the interior (*26*) in which there is typical starch granule initiation and enlargement for a few cell layers, followed by an abrupt reduction in number and size of starch granules. The reduction in starch is accompanied by an increase in phytoglycogen containing plastids. All multiple mutants homozygous for *du,* but none of the others examined, had the type II minor gradient, and Saussy (*26*) suggested that this property was a specific effect of the *du* gene.

Phytoglycogen branching enzyme has been found in *du;* however, no phytoglycogen was isolated by Black and co-workers (*13*). Priess and Boyer (*275*) reported that the *du* mutation lowered the starch synthase II activity and also lowered branching enzyme IIa activity. Because the activities of two enzymes are diminished by *du,* it is possible that *du* may be a regulatory type gene, but a specific genetic lesion has not been associated with it.

6. Amylose-Extender Waxy

The *ae wx* mature kernel phenotype is reduced in size compared to *normal* (Table IV). Similarly, immature and mature kernel dry weights and starch contents are reduced almost 50% (*96, 204, 205*); however, sugar contents are increased (*204, 205, 276, 277*). Only small amounts of material are recovered in the WSP fraction (Table VI) (*276, 277*).

Apparent amylose percentages of 15–26% have been determined for *ae wx* using the blue-value procedure (Table VII), and it was originally thought that *ae wx* is the only genotype producing a significant quantity of amylose when *wx* is homozygous (*208*). However, using potentiometric titration, only 1% amylose was observed, indicating little linear material was present (*223*). This finding was confirmed by chromatographic separations on Sepharose and fine structure analyses which showed that *ae wx* starch consisted solely of loosely branched amylopectin with long external chains (*69, 70, 214*). A similar loosely branched polysaccharide of lower molecular weight also is found in *ae wx* starch in a sweet corn background (Table VIII). Thus, in this double mutant, the *wx* gene is blocking all accumulation of linear polymer, while the *ae* gene is interfering with typical branching. The enzymic reactions discussed under the respective single mutants apparently are both functioning independently in the double mutant.

Increasing doses of *ae* and *wx* effect kernel phenotype (*278*) and amylose content (*96, 223*). Two or 3 doses of the *wx* allele significantly decrease apparent amylose, indicating tighter branching, while 2 or 3 doses of the *ae* allele significantly increase apparent amylose content, indicating looser branching, regardless of the gene dosage at the other locus (*96, 223*). Apparent amylose content of *ae wx* starch decreases with increasing kernel age (*39, 96*), indicating tighter branching. Different *ae* alleles combined with *wx* may also affect the degree of branching, since pollen from different *ae wx* combinations differs in iodine staining (*279*).

Kernels of *ae wx* have the major and minor (type I) developmental gradients characteristic of *normal* (*26, 27*). Starch granules are smaller than *normal* (Table V) and increase somewhat in size with increasing kernel age (*39, 199*). Considerable differences relative to starch granule and plastid development have been observed between dent and sweet corn backgrounds (*26, 27*). In a dent background, no secondary granule initiation, characteristic of *ae*, is observed, but most granules within a cell seem to develop extensions simultaneously (*27*). These granules remain highly birefringent (*27*). In contrast, Brown and coworkers (*199*) reported that the spherical *ae wx* granules have a polarization cross, while the irregular granules only have birefringence on the outer periphery. No phytoglycogen containing amyloplasts are observed in *ae wx* kernels in a dent background (*27*). In a sweet corn background, secondary granule initiation is observed, and many amyloplasts contain a starch granule surrounded by a noncrystalline polysaccharide (*26*). Staining properties of this polysaccharide are similar to those of phytoglycogen. "Phytoglycogen" containing plastids of *ae wx* persist to maturity and, unlike the phytoglycogen plastids in *su* kernels, many of the purified starch granules are still surrounded with the "phytoglycogen-like" polysaccharide (*71*). The nature of this polysaccharide is unknown, but it may be similar to that observed in the triple mutant *ae du wx* to be described later.

7. Amylose-Extender Sugary

Mature kernels of *ae su* are not as full as *ae*, but are fuller than *su* (Table IV); and their phenotype varies with genetic background (Table IV) (*280*). Kernel dry weight and starch concentration are reduced relative to *normal* (*204, 205*). Sugar concentrations are slightly higher than those of *normal* in immature (*204*), but not in mature (*205*), kernels. Minimal WSP levels have been reported in *ae su* and were similar to those in *normal* (Table VI); however, in subsequent studies, significant amounts of phytoglycogen were found (*26, 281*). Specifically, in a dent background *ae su* endosperm contains 11% as much phytoglycogen as *su* endosperm at 20 days post-pollination. Increasing doses of *ae* in a homozygous *su* genotype results in reduced levels of phytoglycogen (*281*). Kernels of *ae su* in a sweet corn background have a large area of cells containing plastids with starch granules surrounded by a non-crystalline "phytoglycogen-like" polysaccharide (*26*). Only a few such plastids were observed in a dent background (*27*). Thus, background is important in the degree of *ae* epistasis relative to the accumulation of "phytoglycogen-like" polymers.

Starch from *ae su* kernels in a dent background consists of 51–60% amylose as determined by the blue-value procedure (Table VII), with the amylose percentage increasing with increasing kernel age (*39*). Yeh and co-workers (*71*), in contrast, reported that *ae su* reduced amylose concentration from 65% for *su* to 28% for *ae su*, based on Sepharose separation of starch polysaccharides isolated from kernels in a sweet corn background. Three fractions were obtained (Table VIII). The first two were loosely branched similar to the amylopectin fractions in *ae*. Amylose from the third peak fraction was similar in iodine staining to that from *normal*; however, some short-chain-length amylose was present as found in *ae*. The second fraction from the Sepharose column eluted in the same position as phytoglycogen and may have been the noncrystalline "phytoglycogen-like" polysaccharide shown to be present with some of the "purified" starch granules (*71*). However, the iodine complex absorption maximum of this lower-molecular-weight branched component was the same as that of the first component and similar to the branched components of *ae wx* (Table VIII). Neither branched component from *ae su*, when complexed with iodine, had an absorption maximum even close to that for *su* phytoglycogen (Table VIII). A similar lower-molecular-weight loosely branched component comprising 7.5% of *ae su* starch has been isolated (*79*).

Kernels of *ae su* have the major and minor (type I) developmental gradients characteristic of *normal* (*26, 27*). Starch granules are smaller than *normal* (Table V) and increase in size with increasing kernel age (*39, 199*). Secondary granule initiation occurs in *ae su* kernels similar to that in *ae* (*26, 27*). Within some cells in a dent background, granules are transformed during development into an amorphous nonbirefringent form (*27, 260*). Badenhuizen (*9*) reported that spherical granules from young kernels have an A-type x-ray pattern; but with development, irregular granules with a B-type x-ray pattern are found. Starch granules and plastid development in *ae su* kernels in a sweet corn background vary considerably from cell to cell (*26*). Some cells contain irregular granules; others contain granules surrounded by "phytoglycogen-like" polysaccharide, and others have plastids with granules in various stages of fragmentation.

The effects of both genes can be seen in the double mutant. Phytoglycogen, as found in *su,* is produced; however, amounts are reduced in *ae su,* although to a lesser degree in a sweet corn background. The *ae* gene reduces branching, which is reflected in the two loosely branched starch fractions obtained by Sepharose chromatography (Table VIII). Furthermore, *ae* probably interferes with phytoglycogen branching, for *ae su* phytoglycogen is degraded more by β-amylase than is the *su* phytoglycogen (*281*). In *su,* the initially formed starch granules are broken down and are thought to be utilized in the production of phytoglycogen. In *ae su,* the *su* gene may be responsible for causing the partial breakdown of the initially formed starch, but *ae* interferes with branching and an amorphous irregular granule is formed in a dent corn background along with a small amount of phytoglycogen (*27*). In the sweet corn background, more "phytoglycogen-like" polysaccharides are formed, apparently because of modifier genes (*26*).

8. Amylose-Extender Sugary-2

The mature kernel phenotype of *ae su2* is similar to that of *ae* (Table IV). Dry weight per kernel is similar to that of *su2* and *normal,* while starch concentration is less than that of *su2* and similar to that of *ae* (*204, 205*). Sugar concentrations in immature (*204*) and mature (*205*) kernels are higher than those in either *ae* or *su2*. Amylose percentage, based on blue-value determinations, is similar to that in *ae* (Table VII). Amylose percentage varies between *ae su2* ears (*230*), although *su2* or *ae* alleles have little effect on *ae su2* amylose percentage (*223*). No dosage effects are observed (*223*). Starch granule sizes and x-ray diffraction patterns are similar to those in *ae,* and the BEPT approaches that of *ae* (Table V) (*206*).

9. Amylose-Extender Dull

The phenotype of mature *ae du* kernels differs from that of both *du* and *ae* (Table IV). Compared to *normal,* dry weight and starch concentration are reduced, while sugar concentration is higher in immature (*204*) and mature (*205*) kernels. The amylose percentage of *ae du,* based on blue-value measurements of starch from kernels in a dent background, is similar to that in *ae* (Table VII). With *ae* homozygous, the apparent amylose percentage decreases with increasing doses of *du* (*223*). The 47% amylose determined by the Sepharose separation of starch from *ae du* kernels in a sweet corn background is intermediate between the amount in *ae* and *du* (Table VIII). The maximum absorption of the iodine–amylopectin complex of *ae du* is similar to that of *ae,* while the amylose component is closer to that of *du* and *normal* (Table VIII). Thus, also in *ae du,* the *ae* gene appears to be interfering with the typical branching of amylopectin resulting in the production of more loosely branched polymers.

Although low levels of WSP have been reported in *ae du* kernels (Table VI), Black and co-workers (*13*) concluded that no phytoglycogen accumulates in *ae du* kernels in a dent background. In contrast, kernels of *ae du* in a sweet corn background produce numerous plastids with one or two starch granules surrounded by a thick layer of noncrystalline ''phytoglycogen-like'' polysaccharide (*26*).

Kernels of *ae du* in a sweet corn background are slightly delayed in development, but have the *normal* major gradient of kernel development (*26*). The type II minor gradient characteristic of *du* is observed in *ae du* (*26*). Saussy (*26*) also reported that secondary starch granule initiation occurs and that granules assume a blocky, elongated irregular shape later in development.

In a dent background, the greatest increase in granule size occurs between 12 and 18 days post-pollination (*199*). Granule size is similar to that of *ae* and *du* granules, but less than that of *normal* granules (Table V). The *ae du* starch granules have a B-type x-ray defraction pattern similar to that of *ae* (Table V). In contrast, the *ae du* BEPT is similar to that of *du* (Table V) (*206, 273*). In *ae du,* the *ae* and *du* genes appear to be functioning independently with *ae* interfering with typical branching, and *du* causing the expression of the type II minor gradient. In *ae du,* branching enzyme fractions IIa and IIb and starch synthase fraction II are considerably reduced (*275*). Thus, the double mutant expresses the enzyme reductions of both individual mutants.

10. Dull Sugary

The mature kernel phenotype of *du su* is similar to that of *su,* although *du su* kernels are often more wrinkled (Table IV). This genotype has been extensively studied to evaluate its potential for improving sweet corn quality (*282, 283*).

Compared to *normal, du su* kernels have reduced dry weight and starch concentration and increased sugar and WSP concentrations (*204, 205*). Sugar (*204, 205, 261, 270, 282, 283*) and WSP (Table VI) (*261, 282*) levels are similar to those in *su,* although starch (*205, 261, 282*) concentration is lower. Thus, *su* is epistatic to *du* relative to phytoglycogen accumulation.

Widely varying amylose percentages have been reported for *du su* starch samples in a dent background when measured by the blue-value procedure (Table VII). In four other reports, *du su* amylose content ranged from 51% to 66% (*222, 224, 256, 261*). Yeh and co-workers (*71*) (Table VIII) reported 70% amylose using Sepharose column chromatography. Dvonch and co-workers (*256*) stated that *du su* amylopectin is intermediate in branching between glycogen and *normal* amylopectin. However, based on the absorption maximum of the iodine complex and the extinction coefficient, the high-molecular-weight branched fraction from *du su* in a sweet corn background appears to be loosely branched with long external chains (Table VIII). The *du su* amylose fraction is similar to that of *du* and *su* alone. No dosage effects have been observed on amylose percentage (*224*).

The overall kernel and plastid development pattern in *du su* is similar to that in *su* except that *du* causes the type II minor gradient (*26*). Compound or semicompound granules are initially formed, followed by slight enlargement, fragmentation, and accumulation of phytoglycogen. At later stages of development in some cells, the phytoglycogen plastid membrane ruptures as in *su,* and the phytoglycogen mixes with the cytosol. Saussy (*26*) also reported a lack of secondary granule initiation. No increase in the average size of *du su* starch granules is observed between 12 and 24 days post-pollination (*199*). The *du su* granules are similar in size to those of *su* (Table V). This lack of size increase is probably due to the observed granule fragmentation (*26*). Granules isolated from mature kernels show weak or no birefrigence (*71*), and those from 24-day-old kernels have a weak A-type x-ray diffraction pattern (*199*).

11. Dull Sugary-2

The mature kernel phenotype of *du su2* differs from both *du* and *su2* (Table IV). The sugar and WSP (Table VI) concentrations in immature (*204*) and mature (*205*) *du su2* kernels are similar to those in *du* and *su2,* except that, in immature *du su2* kernels, the sugar concentration is higher than that in *su2*. Starch concentration is lower than that in either *du* or *su2* (*204, 205*); and the amylose percentage, as measured by the blue-value test (Table VII) or by potentiometic titration (*256*), is higher than that in either *su2* or *du*. No dosage effects on amylose percentage have been observed (*224*). Isolated *du su2* amylose and amylopectin have properties similar to those of *normal* (*256*); however, Whistler and Doane (*79*) isolated 8.7% of *du su2* starch in a loosely branched amylopectin fraction. Average size of *du su2* starch granules is similar to that of the single mutants (Table V). The BEPT of *du su2* starch granules is similar to that of *su2* granules (Table V) (*273*) and *du su2* granules have an A-type x-ray diffraction pattern (Table V).

12. Dull Waxy

The mature *du wx* kernel phenotype differs from that of either *du* or *wx* (Table IV). Dry weights of mature kernels are similar to those for *du* and *wx* and slightly less than those of *normal* (*205*). Sugar concentrations are higher, and starch concentration is lower than in either *normal, du,* or *wx* in immature (*204*) and mature (*205*) kernels.

Starch in the double mutant *du wx* is essentially 100% amylopectin; thus, the *wx* mutant is epistatic to *du*. The absorption maximum and extinction coefficient of the *du wx* branched polysaccharide–iodine complex are the same as for *wx* (Table VIII). When the *wx-a* allele is combined with *du, du wx-a* starch contains 9% amylose, reflecting the increased amylose conditioned by the *wx-a* allele alone (*221*).

Neither *du* nor *wx* in a dent background accumulates phytoglycogen (Table VI), but they both contain phytoglycogen branching enzyme (*13*). However, when combined in the double mutant *du wx,* immature kernels contain up to 11% phytoglycogen (Table VI). Although not quantatively determined, Saussy (*26*) reported numerous phytoglycogen-containing plastids in endosperm cells of *du wx* in a sweet corn background.

Starch granule size of *du wx* at 18 and 24 days postpollination is intermediate between *du* and *wx* (Table V). The mean BEPT and A-type x-ray diffraction pattern of *du wx* starch are the same as for *normal* and the component single mutants (Table V) (*206*).

Kernels of *du wx* in a sweet corn background have the major developmental gradient typical of *normal* and a type II minor gradient characteristic of du (*26*). Secondary granule initiation is observed in many cells. Granule shapes vary from spherical to irregular-blocky. Plastids containing starch granules surrounded by phytoglycogen are generally located in the more mature cells of the central endosperm region (*26*).

13. Sugary Waxy

The *su wx* mature kernel phenotype is similar to that of *su* (Table IV). Immature (*204, 250*) and mature (*205*) kernel carbohydrate composition is similar to that in *su,* except that *su wx* starch is composed of 100% amylopectin (Table VII). The starch component in *su wx* has properties similar to those of both *wx* starch and the amylopectin component of *su* starch (Table VIII) (*256*). With *su* homozygous, increasing doses of *wx* reduce amylose concentration (*222*). The

WSP content (Table VI), β-amylolysis limits, and chain lengths of *su wx* phytoglycogen are the same as those from *su* (*13*). Immature kernels contain phytoglycogen branching enzyme (*13*). Starch granules isolated from *su wx* are small, aggregated, and compound (similar to *su* granules), and phytoglycogen is completely removed from the starch granules during isolation (*71*). The granules are strongly birefringent (similar to those from *wx*) (*71*). Thus, *wx* is epistatic to *su* relative to the absence of amylose in the starch, and *su* is epistatic to *wx* relative to soluble sugar and phytoglycogen concentrations, kernel phenotype, and starch granule size.

14. Sugary-2 Waxy

The mature kernel phenotype for *su2 wx* is similar to that for *wx* (Table IV). Mature (*205*) and immature (*204*) kernel dry weight and carbohydrate composition are similar to those of the single mutants. The *wx* mutant is epistatic to *su2*, resulting in starch with approximately 100% amylopectin (Table VII).

15. Sugary Sugary-2

The *su su2* mature kernel phenotype is similar to that for *su* (Table IV). This genotype has been extensively evaluated for its sweet corn improvement potential (*282, 283*). Immature (*204, 282*) and mature (*205*) kernel carbohydrate composition approaches that of *su* kernels; however, starch accumulation is reduced (*204, 282*). The apparent amylose percentage is similar to that of *su2* starch (Table VII). Amylose concentration increases with increasing doses of *su2* when *su* is homozygous and with increasing doses of *su*, when *su2* is homozygous (*222*). Starch granule size (Table V) (*256*) and BEPT (Table V) (*206*) are similar to *su*. Thus, *su2* is epistatic to *su* for apparent amylose concentration, while *su* is epistatic to *su2* for starch granule size, carbohydrate composition, and mature kernel phenotype.

16. Amylose-Extender Dull Sugary

The mature kernel phenotype for the triple mutant *ae du su* is similar to that for *su* (Table IV). Sugar concentrations of mature (*205*) and immature (*204*) *ae du su* kernels are higher than those of either of the single mutants or the double mutants *ae du* or *ae su*, while starch concentration is similar to that in *su* and the two double mutants. The amylose percentage is near *normal* when measured by either the blue-value test (Table VII) or the Sepharose separation technique (Table VIII). However, in contrast with *normal*, a major proportion of the branched polysaccharide is smaller than typical amylopectin (as is that of *ae su*), and it elutes from a Sepharose column at an intermediate position (Table VIII). The absorption maximum and extinction coefficient of the branched polysaccharide–iodine complexes are similar to those for *ae* and *ae su* and are characteristic of loosely branched polymers. The absorption maximum of the amylose–iodine complex is similar to that for *normal, du,* or *su;* but the extinction coefficient is lower than for either (Table VIII). No short chain amylose has been found in *ae du su* (*71*).

Phytoglycogen accumulates in *su* and *du su* kernels, but not in *ae* or *du* (Table VI). In the double mutant *ae su, ae* is epistatic to *su,* but the addition of *du* allows a somewhat larger amount of phytoglycogen to accumulate (Table VI). Phytoglycogen branching enzyme has been reported in *su* and *du,* but not in *ae* or *ae su* (*13*). Apparently, the branching enzyme activity resulting from the addition of *du* to *ae su* is sufficient to partially overcome the inhibitory effect of *ae* on phytoglycogen accumulation.

Endosperms of *ae du su* in a sweet corn background have the *normal* major developmental gradient and a type II minor gradient characteristic of *du* (*26*). Secondary starch granule initiation has been observed. Starch granules from *ae du su* are similar to those from *du su* and are weakly birefringent (*26, 71*). Various starch granule shapes from simple spherical to irregular are observed in granules from immature (*26*) and mature kernels (*71*). Starch granule fragmentation and disappearance concomitant with increased phytoglycogen in plastids, characteristic of *su,* also are common in *ae du su* (*26*). Thus, *su* is epistatic to *ae du* relative to plastid type.

17. Amylose-Extender Dull Sugary-2

The mature kernel phenotype of *ae du su2* differs from each of the component single or double mutants (Table IV). The sugar and starch concentrations of mature (*205*) and immature (*204*) kernels of *ae du su2* are similar to those of *ae su2* kernels. Sugar concentrations are higher than in the single mutants or other double mutants in this combination, while starch concentration is lower (*204, 205*). Quantity of WSP is higher in mature and immature kernels of *ae du su2* than in *normal* or any of the single and double mutants in this combination (Table VI). However, Black and co-workers (*13*) did not detect phytoglycogen in *ae du su2,* and the nature of the WSP has not been determined. Apparent amylose percentage of *ae du su2* starch is similar to that in *du* and *su2,* but is lower than that in *ae* starch (Table VII).

18. Amylose-Extender Dull Waxy

The mature kernel phenotype of the triple mutant *ae du wx* differs from any of the single mutants (Table IV). Starch concentration is low compared with the component single and double mutants, while sugar concentrations are severalfold higher (*204, 205*). WSP concentration in *ae du wx* kernels in a dent background is lower than in *du wx,* but is similar to the quantity in the single and other double mutants (Table VI). Little if any of this WSP is phytoglycogen (*13*). In contrast, amyloplasts from *ae du wx* in a sweet corn background frequently contain one or two starch granules surrounded by a noncrystalline polysaccharide (*26, 71, 131*). The structure of the noncrystalline polysaccharide has not been determined, but the iodine-staining property appears similar to that of phytoglycogen *in situ*. However, in contrast with phytoglycogen in *su* kernels, it is not readily removed from the granules during aqueous granule isolation (*71, 131*). Extraction of the isolated granules with 10% ethanol removes some polysaccharide (*71*). This extracted material is largely composed of branched polysaccharides of intermediate size having the same polysaccharide–iodine complex absorption maximum as the 10% ethanol residual granules, with this maximum higher than that of *su* phytoglycogen (*71*).

These genes have been incorporated into sweet corn inbreds, and a new type of vegetable corn has been introduced which is intermediate in sweetness between standard sweet corn (*su*) and sweet corns based on the *sh2* mutation (*284*). This hybrid, 'Pennfresh ADX,' has the advantage of extra sweetness at harvest and sugar retention for an extended time in storage (*277*).

Starch from *ae du wx* kernels is composed entirely of branched polysaccharides which are largely of intermediate size between amylopectin and amylose (Table VIII). The absorption maximum and extinction coefficients of the polysaccharide–iodine complexes are similar to those of amylopectin from *wx* and *du* rather than those of the loosely branched polysaccharides in *ae*, the *ae* double, and other *ae* containing triple mutants (Table VIII). Both *du* and *wx* kernels contain phytoglycogen branching enzyme (*13*). In combination, they apparently overcome the effect of *ae*, resulting in the production of polysaccharides, both granular and nongranular, which are more highly branched than those of *ae* amylopectin (Table VIII).

Endosperm of *ae du wx* in a sweet corn background has a major developmental gradient typical of *normal* and a type II minor gradient characteristic of *du* (*26*). Saussy (*26*) reports that starch granule and plastid development in *ae du wx* is similar to that observed in *du wx*.

19. Amylose-Extender Sugary Sugary-2

The mature kernel phenotype of the triple mutant *ae su su2* differs from that of any of the component mutants (Table IV). Mature kernel dry weight is similar to that of *ae su;* and sugar and starch concentrations are intermediate between those in *su* and *su su2* and those in *ae, su2, ae su,* and *ae su2* (*205*). Mature and immature kernels contain intermediate levels of WSP (Table VI). This WSP has not been characterized and may or may not be similar to the phytoglycogen accumulating in *su* kernels. Starch from *ae su su2* kernels has been reported to contain 31–54% apparent amylose (Table VII). Starches from *ae su su2* have not been separated by Sepharose chromatography, and thus the relative sizes of the polysaccharides and degree of branching have not been established.

20. Amylose-Extender Sugary Waxy

The mature kernel phenotype of the triple mutant *ae su wx* differs from that of any of the component mutants (Table IV). The dry weight per mature kernel is similar to that of *ae su* and is higher than that of *su, ae wx,* and *su wx* (*205*). Quantities of sugars and WSP (Table VI) in mature (*205*) and immature (*204*) kernels are intermediate among the component single and double combinations. Starch content is relatively low, but higher than that of *su* (*205*). The WSP fraction is shown to contain phytoglycogen (Table VI) with characteristics similar to *su* phytoglycogen (*13*). Kernels of *ae su wx* also contain phytoglycogen branching enzyme (*13*).

Starches of *ae su wx* are reported to contain 13–14% apparent amylose when measured by blue-value tests (Table VII), but Yeh and co-workers (*71*) (Table VIII) showed that the apparent amylose is due to the loosely branched nature of the starch polysaccharides. Thus, *wx* blocks amylose accumulation, and *ae* influences the degree of branching. The maximum absorption and extinction coefficient of the polysaccharide–iodine complex is similar to that for *ae wx* (Table VIII). Starch granules isolated from mature *ae su wx* kernels vary from large spherical granules to small aggregated and compound granules (*71*). Most granules are strongly birefringent, but occasional phytoglycogen containing plastids and non-iodine-staining and non-birefringent granule particles are present in the starch granule preparation (*71*).

21. Amylose-Extender Sugary-2 Waxy

The *ae su2 wx* kernel phenotype differs from each of the component mutants (Table IV). Mature *ae su2 wx* kernel dry weight is intermediate between that of the lighter *ae* and *ae wx* kernels and the heavier *su2, wx, ae su2,* and *su2 wx* kernels (*205*). The quantities of sugar in mature (*205*) and immature (*204*) *ae su2 wx* kernels are similar to those in *ae wx,* while WSP and starch concentrations are somewhat higher. The small amount of WSP present (Table VI) has not been characterized, and its similarity to phytoglycogen is unknown. Starch of *ae su2 wx,* based on blue-value tests, has been reported to contain 28% amylose (Table VII). Although not yet determined, this apparent amylose is most likely due to the presence of a loosely branched amylopectin similar to that present in *ae wx* (*69, 214*).

22. Dull Sugary Sugary-2

The mature kernel phenotype of the triple mutant *du su su2* is similar to that of *su* (Table IV). Also, sugar concentrations in mature (*205*) and immature (*204*) *du su su2* kernels are similar to those in *su,* but WSP and starch are higher and lower, respectively, in *du su su2*. Various amylose percentages have been reported (Table VII). The 77% amylose observed in one study (*224*) is the highest percentage observed in genotypes lacking *ae;* however, because of the low starch content, this genotype has little or no commercial value. As observed with other genotypes, amylose percentage varies with year of production (*224*). Starch granules of *du su su2* are small, similar to *su* (*256*), and exhibit little or no birefringence (*206, 256*). Starch components from *du su su2* have not been separated by Sepharose chromatography, so the precise nature of the polysaccharides is unknown.

23. Dull Sugary Waxy

The mature kernel phenotype of *du su wx* is similar to that of *su* (Table IV). The quantity of sugars is similar to that in *su* (*204, 205*). The addition of *du* to *su wx* causes an increase in WSP (Table VI) and a decrease in starch (*205*). The phytoglycogen from *du su wx* has a β-amylolysis limit and chain length similar to that of *su*. The enhanced phytoglycogen accumulation may result from the additive effect of the branching enzymes present in each of the component single mutants (*13*).

Starch from *du su wx* is approximately 100% amylopectin (Table VII) and consists of large and intermediate size polymers (Table VIII). The absorption maximum and extinction coefficient of the amylopectin–iodine complex are similar to those for *wx* and *du wx* amylopectins (Table VIII). Starch granules isolated from mature *du su wx* kernels vary from small spherical to aggregated and compound granules (*71*). Although most granules are strongly birefringent, noniodine staining and nonbirefringent granular particles are also observed in the starch granule preparation (*71*). The granular particles probably are the same as the ultra-fine starch granule fragments reported by Saussy (*26*).

Young kernels of *du su wx* have the major gradient in endosperm development characteristic of normal and the type II minor gradient characteristic of *du* (*26*). However, later in development, much of the central endosperm consists of a noncellular cavity containing starch granules and "phytoglycogen" plastids (*26*). Cells near the pericarp contain amyloplasts with small, compound granules, while more interior cells are filled with large "phytoglycogen" plastids void of starch, which appear unique in that the plastid contents are essentially unstained by iodine (*26*). Since the β-amylolysis limit and mean chain lengths of phytoglycogen from *du su wx* are similar to those for *su* (*13*), the reason for the difference in staining properties of phytoglycogen plastids in *du su wx* and *su* is unknown.

24. Dull Sugary-2 Waxy

The mature kernel phenotype of the triple mutant *du su2 wx* differs from that of any of the component mutants (Table IV). Mature kernel dry weight of *du su2 wx* kernels is similar to that of the component mutants (*205*). Sugar concentrations in mature (*205*) and immature (*204*) kernels are similar to those in *du wx* kernels. WSP is slightly higher and starch lower in *du su2 wx* compared to *du wx* (Table VI). The WSP has not been characterized, and its similarity to phytoglycogen is unknown. Starch from *du su2 wx* kernels is 100% amylopectin (Table VII), reflecting the effect of *wx*. The BEPT of *du su2 wx* starch granules is low, reflecting the influence of *su2* (*206*).

25. Sugary Sugary-2 Waxy

The mature kernel phenotype of the triple mutant *su su2 wx* is similar to that of *su* (Table IV). Kernel dry weight and carbohydrate concentrations in mature (*205*) and immature (*204*) *su su2 wx* kernels are similar to those in *su su2* kernels. The elevated concentration of WSP (Table VI) is assumed to be phytoglycogen, although it has not been characterized. The *wx* gene is epistatic to *su su2*, resulting in the accumulation of starch composed of 100% amylopectin (Table VII). Starch granules show little birefringence (*206*). The BEPT is low, similar to that for *su2* (*206*).

26. Amylose-Extender Dull Sugary Waxy

The mature kernel phenotype of the quadruple mutant *ae du su wx* differs from each of the component mutants (Table IV) and varies depending on the sweet corn inbred background (Garwood, unpublished). Mature kernel dry weight is similar to that of *su* kernels (*213*). Starch from *ae du su wx* consists of 100% amylopectin (Table VIII), with most of the polysaccharides of intermediate size (*71*). The degree of branching of the major component (intermediate size) is similar to that of *wx* amylopectin (Table VIII). Aqueously isolated granules contain starch granules with associated nonbirefringent polysaccharides similar to those in *ae du wx*, and extraction of the granule preparation with 10% ethanol removes 27% of the total polysaccharide (*71*). The addition of *su* to *ae du wx* increases the occurrence of small, aggregated, and compound granules (*71*).

Endosperm development in *ae du su wx* is similar to that in *du su wx*, with the type II minor gradient observed and the central endosperm cavity being present by 27 days post-pollination (*26*). Starch granule and phytoglycogen plastid development in *ae du su wx* is similar to that in *su*, except that the quadruple mutant has greater apparent phytoglycogen content at 16 days post-pollination than does *su* or any other mutant combination (*26*). However, with development, there is increasing deterioration of the plastids and central endosperm cells (*26*).

VIII. Conclusions

By using mutants of maize and other species, progress has been made in understanding the pathways and enzymes involved in starch biosynthesis and the fine structure of starch polysaccharides. However, starch biosynthesis and granule formation is still not completely understood. Thus, integration of the information on polysaccharide biosynthesis (Section VI) with that on mutant effects (Section VII) is necessary to evaluate current understanding of polysaccharide biosynthesis and to delineate the limits of this knowledge.

A number of maize endosperm carbohydrate mutants have been shown to influence the *in vitro* activity of particular enzymes (Table X). To date, modification of specific enzyme activities has not been related to endosperm mutants such as *bt* and *su2*. Effects shown in Table X need not necessarily be the primary effect of a mutant, but are the ones known at this writing. Screening for enzyme activities by earlier workers probably would not have detected changes in isozyme activities involving the multiple forms of phosphorylase, starch synthase, and branching enzyme that exist in plants. Thus, more careful examinations will be needed to identify additional enzyme lesions. Also, some mutations may modify the *in vivo* activity of specific enzymes by regulating effector metabolites (*96*). It is possible that current enzyme isolation techniques have not allowed the *in vitro* measurement of certain polysaccharide synthesizing enzymes active in *normal* or mutant tissues. For example, amylosucrase, a bacterial α-D-glucosylase that directly converts sucrose to a glycogen-like α-glucan (*18*), has not been measured in higher plants. Could higher plants have a similar enzyme which is unstable to traditional isolation techniques? Other experimental approaches may be needed to gain information on the precise pathway of starch biosynthesis in the intact, compartmented plant cell.

Mutations such as *sh*, *sh2*, and *bt2* cause major blocks in the conversion of

Table X

Summary of Mutant Effects in Maize Where an Associated Enzyme Lesion Has Been Reported

Genotype	*Major biochemical changes*[a]	*Enzyme affected*
sh	↑ sugars ↓ starch	↓ sucrose synthase
sh2	↑ sugars ↓ starch	↓ ADPG-pyrophosphorylase
		↑ hexokinase
bt2	↑ sugars ↓ starch	↓ ADPG-pyrophosphorylase
sh4	↑ sugars ↓ starch	↓ pyridoxal phosphate
su	↑ sugars ↑ phytoglycogen ↓ starch	↑ phytoglycogen branching enzyme
wx	$\simeq$ 100% amylopectin	↓ starch-bound starch synthase
		↑ phytoglycogen branching enzyme
ae	↑ loosely branched polysaccharide	↓ branching enzyme IIb
	↑ apparent amylose %	
du	↑ apparent amylose %	↓ starch synthase II
		↓ branching enzyme IIa
		↑ phytoglycogen branching enzyme

[a] Changes relative to *normal*, ↑, ↓ = increase or decrease, respectively. Sugars = the alcohol-soluble sugars.

sucrose to the sugar nucleotides UDPG and ADPG (Table X), indicating the key *in vivo* roles of sucrose synthase and ADPG pyrophosphorylase in starch synthesis. The *su* mutant allows the accumulation of phytoglycogen due to the activity of phytoglycogen branching enzyme (*263–265*). Phytoglycogen branching enzyme activity was also found in *wx* and *du* (*13*), but these mutant kernels did not produce phytoglycogen except when they were incorporated into a sweet corn background (*26*). The double mutant *du wx* contains phytoglycogen branching enzyme and also accumulates phytoglycogen (*13*). Approximately 100% amylopectin is produced in kernels homozygous for *wx* (Table VII). In *wx*, the major starch granule bound starch synthase is missing, but the two soluble starch synthase activities are unaffected (*5*). The *ae* mutant interferes with typical branching causing accumulation of a loosely branched polysaccharide (Table VIII). The presence of this polymer causes an increase in "apparent" amylose percentage when measured by iodine binding methods (Table VII). The branching enzyme IIb, which coelutes with starch synthase I from DEAE cellulose columns, is missing in *ae;* but branching enzymes I and IIa are unaffected (*171*). The *du* mutant causes an increase in apparent amylose content through its effects on starch synthase II (the starch synthase which requires primer) and branching enzyme IIa which coelutes with it from DEAE cellulose columns (*275*).

Interaction of these mutants further clarifies the biosynthetic pathway. For example, the *wx* mutant is epistatic to all other known maize endosperm mutants and no amylose accumulates (Table VII). Mutants such as *sh2*, *bt2*, and *su* cause major reductions in starch accumulation, but in combination with *wx* the starch which is produced is all amylopectin (*208*). In the double mutant *ae wx*, *wx* prevents the production of amylose, and *ae* reduces the degree of branching, resulting in the accumulation of a loosely branched polysaccharide (*69*). The *su* mutant is epistatic to *du*, *su2*, and *wx* relative to accumulation of phytoglycogen; but *ae* and *sh2* are partially epistatic to *su*, causing a marked reduction in the *su* stimulated phytoglycogen accumulation (Table VI). The addition of *du* or *wx* to *ae su* partially overcomes the *ae* inhibitory effect, and phytoglycogen accumulates.

Obviously, our understanding of starch biosynthesis is still incomplete, since mutants occur for which the primary metabolic effect has not been determined. Subsequent evaluation of isozymes and effector compounds and studies of the *in vivo* pattern and rate of ^{14}C labeling of intermediates of starch biosynthesis of *normal*, mutants, and mutant combinations should aid in clarifying the nature of the mutations and the pathways of starch biosynthesis. Other aspects of starch formation also remain to be explained. For example, how are starch granules formed? How do reserve starch granules develop species specific shapes? Is a primer needed for starch formation *in vivo?*

In spite of these limitations, the pathway of starch biosynthesis determined using maize mutants can probably be generalized to other plant species because related mutants have occurred in peas, sorghum, barley, and rice and because the same enzymes are found in starch-synthesizing tissues in other plant species. With the existence of isozymes, however, it is possible that the pathway of starch biosynthesis may differ slightly when other species are examined to the same extent as maize.

IX. References

(*1*) N. P. Badenhuizen, "The Biogenesis of Starch Granules in Higher Plants," Appleton-Century-Crofts, New York, 1969.

(*2*) J. A. Radley, "Starch and Its Derivatives," Chapman and Hall, London, 4th Ed. 1968.

(*3*) W. Banks and C. T. Greenwood, "Starch and Its Components," Edinburgh University Press, Edinburgh, 1975.

(*4*) R. G. Creech, *Adv. Agron.*, **20,** 275 (1968).

(*5*) O. E. Nelson, *Adv. Cereal Sci. Technol.*, Vol. III, 1980, Chap. 2, pp. 41–71.

(*6*) J. Preiss and C. Levi, *in* "Photosynthesis II, Encyclopedia of Plant Physiology, New Series," M. Gibbs and E. Latzko, eds., Springer-Verlag, Berlin, 1979, Vol. 6, p. 282.

(*7*) B. O. Juliano, *in* "Rice Chemistry and Technology," D. F. Houston, ed., Amer. Assn. Cereal Chemists, St. Paul, Minnesota, 1972, p. 16.

(*8*) J. J. Marshall, *Wallerstein Lab. Commun.*, **35,** 49 (1972).

(*9*) N. P. Badenhuizen, *in* "Starch: Chemistry and Technology," R. L. Whistler and E. F. Paschall, eds., Academic Press, New York, 1st Ed., 1965, Vol. 1, p. 65.

(*10*) J. S. Craigie, *in* "Algal Physiology and Biochemistry," W. D. P. Stewart, ed., Blackwell Scientific Publications, Oxford, 1974, p. 206.

(*11*) J. F. Frederick, *Ann. N.Y. Acad. Sci.*, **210,** 254 (1973).

(*12*) M. Stacey and S. A. Barker, "Polysaccharides of Micro-organisms," Oxford University Press, Oxford, 1960.

(*13*) R. C. Black, J. D. Loerch, F. J. McArdle and R. G. Creech, *Genetics*, **53,** 661 (1966).

(*14*) F. R. Rickson, *Ann. N.Y. Acad. Sci.*, **210,** 104 (1973).

(*15*) S. A. Barker and E. J. Bourne, *in* "Biochemistry and Physiology of Protozoa," S. H. Hunter and A. Lwoff, eds., Academic Press, New York, 1955, Vol. 2, p. 45.

(*16*) E. Percival and R. H. McDowell, "Chemistry and Enzymology of Marine Algal Polysaccharides," Academic Press, London, 1967, p. 73.

(*17*) E. J. Hehre, *Adv. Enzymol.*, **11,** 297 (1951).

(*18*) G. Okada and E. J. Hehre, *J. Biol. Chem.*, **249,** 126 (1974).

(*19*) J. D. Dodge, "The Fine Structure of Algal Cells," Academic Press, London, 1973.

(*20*) A. R. Archibald, E. L. Hirst, D. J. Manners, and J. F. Ryley, *J. Chem. Soc.*, 556 (1960).

(*21*) D. B. Dickinson, *Plant Physiol.*, **43,** 1 (1968).

(*22*) E. F. Artschwager, *J. Agric. Res.*, **27,** 809 (1924).

(*23*) L. Lampe, *Bot. Gaz.*, **91,** 337 (1931).

(*24*) T. A. Kiesselbach, *Univ. Nebr. Coll. Agric. Exp. Sta. Res. Bull.*, 161 (1948).

(*25*) J. C. Shannon, *Cereal Chem.*, **51,** 798 (1974).

(*26*) L. A. Saussy, M. S. Thesis, Pennsylvania State University, University Park, Pennsylvania, 1978.

(*27*) C. D. Boyer, R. R. Daniels, and J. C. Shannon, *Am. J. Bot.*, **64,** 50 (1977).

(*28*) E. H. Sanders, *Cereal Chem.*, **32,** 12 (1955).

(*29*) L. W. Rooney, *in* "Industrial Uses of Cereals," Y. Pomeranz, ed., Am. Assn. Cereal Chemists, St. Paul, Minnesota, 1973, p. 316.

(*30*) J. E. Freeman, *in* "Sorghum Production and Utilization," J. S. Wall and W. M. Ross, eds., Avi Publishing Co., Westport, Connecticut, 1970, p. 28.
(*31*) R. M. Sandstedt, *Cereal Chem.,* **23,** 337 (1946).
(*32*) B. L. D'Appolonia, K. A. Gilles, E. M. Osman, and Y. Pomeranz, *in* "Wheat Chemistry and Technology," Y. Pomeranz, ed., Am. Assn. Cereal Chemists, St. Paul, Minnesota, 1971, p. 301.
(*33*) D. H. Simmonds and W. P. Campbell, *in* "Rye: Production, Chemistry, and Technology," W. Bushuk, ed., Am. Assn. Cereal Chemists, Inc., St. Paul, Minnesota, 1976, p. 63.
(*34*) D. H. Simmonds, *in* "Triticale: First Man-Made Cereal," C. C. Tsen, ed., Am. Assn. Cereal Chemists, St. Paul, Minnesota, 1974, p. 105.
(*35*) D. E. Briggs, "Barley," Chapman and Hall, London, 1978.
(*36*) C. D. Boyer, R. R. Daniels, and J. C. Shannon, *Crop Sci.,* **16,** 298 (1976).
(*37*) D. Bradbury, M. M. MacMasters, and I. M. Cull, *Cereal Chem.,* **33,** 361 (1956).
(*38*) J. C. Shannon, *Plant Physiol.,* **49,** 198 (1972).
(*39*) C. D. Boyer, J. C. Shannon, D. L. Garwood, and R. G. Creech, *Cereal Chem.,* **53,** 327 (1976).
(*40*) R. Geddes, C. T. Greenwood, and S. Mackenzie, *Carbohydr. Res.,* **1,** 71 (1965).
(*41*) M. J. Wolf, M. M. MacMasters, J. E. Hubbard, and C. E. Rist, *Cereal Chem.,* **25,** 312 (1948).
(*42*) E. B. Earley, *Plant Physiol.,* **27,** 184 (1952).
(*43*) A. W. MacGregor, D. E. LaBerge, and W. O. S. Meredith, *Cereal Chem.,* **48,** 255 (1971).
(*44*) R. Singh and B. O. Juliano, *Plant Physiol.,* **59,** 417 (1977).
(*45*) C. W. Bice, M. M. MacMasters, and G. E. Hilbert, *Cereal Chem.,* **22,** 463 (1945).
(*46*) C. T. Greenwood and J. Thompson, *Biochem. J.,* **82,** 156 (1962).
(*47*) A. K. Paul, S. Mukherji, and S. M. Sircar, *Physiol. Plant.,* **24,** 342 (1971).
(*48*) J. W. Williams, *in* "Starch and Its Derivatives," J. A. Radley, ed., Chapman and Hall, London, 4th Ed., 1968, p. 91.
(*49*) C. T. Greenwood, *Adv. Cereal Sci. Technol.,* **1,** 119 (1976).
(*50*) C. T. Greenwood, *in* "The Carbohydrates," W. Pigman and D. Horton, ed., Academic Press, New York, 2nd Ed., 1970, Vol. IIB, p. 471; L. M. Gilbert, G. A. Gilbert, and S. P. Spragg, *Methods Carbohydr. Chem.,* **4,** 25 (1964); R. L. Whistler, *ibid.,* **4,** 28 (1964).
(*51*) C. T. Greenwood, *Adv. Carbohydr. Chem.,* **11,** 335 (1956).
(*52*) J. Muetgeert, *Adv. Carbohydr. Chem.,* **16,** 299 (1961).
(*53*) T. J. Schoch, *Adv. Carbohydr. Chem.,* **1,** 247 (1945).
(*54*) W. Banks, C. T. Greenwood, and D. D. Muir, *Staerke,* **26,** 73 (1974).
(*55*) B. O. Juliano, *Cereal Sci. Today,* **16,** 334 (1971).
(*56*) R. W. Kerr and O. R. Trubell, *Paper Trade J.,* **117,** 25 (1943).
(*57*) R. M. McCready and W. Z. Hassid, *J. Am. Chem. Soc.,* **65,** 1154 (1943).
(*58*) C. A. Shuman and R. A. Plunkett, *Method. Carbohydr. Chem.,* **4,** 174 (1964).
(*59*) C. M. Sowbhagya and K. R. Bhattacharya, *Starch/Staerke,* **31,** 159 (1979).
(*60*) P. C. Williams, F. D. Kuzina, and I. Hlynka, *Cereal Chem.,* **47,** 411 (1970).
(*61*) M. J. Wolf, E. H. Melvin, W. J. Garcia, R. J. Dimler, and W. F. Kwolek, *Cereal Chem.,* **47,** 437 (1970).
(*62*) D. M. W. Anderson and C. T. Greenwood, *J. Chem. Soc.,* 3016 (1955).
(*63*) F. L. Bates, D. French, and R. E. Rundle, *J. Am. Chem. Soc.,* **65,** 142 (1943).
(*64*) S. Lansky, M. Kooi, and T. J. Schoch, *J. Am. Chem. Soc.,* **71,** 4066 (1949).
(*65*) E. M. Montgomery, K. R. Sexton, and F. R. Senti, *Staerke,* **13,** 215 (1961).
(*66*) W. Banks, C. T. Greenwood, and K. M. Khan, *Staerke,* **22,** 292 (1970).
(*67*) W. Banks, C. T. Greenwood, and K. M. Khan, *Carbohydr. Res.,* **17,** 25 (1971).
(*68*) A. S. Perlin, *Can. J. Chem.,* **36,** 810 (1958).

(69) C. D. Boyer, D. L. Garwood, and J. C. Shannon, *Staerke,* **28,** 405 (1976).
(70) T. Yamada and M. Taki, *Staerke,* **28,** 374 (1976).
(71) J. Y. Yeh, D. L. Garwood, and J. C. Shannon, *Starch/Staerke,* **33,** 222 (1981).
(72) W. Banks and C. T. Greenwood, *Carbohydr. Res.,* **6,** 171 (1968).
(73) D. B. Wankhede, A. Shehnaz, and M. R. Raghavendra Rao, *Starch/Staerke,* **31,** 153 (1979).
(74) S. R. Erlander and D. French, *J. Polymer Sci.,* **32,** 291 (1958).
(75) S. R. Erlander and D. French, *J. Am. Chem. Soc.,* **80,** 4413 (1958).
(76) S. R. Erlander, J. P. McGuire, and R. J. Dimler, *Cereal Chem.,* **42,** 175 (1965).
(77) W. Banks and C. T. Greenwood, *J. Chem. Soc.,* 3486 (1959).
(78) W. Banks and C. T. Greenwood, *Staerke,* **19,** 197 (1967).
(79) R. L. Whistler and W. M. Doane, *Cereal Chem.,* **38,** 251 (1961).
(80) D. Paton, *Starch/Staerke,* **31,** 184 (1979).
(81) G. K. Adkins and C. T. Greenwood, *Carbohydr. Res.,* **11,** 217 (1969).
(82) W. L. Deatherage, M. M. MacMasters and C. E. Rist, *Trans. Am. Assn. Cereal Chemists,* **13,** 31 (1955).
(83) T. J. Schoch and E. C. Maywald, *Cereal Chem.,* **45,** 564 (1968).
(84) R. L. Whistler and P. Weatherwax, *Cereal Chem.,* **25,** 71 (1948).
(85) J. Simek, *Zesz. Probl. Postepow Nauk Roln.,* **159,** 87 (1974).
(86) O. H. Miller and E. E. Burns, *J. Food Sci.,* **35,** 666 (1970).
(87) F. E. Horan and M. F. Heider, *Cereal Chem.,* **23,** 492 (1946).
(88) D. G. Medcalf and K. A. Gilles, *Cereal Chem.,* **42,** 558 (1965).
(89) K. J. Goering, R. F. Eslick, and C. A. Ryan, Jr., *Cereal Chem.,* **34,** 437 (1957).
(90) K. J. Goering, R. Eslick, and B. DeHaas, *Cereal Chem.,* **47,** 592 (1970).
(91) B. O. Juliano, E. L. Albano, and G. B. Cagampang, *Philippine Agriculturalist,* **48,** 234 (1964).
(92) N. Kongseree and B. O. Juliano, *J. Agr. Food Chem.,* **20,** 714 (1972).
(93) S. N. Raghavendra Rao and B. O. Juliano, *J. Agr. Food Chem.,* **18,** 289 (1970).
(94) A. C. Reyes, E. L. Albano, V. P. Broines, and B. O. Juliano, *J. Agr. Food Chem.,* **13,** 438 (1965).
(95) C.-Y. Lii and D. R. Lineback, *Cereal Chem.,* **54,** 138 (1977).
(96) C. D. Boyer, D. L. Garwood, and J. C. Shannon, *J. Hered.,* **67,** 209 (1976).
(97) C. Mercier, R. Charbonnière, D. Gallant, and A. Guilbot, *Staerke,* **22,** 9 (1970).
(98) C. Y. Tsai, F. Salamini, and O. E. Nelson, *Plant Physiol.,* **46,** 299 (1970).
(99) A. O. Ketiku and V. A. Oyenuga, *J. Sci. Food Agr.,* **23,** 1451 (1972).
(100) V. P. Briones, L. G. Magbanua, and B. O. Juliano, *Cereal Chem.,* **45,** 351 (1968).
(101) M. Abou-Guendia and B. L. D'Appolonia, *Cereal Chem.,* **50,** 723 (1973).
(102) L. D. Jenkins, D. P. Loney, P. Meredith, and B. A. Fineran, *Cereal Chem.,* **51,** 718 (1974).
(103) K. Kulp and P. J. Mattern, *Cereal Chem.,* **50,** 496 (1973).
(104) N. K. Matheson, *Phytochem.,* **10,** 3213 (1971).
(105) H. L. Wood, *Aust. J. Agr. Res.,* **11,** 673 (1960).
(106) W. Banks, C. T. Greenwood and D. D. Muir, *Staerke,* **25,** 153 (1973).
(107) G. Harris and I. C. MacWilliam, *Cereal Chem.,* **35,** 82 (1958).
(108) W. Banks and C. T. Greenwood, *Ann. N.Y. Acad. Sci.,* **210,** 17 (1973).
(109) K. J. Goering and B. DeHaas, *Cereal Chem.,* **51,** 573 (1974).
(110) O. Inatsu, K. Watanabe, I. Maeda, K. Ito, and S. Osanai, *J. Japan Soc. Starch Sci.,* **21,** 115 (1974).
(111) B. O. Juliano, *J. Japan Soc. Starch Sci.,* **18,** 35 (1970).
(112) V. L. Fergason and M. S. Zuber, *Crop Sci.,* **5,** 169 (1965).
(113) K. J. Goering, *Cereal Chem.,* **44,** 245 (1967).
(114) K. J. Goering, P. V. Subba Rao, D. H. Fritts, and T. Carroll, *Staerke,* **22,** 217 (1970).

(*115*) M. S. Buttrose, *Aust. J. Biol. Sci.*, **16,** 305 (1963).
(*116*) O. E. Stamberg, *Cereal Chem.*, **16,** 769 (1939).
(*117*) L. H. May and M. S. Buttrose, *Aust. J. Biol. Sci.*, **12,** 146 (1959).
(*118*) M. S. Buttrose, *J. Ultrastructure Res.*, **4,** 231 (1960).
(*119*) D. M. Hall and J. G. Sayre, *Textile Res. J.*, **40,** 256 (1970).
(*120*) E. T. Reichert, "The Differentiation and Specificity of Starches in Relation to Genera, Species, etc.," Carnegie Institution of Washington, Washington, D.C., 1913, Parts I and II.
(*121*) G. E. Moss, *in* "Examination and Analysis of Starch and Starch Products," J. A. Radley, ed., Applied Science Publishers, Ltd., London, 1976, p. 1.
(*122*) N. L. Kent, "Technology of Cereals with Special Reference to Wheat," Pergamon Press, Oxford, 2nd Ed., 1975.
(*123*) D. M. Hall and J. G. Sayre, *Textile Res. J.*, **39,** 1044 (1969).
(*124*) D. M. Hall and J. G. Sayre, *Textile Res. J.*, **41,** 880 (1971).
(*125*) A. J. Klassen and R. D. Hill, *Cereal Chem.*, **48,** 647 (1971).
(*126*) T. Lempiäinen and H. Henriksnäs, *Starch/Staerke*, **31,** 45 (1979).
(*127*) D. N. Duvick, *Am. J. Bot.*, **42,** 717 (1955).
(*128*) B. R. Williams, Ph.D. Thesis, Pennsylvania State University, University Park, Pennsylvania, 1971.
(*129*) N. P. Badenhuizen, *K. Ned. Akad. Wet. Proc. Ser. C*, **65,** 123 (1962).
(*130*) H. W. Heldt and F. Sauer, *Biochim. Biophys. Acta*, **234,** 83 (1971).
(*131*) T.-T.Y. Liu, and J. C. Shannon, *Plant Physiol.*, **67,** 525 (1981).
(*132*) J. C. Shannon, R. G. Creech, and J. D. Loerch, *Plant Physiol.*, **45,** 163 (1970).
(*133*) R. L. Whistler and J. R. Young, *Cereal Chem.*, **37,** 204 (1960).
(*134*) W. B. McConnell, A. K. Mitra, and A. S. Perlin, *Can. J. Biochem. Physiol.*, **36,** 985 (1958).
(*135*) T. Akazawa, T. Minamikawa, and T. Murata, *Plant Physiol.*, **39,** 371 (1964).
(*136*) J. H. Pazur, *in* "Starch Chemistry and Technology," R. L. Whistler and E. F. Paschall, eds., Academic Press, New York, 1965, Vol. I, p. 133.
(*137*) J. C. Shannon and R. G. Creech, *Ann. N.Y. Acad. Sci.*, **210,** 279 (1973).
(*138*) J. F. Turner and D. H. Turner, *Annu. Rev. Plant Physiol.*, **26,** 159 (1975).
(*139*) P. S. Chourey and O. E. Nelson, *Biochem. Genetics*, **14,** 1041 (1976).
(*140*) C. S. Hanes, *Proc. Roy. Soc. B*, **129,** 174 (1940).
(*141*) L. F. Leloir, *Science*, **172,** 1299 (1971).
(*142*) H. K. Porter, *Annu. Rev. Plant Physiol.*, **13,** 303 (1962).
(*143*) C. Y. Tsai and O. E. Nelson, *Plant Physiol.*, **43,** 103 (1968).
(*144*) C. Y. Tsai and O. E. Nelson, *Plant Physiol.*, **44,** 159 (1969).
(*145*) E. Slabnik and R. B. Frydman, *Biochem. Biophys. Res. Commun.*, **38,** 709 (1970).
(*146*) S. J. Gerbrandy and J. D. Verleur, *Phytochem.*, **10,** 261 (1971).
(*147*) M. A. R. deFekete, *Planta*, **79,** 208 (1968).
(*148*) M. A. R. deFekete, *Planta*, **87,** 311 (1969).
(*149*) N. K. Matheson and R. H. Richardson, *Phytochem.*, **15,** 887 (1976).
(*150*) T-T. Y. Liu and J. C. Shannon, *Plant Physiol.*, **67,** 518 (1981).
(*151*) T. Akazawa, *in* "Plant Biochemistry," J. Bonner and J. E. Varner, eds., Academic Press, New York, 3rd Ed., 1976, p. 381.
(*152*) W. Z. Hassid, *in* "The Carbohydrates—Chemistry and Biochemistry," W. Pigman and D. Horton, eds., Academic Press, New York, 1970, Vol. IIA, p. 301.
(*153*) C. E. Cardini and R. B. Frydman, *Methods Enzymol.*, **8,** 387 (1966).
(*154*) E. Recondo and L. F. Leloir, *Biochem. Biophys. Res. Commun.*, **6,** 85 (1961).
(*155*) C. Y. Tsai, *Bot. Bull. Acad. Sinica*, **14,** 125 (1973).
(*156*) C. Y. Tsai, *Biochem. Genetics*, **11,** 83 (1974).

(*157*) T. Murata, T. Sugiyama, and T. Akazawa, *Biochem. Biophys. Res. Commun.* **18,** 371 (1965).
(*158*) O. E. Nelson and H. W. Rines, *Biochem. Biophys. Res, Commun.,* **9,** 297 (1962).
(*159*) O. E. Nelson, P. S. Chourey, and M. T. Chang, *Plant Physiol.,* **62,** 383 (1978).
(*160*) T. Akatsuka and O. E. Nelson, *J. Biol. Chem.,* **241,** 2280 (1966).
(*161*) J. L. Ozbun, J. S. Hawker, and J. Preiss, *Plant Physiol.,* **48,** 765 (1971).
(*162*) C.-Y. Tsai and O. E. Nelson, *Science,* **151,** 341 (1966).
(*163*) D. B. Dickinson and J. Preiss, *Plant Physiol.,* **44,** 1058 (1969).
(*164*) J. D. Vidra and J. D. Loerch, *Biochim. Biophys. Acta,* **159,** 551 (1968).
(*165*) W. N. Haworth, S. Peat, and E. J. Bourne, *Nature,* **154,** 236 (1944).
(*166*) E. J. Bourne and S. Peat, *J. Chem. Soc.,* 877 (1945).
(*167*) D. Borovsky, E. E. Smith, and W. J. Whelan, *Eur. J. Biochem.,* **62,** 307 (1976).
(*168*) J. S. Hawker, J. L. Ozbun, H. Ozaki, E. Greenberg, and J. Preiss, *Arch. Biochem. Biophys.,* **160,** 530 (1974).
(*169*) C. D. Boyer and J. Preiss, *Carbohydr. Res.,* **61,** 321 (1978).
(*170*) S. Schiefer, E. Y. C. Lee, and W. J. Whelan, *FEBS Letters,* **30,** 129 (1973).
(*171*) C. D. Boyer and J. Preiss, *Biochem. Biophys. Res. Commun.,* **80,** 169 (1978).
(*172*) S. Peat, W. J. Whelan, and W. R. Rees, *J. Chem. Soc.,* 44 (1956).
(*173*) E. Y. C. Lee, J. J. Marshall, and W. J. Whelan, *Arch. Biochem. Biophys.,* **143,** 365 (1971).
(*174*) S. R. Erlander, *Enzymologia,* **19,** 273 (1958).
(*175*) H. W. Heldt and L. Rapley, *FEBS Letters,* **10,** 143 (1970).
(*176*) H. W. Heldt, *FEBS Letters,* **5,** 11 (1969).
(*177*) P. S. Nobel and Y. N. S. Cheung, *Nature New Biol.,* **237,** 207 (1972).
(*178*) D. A. Walker, *in* "Transport in Plants III, Encyclopedia of Plant Physiology, New Series," C. R. Stocking and U. Heber, eds., Springer-Verlag, Berlin, 1976, Vol. 3, p. 85.
(*179*) U. Heber, *Annu. Rev. Plant Physiol.,* **25,** 393 (1974).
(*180*) H. W. Heldt, *in* "Transport in Plants III, Encyclopedia of Plant Physiology, New Series," C. R. Stocking and U. Heber, eds., Springer-Verlag, Berlin, 1976, Vol. 3, p. 137.
(*181*) M. D. Hatch, *in* "Plant Biochemistry," J. Bonner and J. E. Varner, eds., Academic Press, New York, 3rd Ed., 1976, p. 797.
(*182*) M. M. Rhoades and A. Carvalho, *Bull. Torrey Botan. Club,* **71,** 335 (1944).
(*183*) W. J. S. Downton and E. B. Tregunna, *Can. J. Bot.,* **46** 207 (1968).
(*184*) C. Levi and J. Preiss, *Plant Physiol.,* **61,** 218 (1978).
(*185*) E. Y. C. Lee, and W. J. Whelan, *in* "The Enzymes," P. D. Boyer, ed., Academic Press, New York, 3rd Ed., Vol. 5, 1971, p. 191.
(*186*) J. Preiss and C. Levi, *in* "4th International Congress on Photosynthesis, Proc." D. B. Hall, J. Coombs, and T. W. Goodwind, eds., Biochemical Society, London, 1978, p. 457.
(*187*) J. Preiss and T. Kosuge, *Annu. Rev. Plant Physiol.,* **21,** 433 (1970).
(*188*) J. Preiss and T. Kosuge, *in* "Plant Biochemistry," J. Bonner and J. E. Varner, eds., Academic Press, New York, 3rd Ed., 1976, p. 278.
(*189*) K. Muhlethaler, *in* "The Structure and Function of Chloroplasts," M. Gibbs, ed., Springer-Verlag, Berlin, 1971, p. 7.
(*190*) C. F. Jenner, *in* "Transport and Transfer Processes in Plants," I. F. Wardlow and J. B. Passioura, eds., Academic Press, New York, 1976, p. 73.
(*191*) P. N. Viswanathan, *Indian J. Biochem.,* **6,** 124 (1969).
(*192*) P. N. Viswanathan and P.S. Krishnan, *Indian J. Biochem.,* **2,** 69 (1965).
(*193*) P. N. Viswanathan and P.S. Krishnan, *Indian J. Biochem.,* **3,** 228 (1966).
(*194*) J. M. Williams and C. M. Duffus, *Plant Physiol.,* **59,** 189 (1977).
(*195*) D. B. Dickinson and J. Preiss, *Arch. Biochem. Biophys.,* **130,** 119 (1969).
(*196*) J. L. Ozbun, J. S. Hawker, and J. Preiss, *Biochem. J.,* **126,** 953 (1972).
(*197*) C. D. Boyer and J. Preiss, *Plant Physiol.,* **64,** 1039 (1979).

(*198*) D. L. Garwood and R. G. Creech, *Crop Sci.*, **12,** 119 (1972).
(*199*) R. P. Brown, R. G. Creech, and L. J. Johnson, *Crop Sci.*, **11,** 297 (1971).
(*200*) C. Y. Tsai and O. E. Nelson, *Genetics*, **61,** 813 (1969).
(*201*) B. Burr and O. E. Nelson, *Ann. N. Y. Acad. Sci.*, **210,** 129 (1973).
(*202*) Y. Ma and O. E. Nelson, *Cereal Chem.*, **52,** 412 (1975).
(*203*) H. Fuwa, Y. Sugimoto, M. Tanaka, and D. V. Glover, *Starch/Staerke*, **30,** 186 (1978).
(*204*) R. G. Creech, *Genetics*, **52,** 1175 (1965).
(*205*) R. G. Creech and F. J. McArdle, *Crop Sci.*, **6,** 192 (1966).
(*206*) H. H. Kramer, P. L. Pfahler, and R. L. Whistler, *Agron. J.*, **50,** 207 (1958).
(*207*) H. L. Seckinger and M. J. Wolf, *Staerke*, **18,** 1 (1966).
(*208*) D. G. Holder, D. V. Glover, and J. C. Shannon, *Crop Sci.*, **14,** 643 (1974).
(*209*) G. Eriksson, *Hereditas*, **63,** 180 (1969).
(*210*) R. M. Hixon and B. Brimhall, *in* "Starch and Its Derivatives," J. A. Radley, ed., Chapman and Hall, London, 4th Ed., 1968, p. 247.
(*211*) D. W. Gorbet and D. E. Weibel, *Crop Sci.*, **12,** 378 (1972).
(*212*) H. G. Nass and P. L. Crane, *Crop Sci.*, **10,** 276 (1970).
(*213*) D. E. Rowe and D. L. Garwood, *Crop Sci.*, **18,** 709 (1978).
(*214*) T. Yamada, T. Komiya, M. Akaki, and M. Taki, *Starch/Staerke*, **30,** 145 (1978).
(*215*) B. O. Juliano, M. B. Nazareno, and N. B. Ramos, *J. Agr. Food Chem.*, **17,** 1364 (1969).
(*216*) Y. Ikawa, D. V. Glover, Y. Sugimoto, and H. Fuwa, *Carbohydr. Res.*, **61,** 211 (1978).
(*217*) B. Brimhall, G. F. Sprague, and J. E. Sass, *J. Am. Soc. Agron.*, **37,** 937 (1945).
(*218*) O. E. Nelson, *Science*, **130,** 794 (1959).
(*219*) E. P. Palmiano and B. O. Juliano, *Agr. Biol. Chem.*, **36,** 157 (1972).
(*220*) A. J. Vidal and B. O. Juliano, *Cereal Chem.*, **44,** 86 (1967).
(*221*) J. L. Helm, V. L. Fergason, and M. S. Zuber, *J. Hered.*, **60,** 259 (1969).
(*222*) H. H. Kramer and R. L. Whistler, *Agron. J.*, **41,** 409 (1949).
(*223*) M. L. Vineyard, R. P. Bear, M. M. MacMasters, and W. L. Deatherage, *Agron. J.*, **50,** 595 (1958).
(*224*) G. M. Dunn, H. H. Kramer, and R. L. Whistler, *Agron. J.*, **45,** 101 (1953).
(*225*) W. Banks, C. T. Greenwood and D. D. Muir, *Staerke*, **26,** 289 (1974).
(*226*) S. Blixt, *in* "Handbook of Genetics," R. C. King, ed., Plenum Press, New York, 1974, Vol. 2, p. 181.
(*227*) V. L. Fergason, J. D. Helm and M. S. Zuber, *J. Hered.*, **57,** 90 (1966).
(*228*) J. T. Walker and N. R. Merritt, *Nature*, **221,** 482 (1969).
(*229*) M. S. Zuber, C. O. Grogan, W. L. Deatherage, J. W. Hubbard, W. Schulze, and M. M. MacMasters, *Agron. J.*, **50,** 9 (1958).
(*230*) R. P. Bear, M. L. Vineyard, M. M. MacMasters, and W. L. Deatherage, *Agron. J.*, **50,** 598 (1958).
(*231*) A. Haunold and M. F. Lindsey, *Crop Sci.*, **4,** 58 (1964).
(*232*) P. J. Loesch, Jr., and M. S. Zuber, *Crop Sci.*, **4,** 526 (1964).
(*233*) J. P. Thomas, Ph.D. Thesis, University of Missouri, Columbia, Missouri, 1968.
(*234*) D. L. Garwood, J. C. Shannon, and R. G. Creech, *Cereal Chem.*, **53,** 355 (1976).
(*235*) J. L. Helm, V. L. Fergason, and M. S. Zuber, *Crop Sci.*, **7,** 659 (1967).
(*236*) V. L. Fergason and M. S. Zuber, *Crop Sci.*, **2,** 209 (1962).
(*237*) J. L. Helm, V. L. Fergason, and M. S. Zuber, *Agron. J.*, **60,** 530 (1968).
(*238*) J. L. Helm, V. L. Fergason, J. P. Thomas, and M. S. Zuber, *Agron. J.*, **59,** 257 (1967).
(*239*) V. L. Fergason, J. L. Helm, and M. S. Zuber, *Crop Sci.*, **6,** 273 (1966).
(*240*) C. Mercier, *Staerke*, **25,** 78 (1973).
(*241*) I. A. Wolff, B. T. Hofreiter, P. R. Watson, W. L. Deatherage, and M. M. MacMasters, *J. Am. Chem. Soc.*, **77,** 1654 (1955).
(*242*) M. J. Wolf, H. L. Seckinger, and R. J. Dimler, *Staerke*, **16,** 375 (1964).

(243) W. C. Mussulman and J. A. Wagoner, *Cereal Chem.*, **45,** 162 (1968).
(244) R. M. Sandstedt, *Cereal Sci. Today,* **10,** 305 (1965).
(245) J. S. Wall and C. W. Blessin, *in* "Sorghum Production and Utilization," J. S. Wall and W. M. Ross, eds., Avi Publishing Co., Inc., Westport, Connecticut, 1970, p. 118.
(246) S. A. Watson and Y. Hirata, *Sorghum Newsletter,* **3,** 6 (1960).
(247) C. W. Culpepper and C. A. Magoon, *J. Agr. Res.*, **28,** 403 (1924).
(248) C. T. Greenwood and P. C. Das Gupta, *J. Chem. Soc.*, 703 (1958).
(249) R. E. Harper and T. R. Quinby, *J. Hered.*, **54,** 121 (1963).
(250) R. H. Andrew, R. A. Brink, and N. P. Neal, *J. Agr. Res.*, **69,** 355 (1944).
(251) P. H. Jennings and C. L. McCombs, *Phytochem.*, **8,** 1357 (1969).
(252) J. R. Laughnan, *Genetics,* **38,** 485 (1953).
(253) R. Singh, Ph.D. Thesis, Purdue University, West Lafayette, Indiana, 1973.
(254) J. W. Gonzales, A. M. Rhodes, and D. B. Dickinson, *Plant Physiol.*, **58,** 28 (1976).
(255) C. M. Duffus and P. H. Jennings, *Starch/Staerke,* **30,** 371 (1978).
(256) W. Dvonch, H. H. Kramer, and R. L. Whistler, *Cereal Chem.*, **28,** 270 (1951).
(257) C. T. Greenwood and P.C. Das Gupta, *J. Chem. Soc.*, 707 (1958).
(258) N. K. Matheson, *Phytochem.*, **14,** 2017 (1975).
(259) J. C. Wanson and P. Drochmans, *J. Cell Biol.*, **38,** 130 (1968).
(260) R. M. Sandstedt, B. D. Hites, and H. Schroeder, *Cereal Sci. Today,* **13,** 82 (1968).
(261) J. W. Cameron, *Genetics,* **32,** 459 (1947).
(262) S. Peat, W. J. Whelan, and J. R. Turvey, *J. Chem. Soc.*, 2317 (1956).
(263) H. F. Hodges, R. G. Creech, and J. D. Loerch, *Biochim. Biophys. Acta,* **185,** 70 (1969).
(264) N. Lavintman, *Arch. Biochem. Biophys.*, **116,** 1 (1966).
(265) D. J. Manners, J. J. M. Rowe, and K. L. Rowe, *Carbohydr. Res.*, **8,** 72 (1968).
(266) D. L. Garwood and R. G. Creech, *Agron. Abstr.*, 7 (1972).
(267) P. C. Mangelsdorf, *Genetics,* **32,** 448 (1947).
(268) D. L. Garwood, *Maize Genet. Coop. News Letter,* **49,** 140 (1975).
(269) D. E. Dahlstrom and J. H. Lonnquist, *J. Hered.*, **55,** 242 (1964).
(270) J. E. Ferguson, A. M. Rhodes, and D. B. Dickinson, *J. Hered.*, **69,** 377 (1978).
(271) J. E. Ferguson, D. B. Dickinson, and A. M. Rhodes, *Plant Physiol.*, **63,** 416 (1979).
(272) N. P. Badenhuizen, *Protoplasmalogia,* **2,** B/26/δ (1959).
(273) P. L. Pfahler, H. H. Kramer, and R. L. Whistler, *Science,* **125,** 441 (1957).
(274) J. H. Davis, H. H. Kramer, and R. L. Whistler, *Agron. J.*, **47,** 232 (1955).
(275) J. Preiss and C. D. Boyer, *in* "Mechanisms of Polysaccharide Polymerization and Deploymerization," J. J. Marshall, ed., Academic Press, New York, 1979.
(276) E. V. Wann, G. B. Brown, and W. A. Hills, *J. Am. Soc. Hort. Sci.*, **96,** 441 (1971).
(277) D. L. Garwood, F. J. McArdle, S. F. Vanderslice, and J. C. Shannon, *J. Am. Soc. Hort. Sci.*, **101,** 400 (1976).
(278) D. L. Garwood, C. D. Boyer, and J. C. Shannon, *Maize Genet. Coop. News Letter,* **50,** 99 (1976).
(279) R. G. Creech and H. H. Kramer, *Am. Naturalist,* **95,** 326 (1961).
(280) H. H. Kramer, R. L. Whistler, and E. G. Anderson, *Agron. J.*, **48,** 170 (1956).
(281) J. E. Ayers and R. G. Creech, *Crop Sci.*, **9,** 739 (1969).
(282) J. W. Cameron and D. A. Cole, Jr., *Agron. J.*, **51,** 424 (1959).
(283) R. M. Soberalske and R. H. Andrew, *Crop Sci.*, **18,** 743 (1978).
(284) D. L. Garwood and R. G. Creech, *HortScience,* **14,** 645 (1979).
(285) J. Preiss, *Ann. Rev. Plant Physiol.*, **33,** 431 (1982).
(286) J. Preiss and C. Levi, *in* "The Biochemistry of Plants, A Comprehensive Treatise; Carbohydrates: Structure and Function," J. Preiss, ed., Academic Press, New York, 1980, Vol. 3, p. 371.
(287) W. Banks and D. D. Muir, *in* "The Biochemistry of Plants, A Comprehensive Treatise;

Carbohydrates: Structure and Function,'' J. Preiss, ed., Academic Press, New York, 1980, Vol. 3, p. 321.
(*288*) L. C. Hannah, D. M. Tuschall, and R. J. Mans, *Genetics,* **95,** 961 (1980).
(*289*) G. L. Matters and C. D. Boyer, *Biochem. Genetics,* **20,** 833 (1982).
(*290*) M. J. Fishwick and A. J. Wright, *Phytochem.,* **19,** 55 (1980).
(*291*) C. D. Boyer, P. A. Damewood, and G. L. Matters, *Starch/Staerke,* **32,** 217 (1980).
(*292*) K. D. Hedman and C. D. Boyer, *Biochem. Genetics,* **20,** 483 (1982).
(*293*) C. D. Boyer, P. A. Damewood, and E. K. G. Simpson, *Starch/Staerke,* **33,** 125 (1981).
(*294*) C. D. Boyer, E. K. G. Simpson, and P. A. Damewood, *Starch/Staerke,* **34,** 81 (1982).
(*295*) D. L. Garwood and S. F. Vanderslice, *Crop Sci.,* **22,** 367 (1982).
(*296*) C. D. Boyer and J. C. Shannon, *Plant Breeding Rev.,* **1,** 139 (1983).
(*297*) J. C. Shannon, *Iowa State J. Res.,* **56,** 307 (1982).

CHAPTER IV

ENZYMES IN THE HYDROLYSIS AND SYNTHESIS OF STARCH

By John F. Robyt

Department of Biochemistry and Biophysics, Iowa State University, Ames, Iowa

I. Introduction and Classification of Starch Hydrolases

Starch, a mixture of amylose and amylopectin, represents a link with the energy of the sun, which is partially captured during photosynthesis. Starch serves as a food reserve for plants and provides a mechanism by which non-photosynthesizing organisms, such as man, can utilize the energy provided by the sun. To utilize starch, the organisms must have enzymes that catalyze the hydrolysis of the (1→4) glycosidic bonds found between the α-D-glucopyranose residues. Enzymes that are capable of catalyzing the hydrolysis of the α-D-(1→4) linkages are called amylases and are widely produced by plants, bacteria, and animals. Several different types of amylases have evolved.

Amylases have been classified by different criteria: (a) the configuration of the anomeric carbon atom of their products; (b) the biological source; (c) the type of

STARCH, 2nd ed.

ISBN 0-12-746270-8

attack on the polymeric substrate, i.e., endo or exo attack; (d) whether they produce a rapid drop in the viscosity of the substrate (liquefying) or a slow drop (saccharifying); (e) the type of product(s) produced, *e.g.*, D-glucose, maltose, and maltotriose; and (f) the nature of their protein structure. These criteria are obviously based on the different attributes of the enzymes and are not necessarily mutually exclusive.

One of the older and more common classifications is the α and β designation which is based on the anomeric configuration of the products released; thus, α-amylases release products with the α-D-configuration and β-amylases release products with the β-D-configuration. In general, α-amylases are endoglycosidases, attacking glucans somewhere away from the chain ends at an internal glycosidic bond and producing a rapid drop in the viscosity of the substrate. α-Amylases also produce varying types of oligosaccharides, characteristic of the particular α-amylase. The α-amylases are probably the most widely distributed of the amylases; they are produced by many different types of bacteria, fungi, animals, and some plants. In mammals, they are produced mostly by two types of organs, the salivary glands and the pancreas. Mammalian α-amylases are frequently referred to according to the organ and the species, *e.g.*, human salivary α-amylase and porcine (pig) pancreatic α-amylase. Bacterial and fungal α-amylases are designated by the genus and species, *e.g.*, *Bacillus subtilis* α-amylase and *Aspergillus oryzae* α-amylase. For some of the more recently discovered amylases that produce a specific type of product, an indication of their product specificity is also given, *e.g.*, *Aerobacter aerogenes* maltohexaose-producing amylase (*1*, *2*). Several α-amylases have been highly purified and crystallized (see Table I).

In general, β-amylases attack glucans in an exo fashion from the nonreducing end to produce a single type of low-molecular-weight product with the β-D-configuration. Most of the β-amylases are of plant origin, and produce β-maltose and a high-molecular-weight limit dextrin. β-Amylase has been found to be produced by at least one bacterium, *Bacillus polymyxa* (*3*). Limit dextrins result when the enzyme reaches a branch point in amylopectin. The β-amylases cannot bypass an α-D-(1→6) branch linkage to attack α-D-(1→4) linkages on the other side of the branch point. Several β-amylases have been crystallized (see Table I).

Another exo-acting amylase is glucoamylase which releases β-D-glucopyranose from the nonreducing end of the starch chain. Crystalline glucoamylases have been prepared from *Aspergillus oryzae, A. niger,* and *Rhizopus delemar* (*28, 30*). These enzymes differ from β-amylases in that they do not produce limit dextrins. They catalyze the hydrolysis of both the α-D-(1→4) and α-D-(1→6) linkages, although at widely different rates, and, therefore, can completely convert starch to D-glucose.

In addition to the amylases, there are the so-called isoamylases or debranching

Table I

Crystalline Amylases

Source and type	Reference for isolation and crystallization
α-Amylases	
Human saliva	*4*
Human pancreas	*5*
Porcine pancreas[a]	*6,7*
Baboon saliva	*8*
Rat pancreas	*9,10*
Bacillus amyloliquefaciens[a,b,c]	*11,12*
Bacillus subtilis var. *amylosaccharitikus*[a,c]	*17*
Bacillus coagulans	*18*
Pseudomonas saccharophila	*19*
Aspergillus oryzae[a]	*20*
Aspergillus candidus	*21*
Barley malt	*22,23*
β-Amylases	
Barley malt	*24*
Soybean	*25*
Sweet potato[a]	*26*
Wheat	*27*
Glucoamylases	
Aspergillus niger[a]	*28*
Rhizopus delemar[a]	*29*
Rhizopus nivens[a]	*30*

[a] Commercially available in crystalline form.

[b] The original designation of this organism was *Bacillus subtilis*, but a reinvestigation produced a new classification as *Bacillus amyloliquefaciens* (*13*).

[c] The two amylases from *Bacillus amyloliquefaciens* and *Bacillus subtilus* var. *amylosaccharitikus* are quite distinct in amino acid composition, molecular weight, and action pattern (*14–16*).

enzymes that hydrolyze the α-D-(1→6) linkages of amylopectin. R-enzyme from broad beans (*31*) was the first recognized enzyme of this type. Another is pullulanase, an enzyme elaborated by *Aerobacter aerogenes* (*32*). This enzyme was originally observed to hydrolyze the α-D-(1→6) linkages of pullulan, a linear polysaccharide composed of maltotriose units linked by α-D-(1→6) bonds, but it was also found to hydrolyze the α-D-(1→6) linkages of amylopectin (*33*). Isoamylases also have been obtained from yeast (*34*) and bacterial species, *Pseudomonas* (*35–37*) and *Cytophaga* (*38*).

II. Assay Methods for Amylases

A number of analytical techniques have been devised for measuring the hydrolysis of glycosidic bonds found in starch. The most quantitative procedures involve the measurement of the formation of new reducing groups, hemiacetal or aldehyde groups, that result from the hydrolysis of the glycosidic, acetal, linkage. The International Committee on Enzymes has recommended that enzyme activity be standardized to the number of micromoles of substrate transformed or product produced per minute under defined conditions. However, when the substrate is a polysaccharide of indefinite molecular weight, as is the case for starch, there are many sites of reaction, and there are produced a mixture of oligosaccharides as products, the International Unit is best defined as the number of micromoles of bond hydrolyzed per minute (*39*).

The colorimetric measurement of the formation of reducing groups has been accomplished primarily by the use of three reagents: alkaline copper (*40*), alkaline ferricyanide (*41*), and alkaline 3,5-dinitrosalicylate (*42, 43*). Each of these reagents have their advantages and disadvantages. All have to be standardized. The most common standard is maltose, especially if oligosaccharides or dextrins are the products; however, care must be taken to obtain highly pure, crystalline maltose. D-Glucose is not an appropriate standard unless it is the only product formed, as it responds differently than maltooligosaccharides to the reagents. The copper and ferricyanide procedures give equimolar reducing values for equimolar reducing ends of maltooligosaccharides, whereas the dinitrosalicylate reagent gives an increasing reducing value with an increase in the number of D-glucose units in the oligosaccharide chain (*41, 44*). Further, the reducing value determined with the dinitrosalicylate reagent is decreased by calcium ions which are frequently present in amylase preparations. Although the dinitrosalicylate reagent is simple to use, it is the least sensitive and the least accurate and can give at best only a qualitative estimate of the number of reducing ends produced. The ferricyanide reagent has the simplest chemistry; it involves measurement of the loss of the yellow color of the ferricyanide ion. Its principal disadvantage is that the plot of absorbance vs. concentration produces a line with a negative slope. A line with a positive slope can be obtained, however, by plotting the percentage transmittance on semi-log paper. The ferricyanide reagent is the most sensitive of the three. An even more sensitive procedure is the modification of the copper procedure using neocuproine (*45*).

A semiquantitative determination of starch hydrolysis by α-amylases involves measurement of the decrease in the blue iodine color. This method is especially useful in surveys of biological samples for α-amylolytic activity (*46, 47*). The procedure reflects the endo cleavage of relatively large starch chains and cannot be used to assay exo-acting amylases. The procedure is sensitive and easy to perform but not adaptable to a quantitative International Unit. A modification of the procedure using starch–agar plates and antibiotic paper discs (*48, 49*) is convenient, extremely sensitive, and can detect down to 10^{-10} molar α-amylase (*50*). The method is quantitated by plotting the diameter of the clear zone against the concentration of the α-amylase.

Measurement of the decrease in the viscosity of a starch solution has also been used to measure α-amylase action (*51, 52*). This method, like the iodine procedure, only measures α-amylase activity, and the results cannot be readily expressed in International Units. Because these procedures only measure endo activity, they can be used to detect α-amylase contaminants in exoamylase preparations. Insoluble chromogenic substrates have been developed for α-amylase assays. These have found some use in clinical laboratories. A dye is covalently linked to starch or one of its components to give an insoluble material (*53–55*). When these substrates are acted on by α-amylase, fragments containing dye are solubilized. The remaining insoluble substrate is removed by centrifugation and the absorbance of the supernatant is measured and taken as an indication of α-amylase activity. Several chromogenic substrates are commerically available. The method is simple and fast, like the iodine and viscosity methods, but it cannot be related to the number of bonds hydrolyzed.

The isoamylases are assayed by measuring the increase in the iodine color at 610 nm using amylopectin as a substrate (*36*) or by measuring the increase in the reducing value using the same substrate (*38*). The latter method can be converted into International Units.

III. Structure and Properties of the Amylases

Most of the α-amylases have molecular weights of about 50,000 (*56, 57*), the exception being *Bacillus amyloliquefaciens* which is a zinc-dimer of two 48,000-dalton subunits (*58*). All α-amylases investigated so far are calcium-metallo enzymes. The calcium ion(s) apparently is chelated by various groups on the enzyme and contributes to the stability of the tertiary structure. The calcium ion(s) imparts resistance against effects of pH, temperature, and proteolysis (*59*). The number of calcium ions per molecule of enzyme varies. Usually there are one or two gram-atoms per mole of enzyme that are very tightly bound and a number (1 to 10) more loosely bound (*60, 61*). The pH-activity curves are usually broad with optima between 6 and 8. In addition to calcium, mammalian α-amylases, pancreatic and salivary, also require chloride ion for maximum activity (*62–64*).

As already mentioned, *Bacillus amyloliquefaciens* α-amylase contains, in addition to calcium, one gram-atom of zinc ion per 96,000 daltons. The removal of zinc with EDTA produces a zinc-free form that has a molecular weight of 48,000 (*58, 65, 66*). By various treatments with EDTA, sodium chloride, and pH values above 8, different active molecular weight forms were obtained (*67*). Among these active forms were those with discrete molecular weights of 24,000, 48,000, and 72,000 both with and without zinc ions, and molecular weight aggregates with molecular weights between 125,000 and 400,000 (*67*).

Porcine pancreatic α-amylase has been shown to exist in two isozyme forms, each with a molecular weight of 53,000 (*76, 190*). Cozzone and co-workers have shown from C- and N-terminal analysis (*191*) and by the ordering of nine peptides from cyanogen bromide cleavage of a pulse-labeled enzyme (*192, 193*) that the isozymes exist as single peptide chains. Fabian has also proposed a single polypeptide chain based on the sequence analysis of ten carboxymethylcysteine containing peptides isolated from the tryptic digestion of the reduced and carboxymethylated enzyme (*194*). Contrary to this, reduction of the enzyme has produced two very similar, amylase-active peptides of 25,000 molecular weight, indicating that the enzyme is composed of two subunits held together by disulfide linkages (*195*). Others have reported results which indicate a two subunit protein, with molecular weights of the two peptides being 31,000 and 22,000 (*196*). Part of this dilemma has been resolved by Corrigan and Robyt (*197*) who isolated porcine pancreatic α-amylase under conditions that prevented proteolysis. The isolated enzyme was a single peptide of 53,000 molecular weight. When this enzyme was subjected to the action of α-chymotrypsin, limited hydrolysis occurred to give four distinct proteins with molecular weights of 53,000, 39,000, 27,000, and 13,000 and with very little decrease in amylase activity. Reduction and carboxymethylation of the α-chymotrypsin-hydrolyzed amylase converted the four proteins into peptide fragments with molecular weights between 27,000 and 13,000. It was concluded that α-chymotrypsin produces limited hydrolysis of the single peptide α-amylase molecule into peptide fragments of 27,000 and 13,000 molecular weight and that these fragments are held together by disulfide linkages. Thus, limited chymotrypsin hydrolysis could convert the single amylase peptide into two 25,000-molecular-weight peptides which are held together by disulfide linkages to give a 50,000-molecular-weight α-amylase molecule.

The second part of the dilemma involves the number of active sites per 53,000 molecular weight. Several lines of evidence suggest two active sites: (a) the formation of two 25,000 MW active proteins after reduction (*195*); (b) equilibrium dialysis studies indicating the binding of two maltotriose molecules per 53,000 daltons (*107*); (c) two binding sites for cyclomaltoheptaose (*198*); (d) hydrodynamic and low angle x-ray scattering studies indicating that the 53,000-MW α-amylase molecule contains two structural domains separated by a large crevice or channel (*199*); and (e) x-ray diffraction studies showing that the enzyme has an internal dyad axis with two identical structural domains (*200*) and two binding sites for maltotriose (*201*). Thus, the data indicate that porcine pancreatic α-amylase is synthesized as a continuous polypeptide chain of 496 amino acids (*202*) containing a tandemly repeated primary sequence that is organized into two identical structural domains (*200*). Limited proteolysis *in vivo* and *in vitro* of this single-peptide molecule gives two peptide fragments of approximately 25,000 molecular weight held together by disulfide bonds, both of which have α-amylase activity.

Many of the α-amylases exist as slightly different isozymic forms that are separable by polyacrylamide gel electrophoresis (*68–72*) and ion exchange chromatography (*73–76*). The formation of the different isozymes found in saliva and pancreatic juice is apparently under genetic control. Differences in the structure of these amylases can be attributed to slight differences in the amino acid composition, the presence or absence of amide groups, and a variable percentage of covalently linked carbohydrate (*65, 77, 78*). The enzymic properties of the isozymes are either very similar or identical (*75*).

An examination of the amino acid composition of the α-amylases shows that they do not have any unusual features except for *B. amyloliquefaciens* α-amylase which contains neither cysteine nor cystine and hence has neither sulfhydryl (thiol) groups nor disulfide bonds (*79*). The α-amylases that have been examined are all high in aspartic and glutamic acids and have acidic isoelectric pH values around 4. *Aspergillus oryzae* and porcine pancreatic α-amylases both are glycoproteins containing a small percentage of covalently linked carbohydrate (*80–82*). Compilations of the number and kind of amino acids in α-amylases have appeared (*39, 57, 83, 84, 204*).

The molecular weight of crystalline sweet potato β-amylase is 197,000 (*85*). Dissociation studies suggest that the molecule is a tetramer made up of subunits of about 48,000 daltons (*85*). Equilibrium dialysis measurements with cyclohexaamylose further indicate that there are four binding sites per 197,000 daltons (*86*). Alkylation of the sulfhydryl groups of sweet potato β-amylase destroys its activity (*85*). The sulfhydryl groups, however, are not required for enzymic activity. It has been speculated that the sulfhydryl groups are involved in regulatory functions *in vivo* (*85*) and that β-amylases can be reversibly inactivated by disulfide interchanges. The pH optima of the β-amylases are usually lower than those of the α-amylases and are between 4 and 5 (*26, 87*).

The molecular weight of *A. niger* glucoamylase is 97,000 and its pH optimum is 5 (*88*). As with the α-amylases, the amino acid compositions of β-amylase and glucoamylase do not have any unusual features (*89, 90*). The fungal glucoamylases have been found to be glycoproteins. *A. niger* glucoamylase contains a significant amount of D-mannose and smaller amounts of D-galactose and D-glucose (*91*). Most glucoamylase preparations have isozymic forms (*92, 93*).

IV. Action of Amylases (see also Chapter V)

1. General

The action of α-amylases is most frequently considered to be a random process in which the enzyme molecules show equal preference for all α-D-(1→4) linkages except those adjacent to the two ends of the substrate chain and those in the vicinity of branch points (*94, 95*). This concept was inferred from the observation that α-amylases produce a very rapid decrease in the iodine color and a rapid drop in the viscosity of starch solutions when compared with a slow decrease in iodine color and a slow drop in viscosity when starch is hydrolyzed with β-amylase or glucoamylase. However, it is now recognized that these differences reflect the difference in the endo vs. exo mechanism by which an amylase encounters the polysaccharide.

A comparison of the rate of decrease in blue iodine color (blue value) with the rate of increase in the reducing value for α-amylases from different sources gives different curves (Fig. 1) and all are different from the curve produced by acid-catalyzed hydrolysis (*96*), which was expected to be a rather random process. Further, during early stages of α-amylase action, a set of specific, low-molecular-weight oligosaccharides are found (*96–101*). Therefore, it was concluded (*96*) that the action of α-amylases is not random as was previously believed.

For enzymic action on polymeric substrates, three distinct mechanisms are possible: single chain, multichain or single attack, and multiple attack (see Fig. 2). In the single-chain mechanism, once the enzyme forms an active complex

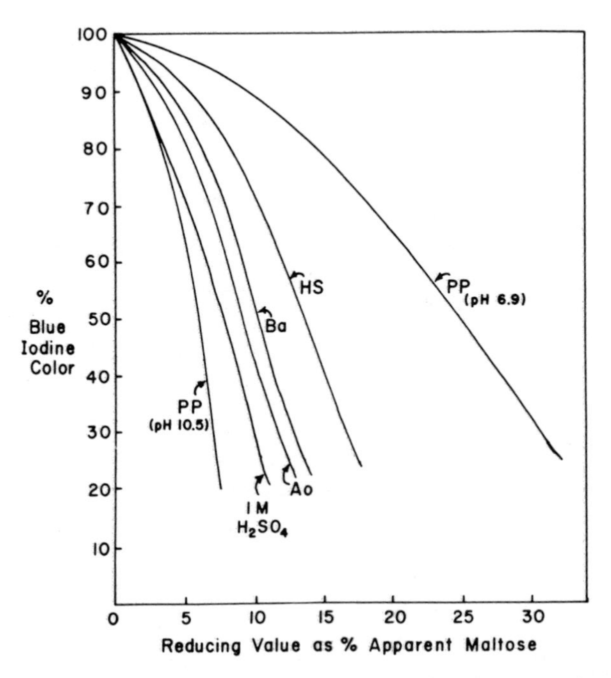

FIG. 1.—Blue iodine color vs. reducing value for the action of porcine pancreatic α-amylase (PP) at pH 6.9, human salivary α-amylase (HS) at pH 6.9, *Bacillus amyloliquefaciens* α-amylase (Ba) at pH 6, *Aspergillus oryzae* α-amylase (Ao) at pH 5.5, 1 *M* sulfuric acid at 60°, and (PP) at pH 10.5 on amylose. Data in part from Robyt and French (*96*).

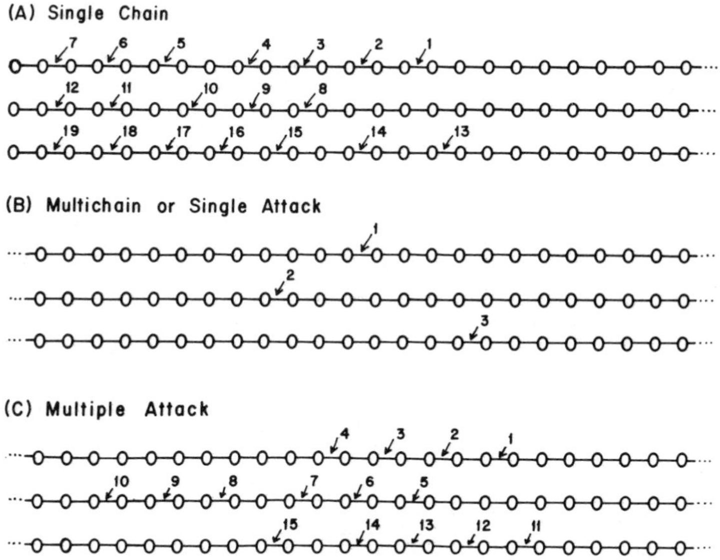

(A) Single Chain

(B) Multichain or Single Attack

(C) Multiple Attack

FIG. 2.—Types of attack patterns for endo amylases. Each type represents the action of a single enzyme molecule. The arrows represent the catalytic hydrolysis of a glycosidic bond; the numbers indicate the sequence of each catalytic event. The direction of multiple attack is toward the nonreducing end as determined for porcine pancreatic α-amylase by Robyt and French (*102*).

with the substrate, it catalyzes a reaction in a "zipper" fashion toward one end of the substrate and does not form an active complex with another substrate until it comes to the end of the first substrate. The multichain process is the classical random attack in which the enzyme catalyzes the hydrolysis of only one bond per encounter. In the multiple-attack mechanism, once the enzyme forms a complex with the substrate and produces the first cleavage, the enzyme remains with one of the fragments of the original substrate and catalyzes the hydrolysis of several bonds before it dissociates and forms a new active complex with another substrate molecule.

It was proposed that α-amylases have a multiple attack mechanism (*96*). This mechanism is consistent with the observed rapid decrease in iodine color and the formation of low-molecular-weight maltodextrins. For an equal drop in blue value (e.g., 50%), the different α-amylases give different percents of conversion to apparent maltose (Fig. 1). These differences in reducing value may be interpreted as differences in the amount of multiple attack that occurs per encounter of enzyme and substrate. Qualitatively, the blue value vs. reducing value curve shows that porcine pancreatic α-amylase has the highest degree of multiple attack at pH 6.9 and the lowest degree at pH 10.5. Human salivary, *Bacillus amyloliquefaciens,* and *Aspergillus oryzae* α-amylases have decreasing, intermediate degrees of multiple attack.

The quantitative amount of multiple attack for α-amylases and for acid-catalyzed hydrolysis is determined by the addition of two volumes of ethanol to the digests to fractionate them into high- and low-molecular-weight products (*96*). The reducing value before fractionation and the reducing value of the ethanol precipitate were measured. The degree of multiple attack was calculated from the ratio (r) of the reducing value before fractionation and the reducing value of the precipitate (*96*). The rationale was as follows. The reducing value before fractionation (total reducing value) represents the sum of two types of bond cleavages, a primary attack on a polysaccharide molecule to give two polysaccharide fragments and secondary attacks on the newly formed end of one of the fragments to give low-molecular-weight oligosaccharides. The reducing value of the ethanol precipitate is the reducing value of the polysaccharide fragments and represents the number of effective encounters. Thus, the ratio (r) of the total reducing value divided by the reducing value of the precipitate is a measure of the total number of bonds broken, divided by the number of effective encounters. Because the first bond broken releases a polymer fragment, the average number of subsequent bonds broken, the quotient minus one ($r - 1$), is equal to the average degree of multiple attack. The results of these experiments are summarized in Table II. For porcine pancreatic α-amylase, the degree of multiple attack ($r - 1$) at optimum pH was 6, *i.e.*, for every effective encounter, this α-amylase on the average breaks 6 bonds and releases maltose, maltotriose, or maltotetraose. The values of ($r - 1$) for porcine pancreatic α-amylase at pH 10.5 and for acid-catalyzed hydrolysis are 0.7 and 0.9, respectively, indicating that, at pH 10.5, the amylase has a random, multichain mechanism.

The direction of the multiple attack was determined for porcine pancreatic α-amylase using reducing-end and non-reducing-end ^{14}C-labeled maltooctaose at the optimum pH of 6.9, which favors multiple attack, and at the unfavorable pH

Table II

Degree of Multiple Attack of α-Amylases and Acid from the Ratio of the Total Reducing Value and the Reducing Value of the Ethanol Precipitate[a]

Catalyst	*Average number of attacks per encounter, (r − 1)*
Porcine pancreatic α-amylase at pH 6.9	6
Human salivary α-amylase at pH 6.9	2
Aspergillus oryzae α-amylase at pH 5.5	1.9
1 *M* sulfuric acid at 60°C	0.9
Porcine pancreatic α-amylase at pH 10.5	0.7

[a] Data from Robyt and French (*96*).

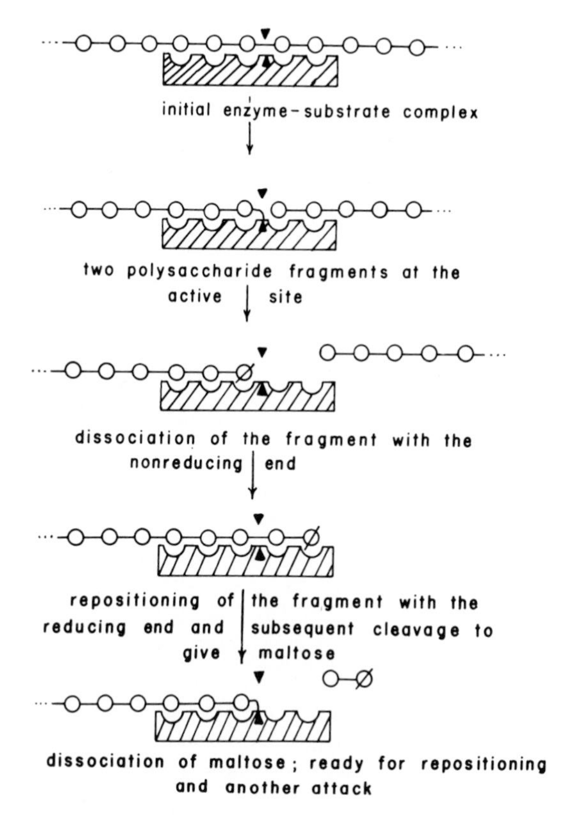

FIG. 3.—Sequence of events at the active site for multiple attack by an endo-acting enzyme. The active site is pictured here with five binding subsites and the catalytic groups located between the second and third subsites; ▲ and ▼ represent the catalytic groups; ○ represents a glucosyl unit; ∅, a reducing glucose unit; and —, an α-D-(1→4) glucosidic bond.

of 10.5, which favors multichain or single attack (*102*). By comparing the radioactive product distributions at the two pH values using the two types of labeled substrates, it was established that the direction of multiple attack is from the reducing end toward the nonreducing end: i.e., after the first cleavage, the fragment with the new nonreducing end dissociates from the active site while the fragment with the newly formed hemiacetal reducing end remains associated with the active site and repositions itself to give another cleavage and the formation of maltose or maltotriose (see Fig. 3).

2. Action of Porcine Pancreatic α-Amylase

A chromatographic study of the types of low-molecular-weight products produced by porcine pancreatic and human salivary α-amylases (*96, 97*) shows that the products are maltose, maltotriose, and maltotetraose. Contrary to previous views, D-glucose is not a product from amylose or amylopectin by the action of these enzymes. A study of the types of labeled products released from reducing-end labeled maltotriose through maltooctaose defined the bond specificity of porcine pancreatic α-amylase (*103*) (see Fig. 4). Only maltotriose and maltotetraose give D-glucose as a product. The former gives nearly three times as much D-glucose as the latter. Neither of these oligosaccharides are good substrates. The first substrate to undergo hydrolysis at an appreciable rate is maltopentaose which is exclusively cleaved at bond two to give labeled maltose. From this investigation, it was postulated that porcine pancreatic α-amylase has five D-glucose subsites and that the catalytic groups are located between the second and third subsite from the reducing-end subsite. The productive complexes with maltotriose, maltotetraose, and maltopentaose to give the observed product specificity is illustrated in Figure 5. The explanation of the product distribution and the relative rates is that the enzyme acts optimally when all subsites are filled.

The distribution of products from the action of porcine pancreatic α-amylase with the poor substrates, maltotriose and maltotetraose, was found to be concentration dependent (*103*). As the concentration of maltotriose increases, the apparent frequency of bond cleavage changes so that hydrolysis of bond two is preferred over hydrolysis of bond one. This apparent change in the distribution of products was shown to be due to a bimolecular reaction to give the condensation of two molecules of maltotriose. The resulting maltohexaose is then rapidly hydrolyzed to give the distribution of products characteristic of maltohexaose. A similar kind of condensation to produce maltooctaose from maltotetraose was also observed (*103*).

The action of porcine pancreatic α-amylase on amylopectin or glycogen eventually gives a series of dextrins that contain α-D-(1→6) linkages. The smallest is a tetrasaccharide, 6^3-α-D-glucopyranosylmaltotriose (B_4) (see Fig. 6) (*104*). In addition, a pentasaccharide, 6^3-α-D-glucopyranosylmaltotetraose (B_5) (see Fig. 6) and higher singly and doubly branched oligosaccharides were obtained by Kainuma and French (*104, 105*). They found that dextrins C–J of Figure 6 undergo further hydrolysis. Dextrin C gives D-glucose and B_4; dextrin D gives maltose and B_4; dextrin E also gives maltose and B_4; dextrin F gives maltotriose and B_4; dextrin G gives maltotriose and B_5; dextrin H gives D-glucose and a resistant heptasaccharide with the B_4 structure at its reducing end; dextrin I gives maltose and the same heptasaccharide with the B_4 structure at its reducing end; and dextrin J gives two B_4 molecules.

Schramm (*106*) found that the incubation of a 1:2 weight ratio of porcine pancreatic α-amylase to glycogen gives a precipitated macrodextrin–amylase complex. When either excess amylase or excess glycogen is added, the precipitate dissolves. The formation of the precipitated complex is compared with the formation of an antigen–antibody complex (*107*). The mechanism for the forma-

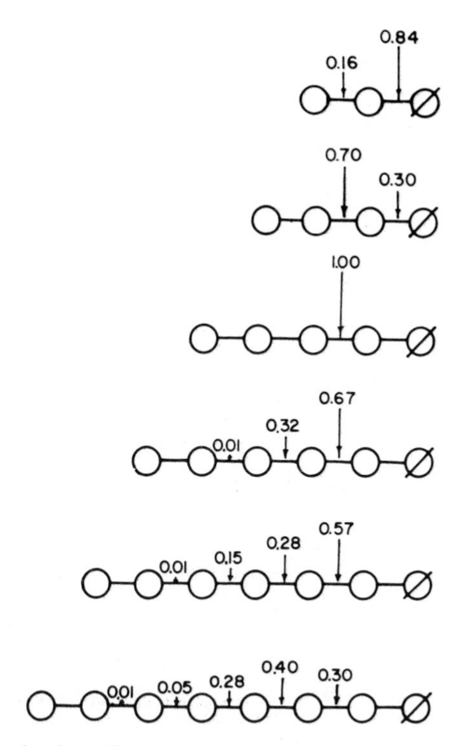

FIG. 4.—Quantitative bond specificity of porcine pancreatic α-amylase acting on pure maltosaccharides, maltotriose through maltooctaose. Under the conditions of the experiment, 84% of the cleavages of maltotriose occurred at bond one and 16% occurred at bond two, etc. From Robyt and French (*103*).

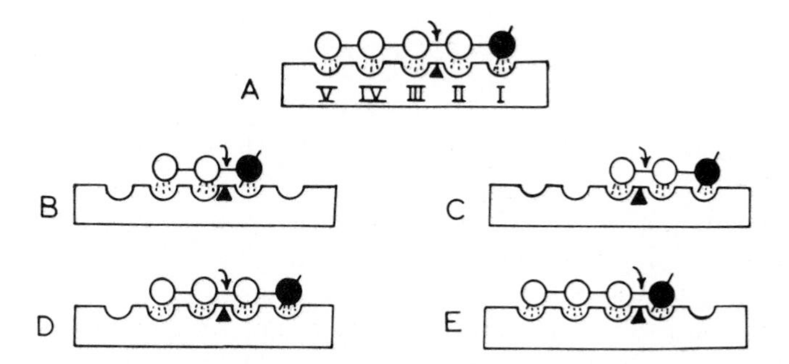

FIG. 5.—Productive complexes of porcine pancreatic α-amylase with maltotriose (B and C), maltotetraose (D and E), and maltopentaose (A). The enzyme is shown as a five-unit subsite with catalytic groups between subsites two and three. From Robyt and French (*103*).

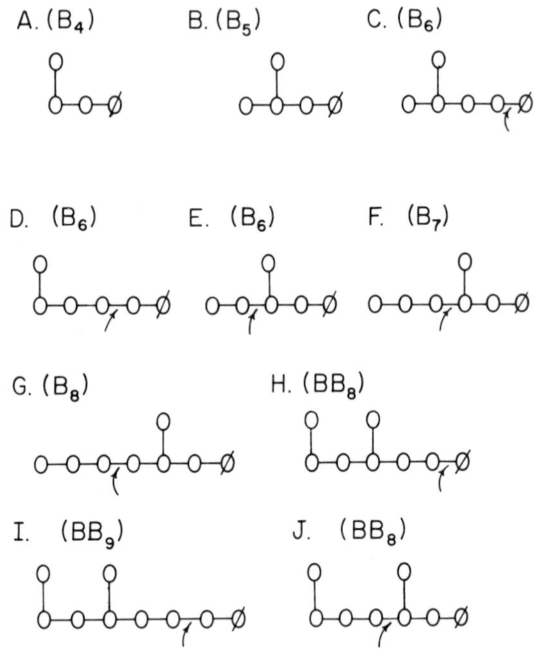

FIG. 6.—Branched dextrins resulting from the action of porcine pancreatic α-amylase on amylopectin. A and B represent B_4, a tetrasaccharide, and B_5, a pentasaccharide α-limit dextrins that cannot be cleaved further. C–G represent singly branched dextrins that undergo a single additional reaction indicated by the arrow. H–J represent doubly branched dextrins that undergo a single additional reaction indicated by the arrow. ○ represents a glucosyl unit, ∅ a reducing glucose unit, — an α-D-(1→4) glucosidic bond, and | an α-D-(1→6) glucosidic bond. From Kainuma and French (*104, 105*).

tion of this type of multimolecular complex requires the amylase to have a minimum of two binding sites per molecule. The precipitation of pancreatic α-amylase with glycogen has been used to purify the enzyme (*108*).

3. Action of *Bacillus amyloliquefaciens* α-Amylase

The action pattern of *Bacillus amyloliquefaciens* α-amylase was shown to be complex (*101*). Reaction with amylose and amylopectin gives predominantly two sets of products, maltose plus maltotriose and maltohexaose plus maltoheptaose. In the early and intermediate stages, only small amounts of D-glucose, maltotetraose, and maltopentaose are formed. Studies with individual oligosaccharides show that maltooctaose is cleaved primarily in two ways: (a) the second bond from the reducing end is hydrolyzed to give maltose and maltohexaose, and (b) the third bond from the reducing end is hydrolyzed to give maltotriose and maltopentaose. Maltoheptaose is likewise cleaved in two ways: (a) the first bond from the reducing end is cleaved to give D-glucose and maltohexaose and (b) the second bond is cleaved to give maltose and maltopentaose. Maltohexaose is hydrolyzed very slowly at bond one to give D-glucose and maltopentaose. Maltopentaose and smaller homologues are hydrolyzed extremely slowly, if at all (*101*).

The action of this α-amylase with the β-amylase limit dextrin of amylopectin produces an apparent change in action pattern from that obtained from amylopectin. The predominant products from the limit dextrin are D-glucose, maltose, and maltotriose. Because of the structure of this substrate, these products must come exclusively from the segments between the α-D-(1→6) branch points. This finding further indicates that the formation of maltohexaose and maltoheptaose from the original amylopectin comes exclusively from the exterior, non-branched chains of the substrate. The distribution of the products from shellfish glycogen shows yet another change. The predominant products are maltopentaose, maltohexaose, and maltoheptaose. In the very late stages of hydrolysis of shellfish glycogen, small amounts of D-glucose and maltose appear, owing to the subsequent hydrolysis of the primary products, maltohexaose and maltoheptaose. No low-molecular-weight products were observed from the β-amylase limit dextrin of shellfish glycogen, although a limited reaction does occur as judged by a small increase in the reducing value. This indicated the maltopentaose, maltohexaose, and maltoheptaose formed from the original glycogen substrate come exclusively from the exterior chains, and that the number of α-D-(1→4) linked D-glucosyl residues between the α-D-(1→6) branch points is not sufficient to give low-molecular-weight products. The smallest branched α-limit dextrin that results from an exhaustive reaction with amylopectin is a pentasaccharide having a maltosyl residue attached to the second D-glucosyl residue of maltotriose by an α-D-(1→6) linkage (*109, 110*).

The action patterns of porcine pancreatic and *Bacillus amyloliquefaciens* α-amylases illustrate the point that the types and amounts of products produced are different for the different α-amylases and are determined by the number of subsites at the active site of the enzyme and the particular structural properties of the substrate. The action of *B. amyloliquefaciens* α-amylase was interpreted in terms of a dual multiple attack along an enzyme binding surface having nine D-glucosyl residues in which the catalytic groups are positioned between the third and fourth subsites from the reducing-end subsite (*101*) (see Fig. 7A).

Thoma and co-workers (*111, 112*) confirmed the nine subsites proposed by Robyt and French (*101*) by studying the product distributions and Michaelis parameters of various maltodextrins containing three to twelve D-glucosyl residues. They applied the concept of the unitary free energy of subsite association with the individual D-glucosyl residues in the maltodextrins and developed a subsite energy profile (see Fig. 7B). The profile indicates that subsite IV is an

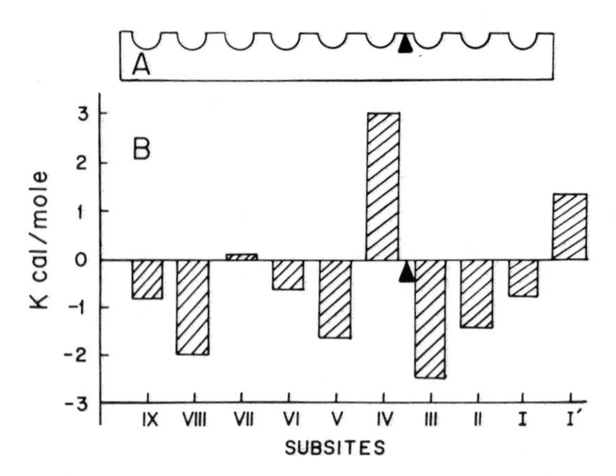

FIG. 7.—Active site of *Bacillus amyloliquefaciens* α-amylase. A, diagrammatic representation of the nine unit subsite with the catalytic groups between subsites III and IV. After Robyt and French (*101*). B, histogram showing the subsite–glucose interaction energies for the enzyme. From Thoma and co-workers (*112*).

anti-binding site that requires energy to force the substrate to occupy it. This anti-binding subsite is located where the glycosidic bond is cleaved and suggests that the D-glucosyl residue at this subsite must be distorted to be bound. This distortion could render the substrate more reactive and thus contribute to the catalytic process.

4. β-Amylase

The β-amylases attack the next to last glycosidic bond from the non-reducing end of a starch chain to specifically release β-maltose. The action of the enzyme is blocked by α-D-(1→6) branch linkages. Amylose, the linear component of starch, should be completely hydrolyzed, whereas only 55% of amylopectin, the branched component, is converted to β-maltose. The remaining 45% is a high-molecular-weight limit dextrin that contains all the branching of the original amylopectin molecule. With extensive reaction with β-amylase, the structure of the resulting limit dextrin depends on whether the chains contain an even or odd number of D-glucosyl residues. The amylopectin molecule can be designated as having two types of chains: A-chains that do not carry branch linkages and B-chains that carry the branch linkage. If both the A and B chains have an even number of D-glucosyl residues, structure I of Figure 8 will result; if both the A and B chains have an odd number of D-glucosyl residues, structure II of Figure 8 results; and if the A and B chains have either an even or an odd number of D-glucosyl residues, structures III and IV result (*113*).

β-Amylase was the first enzyme to be shown definitively to have a multiple

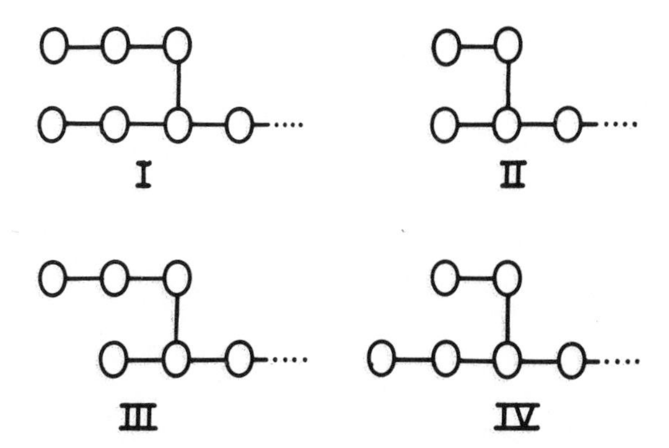

FIG. 8.—Structures of β-amylase limit dextrins resulting from the action of β-amylase on amylopectin. Structure I occurs when both A and B chains have an even number of glucose units to the branch point; II occurs when both A and B chains have an odd number of glucose units; III occurs when the A chain has an even number of glucose units and the B chain has an odd number of glucose units; and IV occurs when the A chain has an odd number of glucose units and the B chain has an even number. A chains are defined as those chains that do not carry branch linkages and B chains are those that carry at least one branch linkage.

attack mechanism (*114*). The enzyme repetitively cleaves a given substrate chain several times before dissociating. For low-molecular-weight amyloses, the average number of such cleavages is four (*114, 115*); but with higher-molecular-weight amyloses, the number of repetitive attacks appears to be much higher. The direction of multiple attack is opposite to that of porcine pancreatic α-amylase, i.e., it is from the non-reducing end toward the reducing end. This, of course, is what would be expected for an exo-enzyme acting from the non-reducing end.

5. New Amylases with Unique Product Specificities

In the last fifteen years or so, several new amylases have been discovered. Interest in them lies in the specific types of products that they produce. The first β-amylase from a source other than plants was reported from *Bacillus polymyxa* (*3*). This enzyme produces β-maltose from starch and β-amylase limit dextrin from amylopectin and is capable of opening the cyclic structure of cyclomaltooctaose. The latter, an unusual reaction, has been attributed to a small contaminant of α-amylase (*116*). Since this discovery, several microbial β-amylases from other species have been obtained (*117–119*).

Two new α-amylases have been reported from *Streptomyces* species (*121, 122*). Both enzymes give between 75% and 80% conversion to α-maltose by an endo mechanism. Their reaction with maltotriose is unusual; one of them gives a three-times higher molar ratio of maltose to D-glucose from maltotriose than do other α-amylases and the other produces 100% maltose from maltotriose. The latter result has been examined using reducing end labeled maltotriose and has been explained as arising from a transfer reaction followed by a hydrolysis reaction (*123*).

A maltotriose-forming amylase with an exomechanism has been obtained from *Streptomyces griseus* (*122*). A maltotetraose-forming amylase with an exomechanism has been obtained from *Pseudomonas stutzeri* (*124*). The latter purified enzyme exclusively cleaves the fourth glycosidic bond from the nonreducing end to give β-maltotetraose; 52% of amylopectin is converted to maltotetraose leaving a high-molecular-weight limit dextrin. A maltohexaose-forming exoamylase is obtained from *Aerobacter aerogenes* (*1, 2*). This enzyme cleaves the sixth glucosidic bond from the nonreducing end (*1*). It was found that this exo amylase would act on the β-amylase limit dextrin of amylopectin. The binding site of this enzyme could accommodate branch linkages and hence could bypass branch points to give branched oligosaccharides (*2*).

An α-amylase from *Bacillus licheniformis* was reported to give high yields of maltopentaose (*125*). Although several oligosaccharides are formed, the starch is converted to 33% maltopentaose by an endo mechanism. The enzyme has a high temperature optimum of 76° and a broad pH optimum from 5 to 8 with considerable activity in the alkaline pH range of 9–10.

6. Mechanism of Glycosidic Bond Hydrolysis

Exoenzymes may be thought of as having active sites that are "pockets" (see Fig. 9). The non-reducing ends of the starch chains fit into the pocket to specifically give a single low-molecular-weight product, *e.g.*, β-maltose in the case of β-amylase, β-D-glucopyranose in the case of glucoamylase, maltotetraose in the case of *P. stutzeri* amylase, etc. The endo enzymes, on the other hand, may be thought of as having active sites that are clefts or crevices. Here, the internal part of a starch chain binds to give endo cleavage; subsequent multiple attack yields one or more low-molecular-weight products. The active site, whether a pocket or cleft, may be conveniently divided into two parts: (a) the binding site made up of some number of subsites, and (b) the catalytic site made up of two or three groups that are proton donors (electrophiles) and proton acceptors (nucleophiles) (*126*). The number of subsites and their arrangement in conjunction with the catalytic groups determines the types of products formed. The catalytic groups essentially promote catalysis by an acid-base mechanism. These groups are specifically positioned in relationship to the bound substrate so that they efficiently catalyze hydrolysis of a specific glycosidic bond.

The further consideration of the quantitative retention or inversion of the cleaved bond for α- and β-amylase action must be made (*127*). The absence of

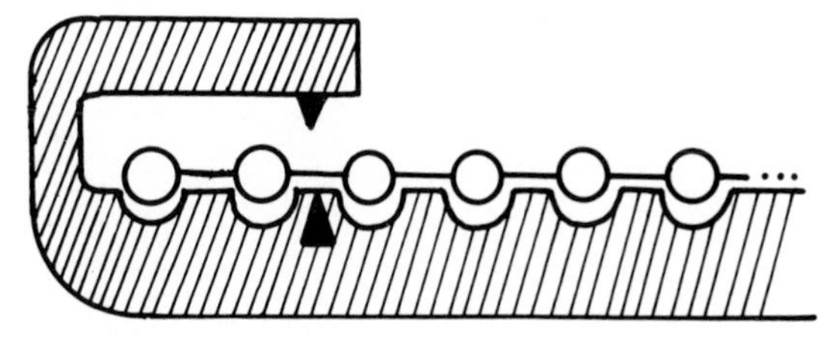

FIG. 9.—The "pocket" active site of β-amylase. Six subsites are shown with the catalytic groups between the second and third subsites from the nonreducing end.

racemization during catalysis by both of these enzymes suggests that a specific mechanism is in operation. In this mechanism, there must be a reaction of a reactive intermediate with water in a sterospecific manner before the product is released from the active site.

Studies of the nature of the catalytic groups of both α- and β-amylases have implied that carboxylate anions act as the nucleophiles and imidazolium cations act as the electrophiles (*128–130*). The retention of configuration occurring during α-amylase action suggests a double displacement mechanism involving a covalent intermediate. Such a mechanism involving the catalytic groups is shown in Figure 10. It would be expected that any covalent intermediate that would occur during amylase action would be unstable and readily hydrolyzed. The acetal–ester intermediate of Figure 10 fulfills this criterion. Although covalent intermediates have never been demonstrated for amylases, they are nevertheless attractive when considering the types of reactions that occur, *e.g.*, hydrolysis, reversion, and condensation (*110, 131, 132*). Furthermore, evidence is mounting that the formation of covalent intermediates is a relatively common phenomenon in enzyme catalysis (*133–136*).

For actions of β-amylase and glucoamylase, in which inversion of configuration occur, a double displacement mechanism on the anomeric carbon is difficult

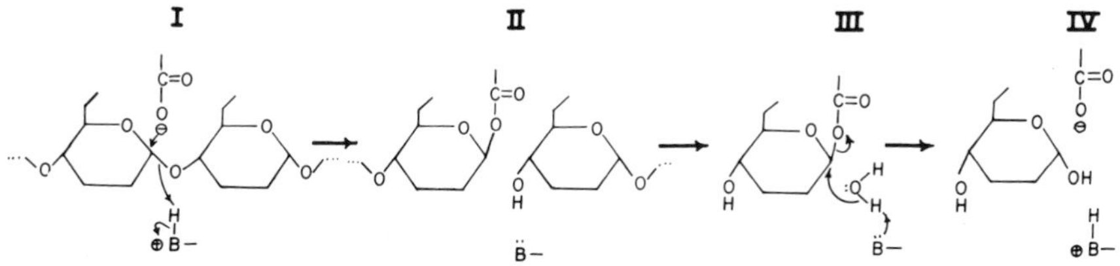

FIG. 10.—Proposed mechanism for α-amylase catalysis. $-BH^+$ represents an imidazolium group at the active site. I, attack of the catalytic carboxylate anion to cleave the glycosidic bond; II, formation of covalent enzyme carboxyl–acetal ester; III, dissociation of non-reducing-end fragment and attack of water on the anomeric carbon atom to hydrolyze the ester; IV, formation of α-product, regeneration of the active ionic forms of the catalytic groups, and ready for the repositioning of the reducing-end fragment for multiple attack.

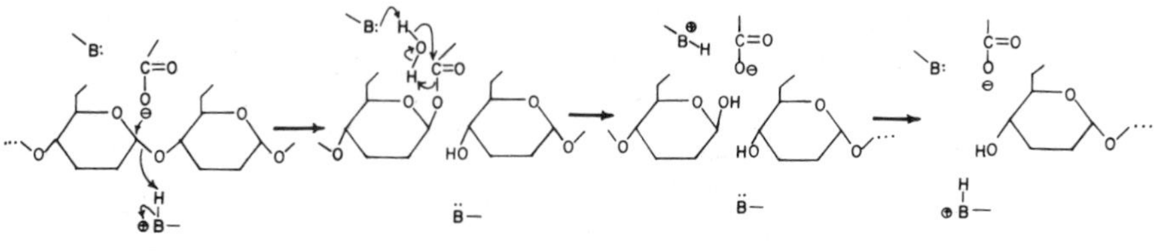

FIG. 11.—Proposed mechanism for β-amylase catalysis. $-BH^+$ represents an imidazolium group and −B: represents an imidazole group at the active site. I, attack of the catalytic carboxylate anion to cleave the glycosidic bond; II, hydrolysis of the covalent enzyme carboxyl–acetal ester by attack of water on the carboxyl carbon; III, formation of the β-product; IV, dissociation of the β-product, regeneration of the active ionic forms of the catalytic groups, and making ready for the repositioning of the non-reducing-end fragment for multiple attack.

to envision. Inversion by β-amylase has been suggested to occur via the enzyme specifically directing the approach of the water molecule to the reaction center in such a way that there is a backside attack onto an enzyme-stabilized, D-glucosyl–oxycarbonium ion (*129, 130*). French (*137*), however, has proposed a very possible displacement mechanism that gives inversion of configuration and is analogous to the double displacement mechanism of α-amylase (see Fig. 11). In contrast to the α-amylase mechanism, the water molecule is not directed to the anomeric carbon atom of the substrate, but is directed to the carbon atom of the carboxyl group that makes a covalent acetal–ester with the anomeric carbon atom. This mechanism is particularly attractive in that the same covalent intermediate, a carboxyl–acetal ester, occurs for both α- and β-amylase mechanisms. Many years ago, Thoma and co-workers (*127*) suggested that the mechanisms of hydrolysis for α- and β-amylases may be very similar. Thoma and co-workers (*84*) have discussed in some detail the stereochemistry, ring distortion, energetic requirements, and polyfunctional amino acid side-chain catalysis for amylase action.

7. *In Vivo* Conversion of Starch to D-Glucose

In the mammal, two α-amylases are produced, one by the salivary glands and the other by the pancreas. Salivary α-amylase is the first to encounter food starch in the mouth. As we have seen, it catalyzes the hydrolysis of amylose to maltose, maltotriose, and maltotetraose and the hydrolysis of amylopectin to the same products plus two α-limit dextrins, a branched tetrasaccharide (B_4, see Fig. 6A) and a branched pentasaccharide (B_5, see Fig. 6B). The food starch and the amylase quickly pass from the mouth into the stomach where the pH is ~ 2, and the action of the amylase ceases. At this time, the starch is only partially hydrolyzed and, therefore, there is an appreciable amount of incompletely hydrolyzed dextrins and unreacted starch present in the stomach. Very little, if any, hydrolysis of starch takes place in the stomach. After some time in the stomach, the food material passes into the small intestine and is neutralized. Pancreatic α-amylase is secreted into the small intestine from the pancreatic duct and completes the hydrolysis of the starch. The products are essentially the same as those that would be produced by the prolonged action of salivary α-amylase, with the possible quantitative exception of the formation of higher amounts of maltose. Neither salivary nor pancreatic α-amylases produce appreciable amounts of D-glucose from starch, as D-glucose is only produced by these amylases from a very slow secondary reaction with maltotriose and maltotetraose (*103*). Neither enzyme has any action on maltose or the α-limit dextrins. Yet the ultimate goal of the organism is to convert starch into D-glucose which can be actively transported across the small intestine membrane into the blood and thereafter be converted into liver or muscle glycogen, energy, or fat. To convert the α-amylase products into D-glucose, two other enzymes are required, an α-1,4-glucosidase and an α-1,6-glucosidase. Both enzymes are secreted by the brush border cells of the lining of the small intestine and by the pancreas (*138, 139*).

The α-1,4-glucosidases convert maltose, maltotriose, and maltotetraose into

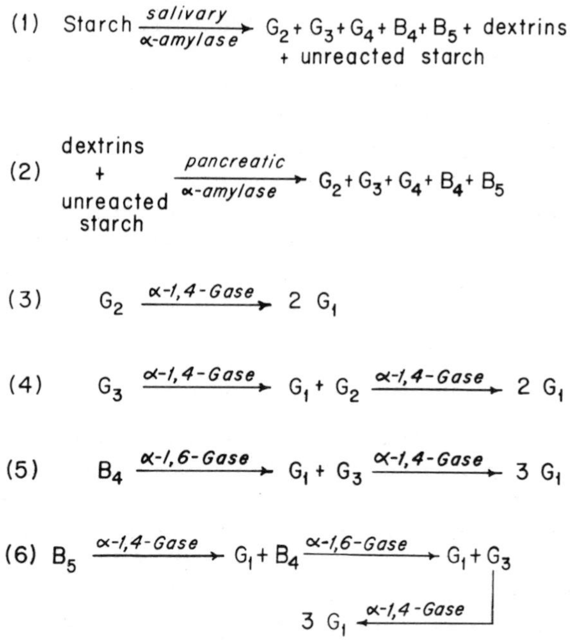

$$(1)\quad \text{Starch} \xrightarrow[\alpha\text{-amylase}]{\text{salivary}} G_2 + G_3 + G_4 + B_4 + B_5 + \text{dextrins} + \text{unreacted starch}$$

$$(2)\quad \text{dextrins} + \text{unreacted starch} \xrightarrow[\alpha\text{-amylase}]{\text{pancreatic}} G_2 + G_3 + G_4 + B_4 + B_5$$

$$(3)\quad G_2 \xrightarrow{\alpha\text{-}1,4\text{-Gase}} 2\ G_1$$

$$(4)\quad G_3 \xrightarrow{\alpha\text{-}1,4\text{-Gase}} G_1 + G_2 \xrightarrow{\alpha\text{-}1,4\text{-Gase}} 2\ G_1$$

$$(5)\quad B_4 \xrightarrow{\alpha\text{-}1,6\text{-Gase}} G_1 + G_3 \xrightarrow{\alpha\text{-}1,4\text{-Gase}} 3\ G_1$$

$$(6)\quad B_5 \xrightarrow{\alpha\text{-}1,4\text{-Gase}} G_1 + B_4 \xrightarrow{\alpha\text{-}1,6\text{-Gase}} G_1 + G_3 \xrightarrow{\alpha\text{-}1,4\text{-Gase}} 3\ G_1$$

FIG. 12.—Summary of the reactions involved in the mammalian *in vivo* conversion of starch into glucose. Gase is an abbreviation for glucosidase.

D-glucose by successive action from the non-reducing end. This enzyme also hydrolyzes B_5 to D-glucose plus B_4. The α-D-(1→6) linkage of B_4 is then hydrolyzed by the α-1,6-glucosidase to D-glucose plus maltotriose, which is then converted to D-glucose by the α-1,4-glucosidase. These two enzymes do not have any action on starch per se, but their combined action completes the job of converting starch into utilizable D-glucose by acting on the low-molecular-weight, primary products, of α-amylase action. See Figure 12 for a summary of these reactions.

8. Amylase Inhibitors

Amylase inhibitors may be divided into two types: (a) those that occur in plants and are primarily proteins, and (b) those that are chemical modifications of monosaccharides or oligosaccharides and that either occur naturally or are produced in the laboratory and are substrate analogs or are thought to mimic the transition state in the hydrolytic process. Of the former, the oldest recognized α-amylase inhibitor was found in a water extract of wheat flour and is known as Kneen's inhibitor (*203, 204*). The inhibitor has protein properties: it is precipitated by ammonium sulfate and ethanol and is hydrolyzed by pepsin; it is relatively heat stable, not being affected by long storage at 70° or a short heating to 100°, but is completely inactivated by autoclaving at 121° for 30 min; it is water-soluble and is inactivated by reducing agents (*205*). The inhibition is non-competitive (*206*). Purification on DEAE- and CM- cellulose resolved two forms of the inhibitor with molecular weights of 18,000 and 26,000 (*207*). Inhibition of α-amylase is much greater after preincubation with inhibitor, followed by addition of substrate, rather than preincubation of substrate with inhibitor, followed by addition of α-amylase (*207*). The inhibition is relatively specific for the mammalian salivary and pancreatic α-amylases; the inhibitor has no effect on fungal (*A. oryzae*) and bacterial (*B. amyloliquefaciens*) amylases and plant α-amylases (*203, 204, 208*). It does inhibit *B. subtilis* var. *amylosaccharitikus* α-amylase (*203, 204*).

A protein α-amylase inhibitor from rye (*203, 204, 209*) is similar to the wheat inhibitor and inhibits salivary and pancreatic α-amylases and has no effect on *B. amyloliquefaciens* and plant α-amylases (*210*). The rye inhibitor is different in that it inhibits salivary α-amylase about 10 times faster than pancreatic α-amylase.

An α-amylase inhibitor from kidney beans, *Phaseolus vulgaris,* was reported in 1945 (*211*) and again reported in 1968 (*212*) when it was observed that rats fed raw kidney beans excreted undigested starch. This inhibitor was also shown to be a protein. Similar α-amylase inhibitors are present in other legumes (*213*). *Phaseolus vulgaris* α-amylase inhibitor has been purified to homogeneity and named phaseolamin (*214*). The inhibition is noncompetitive and occurs optimally at pH 5.5 and 37° with a 1:1 complex of α-amylase and phaseolamin being formed. The inhibition is specific for mammalian, salivary, and pancreatic α-amylases; the inhibitor has no effect on fungal (*A. oryzae*) or bacterial (*B. amyloliquefaciens* and *B. licheniformis*) α-amylases or plant (barley or rye) α-amylases (*214*). Phaseolamin is a glycoprotein of molecular weight between 45,000 and 50,000. Like the wheat inhibitor, phaseolamin inhibits best when preincubated with α-amylase and does not cause any measurable inhibition when added with the substrate. Addition of substrate to α-amylase preincubated with inhibitor, however, does not reverse the inhibition (*214*).

A protein amylase inhibitor from oats inhibits both α- and β-amylases. The inhibition is reversed by the addition of sulfhydryl (thiol) reagents, which suggests that the inhibitor combines with sulfhydryl group(s) on the enzyme to give mixed protein disulfides (*215*).

Two nonprotein α-amylase inhibitors have been found in potato tubers (*216*) and sorghum (*217, 218*). These inhibitors are very heat stable (autoclave temperature), dialyzable, and organic solvent soluble. The inhibitors appeared to be tannins and nonspecific, for they inhibited many enzymes besides α-amylases.

The assay of the α-amylase inhibitor involves preincubation (usually 20 min) at 37°, pH 5.5, and then measurement of amylase activity in the usual manner by adding a starch solution (10 mg/mL) buffered at the optimum pH of the α-amylase. The amylase activity is measured by determining the increase in the reducing value with time or the decrease in iodine–iodide blue color with time. A comparison with α-amylase without inhibitor is made, and 1 unit of inhibitor is the amount that produces 50% inhibition in 20 min (*210*).

A relatively rapid screening procedure for the detection of amylase inhibitors in biological extracts has been described (*219*). A cellulose strip is saturated with the extract by dipping and placed onto a buffered starch–agar plate for 2 h at 37°. The strip is removed and another strip, saturated with amylase, is placed on the gel at right angles across the position which had the strip with the extract. After incubation at 37°, the second strip is removed and the slab flooded with iodine solution. The presence of an inhibitor in the extract is shown by the blue coloration of the area where the inhibitor strip and amylase strip cross.

A variation of this method involves the addition of amylase-extract mixtures to wells in the starch–agar and the measurement of the diameter of the cleared zones. Another, even simpler variation, is the use of antibiotic paper discs, which are dipped into amylase–inhibitor mixtures and then placed onto starch–agar, which is flooded with an iodine solution after incubation. The latter methods offer the advantages of a quantitative determination made by comparing zone diameters and the use of a minimum of manipulative steps with very simple equipment.

In late 1981 and early 1982, phaseolamin appeared in commercial, ethical pharmaceuticals called "starch blockers." These pills were purported to reduce starch digestion and hence reduce the caloric value of starchy foods. The efficacy of these products was immediately questioned. A dichotomy arose regarding the mechanism of their action: on the one hand, it was argued that their effect would be minimal in that the inhibitor has to be preincubated with the amylase and that salivary α-amylase is continuously secreted and a fresh supply of enzyme would be present during mastication of the food in the mouth, well after the starch blocker pills had been taken and swallowed, and that the phaseolamin would be inactivated in the stomach by pepsin before it could inhibit pancreatic α-amylase; on the other hand, however, symptoms of undigested starch in the large intestines, such as flatulence, nausea, vomiting, cramps, and diarrhea, were observed. Because of these undesirable symptoms and an inadequate knowledge of side effects, the Food and Drug Administration, in the summer of 1982, banned the sale of starch blockers and ordered their removal from health and drugstore shelves (*220*), and then in late 1982 concluded that starch blockers were not effective agents for the reduction of the caloric value of starchy foods (*221*). Reports of laboratory studies seemed to substantiate the FDA position (*222*, *223*).

With regard to the second type of inhibitor, it has been postulated that the enzyme-catalyzed hydrolysis of the α-D-(1→4) glucosidic bonds of the starch molecule involves enzyme-induced ring distortion of one of the D-glucosyl residues from the $^4C_1$ chair conformation to a "half-chair" conformation (*129*). It is argued that ring distortion decreases the enthalpy of activation and increases the susceptibility of the glucosyl residue to nucleophilic attack by functional groups on the enzyme and water. X-Ray diffraction studies indicate that D-glucono-1,5-lactone (Fig. 13) is close to a half-chair conformation (*224*). Reese and co-workers (*225*) found that 17 m*M* D-glucono-1,5-lactone produced 27% inhibition of *A. oryzae* α-amylase and 7% inhibition of β-amylase, while 6 mM lactone produced 50% inhibition of *A. niger* glucoamylase. László and co-workers (*226*) also found substantial inhibition of porcine pancreatic α-amylase, sweet potato β-amylase, and *A. niger* glucoamylase. They found that D-glucono-1,5-lactone inhibited α-amylase noncompetitively with a K_i of 0.81 m*M*, and maltobionolactone inhibited competitively with a K_i of 0.31 m*M*; β-amylase was inhibited competitively by maltobionolactone with a K_i of 0.11 m*M*. D-Gluconolactone was a poor inhibitor of β-amylase with a K_i of 21 m*M*. As a comparison, maltobionolactone is a much better inhibitor for α- and β-amylases than is maltose, which has K_i values of 12.5 and 5 m*M* respectively. Maltobionolactone was not an inhibitor for glucoamylase, but D-gluconolactone was with a K_i of 1.3 m*M*.

From these studies, it was concluded by László and co-workers (*226*) that ring distortion or a half-chair conformation is involved in the transition state of α-, β-, and glucoamylases. It must be cautioned, however, that a requirement for ring distortion or half-chair conformation in the transition state does not in itself

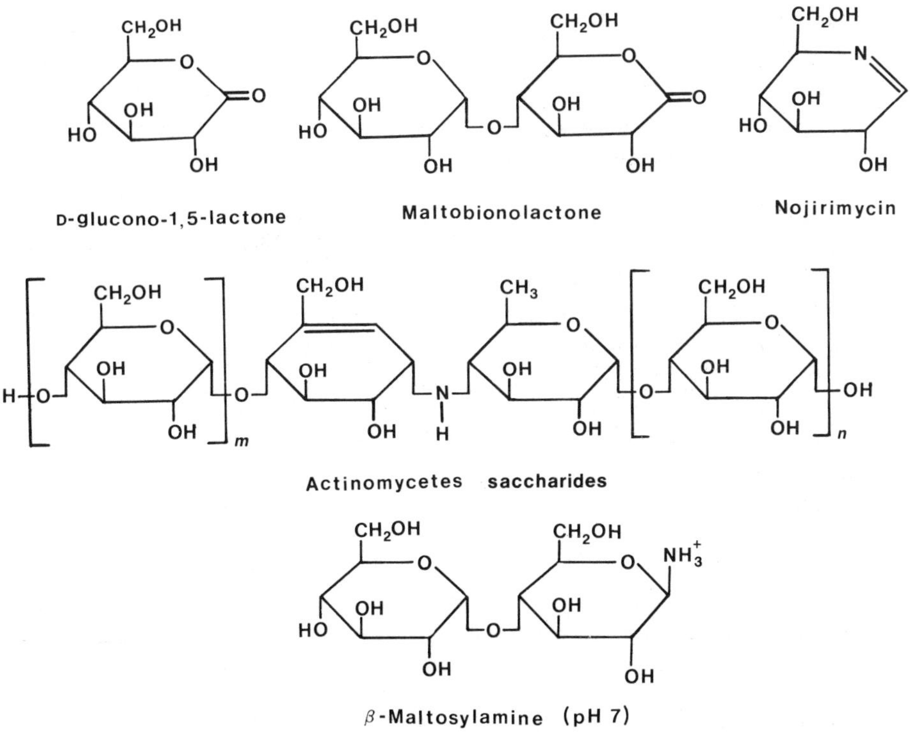

D-glucono-1,5-lactone   Maltobionolactone   Nojirimycin

Actinomycetes saccharides

β-Maltosylamine (pH 7)

FIG. 13.—Modified monosaccharide and saccharide amylase inhibitors.

constitute evidence for the type of mechanism involved, *i.e.*, whether the enzyme-catalyzed hydrolysis goes by a carbonium ion mechanism or by a high-energy covalent intermediate, as both mechanisms could involve a half-chair conformation transition state (*227*).

Another modified D-glucopyranose, nojirimycin, 5-amino-5-deoxy-D-glucopyranose (Fig. 13) gave some inhibition of α- and β-amylase, but inhibited glucoamylase much better (*225*). A possible reason that nojirimycin was not very effective as an inhibitor of α- and β-amylases is that they have several subsites, and nojirimycin, being a monosaccharide, could only occupy one of the subsites at a time. This may also explain why maltobionolactone is a much better inhibitor for these amylases than is D-glucono-1,5-lactone. Other nitrogen-containing α-amylase inhibitors are found in a family of compounds isolated from several *Actinomycetes* species (*228, 229*). The main structural feature of these compounds is α-(1→4)-linked D-glucopyranose units linked to an unsaturated cyclitol which in turn is linked to an amino sugar, 4-amino-4,6-dideoxy-D-glucopyranose, which is linked to another α-(1→4)-linked maltodextrin chain (see Fig. 13). The inhibitory saccharides have one of the glycosidic oxygen atoms replaced by a nitrogen atom. Maximum specific inhibition of α-amylase occurs when $n = 2$ and $m = 2$ or 3. The inhibitor is not effective against β-amylase.

A third nitrogen-containing inhibitor was β-maltosylamine, which is an effective inhibitor for sweet potato β-amylase and glucoamylase (*230*). According to the report, the latter is not inhibited as effectively as the former, and pancreatic and *B. amyloliquefaciens* α-amylases are poorly inhibited, although comparative K_i values were not given. The β-amylase inhibition was uncompetitive and reversible and appeared to be directed to the active site as it protected the essential sulfhydryl group from inactivation by *N*-ethylmaleimide. The uncompetitive inhibition might suggest that the substrate binds to the enzyme first, perhaps inducing a conformational change, and then the inhibitor is bound (*230*).

The Schardinger dextrins, cyclomaltohexaose and cyclomaltoheptaose, are competitive inhibitors of β-amylase, each with K_i of 5×10^{-4} *M* (*86, 231*).

V. Biosynthesis of Starch (see also Chapter III, Section VI)

1. Phosphorylase and Starch Synthetase

The first report of the biosynthesis of a polysaccharide was made in 1939 by Cori and Cori (*140*) who were examining the action of muscle phosphorylase. Almost simultaneously, Hanes (*141*) reported his investigations of the biosynthesis of starch by plant phosphorylase. These investigators observed that chains of polysaccharide are elongated *in vitro* by the transfer of D-glucosyl residues from α-D-glucopyranosyl phosphate (G-1-P) to the nonreducing end of a growing primer chain (*141, 142*). A primer is defined as a necessary molecule (either a polymer or oligomer) whose ends (usually, if not always, the nonreducing end) are *required* and are the sites where new monomers are added. The following reaction may be written for the action of phosphorylase:

$$m\ \text{G-1-P} + \underset{\text{primer}}{G_n} \rightleftharpoons \underset{\substack{\text{elongated}\\ \text{polymer}}}{G_{(n+m)}} + \underset{\substack{\text{inorganic}\\ \text{phosphate}}}{m\ P_i}$$

Plant or starch phosphorylase has been prepared from potatoes, peas, and several cereal grains, and it has been identified in a large variety of other plants. It has been highly purified and crystallized from potato extracts (*143–145*). There are significant physical and chemical differences between animal and plant phosphorylases. Animal muscle phosphorylase is subject to physiological control by phosphorylation and dephosphorylation, dimerization and tetramerization, and AMP, while plant phosphorylases are not. Besides these differences, muscle phosphorylase and plant phosphorylases also differ in their specificities. Muscle phosphorylase utilizes branched primers best and elongates linear oligomers, such as maltoheptaose, poorly (*146*). Potato phosphorylase, on the other hand, can rapidly elongate maltotetraose and can even utilize maltotriose as a primer (*147*).

The equilibrium constant for the phosphorylase reaction is a function of pH, temperature, and ionic strength. At pH 6 and 30°, the equilibrium ratio of P_i to G-1-P is about 7.5 (*148*). This ratio is not particularly favorable for synthesis. Although polysaccharide can be synthesized *in vitro,* the accumulation of inorganic phosphate gradually slows the reaction. Further, unless the phosphorylase concentration is relatively high and the concentrations of primer and G-1-P are very dilute, polymerization is stopped by the precipitation of the synthesized amylose when it is only 150–200 D-glucosyl residues long (*137*). *In vivo,* the situation is even worse. The cytoplasmic ratio of P_i to G-1-P is much higher than 7.5 and the reaction would be expected to proceed from right to left in the above equation to give polysaccharide degradation rather than synthesis (*149*). It is now widely thought that phosphorylase acts as a degradative enzyme and provides the organism with a high-energy form of glucose from the stored polysaccharide. There still is some controversy concerning this view, however. Greenwood and Banks (*150*) have recently stated that, in their opinion, phosphorylase acts in a synthetic manner and that starch is degraded by the action of amylases. This concept has been championed by Badenhuizen (*151–153*) for a number of years.

In 1959, a new pathway for polysaccharide biosynthesis was discovered by Leloir and Cardini (*154*). They found that a nucleoside diphosphate sugar, uridine diphosphate glucose (UDPG), serves as a high-energy donor of D-glucosyl residues in glycogen biosynthesis. Closely following this finding, a starch synthetase was discovered that utilizes UDPG (*155, 156*) or adenosine diphosphate glucose (ADPG) (*157, 158*) as the D-glucosyl donor. The reaction may be formulated as follows:

$$m\ \text{ADPG (or UDPG)} + \underset{\text{primer}}{G_n} \rightarrow \underset{\substack{\text{synthesized}\\\text{polysaccharide}\\\text{chain}}}{G_{(n+m)}} + m\ \text{ADP (or UDP)}$$

The equilibrium constant for the reaction is very favorable for synthesis ($K \simeq 260$).

Starch synthetase, as originally reported (*156*), is a preparation of enzyme attached to starch granules which could utilize either UDPG or ADPG (*159*). It turned out that ADPG is a far better donor than UDPG and that waxy maize starch synthetase would only use ADPG (*158*). A soluble form of starch synthetase has been prepared from extracts of sweet corn (*160*), potatoes (*161*), and spinach leaf (*162, 163*). In the latter case, the synthetase was purified and separated from two types of branching enzymes (*164, 165*). The soluble enzymes exclusively use ADPG as the D-glucosyl donor. The grinding of a number of types of starch granules or the treatment of the granules with α-amylase or 8 *M* urea led to an increase in synthetase activity using ADPG and a decrease in the activity using UDPG (*166*). Further, the addition of ADPG to radioactive UDPG led to a decrease in the incorporation of radioactivity into starch. But the inverse, that is, the addition of UDPG to labeled ADPG, does not produce any change (*157*). Thus, the relationship between the activities using ADPG and UDPG is not clear. Because waxy maize starch, which is exclusively amylopectin, has a synthetase that will only use ADPG, it is tempting to speculate (*159*) that ADPG is used to synthesize amylopectin and UDPG is used to synthesize amylose and that there are two distinct enzymes, with distinct specificities, that synthesize amylopectin and amylose. Experimental support for this concept comes from the findings of Nelson and Rines (*167*) who found that [^{14}C]glucose from [^{14}C]UDPG is incorporated into amylose and amylopectin in about equal amounts. This suggests that there are separate enzymes for amylose and amylopectin and that UDPG favors amylose synthesis, for there is only 20% as much amylose as amylopectin in ordinary starch and the proportion of non-reducing ends, the apparent acceptor sites for glucose transfer, is far less in amylose (*ca.* 1:100) than in amylopectin. Thus in a random system, the chances of the amylose becoming labeled are very much less than the chances of amylopectin becoming labeled; because they were equally labeled, a preference for the use of UDPG for amylose synthesis is indicated.

2. Branching Enzyme

Besides synthetase, another enzyme, branching enzyme or Q-enzyme, is required for the synthesis of the α-D-(1→6) branch linkages found in amylopectin. Such an enzyme was first identified in potatoes (*168, 169*) and then in broad beans (*170*). This enzyme was observed to convert amylose into amylopectin by the synthesis of α-D-(1→6) branch points. Two types of branching enzymes have been found in some plants, one that converts amylose into amylopectin and another that converts amylopectin into a glycogen-like product (*171, 172*). The difference between the two enzymes apparently is in the specificity of the number of D-glucosyl residues between branch points that each enzyme permits. This is of interest, as some plants, sweet corn for instance, produce normal amylose and amylopectin and relatively large quantities of a glycogen-like material called phytoglycogen.

Specificity studies with Q-enzyme, the branching enzyme of potatoes, have shown that branching will not occur on amylose with chain lengths less than 35–40 D-glucosyl residues (*165*). It is capable, however, of introducing additional 1→6 branch linkages into amylopectin. Branching enzyme catalyzes the transfer of a part of one amylose chain to that of another chain (*173*). It is, therefore, an interchain α-1,4–α-1,6 transferase. For a plausible interchain transfer mechanism, see Figure 14. Intrachain transfer does not occur. The minimum chain length that can be transfered is six glucose units. It has been speculated that branching enzyme catalyzes the transfer of one chain to that of another chain, both of which are part of a double helix (*173, 174, 232*). This mechanism would account for the requirement of relatively long amylose chains as short amylose chains would not be expected to form double helices. If the substrate is already a branched molecule, the branch points could help hold the two chains of the double helix together. The idea that these chains would not have to be as long to form a double helix could explain the further branching of amylopectin (see Fig. 15). As stated previously, at 30° branching enzyme requires a minimum chain length of 40 D-glucosyl residues, while, at low tempera-

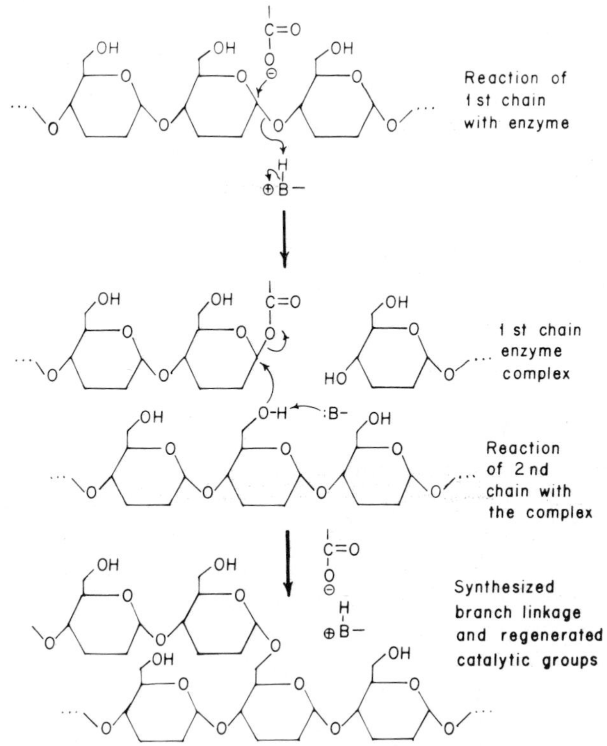

Reaction of 1 st chain with enzyme

1 st chain enzyme complex

Reaction of 2 nd chain with the complex

Synthesized branch linkage and regenerated catalytic groups

FIG. 14.—Proposed interchain mechanism for the formation of α-D-(1→6) branch linkages by the action of branching enzyme. A nucleophile (carboxylate anion) cleaves an α-D-(1→4) glucosidic bond to form two fragments, a covalent carboxyl–acetal ester and a nonreducing end. A hydroxyl group at C-6 of a second chain attacks the anomeric carbon atom of the covalent enzyme complex to form an α-D-(1→6) branch linkage and regenerate the active site catalytic groups. $-BH^+$ represents an electrophile such as an imidazolium ion.

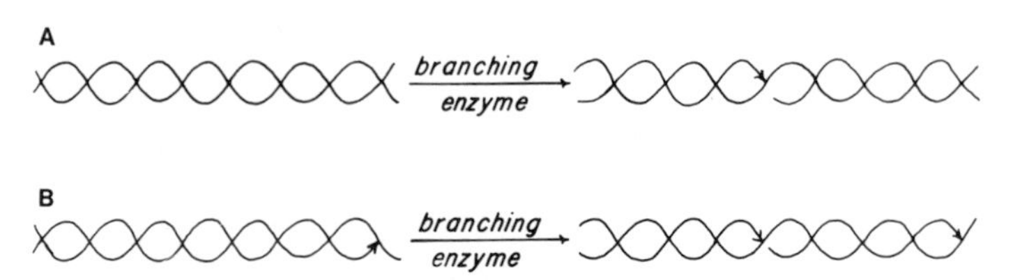

FIG. 15.—Action of branching enzyme (Q-enzyme) with double helices. A, action on an amylose double helix; B, action on an amylopectin double helix. ↘ represents an α-D-(1→6) branch linkage.

tures (14°), much shorter chain lengths of 15–20 D-glucosyl residues will work (*175, 233*). This temperature effect adds support to the hypothesis that branching enzyme requires a double helix; for increasing the temperature would promote dissociation of the helix and, hence, longer chains would be required to give a double helical arrangement.

3. *In Vivo* Synthesis of Starch

The ultimate in the understanding of any biochemical problem is the knowledge of the modes and mechanisms involved in the biosynthesis of the particular product in the cell. The *in vitro* synthesis of amylose may be obtained with phosphorylase or synthetase, and the *in vitro* synthesis of amylopectin may be obtained by the combination of synthetase (or phosphorylase) and branching enzyme (*176*). But, *in vivo,* both amylose and amylopectin are produced and occur side-by-side in a relatively constant ratio of 1:4, respectively, for ordinary starches. Thus, an understanding of the *in vivo* biosynthesis of starch requires an explanation for this side-by-side occurrence of amylose and amylopectin in the starch granule. A further problem is the relatively constant and characteristic morphology of the granule that occurs for each species of plant (*177*).

Various hypotheses have been developed to explain the side-by-side occurrence of amylose and amylopectin in the starch granule. Erlander (*178*) proposed that amylose and amylopectin are formed from a common glycogen precursor. He proposed that low-molecular-weight linear chains are produced by the action of an isoamylase or debranching enzyme and are joined end-to-end by some yet undiscovered enzyme to form amylose. Amylopectin was proposed to result from the partially debranched glycogen. This hypothesis has received little support because of the requirement of the undiscovered "joining" enzyme and the demonstration that glycogen cannot be debranched to give the required increase in the ratio of A-chains to B-chains found in amylopectin (*179*).

Whelan (*180, 181*) proposed that amylose and amylopectin might be synthesized on opposite sides of a semi-permeable barrier that would prevent the access of the branching enzyme to amylose but allow the diffusion of the nucleotide diphosphate glucose to the amylose synthetase and amylose acceptor. He further suggests that the barrier might even be amylopectin, and that the enzyme that synthesizes amylose becomes surrounded by amylopectin, preventing the branching enzyme from penetrating to the amylose but allowing the much smaller nucleotide diphosphate sugar to reach the amylose synthetase. Corroboration of this concept results from short-term labeling experiments with wheat starch in which it has been shown that amylose is not extensively converted into amylopectin once it is laid down in the granule (*182*).

Geddes and Greenwood (*183*) have proposed that a multipathway biosynthetic cycle is operable in the biosynthesis of starch in the amyloplast. They suggest that the enzymes involved are adsorbed at the surface of a nucleation site as a monolayer and that this protein-enzyme layer remains associated with the growing granule. The growth of the granule is by apposition at the granule surface where stereochemical control of the polysaccharide structures would occur. They explain the side-by-side synthesis of amylose and amylopectin as an intense synthesis of linear material at or near the surface which would completely saturate branching enzyme activity in that environment and permit a portion of the linear molecules to remain unbranched. This hypothesis is dependent on the relative immobilization of the enzymes in relationship to each other by adsorption to the starch granule surface and on a relatively larger number of synthetase molecules than branching enzyme molecules, which would give the intense synthesis required to obtain amylose in the presence of amylopectin. It is also dependent on the continuous adsorption of new enzymes at the surface of the granule as it grows.

French (*184*) has also postulated that starch biosynthesis occurs at the granule surface but that the amylose and amylopectin molecules are oriented in opposite directions, the amylose molecules having their reducing ends at the granule surface and the amylopectin molecules having their non-reducing ends at the granule surface. Amylopectin could then be synthesized by the addition of D-glucosyl residues from ADPG and branching of the extended amylopectin chains by branching enzyme. Amylose on the other hand could be synthesized by elongation at the reducing end by an insertion mechanism similar to the mechanisms proposed for the biosynthesis of *Salmonella* O-antigen polysaccharide (*185*), bacterial cell wall polysaccharide (*186*) and dextran (*136*). The opposite orientation of the two polysaccharides thus prevents the action of the branching enzyme on the amylose molecules, which are stereochemically fixed in relation to the branching enzyme and the amylopectin molecules. The relatively constant ratio of amylose and amylopectin would then be dependent on the relative ratio of the amylose synthetase and the amylopectin synthesizing enzymes (amylopectin elongation enzyme and branching enzyme). Evidence has been provided that amylopectin is synthesized by a synthetase-branching enzyme complex (*176*).

Another hypothesis might involve the proposed specificity of branching enzyme for a double helix. Amylose chains could be and probably are isolated from other amylose chains in the starch granule by being interspersed with amylopectin molecules. The "isolated" amylose molecules would be unable to form double helicies and, therefore, would not be acted on by branching enzyme. Amylopectin chains, being shorter and fixed to each other, could form double helices and hence be substrates for branching enzyme.

Radiochemical experiments (*187, 188*) do indeed indicate that starch granules grow by apposition, that is, by the gradual addition of D-glucosyl residues to the surface of the starch granule rather than by intersusception or the addition of new material at the center of the granule or by the uniform addition of material throughout the granule. However, there is little evidence as to how new granules begin. Badenhuizen (*177*) has proposed that the first step may be the formation of a coacervate (*189*), which is a hydrophilic sol that under certain conditions separates into a colloid droplet or particle. The coacervate could then form a gel, and the gel would solidify to give a microgranule. The microgranule could adsorb new synthesizing enzymes and growth would continue by apposition.

Perhaps within the writing of the next edition of this volume, experiments will have been performed that will prove or disprove the various hypotheses and suggest new hypotheses, and we will have a better understanding of the biosynthesis of starch.

VI. References

(*1*) K. Kainuma, S. Kobayashi, T. Ito, and S. Suzuki, *FEBS Letters,* **26,** 281 (1972).
(*2*) K. Kainuma, K. Wako, S. Kobayashi, A. Nogami, and S. Suzuki, *Biochim. Biophys. Acta,* **410,** 333 (1975).
(*3*) J. F. Robyt and D. French, *Arch. Biochem. Biophys.,* **104,** 338 (1964).
(*4*) E. H. Fischer and E. A. Stein, *Biochem. Preps.,* **8,** 27 (1961).
(*5*) E. H. Fischer, F. Duckert, and P. Bernfeld, *Helv. Chim. Acta,* **33,** 1060 (1950).
(*6*) K. H. Meyer, E. H. Fisher, and P. Bernfeld, *Helv. Chim. Acta,* **30,** 64 (1947).
(*7*) M. L. Caldwell, M. Adams, J. T. Kung, and G. C. Toralballa, *J. Amer. Chem. Soc.,* **74,** 4033 (1952).
(*8*) B. L. Williams and P. J. Keller, *Comp. Biochem. Physiol.,* **44A,** 393 (1973).
(*9*) N. G. Heatley, *Nature,* **181,** 1069 (1958).
(*10*) J. Olavarris and M. Torres, *J. Biol. Chem.,* **237,** 1746 (1962).
(*11*) E. A. Stein and E. H. Fischer, *Biochem. Preps.,* **8,** 34 (1961).
(*12*) N. E. Welker and L. L. Campbell, *Biochem.,* **6,** 3681 (1967).
(*13*) N. E. Welker and L. L. Campbell, *J. Bacteriol.,* **94,** 1124 (1967).
(*14*) Y. Nitta, M. Mizushima, and S. Ono, *J. Biochem. (Tokyo),* **22,** 1219 (1972).
(*15*) A. Nishida, J. Fukumoto, and T. Yamamoto, *Agr. Biol. Chem.,* **31,** 682 (1969).
(*16*) H. Fujimori, K. Hiromi, S. Koyama, Y. Momotani, Y. Nakae, M. Ohnishi, T. Shilka, N. Suetsugu, K. Umeki, and T. Yamamoto, *J. Biochem. (Tokyo),* **74,** 1267 (1973).
(*17*) J. Fukumoto, T. Yamamoto, and K. Ichikawa, *Proc. Japan Acad.,* **27,** 352 (1951).
(*18*) L. L. Campbell, *J. Amer. Chem. Soc.,* **76,** 5256 (1954).
(*19*) A. Markovitz, H. P. Klein, and E. H. Fischer, *Biochim. Biophys. Acta,* **19,** 267 (1956).
(*20*) E. H. Fischer and R. DeMontmollin, *Helv. Chim. Acta,* **34,** 1987 (1951).
(*21*) K. Zokaoka, H. Fuwa, and Z. Nikum, *Mem. Inst. Sci. Ind. Res. Osaka Univ.,* **10,** 199 (1952).

(22) S. Schwimmer and A. K. Balls, *J. Biol. Chem.*, **179**, 1036 (1949).
(23) E. H. Fischer and C. H. Haselbach, *Helv. Chim. Acta,* **34**, 325 (1951).
(24) A. Piguet and E. H. Fischer, *Helv. Chim. Acta,* **35**, 257 (1952).
(25) J. Fukumoto and Y. Tsiyisaka, *Kagaku to Kogyo (Osaka),* **29**, 124 (1955).
(26) A. K. Balls, M. K. Walden, and R. R. Thompson, *J. Biol. Chem.*, **173**, 9 (1948).
(27) K. H. Meyer, P. F. Spahr, and E. H. Fischer, *Helv. Chim. Acta,* **36**, 1924 (1953).
(28) L. A. Underkofter and D. K. Roy, *Cereal Chem.*, **28**, 18 (1951).
(29) J. Fukumoto and Y. Tsujisaka, *Kagaku to Kogyo (Osaka),* **28**, 92 (1954).
(30) Y. Tsujisaka, J. Fukumoto, and T. Yamamoto, *Nature,* **181**, 94 (1958).
(31) P. N. Hobson, W. J. Whelan, and S. Peat, *J. Chem. Soc.*, 1451 (1951).
(32) H. Bender and K. Wallenfels, *Biochem. Z.*, **334**, 79 (1961).
(33) M. Abdullah, B. J. Catley, E. Y. C. Lee, J. F. Robyt, K. Wallenfels, and W. J. Whelan, *Cereal Chem.*, **43**, 111 (1966).
(34) Z. H. Gunja, D. J. Manners, and K. Maung, *Biochem. J.*, **81**, 392 (1961).
(35) T. Harda, K. Yokobayashi, and A. Misaki, *Appl. Microbiol.*, **16**, 1493 (1968).
(36) K. Yokobayashi, A. Misaki, and T. Harada, *Agr. Biol. Chem.*, **33**, 625 (1969).
(37) K. Yokobayashi, A. Misaki, and T. Harada, *Biochim. Biophys. Acta,* **212**, 458 (1970).
(38) Z. Gunja-Smith, J. J. Marshall, E. E. Smith, and W. J. Whelan, *FEBS Letters,* **12**, 96 (1970).
(39) J. F. Robyt and W. J. Whelan, *in* "Starch and Its Derivatives," J. A. Radley, ed., Chapman and Hall, London, 4th ed., 1968, p. 439.
(40) N. Nelson, *J. Biol. Chem.*, **153**, 375 (1944).
(41) J. F. Robyt, R. J. Ackerman, and J. G. Keng, *Anal. Biochem.*, **45**, 517 (1972).
(42) G. Noelting and P. Bernfeld, *Helv. Chim. Acta,* **31**, 286 (1948).
(43) G. N. Smith and C. Stocker, *Arch. Biochem.*, **21**, 95 (1949).
(44) J. F. Robyt and W. J. Whelan, *Anal. Biochem.*, **45**, 510 (1972).
(45) S. Dygert, L. H. Li, D. Florida, and J. A. Thoma, *Anal. Biochem.*, **13**, 367 (1965).
(46) R. H. Hopkins and R. Bird, *Biochem. J.*, **56**, 86 (1954).
(47) J. W. van Dyk and M. L. Caldwell, *J. Amer. Chem. Soc.*, **78**, 3345 (1956).
(48) E. Stark, R. Wellerson, Jr., P. A. Tetrault, and C. F. Kossack, *Appl. Microbiol.*, **1**, 236 (1953).
(49) J. Mestecky, F. W. Kraus, D. C. Hurst, and S. A. Voight, *Anal. Biochem.*, **30**, 190 (1969).
(50) J. F. Robyt, unpublished results.
(51) C. T. Greenwood, A. W. MacGregor, and E. A. Milne, *Carbohydr. Res.*, **1**, 303 (1965).
(52) C. T. Greenwood, A. W. MacGregor, and E. A. Milne, *Arch. Biochem. Biophys.*, **112**, 466 (1965).
(53) H. Rinderknecht, P. Wilding, and B. J. Haverbach, *Experientia,* **23**, 805 (1967).
(54) L. Babson, M. Kleinman, and R. E. Megraw, *Clin. Chim. Acta,* **14**, 803 (1968).
(55) L. Fridhandler, J. E. Berk, and S. Take, *Proc. Soc. Exp. Biol. Med.*, **133**, 1212 (1970).
(56) E. H. Fischer, W. N. Summerwell, J. Junge, and E. A. Stein, *Proc. Int. Congr. Biochem., 4th,* **8**, 124 (1958).
(57) E. H. Fischer and E. A. Stein, *in* "The Enzymes," P. D. Boyer, H. Lardy, and K. Myrbäch, eds., Academic Press, New York, 1960, Vol. 4, p. 313.
(58) E. A. Stein and E. H. Fischer, *Biochim. Biophys. Acta,* **39**, 287 (1960).
(59) E. A. Stein and E. H. Fischer, *J. Biol. Chem.*, **232**, 867 (1958).
(60) E. A. Stein, J. Hsiu, and E. H. Fischer, *Biochemistry* **3**, 56 (1964).
(61) J. Hsiu, E. H. Fischer, and E. A. Stein, *Biochemistry* **3**, 61 (1964).
(62) K. Myrbäck, *Z. Physiol. Chem.*, **159**, 1 (1926).
(63) J. Muus, F. P. Brockett, and C. C. Connelly, *Arch. Biochem. Biophys.*, **65**, 268 (1956).
(64) G. J. Walker and W. J. Whelan, *Biochem. J.*, **76**, 257 (1960).
(65) E. A. Stein, *Fed. Proc.*, **16**, 254 (1957).

(66) R. Menzi, E. A. Stein, and E. H. Fischer, *Helv. Chim. Acta,* **40,** 534 (1957).
(67) J. F. Robyt and R. J. Ackerman, *Arch. Biochem. Biophys.,* **155,** 445 (1973).
(68) G. M. Malacinski and W. J. Rutter, *Biochemistry* **8,** 4382 (1969).
(69) S. Meiter and S. Rogols, *Clin. Chem.,* **14,** 1176 (1968).
(70) J. Muus and J. M. Vnenchak, *Nature,* **204,** 283 (1964).
(71) D. L. Kauffman, N. I. Zager, E. Cohen, and P. J. Keller, *Arch. Biochem. Biophys.,* **137,** 325 (1970).
(72) A. D. Merritt, M. L. Rivas, D. Bixler, and R. Newell, *Amer. J. Hum. Genet.,* **25,** 510 (1973).
(73) M. T. Szabo and F. B. Staub, *Acta Biochim. Biophys. Acad. Sci. Hung.,* **1,** 397 (1966).
(74) P. Juhaz and M. T. Szabo, *Acta Biochim. Biophys. Acad. Sci. Hung.,* **2,** 217 (1967).
(75) G. Marchis-Mouren and L. Paséro, *Biochim. Biophys. Acta,* **140,** 366 (1967).
(76) J. J. M. Rowe, J. Wakim, and J. A. Thoma, *Anal. Biochem.,* **25,** 206 (1968).
(77) P. Cozzone, L. Paséro, and G. Marchis-Mouren, *Biochim. Biophys. Acta,* **200,** 590 (1970).
(78) P. J. Keller, D. L. Kauffman, B. J. Allen, and B. L. Williams, *Biochemistry* **10,** 4867 (1971).
(79) J. M. Junge, E. A. Stein, H. Neurath, and E. H. Fischer, *J. Biol. Chem.,* **234,** 556 (1959).
(80) H. Hanafusa, T. Ikenaka, and S. Akabori, *J. Biochem. (Tokyo),* **42,** 55 (1955).
(81) A. Tsugita and S. Akabori, *J. Biochem. (Tokyo),* **46,** 695 (1959).
(82) B. Beaupoil-Abadie, M. Raffalli, P. Cozzone, and G. Marchis-Mouren, *Biochim. Biophys. Acta,* **297,** 436 (1973).
(83) J. H. Pazur, *in* "Starch: Chemistry and Technology," R. L. Whistler and E. F. Paschall, eds. Academic Press, New York, Vol. 1, 1965, p. 140.
(84) J. A. Thoma, J. E. Spradlin, and S. Dygert, *in* "The Enzymes," P. D. Boyer, ed. Academic Press, New York, 1971, Vol. 5, p. 115.
(85) J. E. Spradlin and J. A. Thoma, *J. Biol. Chem.,* **245,** 117 (1970).
(86) J. A. Thoma, D. E. Koshland, J. Ruscica, and R. Baldwin, *Biochem. Biophys. Res. Commun.,* **12,** 184 (1963).
(87) K. H. Meyer, E. H. Fischer, and A. Piquet, *Helv. Chim. Acta,* **34,** 316 (1951).
(88) J. H. Pazur and K. Kleppe, *J. Biol. Chem.,* **237,** 1002 (1962).
(89) J. A. Thoma, D. E. Koshland, R. Shinke, and J. Ruscica, *Biochemistry,* **4,** 714 (1965).
(90) Y. Minoda, M. Arai, and K. Yamada, *Agr. Biol. Chem.,* **33,** 572 (1969).
(91) J. H. Pazur, K. Kleppe, and E. M. Ball, *Arch. Biochem. Biophys.,* **103,** 515 (1963).
(92) J. H. Pazur and T. Ando, *J. Biol. Chem.,* **234,** 1966 (1959).
(93) D. R. Lineback, I. J. Russell and C. Rasmussen, *Arch. Biochem. Biophys.,* **134,** 539 (1969).
(94) K. H. Meyer and P. Bernfeld, *Helv. Chim. Acta,* **24,** 359E (1941).
(95) K. H. Meyer and W. F. Gonon, *Helv. Chim. Acta,* **34,** 294 (1951).
(96) J. F. Robyt and D. French, *Arch. Biochem. Biophys.,* **122,** 8 (1967).
(97) J. H. Pazur, D. French, and D. Knapp, *Proc. Iowa Acad. Sci.,* **57,** 203 (1950).
(98) S. K. Dube and P. Nordin, *Arch. Biochem. Biophys.,* **99,** 105 (1962).
(99) G. G. Freeman and R. H. Hopkins, *Biochem. J.,* **30,** 442 (1936).
(100) G. J. Walker, *Biochem. J.,* **94,** 289 (1965).
(101) J. F. Robyt and D. French, *Arch. Biochem. Biophys.,* **100,** 451 (1963).
(102) J. F. Robyt and D. French, *Arch. Biochem. Biophys.,* **138,** 662 (1970).
(103) J. F. Robyt and D. French, *J. Biol. Chem.,* **245,** 3917 (1970).
(104) K. Kainuma and D. French, *FEBS Letters,* **5,** 257 (1969).
(105) K. Kainuma and D. French, *FEBS Letters,* **6,** 182 (1970).
(106) A. Levitzki, J. Heller, and M. Schramm, *Biochim. Biophys. Acta,* **81,** 101 (1964).
(107) A. Loyter and M. Schramm, *J. Biol. Chem.,* **241,** 2611 (1966).
(108) A. McPherson and A. Rich, *Biochim. Biophys. Acta,* **285,** 493 (1972).
(109) R. C. Hughes, E. E. Smith, and W. J. Whelan, *Biochem. J.,* **88,** 63P (1963).
(110) D. French, E. E. Smith, and W. J. Whelan, *Carbohydr. Res.,* **22,** 123 (1972).

(*111*) J. A. Thoma, C. Brothers, and J. Spradlin, *Biochemistry,* **9,** 1768 (1970).
(*112*) J. A. Thoma, G. V. K. Rao, C. Brothers, and J. Spradlin, *J. Biol. Chem.,* **246,** 5621 (1971).
(*113*) R. Summer and D. French, *J. Biol. Chem.,* **232,** 469 (1956).
(*114*) J. M. Bailey and D. French, *J. Biol. Chem.,* **266,** 1 (1956).
(*115*) D. French and R. W. Youngquist, *Staerke,* **12,** 425 (1963).
(*116*) J. J. Marshall, *FEBS Letters,* **45,** 1 (1974).
(*117*) M. Higashihara and S. Okada, *Agr. Biol. Chem.,* **38,** 1023 (1974).
(*118*) R. Shinke, Y. Kunimi, and H. Nishira, *J. Ferment. Technol.,* **53,** 693 (1975).
(*119*) R. Shinke, Y. Kunimi, and H. Nishira, *J. Ferment. Technol.,* **53,** 698 (1975).
(*120*) H. Hidaka, Y. Koaze, K. Yoshida, T. Niwa, T. Shomura, and T. Niida, *Staerke,* **26,** 413 (1974).
(*121*) H. Hidaka, T. Adachi, K. Yoshida and T. Niwa, *Denpun Kagaku,* **25,** 148 (1978).
(*122*) K. Wako, C. Takahashi, S. Hashimoto, and J. Kanaeda, *Denpun Kagaku,* **25,** 155 (1978).
(*123*) H. Fujimori, T. Suganuma, T. Minakami, M. Sakoda, M. Ohnishi, R. Matsuno, K. Hiromi, J. Kanaeda, K. Wako, and C. Takahashi, *Denpun Kagaku* **24,** 86 (1977).
(*124*) J. F. Robyt and R. J. Ackerman, *Arch. Biochem. Biophys.,* **145,** 105 (1971).
(*125*) N. Saito, *Arch. Biochem. Biophys.,* **155,** 290 (1973).
(*126*) H. Gutfreund, *in* "The Enzymes," P. D. Boyer, H. Lardy, and K. Myrbäch, eds. Academic Press, New York, 2nd Ed., 1960, Vol. 1, p. 233.
(*127*) J. A. Thoma, J. Wakim, and L. Stewart, *Biochem. Biophys. Res. Commun.,* **12,** 350 (1963).
(*128*) J. A. Thoma and D. E. Koshland, Jr., *J. Mol. Biol.,* **2,** 169 (1960).
(*129*) J. A. Thoma, *J. Theoret. Biol.,* **19,** 297 (1968).
(*130*) J. Wakim, M. Robinson, and J. A. Thoma, *Carbohydr. Res.,* **10,** 487 (1969).
(*131*) E. J. Hehre, D. S. Genghof, and G. Okada, *Arch. Biochem. Biophys.,* **142,** 382 (1971).
(*132*) E. J. Hehre, G. Okada, and D. S. Genghof, *Advan. Chem. Ser.,* **117,** 309 (1973).
(*133*) L. B. Spector, *Bioorganic Chem.,* **2,** 311 (1973).
(*134*) R. Silverstein, J. G. Voet, D. Reed, and R. H. Abeles, *J. Biol. Chem.,* **242,** 1338 (1967).
(*135*) J. G. Voet and R. H. Abeles, *J. Biol. Chem.,* **245,** 1020 (1970).
(*136*) J. F. Robyt, B. J. Kimble, and T. F. Walseth, *Arch. Biochem. Biophys.,* **165,** 634 (1974).
(*137*) D. French, "Chemistry and Biochemistry of Starch," *in MTP International Review of Science,* W. J. Whelan, ed. University Park Press, Baltimore, Maryland, 1975, vol. 5, p. 319.
(*138*) G. M. Gray, B. C. Lally, and K. A. Conklin, *J. Biol. Chem.,* **254,** 6038 (1978).
(*139*) R. Datema, J. J. Marshall, C. J. Partington, C. M. Sturgeon, W. J. Whelan, and W. Woloszczuk, *Abstr. Papers Amer. Chem. Soc.,* **176,** Carb 36 (1978).
(*140*) G. T. Cori and C. F. Cori, *J. Biol. Chem.,* **131,** 397 (1939).
(*141*) C. S. Hanes, *Proc. Roy. Soc.,* **B129,** 174 (1940).
(*142*) M. A. Swanson and C. F. Cori, *J. Biol. Chem.,* **172,** 815 (1948).
(*143*) E. H. Fischer and H. M. Hilpert, *Experientia,* **9,** 176 (1953).
(*144*) H. Baum and G. A. Gilbert, *Nature,* **171,** 983 (1953).
(*145*) A. Kamogawa, T. Fukui, and Z. Nikuni, *J. Biochem. (Tokyo),* **63,** 361 (1967).
(*146*) D. H. Brown and C. F. Cori, *in* "The Enzymes," P. Boyer, H. Lardy, and K. Myrbäck, eds. Academic Press, New York, 1961, Vol. 5, p. 222.
(*147*) D. French and G. M. Wild, *J. Amer. Chem. Soc.,* **75,** 4490 (1953).
(*148*) W. E. Trevelyan, P. F. E. Mann, and J. S. Harrison, *Arch. Biochem. Biophys.,* **39,** 419 (1952).
(*149*) D. Stetten, Jr., and M. R. Stetten, *Physiol. Revs.,* **40,** 513 (1960).
(*150*) W. Banks and C. T. Greenwood, *in* "Starch and Its Components," Edinburgh University Press, Edinburgh, Scotland, 1975, p. 296.
(*151*) N. P. Badenhuizen and K. R. Chandorkar, *Cereal Chem.,* **42,** 44 (1965).

(*152*) K. R. Chandorkar and N. P. Badenhuizen, *Cereal Chem.*, **44,** 27 (1967).
(*153*) K. R. Chandorkar and N. P. Badenhuizen, *Staerke*, **18,** 91 (1966).
(*154*) L. F. Leloir and C. E. Cardini, *J. Amer. Chem. Soc.*, **79,** 6340 (1957).
(*155*) M. A. R. DeFekete, L. F. Leloir, and C. E. Cardini, *Nature*, **187,** 918 (1960).
(*156*) L. F. Leloir, M. A. R. DeFekete, and C. E. Cardini, *J. Biol. Chem.*, **236,** 636 (1961).
(*157*) E. Recondo and L. F. Leloir, *Biochem. Biophys. Res. Commun.*, **6,**85 (1961).
(*158*) R. B. Frydman, *Arch. Biochem. Biophys.*, **102,** 242 (1963).
(*159*) D. French, Iowa State University, Ames, Iowa (personal communication).
(*160*) R. B. Frydman and C. E. Cardini, *Biochem. Biophys. Res. Commun.*, **14,** 353 (1964).
(*161*) R. B. Frydman and C. E. Cardini, *Biochem. Biophys. Res. Commun.*, **17,** 407 (1964).
(*162*) A. Doi, K. Doi, and Z. Nikuni, *Biochim. Biophys. Acta*, **92,** 628 (1964).
(*163*) H. P. Ghosh and J. Preiss, *Biochemistry*, **4,** 1354 (1965).
(*164*) J. S. Hawker, J. L. Ozbun, H. Ozaki, E. Greenberg, and J. Preiss, *Arch. Biochem. Biophys.*, **160,** 530 (1974).
(*165*) W. J. Whelan, *Biochem. J.*, **122,** 609 (1971).
(*166*) R. B. Frydman and C. E. Cardini, *J. Biol. Chem.*, **242,** 312 (1967).
(*167*) O. E. Nelson and H. W. Rines, *Biochem. Biophys. Res. Commun.*, **9,** 297 (1961).
(*168*) W. N. Haworth, S. Peat, and E. J. Bourne, *Nature*, **154,** 236 (1944).
(*169*) S. A. Barker, E. J. Bourne, and S. Peat, *J. Chem. Soc.*, 1705 (1949).
(*170*) P. N. Hobson, W. J. Whelan, and S. Peat, *J. Chem. Soc.*, 3566 (1950).
(*171*) N. Lavintman, *Arch. Biochem. Biophys.*, **166,** 1 (1966).
(*172*) D. J. Manners, J. J. M. Rowe, and K. L. Rowe, *Carbohydr. Res.*, **8,** 72 (1968).
(*173*) D. Borovsky and W. J. Whelan, *Fed. Proc.*, **31,** 477 (1972).
(*174*) D. Kainuma and D. French, *Biopolymers*, **11,** 2241 (1972).
(*175*) D. Borovsky and E. E. Smith, *Fed. Proc.*, **33,** 1559 (1974).
(*176*) S. Schiefer, E. Y. C. Lee, and W. J. Whelan, *FEBS Lett.*, **30,** 129 (1973).
(*177*) N. P. Badenhuizen, *Nature*, **197,** 464 (1963).
(*178*) S. R. Erlander, *Enzymologia*, **19,** 273 (1958).
(*179*) J. J. Marshall and W. J. Whelan, *FEBS Letters*, **9,** 85 (1970).
(*180*) W. J. Whelan, "Handbuch der Pflanzenphysiologie," Springer-Verlag, Berlin, 1958, Vol. 6, p. 154.
(*181*) W. J. Whelan, *Staerke*, **15,** 247 (1963).
(*182*) R. L. Whistler and J. R. Young, *Cereal Chem.*, **37,** 204 (1960).
(*183*) R. Geddes and C. T. Greenwood, *Staerke*, **21,** 148 (1969).
(*184*) D. French, *Denpun Kagaku*, **19,** 8 (1972).
(*185*) D. Bray and P. W. Robbins, *Biochem. Biophys. Res. Commun.*, **28,** 334 (1967).
(*186*) J. B. Ward and H. Perkins, *Biochem. J.*, **135,** 721 (1973).
(*187*) N. P. Badenhuizen and R. W. Dutton, *Protoplasmalogia*, **47,** 156 (1956).
(*188*) M. Yoshida, M. Fujii, Z. Nikuni, and B. Maruo, *Bull. Agr. Chem. Soc. Japan*, **21,** 127 (1958).
(*189*) H. R. Kruyt, "Colloid Science," Elsevier Publishing, Amsterdam, 1949, Vol. 2, p. 243.
(*190*) G. Marchis-Mouren and L. Paséro, *Biochim. Biophys. Acta*, **140,** 366 (1967).
(*191*) P. Cozzone, L. Paséro, B. Beaupoil, and G. Marchis-Mouren, *Biochim. Biophys. Acta*, **207,** 490 (1970).
(*192*) P. Cozzone, L. Paséro, B. Beaupoil, and G. Marchis-Mouren, *Biochemie*, **53,** 957 (1971).
(*193*) P. Cozzone and G. Marchis-Mouren, *Biochem. Biophys. Acta*, **257,** 222 (1972).
(*194*) F. Fabian, *Acta Biochim. Biophys., Acad. Sci. Hung.*, **12,** 31 (1977).
(*195*) J. F. Robyt, C. G. Chittenden, and C. T. Lee, *Arch. Biochem. Biophys.*, **144,** 160 (1971).
(*196*) F. B. Staub, M. T. Szabó, and T. Dévényi, *FEBS Symp.*, **18,** 257 (1970).
(*197*) A. J. Corrigan and J. F. Robyt, *Denpun Kagaku*, **30,** 197 (1983).

(*198*) S. Mora, I. Simon, and P. Elodi, *Mol. Cell. Biochem.*, **4,** 205 (1974).
(*199*) I. Simon, S. Mora, and P. Elodi, *Mol. Cell. Biochem.*, **4,** 211 (1974).
(*200*) P. M. D. Fitzgerald, P. J. Stankiewicz, S. C. Smith, and A. McPherson, *J. Mol. Biol.*, **135,** 753 (1979).
(*201*) F. Payan, R. Haser, M. Pierrot, M. Frey, and J. P. Astier, *Acta Cryst.*, *B*, **36,** 416 (1980).
(*202*) I. Kluh, *FEBS Letters*, **136,** 231 (1981).
(*203*) E. Kneen and R. M. Sandstedt, *J. Amer. Chem. Soc.*, **65,** 1247 (1943).
(*204*) E. Kneen and R. M. Sandstedt, *Arch. Biochem.*, **9,** 235 (1946).
(*205*) W. Militzer, C. Ikeda, and E. Kneen, *Arch. Biochem.*, **15,** 309 (1947).
(*206*) W. Militzer, C. Ikeda, and E. Kneen, *Arch. Biochem.*, **15,** 321 (1947).
(*207*) R. Shainkin and Y. Birk, *Biochim. Biophys. Acta*, **221,** 502 (1970).
(*208*) V. Silano, F. Pocchiari, and D. D. Kasarda, *Biochim. Biophys. Acta*, **317,** 139 (1973).
(*209*) D. H. Strumeyer, *Nutr. Rep. Intern.*, **5,** 45 (1972).
(*210*) J. J. Marshall, *Amer. Chem. Soc., Symp. Series*, **15,** 253 (1975).
(*211*) D. E. Bowman, *Science*, **102,** 358 (1945).
(*212*) W. G. Jaffé and C. L. V. Lette, *J. Nutr.*, **94,** 203 (1968).
(*213*) W. G. Jaffé, R. Moreno, and V. Wallis, *Nutr. Rep. Intern.*, **7,** 169 (1973).
(*214*) J. J. Marshall and C. M. Lauda, *J. Biol. Chem.*, **250,** 8030 (1975).
(*215*) B. B. Elliott and A. C. Leopold, *Physiol. Plant.*, **6,** 65 (1953).
(*216*) T. Hemberg and I. Larsson, *Physiol. Plant.*, **14,** 861 (1961).
(*217*) B. S. Miller and E. Kneen, *Arch. Biochem.*, **15,** 251 (1947).
(*218*) D. H. Strumeyer and M. J. Malin, *Biochem. Biophys. Acta*, **184,** 643 (1969).
(*219*) K. Fossum and J. R. Whitaker, *J. Nutr.*, **104,** 930 (1974).
(*220*) FDA Consumer, **16,** No. 7, 2 (1982).
(*221*) FDA Consumer, **16,** No. 10, 3 (1983).
(*222*) G. W. Bo-Linn, C. A. Santa Ana, S. D. Morawski, and J. S. Fordtran, *New England J. Med.*, **307,** 1413 (1982).
(*223*) I. H. Rosenberg, *New England J. Med.*, **307,** 1444 (1982).
(*224*) M. L. Hackert and R. A. Jacobson, *J. Chem. Soc., Chem. Commun.*, 1179 (1969).
(*225*) E. T. Reese, F. W. Parrish, and M. Ettlinger, *Carbohydr. Res.*, **18,** 381 (1971).
(*226*) E. László, J. Holló, A. Hoschke, and G. Sárosi, *Carbohydr. Res.*, **61,** 387 (1978).
(*227*) W. P. Jencks, "Catalysis in Chemistry and Enzymology," McGraw-Hill, New York, 1969, p. 226.
(*228*) D. D. Schmidt, W. Frommer, B. Junge, L. Müller, W. Wingender, E. Truscheit, and D. Schäfer, *Naturwissenschaften*, **64,** 535 (1977).
(*229*) S. Murao and K. Ohyama, *Agric. Biol. Chem.*, **43,** 679 (1979).
(*230*) D. E. Walker and B. Axelrod, *Arch. Biochem. Biophys.*, **195,** 392 (1979).
(*231*) J. A. Thoma and D. E. Koshland, *J. Amer. Chem. Soc.*, **82,** 3329 (1960).
(*232*) D. Borovsky, E. E. Smith, and W. J. Whelan, *Eur. J. Biochem.*, **62,** 307 (1976).
(*233*) D. Borovsky, E. E. Smith, and W. J. Whelan, *FEBS Letters*, **54,** 201 (1975).

CHAPTER V

STARCH OLIGOSACCHARIDES: LINEAR, BRANCHED, AND CYCLIC

By Keiji Kainuma

National Food Research Institute, Tsukuba-Gun, Ibaraki-Ken, Japan

I. Introduction

Starch oligosaccharides, which represent fragments of the original polysaccharide, are composed of α-D-glucopyranosyl units linked by (1 → 4) and (1 → 6) bonds.

The nomenclature of oligosaccharides follows the rules of the Committee on Nomenclature, Spelling and Pronunciation of the American Chemical Society (*1*). However, an alternative nomenclature system suggested by Whelan (*2*) is used frequently because it clearly shows the size and branched structure of oligosaccharides. The Whelan nomenclature is employed here. The generic term "oligosaccharide" is customarily used for saccharides containing fewer than ten monosaccharide units. Thoma and co-workers (*3*) proposed the term "mega-

STARCH, 2nd ed.

ISBN 0-12-746270-8

losaccharide'' for saccharides having ten or more monosaccharide units and this nomenclature is used also.

Structural analysis of starch fragmentation products is of interest for characterization of the native starch molecule. Then, too, interest in the pure oligo- and megalosaccharides is expanding because of their usefulness in the probing of the amylase subsite, elucidation of amylase action, and in clinical amylase assay.

Cyclic oligosaccharides, referred to as cycloamyloses, cyclodextrins or Schardinger dextrins, are formed from starch polysaccharides and are nonreducing D-glucopyranosyl polymers containing 6 or more units linked by α-D-(1 → 4) bonds. Cycloamyloses are rather anomalous structures with interesting physicochemical properties when compared with the linear oligosaccharides. Research on cycloamyloses before 1963 was described in an excellent review by Thoma and Stewart (*4*).

The earliest period of cycloamylose work extends from their discovery by Villers (*5*) in 1891 to approximately the mid-1930s. During this period, the physical and chemical properties of cycloamyloses were investigated. However, the literature of this period is regarded as of little consequence because of work on impure material and the use of incorrect structures. The second period of the cycloamylose research, during which the cyclic nature of the cyclodextrins was recognized and the α-D-(1 → 4) linkages established, was opened by their careful fractionation by Freudenberg and Jacobi (*6*) in 1935. The smallest cyclodextrins were shown to contain 6, 7, or 8 sugar units. French called this period of 1935–1950 the maturation period (*7*). In the third period, work on the *Bacillus macerans* enzyme led to a new method of purification, development of the action pattern, and the cyclization of linear starch chains. Since 1950, attention has been focused on formation of the cycloamylose inclusion complexes and on the use of the dextrins as enzyme models. Since 1970, effort has been directed to find new microorganisms that produce cycloamylose glucanotransferases, to purify the enzymes using modern enzymological techniques, to examine the chemical properties of the enzymes, and to produce cycloamyloses on a large scale to meet needs for industrial applications.

II. Linear and Branched Starch Oligosaccharides

1. Enzymic Methods for Preparation (see also Chapter IV, Section IV)

Maltooligosaccharides are the products of partial hydrolysis of starch by acids or amylases. Prior to 1970, it had been difficult to prepare maltooligosaccharides with definite degrees of polymerization by simple chemical or enzymic hydrolysis. Since 1970, several novel microbial amylases producing specific oligosaccharides have been discovered (Table I). By use of these amylases, pure malto-

Table I

Oligosaccharide-Forming Amylases

Enzyme	*Product*	*Origin*	*Reference*
β-Amylase (EC 3.2.1.2)	β-Maltose	Plant	*8*
		B. polymyxa	*9*
		B. megaterium	*10*
		Bacillus sp. *BQ 10*	*11*
		Pseudomonas sp. *BQ 6*	*12*
		B. cereus	
α-Amylase	α-Maltose	*S. hygroscopicus*	*13*
		S. praecox	*14*
Maltotriose-forming amylase	α-Maltotriose	*S. griseus*	*16*
Exo-maltotetraohydrolase (EC 3.2.1.60)	α-Maltotetraose	*Ps. stutzeri*	*17*
α-Amylase	Mainly α-maltopentaose	*B. licheniformis*	*18*
Exo-maltohexaohydrolase (EC 3.2.1.98)	α-Maltohexaose	*A. aerogenes*	*19*

oligosaccharides can be produced on a large scale, and their possible industrial applications can be examined.

a. *Maltose-Forming Amylases*

β-Amylase of plant origin has been well known to form specifically β-maltose from starch. This enzyme is extracted from sweet potato, barley, soybean, and other sources and used for maltose production on a large scale. Though β-amylase is mainly a plant enzyme, several microbial β-amylases were discovered recently in bacteria such as *Bacillus polymyxa* (*8*), *Bacillus megaterium* (*9*), *Bacillus* sp. BQ 10 (*10*), *Pseudomonas* sp. BQ 6 (*11*), and *Bacillus cereus*. (*12*).

Another group of amylases produce α-maltose from starch and are classified as α-amylases. One of the maltose forming α-amylases, discovered by Hidaka and co-workers (*13*) and designated *Streptomyces hygroscopicus* SF-1084 amylase, converts starch, amylose, and amylopectin to α-maltose in yields of 77%, 86%, and 77%, respectively. The time course of the action of the amylase on potato starch is shown in Figure 1. The *S. hygroscopicus* amylase does not act on pullulan, α-cycloamylose, nor β-cycloamylose. Another maltose-forming α-amylase, obtained from *S. praecox* by Wako and co-workers (*14*), produces 80% α-maltose from starch. An unusual characteristic of the *S. praecox* amylase is that it converts maltotriose into maltose under specific condition. This reaction is explained by the following glucose transfer mechanism (*15*).

$$0\text{-}0\text{-}\emptyset + \text{Enzyme} \rightarrow 0\text{-}\emptyset + 0\text{-Enzyme}$$

$$0\text{-Enzyme} + 0\text{-}0\text{-}\emptyset \rightarrow 0\text{-}0\text{-}0\text{-}\emptyset + \text{Enzyme}$$

$$0\text{-}0\text{-}0\text{-}\emptyset \rightarrow 0\text{-}\emptyset + 0\text{-}\emptyset$$

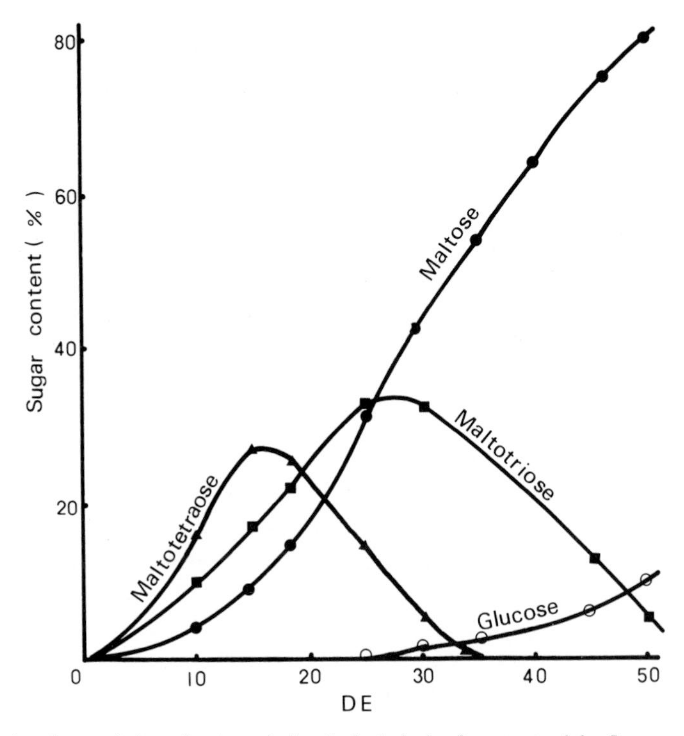

FIG. 1.—Accumulation of maltose during the hydrolysis of potato starch by *Streptomyces hygroscopicus* SF-1084 amylase (*13*).

At the higher substrate concentration, this reaction proceeds quantitatively; for example, when the enzyme acts on 1% maltotriose solution as substrate, the ratio of the formation of maltose-glucose is about 80. The transfer mechanism was also reported for the *hygroscopicus* amylase.

Maltose, the smallest maltooligosaccharide, has been produced industrially using β-amylase from wheat bran, soybean, and sweet potato. The discoveries of bacteria that produce microbial β-amylase or maltose-forming α-amylase will enable the production of large quantities of enzymes in short-time cultivation. These enzymes are considered good for large scale pure maltose production.

b. *Maltotriose-Forming Amylase*

Maltotriose is easily prepared by the action of pullulanase on pullulan, a polysaccharide produced by *Pullularia pullulans*. Pullulan is an α-D-(1 → 6)-linked polymer of α-maltotriose. Though the action of pullulanase on pullulan seems to be a good system to produce pure maltotriose, pullulan frequently contains a small amount of maltotetraose units in place of maltotriose in the structure. This anomalous structure causes slight contamination with maltotetraose in the hydrolyzate. Recently, a new exo-amylase, which produces maltotriose from starch and relating glucans was discovered from *Streptomyces griseus* (*16*). This amylase forms maltotriose from the non-reducing end of the substrate. Wako and co-workers (*16*) purified the enzyme and examined its chemical and biochemical properties. This amylase acts on soluble starch, waxy corn starch, oyster glycogen, and phytoglycogen to form maltotriose in yields of 55%, 51%, 40%, and 20%, respectively. Amylose is hydrolyzed completely into maltotriose and a small amount of maltose and D-glucose (Fig. 2).

In combination with pullulanase, the *S. griseus* amylase converts starch into maltotriose in yields of 90% or higher. The two systems, pullulanase on pullulan and *S. griseus* amylase with pullulanase on starch are promising ways to prepare maltotriose.

c. *Maltotetraose-Forming Amylases (Exo-maltotetraohydrolase, EC 3.2.1.60)*

Robyt and Ackerman (*17*) discovered a novel microbial maltotetraose-forming amylase as an extracellular enzyme from a strain of *Pseudomonas stutzeri*. This enzyme, the next amylase after β-amylase to be classified as a specific oligosac-

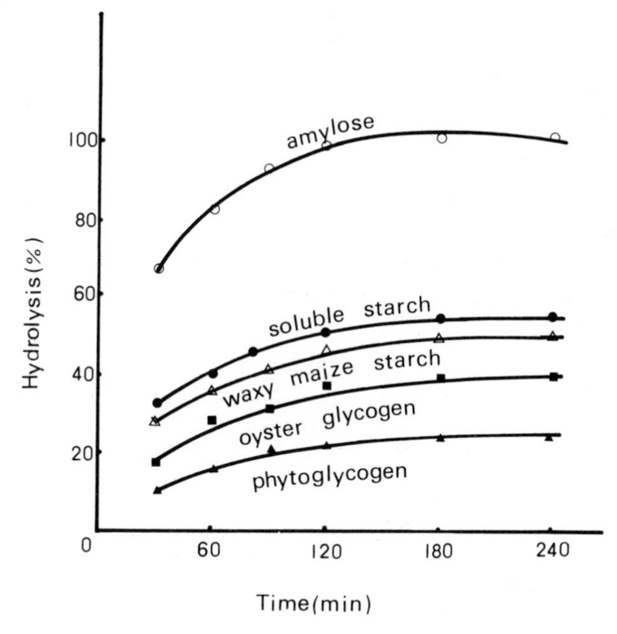

FIG. 2.—Hydrolysis of various substrates by *Streptomyces griseus* maltotriose-forming amylase (*16*).

charide-forming amylase, was purified by ammonium sulfate precipitation and acetone precipitation to 2600 IU/mg protein, the highest activity among the amylases of microbial, plant, and animal sources. The *P. stutzeri* amylase is an exoamylase that attacks only the fourth glucosidic bond from the non-reducing end of the substrate and acts on amylopectin and glycogen to form maltotetraose plus 58% and 62%, respectively, of higher-molecular-weight limit dextrins. The enzyme has been studied further by several groups (*114–116*).

The finding of the *P. stutzeri* amylase has stimulated the search for other amylases that may produce specific oligosaccharides.

d. *Maltopentaose-Forming Amylase*

A thermostable α-amylase obtained from *Bacillus licheniformis* (*18*) can be used for the production of maltopentaose. The enzyme causes accumulation of maltopentaose in the hydrolyzate at an early stage of the reaction. One of the examples of the oligosaccharide composition of the *B. licheniformis* amylase hydrolyzate is shown in Table II.

e. *Maltohexaose-Forming Amylase (Exo-maltohexaohydrolase, EC 3.2.1.98)*

A maltohexaose-forming amylase was discovered by Kainuma and co-workers (*19*) as the fourth exoamylase following β-amylase, glucoamylase, and *Ps. stutzeri* maltotetraose-forming amylase. This unusual enzyme was discovered as a contaminant of crude pullulanase from *Aerobacter aerogenes*. The action pattern of the enzyme was proved to be exo. Maltooligosaccharides are hydrolyzed exclusively at the sixth glucosidic linkage from the non-reducing end by the enzyme.

The enzyme was purified 80-fold by ammonium sulfate precipitation and DEAE-Sephadex and Sephadex G-100 gel filtration chromatography. The *Aerobacter* amylase produces maltohexaose from starch, amylose, and amylopectin by exo-attack, but does not act on α- or β-cycloamylose, pullulan, or maltohexaitol. In an unusual reaction, it produces branched saccharides from the β-limit

Table II

Oligosaccharides in the Hydrolyzate of B. licheniformis (18)

Saccharide	*Percent, %*
G_1	6.8
G_2	8.2
G_3	15.8
G_4	7.7
G_5	33.3
G_{6-12}	37.5

dextrin of amylopectin (*20*). The *Aerobacter* amylase slowly hydrolyzes maltohexaose into mainly maltotetraose and maltose. The crude enzyme forms maltohexaose from soluble starch in a yield of more than 30%. Crystalline maltohexaose may be obtained in this way (*21*). Kainuma and co-workers (*117–119*) obtained a mutant of *A. aerogenes* which produces high levels of an extracellular, maltohexaose-forming amylase. They suggest an application of the enzyme for large-scale production of maltohexaose (*120*). Recently, Takasaki (*121*) and Taniguchi (*122, 123*) discovered a maltohexaose-forming amylase in *Bacillus circulans*.

f. *Cycloamylose Glucanotransferase (EC 2.4.1.19)*

Cycloamylose glucanotransferase is produced by bacteria such as *Bacillus macerans* (*22*), *Bacillus megaterium* (*23*), *Bacillus circulans* (*24*), and *Bacillus stearothermophilus* (*25*). This enzyme catalyzes the formation of cycloamylose from starch, as well as the synthesis of a series of oligosaccharides which are labeled at the reducing end by a specific sugar.

French and collaborators (*26*) used *B. macerans* cycloamylose glucanotransferase to synthesize reducing end-labeled maltosaccharides. This reaction is explained by the following mechanism.

(cyclic O₆: six O linked in a ring) + Ø* —coupling reaction→ O—O—O—O—O—O—Ø*

O—O—O—O—O—O—Ø* + O—O—O—O—O—O—Ø* —disproportionation reaction→ O*, O—Ø*, O—O—Ø*, etc.

The coupling reaction of cycloamylose with radioactive D-glucose produces maltoheptaose labeled exclusively at the reducing end. The same enzyme then catalyzes disproportionation of reducing and labeled maltoheptaose to produce a homologous series of end labeled maltooligosaccharides. Oligosaccharides labeled at specific positions are useful for examining the action pattern of amylases. The action pattern of the *P. stutzeri* maltotetraose-forming amylase (*17*), the *Aerobacter aerogenes* maltohexaose-forming amylase (*19*), porcine pancreatic α-amylase, and various other α-amylases were determined from use of these labeled maltosaccharides.

Kainuma and French (*27*) synthesized maltooligosaccharides stubbed with D-glucosyl units at a specific position by the combined action of *B. macerans* enzyme and pullulanase. Panose was used as the cosubstrate to synthesize 6^3-α-glucosyl maltooligosaccharides; a mixture of 6^3-α-D-glucosyl maltotetraose and 6^3-α-maltosyl maltotriose was used to synthesize 6^3-α-D-glucosyl malto-oligosaccharides. The successive action of pullulanase removed maltosyl or larger branches; consequently, maltooligosaccharides with single D-glucosyl unit branches attached at the second or third unit from the reducing end were obtained. In the first reaction, cyclohexaamylose was coupled with panose.

```
  O                O                                  ···O—O
O   O              |       cycloamylose                    |
|   |   +          |     ──────────────────►               ↓
O   O              ↓     glucanotransferase     O—O—O—O—O—O—O—Ø
  O                O—Ø
```

The resulting oligosaccharide was then disproportionated to form a series of oligosaccharides terminated by panose at the reducing end after debranching by pullulanase.

```
O         O           O             O
↓         ↓           ↓             ↓
O—Ø    O—O—Ø    O—O—O—Ø    O—O—O—O—Ø
```

Cycloamylose glucanotransferase action is a useful tool for synthesizing new oligosaccharides which are essential for examining the action pattern of starch-hydrolyzing enzymes. French and co-workers (*26*) studied cosubstrate specificity in the coupling reaction of *B. macerans* cycloamylose glucanotransferase. D-Glucose, D-*gluco*-heptulose, 1,5-anhydro-D-glucitol (polygalitol), maltose, maltobionic acid, cellobiose, turanose, sucrose, planteose, melezitose, panose, isomaltose, 6^1-α-D-glucopyranosyl maltose, and some glucosides such as phenyl α-D-glucopyranoside, accubin, and phlorizin can be used as cosubstrates. Recently, Kitahata and Okada (*124, 125*) also studied cosubstrate specificities of cycloamylose glucanotransferase of *B. megaterium* and reported that the essential structural requirement for the cosubstrate of the enzyme is a pyranose ring with equatorial hydroxyl groups at C-2, C-3, and C-4.

g. *Branched Oligosaccharides from Amylopectin*

α-D-(1 → 6)-Branched maltooligosaccharides are best prepared from amylopectin and glycogen. Various starch-hydrolyzing enzymes convert them to branched limit dextrins. The following are branched oligosaccharides formed by the action of α-amylase on amylopectin (*28*).

```
O           O—O          O          O—O          O—O—O
↓             ↓          ↓            ↓              ↓
O—O—Ø         O—O—Ø   O—O—O—Ø     O—O—O—Ø        O—O—O—Ø

O           O            O   O            O
↓           ↓            ↓   ↓            ↓
O—O         O—O—O        O—O—O—O—Ø    O  O—O
  ↓             ↓                     ↓    ↓
  O—O—Ø         O—O—Ø                 O—O—O—O—Ø
```

Frequently, porcine pancreatic α-amylase (*28*) and bacterial saccharifying α-amylase (*29*) are used for this purpose. Bacterial saccharifying α-amylase removes D-glucosyl residues from non-reducing terminals of α-D-(1 → 6) branch points so that singly-branched oligosaccharides of the type

```
O
↓
O—O—Ø
```

and multi-branched oligosaccharides of the type

```
O                O
↓                ↓
O—O              O—O—O
  ↓                  ↓
  O—O—Ø              O—O—Ø
```

```
O    O
↓    ↓
O—O—O—O—Ø
```

are formed; those obtained from porcine pancreatic α-amylase limit dextrin have one or two D-glucosyl residues at the terminus. By a combination of the action of acids, glucoamylase, and pullulanase, Kainuma and French (*30*) obtained a series of maltooligosaccharides stubbed with D-glucosyl units at the non-reducing end.

```
O—O ....... O—O                 acid-catalyzed        O—O ....... O—O
              ↓                ──────────────→                      ↓
O—O ....... O—O—O ......         hydrolysis           O—O ....... O—O—O ... Ø

                     O              O               O ... O
glucoamylase         ↓              ↓                     ↓
──────────────→      O—O ... Ø  +  O—O—O ... Ø  +   O ... O ... Ø

                     O              O               O ... Ø
pullulanase          ↓              ↓                  +
──────────────→      O—O ... Ø  +  O—O—O ... Ø  +   O ... O—O ... Ø

                     O                   O
glucoamylase         ↓                   ↓
──────────────→      O—O ... Ø       +   O—O ... Ø  +  nØ
```

Branched saccharides of various structures are important substrates for examination of the action pattern of debranching enzymes (*31, 32*).

2. Chemical Methods for Preparation

Acid-catalyzed hydrolysis of starch is most often used to prepare oligosaccharides. Chemical synthesis is generally limited to the preparation of disaccharides (*33*). The major deficiencies of chemical synthesis are lower specificity of the reaction, formation of large quantities of unwanted side-products, attendant purification problems, synthetic difficulties, and low yield.

a. *Acid-Catalyzed Hydrolysis of Amylose and Starch*

Controlled acid-catalyzed hydrolysis with chromatographic separation is useful for the preparation of a series of maltooligosaccharides. Commercial amylose is a suitable material. Hydrolytic conditions should be carefully controlled to avoid undesired acid-catalyzed reversion. Hydrolysis is generally achieved with dilute hydrochloric or sulfuric acids. Whelan and co-workers (*34*) used 0.33 *N* sulfuric acid at 100° for 2 h to obtain maltooligosaccharides from amylose. Pazur and Budovich (*35*) employed 0.1 *N* hydrochloric acid at 100° for 4 h. Like others, they found that the D-glucosidic bonds in starch or amylose are hydrolyzed at different rates. The rate constant for acid hydrolysis of bonds in (1 → 4)-α-D-glucan was determined also by Weintraub and French (*36*). They found the rate constant for acid-catalyzed hydrolysis of the bond at the non-reducing end of maltotriose and maltohexaose chain to be 1.8 times higher than that of the other glucosidic bonds.

b. *Reversion*

Acid- or enzyme-catalyzed hydrolysis of D-glucosidic bonds is a reversible reaction, and D-glucose recombines to form oligosaccharides. Thompson and co-workers (*37*) report the products from acid reversion of D-glucose as shown in Table III.

c. *Isomerization and Epimerization*

Alkaline isomerization and epimerization of oligosaccharides have been employed to prepare oligosaccharides and keto monosaccharides. Even though these reactions are rather difficult to control, the method is useful to prepare, for

Table III

Reversion Products from D-Glucose (37)[a]

Products	*Linkage*	*Yield, g*[b]
β,β-Trehalose	β,β-(1 → 1)	0.072
β-Sophorose	β-(1 → 2)	0.17
β-Maltose	α-(1 → 4)	0.40
α-Cellobiose	β-(1 → 4)	0.069
β-Cellobiose	β-(1 → 4)	0.25
β-Isomaltose	α-(1 → 6)	4.2
α-Gentiobiose	β-(1 → 6)	0.063
β-Gentiobiose	β-(1 → 6)	3.40
β-Nigerose	α-(1 → 3)	0.22
Levoglucosan[c]		0.30

[a] 100 g of D-glucose was heated in 300 g of 0.082 *N* HCl for 10 h at 98°. Then the products were isolated chromatographically and identified as crystalline octaacetates.

[b] Basis 100 g of initial D-glucose.

[c] Identified as its crystalline triacetate.

example, lactulose (*38*) from lactose, maltulose (*39*) from maltose, 4-*O*-β-D-glucopyranosyl-D-mannose (*40*) from cellobiose, panulose (*41*) [*O*-α-D-glucopyranosyl-(1 → 6)-*O*-α-D-glucopyranosyl-(1 → 4)-D-fructose] from panose, and 2-*O*-α-D-glucosyl-D-mannose (*42*) from kojibiose. Aluminum complex formation of ketoses can be used to give a higher yield of ketoses.

3. Chromatographic Separation of Starch Oligosaccharides

a. *Charcoal Column Chromatography*

Since charcoal column chromatography was first introduced to carbohydrate research by Whistler and Durso (*43*), the technique has been widely used for the preparation of oligosaccharides. In the original method, stepwise elution by ethanol was employed, but later gradient elution with *n*- and *tert*-butanol or *n*-propanol (*44, 45*) was found to improve resolution. The basic technique of the charcoal column chromatography is described elsewhere (*46*). Charcoal column chromatography is especially useful because of the high loading capacity of the column. The amount of carbohydrate that can be loaded on the column is roughly several percent of the weight of the packed charcoal, but for sharp separations, about 0.5% is recommended.

b. *Paper Chromatography*

Paper chromatography is one of the best methods to prepare purified oligosaccharides, and is frequently used as the final purification step for oligosaccharides after preliminary separation by charcoal column chromatography or by gel permeation chromatography.

High-temperature paper chromatography at 85° was developed (*47*) to improve the speed of separation and resolution of maltooligosaccharides of up to ten D-glucosyl units using three ascents. Multiple descending paper chromatography enables the resolution of megalosaccharides up to DP 25. Even though this is a time consuming technique, it is an effective method for separation of oligosaccharides of high DP (*48*).

Two-dimensional paper chromatography is useful for examining the action pattern of starch-degrading enzymes on maltosaccharides (*49*). For analysis, the chromatogram is first dipped in an 80% aqueous acetone solution of glucoamylase to convert weakly reducing higher oligosaccharides into D-glucose (*30*), thereby greatly enhancing the extent of silver reduction (*50*).

French and co-workers (*51*) propose a correlation of structure with mobility on a paper chromatogram. Paper chromatography is also used for preparation of pure maltooligosaccharides (*52*).

c. *Gel Permeation Chromatography*

For gel permeation chromatography, polyacrylamide (BioGel, Toyopearl) gels, crosslinked dextran (Sephadex) gels or agarose (Sepharose) gels are most frequently used. The technique has been reviewed (*53*). BioGel P-2, first applied in 1969 (*54*), has been used frequently for the preparative separation of oligosaccharides. A hot-water-jacketed column can improve separation. Because the column is normally eluted with degassed, deionized water, the fractionated oligosaccharides are easily concentrated or lyophilized without deionization or any further purification (*55, 126*).

d. *High-Pressure Liquid Chromatography (HPLC)*

High-pressure liquid chromatography was introduced to carbohydrate research in 1970. It offers rapid analysis of a large spectrum of saccharides, requires a minimum of sample preparation, and gives accuracy of quantitation equal to other methods.

To achieve good resolution, prepacked columns such as Whatman PXS-1025 pac, Waters Microbondapak/carbohydrate, and Shoden Shodex Ionpak S-801, S-614, may be used. Columns are expensive and their life is relatively short. Schwarzenbach (*56*) describes the bonding of an aminoalkyl substituent to the gel surface and discusses the separation of carbohydrates on the column. He obtained G_1 and G_2 as sharp peaks and G_4–G_7 as relatively broad peaks. Kainuma and co-workers (*127*) successfully fractionated maltooligosaccharides up to DP 15 by HPLC using a JASCOPAK SN-01 column.

e. *Thin-Layer Chromatography (TLC)*

Thin-layer chromatography is another useful technique for separation of maltooligosaccharides. It is as simple as paper chromatography, and can be done in a much shorter time. The principles of thin-layer chromatography are described elsewhere (*57*), but the use of lactic acid in the solvent system to obtain better separation of oligosaccharides is worthy of consideration (*58*). Lactic acid causes almost all monosaccharides to move rapidly, providing a better resolution of oligosaccharides than is obtained with conventional solvents.

4. Chemical and Physical Properties

Only very limited work has been done on the chemical and physical properties of maltooligosaccharides due to the difficulty of large scale preparation in pure form. Various properties were summarized by Thoma (*33*) in the previous edition of this book. More recent results obtained after 1965 are described here. Scallet and his collaborators (*59–61*) have worked extensively on maltooligosaccharides and reported dextrose equivalent, hygroscopicity, and fermentability of fractionated maltooligosaccharides. Johnson and Srisuthep (*62*) separated maltooligosaccharides to DP 12 by a large-scale carbon column chromatography and preparative paper chromatography and carefully determined their chemical and physical properties.

a. *Dextrose Equivalent (DE) of Maltooligosaccharides*

Johnson and Srisuthep (*62*) determined the dextrose equivalent of each oligosaccharide. Their values agreed well with the theoretical values and those reported by others (*59, 63*).

b. *Specific Gravity of Solution*

Specific gravities (*62*) measured by the picnometric method at 20° for 1%, 2%, 4%, 8%, and 10% aqueous solution of maltooligosaccharides are shown in Table IV. There is a general linear relationship between specific gravity and degree of polymerization.

c. *Refractive Index*

Refractive indices of maltooligosaccharides of 1–10% solutions of G_1 to G_{10} at 20° were also determined by Johnson and co-workers (*62*). Refractive index does not increase as the size of the polymer increases, but increases with the concentration of the solution.

d. *Specific Viscosity*

Specific viscosity (*62*) frequently is used to express the relationship of polymer size to viscosity, because specific viscosity depends on the volume occupied by the polymers. The specific viscosities of the maltooligosaccharides of G_1 through G_{10} are shown in Figure 3.

Table IV

True Specific Gravity[a] of Solutions of Maltooligosaccharides at 20° (62)

Degree of polymerization	True specific gravity				
	1%	2%	4%	8%	10%
G_1	1.0018	1.0050	1.0131	1.0282	1.0348
G_2	1.0011	1.0047	1.0126	1.0266	1.0337
G_3	1.0013	1.0049	1.0129	1.0287	1.0364
G_4	1.0014	1.0051	1.0130	1.0289	1.0369
G_5	1.0015	1.0053	1.0130	1.0290	1.0369
G_6	1.0016	1.0057	1.0136	1.0286	1.0386
G_7	1.0019	1.0058	1.0138	1.0287	1.0386
G_8	1.0020	1.0062	1.0140	1.0296	—
G_9	1.0021	1.0057	1.0140	—	—
G_{10}	1.0020	1.0056	1.0139	—	—

[a] Specific gravity *in vacuo*.

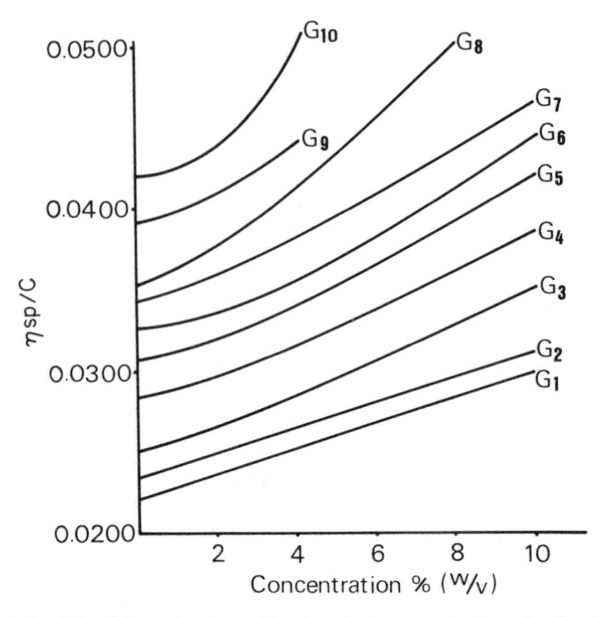

FIG. 3.—Relationship of the ratio of specific viscosity to concentration of maltooligosaccharides (*62*).

e. *Hygroscopicity*

Hygroscopicity of oligosaccharides is particularly important in the application of starch hydrolyzates.

Donnelly and co-workers (*60*) measured the hygroscopicities of malto-oligosaccharides of G_3 through G_{11} under the conditions of 25%, 56%, 75%, and 90% RH at 24°, 30°, and 38°. The relationship of moisture absorption of the individual oligosaccharides at 90% RH and at 24°, 30° and 38° was developed. The listing below relates the moisture absorption powers of these carbohydrates.

$$24° \; G_3>G_4>G_5 = G_7>G_{11}>G_2$$
$$30° \; G_3>G_4 = G_7>G_5>G_6>G_{11}>G_2$$
$$38° \; G_3>G_4 = G_5 = G_7>G_6>G_{11}>G_2$$

Donnelly and co-workers (*60*) speculated that the change in position of G_7 at 30° was due to its increased moisture absorbance based upon the fact that the absorbance of G_5 remains approximately the same. The slightly greater ability of G_7 to take up moisture is possibly related to destruction of the helical content of G_7 at 30°, whereas the difference at 38° may be caused by mobility and flexibility of nonhelical chains. Maltose has a much lower ability to absorb moisture than the higher oligosaccharides. This could be a function of intermolecular hydrogen

Table V

Hygroscopicities of Maltooligosaccharides (G_1 to G_{10}) at 60 ± 5% Relative Humidity at 26° (62)

Degree of polymerization	Moisture (%)		
	15 min	90 min	270 min
G_1	1.72	0.52	0.43
G_2	1.45	0.33	0.34
G_3	3.82	7.93	11.27
G_4	4.21	9.22	10.82
G_5	6.21	9.94	10.10
G_6	7.76	9.14	10.96
G_7	4.01	9.67	11.44
G_8	7.48	11.53	13.23
G_9	9.62	12.73	14.32
G_{10}	8.90	12.30	13.92

bond formation yielding aggregates. For G_3 and above, steric hindrance may prevent such aggregation. Johnson and Srisuthep (*62*) determined hygroscopicities of G_1 through G_{10} at 60% RH and 26°. They measured the weight gain at 15, 90, and 270 min as shown in Table V. The slight difference between these two experiments may be due to the condition used. Donnelly and coworkers measured the hygroscopicity near equilibrium after 24 h, whereas Johnson measured it at the initial stage of the moisture absorption.

Both results agree that G_1 and G_2 have extremely lower hygroscopicity when compared with G_3 and above.

f. *Infrared and Raman Spectra*

Srisuthep and co-workers (*64*) observed the infrared and Raman spectra of G_1 through G_{12}. They found G_3 and higher oligosaccharides could not be readily distinguished by their ir spectra. All the oligosaccharides absorb infrared radiation in approximately the same general frequency ranges. The characteristic absorption band at 1640 cm^{-1} for the bending vibration in water varies in intensity among the oligosaccharides (G_1 to G_{12}). G_1 and G_2 exhibit distinct absorption bands in the region between 1320 and 1220 cm^{-1}, a region characteristic of O–H in plane and C–H, CH_2 bending modes; but this absorption becomes less distinct as chain length increases. The spectra for the oligosaccharides G_4 to G_{12} resemble those of the G_3 spectrum. D-Glucose and maltose show distinct Raman spectra, but maltotriose and higher oligosaccharides exhibit higher fluorescence throughout this spectral range and no satisfactory Raman spectra can be obtained. The differences in the spectra of the higher malto-oligosaccharides are too obscure to be distinguished qualitatively by visual examination.

g. *Fermentability by Yeast*

It is known that D-glucose and maltose are fermented well by yeast. Shieh and co-workers (*61*) examined the fermentability of higher maltooligosaccharides by various strains of *Saccharomyces carlsbergensis* and *Saccharomyces cerevisiae*. Maltose and maltotriose are utilized by many strains of yeast, but the higher oligosaccharides cannot be utilized by those examined.

5. Possible Uses of Pure Oligosaccharides

Pure maltooligosaccharides have been used to examine the action pattern of starch-degrading enzymes, for the determination of the size of the active site of amylases, and as a primer of carbohydrate synthesizing enzymes. Frequently, the oligosaccharides labeled at a specific glucosyl unit are used for these purposes.

Robyt and French (*65*) examined the position specificity of porcine pancreatic α-amylase on maltotriose through maltooctaose specifically labeled in the reducing D-glucosyl unit with ^{14}C. They determined the frequency of bond cleavage of reducing ends during the initial action of porcine pancreatic α-amylase as shown in Figure 4.

Reducing end-labeled maltooligosaccharides also were used for determination of action mechanism of exohydrolases such as *P. stutzeri* maltotetraose-forming amylase (*17*) and *Aerobacter aerogenes* maltohexaose-forming amylase (*19, 20*). D-Glucosyl stubbed maltooligosaccharides were employed to investigate the steric effect of D-glucosyl stubs on the binding of substrates by porcine pancreatic α-amylase or β-amylase (Fig. 5).

The affinities between substrates and subsite of various amylases such as glucoamylase, β-amylase, Taka-amylase A, and bacterial α-amylase were examined using various maltooligosaccharides (*66, 67*) (Fig. 6). The procedure used for the determination of blood amylase activity was that introduced by the Dupont Company for their Automatic Clinical Analyzer, which is based on the amylase-catalyzed hydrolysis of maltopentaose (*68*). The hydrolysis of maltopentaose is coupled to the reduction of NAD^+, which is followed kinetically at 340 nm.

$$\text{Maltopentaose} \xrightarrow{\alpha\text{-amylase}} \text{maltotriose} + \text{maltose}$$

$$\text{Maltotriose} + \text{maltose} \xrightarrow{\alpha\text{-glucosidase}} \alpha\text{-D-glucose}$$

$$\text{D-Glucose} + \text{ATP} \rightarrow \text{D-glucose-6-phosphate} + \text{ADP}$$

$$\text{D-Glucose 6-phosphate} + \text{NAD}^+ \xrightarrow{\text{D-glucose 6-phosphate dehydrogenase}} \text{D-glucono-1,5-lactone 6 phosphate} + \text{NADH} + \text{H}^+$$

The indicator is the reduction of NAD^+ to NADH, followed by the determination of absorption at 340 nm.

Maltooligosaccharides of lower DP can be used as sweeteners. Relatively higher-molecular-weight oligosaccharides can be introduced to improve food quality and can be used in biochemical, clinical, and pharmaceutical applications. Also, oligosaccharides of narrower molecular weight distribution are interesting as raw materials of fine chemicals (*69*).

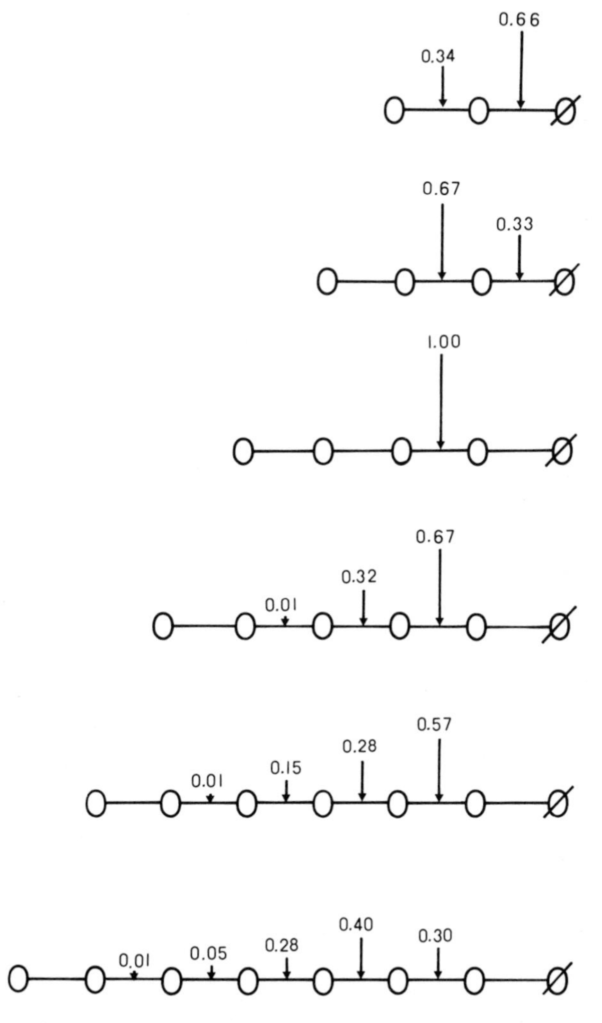

FIG. 4.—The frequency distribution of bond cleavage during initial action of porcine pancreatic α-amylase on reducing end-labeled oligosaccharides (*65*).

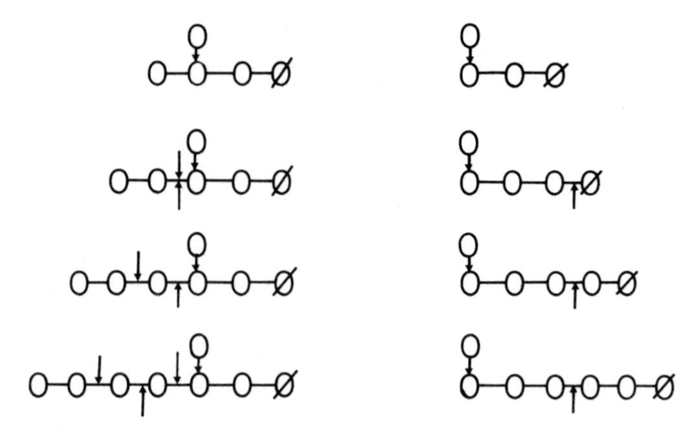

FIG. 5.—Position specificity of porcine pancreatic α-amylase (↑) and sweet potato β-amylase (↓) on glucosyl stubbed maltooligosaccharides (*27, 30*).

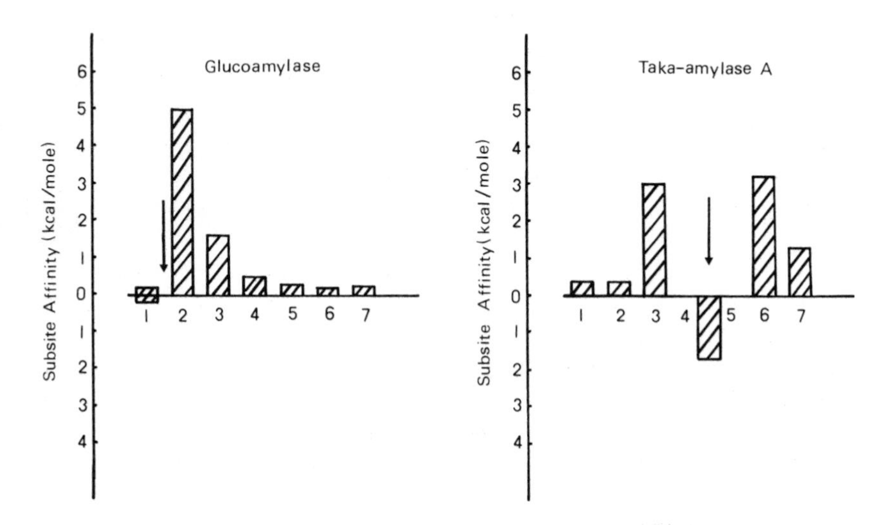

FIG. 6.—Histograms showing the characteristic arrangement of subsite affinities of *Rhizopus delemar* glucoamylase and Taka-amylase A. The subsites are numbered from the terminal where the nonreducing end of the substrate is situated in the productive binding mode. The arrows show the position of the catalytic site at which the substrate is cleaved (*66, 67*).

III. Cycloamyloses

Cycloamyloses are often referred to as cyclodextrins, cycloglucans, and Schardinger dextrins. (See also Chapter IV.)

1. Cycloamylose Glucanotransferase (Cyclomaltodextrin Glucanotransferase, EC 2.4.1.19)

Cycloamylose glucanotransferase is referred to as *Bacillus macerans* amylase, cyclodextrin glucanotransferase, or cyclomaltodextrin glucanotransferase in the literature.

In 1939, Tilden and Hudson (*70*) isolated a cell-free enzyme preparation from *B. macerans* that had the ability to convert starch into cycloamyloses. Prior to this discovery, cycloamyloses were made using live cultures of *B. macerans*. Since Tilden and Hudson's discovery of *B. macerans* cycloamylose glucanotransferase, effort has been given to working out methods for cycloamylose production and details of the mechanism.

The glucanotransferase was purified by several investigators such as Hale and Rawlins (*71*), Depinto and Campbell (*72*), and Kitahata and Okada (*73*) to give a single band by disc gel electrophoresis. More recently, Kobayashi and Kainuma (*74*) successfully crystallized the *B. macerans* enzyme. The enzyme of molecular weight 145,000 was dissociated to subunits of 74,000 by SDS disk gel electrophoresis. The enzyme is stable between pH 6.0 and 9.5 and below 50°. The optimum pH is 6.0 to 6.5, and the optimum temperature is 60°.

Recently, cycloamylose glucanotransferases have been discovered in various microorganisms such as *B. megaterium* (*75*), *B. circulans* (*76*), *B. stearothermohiilus* (*77*), *Bacillus* species (*alkalophilic*) (*78*), and *Klebsiella pneumoniae* (*79*). The enzymes have slightly different properties. The characteristic properties of cycloamylose glucanotransferases obtained from various microorganisms are shown in Table VI. The product specificity of the enzymes are also different. For example, the enzymes obtained from *B. macerans* and *B. stearothermophilus* produce cyclohexaamylose predominantly, while the enzymes obtained from *B. megaterium* and *Bacillus* sp. (*alkalophilic*) produce cycloheptaamylose as the major product.

The cycloamylose glucanotransferase catalyzes three reactions: (a) formation of cycloamyloses from starch; (b) hydrolysis of cycloamylose and transfer of the linear maltooligosaccharide to a cosubstrate; and (c) disproportionation of small oligosaccharides. The details of the reactions are described in Section II.1.f.

2. Preparation and Isolation of Cycloamyloses

The standard method for the preparation of cycloamyloses involves selective precipitation from a cycloamylose glucanotransferase digest of starch. Cramer

Table VI

Properties of Cycloamylose Glucanotransferases Obtained from Various Microorganisms

Microorganisms	*Optimum pH*	*pH Stability*	*Optimum temperature, °C*	*Thermostability*		*Main product*	*Cycloamylose formation, %*	*Ref.*
				$+Ca^+$	$-Ca^+$			
B. macerans	5.0–5.7	8.0–10.0	55	60	—	α	—	*73*
B. macerans	6.0	6.0–9.5	60	—	50	α	50–60	*74*
B. macerans	5.1–6.2	—	—	—	—	—	—	*72*
K. pneumoniae	5.2	6.0–7.5	—	—	45	α	50	*79*
B. stearothermophilus	5.0–5.5	5.0–8.8	70–75	65	70	α	62	*77*
B. megaterium	5.0–5.7	7.0–10.0	55	55	—	β	—	*73*
B. circulans	5.2–6.1	7.5–9.0	55	50	—	β	—	*76*
Bacillus species (alkalophilic)	4.5–4.7	6–10	45	65	65	β	73	*78*

(*80*) developed methods to fractionate cyclohexa-, cyclohepta-, and cyclooctaamylose. Selective precipitation is possible because of the variation in cavity diameter. The selective precipitation method was improved by adding the selective precipitants to the mixture after the starch had been degraded to a point when it no longer formed a precipitate with an organic phase (see Chap. VIII).

Pulley and French (*81*) developed a fractionation method for isolation of larger homologs of cycloamyloses after extensive β-amylase digestion to hydrolyze maltooligosaccharides. The work before 1960 is summarized in Thoma's review (*4*).

More recently, Kobayashi and co-workers (*82*) developed a new method to precipitate cycloamyloses from the *B. macerans* enzyme digest of starch. Glucoamylase is added to hydrolyze the remaining linear and branched oligosaccharides, permitting cycloamyloses to be precipitated upon addition of acetone. Then cyclohexa- and cycloheptaamylose are crystallized from 1-propanol and water. They also investigated the use of complex-forming agents with amylose and the product specificity of the *B. macerans* cycloamylose glucanotransferase (*83*). Cyclohexaamylose is the major product when 1-butanol and sodium dodecyl sulfate are present in the reaction mixture. On the contrary, the presence of *tert*-butyl alcohol and Triton X100 leads to the accumulation of cycloheptaamylose.

3. Chemical and Physical Properties of Cycloamyloses

Cycloamyloses are cyclic nonreducing D-glucopyranosyl polymers containing six, seven, eight, or more residues joined by α-D-(1→4)-linkages and as a result have neither reducing nor non-reducing end groups. Two systems of nomenclature exist. The first indicates the number of residues in the cyclic polymers by prefixing a Greek letter. Because the smallest known cycloamylose is a hexamer, it is assigned the prefix α. The heptamer and octamer are referred to as β-, and γ-cycloamylose. In the other system of nomenclature, the cycloamyloses are designated as cyclohexa-, cyclohepta-, and cyclooctaamylose. Although the latter system is recommended because it is more descriptive, the former nomenclature has been used more frequently.

Although the existence of α, β, γ, δ, ε, ζ, and η cycloamyloses are known, the physical and chemical constant are known well only for α, β, and γ dextrins (Table VII) (*84–88*). The higher-molecular-weight cycloamyloses have not yet been prepared in large scale, and they have not been found to form insoluble complexes with hydrophobic compounds. Specific rotation at the sodium D line of δ, ε, ζ, η are $+191 \pm 3°$, $+197 \pm 3°$, $+200 \pm 2°$, and $+201 \pm 3°$, respectively (*87*).

The cavity of the cycloamyloses has been referred to as hydrocarbon in nature. This theory is attributed to Freudenberg (*89, 90*) who originally proposed that the

Table VII

Chemical and Physical Properties of Cycloamylose (84–88)

	α	β	γ
Specific rotation at sodium D line	+150.5 ± 0.5°	+162.0 ± 0.5°	+177.4 ± 0.5°
Specific rotation in presence of organic acids or salts, $[\alpha]^{35}{}_{D}$			
Sorbic acid	+183°		
Benzoic acid 0.278%	+165°		
1*M* KCl	+150°		
1*M* KBr	+146°		
1*M* KI	+126°		
Molecular weight	972	1135	1297
Crystal character	Hexagonal plates or blade shaped needles	Parallelograms	Square plates or rectangular rods
Cavity diameter	6 Å	8 Å	10 Å
Solubility in water (g/100mL)	14.5	1.85	23.2
Solubility in water in presence of selected addenda (g/100mL)			
Benzene	1.24, 0.714[a]	0.075	1.45
Fluorobenzene	1.14, 1.374[a]	0.00	1.60
Chlorobenzene	1.32	0.00	0.36
Bromobenzene	1.62	0.01	1.64
Iodobenzene	3.12	0.01	1.67
Aniline	2.20	1.215	3.065
N-Methylaniline	1.218	0.175	0.905
p-Chloroaniline	4.7	0.625	
o-Chlorophenol	0.813	0.168	0.845
p-Chlorophenol		0.73	
p-Cymene	2.92	0.13	1.63
Methylcyclohexane	1.36[a]		
Cyclohexanol	1.85	0.987	4.77
Cyclohexylamine	>3	0.37	
2,2-Dimethylbutane	0.172[a]		
2,3-Dimethylbutane	0.513[a]		
1,1,2,2-Tetrachloroethane	0.57[a]	0.12[a]	
Tetrachloroethylene	0.7[a]	0.004[a]	
Carbon tetrachloride	0.212[a]	0.1	
Chloroform	1.383[a]		
1,2-Dichloropropane	0.54[a]		
Ethyl bromide	0.58	0.32	0.5
Butyric acid	1.74	0.5	>7.5
Valeric acid		0.43	
Caproic acid	2.05	0.67	

[a] Temperature at the determination was 25°.

interior of both cycloamyloses and the amylose helix were lined with hydrocarbon-like structures. This proposal was ostensibly supported by the affinity of cycloamyloses for hydrophobic reagents. Thoma (*4*) discusses the nature of the cavity. He believes that the hydrocarbon interior theory is incorrect, arising from construction of molecular models with monomer units in a boat rather than $^4C_1$ chair conformation. Thoma criticized the hydrocarbon interior theory from the standpoint of thermodynamics of complex formation and accurate space filling models of cyclohexaamylose with the anhydro-D-glucopyranosyl rings in the $^4C_1$ conformation. This model indicates that the bridging acetal oxygens are symmetrically distributed around the interior of the cavity with their lone π orbitals orthogonal to the cylindrical axis of the molecule. The model suggests that the "inside" of the molecule contains a region of high electron density (*91*) and might behave as a Lewis base (*92*). Thoma and Stewart (*4*) reviewed the evidences obtained by themselves and others (*86, 93, 94*) on the thermodynamics of cycloamylose complexes and of the transfer of hydrophobic molecules from water to a hydrocarbon solvent and concluded that the thermodynamic evidence is against the hydrocarbon interior hypothesis.

Structural features of crystalline complexes of cycloamyloses have been examined in detail (*95–98*).

4. Biochemical Properties of Cycloamyloses

Cycloamyloses are less susceptible to amylase action than are linear oligosaccharides. As cycloamyloses do not have a non-reducing end, they are stable to the action of exoamylases. Some microbial enzymes hydrolyze cycloamyloses (*99–108*) (Table VIII). Cyclooctaamylose is more susceptible to α-amylase action than is cyclohexa- and cycloheptaamylose and is hydrolyzed even by pancreatic and salivary α-amylase. Suetsugu and co-workers (*109*) examined the action of Taka-amylase A on cyclohexa-, cyclohepta-, and cyclooctaamylose. The rate constants for the cleavage of the cyclic structures of cyclohexa-, cyclohepta-, and cyclooctaamyloses by Taka-amylase A were $3.3–5 \times 10^{-4}$, $2.5–4 \times 10^{-3}$, $5.3–5 \times 10^{-1}$ of those of linear oligosaccharides. In acid-catalyzed hydrolysis, cycloamyloses are hydrolyzed at $2–3.3 \times 10^{-1}$ the rate of hydrolysis of linear maltooligosaccharides, though there is no significant difference between cyclohexa- and cycloheptaamylose. The digestion of cycloamyloses in animal digestive organs has been reviewed (*7*).

Von Hoesslin and Pringsheim (*110*) could not detect synthesis of glycogen when cycloamyloses were administered to a fasted rabbit or guinea pig. With diabetic patients, 50 g of cyclohexaamylose does not give rise to noticeable increase in urinary sugar. Because tests for fecal cyclohexaamylose were negative, Von Hoesslin and Pringsheim concluded that cycloamylose was "directly utilized."

French (*108*) reported that animals utilize cycloamyloses. However, rats fed a

Table VIII

Cycloamylose Hydrolyzing Enzymes

Cycloamylose	*Enzyme*	*Reference*
Cyclohexaamylose	*Pseudomonas amylodermosa* α-amylase	*99*
	Taka-amylase A from *As. oryzae*	*100–102*
	B. macerans cycloamylose glucanotransferase	*103*
Cycloheptaamylose	*Pseudomonas amylodermosa* α-amylase	*99*
	Taka-amylase A from *As. oryzae*	*100–102*
	B. subtilis liquefying α-amylase	*104*
	B. subtilis saccharifying α-amylase	*105,106*
Cyclooctaamylose	*Pseudomonas amylodermosa* α-amylase	*99*
	Taka-amylase A from *As. oryzae*	*102*
	Porcine pancreatic α-amylase	*107*
	Saliva α-amylase	*108*

diet in which a part of the carbohydrate was replaced by cycloheptaamylose refused to eat, except for small amounts, and within a week all animals on the ration died.

Makita and co-workers (*111*) also examined the chronic toxicity of cycloheptaamylose. Cycloheptaamylose was given by stomach tube to rats at daily doses of 0.1, 0.4, and 1.6 g/kg for 6 months. Twelve male and twelve female rats were used. These investigators did not find adverse effects in rat body weight, urine analysis, hematology, clinic chemistry, or organ weight. Autopsy and microscopic examination showed no adverse effects. Because no toxicological symptoms were observed during administration, the workers concluded that cycloheptaamylose is nontoxic to Wistar strain rats at the dose levels used.

Suzuki (*112*) studied the effect of cyclodextrins on the body and organ weights of rats. He replaced pregelatinized starch in the diet with a cyclohexaamylose-containing syrup composed of 30% cyclohexa-, 15% cyclohepta-, and 5% cyclooctaamylose, and 50% low-DE starch hydrolyzate. He checked the increase of body weight daily until the eighteenth day and then measured the weight of the liver, heart, kidney, spleen, lung, testicles, and epididymis. No difference was found in body weight and weight of organs between the group fed with pregelatinized starch and the groups fed with the cyclodextrin–starch hydrolyzate syrup. It was concluded that there is no significant effect of daily administration of cyclodextrins on the development of organs and tissues in rats.

5. Possible Uses of Cycloamyloses

Extensive efforts to find cycloamylose glucanotransferases in various microorganisms and to improve enzyme purification and cycloamylose preparation have been made mainly in Japan. The results suggest the feasibility of wider uses of cycloamyloses in pharmaceuticals, foods, chemicals, cosmetics, toiletries, and pesticides by utilizing the property of cycloamyloses to include compounds or molecular parts in the cavity.

It is now obvious that the inclusion compound does not invariably enter, or is not completely contained in, the cavity. In many instances, adducts are formed, even when the molecular size of the inclusion compound would prohibit its complete insertion into the central cavity. These phenomena are explained as a formation of a "cap" or "lid" to the hole, or an insertion of some functional group into the cavity. By forming the inclusion compound, cycloamyloses can stabilize labile compounds, emulsify oils, mask tastes, flavors or odors, increase solubility, and convert viscous or oily compounds into powders (*113*).

IV. Recent Publications Regarding Maltooligosccharide Preparation and Utilization Not Mentioned in the Text (see also Chapter IV)

H. Kondo, H. Nakatani, and K. Hiromi, *Agric. Biol. Chem.*, **45,** 2369 (1981).
A. Heyraud and M. R. Naudo, *J. Liquid Chromatogr.*, **4,** S2, 175 (1981).
H. David, R. McCroske, and B. Muller, *J. Clin. Chem.*, **19,** 645 (1981).
T. Chang, H. David, and R. McCroske, *Clin. Chem.*, **27,** 1047 (1981).

V. Recent Publications Regarding Cycloamyloses Not Mentioned in the Text (see also Chapter IV)

M. Misaki, A. Konno, M. Miyawaki, and K. Yasumatsu, *Nippon Shokuhin Kogyo Gakkaishi*, **24,** 228 (1982).
A. Konno, M. Misaki, M. Miyawaki, and K. Yasumatsu, *Nippon Shokuhin Kogyo Gakkaishi*, **29,** 255 (1982).
E. Smolkova, *J. Chromatogr.*, **251,** 17 (1982).
K. Hirotsu, T. Higuchi, K. Fujita, T. Ueda, A. Shinoda, T. Imoto, and I. Tabushi, *J. Org. Chem.*, **47,** 1143 (1982).
Y. Ikeya, K. Matsumoto, K. Kunihiro, T. Fuwa, and K. Uekama, *Yakugaku Zasshi*, **102,** 83 (1982).
T. Fujinaga, Y. Uemura, F. Hirayama, M. Otagiri, and K. Uekama, *J. Pharmacobio-Dynamics*, **5,** S11 (1982).
J. Szejtli, *Stärke*, **33,** 387 (1981).
A. Harada, M. Furue, and S. Nozakura, *Polym. J.*, **13,** 777 (1981).
K. K. Chacko and W. Saenger, *J. Amer. Chem. Soc.*, **103,** 1708 (1981).
S. Kobayashi and K. Kainuma, *Kagaku to Seibutsu*, **20,** 453 (1982).
S. Kobayashi and K. Kainuma, *Denpun Kagaku*, **28,** 132 (1981).
K. Uekama, *Yakugaku Zasshi*, **101,** 857 (1981).
K. Uekama, T. Irie, M. Sunada, M. Otagiri, K. Iwasaki, Y. Okano, T. Miyata, and Y. Kase, *J. Pharm. Pharmacol.*, **33,** 707 (1981).

VI. References

(*1*) Anon., *J. Org. Chem.*, **28,** 281 (1963).
(*2*) W. J. Whelan, *Annu. Rev. Biochem.*, **29,** 105 (1960).
(*3*) J. A. Thoma, H. B. Wright, and D. French, *Arch. Biochem. Biophys.*, **85,** 452 (1959).
(*4*) J. A. Thoma and L. Stewart, in "Starch: Chemistry and Technology," R. L. Whistler and E. F. Paschall, eds., Academic Press, New York, Vol. I, 1965, p. 209.
(*5*) A. Villiers, *Compt. Rend.*, **112,** 536 (1891).
(*6*) K. Freudenberg and R. Jacobi, *Ann.*, **518,** 102 (1935).
(*7*) D. French, *Adv. Carbohydr. Chem.*, **12,** 190 (1957).
(*8*) J. Robyt and D. French, *Arch. Biochem. Biophys.*, **104,** 338 (1964).
(*9*) M. Higashihara and S. Okada, *Agr. Biol. Chem.*, **38,** 1023 (1974).
(*10*) R. Shinke, Y. Kunimi, and H. Nishira, *J. Ferment. Technol.*, **53,** 693 (1975).
(*11*) R. Shinke, Y. Kunimi, and H. Nishira, *J. Ferment. Technol.*, **53,** 698 (1975).
(*12*) Y. Takasaki, *Agr. Biol. Chem.*, **40,** 1515, 1523 (1976).
(*13*) H. Hidaka, Y. Koaze, K. Yoshida, T. Niwa, T. Shomura, and T. Niida, *Staerke*, **26,** 413 (1974).
(*14*) K. Wako, C. Takahashi, S. Hashimoto, and J. Kanaeda, *Denpun Kagaku*, **25,** 155 (1978).
(*15*) T. Suganuma, Ph.D. Dissertation, Kyoto University, Kyoto, Japan, 1978.
(*16*) K. Wako, S. Hashimoto, S. Kubomura, K. Yokota, K. Aikawa, and J. Kanaeda, *Denpun Kagaku*, **26,** 175 (1979).
(*17*) J. F. Robyt and R. J. Ackerman, *Arch. Biochem. Biophys.*, **145,** 105 (1971).
(*18*) N. Saito, *Arch. Biochem. Biophys.*, **155,** 290 (1973).
(*19*) K. Kainuma, S. Kobayashi, T. Ito, and S. Suzuki, *FEBS Lett.*, **26,** 281 (1972).
(*20*) K. Kainuma, K. Wako, S. Kobayashi, A. Nogami, and S. Suzuki, *Biochim. Biophys. Acta*, **410,** 333 (1975).
(*21*) K. Kainuma, *Kagaku to Seibutsu*, **11,** 146 (1973).
(*22*) E. B. Tilden and C. S. Hudson, *J. Bacteriol.*, **43,** 527 (1942).
(*23*) S. Kitahata and S. Okada, *Agr. Biol. Chem.*, **38,** 2413 (1974).
(*24*) S. Okada and S. Kitahata, *Proc. Amylase Symp.* (Osaka), **8,** 21 (1973).
(*25*) M. Shiosaka and H. Bunya, *Proc. Amylase Symp.* (Osaka), **8,** 43 (1973).
(*26*) D. French, M. L. Levine, E. Norberg, P. Nordin, J. H. Pazur, and G. M. Wild, *J. Amer. Chem. Soc.*, **76,** 2387 (1954).
(*27*) K. Kainuma and D. French, *FEBS Lett.*, **6,** 182 (1970).
(*28*) K. Kainuma and D. French, *Proc. Amylase Symp.* (Osaka), **5,** 35 (1970).
(*29*) K. Umeki and T. Yamamoto, *J. Biochem.*, **72,** 1219 (1972).
(*30*) K. Kainuma and D. French, *FEBS Lett.*, **5,** 257 (1969).
(*31*) M. Abdullah, B. J. Catley, E. Y. C. Lee, J. Robyt, K. Wallenfels, and W. J. Whelan, *Cereal Chem.*, **43,** 111 (1966).
(*32*) K. Kainuma, S. Kobayashi, and T. Harada, *Carbohydr. Res.*, **61,** 345 (1978).
(*33*) J. A. Thoma, in "Starch: Chemistry and Technology," R. L. Whistler and E. F. Paschall, eds., Academic Press, New York, Vol. I, 1965, p. 177.
(*34*) W. J. Whelan, J. M. Bailey, and P. J. P. Roberts, *J. Chem. Soc.*, 1293 (1953).
(*35*) J. H. Pazur and T. Budovich, *J. Biol. Chem.*, **220,** 25 (1956).
(*36*) M. Weintraub and D. French, *Carbohydr. Res.*, **15,** 251 (1970).
(*37*) A. Thompson, K. Anno, M. L. Wolfrom, and M. Inatome, *J. Amer. Chem. Soc.*, **76,** 1309 (1954).
(*38*) E. M. Montgomery and C. S. Hudson, *J. Amer. Chem. Soc.*, **52,** 2101 (1930).
(*39*) S. Peat, P. J. P. Roberts, and W. J. Whelan, *Biochem. J.*, **51,** XVII (1952).
(*40*) S. N. Danilov, *Zh. Obshch. Khim.*, **16,** 923 (1946).

(41) K. Kainuma, K. Sugawara-Hata, and S. Suzuki, *Staerke*, **26,** 274 (1974).
(42) S. Kawamura and K. Matsuda, *Nippon Nogei Kagaku Kaishi*, **47,** 285 (1973).
(43) R. L. Whistler and D. F. Durso, *J. Amer. Chem. Soc.*, **72,** 677 (1950).
(44) D. French, J. F. Robyt, M. Weintraub, and P. Knock, *J. Chromatogr.*, **24,** 68 (1966).
(45) S. Kobayashi, T. Saito, K. Kainuma, and S. Suzuki, *Denpun Kagaku*, **18,** 10 (1971).
(46) R. L. Whistler and J. N. BeMiller, *Methods Carbohydr. Chem.*, **1,** 42 (1962).
(47) D. French, N. L. Mancusi, M. Abdullah, and G. Brammer, *J. Chromatogr.* **19,** 445 (1965).
(48) K. Umeki and K. Kainuma, *J. Chromatogr.*, **150,** 242 (1978).
(49) D. French, A. O. Pulley, M. Abdullah, and J. C. Linden, *J. Chromatogr.*, **24,** 241 (1966).
(50) W. E. Trevelyan, O. O. Procter, and S. S. Harrison, *Nature*, **166,** 444 (1950).
(51) D. French and G. M. Wild, *J. Amer. Chem. Soc.*, **75,** 2612 (1953).
(52) J. D. Commerford, G. T. VanDuzee, and B. L. Scallet, *Cereal Chem.*, **40,** 482 (1963).
(53) J. Reiland, *Meth. Enzymol.*, **22,** 287 (1971).
(54) M. John, G. Trenel, and H. Dellweg, *J. Chromatogr.*, **42,** 476 (1969).
(55) K. Kainuma, A. Nogami, and C. Mercier, *J. Chromatogr.*, **121,** 361 (1976).
(56) R. Schwarzenbach, *J. Chromatogr.*, **117,** 206 (1976).
(57) E. Stahl and U. Kaltenbach, *in* "Thin Layer Chromatography," E. Stahl, ed., Academic Press, New York, 1965, p. 461.
(58) S. A. Hansen, *J. Chromatogr.*, **105,** 388 (1975).
(59) J. D. Commerford and B. L. Scallet, *Cereal Chem.*, **46,** 172 (1969).
(60) B. J. Donnelly, J. C. Fruin, and B. L. Scallet, *Cereal Chem.*, **50,** 512 (1973).
(61) K. K. Shieh, B. J. Donnelly, and B. L. Scallet, *Cereal Chem.*, **50,** 169 (1973).
(62) J. A. Johnson and R. Srisuthep, *Cereal Chem.*, **52,** 70 (1975).
(63) W. J. Hoover, Ph.D. Thesis, University of Illinois, Urbana, Illinois, 1963.
(64) R. Srisuthep, R. Brockman, and J. A. Johnson, *Cereal Chem.*, **53,** 110 (1976).
(65) J. F. Robyt and D. French, *J. Biol. Chem.*, **245,** 3917 (1970).
(66) K. Hiromi, *Biochem. Biophys. Res. Commun.*, **40,** 1 (1970).
(67) Y. Nitta, M. Mizushima, K. Hiromi, and S. Ono, *J. Biochem.*, **69,** 567 (1971).
(68) T. H. Adams, U. S. Patent 3,879,283 (1975); *Chem. Abstr.*, **84,** 137986e (1976).
(69) K. Kainuma, "Starch Science Handbook," Z. Nikuni, M. Nakamura, and S. Suzuki, eds., Asakura Publishing Co., Tokyo, Japan, 1977, p. 463.
(70) E. B. Tilden and C. S. Hudson, *J. Bacteriol.*, **43,** 527 (1942).
(71) W. S. Hale and L. C. Rawlins, *Cereal Chem.*, **28,** 49 (1951).
(72) J. A. Depinto and L. L. Campbell, *Biochemistry*, **7,** 114 (1968).
(73) S. Kitahata, N. Tsuyama, and S. Okada, *Agr. Biol. Chem.*, **38,** 387 (1974).
(74) S. Kobayashi, K. Kainuma, and S. Suzuki, *Carbohydr. Res.*, **61,** 229 (1978).
(75) S. Kitahata and S. Okada, *Agr. Biol. Chem.*, **38,** 2413 (1974).
(76) S. Okada and S. Kitahata, *Proc. Amylase Symp.* (Osaka), **8,** 21 (1973).
(77) M. Shiosaka and H. Bunya, *Proc. Amylase Symp.* (Osaka), **8,** 43 (1973).
(78) N. Nakamura and K. Horikoshi, *Agr. Biol. Chem.*, **40,** 935 (1976).
(79) H. Bender, *Arch. Microbiol.*, **111,** 271 (1977).
(80) F. Cramer and F. M. Henglein, *Chem. Ber.*, **91,** 308 (1958).
(81) A. O. Pulley and D. French, *Biochem. Biophys. Res. Commun.*, **5,** 11 (1961).
(82) S. Kobayashi, K. Kainuma, and S. Suzuki, *Denpun Kagaku*, **22,** 6 (1975).
(83) S. Kobayashi, K. Kainuma, and S. Suzuki, *Nippon Nogei Kagaku Kaishi*, **51,** 691 (1977).
(84) D. French, M. L. Levine, J. H. Pazur, and E. Norberg, *J. Am. Chem. Soc.*, **71,** 353 (1949).
(85) W. S. McClenahan, E. B. Tilden, and C. S. Hudson, *J. Amer. Chem. Soc.*, **64,** 2139 (1942).
(86) F. Cramer, *Rev. Pure Appl. Chem.*, **5,** 143 (1955).
(87) A. O. Pulley, Ph.D. Dissertation, Iowa State University, Ames, Iowa, 1962.
(88) F. Cramer and F. M. Henglein, *Chem. Ber.*, **90,** 2561 (1957).

(*89*) K. Freudenberg, E. Schaaf, G. Dumpert, and T. Ploetz, *Naturwiss.*, **27,** 850 (1939).
(*90*) K. Freudenberg, *J. Polymer Sci.*, **23,** 791 (1957).
(*91*) F. Cramer and W. Dietsche, *Chem. Ber.*, **92,** 1739 (1959).
(*92*) F. Cramer, *Rec. Trav. Chim. Pays-Bas,* **75,** 891 (1956).
(*93*) W. F. Claussen and M. F. Polglase, *J. Amer. Chem. Soc.*, **74,** 4817 (1952).
(*94*) R. L. Bohon and W. F. Claussen, *J. Amer. Chem. Soc.*, **73,** 1571 (1951).
(*95*) P. C. Manor and W. Saenger, *Nature,* **237,** 392 (1972).
(*96*) P. C. Manor and W. Saenger, *J. Amer. Chem. Soc.*, **96,** 3630 (1974).
(*97*) K. Harata, *Bull. Chem. Soc. Japan,* **48,** 2409 (1975).
(*98*) K. Harata, *Carbohydr. Res.*, **48,** 265 (1976).
(*99*) K. Kato, T. Sugimoto, A. Amemura, and T. Harada, *Biochim. Biophys. Acta,* **391,** 96 (1975).
(*100*) E. Ben-Gershom, *Nature,* **4457,** 539 (1955).
(*101*) V. M. Hanrahan and M. L. Caldwell, *J. Amer. Chem. Soc.*, **75,** 2191 (1953).
(*102*) N. Suetsugu, S. Koyama, K. Takeo, and T. Kuge, *J. Biochem.*, **76,** 57 (1974).
(*103*) S. Kobayashi, K. Kainuma, and S. Suzuki, *Proc. Amylase Symp.* (Osaka), **8,** 29 (1973).
(*104*) M. Ohnishi, *J. Biochem.*, **69,** 181 (1971).
(*105*) M. H. Moseley and L. Keay, *Biotechnol. Bioeng.*, **12,** 251 (1970).
(*106*) L. Keay, *Staerke,* **22,** 153 (1970).
(*107*) M. Abdullah, D. French, and J. F. Robyt, *Arch. Biochem. Biophys.*, **114,** 595 (1966).
(*108*) D. French, *Advan. Carbohydr. Chem.*, **12,** 189 (1957).
(*109*) N. Suetsugu, S. Koyama, K. Takeo, and T. Kuge, *J. Biochem.*, **76,** 57 (1974).
(*110*) H. von Hoesslin and H. Pringsheim, *Hoppe-Seyler's Z. Physiol. Chem.*, **131,** 168 (1923).
(*111*) T. Makita, N. Ojima, Y. Hashimoto, H. Ide, M. Tsuji, and Y. Fujisaki, *Oyo Yakuri,* **10,** 449 (1975).
(*112*) M. Suzuki, personal communication.
(*113*) A. A. Lawrence, "Edible Gums and Related Substances," Noyes Data Corporation, Park Ridge, New Jersey, 1973, p. 289.
(*114*) J. Schmidt and M. John, *Biochim. Biophys. Acta,* **566,** 88 (1979).
(*115*) Y. Sakano, Y. Kashiwagi, and T. Kobayashi, *Agric. Biol. Chem.*, **46,** 639 (1982).
(*116*) Y. Sakano, Y. Kashiwagi, E. Kashiyama and T. Kobayashi, *Denpun Kagaku,* **29,** 131 (1982).
(*117*) T. Nakakuki, M. Monma, K. Azuma, S. Kobayashi, and K. Kainuma, *Denpun Kagaku,* **29,** 179 (1982).
(*118*) T. Nakakuki, K. Azuma, M. Monma, and K. Kainuma, *Denpun Kagaku,* **29,** 188 (1982).
(*119*) T. Nakakuki and K. Kainuma, *Denpun Kagaku,* **29,** 138 (1982).
(*120*) K. Kainuma, *Denpun Kagaku,* **28,** 92 (1981).
(*121*) Y. Takasaki, *Denpun Kagaku,* **29,** 145 (1982).
(*122*) H. Taniguchi, M. J. Chunng, Y. Maruyama, and M. Nakamura, *Denpun Kagaku,* **29,** 107 (1982).
(*123*) H. Taniguchi, F. Odashima, M. Igarashi, Y. Maruyama, and M. Nakamura, *Agric. Biol. Chem.*, **46,** 2107 (1982).
(*124*) S. Kitahata and S. Okada, *Agric. Biol. Chem.*, **39,** 2185 (1975).
(*125*) S. Kitahata and S. Okada, *J. Biochem.*, **79,** 641 (1976).
(*126*) M. John, T. Schmidt, C. Wandrey, and H. Sahn, *J. Chromatogr.*, **247,** 281 (1982).
(*127*) K. Kainuma, T. Nakakuki, and K. Ogawa, *J. Chromatogr.*, **212,** 126 (1981).

CHAPTER VI

MOLECULAR STRUCTURE OF STARCH

By Roy L. Whistler and James R. Daniel*

Department of Biochemistry, Purdue University, West Lafayette, Indiana

I. General Nature of Starch

Proof that starch is a homoglycan composed of but a single type of sugar unit, regardless of the source of the starch, was initiated by a happenstance during investigations by Kirchoff, a German attached to the Russian Imperial Academy of Sciences at St. Petersburg. Kirchoff was searching for a replacement for gum arabic and was attempting to produce a substitute through the partial acid hydrolysis of wheat and potato starches (*1*). On extended hydrolysis with dilute sulfuric acid, he obtained, unexpectedly, a sweet tasting product. Though it was not as sweet as cane sugar (sucrose), Kirchoff thought it might have a household use, a prediction that took some time to become reality (*2*).

Hydrolysis of potato starch with dilute sulfuric acid was examined in more detail by de Saussure (*3*) who proved that the resulting monosaccharide was the same as the principal monosaccharide in grape sugar. He also found that hydrolysis of 100 parts of starch gave 110 parts of monosaccharide from which Salomon (*4*) calculated that the molecular formula of starch must be $C_6H_{10}O_5$. This formula was verified later by combustion analysis (*5*). Partial hydrolysis products precipitated from aqueous solution by alcohol were dextrorotatory (*6*) and, hence, were named *dextrins*. The dextrins gave colored products on exposure to aqueous iodine, a phenomenon investigated by Bondonneau (*7*).

*Present address: Department of Foods and Nutrition, Purdue University, West Lafayette, Indiana

STARCH, 2nd ed.

ISBN 0-12-746270-8

Correct deduction of the molecular formula for starch at such an early period is impressive, for starches retain moisture extremely well, making quantitative deductions somewhat hazardous. Starch was established as a polymer by the remarkably small freezing point depression observed for its aqueous solution (*8*). It was through the brilliant work of Staudinger (*9*) and Carothers (*10*) that the covalent nature of the molecule was firmly established and the general formula could be written as $(C_6H_{10}O_5)_x$.

Hydrolytic investigations of starch were early taken as proof of its poly(D-glucosyl), or D-glucan, nature. However, obtaining a quantitative yield of D-glucose from even a pure glucan is difficult. Freudenberg and Boppel (*11*) obtained a 97% yield of partially methylated D-glucose by methylation and hydrolysis of starch. Banks and co-workers (*12*) investigated the acid-catalyzed hydrolysis of starch amylose (see Section VI) and amylopectin (see Section VII) and concluded that amylose is quantitatively degraded to D-glucose at 100° in 0.75 *M* sulfuric acid after 4 h, whereas amylopectin yields 96% of the theoretical amount of D-glucose and 4% of the disaccharide isomaltose. The identity and structure of isomaltose, if a true fragment, is of considerable importance in ascertaining the structure of amylopectin, as will be discussed later.

II. Fractionation of Starch (see also Chapter VIII)

While it was established that starch consisted largely, or perhaps entirely, of D-glucosyl units, information on molecular structure was lacking. The size and solubility of starch molecules made separation difficult. Though the terms amylose and amylopectin were employed to designate starch fractions, early preparations were not pure and only since 1941 has the heterogenity been generally understood. Aqueous leaching of swollen granules brought about a fairly good separation of starch into two components (*13–15*). Amylose was incompletely extracted, but was of high purity, while the amylopectin was contaminated with residual amylose.

A widely used procedure for separating starch fractions, devised by Schoch (*16, 17*), involves complete dispersion of starch granules in a hot aqueous medium and treatment with 1-butanol to allow, on cooling, formation of large crystalline rosettes of amylose–1-butanol complex. These could be separated easily by centrifugation or filtration. Pure amylose was obtained on redisolution and reprecipitation as the 1-butanol complex with final removal of 1-butanol by distillation from a hot suspension. This basic procedure of starch fractionation can be modified by pretreating the starch with ammonia to aid dispersion (*18*) or by dispersion of the starch in dimethyl sulfoxide prior to treatment with 1-butanol (*19*). Whistler and Hilbert (*20*) showed that amylose could complex with a large number of organic molecules some of which were needed only in a very low concentration. Haworth and co-workers (*21*) commonly used thymol as a fractionation agent.

By fractionation of starch through complexing of the amylose, the amylose: amylopectin ratio is found to be about 1:3. However, some genetic varieties of corn, barley, and rice have no amylose but only amylopectin. Mutant cultivars of pea, corn, and barley have genotypes characterized by high (60–80%) amylose contents.

Meyer and co-workers (*22*) applied methylation analysis (see Section III) in an attempt to deduce structural information on amylose and amylopectin fractions obtained by aqueous leaching of gelatinized corn starch. They obtained 3.7% of 2,3,4,6-tetra-*O*-methyl-D-glucopyranose from the amylopectin fraction and 0.32% from the amylose fraction. The tetramethyl sugar could only come from non-reducing chain ends and, hence, its amounts signified that amylose consists of very long linear chains while amylopectin, because of its high molecular weight as indicated by its high viscosity, must have numerous branches with concomitant numerous non-reducing chain ends.

Hassid and McCready (*23*) performed similar analysis on fractions from potato starch and obtained 4.67% of the tetra-*O*-methyl-D-glucopyranose ether from amylopectin and 0.32% from amylose. They calculated that the average length of a branch in the amylopectin molecule is 25 D-glucosyl units and that amylose must have 350 D-glucosyl units per chain. Hess and Krajne (*24*) using potato starch fractions separated in a different manner, found 4.9% of non-reducing chain ends in amylopectin molecules and 0.5% in amylose molecules. Methylation analysis of barley amylose separated by 1-butanol fractionation gave a chain length (DP, degree of polymerization) of 400 D-glucosyl units (*25*). As further evidence of the structural distinction between amylose and amylopectin, Meyer and co-workers (*22*) showed that, while methylated amylose formed pliable films, those from amylopectin were brittle. The same observations were independently made for the acetate esters by Meyer and co-workers (*26*) and by Whistler and Hilbert (*27*). It is well established that linear polymers can form films and fibers with great strength and flexibility while highly branched molecules cannot form fibers of quality, and films made from them are brittle. Evidence of the essential linearity of amylose was obtained also by Whistler and Schieltz (*28*) who showed that stretched films of amylose triacetate gave X-ray fiber-pattern structures consistent with oriented linear molecules.

Today, it is known that normal starches are composed of at least amylose and amylopectin while in some starches there is present also a small amount of an intermediate fraction that seems to be a less branched amylopectin or a slightly branched amylose. Supporting evidence for each of these structures was obtained by the work of many investigators and only key points are given here.

III. Methylation Analysis

Karrer (*29*) was the first to apply the Purdie methylation to starch. This method was later combined with the alkaline dimethyl sulfate methylation procedure by Karrer and Nageli (*30*) in an attempt to obtain the theoretical methoxyl content of 45.6%. However, a methoxyl content of only 35% was obtained. Irvine (*31*) reported a methoxyl value of 37%. One of the trimethyl ethers of D-glucose identified by Irvine after hydrolysis of the methylated starch was 2,3,6-tri-*O*-methyl-D-glucose, a sugar obtained also by similar treatment of cellulose. Later, Irvine and Macdonald (*32*) repeatedly subjected starch to 24 Purdie and alkaline dimethyl sulfate methylations to obtain a methoxyl content of 43.75%. In a variation of the usual procedure, Haworth and co-workers (*33*) treated starch acetate with alkaline dimethyl sulfate to obtain a methoxyl content of 44%. On hydrolysis, this methyl ether gave identifiable di- and trimethyl ethers of D-glucose. Identification of 2,3,4,6-tetra-*O*-methyl-D-glucopyranose in an amount of 4.5–5.0% of the total hydrolyzate was made by Hirst and co-workers (*34*) from a very carefully methylated starch. As the tetramethyl ether must represent the non-reducing end groups of the starch molecule, a considerable difference was indicated between starch and cellulose; in the latter case, negligible tetramethyl end groups were found by the methods available. The evidence suggests that starch contained shorter molecules than cellulose or contained more highly branched molecules. At that time, no evidence existed to show that starch was a mixture of two polymers.

IV. Maltose, the Repeating Unit

The disaccharide maltose was obtained by enzymic hydrolysis of starch in 1847 by Dubrunfaut (*35*) and termed *glucose de malt* to relate it with glucose. On translating this work into English in 1872, O'Sullivan (*36*) gave it the name maltose. The sugar was crystallized along with crystalline D-glucose by Ost (*37*) in 1904 after hydrolysis of starch with oxalic acid.

Methylation and hydrolysis of maltose yielded 2,3,4,6-tetra-*O*-methyl-D-glucose (*38, 39*) and 2,3,6-tri-*O*-methyl-D-glucose (*40, 41*), indicating that, in the reducing unit, C-4 and C-5 are involved in ring closure and glycoside formation. Which carbon atom was involved in which function was solved by Haworth and Peat (*42*) who oxidized maltose to eliminate the pyranosidic ring in the reducing end unit and then methylated the maltonic acid to isolate, after hydrolysis, 2,3,4,6-tetra-*O*-methyl-D-glucose (**III**) and 2,3,5,6-tetra-*O*-methyl-D-gluconic acid (**IV**). The location of the methyl groups, especially in 2,3,4,6-tetra-*O*-methyl-D-glucose(**III**), was established by Hirst (*43*). These findings clearly showed that, in maltose, one D-glucopyranosyl unit was linked to the oxygen atom on carbon four of the second sugar. The correctness of this formula-

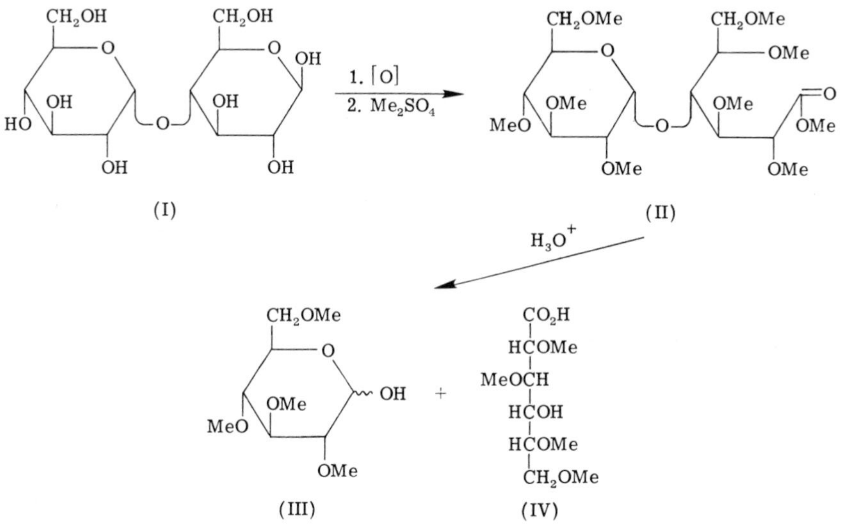

(I) → 1. [O] 2. Me_2SO_4 → (II)

(II) → H_3O^+ → (III) + (IV)

(IV): CO_2H–HCOMe–MeOCH–HCOH–HCOMe–CH_2OMe

tion was later confirmed by Levene and Sobotka (*44*) who showed that the aldonic acid of maltose formed only a 1,5-lactone ring.

To complete the structure of maltose it was only necessary to determine the anomeric (C-1) configuration of the glucosidic linkage joining the two sugar units. The disaccharide is not cleaved by β-glucosidases, which split the anomeric disaccharide cellobiose, but is hydrolyzed by maltase, an α-glucosidase. In the series of D-glucopyranosides, the α-D-glucopyranosides generally exhibit the most positive optical rotations. The equilibrium specific rotation of maltose is +130° while that of cellobiose is +35°, thus suggesting the α-D configuration for the glucosidic linkage in maltose. Application of Hudson's Isorotation Rule (*45*) to the reduction product of maltose (maltitol) and of cellobiose (cellobiitol) also suggests that maltose has the α-D configuration (*46–49*). The molecular rotation of maltitol is +35,200° and that of cellobiitol is −3,000°. Hudson's rule states that, of a pair of anomers of D-sugars, the one possessing the most positive molecular rotation has the α-D configuration. Some doubt about the validity of these optical rotation experiments remained due to the possible optical rotatory effects of the aglycon in maltitol and cellobiitol. This concern was eliminated by Perlin and co-workers (*50, 51*) who performed a series of oxidations and reductions on maltose and cellobiose to yield an isomeric pair of 2-*O*-D-glucopyranosylglycerols. The aglycon in these compounds is optically inactive. The product from maltose showed a specific rotation of +121° whereas that from cellobiose was −30°, again indicating the α-D configuration for maltose.

The α-D-glucosidic nature of maltose, established through these early painstaking chemical investigations, has since been confirmed by such physical and instrumental methods as x-ray diffraction and nuclear magnetic resonance spectroscopy. Hence, crystalline β-maltose can be formulated as 4-*O*-(α-D-glucopyranosyl)-β-D-glucopyranose.

The chemical synthesis of maltose proved somewhat difficult since the conventional Koenigs–Knorr glycoside synthesis (*52, 53*) produces mainly the β-D anomer. There are now useful methods for producing the α-D linkage, but the first synthesis of maltose, as its octaacetate, was by Lemieux (*54*) who reacted acetylated Brigl's anhydride (**V**) with 1,2,3,6-tetra-*O*-acetyl-β-D-glucopyranose (**VI**) without solvent at 120°. After acetylation, the product, the peracetate of β-maltose (**VIII**), was isolated by extrusion chromatography in 8% yield.

(V) + (VI) —120°, 13 h→ (VII) —Ac_2O, NaOAc→ (VIII)

The conformation of the D-glucopyranose rings in maltose has been the subject of much investigation. Early indirect chemical evidence favored a skew boat form (*55–57*). Other evidence indicated the chair form is predominant in maltose and related members of the oligosaccharide series (*58*). Most workers now agree that both rings have the stable $^4C_1$ chair form. Brant and co-workers (*59*) have calculated the conformational energy using structural models based upon the crystal structure of methyl β-maltoside. The potential functions used in these calculations account for intramolecular, van der Waals, coulombic, and hydrogen-bonding interactions. Intramolecular hydrogen-bonding was found to markedly influence the stability of certain conformations. Szejtli (*60*) concluded that, in maltose, the D-glucose residues are in the most stable $^4C_1$ conformation and that the relation between the two rings is such that all equivalent substituents are on the same side of the molecule. This conformation is stabilized by formation of a hydrogen bond between the hydroxyl group on C-2 of the non-reducing ring and that on C-3 of the reducing ring.

V. Maltooligosaccharides

Fragmentation analysis for the determination of polysaccharide structure has been widely used and was early developed by Whistler for aid in the determination of the structure of hemicelluloses. The method has been applied with success in starch structure determinations. Identification of starch fragments as malto-oligosaccharides gives further credence to the belief that the starch molecules are poly(α-D-glucopyranosyl) polymers. Enzymic fragmentation is particularly useful for starch, but partial acid hydrolysis was developed earlier. One difficulty in using acid fragmentation is that acid catalyzes the reverse reaction where new glycosidic bonds are formed and, hence, some isolated oligosaccharides may be artifacts containing glycosidic bonds not present in starch. This process of forming new glycosidic bonds during the principal hydrolysis reaction is known as *acid reversion.* To decrease reversion products, hydrolysis is usually conducted with dilute starch solutions.

Oligomers have been obtained from cellulose up through a degree of polymerization (DP) of 10 and in small yields to higher DP values. While members of the polymer homologous series tend to be crystalline, the property is inversely proportional to the DP (*61, 62*). Maltooligosaccharides crystallize less easily (*63*), but β-maltose crystallizes well as the monohydrate. The trisaccharide can be characterized as its crystalline peracetate (*64, 65*), whereas, the tetra- and pentasaccharides have been crystallized only as their alditol peracetates (*66, 67*).

The earliest methods for isolation of starch hydrolysis products included fractional precipitation. Crude maltotriose was prepared in this way (*68*). The volatile methyl ethers of maltose, maltotriose, and maltotetraose were obtained as syrups by fractional distillation (*69*). Extrusion column chromatography, utilizing a magnesium silicate adsorbent, gave a pure peracetate of β-maltotriose that crystallized (*70–73*). Deacetylation and reduction produced the trisaccharide alditol.

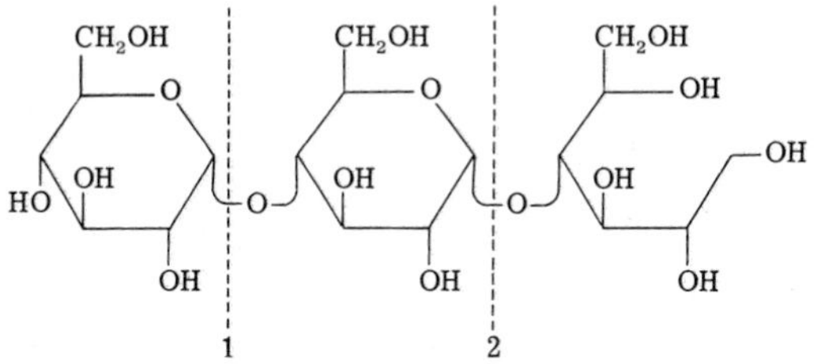

CH_2OH CH_2OH CH_2OH
O O OH
OH
HO OH O OH O OH
OH OH OH
1 2

(IX)

Partial acid-catalyzed hydrolysis of the alditol (**IX**) followed by acetylation yielded D-glucose pentaacetate and maltitol nonaacetate from cleavage at 1 and maltose octaacetate and glucitol hexaacetate from scission at 2 (*65*).

A useful advance in the isolation of starch hydrolyzates was the introduction of carbon column chromatography by Whistler and Durso (*74*). Carbon columns were developed by gradient elution with water–ethanol mixtures. Applying this method, workers isolated linear maltooligosaccharides up to DP 9 (*75, 76*).

Paper chromatography (*77*) represented a considerable advance in the separation of starch hydrolyzates and especially in establishing the homogeneity of fractions obtained by column chromatography. Application of the technique for preparative work is difficult but has been used by Scallet and co-workers (*78*) to obtain maltooligosaccharides to DP 11. The maltooligosaccharides exhibit a linear relationship between R_f and DP (*75*). French and Wild (*79*) reported paper separations of maltooligosaccharides to DP 14. Thin-layer chromatography (*80*) is effective also for separation of starch oligosaccharides on a kieselguhr adsorbent (*81, 82*), and good separation has been reported through DP 9. The methods have been recently supplemented, and in some cases supplanted, by more modern gas chromatographic (*83*) and high performance liquid chromatographic techniques which allow even higher homologs to be separated.

Enzymic techniques have also contributed to the current understanding of the structure of amylose and amylopectin. Lietar and co-workers (*84*) have identified maltotriose, in the form of its per(trimethylsilyl) ether, as a product of the sequential treatment of amylose with α-amylase and β-amylase. Other sugars in the hydrolysis mixture can be removed by fermentation with *Saccharomyces uvarum*. Similarly, a Japanese patent (*85*) indicated that an enzyme (MW 55,000, designated N-A468) from *Streptomyces griseus* is capable of producing maltotriose from amylose, amylopectin, or whole starch. This enzyme preparation is maximally effective at pH 5.0–6.0 and at temperature of 40°–50°. The pentasaccharide, maltopentaose, can be isolated in fairly high yield by dispersing amylose in dimethyl sulfoxide (DMSO), partially hydrolyzing it with acid, and treating it with an amylase from *Bacillus licheniformis* before chromatographic separation of the maltopentaose (*86*). A branched pentasaccharide α-limit dextrin was prepared by the action of α-amylase from *Bacillus subtilis* on amylopectin (*87*). The product, 6^2-maltosylmaltotriose, was structurally characterized by methylation analysis, periodate oxidation, partial acid-catalyzed hydrolysis, enzymic degradation, and enzymic synthesis. The structure supports the view that the branches in amylopectin are of the α-D-(1→6) type.

VI. Nature of Amylose

The essentially linear nature of corn (*22*) and potato (*26*) amylose was proposed in 1940 and confirmed (*23*) three years later. However, evidence is now accumulating that some amyloses may contain a few very long branches.

X-Ray diffraction experiments have been useful in elucidating the structure of both whole starch and amylose (see Chap. VII). Starch is found naturally in three crystalline modifications designated A (cereal), B (tuber), and C (smooth pea and various beans). Starch precipitated from solution, or complexed with various organic molecules, adopts the so-called V structure (Verkleisterung). Amylose also exists in the A, B, C, and V structures. Marchessault and co-workers (*88*) examined the chain conformation of B-amylose and suggested that B-amylose is helical with an integral number of α-D-glucopyranosyl residues per turn. The measured density of B-amylose is consistent with a helical structure containing six α-D-glucopyranosyl residues per unit cell plus 3–4 molecules of water of hydration. Potential energy calculations based on the structure of the repeating unit, maltose, led to the conclusion that, for glycosidic bond angles near 117°, a value suggested by examination of di- and trisaccharides, helices with more than 6 units per turn are geometrically impossible (*89*). Sixfold helices are strongly energetically favored and are characterized by a hydrogen bond between the C-2 hydroxyl group of one α-D-glucopyranosyl unit and the hydroxyl group at C-3 of the succeeding sugar unit. Left-handed helices are slightly favored over right-handed ones, and all other helices corresponding to the observed fiber repeat distance of 10.6 Å are unlikely. Thus, it was concluded that solid-state B-amylose exists most probably as a left-handed sixfold helix. Amylose triacetate, which exists as a left-handed helix, forms B-amylose upon saponification (*90*). Theoretical calculations show that V-amylose, with a fiber repeat of about 8.0 Å is more stable as a left-handed helix than as a right-handed one (*89*). The easy conversion of V-amylose to B-amylose (*58*) can imply that there is no extensive molecular reorganization accompanied by a reversal of chain chirality. Thus, it seems likely that amylose, in the solid state, exists as a left-handed helical polymer with six α-D-glucopyranosyl units per turn. Sarko and Biloski (*197*) report that the amylose–KOH complex is a distorted, left-handed helix with six α-D-glucopyranosyl residues per turn. Amylose triacetate also occurs in a left-handed helix similar to that of V-amylose. A. French (*198*) has reviewed the allowed and preferred shapes of amylose.

A model proposed by Kreger (*91*) in which amylose exists as a threefold helix has been examined (*92*). While this model is consistent with some of the x-ray data on amylose, it has not received wide acceptance. It is worth noting that the model of the starch granule of Frey-Wyssling (*92*) utilizes radially oriented low-molecular-weight amylose with a three-fold helix to account for the granule's positive birefringence and a higher-molecular-weight amylose (DP>280) with a sixfold helix running tangentially, rather than radially, to the granule surface.

A subject just as critical to the understanding of the fine structure of amylose in its solid state is the conformation of the α-D-glucopyranosyl rings as they exist in the polymeric form. Early workers (*58, 93, 94*) suggested that amylose (and amylopectin) contained α-D-glucopyranosyl units in the minimum energy $^4C_1$ (**X**) conformation. However, Sterling (*95*), in order to reconcile x-ray data and the crystallographic densities with a model for B-amylose, proposed that the D-glucopyranosyl rings exist in the $^{O,3}B$ (**XI**) conformation. The ^3B conformation is higher in internal energy than the $^4C_1$ conformation due to eclipsing of

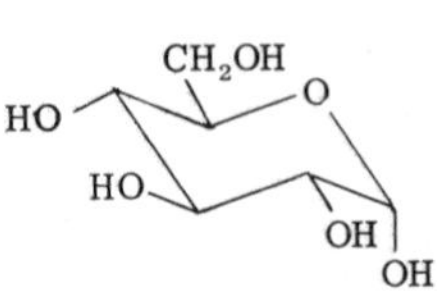

$^{4}C_{1}$

(X)

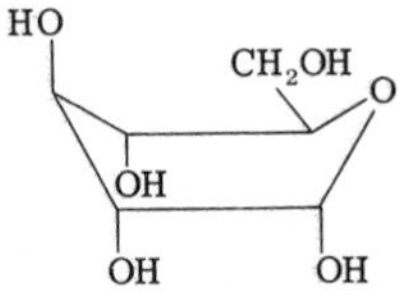

$^{O,3}B$

(XI)

substituents on adjacent carbon atoms. Rao and Foster (*96*) suggested, on the basis of nuclear magnetic resonance data, that the α-D-glucopyranosyl rings in amylose adopt the $^{4}C_{1}$ conformation. Their analysis was based on the similarity of the spectrum of amylose in 2 *M* KOH to those of methyl α-D-glucopyranoside, methyl α-maltoside, and other simple α-D-glucopyranosides known to exist in the $^{4}C_{1}$ conformation in 2 *M* KOH. Subsequent work by Casu and Reggiani (*97*) showed that the chemical shift and splitting of the anomeric hydrogen signal of amylose in dimethyl sulfoxide (DMSO) strongly substantiated the $^{4}C_{1}$ chair conformation. Similar results were reported by Casu (*98*) for amylose and some cyclodextrins based on nmr and infrared experiments. He also suggested that the presence of hydrogen bonds between hydroxyl groups of adjacent chain units would stabilize long helical regions of amylose in solution. This point is discussed later. St. Jacques and co-workers (*99*) have more recently investigated the structure of amylose by nmr at 220 MHz. Their variable temperature experiment in DMSO-d_6 included maltose, cyclohexaamylose, cycloheptaamylose, and amylose. They showed, through a series of very careful experiments, that a hydrogen bond exists between the hydroxyl group at C-2 of one α-D-glucopyranosyl unit and the C-3 hydroxyl group of the adjacent α-D-glucopyranosyl unit (**XII**) with the C-3 hydroxyl group donating the hydrogen atom in the hydrogen

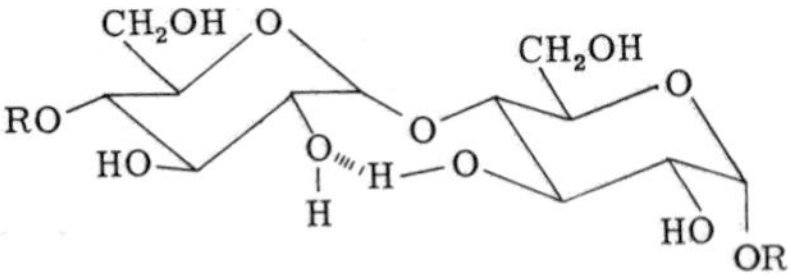

(XII)

bond. The authors made conformation calculations on the amylose chain in DMSO solution and, with data from their nmr experiments, concluded that the glycosidic torsional angles are consistent with loose right-handed helices. However, their data give no hint of helical segments lengths nor of the overall conformation of the dissolved polymer.

^{13}C-Nmr spectra of amylose are shifted downfield in DMSO compared to resonances in sodium hydroxide (*100*). Uniform downfield shift is consistent with a conformation change of the amylose polymer from an expanded random coil in alkaline solution to a more compact coil in DMSO. Similar results were obtained by Colson and co-workers (*101*) who measured the ^{13}C-nmr spectra of amylose in neutral and basic solution. Here also the difference in ^{13}C chemical shifts can be interpreted as a change from expanded random coil in alkaline solution to a more compact coil in neutral solution.

Infrared spectroscopy has been little used to investigate the structure of carbohydrate polymers, largely because of the profusion of functional groups in the polymer and the relative inability of the infrared spectrometer to distinguish chemically identical functional groups in similar, but different, environments. However, Casu and Reggiani (*102*) have obtained infrared spectra of amylose and D-glucose oligomers in the amorphous solid phase, in aqueous (H_2O, D_2O) solution and in DMSO and DMSO-d_6. By consideration of band intensity differences as a function of DP, behavior on deuteration, and comparison with the spectra of some other glucans, most of the spectral transitions can be assigned to specific group vibrations. The spectra suggest that the anomeric hydrogen atom is in an equatorial position and that the complementary C–O bond adopts an axial position, a geometry which is consistent with the $^4C_1$ conformation for the α-D-glucopyranosyl ring in amylose. In related work, Cael and co-workers (*103*) performed a normal coordinate analysis on V-amylose and found good agreement between the calculated and observed frequencies. Their analysis supports the proposed mechanism for the V-amylose → B-amylose conversion, that is, that there are changes in the hydrogen bonding of the —CH_2OH group and rotations of the residues about the glycosidic bonds due to helix extension in the solid state.

Use of β-amylase and the implications of the results have been reviewed by Banks and Greenwood (*104*). β-Amylase acts on amylose by removing maltose units from the non-reducing end of the molecule. Theoretically, this should lead to $n/2$ molecules of maltose from an amylose chain of DP n where n is even, or $n/2$ molecules of maltose plus 1 molecule of D-glucose where n is odd. Crude preparations of β-amylase convert amylose essentially quantitatively to maltose and this was taken as proof of the linearity of amylose. However, Peat and co-workers (*105*) showed that crystalline sweet potato β-amylase caused only 70% conversion of amylose into maltose. Therefore, they suggested that another enzyme (Z-enzyme) must be involved in the degradation of amylose. The incomplete action of β-amylase, producing the so-called β-*amylolysis limit dextrins,* was initially attributed to retrogradation, contamination with amylopectin, or the presence of anomalous residues or linkages in amylose. Amylopectin contamination was easily ruled out by an enzymic technique which showed that the β-amylase resistant samples of amylose were essentially free of amylopectin. Banks and Greenwood argued against retrogradation being a factor on the basis of experiments done to determine the β-amylolysis limit, [β], as a function of the method of dissolving the amylose as its 1-butanol complex. They found that [β] was independent of the solvent, even when DMSO, an excellent solvent for amylose, was used. This evidence indicates that retrogradation of amylose does not contribute significantly to β-amylase resistance.

The indicated β-amylase barrier could be due to a chemical modification of individual α-D-glucopyranosyl units or the presence of an anomalous linkage. Two types of chemical modification suggested were oxidation or the presence of esters, such as phosphate. Anomalous branching at C-2, C-3, or C-6 of the main amylose chain could be effective inhibitors also. Various investigations (*106, 107*) showed that, when oxidized groups were introduced into the amylose molecule, inhibition of β-amylase occurred, but the magnitude of the effect was small, and it was concluded that oxidized groups were not the cause of the barrier. Posternak (*108*) had evidence that phosphate esters obstruct the action of β-amylase, but later workers (*104, 109*) showed that the presence of phosphate ester groups are not the underlying cause of incomplete β-amylolysis.

Support for the view that amylose contains some branch points comes from the observation (*104*) that its hydrodynamic radius is less in those samples producing a β-limit dextrin than for those preparations that are completely degraded. To determine if branches exist in amylose, it was hydrolyzed with a mixture of β-amylase and pullulanase, an enzyme that specifically cleaves α-D-(1 → 6) linkages. The result was an essentially quantitative conversion of the polysaccharide into maltose (*104, 110*). This suggests the presence of some α-D-(1 → 6) branches in addition to the normal (1 → 4)-α-D-glucosidic bonds. The physical properties of the β-amylase resistant fraction are consistent with slightly branched, long-chain molecules wherein the branch points are separated by perhaps hundreds of α-D-glucopyranosyl units. The branched nature of amylose has also been investigated by Hizukuri and co-workers (*199*).

The configuration of amylose in solution has been in controversy for many years. The range of models in solution vary from helical (stiff, rod-like or loosely wound, worm-like) (**XIII**), an interrupted helix (**XIV**), to a random coil (**XV**), as shown below. In a review of this area, Szejtli and Augustat (*111*) conclude that,

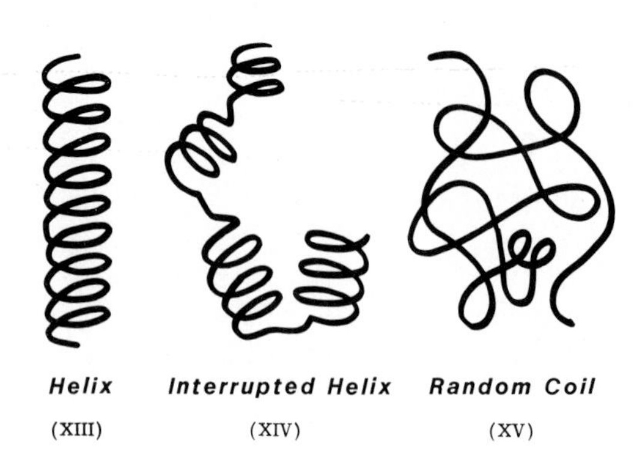

in aqueous solution at room temperature, the rotation about the D-glucosidic bonds is hindered by steric factors and only the helical, or interrupted helical, structure may exist. These helical regions are thought to be relatively stiff, having 10–15 helical turns per region.

Investigation of the viscosity of amylose as a function of molecular weight in aqueous potassium chloride solution gave the relation $[\eta] = 0.115\ \bar{M}_w^{0.5}$ (*112*). The exponent value of $\bar{M}_w$ is that expected for a random coil in a "good" solvent (*113*). If amylose possessed an extended, helical, rod-like structure, an exponent of 1.8 would be expected. However, no distinction is made here between an interrupted helix, which may behave like a random coil, and a pure random coil. Perhaps interrupted helices may not exist because of the lack of stabilizing factors. Banks and Greenwood (*112*) believe that amylose in solution behaves like a random coil and, while there may be some helical character present, it is very loose in structure. They also suggest that the helical nature results not only from intramolecular hydrogen bonding but also as a consequence of the geometry of the (1→4)-α-D-glucopyranosidic linkages. A related investigation (*114*) showed that amylose behaves as a random coil in good solvents such as formamide, dimethyl sulfoxide, and aqueous potassium hydroxide. From a review of its behavior in dilute solution, amylose is likely to have the conformations indicated (*115*).

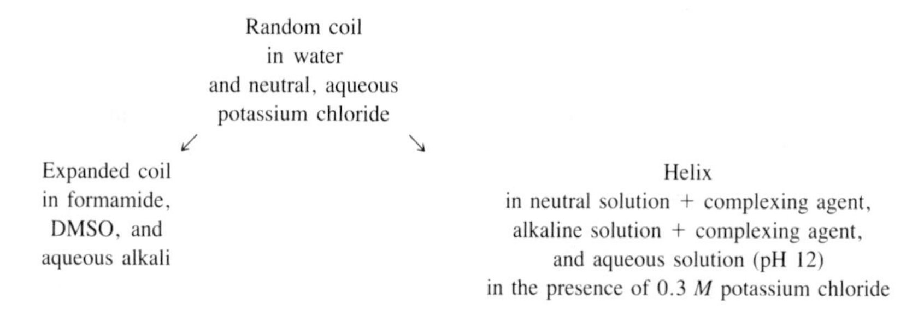

Kodama and co-workers (*116*) report that the solution conformation of amylose depends on its molecular weight. Light scattering–sedimentation equilibrium data shows that, if the molecular weight of amylose is outside the Burchard "dissolving gap" ($6500 < MW < 160{,}000$), the molecules behave as a random coil. Amylose with molecular weights within the "dissolving gap" aggregate to form rigid coils, thought to be double helixes. The investigators believe, unlike Banks and Greenwood, that the β-amylolysis limit is due to retrograded amylose in solution, or incipient retrogradation. However, like others in the field, it is suggested that the random coil may have short stretches of loosely helical regions.

Theoretical calculations on the solution conformation of the amylose molecule have been made by computing the conformational energy of amylose molecules as a function of rotation about the interunit glycosidic bonds (*117*). The characteristic ratio,

$$\frac{[r^2]_o}{n(l_v)^2}$$

(where $[r^2]_o$ = mean square end-to-end distance of the unperturbed chain, n = number of D-glucose residues in the chain, and l_v = the virtual bond length = 4.2 Å), calculated to be 6.9, is in good agreement with the observed value of 6.6 ± 0.2. With free rotation about glycosidic bonds the ratio would be 1 to 1. A conformational map shows that the rotations of the α-D-glucopyranosyl units are highly restricted. From the probability of occurrence of the various α-D-glucopyranosyl conformations, it is concluded that amylose, in neutral aqueous solution, exists as a random coil with short, loosely wound helical segments. Most calculations use a single model compound to establish virtual bond length; that is, the distance from the oxygen at C-4 to that at C-1. French and Murphy (*118*) have pointed out that a range of virtual bond lengths exist in nature and that classification of α-D-glucopyranosyl units by virtual bond length is a useful way of introducing residue geometry as a variable in structural analysis. Geometrically different residues are needed for the construction of satisfactory models of the various amylose polymorphs (*118*). Thus, no single type of α-D-glucopyranosyl unit can be regarded as "typical" of amylose.

A measurement of the optical rotatory dispersion (ORD) spectra of amylose gave data neither excluding nor indicating helical conformation (*119*). Changes in specific rotation of amylose on going from neutral to alkaline solution may be interpreted as due to an increase in the rotational freedom about the (1→4)-α-D-glucosidic bond in alkaline solution (*96*). A similar contribution from a helix-coil transition cannot be ruled out. Dintzis and Tobin (*120*) examined the specific rotation of amylose in dimethyl sulfoxide–water. Refractive index-corrected specific rotations of amylose exhibit a discontinuity in the region of the dimethyl sulfoxide–water system that corresponds to $2H_2O \cdot DMSO$. This discontinuity is dependent on the presence of a number of consecutively (1→4)-linked α-D-glucopyranosyl units and reflects some change in the symmetry of the polymer chain. The optical rotation of amylose in DMSO is invariant from 26.5° to 92.5°C and is only slightly lowered in water at elevated temperatures, a behavior characteristic of a random coil. The data are compatible with the view that amylose molecules have a slight helical twist influenced by removal of strong intrapolymer interactions in a good solvent.

Circular dichroism (CD) spectra have been measured for amylose, maltose, and the cyclodextrins (*121*). CD spectra have been measured in aqueous solution to 1640 Å, and two bands are observed. By examining chromophorically equivalent but conformationally different glucans, it is possible to demonstrate the sensitivity of CD to sugar ring conformation. Conformations of interior maltose residues in amylose appear similar. A comparison of amylose and cycloamylose CD spectra shows that amylose exhibits a substantial chirality bias. The butanol complex of amylose (V-amylose) has a CD spectra similar to that of free amylose in aqueous solution, a result consistent with conformational energy calculations predicting that the maltosyl subunits of amylose are restricted to a small region of the conformational map. The CD spectra is consistent also with the model of amylose as a loosely wound and extended helix which behaves as a random coil in aqueous solution.

VII. Nature of Amylopectin

A branched structure for amylopectin was early inferred from the chemical and physical differences between amylose and amylopectin. Amylopectin does not give a starch–iodine (amylose–iodine) blue color but a purple and sometimes reddish-brown color depending upon its source (*122*).

Methylation and subsequent hydrolysis of whole starch always gives rise to some di-*O*-methyl-D-glucose. These dimethyl ethers were previously regarded as arising from incomplete methylation or from demethylation (Zeisel effect) of higher order ethers. An examination of carefully methylated and hydrolyzed potato starch (*11*), wherein the hydrolyzate was benzoylated (*123*), gave a mixture of benzoylated methyl ethers separable by fractional distillation. The benzoylated dimethyl ether fraction was found to be an isomeric mixture containing 1,4,6-tri-*O*-benzoyl-2,3-di-*O*-methyl-D-glucose as its major component. Thus, Freudenberg assigned the branch point to C-6 of some α-D-glucopyranosyl units. Hirst and co-workers (*124*) also identified 2,3-di-*O*-methyl-D-glucose as a hydrolysis product of tri-*O*-methyl starch. In general it could be shown that the number of moles of 2,3-di-*O*-methyl-D-glucose equated to the number of moles of 2,3,4,6-tetra-*O*-methyl-D-glucose obtained from methylated amylopectin. This indicated the logic that for every branch point there should be a branch end.

If amylopectin is a linear (1→4)-α-D-glucan with chains branched at C-6, then a picture of the chain can be envisioned (**XVI**).

Isolation of a true branched section of the molecule is of importance to proof of structure. Cleavage of the amylopectin at points 2, 3, and 5 of structure **XVI** would yield maltose; cleavage at points 1, 3 and 4 of structure **XVI** would yield the isomeric 6-*O*-(α-D-glucopyranosyl)-D-glucopyranose, known as isomaltose (**XX**). This disaccharide is anomeric with known crystalline gentiobiose (*125*), an acid reversion product of D-glucose (*126, 127*). Isomaltose is generally obtained as a syrup; there is only one report of crystalline isomaltose (*128*).

To obtain more substantial amounts of isomaltose which could be converted to crystalline derivatives for comparison with similar material isolated from amylopectin hydrolyzates, Wolfrom and co-workers (*129, 130*) used a bacterial dextran rich in isomaltosyl units. Partial hydrolysis gave isomaltose charac-

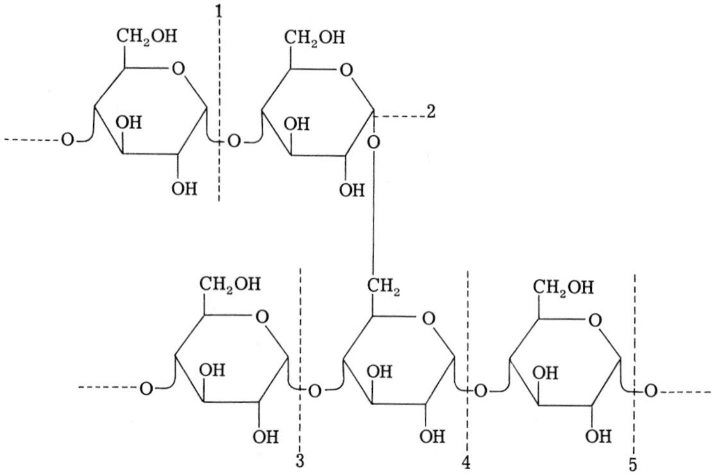

(XVI)

terized as its crystalline β-D-octaacetate, its glycosyl bromide heptaacetate, and as its methyl glycoside heptaacetate. Other crystalline derivatives of isomaltose such as the octa-*p*-nitrobenzoate ester (*128*), isomaltitol, and isomaltitol peracetate (*131*) are known.

Isomaltose has been structurally characterized by polarimetric comparison to its anomer, gentiobiose (*130*), by periodate oxidation of methyl β-isomaltoside (*130*) and isomaltitol (*128*), and by methylation analysis of methyl β-isomaltoside (*132*). Confirmation of the structure of isomaltose (**XX**) is obtained by comparison of natural and synthetic isomaltose, prepared either by acid reversion (*126, 127, 132*) or by the Koenigs–Knorr synthesis (*52, 53*) from 1,2,3,4-tetra-*O*-acetyl-β-D-glucopyranose (**XVIII**) and 3,4,6-tri-*O*-acetyl-2-*O*-nitro-β-D-glucopyranosyl chloride (**XVII**) (*133, 134*). A unique alternative synthesis of isomaltose (*135*) involves the anomerization of gentiobiose octaacetate by titanium tetrachloride (*136*).

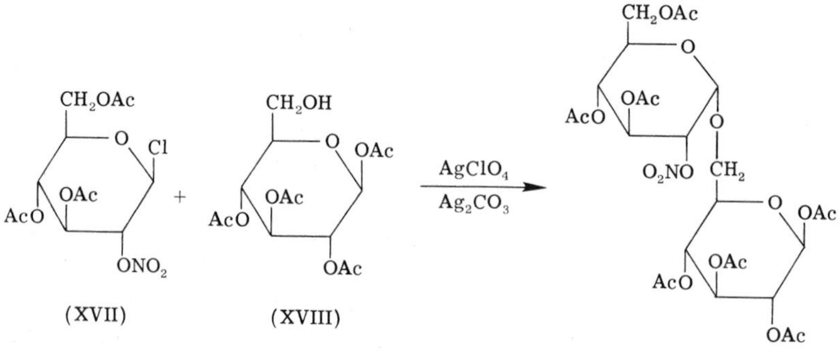

(XVII) + (XVIII) $\xrightarrow[Ag_2CO_3]{AgClO_4}$ (XIX)

XIX → 1. H_2/Pd 2. $Ac_2O/NaOAc$ 3. $Ba(OMe)_2$

(XX)

Once easily identifiable crystalline derivatives of isomaltose were available, it became necessary to isolate such derivatives from amylopectin hydrolyzates. Unexpectedly, initial attempts failed until techniques used with glycogen, a richer source of α-D-(1→6) linkages, were adapted (*137*). Based on the consideration that the rate of hydrolysis of the glycosidic bond of maltose is four times that of the glycosidic bond of isomaltose, the maximum yield of isomaltose should occur at 90% hydrolysis (Fig. 1). At the 90% hydrolysis level, it was possible to isolate β-D-isomaltose, convertable to its peracetate, in 2% yield

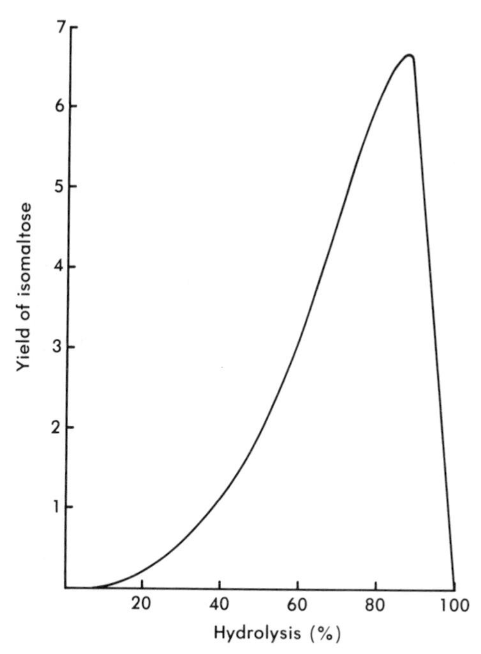

FIG. 1.—Calculated yield of isomaltose from glycogen as a function of percentage hydrolysis.

from glycogen and in 1% yield from waxy maize starch (*72*). Isolation of isomaltose after hydrolysis of amylopectin in dilute (0.4%) solution, where reversion would be minimal, gave significant proof that the isomaltosyl unit is present in amylopectin (*126, 138*).

Three branched trisaccharides isolated from partial hydrolysis of amylopectin were shown to possess the α-D-(1→6) linkage and, thus, gave further proof of its presence.

Reduction of each trisaccharide to its respective alditol followed by partial acid-catalyzed hydrolysis yielded the products shown.

$$\text{panitol} \xrightarrow{H^+} \text{glucose, maltitol, glucitol, isomaltose}$$

$$\text{isopanitol} \xrightarrow{H^+} \text{glucose, isomaltitol, glucitol, maltose}$$

$$\text{DGG alditol} \xrightarrow{H^+} \text{glucose, maltitol, isomaltitol}$$

Panose (**XXI**) was first obtained by Pan and co-workers (*139*) by transferase action on maltose and subsequently was isolated from an amylopectin hydrolyzate (*140*). It crystallizes with remarkable ease, even at elevated temperature, being similar in this property to β-gentiobiose (*125*). Its structure was determined by cleavage of its alditol to D-glucitol, D-glucose, isomaltose, and maltitol, which were separated by column chromatography and identified as their peracetates (*141*). Methylation analysis of panose and separation of products by carbon column chromatography yielded 2,3,6-tri-, 2,3,4-tri-, and 2,3,4,6-tetra-*O*-methyl-D-glucopyranose in the approximate ratio 1:1:1 (*142*). The trisaccharide designated DGG (**XXIII**) has been reported in a partial acid hydrolyzate of floridean starch (*143*) and has been prepared synthetically (*144*) by reaction of 3,4,6-tri-*O*-acetyl-2-*O*-nitro-β-D-glucopyranosyl chloride (**XVII**) with 1,2,3-

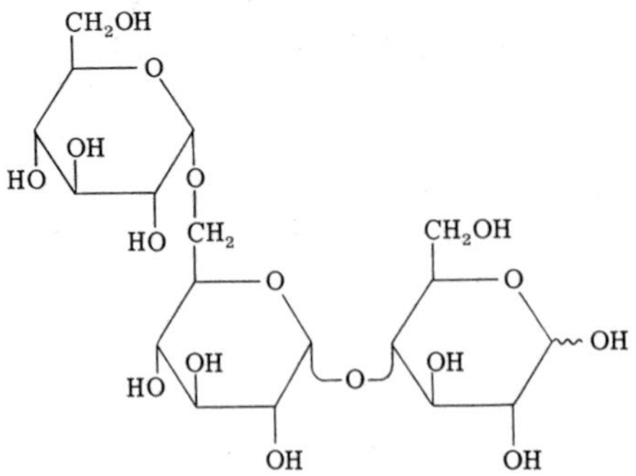

panose

(XXI)

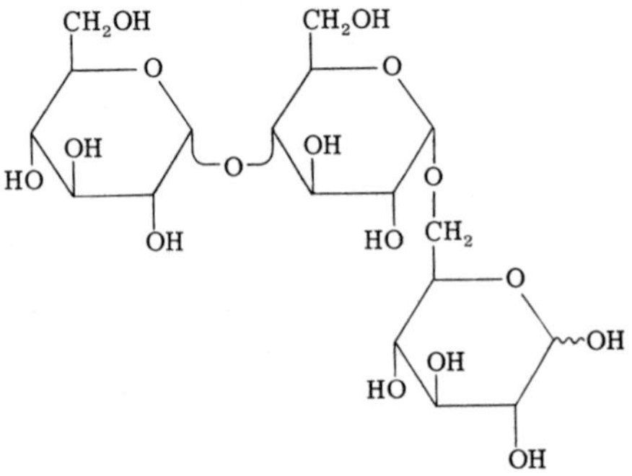

isopanose

(XXII)

4,6-di-O-(α-D-glucopyranosyl)-D-glucopyranose
DGG

(XXIII)

tri-*O*-acetyl-4-*O*-(tetra-*O*-acetyl-α-D-glucopyranosyl)-β-D-glucose. Isopanose (**XXII**) was found in the acid hydrolyzate of a glucan synthesized by *Pullularia pullulans* and characterized as its crystalline alditol peracetate (*145*). It can be synthesized enzymically by the action of *Bacillus macerans* amylase on cyclohexaamylose and isomaltose (*146*).

The possibility of other types of branches, such as α-D-(1→2) or α-D-(1→3), in amylopectin has been sought (*147*). However, such branches, if they do exist, have not been found and can at most represent an exceedingly small percentage of the total number of branch points. A general feature of amylopectin obtained from these chemical results is that of a (1→4)-α-D-glucan with branch linkages of the α-D-(1→6) type.

Several models have been proposed for amylopectin to account for its physicochemical properties. The earliest of these included the laminated structure (**XXIV**) of Haworth and co-workers (*148*) and the so-called "herringbone" model (**XXV**) of Staudinger and Husemann (*149*). Later, Meyer and Bernfeld (*13, 14*) proposed a randomly branched structure (**XXVI**) for amylopectin and most recently a tassel on a string representation has been suggested (**XXVII**) (*150–153*).

The models are characterized by having three distinct types of (1→4)-α-D-glucan chains, except for the herringbone structure which has only two. By

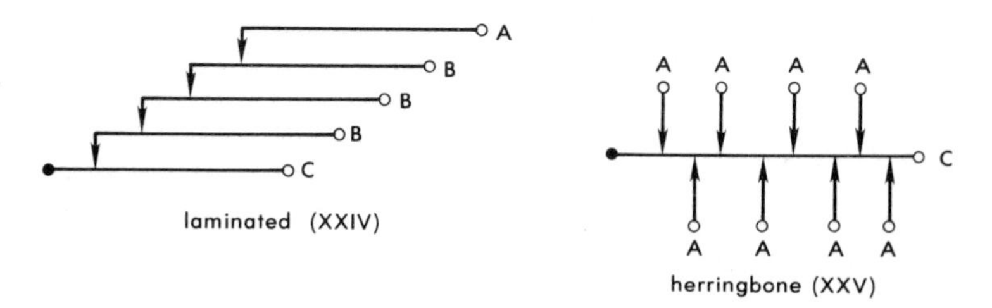

laminated (XXIV)

herringbone (XXV)

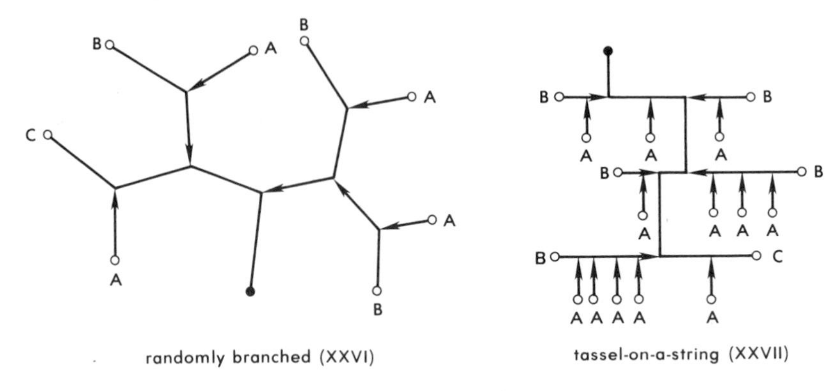

definition, the *A-chains* are those that are linked to the rest of the molecule only through their reducing ends; the *B-chains* are those that are linked to the molecule through their reducing ends but, in addition, are branched at a C-6 position in one or more of their D-glucopyranosyl residues, and the *C-chain* is the one that bears the reducing end group. Examination of these models (Table I) shows that, for relatively large molecular weights, each model has a characteristic ratio of A-type chains to B-type chains. The initial determination of the A:B ratio (*154*) favored the randomly branched structure (Meyer). Other proposed models are mostly variants of the Meyer structure (*92, 155*) but have not achieved general acceptance.

Enzymic methods have been valuable in the determination of the fine structure of amylopectin. A combination of sweet potato β-amylase and *Aerobacter aerogenes* pullulanase has been used in the determination of the average unit chain lengths of amylopectin (*156*). This method assumes that chain segments are equally divided between those with an even and those with an odd number of D-glucopyranosyl units. Bathgate and Manners (*156*) investigated the action of pullulanase on β-amylolysis limit dextrins and obtained results consistent with the major features of the Meyer model. It is interesting to note that the amylopectin β-amylolysis limit dextrin contains side chains that are 3–4 residues long, not

Table I

Comparison of Amylopectin Models

Model	*Chain ratio, A:B*
Laminated	0
Herringbone	∞
Randomly branched	1

Table II

Comparison of Amylopectin Unit Chain Length Determined Chemically and Enzymically

	Av. unit chain length	
Amylopectin	*Enzymic*	*Periodate*
Barley	20	20
Maize	24	27
Oats	19	18
Potato	22	26
Waxy maize	17	23
Wheat	21	20

single α-D-glucopyranosyl units, as was believed earlier (*157*). Lee and Whelan (*158*) utilized a method requiring only 1 mg of polysaccharide for the determination of the segmented chain length involving sequential use of pullulanase and β-amylase. They obtained data in agreement with periodate oxidation results.

Because of difficulties in the periodate method, the enzymic method is preferred for ascertaining branch segment length (*159*). Difficulties in the periodate method are the lack of a suitable correction for over-oxidation and the need to correct for small amounts of contaminating amylose-like material. Comparative data are shown in Table II.

Hydrolysis of starch with isoamylase, a debranching enzyme, and measurement of the molecular weight of the resulting linear chains by gel permeation chromatography can give an excellent profile of branch lengths (*160*). In this way, the distribution patterns of branch lengths is found to be similar for amylopectins of waxy maize, waxy rice, and potato amylopectin, where in each instance two peaks are seen; one for $\overline{DP}$ 50 and one for $\overline{DP}$ 20 (*161*). A comparison of the distribution patterns of unit chains in amylopectin and its β-limit dextrin suggest that in the amylopectin the B-chains are divided into longer (DP = 40–80) and shorter (DP = 20–40) chains. Mercier (*162*), using the same technique, has found amylopectins from waxy and normal starches to be identical, but amylopectin from amylomaize (64% amylose) has longer inner chains (DP $\geq$ 60) than those from waxy or normal starches (DP = 30). This is contrary to earlier findings by Montgomery and co-workers (*163*). While the A:B ratio for glycogen is approximately 1, as required by the Meyer model, the A:B ratio for amylopectin is approximately 2 (*164*). Thus, these related polysaccharides differ in degree of multiple branching as well as average chain segment length and, therefore, amylopectin cannot be formed *in vivo* from a glycogen-like precursor. Results imply that the Meyer model may have to be amended to include a larger proportion of A-chains. Amylopectins from hard wheat, soft wheat, club wheat, durum wheat, triticale, rye, and *Triticum dicoccum* subjected to isoamylase digestion and gel filtration yield two populations of chain unit lengths; DP 11–25 and DP 52–60 (*165*). The average segment chain length ranged from 15 for triticale to 26 for rye, and the ratio of A-chains to B-chains ranged from 1.21 to 2.11. Thus, in certain cases, the Meyer model predicts the presence of too few A-chains. All data point to an average internal chain length or distance between branches of seven D-glucopyranosyl residues. Manners and Matheson (*200*) have re-examined the fine structure of amylopectin and find that the ratio of A chains to B chains is about 1:1, not 2:1 as suggested by others. This is consistent with the "cluster" model of amylopectin suggested by D. French (*150;* this volume, p. 211).

Solution properties measured for amylopectin from potato, tapioca, wheat, barley, waxy maize, smooth-seeded pea, and wrinkled-seeded pea indicate some structural characteristics (*166*). Amylopectin solution values indicate a density of only one-sixth that of glycogen of equivalent molecular weight.

VIII. Structural Indications by Periodate Oxidation

Periodate oxidation, developed by Malaprade (*167*), was applied to corn starch (*168, 169*) with the result that each D-glucosyl unit consumed one mole of periodate. Hydrolysis of the oxidized starch yielded glyoxal (**XXX**) and D-erythrose (**XXXI**), identified as D-erythrono-1,4-lactone.

CH_2OH IO_4^- CH_2OH

(XXVIII) (XXIX)

H_3O^+

glyoxal (XXX) + erythrose (XXXI)

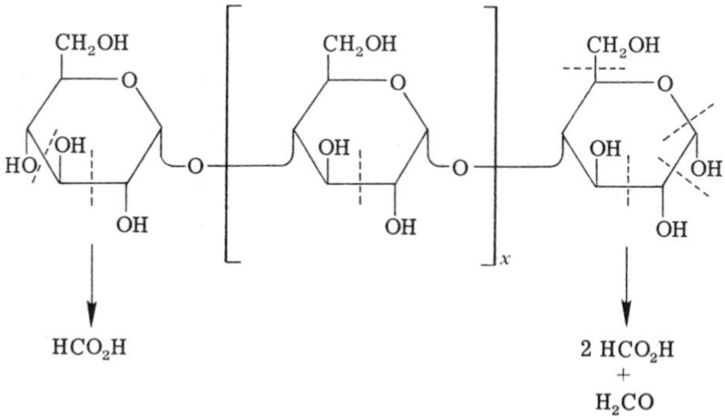

(XXXII)

Hirst and co-workers (*170*) suggested that the formic acid produced from the reducing and nonreducing terminals (**XXXII**) permits a chemical determination of molecular weight. Because most of the formic acid produced by periodate oxidation of amylopectin would come from the nonreducing ends (1 mole per mole of end unit), its amount could constitute a measure of branch lengths (*171*). Correction is required for "non-Malapradian" over-oxidation (*172, 173*). Use of the periodate method to determine molecular weight of purified amylose gave values comparable to osmotic pressure values (*174, 175*). Hirst (*176–177*) and French (*178*) showed also that periodate nonreducing end group measurements agreed with end group numbers obtained by methylation and isolation of 2,3,4,6-tetra-*O*-methyl-D-glucopyranose.

Periodate-oxidized starches, being polyaldehydes (**XXXIX**), are generally unstable and difficult to handle. They can be stabilized by further oxidation to acids (*179*) or by reduction to polyols (**XXXV**) (*180*), as had previously been done in work with periodate-oxidized cellulose (*181*). On reduction and hydrolysis of periodate-oxidized potato starch, erythritol (**XXXVIII**) can be isolated in 54% yield. Improvement of the method gave improved yields of glycerol (**XXXVI**) and erythritol (**XXXVIII**), separable by chromatography (*182, 183*), but the yield of glycerol did not equate well with end-group molecular weight determination. Lead tetraacetate gives similar oxidation results, but the oxidation is slower (*184*).

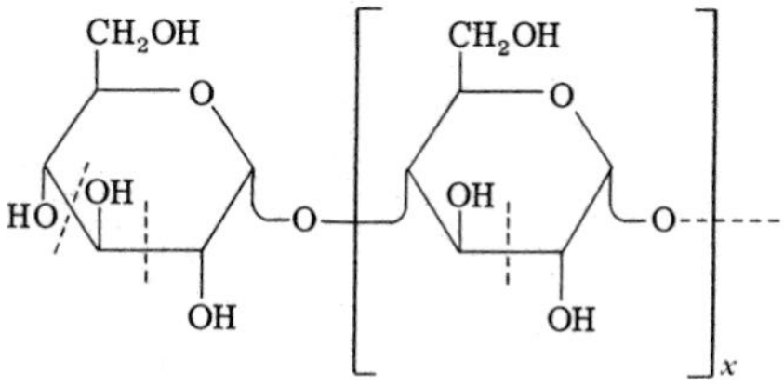

(XXXIII)

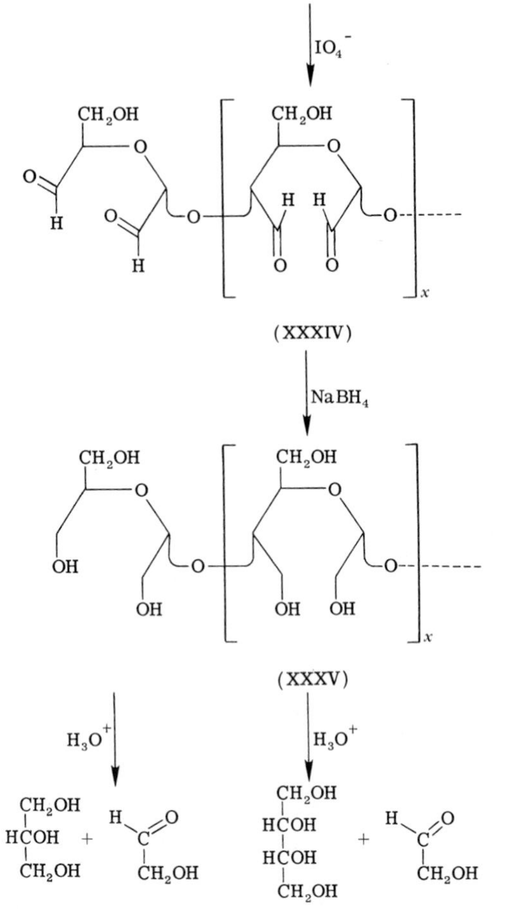

IO_4^-

(XXXIV)

$NaBH_4$

(XXXV)

H_3O^+ H_3O^+

glycerol (XXXVI) glycolaldehyde (XXXVII) erythritol (XXXVIII) glycolaldehyde (XXXVII)

Some of the D-glucopyranosyl residues in starch are resistant to the action of periodate, and this resistance was initially ascribed to the presence of "anomalous" linkages of the α-D-(1→2) or α-D-(1→3) type. To obtain further information on the possible presence of (1→3) linkages, periodate oxidized amylopectin (98.5%) was reduced with borohydride, and the product then was methylated, oxidized and hydrolyzed (*185*). 2,6-Di-*O*-methyl-D-glucose, the product expected if (1→3) linkages had been present in the amylopectin, was detected at a level of 0.02%, a value too low to indicate (1→3) linkages in starch. Later investigators (*186*) showed that some aldehyde groups produced by periodate oxidation can form a six-membered hemicetal ring with a hydroxyl group on a neighboring α-D-glucopyranosyl unit, thereby protecting the second sugar unit from oxidation and that this can account for the low oxidation value observed. Erlander and co-workers (*187*) suggested that periodate oxidation disrupts the helical conformation of amylose in solution.

IX. Starch Hydrolysis

Freudenberg (*188*) noted the similarity between the rates of hydrolysis of the various glycosidic linkages in starch and the linkage of maltose. Only small rate differences occurred between the beginning and end of hydrolysis as compared to larger differences seen with cellulose (*189*). The nearly equal rates of hydrolysis found for starch linkages and their similarity to the rate of maltose hydrolysis was taken as a strong indication for the presence of only one type of linkage (*190*).

Takahashi and co-workers (*191*) have investigated the hydrolysis of starch components and starch oligosaccharides. Hydrolysis values from calorimetric measurements were obtained for starch, panose and isomaltose using glucoamylase (*191*). The heat of hydrolysis ($\Delta H_{1,6}$) of the (1→6) linkage of isomaltose is +1300 cal/mole while that of the similar linkage in panose is +1240 cal/mole. $\Delta H_{1,4}$ for the first step in the cleavage of maltotriose is −1010 cal/mole at 25° while the average $\Delta H_{1,4}$ for amylose is −1030 cal/mole (*192*). The values for (1→4) and (1→6) linkages were later confirmed (*193*). Ono (*193*) found that the heats of hydrolysis of (1→4) and (1→6) glucosidic linkages are approximately equal in magnitude, but opposite in sign. The heat of hydrolysis of oligo- or polysaccharides can be obtained as an additive function of the thermal data of the individual linkages to be hydrolyzed; that is, Hess's law of heat summation is obeyed. Ono showed that by combining the data for linkage content of starches with the data for amylose content of the same starch, it is possible to estimate the average unit chain length for the amylopectin component. Banks and co-workers (*12*) confirmed the work of Takahashi and co-workers (*192*). They demonstrated that amylose was quantitatively converted to D-glucose by 0.75 *M* sulfuric acid at 100° for 4 h, while amylopectin yielded 96% D-glucose and 4% isomaltose, indicating that the (1→6) bonds are less acid labile than the (1→4) bonds.

X. Starch Phosphate Esters

Esterified phosphate groups are present in the amylopectin fraction of potato starch in the amount of 0.1–0.2% (*194*). Samec (*195*) investigated the effect of phosphate esters on the electrophoretic and colloidal properties of starch, and Posternak (*196*) isolated D-glucose 6-phosphate from potato starch after mild hydrolysis, showing that this ester is a natural component. Takeda and Hizukuri (*201*) have studied the distribution of phosphate esters in potato amylopectin and have concluded that about one third of the phosphate groups are located in the inner sections of the B-chains and about two-thirds are located in the outer sections of the B-chains and in the A-chains.

XI. References

(*1*) G. S. C. Kirchoff, *Acad. Imp. Sci. St. Petersbourg, Mem.*, **4,** 27 (1811); R. P. Walton "A Comprehensive Survey of Starch Chemistry," Chemical Catalog Co., New York, 1928, Part 2, p. 1.

(*2*) W. B. Newkirk, U. S. Patent 1,471,347 (1923); *Chem. Abstr.*, **18,** 179 (1924).

(*3*) T. de Saussure, *Bull. Pharm.*, **6,** 499 (1814).

(*4*) F. Salomon, *J. Prakt. Chem.*, **25,** 348 (1882).

(*5*) L. Schulze, *J. Prakt. Chem.*, **28,** 311 (1883).

(*6*) J. B. Biot and J. Persoz, *Ann. Chim. Phys.*, **52,** 72 (1833).

(*7*) L. Bondonneau, *Compt. Rend.*, **81,** 972 (1875).

(*8*) H. T. Brown and G. H. Morris, *J. Chem. Soc.*, **55,** 462 (1889).

(*9*) H. Staudinger, "Die Hochmolekularen Organischen Verbindungen, Kautschuk and Cellulose," Springer-Verlag, Berlin, Germany, 1932.

(*10*) W. H. Carothers, "Collected Papers of W. H. Carothers on High Polymers," Interscience Publishers, New York, 1940.

(*11*) K. Freudenberg and H. Boppel, *Ber.*, **71,** 2505 (1938).

(*12*) W. Banks, C. T. Greenwood, and D. D. Muir, *Staerke*, **25,** 405 (1973).

(*13*) K. H. Meyer and P.Bernfeld, *Helv. Chim. Acta*, **23,** 875 (1940).

(*14*) K. H. Meyer and P. Bernfeld, *Helv. Chim. Acta*, **24,** 359E (1941).

(*15*) K. H. Meyer, P. Bernfeld, R. A. Boissonnas, P. Guertler, and G. Noelting, *J. Phys. Chem.*, **53,** 319 (1949).

(*16*) T. J. Schoch, *Cereal Chem.*, **18,** 121 (1941).

(*17*) T. J. Schoch, *J. Amer. Chem. Soc.*, **64,** 2957 (1942).

(*18*) J. E. Hodge, E. M. Montgomery, and G. E. Hilbert, *Cereal Chem.*, **25,** 19 (1948).

(*19*) P. J. Killion and J. F. Foster, *J. Polymer Sci.*, **46,** 65 (1960).

(*20*) R. L. Whistler and G. E. Hilbert, *J. Amer. Chem. Soc.*, **67,** 1161 (1945).

(*21*) W. N. Haworth, S. Peat, and P. E. Sagrott, *Nature*, **157,** 19 (1946).

(*22*) K. H. Meyer, M. Wertheim, and P. Bernfeld, *Helv. Chim. Acta*, **23,** 865 (1940).

(*23*) W. Z. Hassid and R. M. McCready, *J. Amer. Chem. Soc.*, **65,** 1157 (1943).

(*24*) K. Hess and B. Krajne, *Ber.*, **73,** 976 (1940).

(*25*) I. C. MacWilliam and E. G. V. Percival, *J. Chem. Soc.*, 2259 (1951).

(*26*) K. H. Meyer, P.Bernfeld, and W. Hohenemeser, *Helv. Chim. Acta*, **23,** 885 (1940).

(*27*) R. L. Whistler and G. E. Hilbert, *Ind. Eng. Chem.*, **36,** 796 (1944).

(*28*) R. L. Whistler and N.C. Schieltz, *J. Amer. Chem. Soc.*, **65,** 1436 (1943).

(*29*) P. Karrer, *Helv. Chim. Acta*, **3,** 620 (1920).

(*30*) P. Karrer and C. Nageli, *Helv. Chim. Acta*, **4,** 185 (1921).

(*31*) J. C. Irvine, *Rept. Brit. Assoc. Advan. Sci.*, **90,** 40 (1922).

(*32*) J. C. Irvine and J. MacDonald, *J. Chem. Soc.*, 1502 (1926).

(*33*) W. N. Haworth, E. L. Hirst, and J. I. Webb, *J.Chem. Soc.*, 2681 (1928).

(*34*) E. L. Hirst, M. M. T. Plant, and M. D. Wilkinson, *J. Chem. Soc.*, 2375 (1932).

(*35*) C. Dubrunfaut, *Ann. Chim. Phys.*, **21,** 178 (1847).

(*36*) C. O'Sullivan, *J. Chem. Soc.*, **25,** 579 (1872).

(*37*) H. Ost, *Angew. Chem.*, **17,** 1663 (1904).

(*38*) T. Purdie and J.C. Irvine, *J.Chem. Soc.*, **87,** 1022 (1905).

(*39*) W. N. Haworth and G. C. Leitch, *J. Chem. Soc.*, **115,** 809 (1919).

(*40*) J. C. Irvine and I. M. A. Black, *J. Chem. Soc.*, 862 (1926).
(*41*) C. J. A. Cooper, W. N. Haworth, and S. Peat, *J. Chem. Soc.*, 876 (1926).
(*42*) W. N. Haworth and S. Peat, *J. Chem. Soc.*, 3094 (1926).
(*43*) E. L. Hirst, *J. Chem. Soc.*, 350 (1926).
(*44*) P. A. Levene and H. Sobotka, *J. Biol. Chem.*, **71,** 471 (1926).
(*45*) C. S. Hudson, *J. Amer. Chem. Soc.*, **31,** 66 (1909).
(*46*) P. Karrer and J. Buchi, *Helv. Chim. Acta,* **20,** 86 (1937).
(*47*) M. L. Wolfrom and T. S. Gardner, *J. Amer. Chem. Soc.,* **62,** 2553 (1940).
(*48*) P. A. Levene and M. Kuna, *J. Biol. Chem.*, **127,** 49 (1939).
(*49*) M. L. Wolfrom and D. L. Fields, *TAPPI,* **41,** 204 (1958).
(*50*) A. J. Charlson and A. S. Perlin, *Can. J. Chem.*, **34,** 1200 (1956).
(*51*) A. J. Charlson, P. A. J. Gorin, and A. S. Perlin, *Can. J. Chem.*, **34,** 1811 (1956).
(*52*) W. Koenigs and E. Knorr, *Ber.*, **34,** 957 (1901).
(*53*) W. L. Evans, D. D. Reynolds, and E. A. Talley, *Advan. Carbohydr. Chem.*, **6,** 27 (1951).
(*54*) R. U. Lemieux, *Can. J. Chem.*, **31,** 949 (1953).
(*55*) K. Freudenberg and F. Cramer, *Ber.*, **83,** 296 (1950).
(*56*) R. E. Reeves, *J. Amer. Chem. Soc.*, **76,** 4595 (1954).
(*57*) R. Bentley, *J. Amer. Chem. Soc.,* **81,** 1952 (1959).
(*58*) F. R. Senti and L. P. Witnauer, *J. Amer. Chem. Soc.,* **70,** 1438 (1948).
(*59*) D. A. Brant, C. V. Goebel and W. L. Dimpfl, *Macromolecules,* **3,** 644 (1970).
(*60*) J. Szejtli, *Staerke,* **23,** 295 (1971).
(*61*) M. L. Wolfrom and J. C. Dacons, *J. Amer. Chem. Soc.*, **74,** 5331 (1952).
(*62*) M. L. Wolfrom, J. C. Dacons, and D. L. Fields, *TAPPI,* **39,** 803 (1956).
(*63*) G. Zemplen, *Ber.*, **60,** 1555 (1927).
(*64*) M. L. Wolfrom, L. W. Georges, A. Thompson, and I. L. Miller, *J. Amer. Chem. Soc.*, **71,** 2873 (1949).
(*65*) A. Thompson and M. L. Wolfrom, *J. Amer. Chem. Soc.*, **74,** 3612 (1952).
(*66*) R. L. Whistler and J. L. Hickson, *J. Amer. Chem. Soc.*, **76,** 1671 (1954).
(*67*) R. L. Whistler and J. H. Duffy, *J. Amer. Chem. Soc.,* **77,** 1017 (1955).
(*68*) H. Pringsheim, A. Peiser, K. Wolfsohn, W. Kusenack, and J. Leibowitz, *Ber.*, **57,** 1581 (1924).
(*69*) K. Freudenberg, K. Friedrich, and I. Bumann, *Ann. Chem.*, **494,** 41 (1932).
(*70*) W. H. McNeely, W. W. Binkley, and M. L. Wolfrom, *J. Amer. Chem. Soc.*, **67,** 527 (1945).
(*71*) L. Zechmeister, L. deCholnoky, and E. Ujhely, *Bull. Soc. Chim. Biol.,* **18,** 1885 (1936).
(*72*) M. L. Wolfrom, J. T. Tyree, T. T. Galkowski, and A. N. O'Neill, *J. Amer. Chem. Soc.*, **73,** 4927 (1951).
(*73*) M. L. Wolfrom, R. M. deLederkremer, and L. E. Anderson, *Anal. Chem.*, **35,** 1357 (1963).
(*74*) R. L. Whistler and D. F. Durso, *J. Amer. Chem. Soc.*, **72,** 677 (1950).
(*75*) W. J. Whelan, J. M. Bailey, and P. J. P. Roberts, *J. Chem. Soc.*, 1973 (1953).
(*76*) W. J. Hoover, Ph.D. Dissertation, University of Illinois, Urbana, Illinois, 1963.
(*77*) G. N. Kowkabany, *Advan. Carbohydr. Chem.*, **9,** 303 (1954).
(*78*) J. D. Commerford, G. T. VanDuzee, and B. L. Scallet, *Cereal Chem.*, **40,** 482 (1963).
(*79*) D. French and G. M. Wild, *J.Amer. Chem. Soc.*, **75,** 2612 (1953).
(*80*) J. M. Bobbitt, "Thin Layer Chromatography," Reinhold Publishing Co., New York, 1963.
(*81*) C. E. Weil and P. Hanke, *Anal. Chem.*, **34,** 1736 (1962).
(*82*) J. A. Thoma, *Methods Carbohydr. Chem.*, **4,** 222 (1964).
(*83*) C. T. Bishop, *Advan. Carbohydr. Chem.*, **19,** 95 (1964).
(*84*) C. Lietar, M. Castiau, and A. Devreux, *Rev. Ferment. Ind. Aliment.*, **28,** 233 (1973).
(*85*) S. Hashimoto, J. Kanaeda, S. Kubomura, K. Wako, C. Takahashi, H. Toda, and S. Makino, *Jpn. Kokai* 75 52, 278 (1975).
(*86*) T. J. Pankratz, U. S. Patent 4,039,383 (1977); *Chem. Abstr.*, **87,** 132330p (1977).

(*87*) D. French, E. E. Smith, and W. J. Whelan, *Carbohydr. Res.*, **22,** 123 (1972).
(*88*) J. Blackwell, A. Sarko, and R. H. Marchessault, *J. Mol. Biol.*, **42,** 379 (1969).
(*89*) V. S. R. Rao, P. R. Sundararajan, C. Ramakrishnan, and G. N. Ramachandran *in* "Conformation of Biopolymers," G. N. Ramachandran, ed., Academic Press, London, 1967, Vol. 2, p. 721.
(*90*) A. Sarko and R. H. Marchessault, *J. Amer. Chem. Soc.*, **89,** 6454 (1967).
(*91*) D. Kreger, *Nature*, **158,** 199 (1946).
(*92*) A. Frey-Wyssling, *Amer. J. Bot.*, **56,** 696 (1969).
(*93*) H. Murakami, *J. Chem. Phys.*, **22,** 367 (1954).
(*94*) C. T. Greenwood and H. Rossotti, *J. Polymer Sci.*, **27,** 481 (1958).
(*95*) C. Sterling, *Food Technol. (Chicago)*, **19,** 987 (1965).
(*96*) V. S. R. Rao and J. F. Foster, *J. Phys. Chem.*, **69,** 636 (1965).
(*97*) B. Casu and M. Reggiani, *Tetrahedron*, **22,** 3061 (1966).
(*98*) B. Casu, *Chim. Ind. (Milan)*, **48,** 921 (1966).
(*99*) M. St.-Jacques, P.R. Sundararajan, J. K. Taylor, and R. H. Marchassault, *J. Amer. Chem. Soc.*, **98,** 4386 (1976).
(*100*) K. Takeo, K. Hirose, and T. Kuge, *Chem. Lett.*, 1233 (1973).
(*101*) P. Colson, H. J. Jennings, and I. C. P. Smith, *J. Amer. Chem. Soc.*, **96,** 8081 (1974).
(*102*) B. Casu and M. Reggiani, *Staerke*, **18,** 218 (1966).
(*103*) J. J. Cael, J. L. Koenig, and J. Blackwell, *Biopolymers*, **14,** 1885 (1975).
(*104*) W. Banks and C. T. Greenwood, *Staerke*, **19,** 197 (1967).
(*105*) S. Peat, W. J. Whelan, and S. J. Pirt, *Nature*, **164,** 499 (1949).
(*106*) H. Baum, G. A. Gilbert, and N. D. Scott, *Nature*, **177,** 889 (1956).
(*107*) W. Banks, C. T. Greenwood, and J. Thomson, *Chem. Ind.*, 928 (1959).
(*108*) T. Posternak, *J. Biol. Chem.*, **188,** 317 (1951).
(*109*) S. Peat, S. J. Pirt, and W. J. Whelan, *J. Chem. Soc.*, 714 (1952).
(*110*) W. Banks and C. T. Greenwood, *Arch. Biochem. Biophys.*, **117,** 674 (1966).
(*111*) J. Szejtli and S. Augustat, *Staerke*, **18,** 38 (1966).
(*112*) W. Banks and C. T. Greenwood, *Carbohydr. Res.*, **7,** 349 (1968).
(*113*) P. J. Flory, "Principles of Polymer Chemistry," Cornell University Press, Ithaca, New York, 1953.
(*114*) W. Banks and C. T. Greenwood, *Carbohydr. Res.*, **7,** 414 (1968).
(*115*) W. Banks and C. T. Greenwood, *Staerke*, **23,** 300 (1971).
(*116*) M. Kodama, H. Noda, and T. Kamata, *Biopolymers*, **17,** 985 (1978).
(*117*) V. S. R. Rao, N. Yathindra, and P. R. Sundararajan, *Biopolymers*, **8,** 325 (1969).
(*118*) A. D. French and V. G. Murphy, *Carbohydr. Res.*, **27,** 390 (1973).
(*119*) S. Beychok and E. A. Kabat, *Biochemistry*, **4,** 2565 (1965).
(*120*) F. R. Dintzis and R. Tobin, *Biopolymers*, **7,** 581 (1969).
(*121*) D. G. Lewis and W. C. Johnson, Jr., *Biopolymers*, **17,** 1439 (1978).
(*122*) D. Grebel, *Mikrokosmos*, **57,** 111 (1968).
(*123*) K. Freudenberg and H. Boppel, *Ber.*, **73,** 609 (1940).
(*124*) C. C. Barker, E. L. Hirst, and G. T. Young, *Nature*, **147,** 296 (1941).
(*125*) A. Thompson and M. L. Wolfrom, *J. Amer. Chem. Soc.*, **75,** 3605 (1953).
(*126*) A. Thompson, K. Anno, M. L. Wolfrom, and M. Inatome, *J. Amer. Chem. Soc.*, **76,** 1309 (1954).
(*127*) S. Peat, W. J. Whelan, T. E. Edwards, and O. Owen, *J. Chem. Soc.*, 586 (1958).
(*128*) E. M. Montgomery, F. B. Weakley, and G. E. Hilbert, *J. Amer. Chem. Soc.*, **71,** 1682 (1949).
(*129*) M. L. Wolfrom, L. W. Georges, and I. L.Miller, *J. Amer. Chem. Soc.*, **69,** 473 (1947).
(*130*) M. L. Wolfrom, L. W. Georges, and I. L. Miller, *J. Amer. Chem. Soc.*, **71,** 125 (1949).

(*131*) M. L. Wolfrom, A. Thompson, A. N. O'Neill, and T. T. Galkowski, *J. Amer. Chem. Soc.*, **74,** 1062 (1952).
(*132*) M. L. Wolfrom, A. Thompson, and A. M. Brownstein, *J. Amer. Chem. Soc.*, **80,** 2015 (1958).
(*133*) M. L. Wolfrom, A. O. Pittet, and I. C. Gillan, *Proc. Natl. Acad. Sci. U.S.*, **47,** 700 (1961).
(*134*) M. L. Wolfrom and D. R. Lineback, *Methods Carbohydr. Chem.*, **2,** 341 (1963).
(*135*) B. Lindberg, *Acta Chem. Scand.*, **3,** 1355 (1949).
(*136*) E. Pacsu, *Ber.*, **61,** 1508 (1928).
(*137*) M. L. Wolfrom, E. N. Lassetre, and A. N. O'Neill, *J. Amer. Chem. Soc.*, **73,** 595 (1951).
(*138*) A. Thompson, M. L. Wolfrom, and E. J. Quinn, *J. Amer. Chem. Soc.*, **75,** 3003 (1953).
(*139*) S. C. Pan, L. W. Nicholson, and P. Kolachov, *J. Amer. Chem. Soc.*, **73,** 2547 (1951).
(*140*) A. Thompson and M. L. Wolfrom, *J. Amer. Chem. Soc.*, **73,** 5849 (1951).
(*141*) M. L. Wolfrom, A. Thompson, and T. T. Galkowski, *J. Amer. Chem. Soc.*, **73,** 4093 (1951).
(*142*) E. E. Smith and W. J. Whelan, *J. Chem. Soc.*, 3915 (1963).
(*143*) S. Peat, J. R. Turvey, and J. M. Evans, *J. Chem. Soc.*, 3223 (1959).
(*144*) R. deSouza and I. J. Goldstein, *Tetrahedron Lett.*, 1215 (1964).
(*145*) H. O. Bouveng, H. Kiessling, B. Lindberg, and J. McKay, *Acta Chem. Scand.*, **17,** 797 (1963).
(*146*) J. H. Pazur and T. Ando, *J. Biol. Chem.*, **235,** 297 (1960).
(*147*) M. L. Wolfrom and A. Thompson, *J. Amer. Chem. Soc.*, **78,** 4116 (1956).
(*148*) W. N. Haworth, E. L. Hirst, and F. A. Isherwood, *J. Chem. Soc.*, 577 (1937).
(*149*) H. Staudinger and E. Husemann, *Ann.*, **527,** 195 (1937).
(*150*) D. French, *Denpun Kagaku,* **19,** 8 (1972).
(*151*) D. French, *J. Animal Sci.*, **37,** 1048 (1973).
(*152*) Z. Nikuni, S. Hizukuri, K. Kumagai, H. Hasegawa, Y. Moriuaki, T. Fukui, S. Nara, and I. Maeda, *Memoirs Inst. Sci. Ind. Res. Osaka Univ.*, **26,** 1 (1969).
(*153*) Z. Nikuni, *Denpun Kagaku,* **22,** 78 (1975).
(*154*) S. Peat, W. J. Whelan, and G. J. Thomas, *J.Chem. Soc.*, 4546 (1952).
(*155*) Z. Gunga-Smith, J. J. Marshall, C. Mercier, E. E. Smith, and W. J. Whelan, *FEBS Letters,* **12,** 101 (1970).
(*156*) G. N. Bathgate and D. J. Manners, *Biochem. J.*, **101,** 3C (1966).
(*157*) H. G. Hers and W. Verhue, *Biochem. J.*, **100,** 3P (1966).
(*158*) E. Y. C. Lee and W. J. Whelan, *Arch. Biochem. Biophys.*, **116,** 162 (1966).
(*159*) G. K. Adkins, W. Banks, and C. T. Greenwood, *Carbohydr. Res.*, **2,** 502 (1966).
(*160*) E. Y. C. Lee, C. Mercier, and W. J. Whelan, *Arch. Biochem. Biophys.*, **125,** 1028 (1968).
(*161*) H. Akai, K. Yokobayashi, A. Misaki, and T. Harada, *Biochim. Biophys. Acta,* **252,** 427 (1971).
(*162*) C. Mercier, *Staerke,* **25,** 78 (1973).
(*163*) E. M. Montgomery, K. R. Sexson, R. J. Dimler, and F. R. Senti, *Staerke,* **16,** 345 (1964).
(*164*) J. J. Marshall and W. J. Whelan, *Arch. Biochem. Biophys.*, **161,** 234 (1974).
(*165*) C. Y. Lii and D.R. Lineback, *Cereal Chem.*, **54,** 138 (1977).
(*166*) W. Banks, R. Geddes, C. T. Greenwood, and I. G. Jones, *Staerke,* **24,** 245 (1972).
(*167*) L. Malaprade, *Bull. Soc. Chim. Fr., [5] 1,* 833 (1934).
(*168*) E. L. Jackson and C. S. Hudson, *J. Amer. Chem. Soc.*, **59,** 2049 (1937).
(*169*) E. L. Jackson and C. S. Hudson, *J. Amer. Chem. Soc.*, **60,** 989 (1938).
(*170*) F. Brown, S. Dunstan, T. G. Halsall, E. L. Hirst, and J. K. N. Jones, *Nature,* **156,** 785 (1945).
(*171*) D. M. W. Anderson, C. T. Greenwood, and E. L. Hirst, *J. Chem. Soc.*, 225 (1955).
(*172*) J. M. Bobbitt, *Advan. Carbohydr. Chem.*, **11,** 1 (1956).
(*173*) K. H. Meyer and P. Rathgeb, *Helv. Chim. Acta,* **31,** 1545 (1948).

(*174*) A. L. Potter and W. Z. Hassid, *J. Amer. Chem. Soc.*, **70,** 3488 (1948).
(*175*) A. L. Potter and W. Z. Hassid, *J. Amer. Chem. Soc.*, **70,** 3774 (1948).
(*176*) F. Brown, T. G. Halsall, E. L. Hirst, and J. K. N. Jones, *J. Chem. Soc.*, 27 (1948).
(*177*) T. G. Halsall, E. L. Hirst, and J. K. N. Jones, *J. Chem. Soc.*, 1399 (1947).
(*178*) F. L. Bates, D. French, and R. E. Rundle, *J. Amer. Chem. Soc.*, **65,** 142 (1943).
(*179*) E. L. Jackson and C. S. Hudson, *J. Amer. Chem. Soc.*, **58,** 378 (1936).
(*180*) A. Jeanes and C. S. Hudson, *J. Org. Chem.*, **20,** 1565 (1955).
(*181*) G. Jayme and S. Maris, *Ber.*, **77,** 383 (1944).
(*182*) J. K. Hamilton and F. Smith, *J. Amer. Chem. Soc.*, **78,** 5907 (1956).
(*183*) J. K. Hamilton and F. Smith, *J. Amer. Chem. Soc.*, **78,** 5910 (1956).
(*184*) M. Abdel-Akher, A.-M. Youssef, Y. Ghali, and A. N. Michalinos, *J. Chem. U.A.R.*, **6,** 107 (1963).
(*185*) O. P. Bahl and F. Smith, *J. Org. Chem.*, **31,** 2915 (1966).
(*186*) M. F. Ishak and T. Painter, *Carbohydr. Res.*, **32,** 227 (1974).
(*187*) S. R. Erlander, H. L. Griffin, and F. R. Senti, *Staerke*, **17,** 151 (1965).
(*188*) K. Freudenberg, *Staerke*, **15,** 199 (1963).
(*189*) W. Kuhn, *Ber.*, **63,** 1503 (1930).
(*190*) K. H. Meyer, H. Hopff, and H. Mark, *Ber.*, **62,** 1103 (1929).
(*191*) K. Takahashi, Y. Yoshikawa, K. Hiromi, and S. Ono, *J. Biochem.*, **58,** 251 (1965).
(*192*) K. Takahashi, K. Hiromi, and S. Ono, *J. Biochem.*, **58,** 255 (1965).
(*193*) S. Ono, *Denpun Kogyo Gakkaishi*, **17,** 51 (1969).
(*194*) M. Samec, *Kolloidchem. Beih.*, **6,** 23 (1914).
(*195*) M. Samec, "Kolloidchemie der Starke," T. Steinkopf, Dresden, Germany, 1927.
(*196*) T. Posternak, *Helv. Chim. Acta*, **18,** 1351 (1935).
(*197*) A. Sarko and A. Biloski, *Carbohydr. Res.*, **79,** 11 (1980).
(*198*) A. D. French, *Bakers Digest*, 39 (February, 1979).
(*199*) S. Hizukuri, Y. Takeda, M. Yasuda, and A. Suzuki, *Carbohydr. Res.*, **94,** 205 (1981).
(*200*) D. J. Manners and N. K. Matheson, *Carbohydr. Res.*, **90,** 99 (1981).
(*201*) Y. Takeda and S. Hizukuri, *Carbohydr. Res.*, **102,** 321 (1982).

CHAPTER VII

ORGANIZATION OF STARCH GRANULES

By Dexter French*

Department of Biochemistry and Biophysics, Iowa State University, Ames, Iowa

An exact knowledge of the starch grain is of fundamental importance for the solution of technical problems.

N. P. Badenhuizen (*1*)

*Deceased.

STARCH, 2nd ed.

ISBN 0-12-746270-8

I. Introduction

Nature has chosen the starch granule as an almost universal form for packaging and storing carbohydrate in green plants. In granule form, starch is quasi-crystalline, water-insoluble, and dense. It is hydrated to only a small degree, so that a large amount of carbohydrate is stored in a small volume. In spite of the insolubility of starch, it is adequately convertible to sugar by the enzymes of plant metabolism. Also, raw starch granules are digested by innumerable predators ranging from bacteria and molds to higher animals and man. The abundance of the dense granular form of starch has made it easy to produce starch in essentially pure form by physical processing on a gigantic industrial scale. The unique physical properties and behavior of starch granules are critical in the industrial processing of starch and starch-containing materials and in countless applications.

Starch granules range in size from sub-micron elongated granules of chloroplasts (''leaf starch'') to the relatively huge (frequently over 100 μm) oval granules of potato and canna. The shapes (see Chap. XXII) include nearly perfect spheres typical of small wheat starch granules; discs, as with the large granules of wheat and rye; polyhedral granules, as in rice and maize; ''oystershell'' irregular granules as are often seen with potato starch; elongated cylindrical balloons, often with protuberances, in *Dieffenbachia* starch; needles, as in the milky sap of *Poinsettia;* and highly elongated irregular filamentous granules as in high-amylose maize starch. Moreover, starch granules from many species, such as oats, are compound; that is, they contain many starch particles which appear to have developed simultaneously within a single amyloplast. Semi-compound granules originate as two or more distinct granules, which then fuse together. Pseudo-compound granules, such as pea starch granules, start out as individual granules which then develop several large cracks while remaining a single entity. These differences in size and shape make it possible to recognize most of the ordinary food and commercial starches (see Chap. XXII). While the structure of an individual starch granule is doubtless under strict genetic control, granules isolated from different parts of the same plant may be entirely different, so there are other biological and environmental factors controlling development and regulation of starch formation. Finally, starch granules contain small amounts of non-carbohydrate components, particularly lipids, proteins, phosphate, and ash, that affect the behavior of starch in various applications.

In writing this chapter, selected reports from the literature bearing on the organization, properties and behavior of starch granules have been used. Unsupported statements, except those of an interpretative or speculative nature, are based on a great body of information and folklore which constitutes the general background knowledge of the practicing starch scientist. To appreciate the vast bulk of older literature on starch granules, the reader may wish to peruse the treatises by Naegeli (*2*), which provide a storehouse of knowledge about the botany and morphology of starch; Meyer (*3*), which presents a picture of starch at the end of the 19th century; Samec (*4*), which includes every aspect of the structure and behavior of starch as known in 1928; Reichert (*5*), which contains photomicrographs of hundreds of kinds of starch granules; and Walton (*6*), which contains more than 3000 abstracts of the literature up to 1928.

Many features of starch granule structure and organization are best understood by studying typical photographs and electron micrographs as presented in Chapter XXII.

II. Biological and Biochemical Facets of Starch Granule Structure (see also Chapter III)

The tiny starch granules of chloroplasts are important as a temporary carbohydrate reserve and show diurnal synthesis and degradation (*7*). At the other end of the scale, the starches of fruits, seeds, stems, roots and tubers are relatively long-term reserve plant nutrients and are designed by nature to be compact and stable. Yet these starches are rapidly metabolizable on demands imposed by such physiological crises as ripening of fruit, germination of seeds, or sprouting of roots or tubers.

The morphology of starch granules is dictated by the particular structure and biochemistry of the starch-synthesizing organelle, the chloroplast or amyloplast, as well as by the physiology of the plant (*8*). Different parts of the plant may produce entirely different starches, and even the granules within a given cell may be morphologically different. Although it is easy to observe the anatomical relationship of plastids to starch granules, the means by which plant genetic information is translated into a specific granule morphology is unknown.

In amyloplasts of developing maize endosperm, no starch is visible until the ninth or tenth day after fertilization (*8*). The plastids nearest the nucleus enlarge and begin to accumulate a separate phase or coacervate droplet of amorphous starch (''pro-starch'') (*8*). At a critical moment, the coacervate crystallizes, apparently spontaneously, and becomes the nucleus or hilum of the developing starch granule (*9*). The blue-staining ''cores'' sometimes seen in waxy starch granules, such as waxy sorghum (*10*), may come from amylose in the original coacervate droplet, which, by forming a lipid complex, has escaped action of the branching enzyme. Formation of a separate coacervate phase is the physicochemical consequence of the incompatibility of aqueous solutions of differing high polymers (*11*), in this case the protein of the plastid stroma and the newly forming polysaccharide. It does not require a biological trigger. The crystallization of coacervate droplets can be reproduced artificially, for example, in starch–gelatin mixtures (*12, 13*). At this stage of development, the amyloplast stroma contains enzymes of starch synthesis, such as ADPG–starch glucosyl transferase, branching or Q-enzyme, and phosphorylase (*14–16*).

Starch granules grow by apposition (*17, 18*). Originally the granule is round or ovoid, but as it grows it takes on a shape characteristic of the particular starch variety. Several small granules may grow together to become compound. These may eventually have a single exterior with multiple hyla, or they may simply stick together. The starch-synthesizing organelles are chloroplasts (leaf or transitory starch) or amyloplasts (storage starch). In some tissues, Pellionia stem, for example, the amyloplast may become green (*8*). Some amyloplasts appear to have mitochondrial elements (*19*).

Starch has two major components: amylose and amylopectin. Some, maybe many, starches contain a third component, a short-chain amylose (*20*). Amylose is long-chain of 1000–10,000 D-glucopyranosyl units. The larger molecules, perhaps representing another fraction, may contain a very low degree of branching (*21–23*). Amylopectin is highly branched, but not compact or spherical like glycogen, and some amylopectin molecules may be very lightly branched. It has a relatively high viscosity indicating an elongated structure. Electron microscopy, x-ray difraction, chemical, and biochemical studies show that amylopectin contains clusters of branching at about 70 Å intervals (*24*). The molecules are about 100–150 Å in diameter and 2000–4000 Å long. There would appear to be a rhythmic alternation of chain elongation and branching to give rise to the 70 Å periodicity.

The initial amyloplast may contain more than one starch granule, and it frequently divides into daughter amyloplasts. It is thought to have its characteristic, cytoplasmically inherited DNA, just as with chloroplasts and mitochondria; but there is no known role for this DNA in starch synthesis. As the starch develops, several granules each surrounded by a layer of stroma may develop simultaneously to form compound granules, for example, in rice. Alternatively, initially separate granules may fuse together to produce semi-compound granules with two or more hila, but a single exterior surface. As the starch granules grow they appear to stretch the amyloplast membrane, and often the stroma tends to be pushed to one end of the plastid. This results in asymmetric growth of the granule, distal to the hilum (*3*). With the stem starch of *Dieffenbachia,* after considerable elongation of the distal end of the granule, the pocket of stroma containing starch-synthesizing enzymes appears to retract to one side of the granule. Here it continues to deposit starch on the granule surface, resulting in a protuberance (*3, 10*).

Wrinkle-seeded peas have an amylose content of about 65%; in comparison, smooth-seeded peas contain about 35% amylose (*25*). During development of wrinkled pea starch, the granules develop large cracks or fissures, giving them the appearance of compound granules (*26*).

Growth of granules in the endosperm of *ae* (amylose extender, high-amylose) maize leads to some bizarre, highly elongated or filamentous granules as well as normal-appearing ones (*27–29*). The irregular granules do not show birefringence indicative of a regular pattern of organization or starch deposition, and the whole starch gives a poor B-type X-ray diffraction pattern (*30–32*).

For the first 14 days after pollination, endosperm starch development is the same in normal and *ae* maize (*27*). However, beyond this, some *ae* granules seem to develop a protuberance which eventually leads to the elongated or filamentous granules. With *su* (sugary) gene, after about 18 days, the starch granules gradually disappear, and their place is taken by plastids filled with phytoglycogen. With the double mutant *su ae,* after 18 days, starch granules undergo a metamorphosis and adopt an irregular "snowflake" form.

Starch biosynthesis involves three distinct enzymic processes: initiation, chain elongation and branching. It has been proposed that starch synthesis initiation consists of the elongation of "primer" chains on a glycoprotein (*33*). On the other hand, it is known that starch synthase can elongate starch oligosaccharides, including maltose (*34*), and that these oligosaccharides are present in starch-synthesizing tissues (*35*). Most starch scientists believe that chain elongation of amylose and the outer branches of amylopectin occurs by the action of ADPG- and UDPG-glucosyltransferases (starch synthases) according to the equation (*36*).

$$\mathrm{Glc}_n + \mathrm{ADPG\ (or\ UDPG)} \xrightarrow{\text{starch synthase}} \mathrm{Glc}_{n+1} + \mathrm{ADP\ (or\ UDP)}$$

The ADPG-utilizing enzyme is usually obtained in a granule-bound form, although it is relatively soluble (*37*). By contrast, the UDPG-utilizing enzyme is firmly bound to the granule, presumably to the amylose component (*38*). The waxy cereals, which do not produce amylose, are also lacking in the UDPG-utilizing synthase (*39*). By using radioactive precursors for *in vivo* starch synthesis, it has been shown that newly formed starch is deposited at or near the granule surface; that is, growth is by apposition as opposed to growth by intussusception (*17, 18*). The starch synthases elongate starch chains by joining D-glucopyranosyl units to the non-reducing ends of amylose or amylopectin. When the outer chains of amylopectin become sufficiently long, they undergo branching by action of Q-enzyme. Unlike the synthases, Q-enzyme has little affinity for starch granules, and carefully washed starch has no Q-enzyme activity (*40*).

It has been suggested that starch molecules are synthesized in the amyloplast stroma and the fully formed molecules are then deposited on the growing surface of the starch granule (*8, 10*). However, it seems more likely that the granules grow mainly by elongation of chains already incorporated into the granule structure, with simultaneous elongation and crystallization. This idea is in harmony with the known orientation of the starch chains perpendicular to the granule surface (*41*). For a single spherical granule 15 μm in diameter, with a mass of 2.65×10^{-9}g, there would be about 2.5×10^{9} molecules of amylose (DP 1000, 25% of the total starch), and 7.4×10^{7} molecules of amylopectin (DP 100,000, 75% of the starch). If the starch molecular chains are perpendicular to the surface of the granule, there would be about 1.4×10^{9} molecular chains terminating at the surface of the granule. Of course, 3.5×10^{8} would be amylose molcules, and the rest would be A and B chains of amylopectin.

The role of phosphorylase is controversial. It is thought by some workers (*42*) to be involved in starch chain elongation by a reversal of phosphorolysis. Other workers believe that the phosphorylase is mainly involved in the synthesis and breakdown of starch oligosaccharides, known to be present in starch-synthesizing tissue and possibly involved in carbohydrate transport to the surface of the growing starch granule (*43;* See also *This Volume,* pp. 112–114).

While the concepts of chain elongation and branching are generally accepted, there is no firm theory for the side-by-side synthesis of amylose and amylopectin. *In vitro* experiments with either synthase or phosphorylase lead to amylose (linear) in the absence of Q-enzyme, or if Q-enzyme is present, to an amylopectin or glycogen-like molecule (*44*). It is plausible that synthases can exist as individual molecules or monomers or as multi-enzyme complexes with Q-enzyme. In the former case, only amylose would be produced, provided the synthase maintains its close association with a given amylose molecule. Multi-enzyme complexes of synthase and Q-enzyme could form amylopectin. In spite of a certain amount of evidence for such multi-enzyme complexes of synthase and Q-enzyme (*14, 45*), the issue is far from settled.

During biological utilization of starch, either within plant tissues or by action of the enzymic systems of predators (microorganisms or animals), starch granules undergo erosion typical of the particular type of starch and enzymic system (*46–48*) (Figs. 1–3). In general, α-amylases are the chief agents of starch granule degradation. Debranching enzyme (R-enzyme) accelerates α-amylase attack (*49, 50*). The exterior surface of the starch granule, for example, potato starch, may be extraordinarily resistant to α-amylase attack, but once this layer is penetrated, the enzyme creates channels leading to the less-resistant granule core. Degradation then proceeds laterally revealing the growth layers (*46–48*). These facets of starch granule structure will be discussed further in Section III.5.

III. Ordered Structure of Starch Granules

1. Optical Microscopy

In ordinary light, dry starch granules are clear and colorless; and they have a mean refractive index of 1.5 (*51, 52*). Large hydrated granules, for example, those of potato and canna, show concentric "growth rings," which are shells of

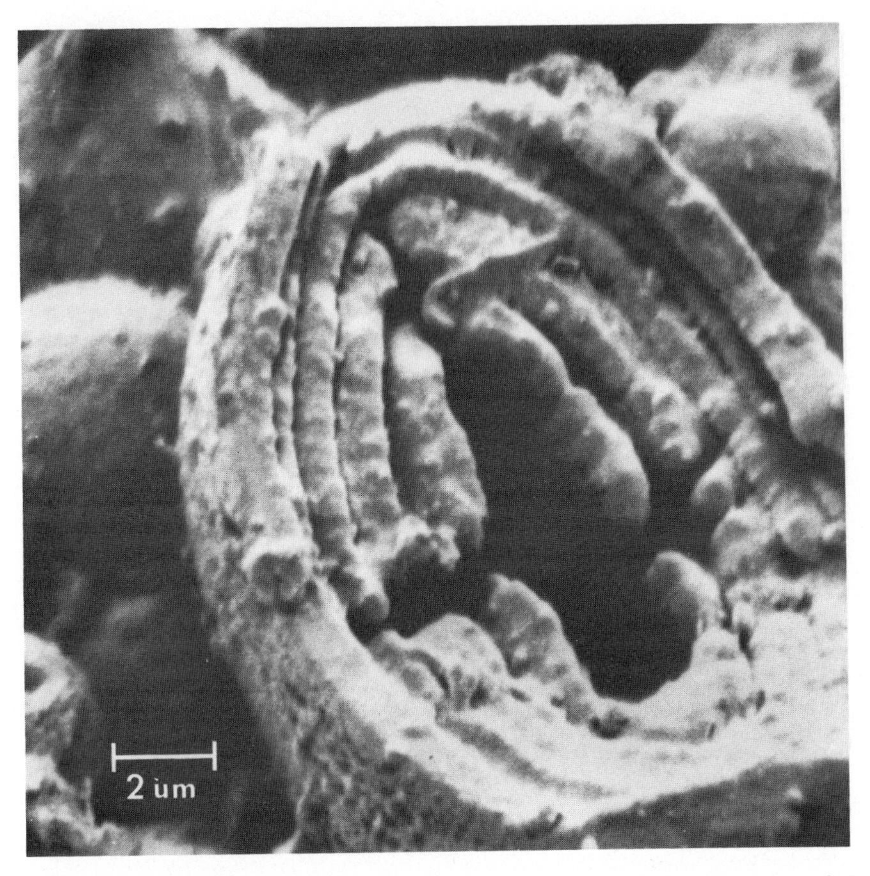

FIG. 1.—Wheat starch granule showing erosion during germination. (Photograph courtesy of W. Bushuk and B. Dronzek.)

alternating high and low refractive index. At the center of the granule is the original growing point or *hilum*. This region is usually less organized than the rest of the granule and may contain non-polysaccharide material from the amyloplast. Frequently, with starch granules that have been dried, there are cracks through the hilum. To get a better view of the three-dimensional arrangement of the microscopically visible features of starch, one should manipulate the granule and view it at various orientations. This is conveniently done by putting a small drop of starch syrup on the underside of a cover glass, and gently pressing this onto a microscope slide which has been lightly dusted with starch granules. By manually sliding the cover glass, one can make the starch granules roll and thereby get a three-dimensional impression of any selected granule. The starch syrup is sufficiently viscous that the objects, once manipulated into a desired

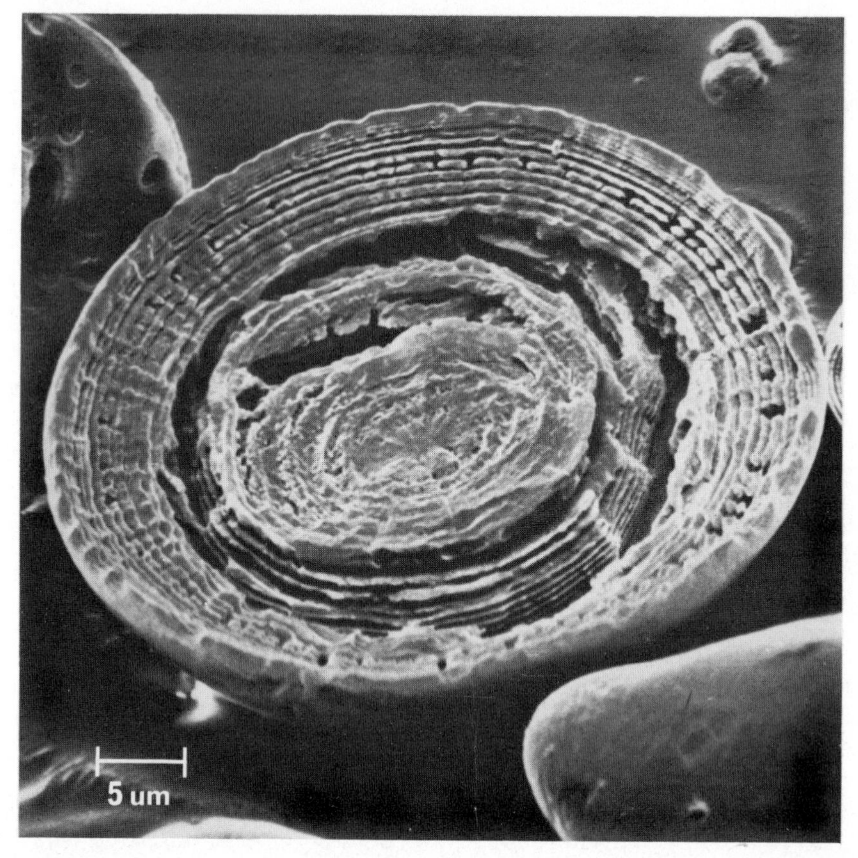

FIG. 2.—Rye starch granule after attack by rye α-amylase. Photograph courtesy of W. Bushuk and B. Dronzek.)

orientation, maintain that orientation long enough to photograph them. In this way, it is easily possible to see that the cracks at the hilum and the concentric shells are interior features, even though in stationary granules they may appear to be at the surface. (See also Chapter XXII.)

The most effective microscopic observation of native starch granules uses the polarizing microscope. Starch granules appear as distorted spherocrystals, with a typical dark cross (Fig. 4). The sign of birefringence is positive with respect to the spherocrystal radius ($n_e - n_o = 0.015$) (*51*). The variety of starch granule polarization patterns can only be appreciated by examining photographs (*53*) or granules under the microscope. The apparent intensity of birefringence is dependent on the thickness of the granule as well as on the degree of crystallinity and orientation of the crystallites. The dark arms of the polarization cross occur along the local average axes of the optic refractive index ellipsoids and intersect at the

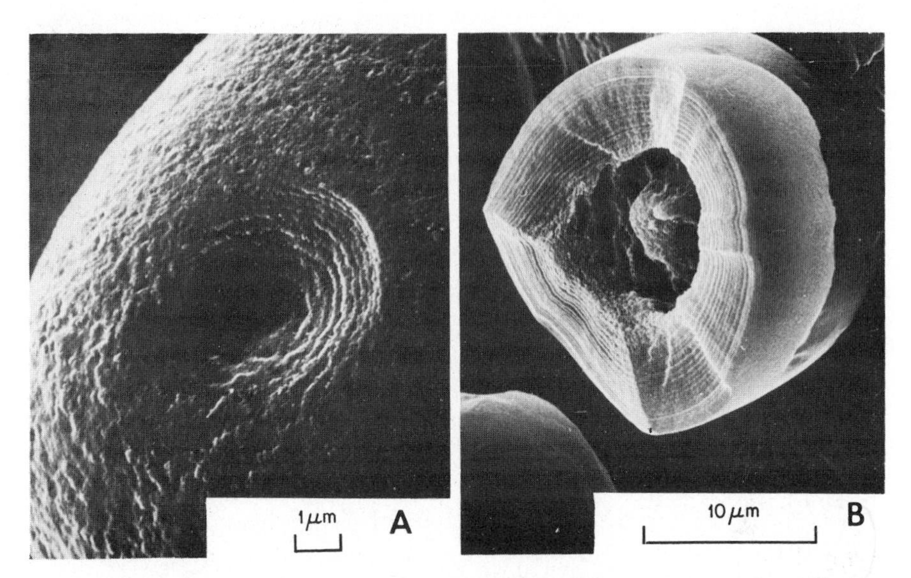

FIG. 3.—Potato starch after attack by *Bacillus subtilis* (*amyloliquefaciens*) α-amylase (*48*). A: conical "bore hole" and pitted surface. B: lightly cross-linked starch with 7.7 cross-links per 100 D-glucosyl residues. Note well-defined lamellae. Even though the central portion is extensively eroded, the outer surface is relatively unattacked. (SEM photograph courtesy of Prof. R. H. Marchessault.)

FIG. 4.—Potato starch granule, crossed polarizers (horizontal and vertical). The granule was mounted in corn syrup so it does not show growth rings. For the view on the right, the granule was rotated 40° from its position on the left. This granule is 53 μm long and 33 μm wide. (Photograph courtesy of Prof. Darryll E. Outka.)

hilum. With a first-order red selenite plate, granules show alternating yellow and blue quadrants. The direction of elongation of the optical ellipsoid (highest refractive index) is along the radius of the spherocrystal. With irregularly shaped starch granules, it is more accurate to say that the optic axis is aligned perpendicular to the growth rings, if they are visible, and to the surface of the granule. It is generally accepted that the optic axis coincides with the chain direction of starch and most other biopolymers (*54*), so that the appearance of starch granules under the polarizing microscope gives a strong clue as to the average orientation of the molecular chains within the various regions of the granule. This characteristic makes it possible to prepare "optical maps" (Fig. 5) showing the average orientation of the molecular chains within an individual granule (*55*). Such optical maps demonstrate that there were wide variations of organization between individual granules from the same source. One should clearly distinguish between starch granules, or portions of starch granules, that are weakly birefringent and those in which there may be a high degree of molecular orientation and birefringence but the molecules or crystallites are oriented along the direction of observation and therefore do not show strong polarization effects. For example, many of the filamentous granules of high-amylose maize starch are only very weakly birefringent, and if one manipulates such granules on a slide, there is no orientation of the granule which shows double refraction. By contrast, many large disk-shaped wheat starch granules show very broad extinction crosses as might be expected for weakly birefringent spherocrystals. However, when such granules are manipulated so that they can be viewed "edge-on," they are brilliantly birefringent. Molecular chains of these granules are aligned nearly perpendicular to the surface of the disc. When one views a granule in a direction

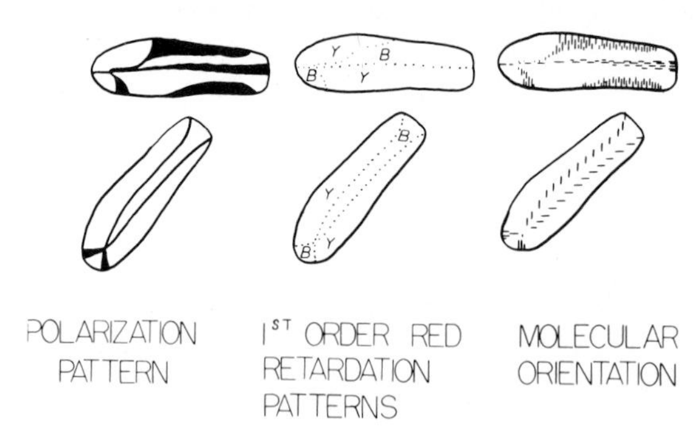

FIG. 5.—Method of obtaining optical maps from birefringence patterns of starch granules (*55*). The granule is photographed between crossed polarizers at different orientations. At the position of the "dark cross" the *average* orientation of the starch chains is either parallel or perpendicular to the plane of light passed by the polarizer. With the first order red plate, starch granules show alternating blue (B) and yellow (Y) patterns or sectors indicating that the highest index of refraction (direction of molecular chains) is perpendicular to the granule surface.

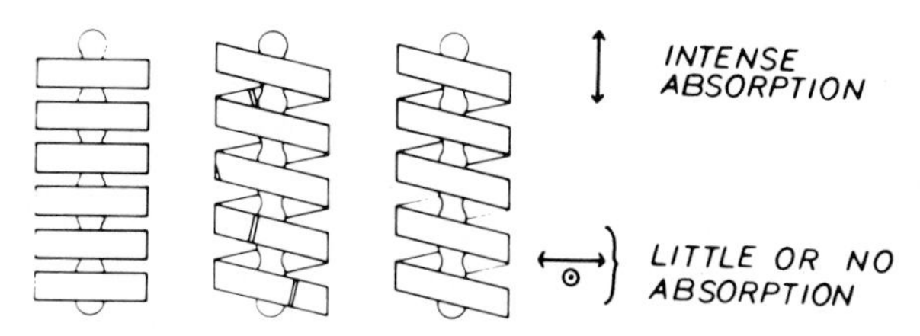

FIG. 6.—Dichroism of iodine complexes. Left: cyclodextrin "channel" complex; center: oligosaccharide interrupted helix complex; right: helical amylose complex. When the plane of polarization (electric vector) is parallel to the axis of the polyiodine array there is intense light absorption. If the electric vector is perpendicular to the polyiodine axis, there is little or no absorption (*55*).

along the axis of the disc, one is looking along the optic axis as well, so the birefringence virtually disappears.

When starch granules are lightly stained with iodine, they may become dichroic; that is, light absorption varies with the plane of polarization of the light (*55*). Crystals of various iodine complexes, as for example, iodine complexes with Schardinger dextrins or starch oligosaccharides, are intensely dichroic (*55*) (Fig. 6). Light is absorbed when the plane of polarization (electric vector) is parallel to the axis of the polyiodine ensemble (*56, 57*). For helical amylose–iodine complexes, light is absorbed when the plane of polarization is parallel to the helix axis (*58, 59*). With potato starch granules, which are strongly birefringent, many iodine-stained granules show little or no dichroism. This may indicate that the more highly oriented amylose molecules are in crystallites where they are inaccessible to iodine or are under other steric constraint so that they are unable to form a complex. Amylose not in crystalline regions is available to form an iodine complex, but it may not have a perceptible degree of orientation. If one stains sections of starch granules, they show a weak degree of dichroism indicative of a more or less radial orientation of the amylose molecules (*55*).

2. Scanning Electron Microscopy (SEM) (see also Chapter XXIII)

SEM of undamaged or unmodified starch granules shows that the granule surface is relatively smooth and free from pores, cracks, or fissures (*60*). With some cereal starches, there are indentations thought to be impressions of spherical protein microbodies which have impeded the local growth of the starch granule. Some starches show a superficial scaliness that may represent remnants of amyloplast membranes or possibly very thin incomplete outer shells of starch.

When a starch granule is damaged or eroded, either by acidic or enzymic treatment, the starch granules are greatly weakened so that many granules crack open and expose a pronounced layer structure (Figs. 1–3). Granules enzymically eroded, as during ripening of fruits or germination of seeds, may show surface terraces which are the exposed edges of layer structures (*46–48*). Strangely, the layering from acid corrosion of potato starch gives thick, 1–3 μm, shells whereas in granules treated with *Bacillus subtilis* liquefying α-amylase, the layers are only 0.3–0.5 μm thick.

SEM reveals that different starches are attacked in patterns that are related to the surface and internal texture of starch granules (*61–69*). For example, maize starch attacked by pancreatic α-amylase shows many surface pits which lead to the granule interior. The granule interior may be highly eroded or solubilized, yet the exterior surface or parts of it are resistant. With banana starch, erosion by pancreatic amylase appears to affect primarily the surface, leaving it roughened and exposing the edges of the growth rings. A typical feature of attack on potato starch is the formation of conical craters. Sometimes a great proportion of the interior of the granule may be dissolved, leaving an "egg-shell"-like resistant outer layer (*48*).

Crushed starch granules often show a radial fibrillar fracture surface, the details of which are too fine to be resolved by SEM (*70*). These fracture surfaces do not show evidence of layering, and there appears to be no tendency of unmodified granules to fracture at the boundaries between growth rings.

3. Transmission Electron Microscopy (TEM)

The method of TEM is one of great power and utility in that it provides a bridge between the gross structural features as seen by optical microscopy or SEM and the finer structural details as revealed by X-ray diffraction and chemical or biochemical structure analysis. With modern equipment, it is possible to obtain granule sections down to about 200 Å thick. The sample may be stained or given chemical or enzymic treatment. Stains with a high electron density and that are noncrystalline, for example, phosphotungstic acid or uranyl acetate, are suitable. The stains may react specifically with certain components of the sample (positive stains) or may merely provide an electron-dense outline of an otherwise electron-transparent object (negative stain).

In freeze fracture methods, a wet sample is rapidly frozen in liquid nitrogen so that the water is in a glassy state. The sample is then impact fractured and some of the excess water is allowed to evaporate at low temperature so that the surface of the fracture reveals topographical details. Then the sample is coated with a thin carbon film, followed or accompanied by a metallic film, and finally the carbon and metal are cleaned by destroying the original object, for example, by using chromic acid, to give a replica of the details of the fracture surface.

For negative staining, a very dilute solution or dispersion of the sample is applied to a film, and excess liquid is drained off. Then a solution of negative stain is applied; the excess is drained off, and the specimen is dried. The electron-opaque negative stain surrounds the less electron-dense particles and outlines them, making them visible.

Work with thin sections of starch was initiated over 20 years ago by Whistler (*71*), Mühlethaler (*72*), and Frey-Wyssling (*54*). These studies showed that starch has a microgranular structure, accentuated by enzymic or acidic erosion (*73, 74*). The microgranules are about 200–300 Å in diameter. If the micro-granules are approximately 50% crystalline, these dimensions correspond more or less to the 145 Å diameter of the crystallites obtained by application of the Scherrer formula to the X-ray diffraction line-width (*75*). Some of the sections, for example with maize starch, show growth rings; but for most sections, rings are not visible. The electron microscopy results, together with the X-ray data, led Nikuni (*76*) to propose a spherical micelle for starch (Fig. 7), in which there are regular chains and chain terminations and branches within the micelle, and starch chains entering and leaving the micelle to form the amorphous phase of the starch granule.

Buttrose (*77–81*) showed that, in the barley endosperm amyloplast, there is a diffuse layer of undefined material, about 100 nm thick, surrounding the growing starch granule (*77*) which could be polysaccharide or possibly biosynthetic enzymes. After acid treatment, growth rings are seen in very young barley granules; and with granules only 3 days old, a maximum of 3 rings is found. Granules formed in barley plants under continuous light and constant temperature do not show growth rings. With normal barley starch, erosion by α-amylase leaves a "saw-tooth" pattern, indicating that the granules have alternating regions of

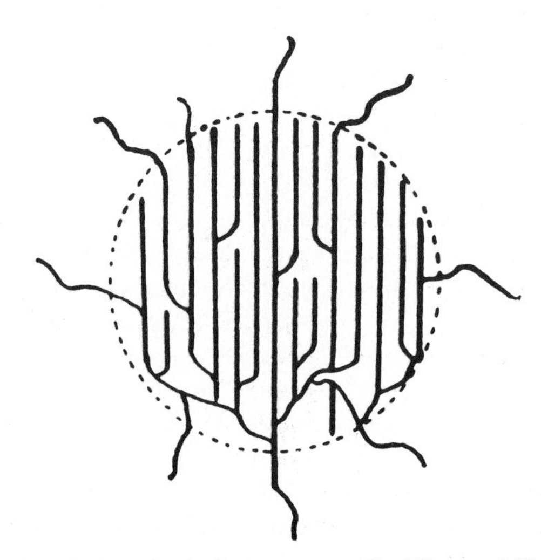

FIG. 7.—Spherical starch micelle, as suggested by Nikuni and Hizukuri (*76*).

greater and lesser susceptibility to amylase action, correlated with the growth rings. But with the constant environment-produced granules, no such saw-tooth patterns are produced. All other starches show growth rings, even the small "granula" (small constituents of a compound granule) (*80*) of oat starch. Frey-Wyssling and Buttrose (*82*) point out that the shell thickness is of the same order as the length of an amylose molecule. Growth shells are seen most distinctly in amylose-free waxy maize starch where the thickness of a growth ring corresponds to the length of an amylopectin molecule (*24*).

Kassenbeck (*83, 84*) stained starch by a very brief reaction with periodate followed by reaction with thiosemicarbazide, then reaction (*fixation*) with silver nitrate. In some cases, the specimens were given a post-fixation staining with phosphotungstic acid or osmium tetraoxide vapor. The thiosemicarbazide reacts mainly with free or hydrated aldehyde groups, as opposed to aldehyde groups involved in hemiacetal formation, and the silver reacts, possibly as a cross-linking reagent, by chelate formation with the thiosemicarbazones. Wheat and maize starches observed in this way show growth rings and a pronounced radial periodicity of 60–70 Å. This period corresponds to the average spacing of clusters in the cluster model of amylopectin (*24, 55*) and supports the idea that amylopectin molecules have a generally radial orientation.

By treating fixed specimens with water vapor and osmium tetraoxide at 60°, the hemiacetal linkages are presumably broken, allowing the freed aldehyde groups to reduce the osmium tetraoxide. With radial (central) sections, this treatment shows a pronounced fibrous texture in a predominantly radial orientation. The staining intensity is higher for normal maize than for waxy maize, suggesting that the fibrous elements are mainly amylose, or possibly that the intercrystalline zones in normal maize are larger than those in waxy maize. In any case, the fibrous elements could not be interpreted as being single amylose molecules, which being less than 10 Å in diameter would not be resolved.

With waxy maize starch, osmium-stained tangential sections show unstained crystalline zones, roughly 1000 × 2000 Å in cross section, separated by an electron-staining matrix 200–300 Å thick.

Kassenbeck distinguishes three types of organization: (*a*) a radial arrangement of amylose molecules, (*b*) amylose in an amorphous arrangement, and (*c*) an arrangement of amylopectin such that the crystallites are in tangential lamellae.

In a study of waxy maize starch, Yamaguchi and co-workers (*24*) examined molecular dispersions, thin (500 Å) sections, and mashed preparations of native and extensively acid-degraded granules. Molecular dispersions of waxy maize starch in dimethyl sulfoxide showed elongated particles and aggregates approximately 100 Å in diameter and 1200–4000 Å long.

Negatively stained, mashed native granules show fibrillar structures and 70 Å ripples similar to those seen by Kassenbeck (Fig. 8). They also show worm-like rippled particles, presumably amylopectin molecules, about 150–200 Å in diam-

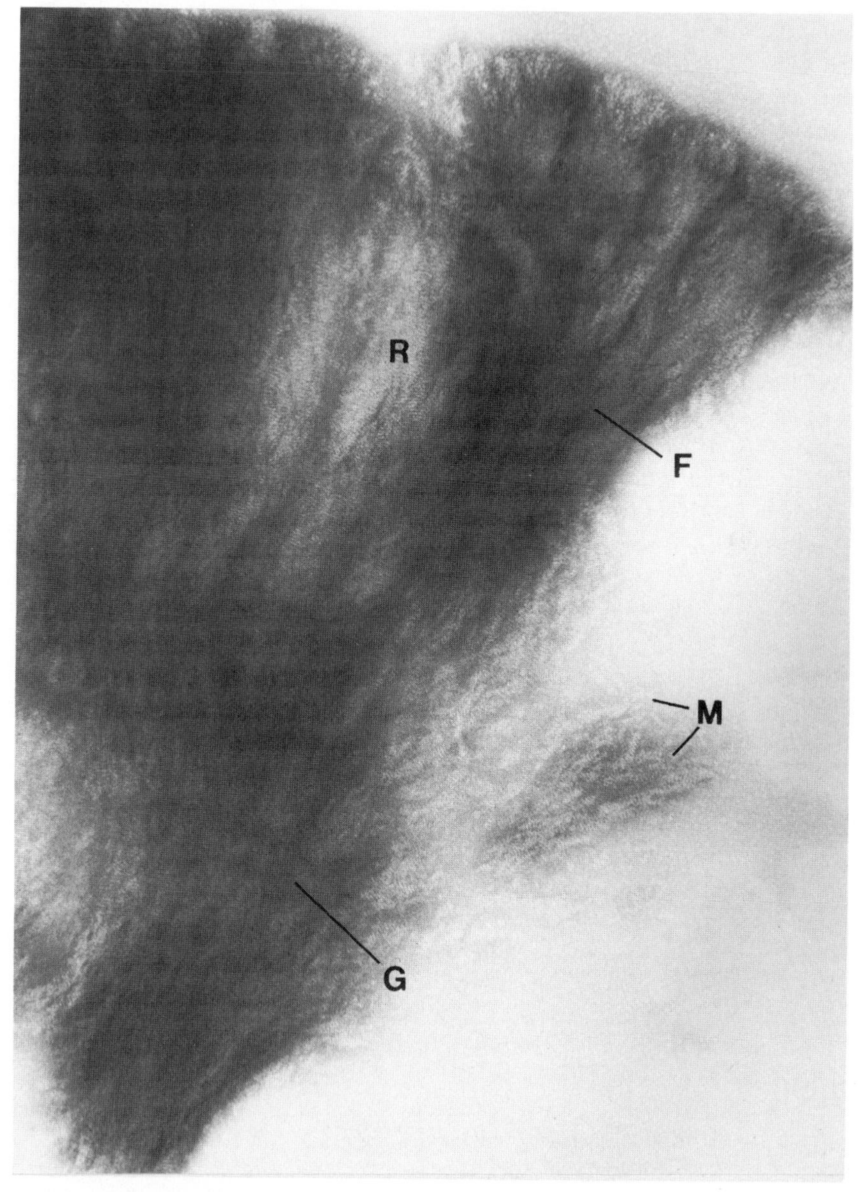

FIG. 8.—Fragment of waxy maize starch granule obtained by wet mashing, negative stain (*24*). F, radial fibrillar structure; R, 70 Å ripples; G, growth rings; M, worm-like rippled structures, presumably amylopectin molecules. ×80,000.

eter and similar in shape to the particles seen in molecular dispersions. Incidentally, the wet mashing procedure does not disrupt seriously the organization at the x-ray level, (*85*), and mashed waxy maize starch retains the A-type x-ray diffraction pattern (*86*).

Extensive acid treatment or "Lintnerization" of native starch granules leads to a relatively acid-resistant material known as Naegeli amylodextrin (*2, 55*). Waxy maize amylodextrin consists of a fairly high-molecular-weight, multiply branched material, a fraction of average DP~25, containing two molecular chains of average DP~11–14 branched at or near the reducing group; and a linear fraction of average DP~14 (*55, 87–90*). Thin sections of the acid-treated granules show pronounced growth rings (*81, 91*) and, at high magnification, stacks of lamellae 50 Å thick and several hundred angstroms broad (*24*) (Fig. 9). The lamellae are approximately parallel to the growth rings. Lamellae in mashed, acid-treated granules are very prominent after negative staining, being about 50 Å thick and stacked at intervals of 70–80 Å. Recrystallized amylodextrin fractions also show regular lamellae 50 Å thick and many hundred Å broad. The 50 Å lamellae in both the acid-treated starch granules and the recrystallized amylodextrin may be two-dimensional crystals, the 50 Å dimension being the

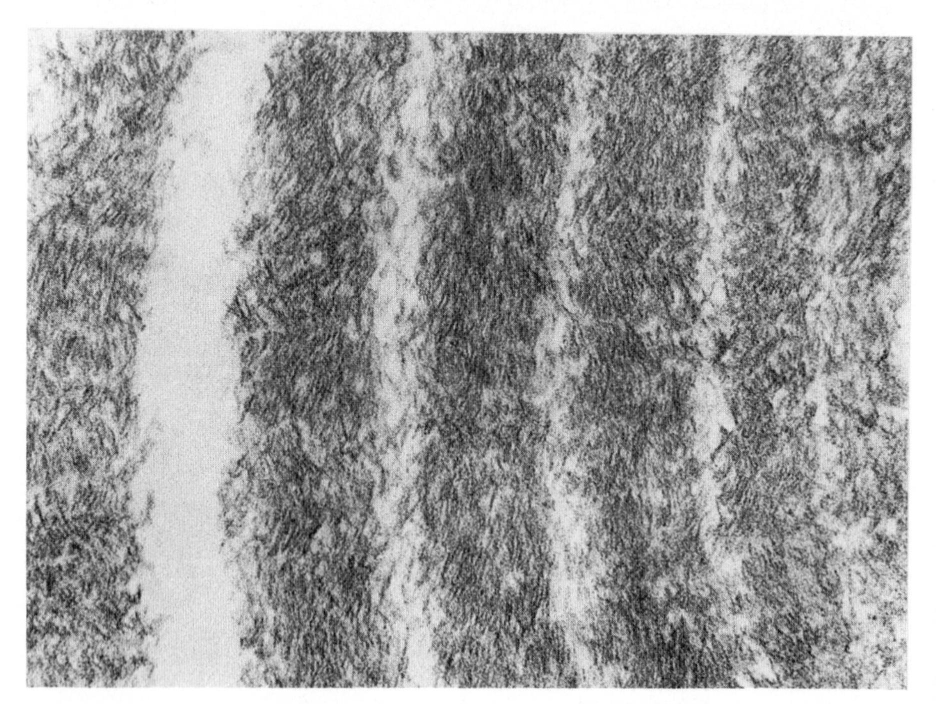

FIG. 9.—Detail from thin section of acid-treated waxy maize starch granule (*24*). Note stacks of lamellae about 50 Å thick. ×110,000.

length of 14-D-glucosyl-unit chains in double helices. The double helices, being roughly cylindrical can pack side-by-side. But the lamellar thickness, only about 50 Å, permits fewer than five 10.5 Å "repeats" along the molecular axis; and the variability in molecular length does not permit a very close or economical stacking of lamellae as would be required for a true three-dimensional crystal. During acid treatment of the granules, the intercrystalline amorphous region is hydrolyzed, leaving the crystallites essentially intact. The crystallites, no longer being separated by amorphous regions, may organize laterally improving the overall crystallinity within the treated granule. This is evident both in the sharpened x-ray diffraction pattern (*32, 92*) and in the high birefringence, as well as in the high resolution TEM micrographs.

In 1958, Sterling and Spit (*93*) in a report of an examination of cereal starch fracture surfaces pointed out that there is very little difference in electron density between the starch crystallites and the amorphous matrix in which the crystallites are imbedded. They fractured starch granules by freezing a suspension in water, hammering the block of ice-embedded starch granules at −20°, and dehydrating the starch with ethanol. Alternatively, starch granules were dehydrated in alcohol and acetone, then placed in xylene, and the suspension mechanically fractured. The fractured granules from either method were shadowed with platinum–carbon to obtain replicas of the surface. The fractures follow lines of least resistance and reveal interfaces between crystallites and less organized areas. Although the micrographs do not show individual molecular strands, they do show crystallite rods or blocks, about 200 Å in diameter, and from 200 to 3000 Å long, as well as more or less cylindrical "microfibrils" 200 Å in diameter and several thousand angstroms long. Many of the small fibrillar strands show a coiled, twisted, or cable-like morphology. Granule surfaces show a papillate texture, each papilla being about 200 Å in diameter, possibly the surface termination of a microfibril.

Sterling and Spit (*93*) emphasize the soft, plastic nature of starch, both the crystallites and the amorphous part. The coiling and twining of the ends of the microfibrils observed at the fracture surface may be an artifact of the fracture process rather than a characteristic of the native starch structure.

Recognizing this softness and plasticity of unmodified starch, to examine details of internal structure of potato starch, Sterling and Pangborn (*94*) first "Lintnerized" the starch by treating it with cold dilute mineral acid for a few days. This treatment made the granules less plastic and more easily fractured. The granules in water were crushed between microscope slides, and the fragments dried, and shadowed.

With crushed, Lintnerized potato starch, bundles of radially oriented microfibrils appear. Each microfibril is about 270 Å in diameter and shows prominent longitudinal ridges and a twined organization.

X-Ray line-width measurements indicate micellar diameters of 77 and 99 Å for dry and wet native starch, and 87 and 126 Å for dry and wet Lintnerized starch. Sterling proposed that the observed microfibrils are cables of seven micelles, each about 80–90 Å in diameter. The micelles are of indefinite length and are separated from each other by intermicellar amorphous starch. During hydration, some of the intermicellar amorphous material may become more highly organized and incorporated into adjacent crystallites, increasing their effective diameters. Sterling suggested that each microfibril may be synthesized by an "organizing body" of the amyloplast, thus maintaining the orientation and organization of the polysaccharide in microfibrils. A similar concept has been applied to biosynthesis of other polysaccharides, notably cellulose and chitin (*95, 96*). The surface papillae seen by Sterling and Spit may then have been the final site of attachment of starch-synthesizing multi-enzyme complexes at completion of starch synthesis.

In none of the replica studies of native starch has there been any sign of the growth rings that are so obvious by optical microscopy or by electron microscopy of chemically or enzymically eroded granules.

4. X-Ray Diffraction

Native starch granules give distinctive x-ray powder diffraction patterns (*97*) of two main types (Fig. 10): A patterns characteristic of cereal grain starches, such as maize, wheat, and rice; and B patterns characteristic of tuber, fruit, and stem starches, such as potato, sago, and banana starches. C patterns, intermediate between A and B patterns, are probably due to mixtures of A- and B-type

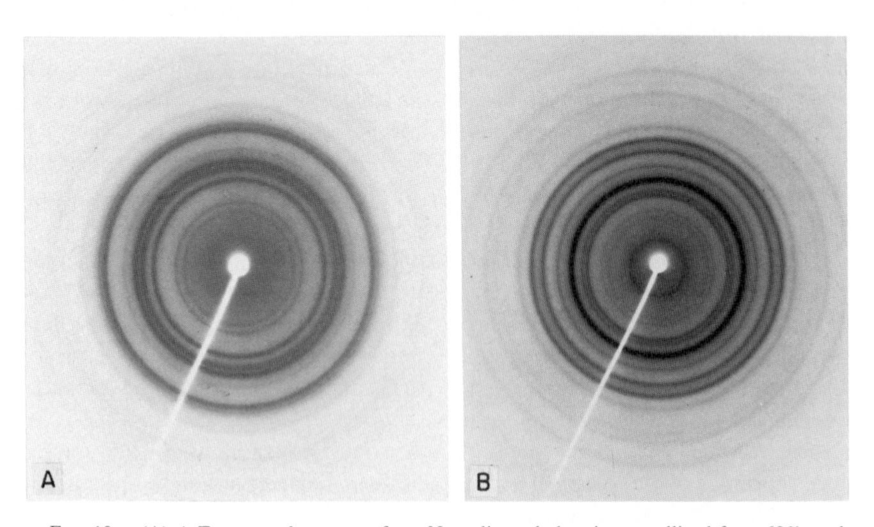

FIG. 10.—(A) A-Type powder pattern from Naegeli amylodextrin crystallized from 60% methanol. (Sample and X-ray pattern by Shoichi Kikumoto.) (B) B-Type powder pattern from retrograded amylose hydrolyzate fraction. (Sample courtesy of H. Zobel.)

crystallites (*98, 99*), either within individual granules, or as mixtures of A- and B-type granules.

Interpretation of x-ray data has been difficult, owing to the inadequate amount of data, the small crystallite size, the imperfection of the crystallites, and the large unit cells. Several unit cells, which have not stood the test of time and are now mainly of historical interest, have been proposed (*100*).

By using a microcamera, Kreger was able to obtain an oriented x-ray pattern from a small region of a single B-type starch granule (*101*). In this pattern, the diffraction spots or arcs were arranged in layer lines, as is characteristic for oriented fibers. The fiber axis was found to be perpendicular to the growth rings, and the fiber period was 10.6 Å. The result is in substantial agreement with the fiber period (10.4–10.6 Å) found by using artificial films or fibers of starch. This spacing represents the repeat distance for three D-glucopyranosyl units in a double-stranded helix. From the rest of the pattern, Kreger proposed a pseudo-hexagonal orthorhombic unit cell with a = 9.0 Å, b (fiber axis) = 10.6 Å, and c = 15.6 Å, essentially identical with that previously proposed on the basis of oriented films and fibers (*102*). Owing to the poor quality of the data obtainable from single starch granules, this approach is not in current use. As yet, no one has published electron diffraction results with single starch granules.

The major progress in understanding the x-ray diffraction properties of starch has come from the use of artificial fibers (103–105). A-Type fibers (*105*) are generally prepared by first acetylating amylose, then casting it into a thin film, cutting narrow strips from the film, orienting the film into a fiber by stretching it at 150°, then deacetylating it in alcoholic alkali, and finally exposing the fiber to high humidity (80% RH). To obtain B-type fibers (*104*), the A-type fibers are exposed to 100% RH then annealed by heating in water at 90°. Both A and B fibers give all the diffractions of A and B powder patterns (*98, 106*), and in addition, many diffractions too weak or too poorly resolved to be visible on the usual powder patterns or diffractometer tracings (Fig. 11).

Using data from the best fiber patterns, Sarko (*106*) has proposed the following unit cells (dimensions in Å). For A, a = 11.90; b = 17.70; c (fiber axis) = 10.52; $\alpha = \beta = \gamma = 90°$; space group, P_1 or $P2_1$(s). For B, a = 18.50, c(fiber axis) = 10.40, space group $P3_221$ (Table I). Sarko has interpreted both A and B structures on the basis of right-handed, parallel-stranded double helices with three D-glucopyranosyl units per 10.4–10.5 Å rise in the helix (Fig. 11). There is an approximate twofold axis of symmetry at the helix axis, so that the repeat distance along a single helical strand corresponds to six D-glucosyl units (21 Å). The three D-glucopyranosyl units in a half turn are more or less in the same conformation, though this is not required by the crystallographic symmetry. The double helices, though parallel stranded, are in antiparallel packing, that is to say, in each unit cell one of the double helices is "up" and the other is "down." From biosynthetic considerations, it would be more reasonable to expect parallel

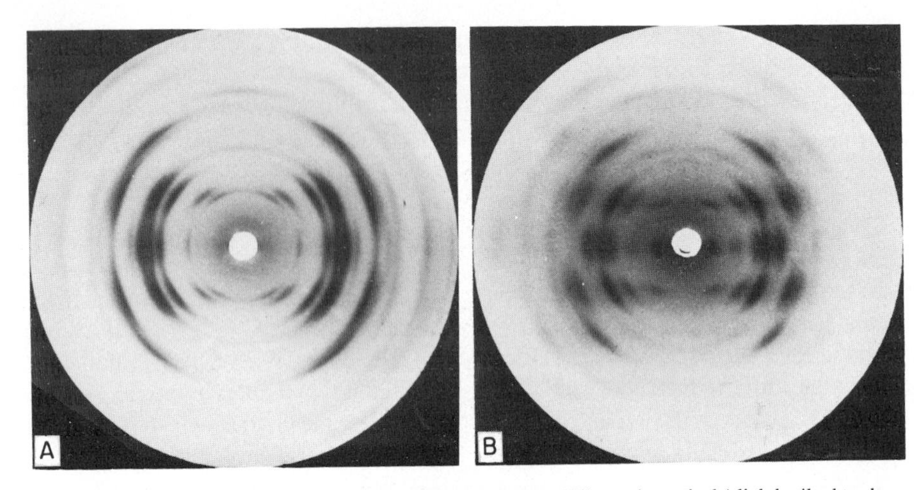

FIG. 11.—(A) A-Type amylose fiber diffraction pattern. Fiber axis vertical (slightly tilted to show [003] diffraction). Fiber spacing is 10.5 Å. (B) B-Type amylose diffraction pattern. The fiber period is 10.4 Å. Fiber axis is vertical (*106*). (Photographs courtesy of Prof. A. Sarko.)

packing of parallel-stranded double helices, so that all the molecular chains have the same polarity. It is conceivable that the A and B patterns obtained from native granular starches and amylodextrins do not reflect precisely the same crystal structures as in the A and B fibers. This enigma and several other puzzling features of the A and B structures remain to be resolved. Recently, Chanzy (*107, 108*) obtained electron diffraction patterns of B-type (retrograded) amylose (Fig. 12). His patterns corroborate the a = 18.5 Å spacing and hexagonal net for B-starch (*104*).

One kind of computer modeling utilizes Φ, ψ maps, in which conformation energy contours are plotted on a two-dimensional map of Φ and ψ, the glycosidic torsion angles (*109*). Only a single molecular geometry for D-glucopyranosyl units can be used on a given map. Usually the geometry chosen is one which seems reasonable and is derived from the average of single crystal structure determinations. For disaccharides, the major interaction between monosaccharide units is around the glycosidic bond, although there may also be hydrogen bonding or interactions elsewhere. Cyclodextrins, in addition to the disaccharide-type interactions, must form a closed loop. Conformational maps of cyclodextrins are based on an n-fold axis of symmetry, though this is contrary to x-ray results (*110*). For long molecular strands there is the constraint that the molecule must be non-intersecting and the additional question as to whether the structure has exact helical symmetry. Finally, multiple helices can be either parallel-stranded or anti-parallel-stranded, and intermolecular as well as intramolecular interactions must be taken into account (*111–113*).

A different approach uses the parameters n, number of monosaccharide units

Table I

Unit Cell Dimensions of Different Polymorphs of Amylose and Amylose Derivatives[a] *(106)*

Structure	*a*	*b*	*c* (*fiber repeat*)	γ (*deg.*)	*Helix symmetry*	*Space group*
Native starch, retrograded starch, and amylose fibers; right-hand double helices:						
A	11.90	17.70	10.52	90	2 × 6/1 in 21.04 Å repeat	$P2_1(S)$
B	18.50	18.50	10.40		2 × 6/1 in 20.8 Å repeat	$P3_121$
Amylose complexes; compact left-handed single helices:						
V_a	12.97	22.46	7.91		$2_1(\sim 6/5)$	$P2_12_12_1$
Intermediate forms	13.30	23.0				
	13.50	23.45				
	13.55	23.50				
V_h	13.7	23.7	8.05		6/5	$P2_12_12_1$
V_{DMSO}	19.17	19.17	24.39		6/5 in 8.13 Å repeat per turn	$P2_12_12_1$
V_h-iodine	13.60	23.42	8.17	90	6/5	$P2_1(S)$
V_a-BuOH	26.4	27.0	7.92			$P2_12_12_1$
V_h-BuOH	13.7	25.8	8.10			
Alkali and salt–amylose complexes and amylose derivatives; extended left-hand helices:						
KOH	8.84	12.31	22.41		6/5	$P2_12_12_1$
KBr	10.88	10.88	16.52		4/3	$P4_32_12$
Amylose triacetate I	10.87	18.83	52.53	90	14/11	$P2_1(S)$
Tri-*O*-methylamylose	17.24	8.70	15.64		$2/1(\sim 4/3)$	$P2_12_12_1$
Tri-*O*-ethylamylose 1	16.13	11.66	15.48		4/3	$P2_12_12_1$
Tri-*O*-ethylamylose 3	15.36	12.18	15.48		4/3	$P2_12_12_1$
Tri-*O*-ethylamylose 1 Chloroform complex 1	16.76	14.28	16.02		4/3	$P2_12_12_1$
Tri-*O*-ethylamylose Dichloromethane complex 1	16.52	13.95	16.02		4/3	$P2_12_12_1$
Tri-*O*-ethylamylose Nitromethane complex	14.70	14.70	15.48		4/3	$P2_12_12_1$

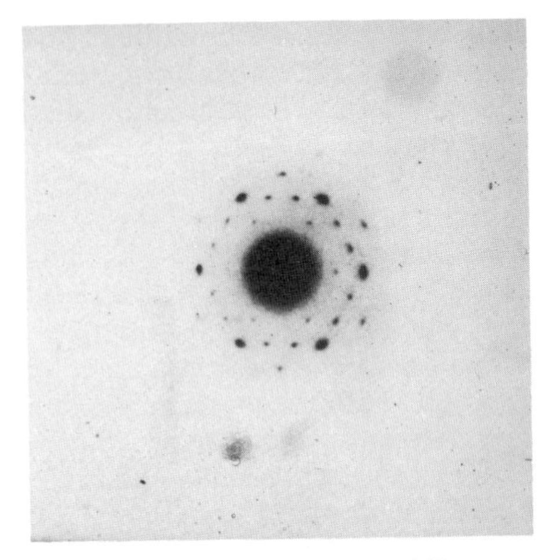

FIG. 12.—Electron diffraction pattern of retrograded (B-type) short-chain amylose; incident beam along "fiber" (hexagonal) axis. The hexagonal net and spacings are in agreement with x-ray results from B-amylose fibers (Photograph courtesy of H. Chanzy.)

per complete turn of the helix, and h, the rise along the helical axis per monomer unit (*114, 115*). Residue geometries are allowed to vary over the range observed in single crystal studies, and a given n-h combination is evaluated as being allowed if there are no nonbonded interatomic distances less than the sum of the van der Waals radii. Criteria for marginally allowed and disallowed conformations are somewhat empirical. For marginally allowed conformations, there may be a few nonbonded interatomic distances somewhat less than the sum of the van der Waals radii. A marginally allowed conformation generally implies that a minor conformational adjustment would relieve the unfavorable nonbonded repulsions. For disallowed conformations, there are one or more nonbonded interatomic distances intolerably less than the sum of van der Waals distances.

With any type of conformation map, it is necessary to use x-ray data, particularly the fiber axis spacing, to arrive at an estimate of the range of allowed chain conformations. With starch, either right- or left-handed double helices are feasible. A 10.5 Å period, as observed for A and B starch, requires a minimum of three D-glucopyranosyl units.

The maximum possible extension of an unstrained starch chain has a rise (h) of ~4.4 Å per sugar unit (*114*). This would occur for $n = 5$, but there is no known example of such a structure. The closest example is KBr amylose (*116*), with $n = 4$ and $h = 4.0$. "Intermediate" amylose has $n = 5$, h = 3.8 (*117*). Various derivatives and alkali and salt complexes exist in the range $h = 3.7$–4.1 Å, $n = 4$–6. In these complexes, economy of packing into a crystal lattice is achieved by stacks of sheets of antiparallel polysaccharide chains. The organic derivative side chains or ions plus water fill in the interstices between the polysaccharide chains.

A small contraction, h = 3.5 Å, n = 6, permits formation of double helices. Starch double helices are stable because the two chains fit together compactly, with the hydrophobic zones of the opposed monomer units in close contact, and the hydroxyl groups located for strong interchain hydrogen bonding (*118*). The chains can dissociate only by unwinding. There is no central cavity or channel. Being roughly cylindrical, they give economical packing in a hexagonal or quasihexagonal lattice, although for parallel stranded double helices having pronounced "crests" and "troughs," a square packing is preferred. Further contractions of the polysaccharide opens up a channel between the chains, destroying interchain hydrophobic bonding. However, when h is about 2.5 Å, the channel becomes large enough to accommodate hydrophobic "guest" molecules (for example, iodine) (communication from H. Zobel). There is no evidence for such a structure, though it seems feasible. A further decrease in h is impossible for double helices. For single six-fold helices, as h decreases below 4 Å, hexagonal packing becomes less and less compact; the overall helix diameter increases, and a central cavity or channel develops. The minimum value of h is $8/n$ Å, corresponding to the totally "collapsed" V-amylose structure.

Single helices in the range ($1.3<h<3.7$ Å) are feasible and probably exist in amylose solutions. However, they are unknown in the solid state, simply because they cannot pack compactly. Interturn hydrogen bonding becomes optimal for h = $8/n$ Å where n = 6, 7, or 8, the well-known V-amylose complexes. A further decrease in h is impossible owing to the intersection of adjacent helical turns.

Cyclodextrins (Schardinger dextrins or cycloamyloses) might be regarded as degenerate helices with $n \geq 6$ and $h = 0$. They would be located at positions on Φ–Ψ or n–h maps (*110*) which are strongly forbidden for amylose, inasmuch as the $(m+n)$th residue of an amylose chain would exactly overlap the mth. Rings with n-fold axes of symmetry are feasible over the range n = 5–10, the ring with n = 7 (cycloheptaamylose or β-cyclodextrin) being most stable. However, the cyclodextrin with n = 5 is unknown, probably because it falls outside the specificity range of enzymes which convert starch into cyclodextrins.

X-Ray data on representative amylose fiber structures are given in Table I; computer drawings are presented in Figures 13 and 14.

In addition to providing information about the crystal structure of the crystallites of starch granules, x-ray diffraction can give information about the relative amounts of crystalline and amorphous phases. If typical diffractometer tracings are examined, one sees peaks from the crystals and a background from the amorphous or gel phase. By integrating the x-ray scattering intensity separately over the peaks and over the background, a number is obtained which can be interpreted as the "x-ray crystallinity" (*119, 120*). Alternatively (*121*), a given diffractometer tracing can be compared with one from a sample which is designated as "maximally crystalline" or "maximally amorphous" (*121*). Because the crystallites in starch granules are very small (~100–150 Å), estimates of the

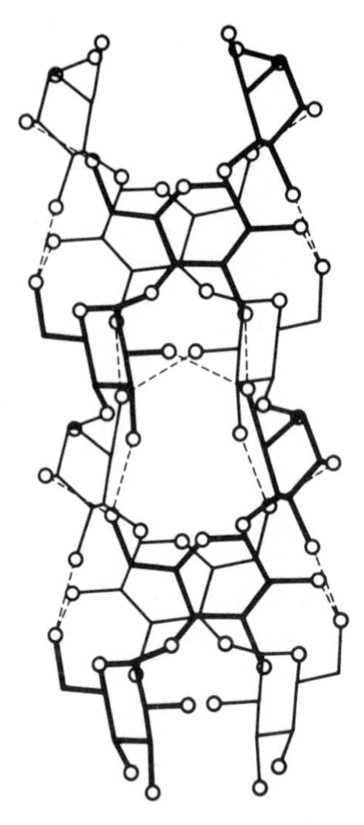

FIG. 13.—Computer drawing of amylose right-handed double helix, one complete turn of each of two strands. The heavy lines indicate the part of the structure nearest the reader; the light lines are away from the reader; the dashed lines represent hydrogen bonds. (Drawing courtesy of A. Sarko, *98*.)

degree of crystallinity cannot be rigorously or unambiguously assigned. At the border of a crystallite, there is a zone in which the crystallinity changes from a high degree of perfection to a totally disordered structure characteristic of a liquid or amorphous arrangement.

The theory of X-ray diffraction indicates that very small or imperfect crystals give broadened diffractions (*122*). For anisodimensional crystallites, line broadening is greatest for those reflections corresponding to the shortest direction of the crystallite, and conversely (*123*). Hizukuri and Nikuni (*75*) used the 16 and 5.22 Å reflections for determining crystallite size of potato starch. These were selected as being both intense and apparently free from overlapping reflections that would give false apparent line broadening. Of these, the B-pattern 16 Å diffraction is less ambiguous. In fiber and single granule patterns, it occurs on the equatorial layer line. The 16 Å line broadening is due to the crystallite size perpendicular to the fiber (molecular) axis, and the corrected line broadening (0.54°) corresponds to a crystallite size of 147 Å. For the 5.22 Å reflection, the corrected line broadening (0.56°) corresponds to a crystallite size of 144 Å.

For waxy maize starch, a periodicity of ~70 Å along the molecular axis has been observed by electron microscopy (*24, 84*).

For the molecular axis, the repeat distance for an individual strand of a double helix is 21 Å or six D-glucopyranosyl units, and the average "outer" chain length is approximately fourteen D-glucopyranosyl units, so that the calculated crystallite dimension along the molecular axis is about 50 Å.

Small angle scattering does not result from the fine structure of the crystallites, but rather from periodic fluctuations in density in the structure. Sterling (*124*) has shown that a variety of native starches give small angle scattering corresponding to Bragg spacings of about 100 Å. This small angle diffraction is fairly sharp, indicating that the fluctuations are rather regular and uniform. In powder diagrams from starch granules, orientation is not possible and it cannot be determined from x-ray powder data alone whether the 100 Å period corresponds to a fluctuation in a radial or tangential direction. Most likely, this long spacing

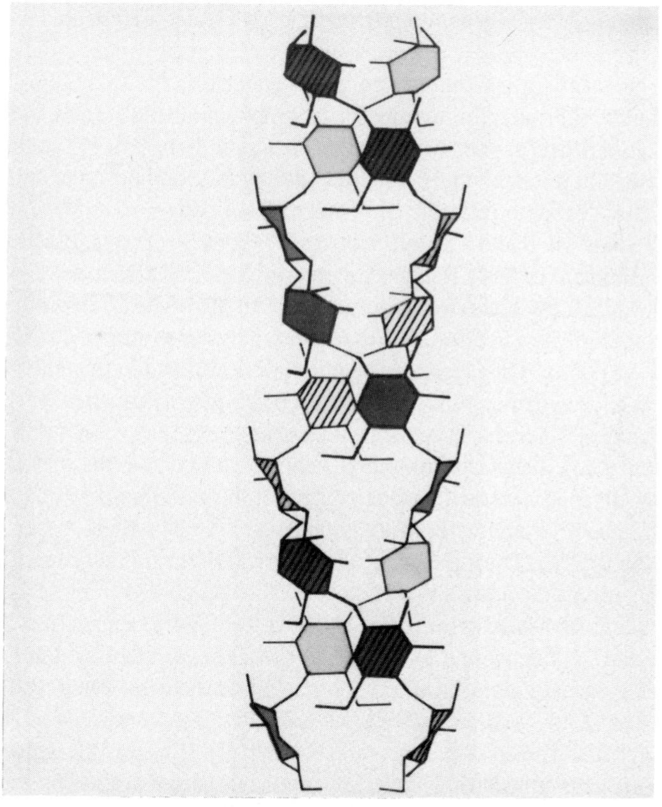

FIG. 14.—Computer drawing of starch left-handed double helix. (Drawing courtesy of Alfred D. French, *117*.)

corresponds to some kind of weighted average of the electron microscopy 70 Å radial spacing and the X-ray 150 Å tangential spacing.

5. Significance of Growth Rings in the Starch Granule

Native starch granules show "growth rings" when observed by optical microscopy, scanning electron microscopy of eroded granules, and transmission electron microscopy of thin sections, especially after chemical treatment. The growth rings represent concentric shells or layers of alternating high and low refractive index, density, crystallinity, and resistance or susceptibility to chemical and enzymic attack. That the rings or shells are internal features is verified by electron microscopy.

Starches that show prominent growth rings, such as potato starch, apparently require high or saturating water levels for ring visibility. When air-dried potato, rye, or pea starches are observed in pentane, rings are not visible (*125*). Cracks and other defects seen when granules are placed in water are not apparent in nonaqueous dispersion media. Growth rings may originate during deposition of layers of increasing organization, retaining a water-rich central region. When the mature starch is dried, this region shrinks more than the rest of the granule, leaving a void or an air pocket. If granules containing these voids are kept under water for several minutes, the voids gradually hydrate and become almost invisible. Why they become invisible in pentane is unknown.

Growth rings are readily visible in granules from fresh, wet, plant tissue, for example, in potato tubers. The rings represent shells of higher and lower starch content produced by fluctuations in the rate or mode of starch deposition. There is a substantial difference in refractive index between water (n_D = 1.333) and starch (n_D = 1.53). Thus, layers of higher and lower starch content would give visible rings, provided there is a sufficient contrast in n_D and the layers are thick enough to be resolved in the optical microscope.

As water is lost, shells shrink radially in such a way that the refractive index differences disappear when the water content of the granule is less than 10%. The refractive index of starch is essentially constant (1.534–1.532) over the range 0–10% water (*126*). During drying, the layers also shrink tangentially, but this has no effect on their visibility.

When starch is heated in air or inert gas at 165–200°, after loss of hydration water the starch is converted into dextrins and gums of practical importance. In the initial stages of dextrinization, starch becomes dehydrated and loses its characteristic X-ray pattern, but retains a polarization cross (*127, 128*). When this slightly dextrinized material is treated with cold water, the granules swell and separate into concentric layers, apparently the original growth layers of the granules (*129*). These layers are particularly noticeable with potato dextrin, but with tapioca dextrin only a single "bag" seems to form. At higher dextrinization temperatures, more and more of the granule becomes soluble (*127, 128*). The less-soluble fraction, presumably from the more crystalline growth rings, has a fairly strong tendency to retrograde, whereas the soluble material does not.

The technique of low-angle light scattering is applicable to individual starch granules, fields containing many granules, and amylose films (*130–136*). In this approach, a narrow beam of monochromatic polarized light from a laser is allowed to pass through the object, and the scattered light is registered on a photographic film or a photoelectric device. With single homogeneous anisotropic spherocrystals, theory predicts a scattering envelope whose geometry is primarily dependent on the radius of the particle. For a mixture of particles having different radii, the bands overlap so that the overall scattering pattern degenerates into a ''four-leaf clover'' pattern. With ringed spherulites or layered spherocrystals, one expects to see a band of scattered light at somewhat higher scattering angle than that from a homogeneous anisotropic spherocrystal. The scattering angle θ is related to the wavelength of the light λ and the distance d between successive shells by $\sin \theta = \lambda d^{-1}$. With V-type amylose films cast from DMSO solution and extracted with methanol, there is a microscopically observable ringed spherulite organization (*131*). Calculation using the above formula gives a ring spacing of 2.2 μm, agreeing with the value 2 μm obtained by direct microscopic measurement. However, attempts to apply this technique to starch granules with visible ring structure have not been particularly fruitful (*132*). Possibly, this is due to the use of dry starch granules which do not show growth rings, and, therefore, do not have significant, periodic fluctuations in refractive index or anisotropy. Ringed spherulitic texture observed in many polymers (*137*) may derive from a spiral pattern of crystallites in which the optic axes of the crystallites are at an angle to the radius of the spherocrystal in a radial, spiral arrangement. The spiral repeat distance is the distance between rings. If one examines the polarization patterns of such ringed spherulites, the polarization cross is seen to have a zigzag appearance, quite unlike that in native starch.

Large-granule starches such as potato and canna show pronounced growth rings when they are fully hydrated. However, as the starch is dried, the growth rings disappear (*125*). With small granule starches, growth rings are seldom if ever seen by optical microscopy. When examined by SEM, starch granules show growth rings only after chemical or particularly enzymic erosion (see Section III.2). Thin sections of native starch examined by TEM show growth rings only weakly or not at all; but after acid or other chemical treatment, the growth rings may be very pronounced (*82*).

With developing wheat starch granules, there is one growth ring per day (*78*). If wheat or barley are grown under continuous illumination in a constant temperature chamber, growth rings may be absent. With potato starch, growth rings are produced even when the potatoes are grown in a constant environment, possibly as a consequence of a rhythmic physiological fluctuation in the potato plant. The number of growth rings visible in the optical microscope is only a fraction of the number found in chemically treated thin sections (*48, 138*).

It is the author's opinion that growth rings represent periodic growth, and with the cereal starches, daily fluctuations in carbohydrate available for starch deposition. The relatively amorphous region between the crystalline, dense rings is more susceptible to acidic and enzymic erosion. It also seems likely that some or all the starch molecules terminate at the boundary of the ring, so that the length of a starch molecule is equal numerically to the thickness of a growth ring (*24, 82*). If starch molecules are glycoproteins, covalently linked at their reducing end to protein, (*33*) the protein should be in the amorphous, interring zone.

The dense, crystalline zones are relatively incapable of radial swelling owing to constraints by the molecular chains which run perpendicular to the growth rings. However, the interring amorphous region is readily penetrated by water, and swelling of this region is accompanied by a local decrease in refractive index, so that the growth rings may become visible. Inspection of polarization crosses of granules showing pronounced growth rings reveals that the interring zones are less birefringent than the crystalline rings. This difference disappears when the granules are dried.

Because the exterior growth rings are always observed to be parallel to the outer surface of the granules, the growth rings provide a visible record of the history of the granule morphology during its development. Finally, it is always observed that the arms of the polarization cross are perpendicular to the growth rings. This shows that the optic axes of the starch crystallites, and hence the molecular axes of the starch molecules, are aligned perpendicular to the growth rings and the granule surface. This permits the construction of optical maps which are of potential value in interpreting various facets of starch granule behavior (see Fig. 5). A possible arrangement of the amylopectin molecules within a growth ring is shown in Figure 15 (*139*).

6. Artificial Starch Granules

There exist in the literature numerous reports of the formation of artificial starch granules. Such "granules" range from the granular, sandy deposits which may form during the slow precipitation of starch or amylopectin from aqueous alcohol (*140*) to the highly oriented spherocrystals produced by synthesis using glucose 1-phosphate and the enzyme system of potato sprout juice or purified potato phosphorylase (*28*). When potato or waxy maize starch granules are used as "primers," potato phosphorylase acting on glucose 1-phosphate produces layers of oriented starch deposited on the periphery of the primer granule (*28;* See Chap. III and IV). These layers apparently consist largely of crystallized or retrograded amylose, as they are resistant to pancreatic amylase action, gelatinization, and staining with iodine.

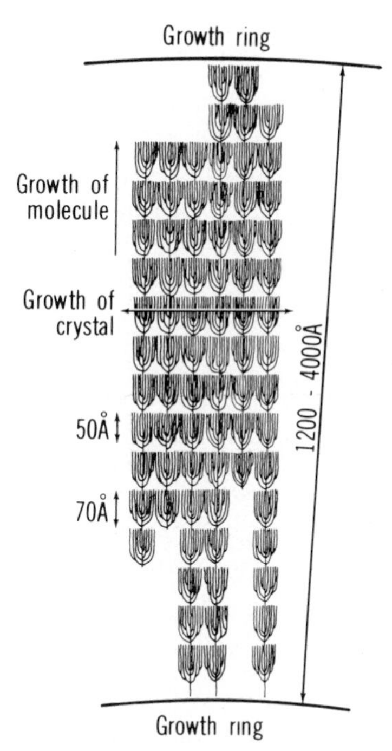

FIG. 15.—Schematic representation of the arrangement of amylopectin molecules within a growth ring. The individual molecules may be entwined. (Adapted from reference *139*).

Coacervate droplets of starch form spontaneously when a dilute starch solution is mixed with an incompatible polymer, for example, gelatin (*12*). The theory of polymer solutions indicates that all polymers of sufficiently high molecular weight are incompatible (*11*), even including very closely related polymers such as amylose and amylopectin. Badenhuizen (*8*) has shown that the immature amyloplast often acculumates a separate phase of polysaccharide as a coacervate drop, or "amylogenic pocket," prior to formation of the starch granule. At a critical time, this phase suddenly condenses or solidifies and becomes the nucleus or hilum of the growing starch granule. However, with artificial starch coacervates formed in gelatin, there is little or no granule type of organization. The spherical particles formed from starch or amylopectin do not show a polarization cross, birefringence, or growth rings. Particles formed at 4°, 20°, or 32° give B-type X-ray diffraction patterns; at higher temperatures, particles giving A-type patterns are formed. Amylose gives conglomerate particles rather than spheres; these give weak, diffuse B-type X-ray patterns. Many conditions give rise to B-type patterns. (*30–32, 98, 99, 102, 104, 131, 141–152*).

When starch sols are treated with various hydrophobic complexing agents, they form crystalline complexes which exhibit a wide variety of morphologies (*153, 154*). Some of these complexes have been called "artificial starch granules,'' but it is only the amylose component of starch which precipitates and crystallizes (*155*). The crystalline organization is also totally different from that of granular starch in that amylose forms helical inclusion complexes which show a V-type X-ray diffraction pattern (*156, 157*) rather than the A–B-type of pattern typical of starch granules. Moreover, electron microscopy and electron diffraction reveals that they are based upon a chain-folded platelet structure typical of the crystallization of many linear high polymers (*108, 158*).

When Naegeli amylodextrins or short-chain amyloses are crystallized from aqueous alcohols (*159*), or when amylose crystallizes as an alcohol, lipid, or non-polar complex at high temperatures (*160*), the products may be spherocrystalline and slightly resemble starch granules.

IV. Amorphous or Gel Phase of Starch Granules

There is no sharp demarcation between the crystalline and amorphous phases of starch granules, and it is generally believed that some or all the starch molecular chains run continuously from one phase to another (*76*). This must be true of the amylopectin component of starch, inasmuch as the individual molecules are far larger than the crystallites. As yet, there is no definitive evidence that amylose participates in starch granule crystallization. Hence, the gel phase may be that region of starch most readily penetrable by water and low-molecular-weight, water-soluble solutes. With water uptake, the gel phase undergoes limited reversible swelling, with consequent swelling of the entire granule. Inasmuch as starch is synthesized in nature in a water medium, the water-saturated granule is most nearly in its native state.

Loss of water from the gel phase leads to formation of interchain and intrachain hydrogen bonds which may generate strain in the polysaccharide structure. During severe drying, the stress induced by hydrogen bond formation distorts the granule structure sufficiently that visible cracks develop (*161*). The X-ray diffraction pattern becomes weaker and more diffuse, presumably because of a decrease in crystallite size and degree of order (*118*). Owing to the essentially hydrophilic nature of the gel phase, most hydrophobic substances have little or no ability to penetrate into or swell the gel phase of dry starch (*52*). Conversely, low-molecular-weight, water-soluble substances readily penetrate the granule if the gel phase is adequately swollen by water or other hydrogen bonding material (*162, 163*). At room temperature, penetration into the gel phase is limited to solutes less than about 1000 daltons. Presumably most chemical reactions with granular starch occur in the gel phase or on the periphery of the crystallites (*164–166*). The total area of the outside surface of a starch granule is practically negligible in comparison with the interior surfaces of crystallites and the ''surfaces'' of molecules in the gel phase. For example, for starch granules 20 μm in diameter, the exterior surface area is $0.2\ m^2g^{-1}$, whereas if one-half the volume of the granule consists of spherical micels 100 Å in diameter, their total surface area would be 200 $m^2\ g^{-1}$, or larger by a factor of 10^3. The surface area for isolated, extended starch molecular chains is about 3000 $m^2\ g^{-1}$ (Section V). If reagents, enzymes, or catalysts can penetrate into the gel phase, there is an enormously greater area available for reaction. Also, the starch chains within the amorphous region are presumably available for reaction, and their total mass is probably more than 50% of the starch. With extensive chemical derivatization of starch in which the granule crystal structure is maintained essentially intact, one may expect to find the bulk of the substituents in those parts of the starch in the amorphous phase. Much of the chemical derivatization practiced commercially is intended to alter the colloidal and crystallization tendencies of starch pastes. To the extent that such derivatization is successful, it indicates that the colloidal properties of gelatinized starch are markedly dependent on the original amorphous phase components. Because amylose is probably the starch component which dominates the initial, rapid stages of association and retrogradation of starch pastes, it would appear that the amylose, or at least significant components of it, are located primarily in the amorphous phase rather than being incorporated into crystallites. This interpretation is also in harmony with the observation that native non-waxy starches stain blue with iodine. However, after extensive acid treatment (extensive "Lintnerization" or Naegeli amylodextrin formation), the granules lose their ability to stain blue with iodine, though they retain crystallite organization (*32*).

In general, low-molecular-weight, water-soluble solutes readily penetrate fully hydrated native starch granules. The distribution coefficient of the solute between the exterior water phase and the interior granule phase is dependent on the molecular size of the solute (Fig. 16). The general form of this relationship indicates that the gel phase of starch granules has a distribution of pore sizes or capillary texture that permits solutes up to about 1000 daltons to be included; that is, the gel phase has an exclusion limit of 1000 daltons. Brown and French (*163*) reported the total water content of fully hydrated potato starch to be in the range 0.48–0.53 g of water per gram of dry starch. Of this internal water, about 70% is available as a solvent for D-glucose or D-mannose (K_d~0.68–0.73), whereas for the tetrasaccharide stachyose, only 8–9% of the water is available (K_d ~0.08–0.09). The K_d values fall on a straight line when the K_d is plotted against the log of the molecular weight, as is typical for gel permeation. These data are compatible with a model in which the pores in the gel phase consist of spaces between parallel plates. It is calculated that the channels in potato starch are approximately 15 Å wide. This dimension is enough to accommodate many low-molecular-weight solutes but not proteins or other biopolymers. Irrespective of the details of the porous channels, the bulk of water in the gel phase can readily accommodate D-glucose (molecular radius = 3.5), but molecules bigger than the cyclodextrins (molecular radius = 7.5) are effectively excluded. BeMiller and

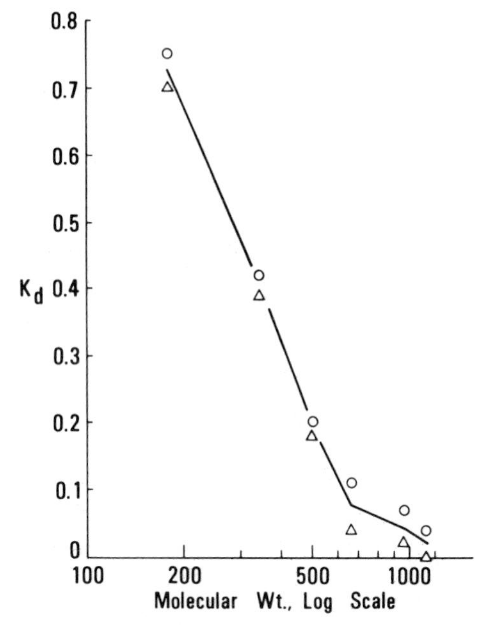

FIG. 16.—Distribution coefficients of sugars (monosaccharides, lactose, raffinose, stachyose, and cyclodextrins) between water and potato starch granules (*163*). The K_d value is an index of the amount of water within a starch granule that is sterically available for dissolving added sugars or other solutes.

Pratt (*293*) reported 33% water in the final starch phase of fully hydrated potato starch (original moisture content of 15.52%) which agrees with the results of Brown and French (*163*).

In examining the behavior of sugars on starch columns, Lathe and Ruthven (*162*) noticed that starch oligosaccharides show "trailing," indicating specific adsorbtion by starch. Brown and French (*163*) measured the adsorption of radioactively labeled starch oligosaccharides and found that maltotriose and higher oligosaccharides, particularly maltotetraose, are specifically adsorbed into the gel phase. The results are interpreted as meaning that the oligosaccharides are able to twist around starch chains in the gel phase to form short double helices. The lesser complexing ability of maltopentaose and higher oligosaccharides might be due to the difficulty of these larger sugars to penetrate the gel structure, or it may be due to the unavailability of sufficiently long unbranched starch segments or chains in the gel phase.

V. ROLE OF WATER IN STARCH GRANULES

Starch is biosynthesized in a water medium; and in its natural form, the starch granule contains much water. Drying starch, even under the gentlest conditions, causes the granule to shrink and crack (*125, 161, 164*), and there is a partial realignment and reorganization of the polysaccharides, so that, on rehydration, the granule does not resume its fully native form. Most of the water bound to starch is in the gel phase or at the surface of crystallites. As seen in Section IV, about 70% of the bound water is accessible as a solvent for D-glucose. The remainder is sterically inaccessible as a solvent, whether it is in the gel or crystalline phase.

The amount and distribution of water within starch granules is critically important with regard to the physical properties and chemical reactions of starch, as well as an understanding of the role of water in industrial processing of starch and starch-containing materials (*243*).

The starch granule is heterogeneous chemically, in that it contains both amylose and amylopectin, and physically, because it has both crystallite and amorphous phases. Consequently, reaction with water cannot be expected to be homogeneous within the granule. Probably the binding of water depends more on the density and regularity of packing of the polysaccharide chains than whether the individual chains are linear or branched. When there is a critical water level in the starch granule, such that the average degree of hydration may be, for example, one molecule of water per D-glucopyranosyl unit, it must be regarded that this is only an average, for not all the D-glucopyranosyl units will bind the same number of water molecules.

The literature abounds with data in which the amount of water bound by starch is expressed as "percent," without any indication as to whether this is percent of the total weight or the add-on weight to dry starch. When moisture values are given, there may be no indication as to how the moisture has been determined. A further problem is that often the moisture content may be only indicated by the statement that the starch is "air-dry," or that it is in equilibrium with an atmosphere having a certain relative humidity.

1. Moisture Sorption Isotherms

Water relationships in starch granules have been extensively studied by measuring the water uptake or loss when starch is exposed to various relative humidities and temperatures (Fig. 17). The moisture isotherms can be analyzed thermodynamically by methods of surface chemistry. However, it is difficult to obtain reproducible equilibrium isotherms necessary for rigorous thermodynamic analysis owing to the slowness with which equilibrium is obtained. Starch–water vapor systems show hysteresis that may not equilibrate for a long period.

The rate of equilibration of starch granules is critically dependent on the thickness of the sample and the residual air pressure (*167*). Water and gas molecules may be effectively trapped within a bulk sample of powdered material, even though the exterior pressure may be very low. To facilitate equilibration, the starch sample should be as thin as possible, and the residual air pressure

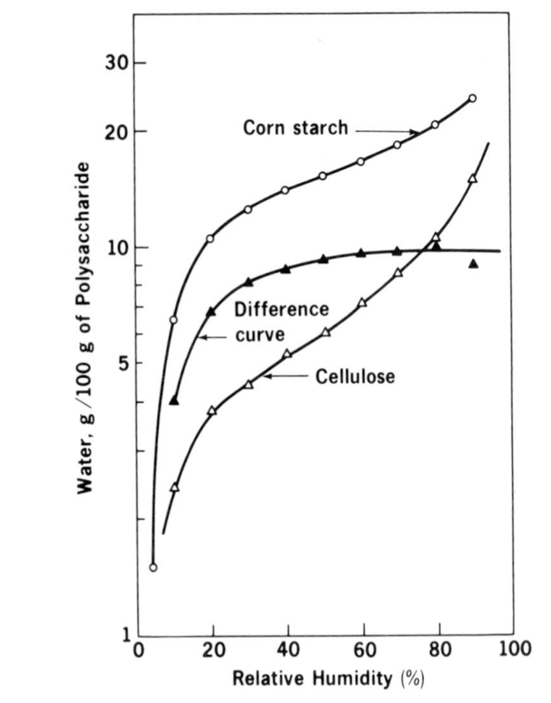

FIG. 17.—Water sorption isotherms of corn starch in comparison with cellulose. The difference curve at relative humidity over 50% corresponds to about one water molecule per anhydroglucose unit.

should be reduced as much as possible, below 5 μm (7×10^{-6} bar). These considerations are important in obtaining the dry weight of starch, but have seldom been applied adequately.

Rates of bulk hydration and dehydration are kinetically biphasic (*4*). For example, dehydration of potato starch initially containing 15.35% water at ~25° over sulfuric acid can be expressed as % water = $11.65\ e^{-0.02t} + 3.7e^{-0.03t}$, where t is in hours. For another sample containing excess water, after removal of the first 14% of water, % water = $15.9e^{-0.03t} + 5\ e^{-0.004t}$. For rehydration over 2% aqueous sulfuric acid of air-dry potato starch containing 15.5% water, after a 24 h initial period, % water = $33 - 8.7\ e^{-0.009t} - 2.3\ e^{-0.001t}$ to give a final water uptake of 33%. These rate expressions indicate that there are two somewhat differing types of water sites in starch, and that there are possibly slow conformational changes within the granule to permit accommodation or release of the slowly exchanging water.

The moisture sorption isotherm can be analyzed according to the Brunauer–Emmett–Teller (BET) equation

$$\frac{p}{v(p_o - p)} = \frac{1}{v_m C} + \frac{C - 1}{v_m C} \cdot \frac{p}{p_o}$$

where v is the amount (volume or other units) of vapor sorbed at the pressure p; p_o is the vapor pressure of water at the temperature of the experiment; v_m is the amount of vapor adsorbed when the surface of the adsorbent is covered with a single layer of molecules; C is a constant, approximately equal to $e(E_A - E_L)/RT$, where E_A is the average heat evolved in the adsorption of the first layer of gas molecules, and E_L is the heat of condensation of the gas to a liquid. By the BET type of analysis, the moisture sorption isotherm can be divided into three regions (*168–172*). In the first, region A, from 0 to about 7% water is sorbed at a water activity between 0 and 0.2 and, according to the BET concept of multilayer adsorption, represents a water monolayer on the surface of the starch molecule. Region B, from 7 to 17% water, requires a water activity from 0.2 to 0.7 and represents further layering of water, but not in direct contact with the carbohydrate. In region C, above 17%, the water is essentially free water absorbed into capillary spaces. The BET plots are linear up to $p/p_o = 0.3$. For sorption of water on starch at 25°, $E_A - E_L$ is 1.74 kcal mole^{-1} and v_m= 75 mg of water per gram of starch or 0.66 molecules of water per anhydroglucose unit.

Incidentally, the "surface" of individual molecular strands of starch (completely dissociated) can be calculated to be 3120 m^2 g^{-1} on the assumption that the molecular strands are cylinders with a specific volume of 0.6 cm^{3}g^{-1} and a length of 3.5 Å per D-glucopyranosyl unit. A monolayer of water (9.7 Å^2 per water molecule) would give a degree of hydration of about nine molecules of water per D-glucopyranosyl unit (100% water, dry starch basis), as compared with 0.66 molecules of water (7.3%) by BET analysis of the moisture sorption isotherms.

Starch sorbs more water per gram than does cellulose. The difference curve between starch and cellulose indicates that about 0.8–1 mole of water per D-glucopyranosyl unit is sorbed by starch in the relative humidity range of up to 40%, thus it is mostly in the monolayer region. This difference remains more or less constant up to a relative humidity of 90%. The so-called monolayer water could be water of crystallization. With native cellulose (cellulose I), it is known that there is no true water of crystallization, but rather that water is avidly sorbed into the intercrystallite regions or onto the crystallite surfaces (*173*).

Moisture sorption isotherm data at two or more temperatures can be analyzed by Clausius-Clapeyron theory to give the enthalpy of water vapor sorption at any moisture level. Bushuk and Winkler (*168*) have applied this approach to the enthalpy of water sorption by flour, starch, and gluten. Sterling and co-workers (*170, 171*) have also used this approach. However, it is difficult to reconcile their enthalpy results with those of Bushuk and Winkler, or with values obtained from the heat of immersion (see Section V.4).

Table II

Swelling of Starches upon Sorption of Water Vapor (181)

Relative humidity (%)	Increase over vacuum-dry diameter (%)							
	Corn		*Potato*		*Tapioca*		*Waxy corn*	
	Absorption	*Desorption*	*Absorption*	*Desorption*	*Absorption*	*Desorption*	*Absorption*	*Desorption*
8		1.5	1.9	2.2	3.6	1.1	1.5	1.8
20	1.9	1.9	4.0	4.1	4.6	2.4	5.2	4.5
31	2.6	2.5	5.4	5.8	5.5	3.9	7.0	6.4
43	3.3	3.3	8.0	8.0	7.6	5.2	9.1	8.9
58	4.1	4.1	9.4	9.4	9.6	7.3	10.7	10.6
75	5.4	5.5	10.3	10.4	12.5	11.4	13.5	13.3
85	—	6.6	—	11.0	16.8	16.1	15.9	15.3
93	7.3	—	11.6	11.7	22.8	21.9	18.5	18.8
100	9.1	—	12.7	—	28.4	—	22.7	—

Table III

Surface Area of Starches (174)

Starch	N_2 sorption[a,b]	Photomicrographic measurement[a]	Water sorption[b,c,d]	
			g $H_2O \cdot g^{-1}$ Dry starch	Surface area[a]
Dasheen	2.62	2.64	0.0801, 0.0941	281, 330
Corn	0.70	0.48	0.0800, 0.0951	280, 334
Tapioca	0.28	0.25	0.0818, 0.0990	287, 348
Potato	0.11	0.15	0.0843, 0.1202	296, 422

[a] Areas expressed as $m^2 \cdot g^{-1}$ of dry starch.
[b] Cross-sectional area of an adsorbed N_2 molecule was taken to be 16.2 Å^2.
[c] Cross-sectional area of adsorbed water molecule was taken to be 10.5 Å^2.
[d] For the water sorption data, the first number is from absorption, the second from desorption.

Hellman and Melvin (*174*) estimated the surface areas of various granular starches at various moisture levels by measuring granule dimensions on photomicrographs (Table II). After drying the starches and degassing them by heating them to 105° for 24 h at 10^{-6} mm, they determined the surface areas by nitrogen adsorption at −195° (Table III). Surface areas measured by nitrogen sorption are essentially the same as obtained by the photomicrographic method. Nitrogen is unable to penetrate significantly into the interior of the granule, or if it does, there are few if any interior surfaces or capillaries available for adsorption. The surface area of starch chains, if they were exposed, would be huge, (3120 m^2 g^{-1}), higher by three or four orders of magnitude than the observed values. Although the data might be accounted for if the granule has an impervious surface film or membrane, it seems more likely that, during the drying preliminary to the gas adsorption analysis, any holes or pores leading to the granule interior are collapsed by hydrogen bonding of adjacent molecular chains during water removal. The general agreement between the nitrogen adsorption and photographic methods also indicates that the exterior of a dried starch granule is relatively smooth on a molecular scale.

2. Density, Specific Volume, and Refractive Index of Starch Granules as Related to Moisture Content

Starch has a variable density depending on its botanical origin, prior treatment, and method of measurement. When air-dried granular maize starch is immersed in water, it increases the volume by 0.611 cm^3 per gram of anhydrous starch (*175*). The measurement is made by adding a weighed amount of starch of known moisture content to a calibrated volumetric flask or pycnometer, making up to volume with water, and obtaining the combined weight of starch plus water. This specific volume corresponds to an apparent density of 1.637. For potato starch, the apparent density is 1.617 (*176*). These density values are higher than the buoyant density (1.50±0.01) of maize or potato starch granules measured by flotation in a mixture of organic solvents (such as carbon tetrachloride plus bromobenzene) or by displacement of a non-swelling organic liquid such as toluene or xylene (*177*). The partial or apparent specific volume (*178–180*) of starch or glycogen dissolved in water is 0.61–0.63 $cm^3\ g^{-1}$. The immersion volumes of starch molecules, amorphous starch, and starch crystallites are experimentally indistinguishable. Air-dry potato (*177*) starch containing 16% water has a buoyant density of only about 1.45. This moisture level corresponds to nearly two molecules of water per D-glucopyranosyl unit. Similarly, air-dry maize starch containing 10% water has a buoyant density of about 1.50. When starch is dried, its buoyant density remains about 1.50, lower than that calculated from the volume of immersion in water (1.63). Presumably during thorough drying, the amorphous part of starch granules shrinks, but as it can not crystallize, the amorphous parts of starch molecules cannot pack together as closely as the more regular, crystalline parts.

When dry starch granules are immersed in water, they undergo a considerable but "limited" swelling, owing presumably to the swelling of the amorphous or gel phase. Potato starch granules can absorb water up to about 50% or more of the weight of the dry starch and expand from 30% to over 100% in volume, due to reversible swelling of the amorphous phase (*181*).

The total volume occupied by fully hydrated starch granules can be measured by application of "negative adsorption" (*163, 164, 182*).

Brown and French found that potato starch absorbed 0.48–0.53 g of water per gram of starch (dry basis) (*163*). BeMiller and Pratt (*243*) found this value to be 0.49 (33% water in the fully hydrated starch); fully hydrated underivatized corn starch had contained about 28% water. Dengate and co-workers (*183*) found the apparent specific volume (immersion volume) of wheat starch to be 0.627±.008 $cm^3\ g^{-1}$, the fully hydrated density (starch plus water) to be about 1.3, the ratio of the hydrated volume to the dry volume to be 1.7, and the water uptake to be roughly 50% (dry basis).

Pycnometric measurements of starch granule density are capable of good accuracy. Either water or an organic liquid may be used. With water, the measurements actually give the immersion volume or the apparent specific volume, that is, the net increase in volume when one gram of starch is added to excess water. The reciprocal of the apparent specific volume is sometimes called the "dry starch density." The differences between starches may reflect differences in internal organization of the granule, such as the relative amounts of crystalline and amorphous components or minor impurities such as ash, lipids, and proteins.

Pycnometric measurements with organic liquids are basically different in that the organic liquid does not swell or penetrate the starch granule significantly. The densities obtained are lower than those obtained in water and depend more on the water content of the starch, the source of the starch, its method of preparation, and the extent to which damaged granules are present.

By using an air comparison pycnometer, it is possible to obtain highly reproducible values for the density of dry or low-moisture starch. The density values average about 1% lower (range 0–3%) than those obtained by xylene displacement (*182*).

In principle, it is possible to analyze systematic data relating moisture content and density to obtain the partial specific volumes of each component, that is, the net increase in volume $\bar{v}$ when one gram of dry starch or, conversely, when one gram of water is added to an infinite amount of moist starch of a certain water content (*184*). The author has been able to find only one example of such data in the literature.

Analysis (*184*) of a curve showing the volume–hydration relationship for potato starch gives $\bar{v}$ starch = 0.677 cm^3 g^{-1} at zero hydration, gradually decreasing to 0.623 for a degree of hydration of 0.2 or beyond, up to saturation where the degree of hydration is 0.43. The corresponding partial or apparent specific volumes for water are $\bar{v}$ water= 0.45 at zero moisture, increasing to 1 at a degree of hydration of 0.2 or higher. Inasmuch as the published curve does not show data points, it is difficult to use it for an exact analysis, and it seems likely that the calculated $\bar{v}$ water value at zero hydration may be too low. Gur-Arieh and co-workers (*186*) have measured the pycnometric density under benzene of a low protein (7%) fraction of wheat flour as a function of the water content. They concluded that the effective density of the sorbed water in the range 0–7% was high, namely 1.48. When their data are graphically examined (*184*), they corroborate the high water density.

These results suggest that, in dry starch, the chains are packed rather inefficiently in comparison with crystalline cellulose (d = 1.6) and that there are very small "voids" into which water can penetrate with a relatively small volume increase, namely, only 0.675 cm^3 per gram of water absorbed (*187*). After uptake of about 10% water, all voids are filled. Further water uptake results in granule volume increases approaching 1 cm^3 per gram of water absorbed.

Speich (*52*) measured the packing volume of dry potato starch after 24-h treatment with various fluids, in comparison with air. For ethanol–water mixtures, dry starch preferentially absorbs water, even from 95% ethanol. Hydrophobic liquids such as benzene and toluene cannot interact with starch through hydrogen bond formation, but may participate in weak hydrophobic interactions. BeMiller and Pratt (*243*) found that dimethoxyethane, a hydrophobic molecule, did not penetrate potato starch granules to a significant extent. Methanol, ethanol, 1-propanol, and 1-butanol, which were preferentially sorbed in comparison to water, appeared to penetrate granules with water and to mix freely with the moisture of the granule.

The author has been unable to find experimental data that adequately gives the true density of the crystalline portion of starch granules.

Where starch is used as an opacifier or "whitener," its performance depends on refractive index differences between the starch and the medium in which it is dispersed. Data analysis of the refractive index of aqueous sugar solutions (*188*) show that there is a nearly linear relationship between refractive index and sugar concentration. To the extent that the starch granule may be considered a solid starch–water solution, the refractive index could describe the relative contributions of starch and water to the specific volumes of hydrated granules.

Wolf and co-workers (*126*) measured refractive indices of wheat starch granules at 0–21% moisture levels, as obtained by interference microscopy. The results suggest that water is tightly bound within the starch granule up to about 10% moisture so that its expected contribution to the refractive index is almost exactly canceled by the increase in granule volume. With higher moisture levels, up to 21%, the refractive index contributions of the water and starch become additive.

Speich (*52*) measured the birefringence of dried starch after equilibration with a number of liquids. For water and most alcohols, at an average wavelength of 550 nm, he found a value of $n_e - n_o \sim 0.0133$ with a range from 0.0131 to 0.0139. For aldehydes, the values were 0.0135–0.0143, while for hydrophobic liquids the values were 0.0134–0.0135. Ethylene glycol and glycerol are exceptions since they react with amylose in native starch granules to form helical complexes (*189*), which in general are negatively birefringent (*190*). Presumably, the decrease in birefringence observed with ethylene glycol and glycerol results from the combination of the positive birefringence, contributed by the amylopectin crystallites, and the negative birefringence of the amylose–ethylene glycol or –glycerol complex.

Speich also tabulated the refractive indices of dry potato starch for the Fraunhofer wavelengths (C = 656 nm, D = 589 nm, and F = 486 nm) and at temperatures from 5° to 55°. For 25° and λ = 589 nm, $n_o = 1.523$ and $n_e = 1.535$.

3. Role of Water in the Gel Phase

The nature of bound water in starch granules has been studied by freezing starch containing excess water, then by differential scanning calorimetry, measuring the endotherms associated with the melting process. By this means, wet wheat starch is found to contain 0.38 g of bound, nonfreezable water per gram of dry starch (*172*). Commercial pregelatinized wheat starch has 0.41 g of bound water. Failure of the bound water to freeze suggests that the water is in more or less fixed positions, rather than as mobile quasi-liquid.

Sterling and co-workers (*170, 171*) used moisture sorption isotherms of maize starch at different temperatures to obtain ΔF, ΔH, and ΔS of sorption. They find that there is a very high binding enthalpy and entropy, higher than that of condensation of water vapor, especially at water levels below 7%. Results are concordant with the integral heat of wetting of rigorously dried starch, namely, 26 cal g^{-1} of dry starch (*191*). Most of this heat is liberated within the first few percent of water uptake; after starch contains about 15% sorbed water, the $-\Delta H$ of water sorption from water vapor drops to 11 kcal $mole^{-1}$ of water, in comparison to 10.5 for simple condensation of water vapor to liquid water (see also Section V.1).

It is clear that dry starch has a great avidity for water. Removal of the last water sets up a high degree of strain, particularly in the gel phase, and this is communicated to the crystal phase as evidenced by loss of intensity and sharpness of its X-ray diffraction pattern (*100, 192*).

Organization in the dried starch granule is such as to give imperfect packing of the starch molecules. There are no significant voids of the size of small molecules such as chloroform or other nonpolar organic liquids, which are incapable of penetrating dry starch. Thus the dried starch granule is a xerogel (dried gel without voids).

It seems likely that the changes in starch granules accompanying drying are due to loss of water from the amorphous or gel phase. However, the crystal structure may have a significant amount of water of crystallization, particularly in the case of B-starch where Wu and Sarko propose a channel of water as large as the double helix (*104*). If such a channel exists and the water is removed from it, non-polar molecules might be unable to penetrate the structure. Such a situation would result in a greatly reduced density for the starch crystallites. If the Wu and Sarko packing arrangement is retained after loss of water, the density calculated for the carbohydrate itself is 1.048, a value incompatible with the density of any known solid form of starch.

4. Heat of Hydration of Starch

When dry starch is immersed in water, heat is evolved (*191, 193*). Winkler and Geddes (*193*) measured the heat of immersion of rice, wheat, and potato starch at levels of hydration of the starch up to 15–18% water (Fig. 18, Table IV). The data show that, for the various starches, as well as for wheat flour, the heat evolved per gram of water adsorbed decreased from about 215–260 cal, with very dry starch or flour, to zero, at about 16–21% moisture for flour and the starches. These heats of adsorption are similiar to those from analysis of the moisture sorption isotherms (see Section V.1). Graphical analysis suggests that when the water content of starch is 20% or more, heat is absorbed in further hydration.

Measurement of the heat of immersion of starches dried to various moisture levels to see whether the method of drying changed the heat of immersion showed no effect, except for starches heated to 140° where decomposition is

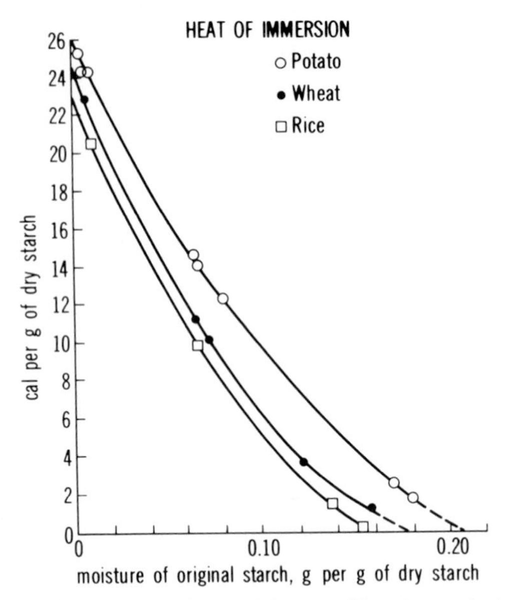

FIG. 18.—Heat of immersion of granular starch in water. Very dry starch gives off 23–26 cal per gram of starch when immersed. The *slopes* of these curves, at any moisture level, give the heat of adsorption per gram of water (approximately 215 for potato, 260 for wheat, and 230 for rice). The point of intersection of the curve with the moisture content axis is called the "zero heat of immersion point." Note that the curves, if extrapolated beyond this point, would give negative heats of immersion at these higher moisture levels (adapted from reference *193*).

significant (*191*). The thermal results for totally dry starch were similar to those of Winkler and Geddes, but were different in regard to the moisture levels for $\Delta H = 0$ (the athermal moisture level) (Table IV).

By an empirical semi-logarithmic plot of ΔH vs percent water a biphasic water uptake is seen with high and low $d(\Delta H)/d(H_2O)$ segments. The break points and athermal points are given in Table IV. By use of an infrared lamp, Schierbaum (*194*) found that, starting with fully wet starch, there are significant breaks in the

Table IV

Heat of Moisture Sorption of Various Starches (191)

Starch	*−ΔH of immersion (cal·g^{-1} dry starch)*	*Water content at ΔH = 0 (%)*	*% Water at discontinuity in semi-log plot*	*Initial slope (cal·g^{-1} of water)*
Potato	27.9	33.8	18.3	~205
Wheat	25.2	30.5	13.2	~215
Maize	24.7	28.0	13.0	—
Rice	24.3	26.5	13.7	—

logarithmic rate of water loss plots. Rate of water loss can also be correlated with the Freundlich-type of analysis of the moisture sorption isotherms of the same starches, suggesting that, with the cereal starches, above about 30% moisture, the water is free. Water loss in the range 12–30% represents loss of "capillary" moisture. From 0% to 12%, the water is strongly bound "adsorption" water. The break points for potato starch are about 5% higher.

These results indicate that, on the average, there is one molecule of water avidly adsorbed per D-glucopyranosyl unit. If all the water-binding sites are of the same affinity as those involved in initial binding, the starch would be saturated after binding about 10% water, that is, about one molecule of water per sugar unit. However, owing to the heterogeneity of the starch, it seems more likely that, in the initial reaction, water first reacts with the small proportion of hydroxyl groups not already hydrogen bonded. As more water is sorbed, there is a tendency to react with weakly hydrogen-bonded hydroxyl groups and thereby partly relieve stresses set up during the original drying of the starch. Additional water enables all polysaccharide chains to seek minimum energy conformations. Finally "capillary" spaces will open to allow the starch granule to assume its native, fully hydrated form.

5. Structure of Hydrated Starch Granules as Indicated by nmr

Lechert (*195*) examined starch at different water contents by pulsed nmr measurement. He reported that, for potato starch, there are two types of bound water with significantly different transverse relaxation times (t_2). From zero to about 0.2 g of water per gram of starch, t_2 increases from zero to about 0.4 ms. This implies that the initially bound water is of low mobility, as one might expect if it is water of crystallization or if the water is "chelated" within the amorphous network. As the amount of water increases, some of it is less tightly bound. Also, there is the possibility of exchange of water between the various types of binding sites. At a water content of about 0.2 g per gram of starch, a second type of water appears with higher t_2. For this water, t_2 increases rapidly with increasing water content.

To look further at the state of water in potato starch, Hennig and Lechert (*196*) exchanged OH and H_2O molecules, converting them into OD groups and D_2O molecules. They found that, at 24% D_2O, the wide-line deuteron spectra shows anisotropy splitting which is more or less constant up to 50°. From 50° to 80°, there is a decrease in anisotropy splitting which might be due either to the "swelling" or melting of crystallites or to a decrease in the number of ordered D_2O molecules. Cooling the sample to 25° permits the splitting to return to its original value. For potato starch containing 39% D_2O, at 0° the excess or "free" D_2O is frozen (the mp for D_2O is 3.8°), and the bound deuterium shows a splitting of 0.88 kc. When the temperature is raised to 5°, a sharp D_2O peak appears together with the 0.88 kc split line. With further heating, the splitting diminishes and disappears between 60° and 80°, only the sharp D_2O peak remaining. On cooling to 25°, the sample does not regain the split peak, indicating an irreversible change in starch granule structure when heated to 80° at 39% D_2O. Wheat starch at 20% D_2O shows deuteron splitting, whereas maize starch and crystallized B-type retrograded amylose do not.

Changes in carbon-bound proton mobility during potato starch gelatinization have been examined using pulsed nmr (*197*). Deuterated starch (3–6% in D_2O) shows little change in the slow decay signal during slow heating up to 55°–60°. At 55°–60°, there is a rapid increase in the amplitude of the slow component which reaches a maximum at about 70°, where it is 100% of the total proton signal. Thus, the amplitude of the slow signal is directly proportional to the degree of gelatinization of the starch.

6. Nature of Sorbed Water from Measurements of Dielectric Absorption

Guilbot and co-workers (*198*) measured the dielectric absorption of wheat starch over the range 25 to 2.4×10^{10} Hz at moisture levels from 0.5% to 51.5%. Below the critical frequency, the dipole can readily orient itself with the electric field. Above the critical frequency, the electric field oscillates so rapidly that the dipole is unable to follow it. At the critical frequency, orientation of the dipole by the electric field is exactly balanced by thermal randomization.

In pilot studies, the authors showed that, in the range 10^3–10^8 Hz, crystalline sugars have very little absorption, whereas glassy or amorphous D-glucose and maltose, as well as dry wheat starch, show absorption with maxima at about 10^6 and 10^8 Hz. Benzoylated sugars or starch have very little absorption, indicating that free hydroxyl groups are essential. Measurement of the peak maxima positions at different temperatures permits calculation of the activation energy by plotting the log of the critical frequency against the reciprocal of the absolute temperature. For the absorption at about 10^6 Hz, the activation energy is 10.1 kcal per mole of dipoles for both anhydrous wheat starch and amorphous maltose.

In the region of dipolar absorption, dry starch shows three overlapping weak maxima with critical frequencies at 2.6, 80, and 4000 MHz. As the water content increases to about 10%, these maxima simultaneously increase, especially the one at 80 MHz. With increasing water, this rapidly increases and tends toward a limit at about 15% water. At this water content, a new maximum at 2000 MHz appears and increases in intensity up to about 30% water. At higher hydration, a new maximum at 7000 MHz increases rapidly, and it is the only peak visible at water contents above 40%. This last peak is essentially a free water peak; its activation energy is about 4.7 kcal $mole^{-1}$.

7. Penetration of Starch Granules by Chemical Reagents

Starch granules are impenetrable by most chemical reagents unless they are hydrated. Small molecules such as acetic anhydride or chromic acid in non-aqueous media have no appreciable action on starch (*199*). When starch is hydrated, aqueous reagents readily penetrate the granule. The simplest example is the interchange of ^{1}H and ^{2}H. Starch contacted with liquid ^2H_2O gives complete exchange of all hydroxyl groups within 1 h (*99*). In fact, potato starch gives complete exchange within 10 min (*197*). When dry, deuterated starch treated with ^1H_2O at 19% relative humidity absorbs 8% of its weight of water. After drying this rehydrated starch, it is found that 92–106% exchange of the ^{2}H for ^{1}H has occurred. Other material including potato starch, high-amylose maize starch, highly crystalline A and B amylose hydrolyzates, crystalline palmitic acid–amylose complex, and various retrograded amyloses, give exchange values in the range of 96–103%. Dry, deuterated wheat starch treated with liquid methanol gives complete exchange in 2–3 h. With ethanol, exchange is incomplete, reaching a plateau of about 20% in 2 h. Samples wet with ethanol require 3 or 4 days of drying to reach constant weight. Samples wet with methanol do not reach constant weight in 5 days.

Acetic anhydride readily penetrates the hydrated starch granule, with surprisingly high efficiency of reaction, as judged by the fraction of the acetic anhydride that reacts with starch in preference to water (*200*). In another example, Hood and Mercier (*165*) examined the structure of manioc (tapioca) starch before and after phosphate cross-linking and etherification with propylene oxide in an aqueous medium to obtain a DS of 0.05–0.1. Substitutents in the modified starch can be located by their effect in blocking action of starch-degrading enzymes, specifically β-amylase, glucoamylase and pullulanase. With amylose, degradation with β-amylase is negligible. This result means that all or most of the amylose molecule has blocking substituents at or near the nonreducing end. In all likelihood, the blocking substituents are spread over the entire amylose molecule, more or less randomly. By contrast, the modified amylopectin component is extensively degraded by β-amylase, indicating that many of the amylopectin outer chains have not been chemically modified. Debranching of the β-amylase limit dextrin by pullulanase is very incomplete, and action of β-amylase on the debranched material gives only 29% maltose as compared with 100% from the unmodified control. Hood and Mercier (*165*) presented a schematic diagram showing that substitution may be very dense around the branching region and essentially negligible in the interior of the outer chains and that there is partial modification at the exposed nonreducing ends of the outer chains.

Periodate readily penetrates and reacts with starch granules, eventually leading to complete conversion to dialdehyde starch (*201*). The reaction kinetics are polyphasic, owing to the formation of internal hemiacetal groups from the aldehyde groups produced by glycol cleavage (*202*).

A 5-min reaction of native potato starch with 0.04 *M* periodate causes initial loss of birefringence in the hilum region (*83*). With reaction time, the loss of birefringence spreads outward from the hilum until after 45 min, only a thin birefringent shell remains at the granule periphery. After 60 min, the granule is no longer birefringent. With Lintnerized starch, loss of birefringence proceeds from the outside inward and after 15–45 min there remains only a small birefringent area around the hilum. Examination of the x-ray patterns of periodate oxidized starch shows a gradual decrease in the intensity up to oxidation of about 50% of the D-glucopyranosyl units, when the crystalline pattern disappears. These results show that the water-saturated starch granule is readily penetrated by periodate and also that the crystallite organization does not present a serious obstacle to formation of the glycol periodate ester.

The reaction of "normal" amylose-containing starch granules with iodine–iodide rapidly gives a blue or opaque stain, indicating rapid reagent penetration and reaction with amylose. Decolorization of such stained granules with thiosulfate is equally rapid. Penetration of iodine–iodide reagent into high-amylose maize starch granules is probably not a limiting factor in the poor granule staining of this starch. Similarly, potato starch Naegeli amylodextrin stains very weakly with dilute iodine, although after it is dissolved in hot water and cooled, it reacts readily (*32*). Prolonged treatment of the cereal starches with acid gives Naegeli amylodextrins with considerable iodine-staining ability (*86*). Survival of amylose under such protracted acid treatment may be due to its existence within the granule as an acid-resistant lipid complex. If the starch is defatted prior to acid treatment, the amylose is not protected, and the starch behaves more like potato starch.

Whereas hydrated starch granules are readily penetrated by aqueous iodine–iodide reagent, air-dry granules are only weakly stained by iodine vapor to give a light brown-red color (*86*). Such weakly stained granules immediately give a blue stain when they are brought in contact with liquid water. The reason for this is not known.

Hizukuri (*203*) showed that large amounts of iodine can be taken up by potato starch granules from sufficiently concentrated iodine–iodide solutions. Although only the amylose component of starch can react with dilute iodine solutions, amylopectin will also react if the concentration of iodine is brought into the range 10^{-4} *M;* amylose is completely reacted at an iodine concentration about 3×10^{-5} *M* (*154*). These minimum concentrations of iodine, though adequate for complete reaction of the dissolved starch components, are not high enough to break the associative forces within starch granules; this requires iodine concentrations higher by one or two orders of magnitude.

At the lower iodine concentrations used by Hizukuri, reaction is primarily with the amylose component to give total iodine uptake up to about 4 mg of iodine per gram of starch. Starch containing this amount of iodine gives a B-type x-ray diffraction pattern only slightly less intense that that of the natural starch, and after removal of the iodine by treating with thiosulfate, the granules appear to have undergone no change. With higher levels of iodine uptake, there is progressive loss of the crystalline B pattern so that with 8.9 mg of iodine per 100 mg of starch only an amorphous pattern remains. Still, removal of the iodine permits the starch to recover its B-type pattern, though it is somewhat less sharp than the original one. Most of the granules show polarization crosses, but appear damaged as judged by the Congo red test. At 12.2 mg of iodine per 100 mg of starch, the deiodinated granules regain only a poor x-ray pattern. Most of the granules have a polarization cross, but many are highly swollen and gelatinized. At 18.4 mg of iodine per 100 mg of starch, the deiodinated granules give only an amorphous x-ray pattern and appear to be totally gelatinized.

These experiments show that at low iodine uptake, up to 5 mg of iodine per 100 mg of starch, the iodine reacts primarily or exclusively with the amylose component. The amylose, though bound to the texture of the intercrystalline regions, is nevertheless able to undergo the conformation change necessary for formation of an iodine complex without seriously disrupting the starch crystallites. Uptake of higher amounts of iodine requires much higher iodine–iodide concentrations. The additional iodine reacts with amylopectin chains, and as it reacts, it irreversibly disrupts the crystallite organization. Essentially total disruption results when about 18 mg of iodine have reacted with 100 mg of starch. Hizukuri's photomicrographs show that, after deiodination, there has been a nonuniform degree of damage to individual granules or even within single granules. In the range 12–14 mg of iodine per mg of starch, the microscopic fields show some totally gelatinized, some partly swollen, and some apparently undamaged granules.

In a study of oxidation of cellulose and starch, Gladding and Purves (*199*) found that chromic acid dissolved in 0.2 *M* sulfuric acid readily penetrates and oxidizes granular starch, whereas chromic anhydride in acetic anyhydride plus acetic acid has no appreciable effect. With the aqueous reagent, a degree of oxidation corresponding to consumption of 0.3 atoms of oxygen per D-glucopyranosyl unit gives an oxystarch containing 0.08 carbonyl groups and 0.06 carboxyl groups per sugar residue.

Reaction of ethylene oxide with starch proceeds readily in aqueous alkaline slurry to give high efficiency of hydroxyethylation (*204*). Hydroxyethylation and hydroxypropylation are effective in reducing the retrograding tendency of starch; thus, it may be concluded that at least part of the reaction occurs on amylose. Positions 2 and 6 of the D-glucopyranosyl units are the main points of substitution (*205, 206*). With cross-linked starch, hydroxypropylation is mainly in the branching region and at the ends of amylopectin outer chains (*165*). To make a hydroxyethylated starch suitable as a blood volume expander, it is necessary to conduct hydroxyethylation on a starch solution, a process far more costly than the slurry reaction (*207*). If hydroxyethylation is done on an aqueous slurry, the derivative is readily hydrolyzed by serum α-amylase and converted to low-molecular-weight products, either metabolizable or excretable by the kidneys. Hydroxyethylation on a starch sol gives a far more enzyme-resistant product.

The location of epichlorohydrin produced cross-links, either within the individual D-glucopyranosyl units or within the total macromolecule, is not known. On the basis of the diminished susceptibility of cross-linked starch to exo-enzymes, Kainuma (*208*) has suggested that at least some of the cross-linking is at or near the non-reducing groups of amylopectin. Up to a degree of cross-linking of 1 per 1000 D-glucopyranosyl units, the susceptibility to β-amylase or pure glucoamylase is altered very little. However, with one cross-link per 120 sugar units, susceptibility to these enzymes drops about one-half. Swollen, lightly cross-linked waxy maize starch is readily degraded by β-amylase to give high yields of maltose without formation of soluble limit dextrins (*209*). When dealing with a giant macromolecule such as amylopectin, it is obvious that only a small degree of cross-linking will convert it into a three-dimensional network. At the same time, it seems likely that most of the cross-linking will be intramolecular, rather than intermolecular, and therefore ineffective in preventing breakdown of the partly gelatinized granule during heating and shearing. The fact that there is such a high degree of β-amylolysis indicates that only a few of the outer chains of amylopectin contain cross-links.

8. Action of Aqueous Acid on Starch Granules: Lintnerization, Thin-Boiling Starch, and Naegeli Amylodextrin

By action of aqueous acid at temperatures below the gelatinization point, starch granules are degraded with cleavage of some or all the macromolecules but little change in the external form or birefringence of the granule (*2, 3, 32, 55, 89, 210–216*). Initial attack is in the gel or amorphous region. This degradation leads to a reduction in the viscosity of starch pastes or sols and an enhancement of crystallinity and gel-forming ability. Presumably, a small degree of macromolecular cleavage permits a much easier realignment and self-association of the molecular chains. No significant amount of low-molecular-weight, water-soluble material is produced in this initial reaction. Typical conditions are, for Lintnerization, treating potato starch granules with 7.5% aqueous hydrochloric acid at room temperature for 7 days or at 40° for 3 days. For production of thin-boiling, reduced-viscosity corn starch, a starch slurry is acidified with hydrochloric or sulfuric acid to 0.1–0.9 *N* and held at 50°–60° for several hours. After the desired degree of treatment, the acid is neutralized and the modified starch is filtered, washed, and dried. For production of Naegeli amylodextrin, native starch granules are suspended in 7.5% hydrochloric or 15% sulfuric acid and

Table V

Treatment of Potato Starch Granules with 7.5% HCl at 37° (212)

Days	Iodine color[a]	Iodine affinity[b]	Reducing groups (%)	Crystallinity index (%)	Micel dimension	
					Peak 1	Peak 4
0	Blue	4.08	0.88	56	148	111
1	Blue-violet, violet red	3.9	1.12	65	137	103
2	Red-violet, red	2.82	1.13	66	137	103
4	Red	2.3	1.48	67	137	103
7	Red	0.8	1.77	67.6	137	103
12	Red	0.3	1.82	70	137	103
22	Red	0.2	3.02	72	137	103
50	Red-brown	0	3.4	75	137	103
110	Maroon	0	n.d.	80	134	103

[a] Acid-treated granules were stained with Lugol solution (1 g of iodine and 2 g of KI in 300 mL of water).

[b] mg iodine bound/100 mg starch.

held either at room temperature or at temperatures up to about 35° for days, weeks, months, or years. During this process, there is a gradual erosion of the starch and conversion to soluble sugars, but the main crystalline granule structure may remain intact, as judged by microscopic appearance and x-ray diffraction properties (Table V).

To avoid starch granule gelatinization, starch can be heated with alcoholic or aqueous alcoholic acid (*213*). In the Small process for making soluble starch, starch is refluxed with ethanolic hydrochloric acid without overt granule destruction (*215*). After filtering and washing, the product is readily soluble in hot water. It has a negligible reducing value owing to the nonaqueous acid-catalyzed formation of nonreducing glycosides, in contrast to aqueous acid-catalyzed hydrolysis. An incidental advantage of the alcoholic acid treatment is that it removes lipids and pigments from the starch.

The course of heterogeneous acid degradation (*90, 212, 216*) can be followed by the decrease in paste viscosity, development of reducing value, conversion of insoluble starch to water-soluble sugars, loss in iodine color, and fractionation of the product on a molecular sieve such as BioGel P6.

Native cereal starches contain lipid, some complexed to amylose. These amylose-lipid complexes are resistant to degradation by cold aqueous acid and therefore persist through extensive acid treatment (*32*). This resistance has given rise to the misconception that amylose is in the protected crystallite phase of the starch granule. However, if the lipids are removed prior to acid treatment, amylose is readily degraded. This probably applies also to the "blue cores" seen in some waxy cereal starches following Lintnerization and iodine staining (*10*).

During extensive heterogeneous acid degradation, waxy maize starch, containing only amylopectin, is converted into low-molecular-weight products. These products can be fractionated by gel permeation chromatography or by use of organic solvent precipitation (*55, 87, 90, 216*). Fraction I, highest molecular weight, is a mixture of branched dextrins containing two or more branches per molecule. Fraction II, $\overline{DP}$ about 25, is mainly a singly branched material, with the branch located close to or at the reducing end of the molecule (*55, 87, 88, 90, 216*). The unit chains average 12–14 D-glucopyranosyl units in length. Fraction III is linear with $\overline{DP}$ 12–14 (*87*). During acid degradation, Fraction I gradually disappears and Fractions II and III become better resolved.

Naegeli amylodextrin is readily soluble in hot water. It does not give a paste or gel, but a hot concentrated solution can be cooled to give a semi-solid crystalline mass. The crude material or the purified fractions can then be crystallized by adding methanol or similar organic liquids to hot aqueous solutions, then cooling. When crystallized from water, it gives a B-type x-ray pattern; the crystals from aqueous alcohol give A-type patterns (*148*).

The great resistance of amylodextrin to acidic hydrolysis, in comparison with the gel or amorphous phase of the starch granule or starch solutions or pastes, and the high retention of crystallinity point to the inaccessibility or unreactivity of the glycosidic linkages in the crystallites. This lack of reactivity may stem from the inability of hydrogen ions to penetrate the crystal lattice. It is well known that deuterium can exchange with all the hydrogen atoms of hydroxyl groups of granular starch (*99*), so the crystal lattice per se is not inpenetrable. In a starch molecule double helix, the glycosidic oxygen is ''buried'' in the interior, where it is protected from contact with hydrated hydrogen ions. On the other hand, the hydroxyl groups are in relatively accessible positions on the outside of the double helix. Also, the actual hydrogen ion concentration within the crystallites may be so small that the reaction velocity is negligible. Alternatively, to go into a transition state as appropriate for the hydrolysis of glucosidic linkages, the D-glucopyranosyl unit involved must undergo a change in conformation, and it seems possible that the crystallite structure so immobilizes the sugar conformation that this transition is sterically improbable. These explanations could also account for the extreme resistance of native cellulose to aqueous acidic hydrolysis (*244*).

VI. Granule Swelling and Gelatinization (see also Chapter IX)

When dry, native starch granules are exposed to water vapor or liquid water at 0°–40°, they absorb water and undergo limited, reversible swelling. Heating starch containing limited water results in melting of the starch crystallites, with loss of x-ray and optical crystallinity. The melting temperature depends on the moisture content; with low moisture the melting temperature may be over 100°. With excess water, melting is accompanied by hydration and profound irreversible swelling; the collective process is known as gelatinization. These processes are of cardinal importance in starch technology, and much effort has gone into trying to understand them.

The crystalline organization of starch granules can also be destroyed by mechanical means, such as by dry grinding, washing, and subjecting to high pressure. Alsberg (*85*) has shown that grinding starch in a ball mill at room temperature results eventually in a total destruction of granule structure as judged by disappearance of the x-ray pattern and optical birefringence. The ground starch becomes water-soluble and eventually is chemically degraded as indicated by an increase in reducing value and loss of ability to form a gel. These changes may be due to locally high temperatures at the moment of impact of the grinding balls on the starch granule.

Starch can be damaged by pressure between two glass surfaces (microscope slides) with loss of birefringence (*85*). Such injured granules swell when brought in contact with cold water.

The effect of high pressure (6000 kg cm^{-2}) on starches at moisture levels of from 1% to 36% (dry basis) has been studied (*217*). The changes are somewhat similar to those with ball-milling. Up to about 18% moisture (dry basis), only a few tenths of 1% of the starch becomes solubilized, but beyond 18% there is a sharp rise in the amount of starch solubilized (7% with potato starch at 30% moisture). Up to 40% or 50% of the soluble carbohydrate is maltose and maltotriose, apparently produced by endogenous amylases entrapped within the granule. Damage to potato starch is least at 19% moisture, as indicated by the retention of about three quarters of the original x-ray pattern intensity; whereas at 0% or 36% moisture, the intensities are less than one-half those of the unpressed sample. Starches cut (*218*) or ground (*86*) under water appear to suffer less damage, but no thorough study has been done.

1. Phase Transitions during Heating Starch Granules at Various Moisture Levels

Three or four distinct processes are involved, namely, insertion of water of crystallization into defined or crystallographic positions in the starch crystallites, limited swelling of the gel or amorphous phases, melting of the starch crystallites, and hydration and swelling of the molten crystallites together with the noncrystalline gel region. As the crystallites melt, their substance becomes available for swelling and hydration. Hydration of newly molten crystallites requires a redistribution of water within the granule. During swelling, strong stresses develop which strain the crystallites and contribute to their melting. Thus, to a degree, swelling and gelatinization are cooperative processes.

Gelatinization of starch is readily observed in a polarizing microscope with a hot stage (*219, 220;* see also Chap XXIII). Loss of birefringence and the polarization cross are accompanied by rapid swelling of the granule. These processes occur in an individual granule over a small temperature interval (0.5°–1.5°). However, in any population of starch granules, there will be some granules that gelatinize at lower and higher temperatures, so that macroscopic gelatinization will often occur over a temperature range of 10°–15°. The birefringence end point is taken as that temperature at which 98% of the granules have lost their birefringence.

Loss of birefringence can be quantitated by using a polarized laser beam as light source ($\lambda = 632.8$ nm), a temperature-controlled sample chamber, a polaroid analyzer, and a photodetector with a scanning system (*134*). Light is scattered in a characteristic four-leaf clover pattern over a scattering angle, Θ, up to about 30°, with maxima at 45° to the plane of polarization of the incident radiation and the analyzer. By rapidly scanning the scattered light, it is possible to obtain the integrated intensity, which is a measure of the birefringence of the sample. Starch samples in the hot chamber can be subjected to controlled temperatures and temperature jumps, and the decrease in birefringence after each temperature jump can be followed with time. Working with wheat starch, Marchant and Blanshard (*134*) have shown that, following a small (~2°) temperature jump in the gelatinization range, there is an immediate loss in birefringence (fast component) with a time constant of less then ~0.5 sec^{-1} ($t_1 = 2$ sec) followed by a relatively slow process with a time constant in the range 0.001–0.005 sec^{-1} ($t_2 = 200$–800 sec). Large temperature jumps (~8°) show only the fast component. Marchant and Blanshard interpret their data as follows:

> The starch granule is a highly concentrated and condensed gel system in which . . . the junction zones are crystallites. . . . The energy relationships governing the stability of the crystallites and the conformation of the chains in the amorphous region are *mutually interdependent*. Furthermore, the polymer chains of the amorphous region are susceptible to rearrangement under appropriate conditions which may significantly affect the gelatinization process of the granule. . . . Gelatinization [is] a *semi*-cooperative process.

If one views gelatinization as a semi-cooperative process, there is no need to infer that the crystallites of a given granule melt within the narrow temperature range characteristic of the gelatinization process. Rather, following the melting of a few crystallites, their hydration and swelling exert an increased stress on the remaining crystallites ("junction zones") so that melting is facilitated. After small (2°) temperature jumps, or presumably during gradual heating, some particularly stressed crystallites melt. There is then opportunity for some annealing of the polymer chains, not necessarily into a crystallite structure. Also, there may be some disentanglement of the chains in the amorphous phase. Both processes would be slow in comparison with melting. To the extent that annealing and disentanglement occur, this would make remaining crystallites more stable. With untreated wheat starch, a sudden jump to 58° gelatinizes all the granules, even though this is only about halfway through the usual gelatinization range. With the annealed granules, small jumps give only small losses in birefringence until a temperature of 65°, where 72% of the total loss in birefringence suddenly occurs. These results imply that, during slow heating of untreated starch, there is a significant degree of reorientation, annealing, and possibly, disentanglement.

Annealing starch granules in water at 50° for periods such as 72 h results in an increase in the gelatinization temperature, a reduced gelatinization range, a reduction in birefringence, a reduction in x-ray crystallinity, and an increase in enthalpy of gelatinization as measured by differential scanning calorimetry (*221*). When annealed wheat starch granules are examined by the stepwise small temperature jump method, the most significant loss of birefringence (72% of the total birefringence) occurs for a jump from 63.7° to 65.5°. By contrast, with untreated wheat starch with a jump from 58.0° to 60.0° (midway through the gelatinization range), the loss of birefringence is only 18% of the total initial birefringence.

Donovan (*222, 223*) examined the gelatinization of starch over water contents ranging from a large excess to volume fractions of less than 0.3. The samples were heated at a rate of 10° per minute from room temperature to temperatures over 130°, the maximum attainable temperature (~150°) being limited by the pressure rupturing of the sample cell. With excess water, this rate of heating gives strong gelatinization (G) endotherms with half-widths of 8°–10°. Peak maxima are at 64° for potato and wheat, 71° for maize, and 73° for waxy maize starch. When starch is heated in a volume fraction of water of less than about 0.7, the intensity of the G endotherm diminishes with decreasing water content, and a melting endotherm (M_1) appears at temperatures up to 130° with decreasing water content. By integrating the area under each of the endotherms, it is possible to find the enthalpy for each of the transitions, as a function of the water to starch ratio. A third endotherm (M_2) at about 117° occurs with maize and wheat starch; M_2 is subsequently shown to be due to dissociation of an amylose–lipid complex (*224*). Results of treatment of the data for M_1 and M_2 by the Flory–Huggins equation are given in Table VI.

Potato starch after extensive acid treatment (Naegeli amylodextrin) gives significantly wider G endotherms at a higher temperature, but with the same specific enthalpy, in comparison with unmodified starch (*223*). Inasmuch as the acid treatment brings about hydrolysis of mainly the amorphous phase, leaving the crystallites disconnected from each other and from a gel network, a cooperative or semi-cooperative melting or swelling is impossible. Hence the G and M_1 transitions occur over a range of temperatures reflecting the range of stabilities of the individual crystallites.

During the G and M_1 transitions, the effective water volume fraction v_1

Table VI

Extrapolated Melting Temperatures and Unitary Enthalpies of Melting for the M_1 and M_2 Transitions (223)

Starch and x-ray type	M_1 transition		M_2 transition[a]	
	T_m, °	ΔH_u, kcal/mol	T_m, °	ΔH_u, kcal/mol
Maize, A	187	13.8	172	23.0
Wheat, A	181	12.6	167	23.4
Waxy maize, A	197	14.6	—	—
Potato, B	168	14.3	—	—

[a] The high temperature M_2 transition is due to an amylose–lipid complex (*224*).

decreases. With the Naegeli amylodextrin, the initial state is a mixture of crystallites with essentially pure water, so that the effective v_1 is nearly 1. As crystallites melt and dissolve, v_1 decreases until after the last crystallite has dissolved, when v_1 becomes that calculated from the proportions of water and carbohydrate. With unmodified starch granules, prior to the onset of gelatinization, the effective v_1 is the ratio of the volume of total water, less whatever may be as water of crystallization in the crystal phase, to the total volume of the hydrated gel phase. As with amylodextrins, only after total melting and hydration of the crystallites does v_1 become that calculated from the proportions of water and carbohydrate added together.

In his interpretation of the architecture of the starch granule, Donovan (*222*) indicates that the data from acid-treated starch show that the starch crystallites have a wide range of stabilities, hence a wide melting range. However, in the untreated starch granule, the crystallites are linked to each other and to the macromolecules in the gel phase. Hydration and swelling of the gel phase induces strain in the crystallites and tears molecular chains away from the crystallites. This further weakens the crystallites so that they tend to melt, cooperatively, at a lower temperature than they would if they were not coupled to the gel phase. Starch granules containing less than about 35% water undergo a melting transition, similar to the melting of hydrated crystalline materials, which on heating, eventually "dissolve" in their own water of crystallization.

The results with amylodextrin, in comparison with the original starch or lightly treated starch show that amylodextrin is more highly crystalline and, at high water levels, has a higher melting temperature than the original starch; but at low moisture levels, the small individual molecular chains of amylodextrin melt or dissolve more readily than the relatively immobile chains in high-molecular-weight starch.

The observed variation in gelatinization temperatures for various starches is not well understood. However, it seems reasonable, and in agreement with observations on the melting of synthetic polymers, that the melting temperature should be directly correlated with the perfection and size of crystallites. In the case of legume starches, Biliaderis (*225*) has measured gelatinization temperatures and average chain lengths and found an almost linear correlation between the average chain length and gelatinization temperature.

2. Heat-Moisture Treatment and Annealing

Sair and Fetzer (*149–151*) showed that heat treatment of moist granular starches alters several properties, especially the gelatinization temperature, moisture sorption capacity, and pasting characteristics; and in the case of potato starch, the x-ray pattern changes from B to A. In this treatment, starch is exposed to steam at 90°–110° for 2–18 h or heated at 27% moisture at 90°–100° up to 16 h. Some effects are shown in Table VII. Analogous products also can be prepared by heating granular starches in concentrated salt solutions (*182*) or in 70% aqueous 4-hydroxy-4-methyl-2-pentanone (*166, 226*). Heat-moisture treatment probably causes rearrangement or annealing and a higher degree of association of starch chains. Conversion of the B to A x-ray pattern, with potato starch, indicates that the crystallites have melted and recrystallized or at least have undergone significant reorientation. Though the exact water content of potato starch at the temperature of treatment is unknown, it is probably of the order of 35–45%. At this water level, potato starch undergoes a phase transition (melting) over the temperature range 80°–110°, even though there is insufficient water for the usual gelatinization process (*222*). Donovan and Mapes (*223*) found that a

Table VII

Properties of Heat–Moisture-Treated Corn and Potato Starches (205)

Property	*Potato starch*		*Corn starch*	
	Control	*Treated 16 h at 100°*	*Control*	*Treated 24 h at 100°*
Water sorption capacity at 92% RH (mg water per g)	324	202	245	234
Swelling volume (mL per g dry starch)	65	12	11	10
Hot-paste viscosity (cP)	350	25	40	25
Alkali viscosity (cP)	750	420	980	275
Pasting temperature	61°	70°	67°	74°
Final translucency (0.1% conc.)	600	322	318	299
Nondialyzable phosphorus (*206*)	0.076	0.069	—	—
Reducing power (*207*)	3.5	3.8	—	—

corn starch sample at 50% water, heated to 100° and then dried at that temperature, has an essentially amorphous x-ray pattern.

Crystallization of starch or amylodextrins at high temperatures and low water contents gives A-type crystallization. Gough and Pybus (*221*) found that the physical properties of wheat starch are markedly altered by treatment for 72 h in excess water at 50°. This temperature is just under the gelatinization range of 52°–61°. Although up to 35% of the granules are damaged by this process as measured by dye adsorption and loss of birefringence, there is a significant increase in the gelatinization temperature and a marked sharpening of the gelatinization range.

It appears that the annealing process at 50° permits a realignment of starch chains in the amorphous or gel phase and some additional crystallization. In the high-temperature moisture treatment, the crystallites melt, at least in part, and during cooling assume a more stable and more highly associated form than in the native granule. A type of heat–moisture treatment for production of 'lump' starch has been patented (*227*).

In examining bread staling, Hellman and co-workers (*152*) heated wheat starch–water mixtures at 95° for 30 min and, after various times of aging at 25°, compared the x-ray patterns with standards. "Gels" containing 29% or less water give an A pattern, both fresh and after aging. With 46–56% water (85–127%, dry starch basis), the fresh gels have only a weak lipid–amylose complex V pattern, characterized by a sharp line at 4.4 Å. On aging, only a B pattern develops. With higher water levels, up to 63% water (170%, dry starch basis), development of the B pattern is very slow; and after 8 days, the pattern is only about 20% as intense as with the 50% aged gel. With gels containing 35–43% water (54–75%, dry starch basis), there is a decrease in intensity of the A pattern, and an increase on aging of the B pattern. These intermediate patterns are similar to the so-called C patterns.

Bear and French (*100*), in reexamining the work of Katz and Derksen (*192*), reported that slowly drying a soluble starch solution at 25° gives a B-type pattern. Drying at increasing temperatures gives a progressive shift towards an A pattern; and at 70°, only a pure A pattern is obtained. The shift with increasing temperature was interpreted at the time as indicating a continuum of crystal structures from B to A. However, it now appears clear that the intermediate C patterns are in reality due to a mixture of A and B types of crystals (*98*).

3. Low-Moisture, High-Pressure Cooking of Starch

In extrusion cooking (*228–232*) of starch-containing materials, such as cereal grains, the starch is heated under pressure (100 atm) at up to 40% moisture (dry starch basis) to 200°–250°. At this temperature and moisture content, starch crystallites undergo melting, although the granules retain their individuality. On the passage of the hot mass through the extruder die, there is a sudden reduction of pressure, and the superheated water expands and evaporates dropping the temperature and leaving the starch as an amorphous solid. Puffing of cereals accomplishes the same transformation of starch into an amorphous form.

In the popping of corn, the individual corn kernel pericarp serves as the pressure container. For corn containing optimal moisture (*233*), 13.5%, the exterior temperature required for popping is about 235°. An additional heating time of 30–80 sec at 235° is needed to cause popping. When the temperature within the kernel reaches 205°, the pericarp ruptures with explosion of the kernel contents. Waxy cereals do not pop. Popping volume is improved with increasing amylose content, up to 34%.

Microscopic and x-ray examination indicate that, in these low-moisture cooking procedures, the starch granules are broken and distorted, and that the starch crystal structure is completely destroyed (*230*).

If starch at similar moisture levels is heated under atmospheric pressure, at temperatures between 140° and 200°, it loses its x-ray diffraction pattern and is converted into a pyrodextrin with altered chemical structure (*127*).

When extruded at 190°, potato starch (*230*) containing 23% moisture becomes soluble (up to 80%) in cold water. A part of the water-soluble material consists of D-glucose and linear starch oligosaccharides up to at least DP 11. During extrusion cooking, the amylose component of potato starch is largely destroyed, whereas the outer chains of amylopectin seem to be essentially unaltered. With the lipid-containing cereal starches, no corresponding breakdown is observed (*229, 232*).

4. Geometry of Swelling and Gelatinization of Starch Granules

Hanssen and co-workers (*125*) show that, up to a diameter of 10–15 μm, almost all potato starch granules are spherical, but with granule diameters larger than 30 μm, all granules are ellipsoidal. When totally dry, elliposoidal potato starch granules are immersed in water, they swell anisotropically, increasing 47% in length but only 29% in diameter. Pea starch is even more extreme; it contracts 2% in length while it expands 35% in diameter. The explanation of these observations is not obvious, and more study is required for their understanding.

Bear and Samsa (*234*) investigated potato starch during swelling and gelatinization in calcium nitrate solution. They find that the cavity often seen at the hilum, during rapid swelling, is in reality a gas bubble which expands during gelatinization owing to a decrease in pressure. When the granule structure is sufficiently weakened by gelatinization, the bubble collapses. Their observations indicate that swelling is tangential rather than isotropic or radial. A radial contraction during swelling cannot be ruled out. From present knowledge of starch granule structure, it seems likely that the tangential swelling is due to the hydration and lateral expansion of amylopectin crystallites. The amylopectin molecule as a whole cannot expand radially, that is, along its molecular axis. Radially oriented amylose molecules, whether in the crystallites or in the amorphous phase, also restrict swelling along their molecular length.

A unique swelling and gelatinization pattern is seen with wheat and related starches. The native starches are a mixture of small spherical granules and large flattened ellipsoidal or discus-shaped granules. The discs on swelling undergo a remarkable transformation into a saddle shape, with very little if any increase in thickness (*129, 235–238*). Such saddles have a much larger surface and perimeter than the original discs. If one examines an optical map of a wheat starch granule, it is apparent that swelling perpendicular to the chain direction will greatly increase the surface area but not the thickness.

Sterling (*239*) investigated the swelling pattern of the highly elongated granules of *Dieffenbachia* starch. The starch granules are long, flat rods, with the proximal (hilum) end rounded and the distal end square. However, the granules observed by the author and by Hall and Sayre (*240*) are essentially cylindrical (round in cross section). To be observed fully, a granule should be manipulated so that one can see different aspects. *Dieffenbachia* ganules from different varieties or in different stages of development may have different shapes. Sterling does not show or comment on the occurrence of granules with large lateral protuberances, as described by A. Meyer (*3*) and observed many times by the author. In the central core, especially at the end distal to the hilum, the growth shells are nearly flat and perpendicular to the long axis of the granule. In this region, the molecular chains are highly oriented along the granule axis. When *Dieffenbachia* granules are heated in water (65°–75°), they undergo an internal longitudinal splitting, usually beginning in the central part. The inner core, where the molecules are aligned along the long axis of the granule, can only swell laterally, so that during swelling, the total length of the swollen granule remains more or less constant. Swelling must therefore occur in a direction perpendicular to the granule axis, and in fact, the diameters of the swollen granules appear to be about twice the diameters of the native granules. Around the periphery there is tendency to expand. A constraint against swelling exists because molecules nearer the granule surface are oriented perpendicular to the granule axis. Swelling exerts a tension on the interior, longitudinally oriented molecules, and tears apart the central structure into longitudinal fibers easily visible in the optical microscope. Banana starch is somewhat intermediate between wheat and potato in its swelling pattern. Canna starch also shows a unique geometry of swelling.

It is the author's opinion that the hot water swelling and gelatinization patterns of the various types of starch granules involve the following general principles (*237*). Swelling begins in the least organized, amorphous, intercrystallite regions of the granule. As this phase swells, it exerts a tension on neighboring crystallites and tends to distort them. Further heating leads to uncoiling or dissociation of double helical regions and break-up of amylopectin crystallite structure. The liberated side chains of amylopectin become hydrated and swell laterally, further disrupting crystallite structure. The starch molecules are unable to stretch longitudinally, and actually may have a tendency to contract to approach a random coil conformation. This provides a constraint against swelling in the chain direction. Increased molecular mobility with further hydration permits a redistribution of molecules; and the smaller, linear amylose molecules diffuse out. Further heating and hydration weaken the granule to the point where it can no longer resist mechanical or thermal shearing, and a sol results.

5. Dissolution of Starch in Dimethyl Sulfoxide

Granular starches, retrograded starch, isolated starch fractions, and most uncrosslinked starch derivatives dissolve in dimethyl sulfoxide (DMSO) (*241, 242*). In anhydrous DMSO at room temperature, dissolution is slow, requiring several days, and may be incomplete. Native potato and canna starches are especially resistant to dissolution in cold DMSO. Heating or adding 5–15% water to the DMSO greatly enhances the rate of solution of starch. Conversely, if the DMSO is diluted by a solvent such as chloroform or by aliphatic alcohols, its solvent action on starch is diminished. Dissolution of starch by DMSO follows an entirely different course from the gelatinization of starch in hot water or aqueous solutions of salt, alkali, or strong acids. Granular starch does not swell in DMSO (*242*). Rather the granules appear to dissolve either by surface erosion or by granule fragmentation. DMSO exerts its solvent action by being a powerful hydrogen bond acceptor, thereby breaking associative hydrogen bonds both in the polysaccharide and in water. In a sense, it converts starch into a temporary organic derivative in which the starch hydroxyl groups have been replaced by hydrogen-bonded DMSO.

$$\text{starch—OH}\cdots\overset{\ominus}{\text{O}}\text{—}\overset{\oplus}{\text{S}}\begin{matrix}\diagup \text{CH}_3\\ \diagdown \text{CH}_3\end{matrix}$$

At the same time, water is similarly converted into an "organic solvent." At the optimal water level, about 85–90% by weight DMSO, the DMSO:water mole ratio is 1.3:2.1. It is not obvious why a small amount of water markedly improves DMSO as a solvent.

In a study of potato and canna starch, Hall and Sayre (*161*) found that granules retain their polarization crosses during dissolution in DMSO. With DMSO–water (85:15 by weight), canna starch appears to extrude curious balloon-like nodules, apparently consisting of the intercrystallite portions of the starch granule.

The nodules readily stain with iodine. Possibly they consist largely of amylose, as amylose is readily soluble in DMSO. Scanning electron microscopy shows a great deal of granule splitting and fragmentation (*161*). With aqueous DMSO, by contrast, the granules appear to undergo peeling, with large pieces of membrane-like layers or lamellae still adhering to the smooth regions of the granule surface; and in other regions deep holes are seen. By optical microscopy, treatment of canna starch with DMSO–water or DMSO–dimethylformamide produces a pronounced radial swelling with the formation of "pie layers" (*161*). These layers appear to be growth shells, particularly from the end of the granule distal to the hilum. They retain their identity even after mechanical separation from the rest of the partly swollen granule.

VII. References

(*1*) N. P. Badenhuizen, *Trans. Faraday Soc.*, **42B,** 255 (1946).

(*2*) C. W. Naegeli, "Die Stärkekörner, Pflanzenphysiologischer Untersuchungen," Zurich, 1858, Heft 2, p. 624.

(*3*) A. Meyer, "Untersuchungen über die Stärkekörner, Wesen und Lebensgeschichte der Stärkekörner der hoheren Pflanze," Gustav Fischer, Jena, 1895.

(*4*) M. Samec, "Kolloidchemie der Stärke," Theodor Steinkopf, Dresden and Leipzig, 1927.

(*5*) E. T. Reichert, "The Differentiation and Specificity of Starches in Relation to Genera, Species, etc. Stereochemistry Applied to Protoplasmic Processes and Products and as a Strictly Scientific Basis for the Classification of Plants and Animals," Carnegie Institution, Washington, 1913.

(*6*) R. P. Walton, ed., "A Comprehensive Survey of Starch Chemistry," Chemical Catalog Co., New York, 1928.

(*7*) N. P. Badenhuizen, *in* "Starch: Chemistry and Technology," R. L. Whistler and E. F. Paschall, eds., Academic Press, New York, 1965, Vol. I, p. 65.

(*8*) N. P. Badenhuizen, "The Biogenesis of Starch Granules in Higher Plants," Appleton Century Crofts, New York, 1969.

(*9*) R. Salema and N. P. Badenhuizen, *J. Ultrastr. Res.*, **20,** 383 (1967).

(*10*) N. P. Badenhuizen, *in* "Protoplasmatologia, Handbuch der Protoplasmaforschung," L. V. Heilbrunn and F. Weber, eds., Springer, Vienna, 1959.

(*11*) P. J. Flory, "Principles of Polymer Chemistry," Cornell Univ. Press, Ithaca, New York, 1953.

(*12*) K. Doi and Z. Nikuni, *Staerke*, **14,** 461 (1962).

(*13*) K. Doi and A. Doi, *Denpun Kagaku*, **17,** 89 (1969).

(*14*) C. D. Boyer and J. Preiss, *Carbohydr. Res.*, **61,** 321 (1978).

(*15*) J. L. Ozbun, J. S. Hawker, E. Greenberg, C. Lammel, J. Preiss, and E. Y. C. Lee, *Plant Physiol.*, **51,** 1 (1973).

(*16*) C. Y. Tsai, F. Salamini, and O. E. Nelson, *Plant Physiol.*, **46,** 299 (1970).

(*17*) N. P. Badenhuizen and R. W. Dutton, *Protoplasma*, **XLVII,** 156 (1956).

(*18*) M. Yoshida, M. Fujii, Z. Nikuni, and B. Maruo, *Bull. Agr. Chem. Soc. Japan*, **21,** 127 (1958).

(*19*) N. P. Badenhuizen, *South African J. Sci.*, **51,** 41 (1954).

(*20*) W. Banks, C. T. Greenwood, and D. D. Muir, *Staerke*, **23,** 199 (1971).

(*21*) O. Kjoølberg and D. J. Manners, *Biochem. J.*, **86,** 258 (1963).

(*22*) S. Kizukuri and Y. Takeda,'' *in Proc. Australian Biochem. Soc. and Royal Australian Chem. Inst. Cereal Chem. Div., Workshop Biochem., Nutritional and Industrial Aspects of Starch and Other Cereal Carbohydrates,* July 2 and 3, Windsor Hotel, Melbourne (1980).
(*23*) W. Banks and C. T. Greenwood, *Staerke,* **19,** 197 (1967).
(*24*) M. Yamaguchi, K. Kainuma, and D. French, *J. Ultrastructure Res.,* **69,** 249 (1979).
(*25*) G. E. Hilbert and M. M. MacMasters, *J. Biol. Chem.,* **162,** 229 (1946).
(*26*) W. Banks, C. T. Greenwood, and D. D. Muir, *Staerke,* **26,** 46 (1974).
(*27*) C. D. Boyer, R. R. Daniels, and J. C. Shannon, *Amer. J. Bot.,* **64,** 50 (1977).
(*28*) R. M. Sandstedt, *Cereal Sci. Today,* **10,** 305 (1965).
(*29*) M. J. Wolf, H. L. Seckinger, and R. J. Dimler, *Staerke,* **12,** 375 (1964).
(*30*) R. Charbonniere, C. Mercier, M. T. Tollier and A. Guilbot, *Staerke,* **20,** 75 (1968).
(*31*) M. Kapp, C. Legrand, and O. Yovanovitch, *Compt. Rend.* **255,** 2967 (1962).
(*32*) K. Kainuma and D. French, *Biopolymers,* **10,** 1673 (1971).
(*33*) J. Tandecarz, N. Lavintman, and C. E. Cardini, *Biochim. Biophys. Acta,* **399,** 345 (1975).
(*34*) L. F. Leloir, M. A. R. de Fekete, and C. E. Cardini, *J. Biol. Chem.,* **235,** 636 (1961).
(*35*) K. Nishida, *Physiol. Plantarum,* **15,** 47 (1962).
(*36*) E. Recondo and L. F. Leloir, *Biochem. Biophys. Res. Commun.,* **6,** 85 (1961).
(*37*) J. S. Hawker, J. L. Ozbun, H. Ozaki, E. Greenberg, and J. Preiss, *Arch. Biochem. Biophys.,* **160,** 530 (1974).
(*38*) A. A. Perdon, E. J. Del Rosario, and B. O. Juliano, *Phytochem.,* **14,** 949 (1975).
(*39*) O. E. Nelson and H. W. Rines, *Biochem. Biophys. Res. Commun.,* **9,** 297 (1962).
(*40*) J. W. Bolcsak, Ph.D. Thesis, Iowa State University, Ames, Iowa, 1979.
(*41*) V. E. Stöckman, *Biopolymers,* **11,** 251 (1972).
(*42*) E. Slabnik and R. B. Frydman, *Biochem. Biophys. Res. Commun.,* **38,** 709 (1970).
(*43*) J. C. Linden, *Diss. Abstr.,* **30,** 4903-B (1970).
(*44*) S. A. Barker, E. J. Bourne, S. Peat, and I. A. Wilkinson, *J. Chem. Soc.,* 3022 (1950).
(*45*) S. Schiefer, E. Y. C. Lee, and W. J. Whelan, *FEBS Lett.,* **30,** 129 (1973).
(*46*) B. L. Dronzek, P. Hwang, and W. Bushuk, *Cereal Chem.,* **49,** 232 (1972).
(*47*) H. Fuwa, Y. Sugimota, and T. Takaya, *Carbohydr. Res.,* **70,** 233 (1979).
(*48*) G. Hollinger and R. H. Marchessault, *Biopolymers,* **14,** 265 (1975).
(*49*) I. Maeda, Z. Nikuni, H. Taniguchi, and M. Nakamura, *Carbohydr. Res.,* **61,** 309 (1978).
(*50*) S. Ueda and J. J. Marshall, *Carbohydr. Res.,* **84,** 196 (1980).
(*51*) A. Frey-Wyssling, *Ber. Schweiz. Bot. Ges.,* **50,** 321 (1940).
(*52*) H. Speich, *Ber. Schweiz. Botan. Ges.,* **52,** 175 (1942).
(*53*) A. T. Czaja, ''Die Mikroskopie der Stärkekörner,'' Paul Parey, Berlin and Hamburg (1969).
(*54*) A. Frey-Wyssling, ''Submicroscopic Morphology of Protoplasm, Elsevier, Amsterdam, 2nd English Ed., 1953.
(*55*) D. French, *Denpun Kagaku,* **19,** 8 (1972).
(*56*) E. H. Land and C. D. West, *in* ''Colloid Chemistry,'' J. Alexander, ed., Reinhold, New York, 1946, Vol. 6, p. 160.
(*57*) C. D. West, *J. Chem. Phys.,* **15,** 689 (1947).
(*58*) R. E. Rundle and R. R. Baldwin, *J. Amer. Chem. Soc.,* **65,** 554 (1943).
(*59*) R. E. Rundle, J. F. Foster, and R. R. Baldwin, *J. Amer. Chem. Soc.,* **66,** 2116 (1944).
(*60*) D. M. Hall and J. G. Sayre, *Textile Res. J.,* **39,** 1044 (1969).
(*61*) Y. Sugimoto, K. Ohnishi, T. Takaya, and H. Fuwa, *Denpun Kagaku,* **26,** 182 (1979).
(*62*) H. Fuwa, *Proc. Symposium Amylase,* **9,** 23 (1974).
(*63*) K. Endo, Y. Sugimoto, T. Takaya, and H. Fuwa, *Denpun Kagaku,* **25,** 24 (1978)
(*64*) H. Fuwa, Y. Sugimoto, M. Tanaka and D. V. Glover, *Starch/Staerke,* **30,** 186 (1978).
(*65*) H. Fuwa, D. V. Glover and Y. Sugimoto, *Denpun Kagaku,* **24,** 99 (1977).

(*66*) H. Fuwa, Y. Sugimoto and T. Takaya, *Denpun Kagaku,* **26,** 105 (1979).
(*67*) H. Fuwa, *Denpun Kagaku,* **24,** 128 (1977).
(*68*) H. Fuwa, D. V. Glover, and Y. Sugimoto, *J. Nutr. Sci. Vitaminol.,* **25,** 103 (1979).
(*69*) A. D. Evers and E. E. McDermott, *Staerke,* **22,** 23 (1970).
(*70*) D. M. Hall and J. G. Sayre, *Textile Res. J.,* **40,** 147 (1970).
(*71*) R. L. Whistler and E. S. Turner, *J. Polymer Sci.,* **18,** 153 (1955).
(*72*) K. Mühlethaler, *Z. Wiss. Mikrosk. Microsk. Tech.,* **62,** 394 (1955).
(*73*) Z. Nikuni, *Starch/Staerke,* **30,** 105 (1978).
(*74*) J. F. Chabott, J. E. Allen, and L. F. Hood, *J. Food Sci.,* **43,** 727 (1978).
(*75*) S. Hizukuri and Z. Nikuni, *Nature,* **180,** 436 (1957).
(*76*) Z. Nikuni and S. Hizukuri, *Mem. Inst. Sci. Ind. Res. (Osaka Univ.),* **14,** 173 (1957).
(*77*) M. S. Buttrose, *J. Ultrastr. Res.,* **4,** 231 (1960).
(*78*) M. S. Buttrose, *J. Cell Biol.,* **14,** 159 (1962).
(*79*) M. S. Buttrose, *Naturwissenschaften,* **12,** 450, (1963).
(*80*) M. S. Buttrose, *Naturwissenschaften,* **49,** 307 (1962).
(*81*) M. S. Buttrose, *Staerke,* **3,** 85 (1963).
(*82*) A. Frey-Wyssling, and M. S. Buttrose, *Makromol. Chem.,* **44/46,** 173 (1961).
(*83*) P. D. Gallant and A. Guilbot, *Staerke,* **21,** 156 (1969).
(*84*) P. Kassenbeck, *Starch/Staerke,* **30,** 40 (1978).
(*85*) C. L. Alsberg, *Plant Physiol.,* **13,** 295 (1938).
(*86*) D. French, unpublished data.
(*87*) T. Watanabe and D. French, *Carbohydr. Res.,* **84,** 115 (1980).
(*88*) R. S. Hall and D. J. Manners, *Carbohydr. Res.,* **83,** 93 (1980).
(*89*) J. P. Robin, C. Mercier, R. Charbonnier, and A. Guilbot, *Cereal Chem.,* **51,** 389 (1974).
(*90*) J. P. Robin, C. Mercier, F. Duprat, R. Charbonniere, and A. Guilbot, *Staerke,* **27,** 36 (1975).
(*91*) W. C. Mussulman and J. A. Wagoner, *Cereal Chem.,* **45,** 162 (1968).
(*92*) F. Duprat, R. Charbonniere, J. P. Robin, and A. Guilbot, *Zesz. Probl. Postepow Nauk Roln.,* **159,** 155 (1974).
(*93*) C. Sterling and B. J. Spit, *J. Exptl. Bot.,* **9,** 75 (1958).
(*94*) C. Sterling and J. Pangborn, *Amer. J. Botany,* **47,** 577 (1960).
(*95*) S. C. Mueller, R. M. Brown, Jr, and T. K. Scott, *Science,* **194,** 949 (1976).
(*96*) J. Ruiz-Herrera, V. O. Sing, W. J. Van der Woude, and S. Bartnicki-Garcia, *Proc. Nat. Acad. Sci.,* **72,** 2706 (1975).
(*97*) M. Samec and M. Blinc, ''Die Neuere Entwicklung der Kolloidchemie der Stärke,'' Steinkopf, Dresden and Leipzig, 1941, p. 477.
(*98*) A. Sarko and H. C. H. Wu, *Starch/Staerke,* **30,** 73 (1978).
(*99*) N. W. Taylor, H. F. Zobel, M. White, and F. R. Senti, *J. Phys. Chem.,* **65,** 1816 (1961).
(*100*) R. S. Bear and D. French, *J. Amer. Chem. Soc.,* **63,** 2298 (1941).
(*101*) D. R. Kreger, *Biochim. Biophys. Acta,* **6,** 406 (1951).
(*102*) R. E. Rundle, L. Daasch and D. French, *J. Amer. Chem. Soc.,* **66,** 130 (1944).
(*103*) F. R. Senti and L. P. Witnauer, *J. Amer. Chem. Soc.,* **68,** 2407 (1946).
(*104*) H. C. Wu and A. Sarko, *Carbohydr. Res.,* **61,** 7 (1978).
(*105*) H. C. Wu and A. Sarko, *Carbohydr. Res.,* **61,** 27 (1978).
(*106*) A. Sarko and P. Zugenmaier, *in* ''Fiber Diffraction Methods,'' A. D. French and K. H. Gardner, eds., ACS Symposium Series, No. 141, 1980, p. 459.
(*107*) H. Chanzy, private communication.
(*108*) F. P. Booy, H. Chanzy, and A. Sarko, *Bioploymers,* **18,** 2261 (1979).
(*109*) G. N. Ramachandran, *in* ''Conformation of Biopolymers,'' G. N. Ramachandran, ed., Academic Press, New York, 1967, p. 721.
(*110*) P. C. Manor and W. Saenger, *J. Amer. Chem. Soc.,* **96,** 3630 (1974).

(*111*) D. A. Brant, *in* "The Biochemistry of Plants," P. K. Stumpf and E. E. Conn, eds., Academic Press, New York, 1980, Vol. 3, p. 425.
(*112*) A. D. French, *Baker's Dig.,* **53,** 39 (1979).
(*113*) C. V. Goebel, W. L. Dimpfl, and D. A. Brant, *Macromolecules,* **3,** 644 (1970).
(*114*) A. D. French, V. G. Murphy, and K. Kainuma, *Denpun Kagaku,* **25,** 171 (1978).
(*115*) A. D. French and W. A. French, *in* "Fiber Diffraction Methods," A. D. French and K. H. Gardner, eds., ACS Symposium Series, No. 141, 1980, p. 239.
(*116*) D. P. Miller and R. C. Brannon in "Fiber Diffraction Methods," A. D. French and K. H. Gardner, eds., ACS Symposium Series, No. 141, 1980, p. 93.
(*117*) A. D. French and V. G. Murphy, *Cereal Foods World,* **22,** 61 (1977).
(*118*) K. Kainuma and D. French, *Biopolymers,* **11,** 2241 (1972).
(*119*) M. J. Richardson, *Brit. Polym. J.,* **1,** 132 (1969).
(*120*) S. Nara, A. Mori, and T. Komiya, *Starch/Staerke,* **30,** 111 (1978).
(*121*) M. Ahmed and J. Lelievre, *Starch/Staerke,* **30,** 78 (1978).
(*122*) A. G. Walton and J. Blackwell, "Biopolymers," Academic Press, New York, 1973, p. 92.
(*123*) A. Guinier, "X-Ray Diffraction in Crystals, Imperfect Crystals, and Amorphous Bodies," W. H. Freeman and Co., San Francisco, 1963, p. 121.
(*124*) C. Sterling, *J. Polymer Sci.,* **56,** S10 (1962).
(*125*) E. Hanssen, E. Dodt, and E. G. Niemann, *Kolloid-Z.,* **130,** 19 (1953).
(*126*) M. J. Wolf, V. J. Ruggles, and M. M. MacMasters, *Biochim. Biophys. Acta,* **57,** 135 (1962).
(*127*) J. R. Katz and A. Weidinger, *Z. Physik. Chem.,* **184,** 100 (1939).
(*128*) N. P. Badenhuizen, Jr., and J. R. Katz, *Z. Physik. Chem.,* **182,** 73 (1938).
(*129*) O. A. Sjostrom, *Ind. Eng. Chem.,* **28,** 63 (1936).
(*130*) J. Borch, Ph.D. Thesis, State University College of Forestry, Syracuse, New York, 1969.
(*131*) J. Borch, R. Muggli, A. Sarko, and R. H. Marchessault, *J. Applied Phys.,* **42,** 4570 (1971).
(*132*) R. S. Finkelstein, and A. Sarko, *Biopolymers,* **11,** 881 (1972).
(*133*) Z. Mencik. R. H. Marchessault, and A. Sarko, *J. Mol. Biol,* **55,** 193 (1971).
(*134*) J. L. Marchant and J. M. V. Blanshard, *Starch/Staerke,* **30,** 257 (1978).
(*135*) Z. H. Bhuiyan and J. M. V. Blanshard, *Cereal Chem.,* **57,** 262 (1980).
(*136*) J. M. V. Blanshard, *in* "Polysaccharides in Food," J. M. V. Blanshard and J. R. Mitchell, eds., Butterworths, London, 1979, p. 139.
(*137*) F. P. Price, *J. Polymer Sci.,* **37,** 71 (1959).
(*138*) A. Frey-Wyssling, *Schweiz. landw. Forsch.,* **13,** 385 (1974).
(*139*) K. Kainuma, *Chori Kagaku,* **13,** 83 (1980).
(*140*) T. J. Schoch and D. French, *Cereal Chem.,* **24,** 231 (1947).
(*141*) J. R. Katz and J. C. Derksen, *Z. Physik. Chem.,* **A165,** 228 (1933).
(*142*) S. Hizukuri, *Agr. Biol. Chem.,* **25,** 45 (1961).
(*143*) S. Hizukuri, M. Fujii, and Z. Nikuni, *Biochem. Biophys. Acta,* **40,** 346 (1960).
(*144*) D. French and R. W. Youngquist, *Abstr. Papers Amer. Assoc. Cereal Chem.,* 45th Meeting, Abstr. 42 (1960).
(*145*) B. Pfannemüller, private communication.
(*146*) S. Kikumoto and D. French, *Archives of Biochem. Biophys.,* **156,** 794 (1973).
(*147*) S. Hizukuri, M. Fujii, and Z. Nikuni, *Nature,* **192,** 239 (1961).
(*148*) S. Hizukuri, Y. Takeda, S. Usami, and Y. Takase, *Carbohydr. Res.,* **83,** 193 (1980).
(*149*) L. Sair and W. R. Fetzer, *Ind. Eng. Chem.,* **36,** 205 (1944).
(*150*) L. Sair, *Cereal Chem.,* **44,** 8 (1967).
(*151*) L. Sair, *Methods Carbohydr. Chem.* **4,** 283 (1964).
(*152*) N. N. Hellmann, B. Fairchild, and F. R. Senti, *Cereal Chem.,* **31,** 495 (1954).
(*153*) E. Wiegel, *Kolloid-Z.,* **102,** 145 (1943).
(*154*) D. French, A. O. Pulley, and W. J. Whelan, *Staerke,* **15,** 349 (1963).

(*155*) T. J. Schoch, *Advan. Carbohydr. Chem.*, **1,** 247 (1945).
(*156*) R. S. Bear, *J. Amer. Chem. Soc.*, **66,** 2122 (1944).
(*157*) F. F. Mikus, R. M. Hixon, and R. E. Rundle, *J. Amer. Chem. Soc.*, **68,** 1115 (1946).
(*158*) R. St. J. Manley, *J. Polymer Sci., Part A,* **2,** 4503 (1964).
(*159*) S. Hizukuri and Z. Nikuni, *Mem. Inst. Sci. Ind. Res. (Osaka Univ.),* **16,** 231 (1959).
(*160*) T. Davies, D. C. Miller, and A. A. Procter, *Starch/Staerke,* **32,** 149 (1980).
(*161*) D. M. Hall and J. G. Sayre, *Textile Res. J.*, **41,** 404 (1971).
(*162*) G. H. Lathe and C. R. J. Ruthven, *Biochem. J.*, **62,** 665 (1956).
(*163*) S. A. Brown and D. French, *Carbohydr. Res.*, **59,** 203 (1977).
(*164*) H. W. Leach, *in* "Starch: Chemistry and Technology," R. L. Whistler and E. F. Paschall, eds., Academic Press, New York, 1965, p. 289.
(*165*) L. F. Hood and C. Mercier, *Carbohydr. Res.*, **61,** 53 (1978).
(*166*) H. W. Leach, L. D. McCowen, and T. J. Schoch, *Cereal Chem.*, **36,** 534 (1959).
(*167*) N. W. Taylor, J. E. Cluskey, and F. R. Senti, *J. Phys. Chem.*, **65,** 1810 (1961).
(*168*) W. Bushuk and C. A. Winkler, *Cereal Chem.*, **34,** 73 (1957).
(*169*) F. Schierbaum, *Staerke,* **12,** 257 (1960).
(*170*) M. Masuzawa and C. Sterling, *J. Appl. Polymer Sci.*, **12,** 2023 (1968).
(*171*) M. K. S. Morsi, C. Sterling, and D. H. Volman, *J. Appl. Polymer Sci.*, **11,** 1217 (1967).
(*172*) M. Wootton, J. Regan, N. Munk and D. Weeden, *Food Technol. Australia,* 24 (January, 1974).
(*173*) A. R. Urquhart, *in* "Recent Advances in the Chemistry of Cellulose and Starch," J. Honeyman, ed., Interscience, New York, 1959, p. 240.
(*174*) N. N. Hellman and E. H. Melvin, *J. Amer. Chem. Soc.*, **72,** 5186 (1950).
(*175*) J. E. Cleland, E. E. Fauser, and W. R. Fetzer, *Ind. Eng. Chem., Anal. Ed.*, **15,** 334 (1943).
(*176*) H. Rodewald, "Untersuchung über die Quellung der Stärke," Kiel and Leipzig, 1896, p. 65.
(*177*) E. Parow, *Z. Spiritusind.*, **30,** 432 (1907).
(*178*) G. L. Brammer, M. A. Rougvie, and D. French, Carbohydr. Res., **24,** 343 (1972).
(*179*) R. Geddes, J. D. Harvey, and P. R. Wills, *Biochem. J.*, **163,** 201 (1976).
(*180*) W. Banks, R. Geddes, C. T. Greenwood, and I. G. Jones, *Staerke,* **24,** 245 (1972).
(*181*) N. N. Hellman, T. F. Boesch, and E. H. Melvin, *J. Amer. Chem. Soc.*, **74,** 348 (1952).
(*182*) H. W. Leach, personal communication.
(*183*) H. N. Dengate, D. W. Baruch, and P. Meredith, *Starch/Staerke,* **30,** 80 (1978).
(*184*) G. N. Lewis and M. Randall, "Thermodynamics," McGraw-Hill, New York, 1923, p. 33.
(*185*) F. Duprat, D. Gallant, A. Guilbot, C. Mercier, and J. P. Robin, *in* "Les Polymeres Vegetaux," B. Monties, ed., Gautheir-Villars, 1980, p. 176.
(*186*) C. Gur-Arieh, A. I. Nelson, and M. P. Steinberg, *J. Food Sci.*, **32,** 442 (1967).
(*187*) H. J. Woods, *in* "Recent Advances in the Chemistry of Cellulose and Starch," J. Honeyman, ed., Interscience, New York, 1959, p. 134.
(*188*) F. J. Bates and associates, "Polarimetry, Saccharimetry and Sugars," Circular of the National Bureau of Standards C440, U.S. Govt. Printing Office, Washington, D.C., 1942.
(*189*) R. S. Bear, *J. Amer. Chem. Soc.*, **64,** 1388 (1942).
(*190*) R. E. Rundle and D. French, *J. Amer. Chem. Soc.*, **65,** 558 (1943).
(*191*) F. Schierbaum and K. Täufel, *Staerke,* **14,** 233 (1962).
(*192*) J. R. Katz and J. C. Derksen, *Z. Physik, Chem.*, **A150,** 100 (1930).
(*193*) C. A. Winkler and W. F. Geddes, *Cereal Chem.*, **8,** 455 (1931).
(*194*) F. Schierbaum, K. Täufel, and M. Ulmann, *Staerke,* **14,** 161 (1962).
(*195*) H. Lechert, *Staerke,* **28,** 369 (1976).
(*196*) H. J. Hennig and H. Lechert, *J. Colloid. Interf. Sci.*, **62,** 199 (1977).
(*197*) H. J. Hennig, H. Lechert, and W. Goemann, *Staerke,* **28,** 10 (1976).
(*198*) A. Guilbot, R. Charbonniere, P. Abadie, and P. Girard, *Staerke,* **12,** 327 (1960).
(*199*) E. K. Gladding and C. B. Purves, *Paper Trade J., TAPPI Sect.*, **116,** 150 (1943).

(*200*) O. B. Wurzburg, *Methods Carbohydr. Chem.*, **4,** 286 (1964).
(*201*) E. L. Jackson and C. S. Hudson, *J. Amer. Chem. Soc.*, **59,** 2049 (1937).
(*202*) T. Painter and B. Larsen, *Acta Chem. Scand.*, **24,** 2724 (1970).
(*203*) S. Hizukuri, *Mem. Inst. Sci. Ind. Res. (Osaka Univ.),* **15,** 215 (1958).
(*204*) C. C. Kesler and E. T. Hjermstad, *Methods Carbohydr. Chem.* **4,** 304 (1964).
(*205*) H. C. Srivastava and K. V. Ramalingam, *Staerke,* **19,** 295 (1967).
(*206*) H. C. Stivastara and K. V. Ramalingam, *Staerke,* **21,** 181 (1969).
(*207*) T. Shiba, German Patent 2,837,067 (1979).
(*208*) K. Kainuma, S. Miyamoto, and S. Suzuki, *Denpun Kagaku,* **22,** 66 (1975).
(*209*) R. A. Harrington, M. S. Thesis, Iowa State University, Ames, Iowa, 1974.
(*210*) W. Naegeli, *Liebig's Ann. Chem.*, **173,** 218 (1874).
(*211*) H. T. Brown and G. H. Morris, *J. Chem. Soc.*, **55, 449 (1889).**
(*212*) S. Ferri and C. Carlozzi, *Atti Soc. Tosc. Sci. Nat. Mem., Serie B,* **86,** 63 (1979).
(*213*) S. Hizukuri, Y. Takeda, and S.Imamura, *J.Agr. Chem. Soc. Japan,* **46,** 119 (1972).
(*214*) C. J. Lintner, *J. Prakt. Chem.*, **34,** 378 (1886).
(*215*) J. C. Small, *J. Amer. Chem. Soc.*, **41,** 113 (1919).
(*216*) S. Kikumoto, N. Nimura, Y. Hiraga, and T. Kinoshita, *Carbohydr. Res.*, **61,** 369 (1978).
(*217*) C. Mercier, R. Charbonniere, and A. Guilbot, *Staerke,* **20,** 6 (1968).
(*218*) W. W. Lepeschkin, *Protoplasma,* **30,** 309 (1938).
(*219*) S. A. Watson, *Methods Carbohydr. Chem.*, **4,** 240 (1964).
(*220*) T. J. Schoch and E. C. Maywald, *Anal. Chem.*, **28,** 382 (1956).
(*221*) B. M. Gough and J. N. Pybus, *Staerke,* **23,** 210 (1971).
(*222*) J. W. Donovan, *Biopolymers,* **18,** 263 (1979).
(*223*) J. W. Donovan and C. J. Mapes, *Starch/Staerke,* **32,** 190 (1980).
(*224*) M. Kugiyama, J. W. Donovan, and R. Y. Wong, *Starch/Staerke,* **32,** 265 (1980).
(*225*) C. G. Biliaderis, Ph.D. Dissertation, University of Saskatchewan, Saskatoon, Saskatchewan, 1980.
(*226*) G. A. Hull and T. J. Schoch, *TAPPI,* **42,** 438 (1959).
(*227*) A. Moffatt, U.S. Patent 541,941 (1895).
(*228*) R. E. Tribelhorn and J. M. Harper, *Cereal Foods World,* **25,** 154 (1980).
(*229*) C. Mercier and P. Feillet, *Cereal Chem.*, **52,** 283 (1975).
(*230*) C. Mercier, *Staerke,* **29,** 48 (1977).
(*231*) C. Mercier, R. Charbonniere, D. Gallant, and A. Guilbot, *in* "Polysaccharides in Food," J. M. V. Blanshard and J. R. Mitchell, eds., Butterworths, London, 1979, p. 153.
(*232*) C. Mercier, R. Charbonniere, J. Grebaut, and J. F. de la Gueriviere, *Cereal Chem.*, **57,** 4 (1980).
(*233*) L. V. Babichenko and E. N. Sorochinskaya, *Izv. Vyssh. Uchebh. Zaved., Pishch. Tekhnol.*, 63, 1972; *Chem. Abstr.* **79,** 17130w (1973).
(*234*) R. S. Bear and E. G. Samsa, *Ind. Eng. Chem.*, **35,** 721 (1943).
(*235*) S. Woodruff and L. R. Webber, *J. Agr. Res.*, **46,** 1099 (1933).
(*236*) R. I. Derby, B. S. Miller, B. F. Miller, and H. B. Trimbo, *Cereal Chem.*, **52,** 702 (1975).
(*237*) D. French, *Abstr. Papers, 2nd Chem. Congress North Amer. Continent, Div. Agr. Food Chem.*, 23 (1980).
(*238*) P. Bowler, M. R. Williams, and R. E. Angold, *Starch/Staerke,* **32,** 186 (1980).
(*239*) C. Sterling, *Staerke,* **29,** 117 (1977).
(*240*) D. M. Hall and J. G. Sayre, *Textile Res. J.*, **41,** 880 (1971).
(*241*) P. J. Killion and J. F. Foster, *J. Polymer Sci.*, **46,** 65 (1960).
(*242*) H. W. Leach and T. J. Schoch, *Cereal Chem.*, **39,** 318 (1962).
(*243*) J. N. BeMiller and G. W. Pratt, *Ceral Chem.*, **58,** 517 (1981).
(*244*) J. N. BeMiller, *in* "Starch: Chemistry and Technology," R. L. Whistler and E. F. Paschall, eds., Academic Press, New York, 1965, Vol 1, Chap. 20.

CHAPTER VIII

FRACTIONATION OF STARCH

By Austin H. Young

A. E. Staley Manufacturing Company, Decatur, Illinois

I. Introduction

1. Starch

Starch is a granular mixture of homopolymers of α-D-glucopyranosyl units (*1–10*). These polymer molecules are hydrogen bonded and aligned radially in the granule (*5*, Chapter VII). The granule can be swollen and dispersed in water as well as other hydrogen bond-breaking solvents (Chapter VII). Methods have been developed to fractionate starch into a mixture of an essentially linear, isotactic polymer of α-D-glucopyranosyl units linked (1→4), called amylose according to Meyer (*2, 11, 12*), and a group of highly branched polymers with α-D-glucopyranosyl units linked (1→4) and (1→6) called amylopectin. The branches of amylopectin occur at O-6, with an average of one for every 18 to 28 α-D-glucopyranosyl units (*13*). Maquenne and Roux (*14*) introduced the term amylopectin in 1905.

STARCH, 2nd ed.

ISBN 0-12-746270-8

All starches examined by Whistler and Doane (*15*) contained an intermediate fraction. The intermediate fraction comprises 4–9% of normal corn starch (*6, 9, 16–18*) and may be considered amylopectin with a lower degree of branching (see Chap. VI). Some of the amylose molecules may be slightly branched with infrequent α-D-(1→6) linkages that do not significantly affect the molecule's physical properties (*19*). Greenwood (*20*) thinks that most amylose contains a small degree of branching (see Chap. VI).

2. Comparison of Amylose and Cellulose

Isotactic amylose and syndiotactic cellulose have the same empirical formula, but in cellulose the D-glucopyranosyl units are linked in the β-D-(1→4) configuration with the D-glucopyranosyl units alternating in 180° rotation.

Cellulose forms strong microcrystalline regions called crystallites. Its β-D-(1→4) glucosidic linkages permit an extended chain enabling molecular alignment in crystallites and provides reduced solubility. Only through solvents or formation of derivatives that disrupt hydrogen bonds is cellulose dispersible. The uniform linear-chain nature of amylose polymers permits them to crystallize from solution as well as in the semi-solid state in films and coatings in a manner similar to that of other linear polymers such as poly(propylene) and cellulose. The α-D-(1→4) bonds of amylose give the linear chains a less extended configuration. The resulting more poorly crystallized amylose can be solubilized in water alone when it is heated for 1–2 min at temperatures above 124°. On cooling, however, the amylose readily crystallizes and precipitates.

3. Amylopectin Molecule (see also Chapter VI)

The branched structure of amylopectin provides greater solution stability than is possessed by amylose. With time and low temperatures or very high concentrations, however, amylopectin can also partially crystallize. The crystalline nature of amylopectin is observed in all starch granules, even those containing only branched amylopectin such as waxy corn (*21*, Chapter VII). Leaching experiments which release amylose preferentially from the granule imply that amylose is not extensively associated within the crystalline regions of the native granule.

Amylopectin is heterogeneous with respect to molecular weight distribution and very likely with respect to frequency of branching (*22*). In corn starch, branches occur an average of once for every 26 α-D-glucopyranosyl units through α-D-(1→6) linkages (Chapter VI). Branches may be close together or be over 40 sugar units apart. The inner chain segments seem to be shorter and more closely spaced, giving the molecule a dense core (*23, 24*). Some amylopectin molecules may have very long outer branches (*15*). There is some evidence that a small amount of branching occurs through α-D-(1→3) linkages (*25*). Fractions which are more highly branched than normal amylopectin are also known (*26–29*). The mean variation in branch length is not known. The average chain length and the β-amylolysis limit vary with increase in maturity of the plant (*20*).

4. Starch from Various Plants (see also Chapter III)

The amylose content of a number of starches is shown in Table I (*10, 13, 30–33*). Starches from different sources, such as wrinkled pea, normal corn, and waxy maize, contain widely different amounts of amylose. Most native starches contain 20–30% by weight of amylose. One variety of wrinkled pea contains a starch having 66% amylose. The ratio of amylose to amylopectin and/or the linearity and frequency of branching can be genetically controlled (*34*). Certain varieties of hybrid corn, such as Amylomaize, have starch with more than 80%

Table I

Amylose Content of Starches

Starch	*Amylose (%)*	*Reference*
Acorn	24	*30*
Apple	19	*13*
Arrowroot	20.5	*31*
Banana	16	*13*
Barley	22	*13*
Waxy	0	*32*
Easter lily	34	*32*
Elm, sapwood	21.5	*33*
Iris tuber	27	*31*
Corn (*Zea mays*)	28	*13*
Hybrid amylomaize Class V	52	*13*
Hybrid amylomaize Class VII	70–75	*10*
Hybrid waxy maize	0.8	*31*
Oat	27	*13*
Pea		
Smooth	35	*13*
Wrinkled	66	*13*
Manioc	15.7	*31*
Parsnip	11	*13*
Potato	20	*31*
Rice	18.5	*31*
Waxy	0	*32*
Sago	25.8	*31*
Sorghum, waxy	0	*32*
Sweet potato	17.8	*31*
Tapioca	16.7	*31*
Wheat	26	*13*

amylose (*35*) when measured by iodine binding capacity (*32*). Waxy maize, waxy sorghum, waxy barley, and waxy rice starches contain no amylose.

5. Considerations of Starch Structure

To disperse and fractionate starch molecules that might possess outer branches of up to 12,000 α-D-glucopyranosyl units, heat, shear, and complex formation have been used, but these methods may cleave the long outer branches, especially if high concentrations of starch are used. Therefore, separation of starch into two or more fractions is not conclusive evidence that the polymers were originally present as a mixture in the starch granule (*36*).

The occurrence of a linear α-D-(1→4)-linked amylose and a branched amylopectin with about 4% α-D-(1→6) linkages as two separate natural polymers is not universally accepted (*8*). Potze (*8*) points out that fractionation procedures are usually preceded by such rigorous dissolution methods that two polymers could be formed from pre-existing large molecules by debranching. Badenhuizen (*37*) supports arguments for the existence of only one constituent substance in starch. Bauer and Pacsu (*38*) conclude that starch amylose and amylopectin are bound together to form native starch molecules. Langlois and Wagoner (*36*) believe that the fractionation of starch into linear and branched components is not proof that a mixture exists in the granule. Richardson and coworkers (*39*) conclude that the differences in the relative proportions of chemically defined "amylose" and "amylopectin" observed by various workers owe partly to differences in pretreatments and in separation methods.

Greenwood (*20*) states that, although there is a vast amount of evidence for the heterogeneous character of starch, the starch granule may be a giant molecule and such a unitarian concept has to be seriously considered. Banks and Greenwood (*40*) claim that, the more that is known about starch, the more it is realized that few generalizations about its structure and behavior are possible. They conclude that the biosynthetic routes and the reasons for deposition of a multiplicity of polysaccharide structures remain matters of conjecture (*41*).

Exclusion chromatography (*42, 43*) produces fractions that raise the possibility that native starch may be, not a mixture of branched and linear molecules of D-glucopyranosyl units, but a mixture of covalently bound branched molecules having some extremely long chains, a broad degree of branching, and broad molecular weight distributions.

Regardless of how amylose occurs in the starch granule, methods have been developed to separate it from the branched amylopectin; and some methods have been used commercially (*7, 44*). The present high cost of energy required for fractionation of starch makes commercial fractionation less attractive, (*45*) especially because developed genetic hybrids of corn provide starches that function commercially as well as commercial amylose (*35, 46*) and amylopectin (*47*).

Except for exclusion chromatography, most analytical methods for starch fractionation are based on differential solubility. Free amylose and low-molecular-weight amylopectin can be leached from the granule, or the granule can be dispersed and the less soluble amylose component complexed or precipitated. Refined techniques can subfractionate amylose into different molecular weight groups.

II. Molecular Weight of Fractions

Standard methods for determination of polymer molecular weight (*48*) can be used. The average molecular weights ($\overline{M}_n$ = number average, $\overline{M}_w$ = weight average, and $\overline{M}_v$ = viscosity average) are defined by the equations:

$$\overline{M}_n = \frac{\Sigma_i n_i M_i}{\Sigma_i n_i} \tag{1}$$

$$\overline{M}_w = \frac{\Sigma_i n_i M_i^2}{\Sigma_i n_i M_i} = \Sigma_i w_i M_i \tag{2}$$

$$\overline{M}_v = \left[\frac{\Sigma_i n_i M_i^{(1+a)}}{\Sigma_i n_i M_i} \right]^{1/a} \tag{3}$$

where n_i = number of each molecule possessing a molecular weight M_i; w_i = weight fraction of each molecule possessing a molecular weight M_i; and a = constant for each type of polymer–solvent system, usually in the range 0.5–1.0.

The number-average molecular weight is the total weight of sample divided by the total number of moles. Any colligative method, such as osmotic pressure or end-group analysis, can give the number of solute molecules.

Weight-average molecular weight, $\overline{M}_w$, is often determined by light scattering. For heterogeneous samples, $\overline{M}_w$ is always $>\overline{M}_n$.

Viscosity-average molecular weight becomes equal to $\overline{M}_w$ if $a = 1$.

Small molecules affect the number-average molecular weight to a greater extent than do large molecules. The number-average molecular weight loses significance for broad molecular weight distributions, since it is not sensitive to larger molecules. The weight-average molecular weight is more sensitive to the contributions of large molecules and not as sensitive to small molecules. Thus, both averages have shortcomings for broad molecular weight distributions and are best employed for molecular characterization of narrow molecular weight fractions. The extent of the molecular weight distribution can be indicated by the ratio of $\overline{M}_w/\overline{M}_n$. For fractions of narrow molecular weight distributions, the ratio can be close to 1.

The following two examples show how average degree of polymerization values can vary. A mixture of molecules composed of α-D-glucopyranosyl units and with degrees of polymerization of 1, 10, 100, 1000, and 10,000 have a weight average $\overline{DP}_w = 9090$ and a number average $\overline{DP}_n = 2222$, with a value of $\overline{DP}_w/\overline{DP}_n = 4.09$. For a mixture where 25% by weight of the molecules are DP 1000 and 75% by weight are DP 25, the $\overline{DP}_w$ is 269 and $\overline{DP}_n$ is 33.1. For the same weight percentage mixture but with the DP of the 25% by weight fraction increased from 1000 to 2000 $\overline{DP}_w = 519$ and $\overline{DP}_n = 33.1$, showing that the value of $\overline{DP}_n$ is rather insensitive to changes in the population of large molecules.

Staudinger's early work showed the practical usefulness of solution viscosity as a measure of polymer molecular weight (*49*). Solution viscosity is a measure of the space swept out by a rotating molecule (*48*). Comparative volumes are most meaningful when the intrinsic value, $[\eta]$, is obtained by extrapolation to zero solution concentration.

The relation is expressed in the Mark–Houwink equation

$$[\eta] = KM^a \tag{4}$$

where

$$[\eta] = \lim_{c \to 0} \frac{\eta_{sp}}{c} \tag{5}$$

$$\frac{\eta_{sp}}{c} = \frac{\eta_{solution} - \eta_{solvent}}{\eta_{solvent}} \times c^{-1} \tag{6}$$

and a = constant for molecular shape.

For most stiff or fully extended molecules in solution, the value of a is nearly 1, so that simple proportionality holds. However, for macromolecules which are not as stiff or fully extended in solution, the value of a usually lies between 0.5 and 1.0. The constants K and a are characteristic for a given species of macromolecule and must be established for each species from an "absolute" method such as osmometry or light scattering procedures. The latter gives results which are typical of the viscosity-average molecular weight (*48*). From the Mark–Houwink equation [Eq. (4)], it is possible to calculate the weight average molecular weight of an amylose sample if the intrinsic viscosity is known. Results are valid only for fractions with narrow molecular weight distributions.

Methods for determining the size and shape of polysaccharides have been reviewed (*22, 50*). Light scattering has been used most often for determination of high-molecular-weight starch fractions. Dimethyl sulfoxide is a preferred solvent. Nuclear magnetic resonance shows that DMSO forms a strong complex with carbohydrates that involves the hydroxyl groups at C-1 and C-2 of the D-glucopyranosyl units (*51*). Organic solvent derivatives of the starch fractions can also be used (*52–54*). Amylose molecular weights are given in Table II.

The molecular weights of amylopectin fractionated from different starches are given in Table II (*55*). The lowest molecular weight (10^7) was obtained on waxy maize amylopectin degraded by shear through vigorous stirring of the starch in a

Table II

Molecular Weight of Amylose and Amylopectin

Source	Molecular weight $\times 10^{-6}$		Reference
	Amylose	Amylopectin	
Anaerobic Fractionation			
Apple	0.24[a]		58
Banana	0.27[a]		58
Broad bean	0.29[a]		58
Barley	0.30[a]		58
Iris (rhizome)	0.29[a]		58
Mango seed	0.29[a]		58
Oat	0.21[a]		58
Parsnip	0.71[a]		58
Pea			
Smooth	0.21[a]		58
Wrinkled	0.16[a]		58
Potato	0.49[b]		58
Rubber seed	0.24[b]		58
Sweet corn (*Z. mays*)	0.18[b]		58
Wheat	0.34[b]		58
Fractionation in DMSO			
Potato	1.9[b]		59
Barley	2.11[b]		60
Oat	2.19[b]		66
Rye	2.50[b]		66
Wheat I	1.33[b]		66
Wheat II	2.65[b]		66
Barley		400[b]	55
Pea			
Smooth		500[b]	55
Wrinkled		500[b]	55
Potato I		440[b]	55
Potato II		65[b]	55
Tapioca		450[b]	55
Waxy maize		400[b]	55
Waxy maize (sheared)		10[b]	55
Wheat		400[b]	55

[a] Determined by viscosity.
[b] Determined by light scattering.

15% by weight dispersion in boiling water. When waxy maize starch was dissolved without shear in dimethyl sulfoxide, the molecular weight was 40 times higher, suggesting that earlier measurements were made on degraded amylopectin. Potato amylopectin at 500 million daltons is one of the largest natural polymers. Foster (*22*) interpreted the wide difference between weight-average and number average molecular weights for amylopectin (ratios of 300 to 500) as evidence of a wide molecular size distribution.

Greenwood and co-workers (*55*) found a variation in the molecular weight of amylopectins in different cultivars of the same plant type; for example, the molecular weights of potato amylopectins of different cultivars were consistently 440×10^6 and 65×10^6. The molecular weight of amylopectin of a given plant type may also vary from year to year. The amount, molecular weight, and fine structure of amylose and amylopectin also change during granule growth. Thus, the molecular weight of amylose and amylopectin increase as granules develop in normal barley and high-amylose barley starch (*56*) and in potato starch (*57*). As the granule size increases in potato starch, the proportion of amylose also increases.

The average length of the branch segments of the amylopectins from different starches has been determined by chemical methods such as methylation (*5*) and periodate oxidation (*20*) which give number average values. If the distribution of branches is large, the $\overline{DP}_w$ value of D-glucopyranosyl units in corn amylopectin branches could be higher with some of the branch segments 100 times larger than the $\overline{DP}_n$ value of 26 D-glucopyranosyl units in corn amylopectin obtained by chemical methods.

The average molecular weight of amylose molecules is orders of magnitude lower than the average molecular weight of the amylopectin molecules. The average molecular weights of the amylose and amylopectin depend upon the history of the starch as well as the method of fractionation. Greenwood (*58*) used anaerobic procedures to separate amylose from different starches and measured molecular weights ranging from 160,000 to 700,000. The results were calculated from the relationship $\overline{DP} = 740 \times [\eta]$, where $[\eta]$ is the intrinsic viscosity in dL g^{-1}. The amylose was heterogeneous with at least two components; one of $\overline{DP}$ $\approx$2000 and a β-amylase limit of 100% and one with a $\overline{DP}$ up to 6000 with a possible random barrier to β-amylase that yielded a limit of 50%.

Killion and Foster (*59*) used neat dimethyl sulfoxide on potato starch to isolate amylose with a weight average molecular weight by light scattering of 1.9×10^6. Other workers (*60*) used dimethyl sulfoxide on different starches to isolate amylose with $\overline{M}_w$ of about 2×10^6.

III. Chromatographic Separation

Since molecular size distribution effects molecular weight averages, Young and Griffin (*61*) have used exclusion chromatography with dextran standards

(*62*) to obtain a better measurement of molecular size. Dextran is mainly a linear α-D-(1→6) glucopyranosyl polymer with some (1→3) linkages and has a $\overline{M}_w/\overline{M}_n$ of the order of 10^2 to 10^3. It can be hydrolyzed and fractionated into rather narrow molecular weight fractions $\overline{M}_w/\overline{M}_n$ = 1.2–1.5 characterized by both chemical end-group analysis ($\overline{M}_n$) and by light scattering ($\overline{M}_w$). Dintzis and Tobin (*63*) found that an amylose and a dextran of about the same intrinsic viscosity, [η], have nearly the same $\overline{M}_w$ and elution volume in exclusion chromatography. The product of the [η] and $\overline{M}_w$ is proportional to the end-to-end length of a polymer through the universal constant Φ.

$$[\eta]\overline{M}_w = \Phi(r^2)^{3/2} \qquad (7)$$

For amylopectin molecules, $\overline{M}_w$ is higher than for amylose or a linear dextran of the same hydrodynamic volume.

Mixtures of two polymer types should be evidenced by two separate molecular size distributions, one for a lower average molecular size of amylose, the other for the much, much larger molecule, amylopectin.

Chromatograms on porous glass at 25° of moisture-free regular dent corn starch dispersed at 0.1% by weight in anhydrous dimethyl sulfoxide for 1 h at 150° were monomodal with no indication of a low-molecular-weight linear fraction. The molecular size distribution of waxy corn starch shows a similar narrow molecular weight distribution. The molecular size distributions for Amylomaize starch of 55–70% amylose are broader. These high-amylose starches contain more low-molecular-weight species than does regular dent corn starch (*61*).

Molecular size distributions of several corn starches are shown in Table III. The $\overline{M}_{50}$, molecular weight median based on the dextran standards, is included since it is less sensitive to the tails of the distribution than are $\overline{M}_n$ on the lower molecular weight side and $\overline{M}_w$ on the higher molecular weight side.

A similar molecular size distribution is seen for waxy maize starch and regular dent corn, lending support to the idea that regular dent corn starch is not a mixture of high-molecular-weight, branched amylopectin and low-molecular-

Table III

Molecular Size Distribution of Corn Starches (61)[a]

	[η] *1 N KOH at 35°, dL·g⁻¹*	$M_{50} \times 10^{-6}$	$\overline{M}_n \times 10^{-6}$	$\overline{M}_w \times 10^{-6}$	$\overline{M}_w/\overline{M}_n$
Regular dent corn starch	1.53	2.2	0.214	14.5	68
Waxy maize	1.30	3.3	0.148	21.8	147
Amylomaize (70–75% amylose)	0.92	0.127	0.048	3.96	82
Amylomaize (52% amylose)	1.01	0.175	0.054	5.75	106

[a] Chromatography conditions: solvent, dimethyl sulfoxide (DMSO); concentration, 0.1% by wt; columns (4 ft × ⅜ in.), Bio-Glas 2500, Bio-Glas 500, Bio-Glas 200 (80–100 mesh); temperature, ambient (25°); flow, 1 mL·min⁻¹.

weight, linear amylose polymers, but rather contains high-molecular-weight branched structures with some branches consisting of long linear chains of (1→4)-linked α-D-glucopyranosyl units. The long branches could be severed from the branched structure when the starch is dispersed in water with shear. In contrast, the mild agitation given at the low (0.1% by weight) concentration for regular dent corn starch solutions in dimethyl sulfoxide does not greatly alter the monomodal molecular size distribution present in the granule, but this is not irrefutable evidence for the amylose fraction being chemically bound to the branched molecules.

It is possible that the regular dent starch contains a mixture of linear and branched polymers with a size distribution that gives a narrow monomodal pattern similar to that of waxy maize starch. The results support the possibility that the linear amylose fraction in regular starch may be chemically bound to the amylopectin before dispersion in water or other solvents.

Yamada and Taki (*64*) used gel chromatography for fractionation of amylose and amylopectin from normal, waxy, amylo-, and amylowaxy-type maize starches. Starch was dispersed in water with the aid of perchloric acid, and the dispersion was passed through a column of agarose gel (Sepharose 2B) capable of separating molecules and particles with molecular particle weights of several million daltons (*65*). By extrapolation the exclusion limit of Sepharose 2B for polysaccharides is estimated at 20–30 × 10^6. Eberman and Schwarz (*66*) claim that agarose beads can exclude amylopectin fractions greater than 20–30 × 10^6 daltons.

Gel chromatography shows an average molecular weight of potato amylose of 400,000, but 100,000–200,000 for amylose from rice and corn starch (*67*). The homogenizer used to disperse the starch in sodium hydroxide solution may have sheared long linear chains from the starch molecule.

Molecular weight distribution of amylose has been measured by gel permeation chromatography with dimethyl sulfoxide and deactivated silica gel columns (*68*). Dintzis and Tobin (*63*) find that dimethyl sulfoxide–water (95:5, v/v) is a satisfactory solvent for separation on Corning porous glass of amylose and dextran of weight-average molecular weights ranging from 2 × 10^4 to 2 × 10^5. Starch amylose and amylopectin can be fractionated on calcium phosphate columns (*69*), and iodine complexes of amylose and amylopectin have been separated chromatographically on paper strips using 40% perchloric acid as the irrigant at 50° (*70*).

Amylose and amylopectin have been separated on paper using 0.2 *M* sodium hydroxide as the eluant (*71*) and on an alumina column eluted with an acid buffer (*72*). Fischer and Settele (*73*) used an alumina column to subfractionate amylose and concluded that potato amylose and corn amylose were homogeneous, but that tapioca amylose contained two fractions.

Structural studies on acid-treated starches have been conducted (*290*) using gel filtration chromatography. Starch molecules have been separated by high-pressure liquid chromatography (*291*). The amylose: amylopectin ratios of starches have been determined (*292*) by solubilization in dimethyl sulfoxide, debranching with isoamylase, and chromatography of the linear components on a column of Sepharose CL-6B. Starch components have been separated by affinity chromatography (*293*).

IV. Aqueous Leaching of Gelatinized Granules

Leaching occurs from swelling starch granules in water at temperatures of 57–100°. Mobile amylose molecules diffuse out of the swollen granule while the granule is intact, while most of the amylopectin remains hydrogen bonded or crystallized in the granule residue. The amylose is collected in the supernatant solution by centrifugation. With successive leaching at higher temperatures, some of the branched amylopectin molecules diffuse out to contaminate the amylose fractions. Early methods of fractionating starch by aqueous leaching have been discussed by Whistler (*3*) and also by Banks and Greenwood (*50*).

Meyer (*11*), and Schoch (*6*) were unable to obtain pure amylose by swelling defatted corn starch granules in water at 80°. Kerr and Severson (*74*) extracted amylose from defatted corn starch at 57° and obtained an amylose purity of 80–85%. Purification of the leached portion was achieved using 1-butanol to form a crystalline amylose–butanol complex following the general procedures of Schoch (*75*). In spite of the purification step, Schoch (*16*) found the amylose obtained by aqueous leaching still contaminated by amylopectin.

Baum and Gilbert (*76*) showed that removal of oxygen from the water reduced degradation of potato starch when extracted for 5 min at 100°. Greenwood and co-workers (*27*), however, claim degradation can occur during dispersion, fractionation, and centrifugation even in a nitrogen atmosphere. Killion and Foster (*59*) repeated and scaled up the anaerobic leaching of potato starch according to the Baum and Gilbert (*76*) procedure, but obtained low yields of amylose that had a molecular weight of 1.46×10^6. Banks, Greenwood, and Thompson (*77*) improved separation by successive extractions under conditions that prevented retrogradation.

Because defatting starch granules by heating in aqueous methanol enhances leaching of the amylose, other pretreatments of starch in hot aqueous mixtures of organic solvents such as glycerol, 1-butanol, Pentasol, Cellosolve, and dioxane have been used (*78*). These techniques are useful for extraction of amylose from high-amylose corn starch which strongly resists swelling, gelatinization, and solubilization (*79*). Pretreatment of starch with hot organic solvents such as dioxane causes some degradation (*77*). Baum and Gilbert (*80, 81*) suggested that the cleavage may occur at anomalous linkages in the amylose.

By a low temperature pretreatment of granules to induce further crystallization, Montgomery and Sexson (*82*) reportedly extracted all the amylose at 100°

without granule rupture and with retention of crystallinity. They froze a slurry (33% by weight) of high-amylose corn starch in distilled water for 1 min at −78° prior to 3 extractions of 2 h each at 100° to obtain amylose of 90% purity (*82*).

Potato starch has been extracted with 0.5 *M* sodium hydroxide solution at 15° for about 30 min with exclusion of oxygen to give amylose with an intrinsic viscosity of 5.5 $dL \cdot g^{-1}$ (*83*). Muetgeert (*7*) used 30% chloral hydrate solution to remove amylose; on dilution of the resulting solution with water, amylose precipitated.

Although leaching can be used for fractionation, complete separation of pure fractions is not achieved. Presumably, granular structure must be ruptured for fractionation to be successful (*84*).

V. Dispersion of the Granule and Fractionation with Complexing Agents

Schoch (*75*), in 1941, pioneered a new method of fractionating starch by dispersing 1–3% by weight of starch in water at 105–109° for 2–3 h and then precipitating the amylose as a complex with butanol followed by centrifugal separation. Defatting procedures and buffering at pH 6.2–6.3 were employed later to reduce degradation (*16*).

Amylose has been reported to form complexes with aliphatic alcohols such as isopropyl, *n*-butyl and isoamyl (*85*), lower aliphatic ketones, fatty acids (*86*), benzenoid derivatives having aldehyde groups (*85*), alkyl halides (*87, 88*), cyclic alcohols, phenols (*89, 90*), and aliphatic and aromatic esters (*18*). Some non-polar molecules, such as benzene (*88*) and cyclic and aliphatic hydrocarbons (*88*), are also capable of forming complexes with amylose. Carbon tetrachloride has been used to recover amylose from starch (*91, 92*). Another reported complexing agent is *l*-menthone (*93*). Whistler and Hilbert (*18*) found that amylose forms microcrystalline organic complexes with nitroethane, nitropropanes, nitrobenzene, 1-pentyl acetate, 2-butanone, butanethiol, and pyridine. For a complex to form with butanol, the DP of the linear chain has to be greater than 20–40 (*28*).

Alcohols are often used at concentrations which saturate water. Muetgeert (*7*) proposed that less than saturation quantities of the alcohol provide a more satisfactory separation; otherwise the complexes may form with the outer branches of the amylopectin and reduce the effectiveness of the separation.

To avoid the autoclave and hydrolytic degradation at elevated temperatures various pretreatments have been employed which lower the gelatinization temperature. Bauer and Pacsu (*38*) dispersed corn starch at 2–3% by weight in 1.0 *N* alkali at 25°. After the dispersion was neutralized, butanol or Pentasol at saturation did not produce a precipitate. However, when such neutral dispersions were heated to 90°, they precipitated as the mixture cooled. Pacsu (*38*) found it necessary to heat the solution to 60° before a complex with butanol formed.

Undried potato starch has been dispersed to a concentration of 0.5% in dimethyl sulfoxide by stirring for one week at 25° (*59*). Addition of 1-butanol produced a gummy precipitate that was redissolved by heating to 70° in water for 10 min and reprecipitated with 1-butanol. [It is necessary to heat the mixture to form the amylose butanol complex (*77*).]

Amylose purified by Killion and Foster (*59*) had a light scattering $\overline{M}_w$ of 1.9×10^6 (Table II) and was 96% converted to maltose by β-amylase. The β-amylase limits of three preparations of amylose prepared: (a) commercially, (b) by autoclaving followed by butanol precipitation, and (c) by aqueous leaching were 65.6%, 74.1%, and 94.9% (*59*). Explanations for the lower β-amylase limits were that harsh methods produce anomalous linkages that sterically hinder enzyme hydrolysis (*94, 95*) or that produced smaller molecules aggregates and thereby resist hydrolysis.

Muetgeert (*7*) attributed the failure to obtain a butanol complex precipitate at room temperature to lack of a critical concentration of butanol. He believes that separation of amylose complexes at 1.5% concentration from solutions saturated with complexing agent requires gravitational fields of 10,000 *g* or higher. Greenwood and co-workers (*96*) used the critical concentration of 1-butanol proposed by Muetgeert (*7*) to obtain potato amylose with a high intrinsic viscosity. Banks and Greenwood (*60*) also report that pretreatment of cereal starches with dimethyl sulfoxide prior to fractionation removes fatty acids that could hydrolyze the starch components. Under anaerobic conditions, they obtained higher molecular weights for amylose fractions than reported for barley, oat, rye, and wheat starches (Table II).

Pacsu and Bauer (*97*) report that fractionation of thin-boiling starches with complexing agents raises the amylose yield because of hydrolysis occurring during the dissolution process.

Amylose has been separated from amylomaize as complexes with alcohols or with acids of 8–10 carbon atoms (*98*). Banks and Greenwood (*99*) claimed that the linear material isolated as a butanol complex from amylomaize amylopectin was contaminated by short linear chains that produced the anomalous properties observed.

Adkins and Greenwood (*10*) separated maize starches into different fractions. Starch granules were defatted, pretreated with dimethyl sulfoxide, precipitated with butanol, dispersed in water, precipitated in acetone, dried, redissolved in dimethyl sulfoxide, diluted with water and butanol to form the amylose butanol complex at 25°, and twice recrystallized as a complex at 50°–55° (Table IV). The isolated amylose behaved normally, but its intrinsic viscosity decreased with its increased proportion in the starches. The yields of amylose are directly proportional to the apparent amylose content of starches, but are smaller than analyzed

Table IV

Fractionation of Corn Starches Using 1-Butanol and Iodine (10)

	Waxy maize	*Regular corn starch*	*High-amylose 57*	*High-amylose 70*
Recrystallized butanol complex (%) (normal amylose)	0	25	51	63
Supernatant from butanol complex recrystallization	0	2	4	5
Iodine complex-intermediate material	0	n.d.[a]	3	4
Not complexed by iodine, normal amylopectin, average chain length 14–22	0	n.d.[a]	1	1
Non-complexing components in supernatant from 1-butanol	99	73	45	32
Iodine complex-intermediate material	2	4	36	27
Material not complexed by iodine (amylopectin, chain length 16–27)	96	68	9	4
Total % of original starch recovered	98%	97%	100%	99%

[a] n.d. = not determined.

values would indicate. Characteristics of the supernatant liquors from recrystallization of amylose shows that the intermediate fraction occluded in the butanol complex is a minor component of the starches. Subfraction with iodine shows that the intermediate material is probably a mixture of short-chain amylose and amylopectin-like material.

In the fractionation of waxy and regular maize starch, the materials in the first supernatant that do not complex with 1-butanol are similar to those obtained in fractionation of regular amylopectin, whereas similar material from high-amylose starches has abnormally high iodine affinities, β-amylolysis limits, and average chain lengths. Subfractionation of these samples by formation of iodine complexes gives short-chain amylose and material having the characteristics of normal amylopectin. The amount of regular amylopectin in high-amylose starches is low and decreases with increases in the analyzed amylose content of the starch.

These results suggest that 57% amylose starch (by iodine affinity), in addition to normal amylose, contains up to 39% of short-chain amylose of DP 90–120 (molecular weight 16,000) and perhaps 10% amylopectin. High-amylose starches seem to contain an unusually high proportion of intermediate material which is neither amylose nor amylopectin (*100*). There is little normal amylopectin present, but rather there is large amounts of an intermediate material with longer outer branches and longer linear segments between branch points than in normal starch. One-half of the intermediate fraction in Amylon 50 may consist of linear amylose with chain lengths between 50 and 90 D-glucopyranosyl units. Percent by weight of molecular fractions of 19,000 daltons or less in amylomaize Class V and amylomaize Class VII are 14.5% and 16.5%, respectively.

Solubility of starches in dimethyl sulfoxide increases with increasing amylose content (*101*). Waxy starches disperse slowly, and their unusually high viscosity requires a low concentration of polysaccharide during treatment. It is difficult to obtain a true molecular dispersion of waxy maize starch in hot water, just as it is with normal dent corn starch. Waxy corn starch and regular corn starch have essentially the same molecular weight distribution. As pointed out earlier, both starches may consist of similar distributions of branched molecules, with the branches of the regular corn starch containing longer linear outer branches.

VI. Fractional Precipitation

Solubility of a homologous series of polymers decreases with increasing molecular weight. Addition of a nonsolvent to a polymer solution causes the molecules to desolvate and precipitate, beginning with the highest molecular weight fraction. In a mixture of two polymers, where one type is less likely to crystallize, the addition of a nonsolvent will precipitate, or salt-out, the polymer having the more isotactic structure and, therefore, a higher tendency to crystallize.

A thorough examination of the fractionation of starch in magnesium sulfate has been made using the phase rule with the approximations cited by Muetgeert (*7*) and the concepts of polymer chemistry (*102*). A commercial process salted out the amylose from potato starch with magnesium sulfate (*44*). This method was based upon the fractional crystallization of amylose from a 10% by weight aqueous solution of potato starch in the presence of 10–13% magnesium sulfate. The starch was solubilized by heating to 160°. It was then cooled to 80°, and the precipitated amylose was removed by centrifugation. Cooling to 20° allows the amylopectin to precipitate. The process can be modified by dilution with water at 90° to reduce the salt concentration to 10%. At the lower salt concentration, the amylose can be separated at 20° by raising the salt concentration to 13%. Potze (*8*) pointed out that, for an efficient process, the energy consumption must be low and inexpensive (*8*). In 1960, the Avebe Company in Holland produced 5.4 tons/day of amylose and 15.4 tons/day of amylopectin (*103*).

Cantor and Wimmer (*104*) found that starch dispersed in calcium chloride solution can be precipitated with calcium hydroxide. Addition of water causes the amylose complex to dissolve, and thus, stepwise addition of water can effect separation.

Corn amylose dissolved in ethylenediamine can be fractionated by gradual addition of diethyl ether (*105*). It can also be fractionated from water–octanol mixtures by cooling (*16*). Amylose in potassium hydroxide solution has been fractionated by gradual addition of isopropanol (*106*) or in dimethyl sulfoxide solution by gradient addition of ethanol (*107*).

VII. Fractionation by Retrogradation and Controlled Polymer Crystallization

When starch is solubilized at low solids, amylose molecules diffuse and, if the temperature is lowered appropriately, orient and crystallize from solution. This readiness to crystallize is characteristic of the isotactic, stereoregular, α-D-(1→4) linkage of the linear amylose molecule. If the concentration is too high or the temperature too low, diffusion decreases and crystallization of amylose is hindered; instead a three-dimensional gel network results, and the starch is said to "retrograde" (*108*). The linear outer branches of amylopectin can also participate in gel formation and make separation difficult or impossible. Early controlled crystallization can be achieved by selective retrogradation (*14, 109*).

Whistler (*110, 111*) and Schoch (*16*) have shown that the rate of retrogradation of corn amylose has a maximum, depending on molecular weight. With high molecular weights, rates of orientation and diffusion are too slow for rapid crystallization of amylose. With low molecular weights, the increased solubility of the fractions prevents crystallization. Maximum rates of crystallization and retrogradation occur at the normal molecular weight of corn amylose (reduced viscosity 0.5 $dL \cdot g^{-1}$).

Meyer and co-workers (*112*) showed that retrogradation leaves less than 0.4% of the starch in solution. Retrograded starch, as well as retrograded amylose, exhibits the B-type x-ray diffraction pattern. Retrogradation is favored by low temperatures, neutral pH, relatively low degrees of polymerization, absence of branching, low degrees of polydispersity, high concentrations, presence of inorganic ions (according to the Hofmeister lyotropic series), presence of other dehydrating substances, and absence of surface-active (wetting) agents (*111, 113–117*).

A semicommercial method was developed using controlled crystallization to separate amylose in fairly high purity (*118*). It was not pursued commercially because of the availability of high-amylose corn starch and waxy corn starch, and because of the increasing cost of energy. Heating a starch slurry at a very high temperature under high shear for a short time produces an effective dispersion of the starch. Starch dispersions at 10% by weight undergo a transition at neutral pH and 125° that lowers the viscosity and permits greater diffusion of amylose molecules. When commercial corn starch is used, the fatty acids normally present complex with a portion of the amylose to form microcrystalline spherulites. On further cooling, the remaining amylose crystallizes as fat-free spherulites leaving the amylopectin in solution. Defatting procedures have been developed to give a pure product (*119*). An advantage of this process is that the amylopectin solution is not contaminated by salt.

Another separation based upon retrogradation has been designated a hydrodynamic process (*120*). A 7–9% slurry of potato starch is gelatinized at 85° and homogenized in a blender. Sufficient energy is given the system to cause starch dissolution. The mixture is then centrifuged and cooled so that the amylose separates in the form of globules from a liquid containing the amylopectin. Heating to 85° and dispersing the starch molecules by high shear is probably no better than heating alone to higher temperatures. At the higher temperatures, the starch can be dispersed in water or salt solutions, without high speed stirring. When conditions are severe, some hydrolytic degradation occurs. Hydrolysis and color formation can be reduced by buffering to neutral pH and by adding sulfite to provide reducing conditions (*121*). Degradation can occur during the solubilization if the slurry is heated under alkaline conditions (*97*).

Amylose can be fractionally precipitated in high yields from normal or high-amylose corn starch after debranching the branched molecules with an amylo-1,6-glucosidase, such as pullulanase or isoamylase (*122–125*).

VIII. Conformation of Amylose in Dilute Solution

Foster (*22*) and Banks and Greenwood (*126*) describe the physical properties and the conformation or shape of amylose molecules in solution. Available evidence indicates that solution behavior of amylose can be explained on the basis of a model in which the α-D-glucopyranosyl units are in the $^4C_1$ conformation (*127*).

The Mark–Houwink constants K and a for amylose are given in Table V (*128–141*). Values from different laboratories do not completely agree but are in the expected range of 0.5–1.0 for a. In neutral aqueous potassium chloride, a has the ideal value 0.5 as expected for an unperturbed polymer in a ''theta'' solvent. In ''theta'' solvents, the polymer–solvent and polymer–polymer interactions are nearly equal, thus the solvent is indifferent (*142*).

Foster (*22*) concludes that amylose exists in solution as a relatively stiff wormlike coil consisting largely of an imperfect or deformed helical backbone. Thus, it is similar to the iodine complex which is helical in character (*143*). The amylose polymer chain is more flexible in alkaline solution than in neutral water. Near pH 12, the helix structure starts to break down as hydroxyl groups ionize, resulting in greater chain flexibility and a shrinking of the hydrodynamic volume. At still higher pH, a progressive increase in the negative charge on the polymer molecule due to increased degree of ionization results in an expansion of the coil from charge repulsion (*22*)

Brant and Dimpfl (*144*) conclude that, in aqueous solution, amylose is a

Table V

Parameters K and a of the Mark–Houwink Equation as Determined for Amylose and Amylose Derivatives in Various Solvent Systems[a]

Polymer			$K \times 10^5$	*a*	*Reference*
Amylose	Water		13.2	0.68	*128*
	0.5 *N* NaOH		1.44	0.93	*129*
	0.5 *N* NaOH		3.64	0.85	*128*
	0.15 *N* KOH		8.36	0.77	*130*
	0.2 *N* KOH		6.92	0.78	*131*
	0.5 *N* KOH		8.5	0.76	*132*
	1.0 *N* KOH		1.18	0.89	*133*
	0.33 *N* KCl		113	0.50	*132*
	0.33 *N* KCl		112	0.50	*134*
	0.33 *N* KCl		115	0.50	*135*
	0.50 *N* KCl		55	0.53	*133*
	0.50 *N* KCl		55	0.53	*136*
	Aqueous KCl (acetate buffer)		59	0.53	*136*
	Dimethyl sulfoxide		1.25	0.87	*133*
			30.6	0.64	*132*
			15.1	0.70	*130*
			3.95	0.82	*128*
	Ethylenediamine		15.5	0.70	*133*
	Formamide		22.6	0.67	*128*
			30.5	0.62	*130*
Amylose triacetate	Chloroform		1.06	0.92	*137*
	Nitromethane		1.10	0.87	*137*
Amylose tricarbanilate	Acetone		0.814	0.90	*138*
	Dioxane		0.906	0.92	*138*
	Pyridine		0.589	0.92	*138*
Sodium carboxymethylamylose	D.S. = 0.8	0.65 *M* NaCl	135	0.5	*139*
	D.S. = 0.55	0.50 *M* NaCl	209	0.53	*140*
	D.S. = 0.29	0.78 *M* NaCl	37	0.61	*141*
Dimethylaminoethylamylose hydrochloride	D.S. = 0.28	0.78 *M* NaCl	83.4	0.55	*141*

[a] Based on concentration units of deciliters per gram.

statistical coil without helical character. Banks and Greenwood (*145*) believe that the conformation of amylose is a random coil in water and neutral potassium chloride, an expanded coil in formamide, dimethyl sulfoxide and aqueous alkali, and a helical coil in neutral or alkaline solution in the presence of a complexing agent and in aqueous solution at pH 12 in the presence of 0.3 *M* potassium chloride.

A transition from helix to coil at pH 12 for amylose, amylopectin, and glycogen is observed in the presence of sodium chloride (*146*). Studies on amylose and amylopectin in dimethyl sulfoxide show that dimethyl sulfoxide and its hydrate stabilize the helical structure (*147*). Marchessault and co-workers (*148*) conclude that amylose in dimethyl sulfoxide exhibits substantial right-handed helical character.

The iodine-induced aggregation of amylose chains by proteins, which is associated with the coil → helix transition of the amylose chain, has been reported (*294*). Near infrared and spin probe studies on the helix stability of amylose and amylopectin have been made (*295*).

IX. Solution Properties of Amylose

Langlois and Wagoner (*36*) have described the physical properties of commercial grade, retrograded amylose and have provided quantitative data on its solution properties. BeMiller (*46*) has reviewed starch amylose and amylose derivatives. The solubility of amylose in water has been discussed (*296, 297*).

Although retrograded amylose is essentially insoluble in cold water, some commercial preparations adsorb approximately four times their weight of water. With less water, a thixotropic gel forms (*36*).

Commercial amylose is retrograded and is insoluble in water unless heated above the 124° transition temperature for a short time which destroys hydrogen bonding at neutral pH (*149*). Once solubilized, amylose remains cold water soluble on rapid drying from the solution on a heated belt or drum because it is thereby immobilized in an amorphous state (*150–152*).

Amylose separated as the butanol complex can be obtained as a free-flowing powder soluble in boiling water to concentrations of up to 15% by weight (*74*). Commercial retrograded amyloses rapidly disperse in water at 150°. Solubilization over a period of only 1–2 min is recommended to avoid hydrolytic degradation. In general, amylose is not seriously degraded in water at neutral pH and 140° for up to 20 min or at 170° for periods up to 2 min (*149*).

Differences in the crystalline–amorphous ratio of amylose can influence viscosity behavior of its dispersions (*149*). Incompletely solubilized crystallites of amylose serve as nuclei to initiate gelation. Highly retrograded corn starch amylose is not dissolved completely when a 10% by weight slurry is held at 150° for 2 min, while a more amorphous amylose is dissolved under these conditions. Most of the amylose and nuclei are dissolved at 160° in 2 min, and the solution exhibits a stable viscosity at 90°.

The rate of amylose gelation or retrogradation is a function of time, temperature, pH, concentration of amylose and the method used to solubilize the amylose. The more completely the amylose is dissolved, the lower the tendency of the amylose to gel. Solution stability is increased by lowering concentration. Gelation rates increase at high solids, 20% by weight being about the upper limit possible for control of gelation.

Defatted amylose gives solutions of high clarity. In the presence of fat, amylose solutions show formation of fatty acid complexes as opaqueness, even at temperatures above the gelation temperature. Rates of gelation increase with increasing reduced viscosity in the range of 1.1 $dL \cdot g^{-1}$ to 1.6 $dL \cdot g^{-1}$, while the rates of precipitation of the retrograded amylose as well as the precipitation of the fatty acid-amylose complex increase with decreasing viscosity. Gel time for commercial corn amylose of 700 DP decreases with increasing concentration, and increases with increasing holding temperature. The relationship does not hold for amylose of other molecular weights. Muetgeert and Bus (*153*) found that potato amylose of 1000 DP has a much lower gelation temperature.

Amylose gels which are opaque and chalk-white may be resolubilized by heating above 124°. Examination of the gels with a microscope reveals a granular or spherulitic structure typical of retrogradation (*153*). The gel strength of corn starch amylose increases with concentration. It also increases with decreasing temperature and age, but usually reaches its maximum in 24 h or less.

Addition of aldehydes, such as formaldehyde or glyoxal, to amylose solutions increases their stability, very likely from hemi-acetal formation which makes the amylose less stereoregular (*75, 154*).

Amylose solutions are stabilized by alkali in the absence of oxygen. Concentrations of alkali greater or less than a critical amount promote gelation. Viscosity is a minimum in 1 *M* potassium hydroxide and sodium hydroxide solutions. Rigid, transparent, colorless gels are reportedly made from alkaline solutions (*155*). The degradative effects of oxygen in the presence of alkali are well known (*80, 81, 111, 298*). Amylose dispersions of limited stability are reported for alkaline solutions that are neutralized (*156*).

Amylose is soluble in aqueous chloral hydrate, formamide, dichloroacetic acid, pyrrolidine, dimethyl sulfoxide, acetamide, ethylenediamine, piperazine, formic acid, and urea. Concentrations in most of these solvents are limited by viscosities. Dimethyl sulfoxide is a solvent of great interest. The anhydrous reagent will dissolve up to 50% by weight of commercial amylose with 10% moisture. The amylose can then be precipitated by addition of water.

X. Solution Properties of Amylopectin

Powell (*47*) has reviewed starch amylopectin. Once aggregation owing to hydrogen bonding has been eliminated, amylopectins give stable solutions that are not readily subject to retrogradation, gelation, or precipitation. Highly concentrated solutions of amylopectin slowly retrograde.

XI. Amylose Films

The physical properties of corn amylose films have been described by Langlois and Wagoner (*36*). Methods of preparing amylose films by casting from aqueous solutions have been reported (*149, 157–168*).

Water is a plasticizer for amylose. Water weakens the intermolecular forces; relative humidity increases and reduces the tensile strength and stiffness (elastic modulus). Tensile strengths of about 13,000 psi under dry conditions for corn amylose (*160*) and potato amylose (*161*) films cast and dried at room temperature have been reported. The tensile strength, elongation, and fold resistance of the films reportedly decrease when the intrinsic viscosity falls below 1.0 dL g^{-1} in accord with the general behavior of polymeric materials. The tensile properties of films are decreased by amylopectin (*149, 160*), because the branched molecules cannot orient and pack as closely as linear molecules. A film cast from corn starch has about the same tensile strength as a film cast from a solution containing 75% by weight amylopectin and 25% by weight amylose.

Protzman and co-workers (*149*) have determined the effects of solubilization temperature and casting temperature on amylose films. Commercial-grade corn amylose was solubilized at 120°–170° and pH 7.0 for 1–7 min. The amylose dispersion was defatted using filter aid supported on micrometallic filters. When an amylose film is prepared by casting and drying on a support at a temperature above the gelation temperature, the resulting film has less tendency to swell on exposure to water and appears to be more dense than amylose films cast and dried at room temperature. Amylose film cast and dried above its gelation temperature has initially a much lower degree of hydrogen bonding than amylose film prepared at room temperature. Amorphous films tend to embrittle on aging, particularly in the presence of certain agents such as a polyhydric alcohol humectant.

In large measure, the final crystallinity of an amylose film is dependent upon the completeness of dissolution of amylose in water and the extent of hydrogen bonding in the film (*149*). Amylose films having excellent properties can be produced by dissolving amylose in water at a temperature of at least 130°, casting the aqueous solution onto a support having a temperature at least 10° less than the gelation temperature of the amylose solution, and drying the amylose gel at a temperature above the gelation temperature.

Rapid drying of the gel inhibits the crystallization that occurs when the amylose gel is dried slowly at room temperature. Rapid drying results in the films being more dense and more water resistant (*149*). Small amounts of fat (1–2%) tend to detract from clarity. Fat also reduces the effectiveness of plasticizers. Suitable methods of defatting amylose are reported (*119, 169*).

Films of dimethyl sulfoxide-pretreated high-amylose starch dried at temperatures above 50° are reported to have an amorphous x-ray pattern, whereas films dried at 25° give the B pattern, typical of retrograded starch (*170*). Zobel (*171*) has described the improved flex-fold resistance of amorphous amylose films relative to those showing crystalline x-ray patterns (*171*). Price and Haskell (*172*) report improved properties of amorphous cellophane produced by a dry, high-temperature casting process or by regeneration of amorphous cellulose acetate in the dry state.

The swelling of amylose films reportedly is decreased by subjecting the film to superheated steam (*160*). Zobel (*173*) points out that starch films precipitated from pastes by evaporation at temperatures below about 50° exhibit the B pattern; above 50°, the A and C patterns result. (See Chap. VII for more on X-ray patterns.)

Evidence strongly suggests that the B pattern is associated with the more open structure, while the A and C patterns are characteristic of the more compact structure. The B pattern includes a medium intensity interplanar d-spacing of 15.8 Å. In contrast, the A pattern gives weak intensities for the interplanar d-spacings for values higher than the strong intensity 5.78 Å d-spacing. The C pattern also has weak interplanar d-spacing for values higher than the strong 5.78 Å spacing. The presence of a weak 15.4 Å d-spacing in the C pattern suggests it is a more open structure than the A structure, but more compact than the B structure. These observations are in agreement with the water absorption properties for different starches (*174*). The B pattern observed for potato starch has the greatest water absorption, followed by the C pattern observed for sago and arrowroot starch, followed by the A pattern observed for corn and wheat starch.

Typical amylose plasticizers are ethylene glycol and glycerol. The solvent or swelling properties of these plasticizers, however, facilitate movement of amorphous chain segments of amylose molecules toward crystallization (*149*).

The elastic modulus (stiffness) of amylose films plasticized with 20% by weight glycerol and solubilized for 2 min at 150° increases with aging. More stable amylose films having excellent properties can be produced by casting at temperatures below the gelation temperature to obtain a dimensionally stable wet film which does not flow on the casting surface and which is then rapidly dried at elevated temperature (*149*).

Crystallization in amylose films reportedly causes decreased swelling in superheated steam (*160*) and conversion of the B x-ray pattern to the A pattern (*173*). Tobolsky (*175*) has examined the viscoelastic properties of amylose films plasticized with dimethyl sulfoxide and found that the modulus-temperature curves at high concentrations are typical for semi-crystalline polymers and transform to those of amorphous polymers when amylose contents below about 60% by weight are used. Amylose seems to exist as a semi-crystalline polymer above 60% by weight amylose with crystalline regions acting as cross-links.

Higher levels of plasticizer as well as high storage temperature can increase the diffusion of molecular chain segments and may account for the higher crystallinity rates observed, especially in glycerol-plasticized films. In fact, films of amylose that have been plasticized with glycerol and cast and aged at room temperature embrittle more and develop lower elongation with age than do unplasticized films. A quick chill and proper plasticizer prevents crystallization of polycarbonate resin (*176*).

Suitable amylose plasticizers include polyhydric alcohols, amino alcohols, hydroxyalkyd amides, and quaternary ammonium compounds (*177–181*). Poly(vinyl alcohol) is a useful polymeric plasticizer for amylose (*182*). Poly(vinyl alcohol) can be used alone or together with one or more other plasticizers. Poly(vinyl alcohol) increases the elongation and tear strength of amylose films. Although there is some decrease in tensile strength, it is not as sensitive to humidity change as is glycerol. Amylose films plasticized with poly(vinyl alcohol) retain their elongation and plasticized condition at low humidities. High-molecular-weight, fully hydrolyzed grades of poly(vinyl alcohol) are compatible with amylose in films and enhance the film properties, probably by lowering intermolecular attraction and interfering with the packing of the isotactic amylose molecules which, in the unplasticized condition, result in excessive crystallinity, similar to isotactic polypropylene.

Both the toughness (area under tensile stress–strain curves) and flex-fold resistance of amylaceous film increase with increasing amounts of the linear isotactic amylose polymer, as well as with increasing amounts of poly(vinyl alcohol). In the absence of poly(vinyl alcohol), regular corn starch containing the greatest amount of branched amylopectin has the lowest toughness and flex-fold resistance. As the poly(vinyl alcohol) level increases, both toughness and flex-fold resistance of all amylaceous films, but especially those of higher amylose content, improve. Tensile strength and elongation of films prepared from different corn starches decrease with decreasing amylose content (*183*).

Amylose films are reportedly equivalent to cellulose films in water vapor permeability and in permeability to organic vapors (*184*). Amylose film is more permeable to polar gases, such as ammonia and sulfur dioxide.

Certain methods for making flat films or tubes from amylose have been reported to be based on the regeneration of an alkaline solution of amylose (*185, 190*). Amylose has been solubilized up to 50% by weight in aqueous solutions of formaldehyde which have been cast or formed into films and fibers (*154*) with or without regeneration using ammonium hydroxide. Wagoner and Protzman (*191*) have extruded plasticized objects in the form of films, filaments, and tubes from amylose. Others (*192–198*) reportedly have also investigated the extrusion of amylose. Film has been reported to be produced from amylose obtained by treating starch with α-1,6-glucosidase (*199*).

Self-supported amylose films reportedly having excellent transparency, flexibility, tensile strength, and water insolubility may be prepared from amylose by

Table VI

Uses of Amylose and High-Amylose Starch

Product	*Amylose function*	*Ref.*
Film		
Water-soluble	Nontoxic, blended with poly(vinyl alcohol)	*201,202*
Biodegradable	Substrate for enzyme attack	*203*
Meat casing	Water soluble for skinless products	*159*
Edible	Food wrapping and coating	*204–207*
Heat sealing	Food packaging film	*208*
Packaging	Impermeable to oil and oxygen	*209*
Plasticized	Enzyme debranched starch	*201,209,210*
Water resistant	Hydrophobic plasticizer and/or coating	*211–214*
Water resistant	Crosslinked by irradiation	*215*
Laminate	Coating on paper, heat sealable	*216*
Coating		
Ink	Marking ink for textiles when complexed with iodine	*217*
Glass fiber	Reinforcement for polyester or epoxy resin	*218*
Paper	Grease resistance	*219,220*
Laminate	Grease resistant, oxygen resistant	*221*
Food		
High-amylose	Thickener	*222*
Gum and jelly confections	Reduced gelation time	*223–225*
Confection	Creme centers	*226*
Fried potatoes	Binding agent for dehydrated potatoes	*227*
Potatoes	Coating to reduce oil absorption during deep fat frying	*228*
Fruits	Coating prevents clumping and sticking together	*229*
Pudding	Cold-water-soluble thickener for instant pudding	*150–152,230*
Pudding	Thickener, aseptically packed	*231*
Gelatin replacement	Gelling agent	*232*
Tomato paste, applesauce	Texturizing agent to produce pulpy texture	*233*
Review—food		*299*
Bread	Quality and anti-staling improvement	*234–236*
Pastry dough	Decreased sogginess, less shrinkage	*236,237*
Encapsulating agent	Coating	*238,239*
Pharmaceutical		
Tablets	Binding agent	*240–243*
Gels and foams		
Air freshener	Gelling agent	*244–246*
Foams	For edible food	*247–249*
Sponge	Structural material	*250–251*
Adhesive		
Corrugating	Carrier solution stabilized by formaldehyde or high temperature	*252–254*
Acoustical tile	Binder	*255*
Cement	Binder	*256*

Table VII

Amylose and High Amylose Starch Derivatives

Derivatives	*Uses*	*Reference*
Esters		
Acetate	Water-soluble film	*257*
	Textile printing paste, hydroxyethylated	*258*
	Coating for French fried potatoes reduces use of cooking oil	*259*
Benzoate	Water-soluble film	*257*
Glutarate	Hydrogels and sponges	*260*
Phosphate	Foundry core binder	*261,262*
Propionate	Water-soluble film	*257*
Succinate	Water-soluble meat casing for skinless frankfurter	*263*
	Hydrogels and sponges	*260*
Sulfonate	Amphoteric coating with tertiary amino groups for textile size	*264–266*
Fatty acids	Waterproofing agents for textiles	*267*
Ethers		
Cyanoethyl	Water-soluble film for dry detergents	*257,268,269*
Hydroxyalkyl	Water-soluble film for dry detergents	*257,268,269*
	Water-soluble, nontoxic filaments for medical sutures, bandages and filters	*270,271*
	Oriented filaments	*169,175,193,195, 196,197,202, 272–274,289*
Hydroxyalkyl	High water binding, good film former	*275*
	Textile size	*276*
	Textile printing of hydrophobic fibers	*258*
	Glass fiber size	*277,278*
	Retort thickener for foods such as pudding	*279–281*
	Hydrogel for slow release of drugs	*282*
	Tobacco binder	*283*
Carboxyalkyl	Coatings or binders	*284*
Carboxymethyl	Hydrogel for slow release of drugs	*282,285*
Miscellaneous		
Cross-linked, grafted toluene diisocyanate	Biodegradable packaging film	*286,287*
Irradiated, 0–4 Mrad	Water resistant film for wrapping food	*215*
Tertiary amino	Amphoteric size with sulfonic acid or sulfonate groups for textiles	*264,265*
Acetals	Water-soluble size for natural and synthetic fibers	*266*
Carbamate	Textile size	*288,300*
Carboxylated (oxidized)	Glass fiber size	*289*
Cross-linked with epichlorohydrin	Retort thickener for foods such as pudding	*279–281*
Graft copolymer with acrylamide	Adhesive tape	*301,302*

chemical modification of the surface of the preformed film with lower alkyl halides, fatty acid anhydrides, ketenes, acetyl halides, benzyl halides, lower alkylene oxides, lactones of lower carboxylic acids, phenyl or lower alkyl isocyanates, organosilicon halides, phenyldichlorophosphine oxide, or dibutylchloroborane (*200*).

XII. Uses for Amylose and Amylopectin

Industrial uses of amylose (*36, 46*) and of amylopectin (*47*) have been described. The high cost of energy for fractionation of starch makes it difficult for the amylose and amylopectin to compete with products made from high-amylose starch and waxy maize starch. To improve solution stability of amylose, it has been hydroxyethylated and hydroxypropylated. In some applications, with proper process control for solution casting and extrusion techniques, advantage has been taken of the water-resistant properties of unmodified amylose and high-amylose starch. Some of these applications are summarized in Tables VI and VII.

XIII. References

(*1*) W. Banks and C. T. Greenwood, "Starch and Its Components," Edinburgh University Press, Edinburgh, Scotland, 1975, p. 247.
(*2*) A. Van Leeuwenhoek, Opera omina, Vol. IV, "Epistolae Physiologicae super compluribus Naturae Arcanus," (1719) p. 232; quoted by A. Payen, *Ann. Sci. Nat. Bot.*, **[2], 10,** 8 (1838).
(*3*) R. L. Whistler, *in* "Starch: Chemistry and Technology," R. L. Whistler and E. F. Paschall, eds., Academic Press, New York, 1965, Vol. I, p. 331.
(*4*) R. W. Kerr, *in* "Chemistry and Industry of Starch," R. W. Kerr, ed., Academic Press, New York, 2nd Ed., 1950, p. 180.
(*5*) K. H. Meyer, "Natural and Synthetic High Polymers," Interscience Publishers, New York, 2nd Ed., 1950, p. 458.
(*6*) T. J. Schoch, *Advan. Carbohyd. Chem.*, **1,** 247 (1945).
(*7*) J. Muetgeert, *Advan. Carbohydr. Chem.*, **16,** 299 (1961).
(*8*) J. Potze, *in* "Starch Production Technology," J. A. Radley, ed., Applied Science Publishers, London, 1976, p. 257.
(*9*) C. T. Greenwood, *Advan. Carbohydr. Chem.*, **11,** 342 (1956).
(*10*) G. K. Adkins and C. T. Greenwood, *Carbohydr. Res.*, **11,** 220 (1969).
(*11*) K. H. Meyer, W. Brentano, and P. Bernfeld, *Helv. Chim. Acta,* **23,** 845 (1940).
(*12*) K. H. Meyer, M. Wertheim, and P. Bernfeld, *Helv. Chim. Acta,* **23,** 865 (1940).
(*13*) C. T. Greenwood and J. Thomson, *J. Chem. Soc.*, 222 (1962).
(*14*) L. Maquenne and E. Roux, *Compt. Rend.*, **140,** 1303 (1905); *Bull. Soc. Chim.*, **[3], 33,** 723 (1905).
(*15*) R. L. Whistler and W. M. Doane, *Cereal Chem.*, **38,** 251 (1961).
(*16*) S. Lansky, M. Kooi, and T. J. Schoch, *J. Amer. Chem. Soc.*, **71,** 4066 (1949).
(*17*) B. A. Lewis and F. Smith, *J. Amer. Chem. Soc.*, **79,** 3929 (1957).
(*18*) R. L. Whistler and G. E. Hilbert, *J. Amer. Chem. Soc.*, **67,** 1161 (1945).

(*19*) R. W. Kerr and F. C. Cleveland, *J. Amer. Chem. Soc.*, **74,** 4036 (1952).
(*20*) C. T. Greenwood, *in* "The Carbohydrates—Chemistry and Biochemistry," W. Pigman and D. Horton, eds., Academic Press, New York, 2nd Ed., 1970, Vol. IIB, p. 481.
(*21*) R. L. Whistler, *in* "Starch: Chemistry and Technology," R. L. Whistler and E. F. Paschall, eds., Academic Press, New York, 1965, Vol. I, p. 332.
(*22*) J. F. Foster, *in* "Starch: Chemistry and Technology," R. L. Whistler and E. F. Paschall, eds., Academic Press, New York, 1965, Vol. I, p. 349.
(*23*) E. Y. C. Lee, C. Mercier, and W. J. Whelan, *Arch. Biochem. Biophys.*, **125,** 1028 (1968).
(*24*) G. K. Adkins, W. Banks, and C. T. Greenwood, *Carbohydr. Res.*, **2,** 502 (1966).
(*25*) M. L. Wolfrom and A. Thompson, *J. Amer. Chem. Soc.*, **78,** 4116 (1956).
(*26*) W. Banks and C. T. Greenwood, *J. Chem. Soc.*, 3436 (1959).
(*27*) J. M. G. Cowie and C. T. Greenwood, *J. Chem. Soc.*, 4640 (1957).
(*28*) W. Dvonch and R. L. Whistler, *J. Biol. Chem.*, **181,** 889 (1949); W. Dvonch, H. J. Yearin and R. L. Whistler, *J. Amer. Chem. Soc.*, **72,** 1748 (1950).
(*29*) A. S. Perlin, *Can. J. Chem.*, **36,** 810 (1958).
(*30*) E. L. Hirst, J. K. N. Jones, and A. J. Roudier, *J. Chem. Soc.*, 1779 (1948).
(*31*) D. M. W. Anderson, C. T. Greenwood, and E. L. Hirst, *J. Chem. Soc.*, 225 (1955).
(*32*) F. L. Bates, D. French, and R. E. Rundle, *J. Amer. Chem. Soc.*, **65,** 142 (1943).
(*33*) W. G. Campbell, J. L. Frahn, E. L. Hirst, D. G. Packman, and E. G. V. Percival, *J. Chem. Soc.*, 3489 (1951).
(*34*) M. S. Zuber, *in* "Starch: Chemistry and Technology," R. L. Whistler and E. F. Paschall, eds., Academic Press, New York, 1965, Vol. I, p. 44.
(*35*) F. Senti, *in* "Starch: Chemistry and Technology," R. L. Whistler and E. F. Paschall, eds., Academic Press, New York, 1967, Vol. II, p. 499.
(*36*) D. P. Langlois and J. A. Wagoner, *in* "Starch: Chemistry and Technology," R. L. Whistler and E. F. Paschall, eds., Academic, Press, New York, 1967, Vol. II, p. 451.
(*37*) N. P. Badenhuizen, *Protoplasma*, **33,** 440 (1939).
(*38*) A. W. Bauer and E. Pacsu, *Text. Res. J.*, **23,** 870 (1953).
(*39*) W. A. Richardson, R. S. Higginbottom, and F. D. Farrow, *Shirley Inst. Mem.*, **14,** 63 (1940).
(*40*) W. Banks and C. T. Greenwood, "Starch and its Components," Edinburgh University Press, Edinburgh, Scotland, 1975, p. 2.
(*41*) W. Banks and C. T. Greenwood, "Starch and its Components," Edinburgh University Press, Edinburgh, Scotland, 1975, p. 306.
(*42*) J. Cazes, *J. Chem. Ed.*, **43,** A567 (1966).
(*43*) J. Cazes, *J. Chem. Ed.*, **43,** A625 (1966).
(*44*) W. C. Bus, J. Muetgeert, and P. Hiemstra, U.S. Patents 2,822,305 (1958), *Chem. Abstr.*, **52,** P9635c (1958); 2,829,987 (1958), *Chem. Abstr.*, **52,** P13295d (1958); 2,829,988 (1958), *Chem. Abstr.*, **52,** P14204i (1958); 2,829,989 (1958), *Chem. Abstr.*, **52,** P13295d (1958); 2,829,990 (1958), *Chem. Abstr.*, **52,** P17768e (1958).
(*45*) J. Potze, *in* "Starch Production Technology," J. A. Radley, ed., Applied Science Publishers, London, 1976, p. 267.
(*46*) J. N. BeMiller, *in* "Industrial Gums: Polysaccharides and Their Derivatives," R. L. Whistler, ed., Academic Press, New York, 2nd Ed., 1973, p. 546.
(*47*) E. L. Powell, *in* "Industrial Gums: Polysaccharides and Their Derivatives," R. L. Whistler, ed., Academic Press, New York, 2nd Ed., 1973, p. 567.
(*48*) F. W. Billmeyer, Jr., "Textbook of Polymer Science," Wiley-Interscience, New York, 2nd Ed., 1971, p. 84.
(*49*) H. Staudinger and W. Heuer, *Ber.*, **63B,** 222–234 (1930).
(*50*) W. Banks and C. T. Greenwood, "Starch and Its Components," Edinburgh University Press, Edinburgh, Scotland, 1975, p. 121.

(*51*) V. S. R. Rao and J. F. Foster, *J. Phys. Chem.*, **69,** 656 (1965).
(*52*) J. M. G. Cowie, *J. Polymer Sci.*, **49,** 455 (1961).
(*53*) E. Husemann, W. Burchard, B. Pfannemüller, and R. Werner, *Staerke*, **13,** 196 (1961).
(*54*) W. Burchard and E. Husemann, *Makromol. Chem.*, **44/46,** 358 (1961).
(*55*) W. Banks, R. Geddes, C. T. Greenwood, and I. G. Jones, *Staerke*, **24,** 245 (1972).
(*56*) W. Banks, C. T. Greenwood, and D. D. Muir, *Staerke*, **25,** 153 (1973).
(*57*) R. Geddes, C. T. Greenwood, and S. MacKenzie, *Carbohydr. Res.*, **1,** 71 (1965).
(*58*) C. T. Greenwood, *Staerke*, **12,** 169 (1960).
(*59*) P. J. Killion and J. F. Foster, *J. Polymer Sci.*, **46,** 65 (1960).
(*60*) W. Banks and C. T. Greenwood, *Staerke*, **19,** 394 (1967).
(*61*) A. H. Young and P. G. Griffin, unpublished work, 1969.
(*62*) Anon., "Dextran Fractions, Dextran Sulphate, DEAE-Dextran, Defined Polymers for Biological Research," Pharmacia Fine Chemicals, Piscataway, N.J., 1971, p. 5.
(*63*) F. R. Dintzis and R. Tobin, *J. Chromatogr.*, **88,** 77 (1974).
(*64*) T. Yamada and M. Taki, *Staerke*, **28,** 374 (1976).
(*65*) Anon., "Sepharose[R] Agarose Gels in Bead Form," Pharmacia Fine Chemicals, Piscataway, N.J.
(*66*) R. Ebermann and R. Schwarz, *Staerke*, **27,** 361 (1975).
(*67*) R. Ebermann and W. Praznik, *Staerke*, **27,** 329 (1975).
(*68*) J. A. P. P. Van Dijk, W. C. M. Henkens, and J. A. M. Smit, *J. Polymer. Sci., Polymer Physics Ed.*, **14,** 1485 (1975).
(*69*) N. A. Smykova and B. N. Stepanenko, *Biokhim Microbiol.*, **5,** 219 (1969).
(*70*) M. Taki, *Agr. Biol. Chem.*, **26,** 1 (1962).
(*71*) M. Ulmann and M. Richter, *Staerke*, **14,** 455 (1962).
(*72*) M. Ulmann, *Kolloid-Z.*, **116,** 10 (1950); **123,** 105 (1951); *Biochem. Z.*, **321,** 377 (1951); *Naturwissenschaften*, **37,** 309 (1950); *Makromol. Chem.*, **9,** 76 (1952).
(*73*) E. H. Fischer and W. Settele, *Helv. Chim. Acta*, **36,** 811 (1953).
(*74*) R. W. Kerr and G. M. Severson, *J. Amer. Chem. Soc.*, **65,** 193 (1943).
(*75*) T. J. Schoch, *J. Amer. Chem. Soc.* **64,** 2957 (1942); *Cereal Chem.*, **18,** 121 (1941).
(*76*) H. Baum and G. A. Gilbert, *Chem. Ind.*, 490 (1954).
(*77*) W. Banks, C. T. Greenwood, and J. Thomson, *Makromol. Chem.*, **31,** 197 (1959).
(*78*) E. M. Montgomery and F. R. Senti, *J. Polymer Sci.*, **28,** 1 (1958).
(*79*) E. M. Montgomery, K. R. Sexson, and F. R. Senti, *Staerke*, **13,** 215 (1961).
(*80*) R. T. Bottle, G. A. Gilbert, C. T. Greenwood, and K. N. Saad, *Chem. Ind.*, 541 (1953).
(*81*) H. Baum and G. A. Gilbert, *Chem. Ind.*, 489 (1954).
(*82*) E. M. Montgomery and K. R. Sexson, U.S. Patent 3,046,161 (1962).
(*83*) H. Baum and G. A. Gilbert, *J. Colloid Sci.*, **11,** 428 (1956).
(*84*) W. Banks and C. T. Greenwood, "Starch and Its Components," Edinburgh University Press, Edinburgh, Scotland, 1975, p. 13.
(*85*) R. S. Bear, *J. Amer. Chem. Soc.*, **66,** 2122 (1944).
(*86*) F. F. Mikus, R. M. Hixon, and R. E. Rundle, *J. Amer. Chem. Soc.*, **68,** 1115 (1946).
(*87*) F. F. Mikus, *Iowa State Coll. J. Science*, **22,** 58 (1947).
(*88*) D. French, A. O. Pulley, and W. J. Whelan, *Staerke*, **15,** 349 (1963).
(*89*) W. N. Haworth, S. Peat, and P. E. Sagrott, *Nature*, **157,** 19 (1946).
(*90*) E. J. Bourne, G. H. Donnison, N. Haworth, and S. Peat, *J. Chem. Soc.*, 1687 (1948).
(*91*) D. P. Marcus and P. R. Shildneck, U.S. Patent 3,313,654 (1967); *Chem. Abstr.*, **67,** P3991t (1967).
(*92*) R. S. Leiser, D. P. Macarus, and J. A. Wagoner, U.S. Patent 3,323,949 (1967); *Chem. Abstr.*, **67,** P55325y (1967).
(*93*) T. Kuge and K. Tenichi, *Agr. Biol. Chem. (Tokyo)*, **32,** 1232 (1968).

(*94*) C. T. Greenwood, *Advan. Carbohydr. Chem.*, **11,** 335 (1956).
(*95*) G. A. Gilbert, *Staerke,* **10,** 95 (1958).
(*96*) R. Geddes, C. T. Greenwood, A. W. MacGregor, A. R. Proctor, and J. Thomson, *Makromol. Chem.*, **79,** 189 (1964).
(*97*) E. Pacsu and A. W. Bauer, U.S. Patent 2,779,694 (1957); *Chem. Abstr.*, **51,** P8459i (1957).
(*98*) R. A. Anderson and C. Vojnovich, U.S. Patent 3,252, 836 (1966); *Chem. Abstr.*, **65,** P2457b (1966).
(*99*) W. Banks and C. T. Greenwood, *Carbohydr. Res.*, **6,** 241 (1968).
(*100*) W. Banks, C. T. Greenwood, and D. D. Muir, *Staerke,* **26,** 289 (1974).
(*101*) W. Banks and C. T. Greenwood, "Starch and Its Components," Edinburgh University Press, Edinburgh, Scotland, 1975, p. 14.
(*102*) P. J. Flory, "Principles of Polymer Chemistry," Cornell University Press, Ithaca, New York, 1953, p. 344.
(*103*) S. Augustat, *Staerke,* **12,** 145 (1960).
(*104*) S. M. Cantor and E. L. Wimmer, U.S. Patent 2,779,692 (1957); *Chem. Abstr.*, **51,** P8460i (1957).
(*105*) R. W. Kerr, *J. Amer. Chem. Soc.*, **67,** 2268 (1945); *Arch. Biochem.*, **7,** 377 (1945).
(*106*) R. Speiser and R. T. Whittenberger, *J. Chem. Phys.*, **13,** 349 (1945).
(*107*) W. W. Everett and J. F. Foster, *J. Amer. Chem. Soc.*, **81,** 3459 (1959).
(*108*) J. Olkku and C. K. Rha, *Food Chem.*, **3,** 293 (1978).
(*109*) M. Samec, C. Nucic, and V. Pirkmaier, *Kolloid. Z,* **94,** 350 (1941).
(*110*) R. L. Whistler, *in* "Starch and Its Derivatives," J. A. Radley, ed., Chapman and Hall, London, England, 1953, Vol. 1, p. 213.
(*111*) R. L. Whistler and C. Johnson, *Cereal Chem.*, **25,** 418 (1948).
(*112*) K. H. Meyer, P. Bernfeld, and E. Wolff, *Helv. Chim. Acta,* **23,** 854 (1940).
(*113*) J. F. Foster and M. D. Sterman, *J. Polymer Sci.*, **21,** 91 (1956).
(*114*) J. Holló, J. Szejtli, and G. Gantner, *Magyar Tudományos Akad. Kém. Tudományos Osztályának Közleményei,* **11,** 465 (1959); *Chem. Abstr.*, **54,** 4009 (1960).
(*115*) J. Hollo, J. Szejtli, and G. Gantner, *Periodica Polytech.*, **3,** 95 (1959); *Chem. Abstr.* **54,** 5135h (1960).
(*116*) J. Hollo, J. Szejtli, and G. S. Gantner, *Staerke,* **12,** 73 (1960).
(*117*) J. Hollo, J. Szetli, and G. Gantner, *Staerke,* **12,** 106 (1960).
(*118*) O. R. Etheridge, J. A. Wagoner, J. M. McDonald, and D. A. Lippincott, U.S. Patent 3,067, 067 (1962); *Chem. Abstr.*, **58,** P4723f (1963).
(*119*) R. A. Schnell and F. Verbanac, U.S. Patent 3,255,042 (1966).
(*120*) Hoffman's Starkfabriken A. G., British Patent 1,014,105 (1965); Belgian Patent 628,265 (1963); Chem. Abstr., **60,** 14724 (1964).
(*121*) W. C. Bus, J. Muetgeert, and P. Hiemstra, U.S. Patent 2,829, 989 (1958); *Chem. Abstr.*, **52,** P13295d (1958).
(*122*) M. Seidman, U.S. Patent 3,532,602 (1970); *Chem. Abstr.*, **73,** P132225u (1970).
(*123*) M. Kurimoto and K. Sugimoto, German Offen. 2,003,335 (1970). *Chem. Abstr.*, **73,** P89423t (1970).
(*124*) M. Yoshida and M. Hirao, South African Patent 69 06,182 (1970); *Chem. Abstr.*, **73,** P111200s (1970).
(*125*) Hayashibara Co., Netherlands Patent 6,906,490 (1969).
(*126*) W. Banks and C. T. Greenwood, *Staerke,* **23,** 300 (1971).
(*127*) W. Banks and C. T. Greenwood, "Starch and Its Components," Edinburgh University Press, Edinburgh, Scotland, 1975, p. 138.
(*128*) W. Burchard, *Makromol. Chem.*, **64,** 110 (1963).
(*129*) W. W. Everett, Ph.D. Thesis, Purdue University, Lafayette, Indiana, 1959.

(*130*) W. Banks and C. T. Greenwood, *Carbohydr. Res.*, **7,** 414 (1968).
(*131*) W. Banks and C. T. Greenwood, *European Polymer J.*, **5,** 649 (1969).
(*132*) W. W. Everett and J. F. Foster, *J. Amer. Chem. Soc.*, **81,** 3464 (1959).
(*133*) J. M. G. Cowie, *Makromol. Chem.*, **42,** 230 (1960).
(*134*) W. Banks and C. T. Greenwood, *Makromol. Chem.*, **67,** 49 (1963).
(*135*) W. Banks and C. T. Greenwood, *Carbohydr. Res.*, **7,** 349 (1968).
(*136*) J. M. G. Cowie, *Makromol. Chem.*, **59,** 189 (1963).
(*137*) J. M. G. Cowie, *J. Polymer Sci.*, **49,** 455 (1961).
(*138*) W. Burchard and E. Husemann, *Makromol. Chem.*, **44/46,** 358 (1961).
(*139*) J. R. Patel, C. K. Patel, and R. D. Patel, *Staerke,* **19,** 330 (1967).
(*140*) D. A. Brant and B. K. Min., *Macromolecules,* **2,** 1 (1969).
(*141*) K. D. Goebel and D. A. Brant, *Macromolecules,* **3,** 634 (1970).
(*142*) B. B. Jorgenson and O. B. Jorgenson, *Acta Chem. Scand.*, **14,** 2135 (1960).
(*143*) S. V. Bhide and N. R. Kale, *Carbohydr. Res.*, **68,** 161 (1979).
(*144*) D. A. Brant and W. Dimpfl. *Macromolecules,* **3,** 655 (1970).
(*145*) W. Banks and C. T. Greenwood, *Carbohydr. Res.*, **21,** 229 (1972).
(*146*) S. R. Erlander, R. M. Purvinas, and H. L. Griffin, *Cereal Chem.*, **45,** 140 (1968).
(*147*) S. R. Erlander and R. Tobin, *Makromol. Chem.*, **111,** 194 (1968).
(*148*) M. St. Jacques, P. R. Sundarajan, K. J. Taylor, and R. H. Marchessault, *J. Amer. Chem. Soc.*, **98,** 4386 (1976).
(*149*) T. F. Protzman, J. A. Wagoner, and A. H. Young, U.S. Patent 3,344,216 (1967).
(*150*) A. Sarko, B. R. Zeitlin, and F. J. Germino, U.S. Patent 3,086,890 (1963); *Chem. Abstr.*, **59,** P1829c (1963).
(*151*) A. Sarko, F. J. Germino, B. R. Zeitlin, and P. F. Dapas, *J. Appl. Polymer Sci.*, **8,** 1343 (1964).
(*152*) F. J. Germino, B. R. Zeitlin, and A. Sarko, U.S. Patent 3,128,209 (1964); *Chem. Abstr.*, **60,** P14724b (1964).
(*153*) J. Muetgeert and W. C. Bus, *J. Chem. Eng. Data,* **7,** 272 (1962).
(*154*) P. Hiemstra, J. M. Muetgeert, and W. C. Bus, *Staerke,* **10,** 213 (1958).
(*155*) H. W. Leach, U.S. Patent 3,265,632 (1966).
(*156*) R. B. Evans, U.S. Patent 3,130,081 (1964). *Chem. Abstr.*, **61,** P2040g (1964).
(*157*) I. A. Wolff, H. A. Davis, J. E. Cluskey, and L. J. Gundrum, U.S. Patent 2,608,723 (1952); *Chem. Abstr.*, **47,** 2523 (1952).
(*158*) H. A. Davis, I. A. Wolff, and J. E. Cluskey, U.S. Patent 2,656,571 (1953); *Chem. Abstr.*, **48,** 1040 (1954).
(*159*) R. E. O'Brian and E. D. O'Brian, U.S. Patent 2,729,565 (1956); *Chem. Abstr.*, **50,** 5945 (1956).
(*160*) I. A. Wolff, H. A. Davis, J. E. Cluskey, L. J. Gundrum, and C. E. Rist, *Ind. Eng. Chem.*, **43,** 915 (1951).
(*161*) J. Muetgeert and P. Hiemstra, U.S. Patent 2,822,581 (1958); *Chem. Abstr.*, **52,** 8602 (1958).
(*162*) D. Kudera, U.S. Patent 2,973,243 (1961); *Chem. Abstr.*, **55,** 17030 (1961).
(*163*) A. Dekker, U.S. Patent 2,999,032 (1961); *Chem. Abstr.*, **56,** 2626 (1962).
(*164*) Cooperative VerKoop-en Productievereniging van Aardappelmeel en Derivaten "Avebe" G. A., Netherlands Patent Appl. 88,996 (1958); *Chem. Abstr.*, **53,** 23020g (1959).
(*165*) American Machine and Foundry Co., Netherlands Patent Appl. 6,501,338 (1965); *Chem. Abstr.*, **64,** 3805 (1966).
(*166*) American Machine and Foundry Co., Netherlands Patent Appl. 6,414,395 (1965); *Chem. Abstr.* **64,** 14404 (1966).
(*167*) Wolff & Co., K. G. auf Aktien, West German Patent 1,127,074 (1962); *Chem. Abstr.* **57,** 2460 (1962).

(*168*) J. N. BeMiller and R. L. Whistler, *in* "Industrial Gums," R. L. Whistler, ed., Academic Press, New York, 1st Ed., 1959, p. 675.
(*169*) D. P. Macarus and C. L. Royal, U.S. Patent 3,238,064 (1966).
(*170*) A. M. Mark, W. B. Roth, H. F. Zobel, and E. L. Mehltretter, *Cereal Chem.*, **42,** 209 (1965).
(*171*) H. F. Zobel, personal communication, Sept. 1965.
(*172*) C. R. Price and V. C. Haskell, *J. Appl. Polymer Sci.*, **5,** 635 (1961).
(*173*) H. F. Zobel, *in* "Methods in Carbohydrate Chemistry," R. L. Whistler, ed., Academic Press, New York, 1964, p. 109.
(*174*) J. Hofstee, *Staerke*, **14,** 318 (1962).
(*175*) S. Nakamura and A. V. Tobolksky, *J. Appl. Polymer Sci.*, **11,** 1371 (1967).
(*176*) J. K. Sears and J. R. Darby, *SPE Journal*, **19,** 623 (1963).
(*177*) H. M. Walton, U.S. Patent 3,312,560 (1967); *Chem. Abstr.*, **67,** 12119s (1967).
(*178*) M. T. Tetenbaum, U.S. Patent 3,318,715 (1967); *Chem. Abstr.*, **67,** 23096h (1967).
(*179*) A. H. Young, U.S. Patent 3,312,559 (1967); *Chem. Abstr.*, **67,** 3994w (1967).
(*180*) A. H. Young, M. T. Tetenbaum, and R. J. Pratt, U.S. Patent 3,320,081 (1967); *Chem. Abstr.*, **67,** 23095g (1967).
(*181*) A. H. Young, U.S. Patent 3,314,810 (1967).
(*182*) A. H. Young, U.S. Patent 3,312,641 (1967).
(*183*) N. E. Lloyd and L. C. Kirst, *Cereal Chem.*, **40,** 154 (1963).
(*184*) J. C. Rankin, I. A. Wolff, H. A. Davis, and E. C. Rist, *Chem. Eng. Data Ser.*, **3,** 120 (1958).
(*185*) P. Hiemstra and J. Muetgeert, U.S. Patent 2,902,336 (1959); *Chem. Abstr.*, **54,** 915 (1960).
(*186*) Etzkorn and Co., West German Patent 1,063,325 (1959); *Chem. Abstr.*, **55,** 15951 (1961).
(*187*) A. G. Kalle, British Patent 847,431 (1960); *Chem. Abstr.*, **55,** 6896 (1961); see also West German Patent 1,097,793 (1961) and French Patent 1,205,436 (1960).
(*188*) W. B. Kunz, U.S. Patent 3,030,667 (1962); *Chem. Abstr.*, **57,** 3684 (1962).
(*189*) F. C. Wohlrabe, A. M. Mark, W. B. Roth, and C. L. Mehltretter, U.S. Patent 3,116,351 (1963); *Chem. Abstr.*, **60,** 9466 (1964).
(*190*) F. E. Carevic, U.S. Patent 3,336,429 (1967); *Chem. Abstr.*, **67,** P101244w (1967).
(*191*) A. E. Staley Mfg. Co., Belgian Patent 641,123 (1964); *Chem. Abstr.*, **64,** 11429 (1966); see also British Patent 1,075,001 (1967).
(*192*) Midwest Research Inst., 16th Annual Report, May 1961.
(*193*) Nebraska, Dept. of Agriculture and Inspection, British Amended Patent 965,349 (1969); *Chem. Abstr.* **76,** 35432 (1972).
(*194*) C. G. Caldwell and O. B. Wurzburg, Paper presented at the 26th Annual Packaging Forum, October 19–21, 1964, New York City Packaging Institute.
(*195*) E. H. Simpson and Polleck, SPE Technical Papers, IX Nineteenth ANTEC, Feb 26–March 1, 1963, p. XVIII-2.
(*196*) A. Asanumu and Y. Fujiwara, Japanese Patent 75/86,557 (1975); *Chem. Abstr.*, **83** 195628e (1975).
(*197*) A. Asanuma and K. Iwasa, Japanese Patent 75/105,766 (1975); *Chem. Abstr.*, **84,** P6833e (1976).
(*198*) A. Asanuma and Y. Fujihara, Japanese Patent 75/122,553 (1975); *Chem. Abstr.*, **84,** 19531e (1976).
(*199*) M. Yoshida and M. Hirao, German Patent 2,055,029 (1971). *Chem. Abstr.*, **75,** P89446u (1971).
(*200*) S. A. Buckler and F. J. Germino, U.S. Patent 3,398,015 (1968); *Chem. Abstr.*, **69,** P97855t (1968).
(*201*) H. Hijiya and M. Shiosaka, German Patent 2,235,991 (1973); *Chem. Abstr.* **78,** 125459y (1973).
(*202*) H. Hijiya and M. Hirao, German Patent 2,206,950 (1972); *Chem. Abstr.*, **78,** 44672k (1973).

(*203*) P. Engler, S. A. Bradley, and S. H. Carr, *Tech. Paper, Reg. Tech. Conf., Soc. Plastics Eng., Chicago Sec., 1972,* p. 78; *Chem. Abstr.*, **78,** 59911m (1973).
(*204*) A. Asanuma and Y. Fujiwara, Japanese Patent 75/75,655 (1975); *Chem. Abstr.*, **83,** 130302z (1975).
(*205*) A. Asanuma and Y. Fujiwara, Japanese Patent 75/75,654 (1975); Chem. Abstr., **83,** 130303a (1975).
(*206*) T. Kokura, *Shokuhin Kogyo,* **14,** 33 *(1971); Chem. Abstr.*, **76,** 32913 (1972).
(*207*) K. Hayashibara, French Patent 2,116,021 (1972); *Chem. Abstr.*, **79,** 114194c (1973).
(*208*) R. Nakatsuka, S. Suzuki, E. Funatsu and S. Tanimoto, Japanese Patent 76/121,063 (1976); *Chem. Abstr.*, **86,** 57104s (1977).
(*209*) H. Hijiya and M. Yoshida, German Patent 2,137,767 (1972); *Chem. Abstr.*, **76,** P155918y (1972).
(*210*) M. Yoshida, German Patent 2,103,620 (1971); *Chem. Abstr.*, **76,** 15964k (1972).
(*211*) A. Asanuma and K. Iwasa, Japanese Patent 75/105,767 (1975); Chem. Abstr., **84,** P6832d (1976).
(*212*) K. K. Teijin, Japanese Patent Appl. NS 119853/75 (1975).
(*213*) P. Sumitomo Bakelite K. K., Japanese Patent Appl. NS 5270/78 (1978).
(*214*) K. Yokoyama, N. Yano, M. Kishi, and K. Ainoya, Japanese Patent 76/44,180 (1976); *Chem. Abstr.*, **85,** P65150y (1976).
(*215*) W. R. Grace and Co., Netherlands Patent Appl. 6,600,774 (1966); *Chem. Abstr.*, **66,** 96462p (1967).
(*216*) O. B. Wurzburg and W. Herbst, U.S. Patent 3,071,485 (1963); *Chem. Abstr.* **58,** 7022d (1963).
(*217*) I. Okada, Japanese Patent 75/132,284 (1975); *Chem. Abstr.*, **84,** 123617y (1975).
(*218*) C. E. Nalley and D. L. Motsinger, German Patent 2,237,902 (1973); *Chem. Abstr.*, **79,** 19695h (1973).
(*219*) R. W. Best and R. M. Powers, U.S. Patent 3,329,523 (1967); *Chem. Abstr.*, **67,** P74685g (1967).
(*220*) R. M. Powers, U.S. Patent 3,329,525 (1967), *Chem. Abstr.*, **67,** P65645s (1967).
(*221*) E. E. Kimmel and A. H. Young, U.S. Patent 3,661,697 (1972); *Chem. Abstr.*, **77,** P49613f (1972).
(*222*) C. H. Hullinger, E. Van Pattern, and J. A. Freck, *Food Technol.*, **27,** 22 (1973).
(*223*) J. W. Robinson and F. H. Brock, U.S. Patent 3,218,177 (1965).
(*224*) O. B. Wurzburg and W. G. Kunze, U.S. Patent 3,265,509 (1966); *Chem. Abstr.*, **65,** P14343d (1966).
(*225*) F. J. Germino and J. R. Carracci, Jr., British Patent 1,259, 368 (1972); *Chem. Abstr.* **76,** 84667c (1972).
(*226*) C. O. Moore, U.S. Patent 3,687,690 (1972).
(*227*) C. W. Cremer, French Patent 2,068,997 (1971); *Chem. Abstr.*, **77,** 33152u (1972).
(*228*) E. M. Van Patten and J. A. Freck, U.S. Patent 3,751,268 (1973); *Chem. Abstr.*, **79,** P124923d (1973).
(*229*) C. O. Moore and J. W. Robinson, U.S. Patent 3,368,909 (1968); *Chem. Abstr.*, **68,** P77068q (1968).
(*230*) J. R. Feldman, R. E. Klose, and R. V. MacAllister, U.S. Patent 3,515,591, (1970); *Chem. Abstr.*, **73,** P65210v (1970).
(*231*) E. P. L. DeMuynck and A. E. Van Brateghem, South African Patent 68/02,001 (1968); *Chem. Abstr.*, **70,** P95600c (1969).
(*232*) E. R. Jensen and J. E. Long, U.S. Patent 3,666,557 (1972); *Chem. Abstr.*, **77,** P60346s (1972).
(*233*) N. G. Morotta, H. Bell, and P. C. Trubiano, U.S. Patent 3,650,770 (1972).

(*234*) M. Yoshida, M. Mitsuhashi, and M. Hirao, South African Patent 700,730 (1970); *Chem. Abstr.*, **75,** P34225r (1971).
(*235*) H. Hiromi and M. Shiosaka, U.S. Patent 3,872,228 (1975); *Chem. Abstr.*, **83,** 26550t (1975).
(*236*) E. M. Van Patten, D. O'Rell, and L. B. Kdondrot, U.S. Patent 3,792,176 (1974); *Chem. Abstr.*, **81,** 36693q (1974).
(*237*) E. M. Van Patten and J. A. Freck, U.S. Patent 3,777,039 (1973); *Chem. Abstr.*, **80,** 69428q (1974).
(*238*) O. B. Wurzburg, P. C. Trubiano, and W. Herbst, U.S. Patent 3,499,962 (1970); *Chem. Abstr.*, **72,** P113051x (1970).
(*239*) Nihon Kayaku K. K. and K. K. Hayashibara Seibutsukagaku Kenkyusho, Japenese Patent Appl. NS 13521/75 (1975).
(*240*) K. C. Kwan and G. Milosovich, *J. Pharm. Sci.*, **55,** 340 (1966).
(*241*) G. K. Nichols and R. W. P. Short, U.S. Patent 3,490,742 (1970); *Chem. Abstr.*, **72,** 103756s (1970).
(*242*) N. Ohgusa and M. Fukui, Japan Patent 74 80,227 (1974); *Chem. Abstr.*, **82,** 77102k (1975).
(*243*) A. Stamm and C. Mathis, *Acta Pharm. Technol. Suppl.*, **1,** 7 (1976) *Chem. Abstr.*, **85,** 166583x (1976).
(*244*) S. C. Johnson & Sons, Inc., British Patent 1,515,630 (1978).
(*245*) S. C. Johnson and Son, Inc., Netherlands Patent Appl. 76 12,909 (1977); *Chem. Abstr.*, **88,** 126188c (1978).
(*246*) D. R. Bloch, U.S. Patent 4,071,616 (1978); *Chem. Abstr.*, **88,** 197456z (1978).
(*247*) A. Asanuma, Y. Fujiwara and K. K. Teijin, Japanese Patent Appl. NS83469/75 (1975); *Chem. Abstr.*, **83,** 207823e (1975).
(*248*) A. Asanuma, Japanese Patent Appl. 122575/75 (1975).
(*249*) A. Asanuma, Japanese Patent 75/122,570 (1974); *Chem. Abstr.*, **84,** 29470y (1976).
(*250*) M. W. Rutenberg and W. Jarowenko, U.S. Patent 3,081,181 (1963).
(*251*) American Cyanamid Co., Japanese Patent 60/16496 (1960).
(*252*) L. J. Hickey, R. H. Williams, and E. D. Mazzarella, U.S. Patent 3,284,381 (1966); *Chem. Abstr.*, **66,** 12114r (1967).
(*253*) D. L. Wilhelm, U.S. Patent 3,532,648 (1970); *Chem. Abstr.*, **73,** 132229y (1970).
(*254*) National Starch and Chemical Corp., Netherlands Patent Appl. 6,608,467 (1966); *Chem. Abstr.*, **67,** 34043g (1967).
(*255*) National Starch and Chemical Co., British Patent 916,710 (1962); Can. Patent 648,053 (1962).
(*256*) M. P. Ptasienski and J. W. Gill, U.S. Patent 3,003,979 (1961); *Chem. Abstr.*, **56,** 1160 (1962).
(*257*) M. W. Rutenberg and W. Jarowenko, U.S. Patent 3,038,895 (1962); *Chem. Abstr.*, **57,** 10086 (1962).
(*258*) M. Kamiyama, K. Nagatsuka, Japanese Patent 74/14,438 (1974); *Chem. Abstr.*, **82,** 32376m (1975).
(*259*) D. G. Murray, N. G. Marotta, and R. M. Boettger, U.S. Patent 3,597,227 (1971); *Chem. Abstr.*, **78,** 56726u (1973).
(*260*) J. H. Manning and J. H. Stark, German Patent 2,533,005 (1976); *Chem. Abstr.*, **84,** 152497g (1976).
(*261*) J. W. Frieders, U.S. Patent 2,974,048 (1961); *Chem. Abstr.* **55,** 14267 (1961).
(*262*) T. A. Hoglan and J. W. Sietsema, U.S. Patent 2,988,453 (1961); *Chem. Abstr.*, **55,** 19726 (1961).
(*263*) J. L. Louis, U.S. Patent 2,627,466 (1953); *Chem. Abstr.*, **47,** 4520 (1953).
(*264*) L. H. Elizer, U.S. Patent 3,650,787 (1972); *Chem. Abstr.*, **77,** P7646p (1972).
(*265*) L. H. Elizer, U.S. Patent 3,751,411 (1973). *Chem. Abstr.*, **79,** P106307b (1973).

(*266*) W. A. Scholten's Chemische Fabrieken N.V., West German Patent 1,118,151 (1961); *Chem. Abstr.*, **56**, 14501 (1962).
(*267*) Rootry Exploitatie Maatschappij N. V., Netherlands Patent 60/93,356 (1960); *Chem. Abstr.*, **55**, 6736 (1961).
(*268*) A. E. Staley Mfg. Co., Belgian Patent 609,702 (1961); *Chem. Abstr.*, **57**, 10086 (1962).
(*269*) E. E. Fisher and J. L. Harper, A. E. Staley Mfg. Co., Belgian Patent 611,052 (1961); *Chem. Abstr.*, **57**, 14042 (1962).
(*270*) J. W. Barger and C. E. Mumma, U.S. Patent 3,499,074 (1970); *Chem. Abstr.*, **72**, P91407p (1970).
(*271*) J. W. Barger and C. E. Mumma, German Patent 2,001,533, (1970); *Chem. Abstr.*, **75**, P152861t (1971).
(*272*) Tefag Etzkorn and Co. Technische Faser G.m.b.H., West German Patent 1,063,325 (1959); *Chem. Abstr.*, **55**, 15951 (1961).
(*273*) Dept. Agr. Inspect., Nebraska, British Patent 965,349 (1964); *Chem. Abstr.*, **61**, 12156 (1964).
(*274*) D. J. Bridgeford and D. M. Gallagher, U.S. Patent 3,497,584 (1970); *Chem. Abstr.*, **72**, 113005k (1970).
(*275*) J. Muetgeert, A. Schors, and P. Hiemstra, U.S. Patent 3,122,534 (1964).
(*276*) J. Lolkema, G. Moes, and W. F. Vogel, U.S. Patent 3,036,935 (1962).
(*277*) C. W. Charon and L. C. Renaud, German Patent 1,283,446 (1968).
(*278*) C. W. Charon and L. C.Renaud, U.S. Patent 3,108,891 (1963).
(*279*) M. M. Tessler, W. Jarowenko, and R. A. Amitrano, German Patent 2,364,056 (1974); *Chem. Abstr.*, **81**, P154960x (1974).
(*280*) M. M. Tessler and W. Jarowenko, U.S. Patent 3,969,340 (1976); *Chem. Abstr.*, **85**, P122115s (1976).
(*281*) M. M. Tessler and W. Jarowenko, U.S. Patent 3,970,767 (1976); *Chem. Abstr.*, **85**, P122140w (1976).
(*282*) K. K. Toray, Japanese Patent Appl. NS54515/74 (1974).
(*283*) M. M. Samfield and M. G. Christy, U.S. Patents 3,009,835; 3,009,836 (1961); *Chem. Abstr.*, **56**, 7566 (1962).
(*284*) M. Nishinohara, T. Niimura, and K. Kamemaru, Japanese Patent 75,156,562 (1975); *Chem. Abstr.*, **84**, 140722w (1976).
(*285*) Toray Industries, Inc., British Patent 1,388,580, (1975); *Chem. Abstr.*, **83**, 152359a (1975).
(*286*) H. S. Primack and S. H. Carr, *Amer. Chem. Soc., Div. Org. Coat. Plastics Chem.*, Paper, **34**, (1), 672-7 (1974); *Chem. Abstr.*, **84**, 60384x (1976).
(*287*) S. A. Bradley, P. Engler, and S. H. Carr, "Applied Polymer Symposium," No. 22, John Wiley and Sons, 1973, p. 269.
(*288*) M. Yoshida and S. Yuen, West German, Patent 2,104,780 (1971); *Chem, Abstr.*, **76**, 26364r (1972).
(*289*) K. Miyakoshi, Japanese Patent 76 23,392 (1976); *Chem. Abstr.*, **85**, 7208n (1976).
(*290*) C. G. Biliaderis, D. R. Grant and J. R. Vose, *Cereal Chem.*, **58**, 502 (1981).
(*291*) Meuser, R. W. Klingler and E. A. Niediek, *Ber. Tag. Getreidechem.*, **30**, 52 (1979); *Chem. Abstr.*, **93**, 28135s (1980).
(*292*) J. G. Sargeant and H. Wycombe, *Starch/Staerke*, **34**, 89 (1982).
(*293*) M. S. Karve, S. V. Bhide, and N. R. Kale, *Am. Chem. Soc.*, Symp. Ser., **150**, 559 (1981).
(*294*) S. V. Bhide and N. R. Kale, *Carbohyd. Res.*, **68**, 161 (1979).
(*295*) J. J. Lindberg and T. Kaila, *Acta Chem. Scand.*, **B34**, 757 (1980).
(*296*) M. M. Green, G. Blankenhorn, and H. Hart, *J. Chem. Ed.*, **53**, 729 (1975).
(*297*) W. A. Mitchell, *J. Chem. Ed.*, **54**, 132 (1977).

(*298*) F. L. Arbin, L. R. Schroeder, N. S. Thompson, and E. W. Malcolm, *Cellul. Chem. Technol.*, **15,** 523 (1981); *Chem. Abstr.*, **96,** 104641h (1982).
(*299*) K. Nishinari, *New Food Ind.*, **24,** 82 (1982); *Chem. Abstr.*, **96,** 102522w (1982).
(*300*) C. L. McCormick, D. K. Lichatowich, J. A. Pelezo, and K. W. Anderson, *Am. Chem. Soc., Symp. Ser.*, **121,** 371 (1980).
(*301*) R. W. Monte, U.S. Patent 4,105,824 (1978); *Chem. Abstr.*, **90,** 24445h (1979).
(*302*) C. L. McCormick and K. C. Lin, *J. Macromol. Sci., Chem.*, **A16,** 1441 (1981); *Chem. Abstr.*, **96,** 7185e (1982).

CHAPTER IX

GELATINIZATION OF STARCH AND MECHANICAL PROPERTIES OF STARCH PASTES

By Henry F. Zobel

Moffett Technical Center, Corn Products, Summit-Argo, Illinois

I. Introduction

Cereal grains, tubers, and roots are the main sources of starches for industrial and consumer applications, although starches are found in the piths and leaves of plants as well as in nuts, seeds, and fruits. Following synthesis, starch is stored in the plant as compact micron-sized (2–150 μm) granules that are partly crystalline and, hence, water-insoluble, facilitating starch isolation and handling. However, a first step in starch utilization generally is one that disrupts the granular structure leading to granule swelling and hydration and solubilization of starch molecules. These events, referred to collectively as starch gelatinization,

STARCH, 2nd ed.

ISBN 0-12-746270-8

are commonly effected by heating granules that are slurried in water. While primary consideration is being given to this aspect of gelatinization, solvents other than water (liquid ammonia, formamide, formic acid, chloroacetic acids, and dimethyl sulfoxide) also effect gelatinization by disrupting hydrogen bonding within the granule or by forming soluble complexes with the starch. Furthermore, alkalies (*1, 2, 146, 147*), salts (*3–10, 147–150, 169, 174, 175*), sugars (*11–18, 148, 149, 151, 173*), lipids (*18–29, 152–156*), alcohols (*30, 148, 169*), and organic acids and their salts (*169*) are examples of chemicals that decrease or increase starch gelatinization temperature and that determine extent of gelatinization. Thus, under alkaline conditions, starch gelatinizes at a lower temperature. This feature is used in adhesives used to make corrugated boxes since it is essential that the starch granules gelatinize rapidly to develop the tack needed to hold the board together as it moves through the process. Salts (sodium chloride, sodium sulfate) are used to raise the gelatinization temperature of starches being derivatized, while sugars and lipids (including surfactants) are combined with starches to provide texture in foods. Anions or cations that cause a decrease or increase in starch gelatinization temperatures also appear to have the same effect on water structure (*31–34, 149*); similarly, sugars increase gel temperature (*11, 13, 15–17*) and show varying degrees of increased water structure and association with water (*35, 36, 151*). However, Evans and Haisman (*174*) suggest a relationship between initial gelatinization temperature, water activity, and the volume fraction of water in starch granules for selected sugars, organic hydroxy compounds, and inorganic salts. Oosten (*175*) views starch as a weak ion-exchanger; cations stabilize starch structures while anions promote rupture of hydrogen bonds.

II. Granule Composition and Structure (see also Chapter VII)

Starch molecules are classified as either linear (amylose) or branched (amylopectin). Some granules are almost entirely amylopectin (98%); others are high in amylose (45–80%), but granules generally are a mixture of amylose and amylopectin with amylose contents of 15–30% (*37*). Amylose is a polymer of (1→4)-linked α-D-glucopyranosyl units with an average molecular weight ($\overline{MW}$) of about 250,000 (typical of corn starch); some potato amylose fractions are of unusually large size, with MW values on the order of a million. Amylose in water is unstable, quickly precipitating to initiate gelation, which is the terminology used here to describe the process of "setback" or rigidity development that occurs as starch gels are cooled or aged. Firm resilient gels are obtained that may require temperatures as high as 115°–120° to reverse gelation. Amylopectin has short branches on about 4% of the D-glucosyl residues. Amylopectin molecules are very large, ranging from 50 million to over a 100 million in MW. Gelation of amylopectin occurs at a much slower rate than does amylose and requires higher concentrations. Amylopectin gels are soft and may be thermally reversed at temperatures of only 50°–85°. These starch molecules deposit to give partly crystalline (10–40%, as determined by x-ray diffraction) granules that are described as spherocrystals. This spherocrystalline characteristic is clearly shown by laser light scattering data obtained on granular starches (*38–40*).

Starches that contain amylose and fatty acids are unique in that amylose–fatty acid complexes are formed that can be detected by x-ray diffraction. Characteristic diffraction lines identify the type (V-structure) and extent of complex formation (*41*). For native high-amylose corn starches, extensive crystallization into a fatty acid–amylose structure can be detected; in native corn and wheat starches, complex formation occurs to a lesser degree (*42*). Furthermore, the first evidence of crystalline structure in starch gels is often that of the complex (*43, 44*). Its formation is commonly enhanced in gels by adding selected long-chain fatty acid derivatives to starch to alter gel texture (*19, 21, 23, 26, 28, 156*).

Several premises underlie this review. (a) Gelatinization is the disruption of molecular order within the granule. (b) Granule organization and gelatinization are controlled by the major molecular component present. (c) Gelation is dominated by the amylose content of the starch regardless of whether or not it is the major fraction. (d) Gelation occurs as molecular association takes place, presumably through hydrogen bonding, forming a network of junction zones between molecules. More often than not, crystalline regions develop from this point and provide additional gel strength (rigidity) and stability.

Reviews by Olkku (*45*) and Sterling (*46*) may be of interest for supplemental information on structure and gelatinization. Olkku (*45*) discusses wheat starch, including the affects of proteins and pentosans on starch gelatinization. Sterling (*46*) covers the textural aspects of food products in relation to starch/water interactions taking place during gelatinization and gelation; see also Hofstee (*157*).

III. Melting Concept for Gelatinization

The term "gelatinization" generally is used to describe the swelling and hydration of granular starches. The process also can be described as the melting of starch crystallites which should be amenable to thermodynamic analysis (*158*). Evidence for this viewpoint comes from calorimetric studies showing gelatinization endotherms (Figs. 1 and 2) that are first-order thermal transitions (*47–49, 159–165*). Such starch properties can be routinely determined using instruments that automatically take samples through programmed temperature cycles and record heat flow as thermal events occur, i.e., differential thermal analyzers (DTA) and differential scanning calorimeters (DSC). Investigations, in which these techniques were used, have shown the dependence between the temperature of melting and starch:water ratios, and have led to a thermodynamic treatment of the gelatinization process. Accordingly, Flory has shown a relationship between the melting point of crystalline polymers and an added diluent (*50*). In the case of starch, water is the diluent. Thus,

$$1/T_m - 1/T_m^\circ = (R/\Delta H_u)\,(V_u/V_1)(v_1 - \chi_1 v_1^2) \qquad (1)$$

where R is the gas constant; ΔH_u is the enthalpy of fusion per repeating unit; V_u/V_1 is the ratio of the molar volumes of the repeating unit and the diluent; v_1 is the volume fraction of the diluent; χ is the Flory–Huggins polymer–diluent interaction parameter; and T_m° is the melting point in the absence of diluent. Rewritten, the equation is

$$(1/T_m - 1/T_m^\circ)/v_1 = (R/\Delta H_u)(V_u/V_1)(1 - \chi_1 v_1) \qquad (2)$$

A plot of the left side of Eq. (2) against v_1 gives ΔH_u from the intercept and the energy of interaction B equal to $\chi\, RT/V_1$. A comprehensive discussion of these concepts and polymer crystallization is presented by Mandelkern (*51*). Using these principles, Zobel (*48*) estimated a ΔH_u for potato starch of 14.9 kcal/mol of D-glucosyl unit; Donovan (*52*) reports 13.5 ± 4 kcal and a B value of 3.5 ± 2.3 cal/mL, indicating that water is a poor solvent for starch. Theoretical melting points of a starch crystal without diluent (T_m°) were reported as 153° and 168°, respectively, by these authors. Donovan (*53*) subsequently reported ΔH_u and T_m° values of 13.8, 187; 12.6, 181; 14.6, 197; and 14.3, 168 for maize, wheat, waxy maize, and potato starches, respectively. Overall, the values are remarkably close, considering the variation in amylose content and x-ray structure of the different starches. Lelievre (*54*) extended the above theory of polymer–diluent interactions to include polymer, diluent, and solute. Based on an idealized model of the complex gelatinization transition, he successfully predicted the effect of sugar-solutes on gelatinization temperature.

Differential thermal analysis of starch–water slurries can be used to determine quantitatively starch gelatinization as an enthalpy ($-\Delta H_G$) of gelatinization (*48, 49, 52, 55–61, 160–164*). Enthalpy measurements also can be used to measure the return of crystallinity in aged starch gels (*161, 166–168*), granular starch damage (*163*), and the effects of heat/moisture treatment on starches (*178*).

Figure 1 shows typical gelatinization patterns for several cereal starches at a 1:1 starch–water ratio. The A-type starches show degrees of crystallinities as measured by x-ray diffraction of 36, 39, and 42% for wheat, corn and waxy maize, and dasheen starches, respectively (*62*). Endotherm excursions show a broad temperature range over which crystal structure is being melted. The relative intial temperatures and peak shapes of dasheen starch and wheat starch gelatinization parallel their differences in crystallinity. The high temperature peak (110°–120°) in wheat starch which was attributed to the melting of fatty acid–amylose complexes has been confirmed by x-ray diffraction patterns taken

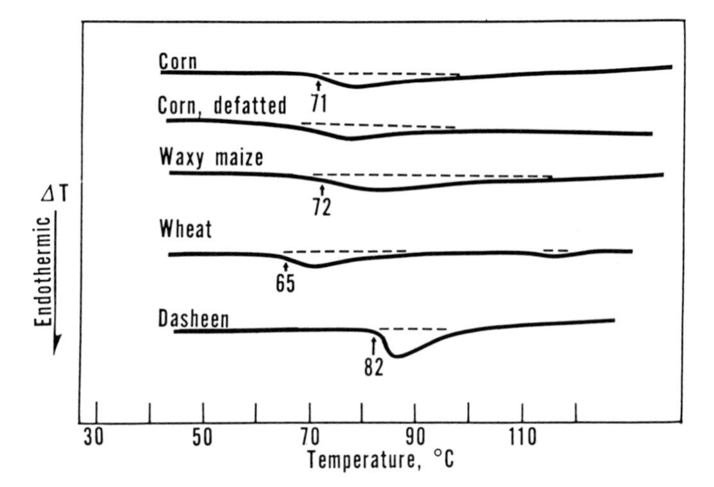

FIG. 1.—Gelatinization endotherms of granular starches with A-type X-ray structures.

by Kugimiya (*60*). Other lipid complexes in gelled starches, identified by V-type x-ray patterns, are reported by Osman and Zobel (*20, 63*). The formation of complexes upon gelatinization of starches with added lysolecithin was demonstrated by Kugimiya and Donovan (*162*) using DSC.

The temperature at which gelatinization is initiated (Fig. 1) is in general agreement with values reported from loss of birefringence measurements (*49, 58*). It is obvious from the DSC tracings, however, that major structural changes occur beyond the birefringence end-point. Hence, the data indicate the temperature range needed to provide maximum disruption of structure to achieve granule swelling and high viscosities in the hot starch paste. This is demonstrated by following the viscosity development in a Brabender Amylograph and noting that the DSC and viscosity maximum are in close correspondence. Takahashi (*64*) reports such a comparison, including photographic data, indicating additional changes in starch after completion of the viscosity increase. The persistence of crystal structure in corn starch upon heating and its loss at temperatures of 85°–92° were confirmed by Zobel and Rocca (*65*) using x-ray diffraction. In their experiments, loss of diffraction lines characteristic of the A structure was recorded as a starch slurry was heated through a programmed temperature increase.

The crystallinity of potato starch is in the order of 20–28%, while high-amylose corn starches show lower levels of crystallinity (10–15%) as detected by x-ray diffraction (*62, 66*). The effect of the amylose extender gene in waxy maize starch (Fig. 2) is to shift the endotherm of the waxy maize starch to a higher temperature and change the crystal structure from the A to the B type, but with little change in overall crystallinity from that of normal waxy maize starch.

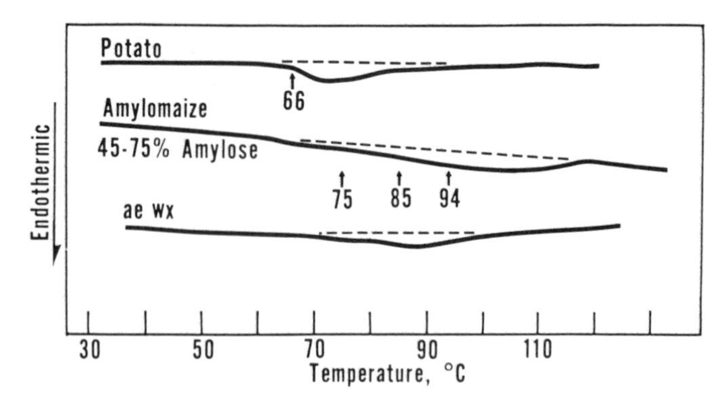

FIG. 2.—Gelatinization endotherms of granular starches with B-type X-ray structures.

The trend of a lower ΔH_G with decreasing amylose content in the B starches (Table I) is interesting, but needs to be confirmed. For a high-amylose corn starch, a starch with a B-type X-ray structure, as is potato starch, the initial (75°), midpoint (85°), and final gel temperature (94°) by loss of birefringence measurement are shown on Figure 2.

Figures 1 and 2 were obtained on samples with excess water to insure complete starch gelatinization and to emphasize the relationship between gelatinization and melting. Variations in the water:starch ratio, however, can result in multiple peaks as demonstrated by several authors (*52, 61, 160, 161*, see also Chap. VII). A simple and consistent explanation of multiple peaks is first that each peak is characteristic of a first-order transition, namely, melting. Second, a decrease in available water results in resistant crystalline regions showing more resistance to melting in accordance with the theory and practice of polymer–diluent interaction. Thus, multiple peaks provide information on the distribution of forces holding a granule together. Third, it is not evident, from DSC or other data, that partial gelatinization of samples containing 40–70% (w/w) water results in the water being bound to such an extent that neighboring crystalline regions (or granules) are inhibited from gelatinizing.

A partial compliation of enthalpy ($-\Delta H_G$) calculations is shown in Table I. Similar results on starches analyzed by both Zobel and co-workers (*48*) and Stevens and Elton (*49*) permit the A-series to be enlarged and a comparison of C-type starches with A and B types. High-amylose corn starch, which is low in crystallinity by X-ray standards, shows a high heat of gelatinization (6.7 cal/g), attributed to extensive association between amylose chains. Other workers report values of 7.6 (*58*), 4.2 (*160*), and 2.7 (*164*) cal/g for high-amylose corn starches. The low ΔH_G value for wheat starch is believed to owe partly to a transformation, during gelatinization, to the higher melting form observed on the DSC scans. Confirmatory evidence for such a transformation is found in X-ray studies on wheat starch gels indicating the formation of V-type structures during gelatinization (*43, 67, 68*). The remaining starches show relatively small differences in ΔH_G, regardless of amylose content or crystal structure. These data reinforce the view that the major molecular component governs organization and gelatinization of the native granule. The low ΔH_G value for compacted corn starch is due to partial destruction of granule crystallinity during the compaction process. Additionally, endotherms are not obtained on pregelatinized starches which show little or no crystallinity.

IV. Methods for Determining Starch Gelatinization

Determination of starch gelatinization requires ways of heating the starch slurry and of sensing the structural changes that occur. Heating may be continuous through a programmed heating cycle or it may be stepwise. The sensor used, its sensitivity to physical change, and starch concentration are factors that affect reported starch gelatinization determinations and that provide different insights into the physics of gelatinization. The term paste is reserved here to describe the character of the gelatinized starch and not the process itself. Methodology, other than the DTA/DSC approach, and selected results are described in the following subsections.

1. Light Microscopy

Loss of birefringence measurements requires an optical microscope with crossed polarizers and a heating stage, such as the Kofler (*69, 70*). Most granules show some degree of birefringence (optical anisotropy or double refraction) owing to a regular orientation of D-glucosyl units in amorphous and crystalline regions. Crystallinity, however, is not essential for birefringence since both

Table I

Enthalpy of Gelatinization ($-\Delta H_G$) for Native Granular Starches

X-ray type A		*X-ray type B*		*X-ray type C*	
Starch[a]	$-\Delta H_G$[b]	*Starch*[c]	$-\Delta H_G$	*Starch*[d]	$-\Delta H_G$
Corn	4.1–4.9	High-amylose corn	6.7	Arrowroot	4.6
Wheat	2.9	Potato	4.5	Tapioca	4.0
Rice	3.4–3.9	ae wx	3.9		
Dasheen	4.3				
Waxy maize	4.8				
Compacted[e] corn	2.2				

[a] Top to bottom, amylose content (%) = 27, 23, 17, 16, 2, 27.
[b] Calories/gram dry starch, ± 0.4 cal.
[c] Amylose content (%) = 50–70, 22, ~2.0.
[d] Amylose content (%) = 21, 18.
[e] Pearl starch passed through compacting rolls.

granules ball-milled to destroy crystallinity and stretched amorphous amylose films can be highly birefringent. Thus, with the release of stress during gelatinization, granule swelling occurs just prior to or simultaneously with loss of granule birefringence. The view that granule birefringence primarily reflects stress or orientation of molecules in amorphous regions is augmented by the DSC/x-ray/maximum swelling and viscosity relationships discussed earlier.

Starch gelatinization is often measured by loss of granule birefringence because of its simplicity in equipment and application. It is limited, however, to dilute granule suspensions. The methodology is described in Chapter XXII. Table III of Chapter XXII shows gelatinization temperatures for unmodified and modified starches alone and in combination with sucrose, sodium hydroxide, sodium chloride, and sodium carbonate. The text indicates also that starch granules within a given species gelatinize over a range of temperatures. Seidmann (*71*) discusses this method at length and gives gelatinization temperatures for many different starches. The Staerke Atlas (*72*) summarizes the gelatinization of starch in water, alkali, acid, salt solutions, and organic compounds. Physical changes that starch granules undergo upon enzymic attack, by heat treatments, and ultrasonic irradiation are also described. A photometric device for recording the loss of birefringence is described by Berry (*73*).

Microscopic examination of granules undergoing gelatinization permits one to observe swelling duration, degree of swelling, and swollen granule integrity and size. Collison (*74*)used a photomicrograph technique to measure the swelling of individual granules directly. He showed that wheat starch containing the surfactant cetyltrimethylammonium bromide exhibited a high degree of swelling, while indirect methods, such as viscosity or volume sedimentation, suggested a low swelling behavior. Bean and coworkers (*17*) measured the swelling diameter of wheat starch and found that D-glucose, D-fructose, and sucrose inhibited granule swelling. Gelatinization temperatures increased with sugar concentration with sucrose > D-glucose > D-fructose in affecting gelatinization temperature. Counting of swollen granules in a series of photomicrographs was used by Freke (*75*) to demonstrate the effects of sucrose concentration and pH on the gelatinization of native and modified waxy corn starches. Since dyes such as chlorazol violet R and congo red stain damaged granules, Collison and Chilton (*76*) counted stained granules to determine gelatinization of wheat and potato starches as a function of their water content.

For very small granules, or those showing weak birefringence, granule swelling alone can be used as a criteria for gelatinization. The process is tedious compared to loss of birefringence measurements.

2. Electron Microscopy

Structural changes occurring during gelatinization have been documented with the scanning electron microscope (SEM) (*77–80*). The technique is most useful under conditions where the swollen granules have maintained some integrity.

The report by Hill and Dronzek (*78*) on 12% slurries illustrates that considerable residual structure remains throughout the gelatinization range as measured by loss of birefringence. Investigations on dilute (1%) systems, or on completely pasted starches, have the problem of artifacts owing to the well-known phenomena that starch molecules aggregate during drying or freezing. These effects were investigated by Berghofer and Klaushofer (*81*) using SEM and by Woodruff and MacMasters (*82*) using light microscopy. The possibility of such artifacts was recognized by Miller and co-workers (*77*), but was not ruled out, in their SEM study of swollen granules.

3. Light Transmission

Changes in light transmission are the basis for several devices for following gelatinization. Morgan (*83*) used the technique to identify native starches, analyze starch mixtures, determine initial gel temperature, and to characterize dextrins and oxidized and thin-boiling starches. Beckford and Sandstedt (*84*) describe an apparatus that uses a spectrophotometer to measure transmitted light. Longley and Miller (*85*) modified the instrument developed by Cook and Axtmayer (*86*) and concluded that amylose precipitation occurs within granules swollen in the presence of added saturated monoglycerides.

4. Viscometry

A Brabender Visco/amylo/Graph provides information, not only on gelatinization, but also on the properties of the cooled paste. This instrument records the torque required to balance the viscosity that develops when a starch slurry is subjected to a programmed heating and cooling cycle. The temperature at which the first major rise in viscosity occurs is highly dependent on starch concentration and is generally higher than gel temperatures obtained by loss of birefringence. Viscosity is measured in arbitrary units that reflect paste consistency; accordingly, Brabender results are described in terms of paste properties and pasting temperatures at given times. The Amylograph Handbook covers the theory, construction, applications, sources of error, troubleshooting, and adjustment of units (*87*). Addition of carboxymethylcellulose (CMC), which binds part of the water, to the starch slurry permits detection of the initial gelatinization or swelling temperature (*88*). Data obtained by this technique are compared with light transmission, water adsorption, and solubility data by Sandstedt (*89*). Reported Amylograph data should include instrument model, torsion spring used, bowl speed, volume of slurry, slurry concentration (including basis), and start, hold, and final temperatures.

Freeman and Verr (*90*) measured gelatinization and paste development by heating starch in a closed bottle submerged in a water bath with a shaking mechanism. At specified times or temperature, samples were removed and viscosities determined with a Brookfield Synchro-Lectric viscometer. Samples were also cooled and viscosity again measured to determine setback. The authors claim the procedure is rapid and that it gives precise viscosity values when measuring the effects of starch concentration, cooking times, temperature range, and pH.

5. Swelling and Solubility Determinations

Gelatinization of corn and wheat starch granules as a function of temperature and degree of swelling are illustrated in a classical paper by MacMasters (*82*). Photographs of native and gelatinized corn, sorghum, waxy corn, waxy sorghum, wheat, barley, rye, and oat starches have also been published by MacMasters (*91*). Photographs of corn and potato starches at intermediate and high levels of swelling were published by Schoch (*92*).

Leach and co-workers (*93*) developed swelling power and solubility curves to characterize potato, corn, waxy and cross-linked (-bonded) waxy and other starches over their gelatinization range (65°–95°). Swelling power, in this case, is a measure of hydration capacity, since the test is a weight measure of swollen granules and their occluded water. At 90°, the swelling power of potato starch is 350 compared to 65 for tapioca, 60 for waxy corn, and 16 for regular corn starch. The same pattern is followed by Brabender peak viscosities using 35 g of starch in 500 mL for analysis; viscosities exhibited are on the order of 2500 units for potato starch, 1400 for tapioca, 1000 for waxy maize, and 500 for regular corn starches.

Hellman and co-workers (*94*) measured granule swelling from the increase in vacuum-dry diameters at 25° when humidified at 100% relative humidity. For tapioca, waxy maize, potato, and corn starches, diameters increased by 28.4, 22.7, 12.7, and 9.1%, respectively, indicating the degree of interaction of water with the granular structure independent of temperature. Except for potato starch, the relative standing of the starches is the same as in the 90° swelling and Brabender viscosity data. The properties shown by potato starch suggest cohesiveness of structure and resistance to disruption when heated in water. The observed swelling and viscosity properties of potato starch are accounted for, in part, by the absence of lipids that may complex with amylose during gelatinization and limit swelling and the presence of a viscous cohesive amylose fraction owing to its high molecular weight.

6. Enzymic Analysis

Although changes in starch properties initiated by gelatinization are easily measured, it is difficult to quantitatively determine the degree of gelatinization that takes place. This is especially true in the analysis of foods and feeds where heterogeneity of the mixture makes it difficult to detect changes in the starch fraction. Selective digestion of the gelled starch granules by enzymes, however, has proven to be an accurate method for measuring extent of gelatinization. Shetty and co-workers (*95*) reviewed enzymic methods and describe one utilizing glucoamylase and D-glucose oxidase in a solvent of dimethyl sulfoxide (DMSO) and water (*95*). Chaing and Johnson (*96*) propose using glucoamylase and reacting the resulting D-glucose with *o*-toluidene to form a green chromagen in glacial acetic acid. This method was used to determine gelatinized starch in extruded products (*97*). Wu (*98*) claims an alkaline digestive procedure for total starch that eliminates the decrease in D-glucose release with prolonged digestion in DMSO. Other than using a different glucoamylase, the methodology is similar to that of Shetty and co-workers (*95*).

7. Nuclear Magnetic Resonance

Starch gelatinization at a molecular level is measured using either high-resolution or wide-line proton magnetic resonance (pmr) techniques (*99–101*). Jaska (*101*) used spectral line-width changes to follow the hydration and solubilization of potato, wheat, and corn starches. Three stages of gelatinization were noted with corn starch that parallelled earlier observations (*94, 67, 102*); a first stage corresponding to reversible hydration and swelling; a second stage, defined by loss of birefringence or swelling measurements (62°–74°), where some granule structure remains; and a third stage (74°–92°) where the last vestige of granule structure is broken giving rise to additional chain mobility (decreased line widths). An abrupt change in line width was observed at the initiation of gelatinization (*101*).

Wide-line techniques are more suitable, however, since they can be used to measure the long relaxation time observed in dilute starch suspensions. The relaxation of a nucleus is the time required for reorientation once it is perturbed by an electromagnetic pulse of the proper frequency and pulse length. Relaxation is characterized by T_1 and T_2 parameters, which are spin-lattice and spin-spin relaxation times, respectively. The spin-spin (T_2) relaxation time is sensitive to changes in structure that are likely to occur during gelatinization and solubilization. Lelièvre (*103*) followed T_2 vs. temperature with wheat starch suspensions ranging from 2 to 40% w/w. Water mobility decreased as hydration of the starch occurred at the onset of gelatinization. As heating progressed, T_2 went through a minimum and then increased as starch chain mobility increased. The transition parallels that of a crystalline to amorphous transition suggesting again that starch gelatinization is a melting process.

Publications of Lechert and co-workers (*104–109, 170*) and by H.-J. Hennig (*110*) illustrate the use of wide-line pmr for determining the rate and extent of water diffusion into gels, free and bound water in granules, starch swelling and retrogradation, and syneresis of water by aged gels.

8. Laser Light Scattering

A relatively new technique for following changes in the internal structures of starch granules during gelatinization is the use of polarized laser-light scattering. As indicated earlier, the anisotropic scattering indicates a spherulic organization within the granule (*38–40*). Thus, changes reflected in this structure during gelatinization are directly related to the supermolecular ordering of the starch macromolecules. Experimentally, the incident laser beam is vertically polarized which, when scattered, is recorded after passing through an analyzer that is either vertical (V) or horizontal (H) with respect to the incident beam. Intensity maxima within the cloverleaf scattering pattern (H_v) shift toward smaller scattering angles, indicating an increase in average spherulite size as the granules swell during gelatinization. The V_v scattering pattern reflects mainly the anisotropic nature, and preferred radial arrangement, of molecular segments within the granule. Theory and methodology for using laser optics as well as other forms of electromagnetic radiation to study the micro- and supermolecular structure of polymeric solids and the origin of birefringence and its characteristic forms are discussed by Wilkes (*111*).

Utilizing a laser small-angle light scattering photometer (SALS), Blanchard and co-workers (*112–116*) have measured the changes in H_v scatter when starch/water slurries subjected to small (2°) and large (8°–10°) temperature jumps. Large temperature jumps were introduced at a temperature just above the initiation of loss of granule birefringence. Total loss of birefringence occurred as fast as the rise in temperature without a significant relaxation time. The 2° temperature jump, however, produced time-dependent changes in birefringence as the initial instantaneous (2–3 sec) change was followed by a much slower process (60–900 sec). The long relaxation times indicate rearrangement and/or disentangling of molecules in the amorphous region of the granule while the fast process reflects crystallite melting. Thus, the loss of crystalline order is mutually interdependent with mobility of the polysaccharide in the non-crystalline portion of the granules.

SALS analysis of the time dependence of birefringence loss occurring in heat or chlorine-treated flours indicated that treated granules have a less rigid molecular network than untreated granules and, hence, showed greater swelling. In contrast, negligible differences were observed in gelatinization temperatures or heats of gelatinization although the two flours varied in baking qualities (*116*). Research using SALS is continuing and should provide new insight into granule structure and gelatinization.

V. Mechanical Properties of Starch Pastes

1. Introduction

For thickening foods, sizing and coating papers, sizing textiles, and use in drilling muds or adhesive formulations as well as in a host of other applications, starch gelatinization in aqueous media is a first step. When starch is cooked, the flow behavior of a granule slurry changes markedly as the suspension becomes a dispersion of swollen granules, partially disintegrated granules, and molecularly dispersed granule contents. The cooked product is called a starch paste. Transition from a suspension of granules to a paste is accompanied by a large increase in apparent viscosity. Although the term viscosity finds general usage, starch pastes are non-Newtonian and a more apt description is that pastes develop resistance to deformation, or simply, consistency.

During cooling, paste consistency increases as molecular association forms a cross-linked network that increases paste resistance to deformation. In a cooled state, pastes may remain fluid or form a semi-solid or solid gel showing considerable strength. Starch suppliers and users both require methodology for judging starch quality and probable flow behavior in its end-use applications. For this purpose, viscosity, gel rigidity, and gel strength measures have been devised.

2. Paste Viscosity

a. *Scott Hot Paste Viscosity*

Flow-through orifice devices are commonly used for hot paste viscosity measurements. Scott viscometer methodology, as modified by CPC International, is used throughout the wetmilling industry to evaluate unmodified, acid-modified, oxidized, and derivatized starches. A specially constructed water bath is maintained at the boiling point with live steam. Samples are cooked in a reproducible manner with fixed time intervals for heating and stirring. Flow characteristics of the cooked pastes are determined at the temperature of the water bath. The accurately machined Scott cup with its polished orifices and an overflow to provide a uniform sample head contributes to reproducible results. Flow of a 100-mL sample is followed in a preferred time span of 40–80 sec. Sample weight is adjusted to meet these conditions for different starch samples. The speed and precision of this relatively simple test makes it an effective instrument for process control and finished product analysis.

b. *Stein-Hall Cup*

This flow-through device is used by corrugated board manufacturers to monitor adhesive viscosity. Glue line tack, penetration into the paper board, and final bond strength are all functions of starch paste viscosity (*177*).

c. *Brookfield Viscometer*

Brookfield Synchro-Lectric viscometers have a rotating cylinder or disc coupled, through a torsion spring, to a constant speed synchronous motor. The spring applies a restoring force to balance the resistance offered by sample viscosity to spindle rotation. The resulting torque, in centipoise (cP), is read directly from a dial. Ease of cleaning, operational simplicity, and instrument portability favor use of this instrument by starch users and manufacturers. Operation is limited to fluid or semi-solid pastes where flow around the spindles can take place; otherwise, the spindle rotates in a hole of its own making. The problem can be partially overcome by using a Helipath stand where the spindle head is advanced into fresh undisturbed gel as measurements are being made.

Brookfield viscometers are available for measuring apparent viscosities ranging from 1 to 64 MM cP. Viscosity measurements at various shear rates permit determination of dilatancy, thixotropy, and other rheological properties of non-Newtonian starch pastes (*118*).

Elder and Schoch (*119*) report Brookfield data that show an order of increasing shear stability as follows: waxy sorghum $\simeq$ tapioca $<$ potato $<$ corn $<<$ cross-bonded waxy sorghum. Sorghum starch showed rapid breakdown at low shear rates while introduction of a few phosphate cross-links into the starch resulted in a sevenfold increase in shear stability.

Brookfield data on corn starch pastes (Table II) illustrate the effects of starch concentration, cooking time and temperature, and shear on viscosity. Microscopic observations on granule breakage are included to delineate the role of swollen granules on paste viscosity. Marked effects on viscosity are apparent owing to concentration and granule breakdown (*120*).

Viscosity data obtained with a Wells–Brookfield cone and plate viscometer on native and partially hydrolyzed waxy maize starch were modeled by the power law $\sigma = KD^a$, where σ (the shear stress) and D (the shear rate) are related through parameters K and a (*121*). A similar model developed on pastes undergoing particle buildup owing to starch retrogradation provided a means to monitor both changes in starch structure with time (*122*). A relationship also was developed between total hydrodynamic volume (polymer plus immobilized liquid) and viscosity indicating a behavior consistent with rigid spheres in dilute solutions and deformable particles in concentrated starch pastes (*121*).

d. *Rheometers*

Schutz and co-workers (*122–126*) have extensively investigated the rheology of starch pastes indicating the need for, and factors to be controlled for, reproducible paste preparations. The physical importance of parameters K and a of the power law ($\sigma = KD^a$) has been examined, and a procedure developed for correlating viscosity data obtained on different coaxial cylindrical rheometers.

Table II

Factors Affecting Corn Starch Paste Viscosity

Starch concentration (%)	*Temperature (°C)*	*Time at temperature (min)*	*Stirring (rpm)*	*Granule breakage (microscopic)*	*Apparent viscosity (cP)*
3	90	30	120	None	15
5	90	30	120	None	1,080
5	90	30	1,800	Appreciable	317
5	100	30	160	Appreciable	754
5	100	30	1,800	Complete	90
10	90	30	120	None	11,800
10	90	30	1,000	Unappreciable	7,290
20	100	30	1,000	Complete	18,500
20	100	60	1,000	Complete	9,480

The authors stress the need for such rheometers if absolute viscosity characterizations are to be made.

Absolute viscosity determinations on starch pastes undergoing swelling and shear are reported by Cěh and Stropnik (*127*) and by Winkler and co-workers (*128*) using a Haake-Rotovisco rheometer. Djaković and Dokić (*129*) used this rheometer not only to develop parameters describing rheological characteristics, but also to develop a thixotropy coefficient for starch pastes. Doublier (*171*) discusses the rheology of wheat starch pastes; interpreting large differences in behavior on the basis of particulates dispersed in a macromolecular medium.

Relaxation modulus (G) and half-relaxation time ($T_{1/2}$) were studied by Eliasson and Bohlin (*177*) using concentrated wheat starch paste prepared at different water levels (80% and 50% by wt) to simulate conditions that give rise to one or more gelatinization endotherms, respectively, as observed in DSC thermograms. Results were interpreted as a model composite system of a viscoelastic continuous phase and a dispersed particle phase. Marked increases in G (6.5×) and $T_{1/2}$ (15×) were observed when the 80% and 50% water slurries were heated to about 80°. The data indicate that the higher temperature DSC endotherm, in the lower water content experiment, reflects introduction of a significant level of swollen granules into the continuous phase. However, because of the lower water level (50%), these granules impart more rigidity and adhesion to the paste than those in a gel with 80% water.

Current commercial rheometer systems include brand names such as Brabender, Contraves, Ferranti–Shirley, Haake, and Rheometrics. In some cases, viscometer designs also are offered for continuous in-process viscosity control. As continuous modes of starch processing and utilization increase, such systems are receiving more attention. For the most part, however, there is little published information on the subject.

An autoviscometer described by Poyser and Wright (*130*), while not an inline system, enables handling one sample every 10 min with viscosity values available in 7 min.

e. *Brabender Visco/amylo/Graph*

The Amylograph, described earlier, finds worldwide usage in the characterization of starch pastes (*87*). Amylograph viscosity curves are widely published in the format developed by Mazurs and co-workers (*131*) and will not be repeated here. (See Chap. XIX for examples.) Six significant points on the curve are generally recognized.

1. Pasting Temperature.—Denotes initiation of paste formation; varies with starch type and modification and with additives present in the slurry.
2. Peak Viscosity.—Cited irrespective of the temperature at which the peak is attained. Generally, cooking must proceed through this stage to obtain a usable paste.
3. Viscosity at 95°.—Reflects ease of cooking the starch.
4. Viscosity after 1 h at 95°.—Indicates paste stability, or lack thereof, during cooking under relatively low shear.
5. Viscosity at 50°.—Measures the setback that occurs on cooling the hot paste.
6. Viscosity after 1 h at 50°.—Indicates stability of the cooked paste under simulated use conditions.

Typical values for these six points are shown in Table III for several common starches all run under the same conditions. Highly swollen potato starch granules give the highest peak viscosity, but undergo the most breakdown as indicated by the viscosity after the 95° hold. Although tapioca and waxy maize starches produce less viscosity and have a lesser degree of swelling than does potato starch, all three starches show four- to sevenfold decreases in viscosity from the peak to after the 95° hold. In contrast, corn and wheat starches show little or no change in viscosity through the 95° hold, while peak viscosities vary almost eightfold. This lack of change in viscosity is attributed to a retardation of swelling caused by a strengthening of the swollen corn and wheat starch granules by complex formation between the amylose fraction and lipids present in these native starches. MacMasters and Wolff (*179*) demonstrated the *in situ* formation of such complexes both within and outside starch granules by tempering pastes for several days at 80°–90° and observing spherocrystal formation microscopically. Priestley (*132–135*) has shown by X-ray diffraction that both solubility and amylograph viscosities for rice starch decrease as complex formation increases.

Data to show the effect of cooking and hold times on viscosities are given in

Table III

Brabender Pasting Temperatures and Viscosities for Several Common Starches[a]

Starch	Paste temperature (°)	Brabender viscosity units[b]				
		Peak	95°	95° hold	50°	50° hold
Potato	61	2500	850	340	600	630
Tapioca	59	1400	520	280	500	510
Tapioca[c]		1400	520	380	610	650
Waxy corn	69	1000	400	250	390	370
Corn	73	470	470	350	830	760
Corn[c]		520	520	445	1050	1040
Wheat	77	65	60	60	300	270

[a] Conditions: 35 g of starch/500 mL; 1.5° per min heating rate; regular 60 min hold at 95° and 50°.

[b] See text for definition of data points.

[c] Data obtained with 30 min holds at 95° and 50°.

Table III for tapioca and corn starches. Both starches show less cook-out (95° hold) and higher viscosities (50° and 50° hold) with shorter times. This behavior seems to contradict the viewpoint that paste viscosity is largely a function of the exudate, since longer cooking times, by this reasoning, should show a higher viscosity (*77*).

Amylographs of starch dispersions at pH 5–7 show essentially the same consistency behavior; corn starch shows substantial breakdown at pH 3.5–4.0. Waxy starches and particularly potato starch show rapid viscosity losses below pH 5. A cross-linked waxy starch, however, maintains a moderate consistency at pH 3.5 (*136*).

Viscosity changes occurring at temperatures lower than 50° or over an extended time period are also of interest in certain industrial applications. For such viscosity changes, one can use the Amylograph for an initial analysis and for paste preparation. Brabender and Brookfield viscosity data of corn starch cooked with and without glycerol monostearate (GMS) are given in Table IV to illustrate changes in gelation for corn starch at 25°.

The pasting temperatures and Brabender viscosities shown in Table IV are typical for many cases when lipids are added to starches. Results vary widely among starches and depend on the polar versus non-polar nature of the additive and level of addition, all of which make predictive generalizations difficult (*21*, *28*).

Brookfield viscosities shown in Table IV were taken from the Amylograph after the final 50° reading. In this way, samples are prepared in a uniform manner and the course of paste setback with time can be followed. As shown, corn starch

Table IV

Brabender and Brookfield Viscosities of Corn Starch Pastes with and without Added Glycerol Monostearate (GMS)[a]

Sample	Paste temperature (°)	Brabender viscosity units[b] Peak	95°	95° hold	50°	50° hold
Corn starch	73	520	520	445	1050	1040
Corn starch + 1.3% GMS	84	640	100	600	890	840
Corn starch + 5.0% GMS	92	1340[c]	140	800	1170	1120

Sample	Brookfield viscosity (cP) 0 h (45°)	3 h (25°)	24 h (25°)
Corn starch	38,000	—Gelled—	
Corn starch + 1.3% GMS	26,000	70,000	75,000
Corn starch + 5.0% GMS	30,000	28,000	28,500

[a] Conditions: 35 g of starch/500 mL, 1.5°/min heating rate; 30 min holds at 95° and 50°.
[b] See text for definition of data points.
[c] Peak occurs at 73° during 95°→50° cooling.

sets to a gel too heavy for viscosity measurement. With added GMS at the 1.3% level, viscosity is higher at 25° than at 45°, but viscosity remains the same with time. With 5% added GMS, viscosity is essentially the same through the 24-h period, showing no effect of temperature on viscosity.

The Ottawa starch viscometer developed by Voisey and co-workers (*137*) is similar operationally to the Visco/amylo/Graph. The developers claim that more information is obtained during the initial stages of granule swelling, and that the viscosity–temperature relationships are more realistic with respect to starch processing since temperature is recorded in the region where the paste is undergoing shear. Versatility is claimed by having several variable testing modes such as bowl and sample size and shear rate determined by rational speed and shear gap size.

A summary of paste properties derived from viscometric measurements and visual observations is given in Table V for native starches.

3. Gel Rigidity

Gelation occurs as the hydrated and dispersed starch molecules reassociate. Amylose-containing starches, as a rule, set up quickly because linear amylose molecules associate more readily than branched amylopectin molecules. Gelation behavior varies among the amylose-containing starches with corn starch proceeding faster and further than potato starch, as evidenced by differences in final gel strengths. The reasons for this behavior are not completely understood, but differences in fat content and amylose molecular weight probably play major roles. Another factor, about which more information is needed, is the extent that the amylose and amylopectin molecules are separated, or associated, in native granules. It is known, for example, that the amylose is more readily leached from corn than from potato starch, and that the amylose in corn starch is partially segregated by the presence of native lipid. Also, corn starch granules are more amenable to complex formation with added adjuncts than are potato starch granules. Hence, amylose and amylopectin are largely separated in corn starch granules; while in potato starch granules, they are more intimately associated. This model accounts for the high degree of swelling and lesser gelation (softer gels) shown by potato starch, that is, the amylose fraction being associated, as postulated in potato starch, strengthens the granules to give large swollen vesicles during gelatinization. Its effectiveness in forming rigid gels, however, is partially obscured by the amylopectin fraction.

As gelation proceeds to form solid gels, other modes of analysis are required for following gel rigidity development. A simple device for this purpose is a modified penetrometer where gel compression is measured as a function of added

Table V

Paste Properties of Thick-Boiling Starches

Property	*Potato*	*Corn*	*Waxy sorghum (white milo)*[a]	*Cross-bonded waxy sorghum*[a]	*Major factors involved*
Ease of cooking	Fast	Slow	Rapid	Rapid	Rate of granule swelling
Viscosity peak	High	Moderate	Very high	None	Extent of granule swelling and solubility
Stability during cooking	Poor	Good	Poor	Very good	Swollen granule fragility and solubility
Set-back on cooling	Low	Very high	Very low	Very low	Retrogradation of linear molecules
Cold-paste consistency	Long, stringy	Short, heavy-bodied, congealing	Long, cohesive, noncongealing	Very short, heavy-bodied, noncongealing	Swollen granule rigidity and retrogradation
Thickening power	High	Moderate	Moderate	High	Extent of granule swelling and integrity
Resistance to shear	Poor	Fair	Poor	Very high	Swollen granule rigidity

[a] Similar properties can be expected from regular or cross-bonded waxy maize starches.

weights. From the slope of the compression curve, a Young's modulus for compressibility can be calculated. This approach and a technique for preparing and storing the gels for analysis is given by Cluskey and co-workers (*138*).

Utility of an Instron Tensile Tester for making gel rigidity measurements is shown in a study on sweetener–starch–water systems. Plotting rigidity (E) versus time (t) indicated first order decay curves which led to a "diminishing growth model" for rigidity development, since starch crystallization occurs at a diminishing rate with time (*139*). Thus,

$$E = A - B \exp(-kt^{n})$$

or

$$E = A - B \exp(-t/T)$$

where E is the rigidity in 10^6 dynes cm^{-1}; A is final rigidity in 10^6 dynes cm^{-2}; B is magnitude of the rigidity change, also in dynes cm^{-2}; k is the rate constant in h^{-1}; T is a time constant in h; n is taken as 1, meaning, instantaneous nucleation and rod-like crystal growth with no lag period.

Basically, this is the Avrami model developed to account for the growth of crystals from an amorphous polymer phase. Maxwell and Zobel (*139*) used non-linear least squares methods for fitting a large number of data points and made no assumption regarding final rigidity. Data analysis showed that the major pure effect on gel rigidity is the starch to water ratio. A sweetener–starch interaction was observed, indicating a marked acceleration in starch crystallization with added fructose. Time constants ($1/k$) for rigidity development of 40, 54, 63, and 64 h are reported for wheat starch gels with added D-fructose, D-glucose, and sucrose, and starch without added sweetener, respectively.

The degree and type of interaction of the various sugars with water appears to account for their reported behavior in concentrated gel systems. D-Fructose, for example, shows greater propensity than the other two sugars to associate with water (*32, 34–36*). Hence, at elevated temperatures, a fructose solution appears to be a better solvent for starch because it results in the least increase in gelatinization temperature over that shown by the wheat starch control (*13, 16, 17, 140*). With aging at lower temperatures, however, the strength of the D-fructose–water bond reduces the level of water available for starch hydration. Consequently, final gel rigidity is highest for D-fructose which is consistent with the findings of Maxwell and Zobel (*139*) and of Hellman and co-workers (*141*) that rigidity increases as the water level in starch gels decreases.

4. Gel Strength

Discussion on the mechanical properties of pastes has focused on gel rigidity, but rupture of gel structure, i.e., gel strength, is measured in some applications such as gum confections. Both the penetrometer described earlier and the Instron

Tensile Tester may be used to measure gel strength. Another instrument is the Bloom Gelometer, with which the weight of accumulated small lead shot is used to measure gel strength (*142*).

5. Viscoelasticity

Discussions up to this point have centered around apparent viscosity measurements that have been correlated with paste mechanical properties defined by food uses and industrial applications. Viscosity is a measurable quantity only if the energy required to impose a stress on the system is dissipated when the stress is relaxed. At 5% or higher concentrations, starch pastes possess a certain amount of rigidity as a result of granule swelling, the binding of solubilized molecules, and the formation of physical cross-links owing to molecular reassociation. Rigidity results from the applied stress being stored in the pastes rather than being dissipated, and such pastes are said to be viscoelastic. Myers and co-workers (*143*) suggest that dynamic measurements in which low frequency shear waves are applied to the paste are best suited to separate viscosity (η) and rigidity (G), or modulus of shear elasticity, effects in starch pastes. Values for each parameter were determined for 5% pastes of white milo starches (waxy sorghum) that were cross-linked to varying degrees with sodium trimetaphosphate. Long stringy or short cohesive pastes at 25° gave η and G values of 1.8 and 110 or 3.0 and 810, respectively. Variation in phosphate cross-linking altered rigidity (G) tenfold while viscosity (η) merely doubled; in addition, η increased linearly with concentration (C), while G increased as C^2. Essentially, the mechanical properties of pastes can be measured by a quantitative determination of two rheological parameters that, in turn, can be controlled to optimize products for specific applications.

Dynamic rigidity measurements on aging wheat starch pastes by Wong and Lelievre (*176*) suggest that observed increases in rigidity owe to a crystallization process. The mechanism of crystallization was interpreted using the Avrami equation that models nucleation and crystal growth. Dynamic viscoelastic measurements also are reported on cassava and sago starches by Kawabata and co-workers (*172*).

Regardless of the perceived value in characterizing starch pastes by determining η and G, the technique has not been widely adapted, in part, due to lack of readily available hardware (*143–145*). Rheometric's Visco-Elastic Tester, however, appears to be a viable unit for testing starch pastes; hopefully, the future will see starch technologists utilizing dynamic rheological measurements.

VI. References

(*1*) H. W. Leach, T. J. Schoch, and E. F. Chessman, *Staerke*, **13**, 200 (1961).
(*2*) E. B. Lancaster and H. F. Conway, *Cereal Sci. Today*, **13** (6), 248 (1968).
(*3*) C. E. Mangels and C. H. Bailey, *J. Amer. Chem. Soc.*, **55**, 1981 (1933).

(*4*) R. M. Sandstedt, W. Kempf, and R. C. Abbott, *Staerke*, **12,** 333 (1960).
(*5*) D. G. Medcalf and K. A. Gilles, *Staerke*, **18,** 101 (1966).
(*6*) B. M. Gough and J. N. Pybus, *Staerke*, **23,** 210 (1971).
(*7*) B. M. Gough and J. N. Pybus, *Staerke*, **25,** 123 (1973).
(*8*) C. F. Ciacco and J. L. A. Fernandes, *Starch/Staerke*, **31,** 51 (1979).
(*9*) I. Lindquist, *Starch/Staerke*, **31,** 195 (1979).
(*10*) A. J. Ganz, *Cereal Chem.*, **42,** 429 (1965).
(*11*) E. E. Hester, A. M. Briant, and C. J. Personius, *Cereal Chem.*, **33,** 91 (1956).
(*12*) A. M. Campbell and A. M. Briant, *Food Research*, **22,** 358 (1957).
(*13*) M. M. Bean and E. M. Osman, *Food Res.*, **24,** 665 (1959).
(*14*) H. Knoch, Manufr. *Confectioner*, **52** (10), 42 (1972).
(*15*) H. L. Savage and E. M. Osman, *Cereal Chem.*, **55,** 447 (1978).
(*16*) M. M. Bean and W. T. Yamazaki, *Cereal Chem.*, **55,** 936 (1978).
(*17*) M. M. Bean, W. T. Yamazaki, and D. H. Donelson, *Cereal Chem.*, **55,** 995 (1978).
(*18*) E. M. Osman, *Food Technol.*, **29** (4), 30 (1975).
(*19*) E. M. Osman and M. R. Dix, *Cereal Chem.*, **37,** 464 (1960).
(*20*) E. M. Osman, S. J. Leith, and M. Flĕs, *Cereal Chem.*, **38,** 449 (1961).
(*21*) V. M. Gray and T. J. Schoch, *Staerke*, **14, 239 (1962).**
(*22*) K. Yasumatsu and S. Moritaka, *J. Food Sci.*, **29,** 198 (1964).
(*23*) G. Y. Brokaw, L. J. Lee, and G. D. Neu, *Food Processing*, **29,** 198 (1964).
(*24*) G. R. Jackson and B. W. Landfried, *Cereal Chem.*, **42,** 323 (1965).
(*25*) T. Yasunaga, W. Bushuk, and G. N. Irvine, *Cereal Chem.*, **45,** 269 (1968).
(*26*) N. Krog and B. N. Jensen, *J. Food Technol.*, **5,** 77 (1970).
(*27*) N. Krog, *Staerke*, **23,** 206 (1971).
(*28*) N. Krog, *Staerke*, **25,** 22 (1973).
(*29*) F. T. Orthoefer, *Cereal Chem.*, **53,** 561 (1976).
(*30*) S. Hizukuri and C. Takeda, *Starch/Staerke*, **30,** 228 (1978).
(*31*) E. C. Bingham, *J. Phys. Chem.*, **45,** 885 (1941).
(*32*) J. B. Taylor and J. S. Rowlinson, *Trans. Faraday Soc.*, **51,** 1183 (1955).
(*33*) E. R. Nightingale, Jr., *J. Phys. Chem.*, **63,** 1381 (1959).
(*34*) A. Suggett, *in* "Water: A Comprehensive Treatise," F. Franks, ed., Plenum Press, New York, 1975, Vol. 4, p. 519.
(*35*) M. Mathlouthi and D. V. Luu, *Carbohydr. Res.*, **81,** 203 (1980).
(*36*) M. Mathlouthi, C. Luu, A. M. Meffroy-Biget, and D. V. Luu, *Carbohydr. Res.*, **81,** 213 (1980).
(*37*) W. L. Deatherage, M. MacMasters, and C. E. Rist, *Trans. Amer. Assoc. Cereal Chem.*, **13,** 31 (1955).
(*38*) J. Borch and R. H. Marchessault, *J. Colloid Interfac. Sci.*, **27,** 355 (1968).
(*39*) J. Borch, A. Sarko, and R. H. Marchessault, *Staerke*, **21,** 279 (1969).
(*40*) J. Borch, A. Sarko, and R. H. Marchessault, *J. Colloid Interfac. Sci.*, **41,** 574 (1972).
(*41*) H. F. Zobel, *Methods Carbohydr. Chem.*, **4,** 109 (1964).
(*42*) H. F. Zobel, *in* "Starch: Chemistry and Technology," R. L. Whistler and E. F. Paschall, eds., Academic Press, New York, 1967, Vol. II, p. 510.
(*43*) N. N. Hellman, B. Fairchild, and F. R. Senti, *Cereal Chem.*, **31,** 495 (1954).
(*44*) H. F. Zobel and F. R. Senti, *Cereal Chem.*, **36,** 441 (1959).
(*45*) J. Olkku, *Food Chem.*, **3,** 293 (1978).
(*46*) C. Sterling, *J. Texture Stud.*, **9,** 225 (1978).
(*47*) A. Kuntzel and K. Doehner, *Kolloid Z.*, **86,** 130 (1939).
(*48*) H. F. Zobel, F. R. Senti, and D. S. Brown, *Abstr., Ann. Meeting Amer. Assoc. Cereal Chem., 50th, 1965*, p. 77.

(*49*) D. J. Stevens and G. A. H. Elton, *Staerke,* **23,** 8 (1971).
(*50*) P. J. Flory, "Principles of Polymer Chemistry," Cornell University Press, Ithaca, New York, 1953, p. 569.
(*51*) L. Mandelkern, "Crystallization of Polymers," McGraw-Hill Book Company, New York, 1964, p. 40.
(*52*) J. W. Donovan, *Biopolymers,* **18,** 263 (1979).
(*53*) J. W. Donovan and C. J. Mapes, *Starch/Staerke,* **32,** 190 (1980).
(*54*) J. Lelievre, *Polymer,* **17,** 854 (1976).
(*55*) M. Ahmed and J. Lelievre, *Staerke,* **30,** 78 (1978).
(*56*) G. Hollinger, L. Kuniak, and R. H. Marchessault, *Biopolymers,* **13,** 879 (1974).
(*57*) K. Wada, K. Takahashi, K. Shirai, and A. Kawamura, *J. Food Sci.,* **44,** 1366 (1979).
(*58*) M. Wooton and B. Baumunarachchi, *Starch/Staerke,* **31,** 201 (1979).
(*59*) M. Wooton and A. Baumunuarachchi, *Starch/Staerke,* **31,** 262 (1979).
(*60*) M. Kugimiya, J. W. Donovan, and R. Y. Wong, *Starch/Staerke,* **32,** 265 (1980).
(*61*) A. C. Eliasson, *Starch/Staerke,* **32,** 270 (1980).
(*62*) H. F. Zobel, unpublished.
(*63*) H. F. Zobel, *Baker's Dig.* **47,** 52 (1973).
(*64*) K. Takahashi, K. Shirai, K. Wada, and A. Kawamura, *Nippon Nogei Kagaku Kaishi,* **52,** 201 (1978).
(*65*) H. F. Zobel and L. A. Rocca, unpublished.
(*66*) C. Sterling, *Staerke,* **12,** 182 (1960).
(*67*) J. R. Katz, *Bakers Weekly,* **81,** 34 (1934).
(*68*) J. R. Katz, *Bakers Weekly,* **83,** 26 (1934).
(*69*) T. J. Schoch and E. C. Maywald, *Anal. Chem.,* **28,** 382 (1956).
(*70*) S. A. Watson, *Methods Carbohydr. Chem.,* **4,** 240 (1964).
(*71*) J. Seidemann, *Staerke,* **15,** 291 (1963).
(*72*) J. Seidemann, "Stärke-Atlas," Paul Parey, Berlin, 1966.
(*73*) G. K. Berry and G. W. White, *J. Food Technol.* **1,** 249 (1966).
(*74*) R. Collison, *Staerke,* **13,** 9 (1961).
(*75*) C. D. Freke, *J. Food Technol.,* **6,** 273 (1971).
(*76*) R. Collison and W. G. Chilton, *J. Food Technol.,* **9,** 309 (1974).
(*77*) B. S. Miller, R. I. Derby, and H. B. Trimbo, *Cereal Chem.,* **50,** 271 (1973).
(*78*) R. D. Hill and B. L. Dronzek, *Staerke,* **25,** 367 (1973).
(*79*) R. C. Hoseney, W. A. Atwell, and D. R. Lineback, *Cereal Foods World,* **22,** 56 (1977).
(*80*) D. R. Lineback and E. Wongsrikasem, *J. Food Sci.,* **45,** *71 (1980).*
(*81*) E. Berghofer and H. Klaushofer, *Staerke,* **28,** 113 (1976).
(*82*) S. Woodruff and M. M. MacMasters, University of Illinois Agricultural Experiment Station, Bulletin 445 (1938).
(*83*) W. L. Morgan, *Ind. Eng. Chem.,* **12,** 313 (1940).
(*84*) O. C. Beckford and R. M. Sandstedt, *Cereal Chem.,* **24,** 250 (1947).
(*85*) R. W. Longley and B. S. Miller, *Cereal Chem.,* **48,** 81 (1971).
(*86*) D. H. Cook and J. H. Axtmayer, *Ind. Eng. Chem., Anal. Ed.,* **19,** 226 (1937).
(*87*) W. C. Shuey and K. H. Tipples, "The Amylograph Handbook," Amer. Assoc. Cereal Chem., St. Paul, Minnesota, 1980.
(*88*) L. B. Crossland and H. H. Favor, *Cereal Chem.,* **25,** 213 (1948).
(*89*) R. M. Sandstedt and R. C. Abbott, *Cereal Sci. Today* **9** (1), 13 (1964).
(*90*) J. E. Freeman and W. J. Verr, *Cereal Sci. Today,* **17** (2), 46 (1972).
(*91*) M. M. MacMasters, M. J. Wolf, and H. L. Seckinger, *J. Agr. Food Chem.,* **5,** 455 (1957).
(*92*) T. F. Schoch, *Baker's Dig.,* **39** (2), 48 (1965).
(*93*) H. W. Leach, L. D. McCowen, and T. J. Schoch, *Cereal Chem.,* **36,** 534 (1959).

(*94*) N. N. Hellman, T. F. Boesch, and E. H. Melvin, *J. Amer. Chem. Soc.*, **74**, 348 (1952).
(*95*) R. M. Shetty, D. R. Lineback, and P. A. Seib, *Cereal Chem.*, **51**, 364 (1974).
(*96*) B. Y. Chaing and J. A. Johnson, *Cereal Chem.*, **54**, 429 (1977).
(*97*) B. Y. Chaing and J. A. Johnson, *Cereal Chem.* **54**, 436 (1977).
(*98*) J. M. Wu, *Li Kung Hsueh Pao (Kuo Li Chung Hsing Ta Hsueh)*, **14**, 271 (1977).
(*99*) A. Collison and M. P. McDonald, *Nature*, **166**, 548 (1960).
(*100*) C. Sterling and M. Masuzawa, *Makromol. Chem.*, **116**, 140 (1968).
(*101*) E. Jaska, *Cereal Chem.*, **48**, 457 (1971).
(*102*) D. French, *in* "The Chemistry and Industry of Starch," R. W. Kerr, ed., Academic Press, New York, 1950, p. 165.
(*103*) J. Lelievre, *Staerke*, **27**, 113 (1975).
(*104*) W. Basler and H. Lechert, *Staerke*, **26**, 39 (1974).
(*105*) H.-J. Hennig and L. Lechert, *Staerke*, **26**, 232 (1974).
(*106*) H.-J. Hennig, H. T. Lechert, and W. Goemann, *Staerke*, **28**, 10 (1976).
(*107*) H. T. Lechert, *Staerke*, **28**, 369 (1976).
(*108*) H.-J. Hennig, H. T. Lechert, and B. Kirsche, *Staerke*, **27**, 151 (1975).
(*109*) H.-J. Hennig and H. T. Lechert, *J. Colloid Interfac. Sci.* **62**, 199 (1977).
(*110*) H.-J. Hennig, *Staerke*, **29**, 1 (1977).
(*111*) *G. L. Wilkes, J. Macromol. Sci. Revs. Macromol. Chem.*, **C10** 149 (1974).
(*112*) J. L. Marchant, M.C.A. Chapman, and J. M. V. Blanshard, *J. Phys. E* **10**, 928 (1977).
(*113*) J. L. Marchant and J. M. V. Blanshard, *Starch/Staerke*, **30**, 257 (1978).
(*114*) J. M. V. Blanshard, *in* "Polysaccharides in Foods," J. M. V. Blanchard and J. R. Mitchell, Butterworths, Boston, 1979, p. 139.
(*115*) J. L. Marchant and J. M. V. Blanshard, *Starch/Staerke*, **32**, 223 (1980).
(*116*) Z. H. Bhuiyan and J. M. V. Blanshard, *Cereal Chem.*, **57**, 262 (1980).
(*117*) "Preparation of Corrugating Adhesives," W. O. Kroeschell, ed., TAPPI Press, Atlanta, Georgia 1977, p. 49.
(*118*) H. Klushofer, E. Berghofer, and E. Neugeschwandtner, *Staerke*, **27**, 185 (1975).
(*119*) A. L. Elder and T. J. Schoch, *Cereal Sci. Today*, **4**, 202 (1959).
(*120*) M. M. MacMasters, personal communication.
(*121*) A. Cruz, W. B. Russell, and D. F. Ollis, *AIChE J.*, **22**, 832 (1976).
(*122*) Y. Nedonchelle and R. A. Schutz, *C. R. Acad. Sci., Paris, Ser. C*, **265**, 16 (1967).
(*123*) R. Schutz, *Staerke*, **15**, 394 (1963).
(*124*) R. A. Schutz and Y. Nedonchelle, *in* "Polymer Systems, Deformation and Flow," Proc. Ann. Conf. Brit. Soc. Rheology, R. E. Wetton, ed., 1966, p. 325.
(*125*) R. A. Schutz, *Staerke*, **23**, 359 (1971).
(*126*) R. A. Schutz and G. Tatin, *Staerke*, **26**, 261 (1972).
(*127*) M. Cěh and Č. Stropnik, *Staerke*, **28**, 172 (1976).
(*128*) S. Winkler, G. Luckow, and H. Donie, *Staerke*, **23**, 325 (1971).
(*129*) Lj. Djaković and P. Dokić, *Staerke*, **24**, 195 (1972).
(*130*) I. R. Poyser and R. G. Wright, *Starch/Staerke*, **31**, 365 (1979).
(*131*) E. G. Mazurs, T. J. Schoch, and F. E. Kite, *Cereal Chem.*, **34**, 141 (1957).
(*132*) R. J. Priestley, *Staerke*, **27**, 416 (1975).
(*133*) R. J. Priestley, *Food Chem.*, **1**, 5 (1975).
(*134*) R. J. Priestley, *Food Chem.*, **1**, 139 (1976).
(*135*) R. J. Priestley, *Food Chem.*, **2**, 43 (1977).
(*136*) F. E. Kite, E. C. Maywald, and T. J. Schoch, *Staerke*, **15**, 131 (1963).
(*137*) P. W. Voisey, D. Paton, and G. E. Timbers, *Cereal Chem.*, **54**, 534 (1977).
(*138*) J. E. Clusky, N. W. Taylor, and F. R. Senti, *Cereal Chem.*, **36**, 236 (1959).
(*139*) J. L. Maxwell and H. F. Zobel, *Cereal Foods World*, **23**, 124 (1978).

(*140*) E. Osman, *in* "Food Theory and Applications," P. C. Paul and H. H. Palmer, eds, John Wiley and Sons, New York, 1972, p. 151.
(*141*) N. N. Hellman, B. Fairchild, and F. R. Senti, *Cereal Chem.*, **31,** 495 (1954).
(*142*) GCA/Precision Scientific Apparatus Supply, Chicago, Illinois.
(*143*) R. R. Meyers, C. J. Knauss, and R. D. Hoffman, *J. Appl. Poly. Sci.*, **6,** 659 (1962).
(*144*) R. R. Meyers and C. J. Knauss, *Methods Carbohydr. Chem.*, **4,** 128 (1964).
(*145*) R. R. Meyers and C. J. Knauss, *in* "Starch: Chemistry and Technology," R. L. Whistler and E. F. Paschall, eds., Academic Press, New York, Vol. I, 1965, p. 393.
(*146*) B. J. Oosten, *Starch/Staerke,* **31,** 228 (1979).
(*147*) B. J. Oosten, *Starch/Staerke,* **32,** 272 (1980).
(*148*) K. Takahashi, S. Kunio, K. Wada, and A. Kawamura, *Nippon Nogei Kagaku Kaishi,* **52,** 441 (1978).
(*149*) K. Takahashi, K. Shirai, K. Wada, and A. Kawamusa, *Denpun Kagaku,* **27,** 22 (1980).
(*150*) K. Takahashi, K. Shirai, and K. Wada, *Denpun Kagaku,* **28,** 1 (1981).
(*151*) M. Wootton and A. Bamunuarachchi, *Starch/Staerke,* **32,** 126 (1980).
(*152*) K. Larsson, *Starch/Staerke,* **32,** 125 (1980).
(*153*) K. Ohashi, G. Goshima, H. Kusada, and H. Tsuse, *Starch/Staerke,* **32,** 54 (1980).
(*154*) E. Varriano-Marston, V. Ke, G. Huana, and J. Ponte, Jr., *Cereal Chem.*, **57,** 242 (1980).
(*155*) K. Ghiasi, R. C. Hoseney, and E. Varriano-Marston, *Cereal Chem.*, **59,** 81 (1982).
(*156*) K. Ghiasi, E. Varriano-Marston, and R. C. Hoseney, *Cereal Chem.*, **59,** 86 (1982).
(*157*) J. Hofstee, *LWT-Edition,* **5,** 31 (1980).
(*158*) J. Lelievre, *J. Appl. Polymer Sci.*, **18,** 293 (1973).
(*159*) J. W. Donovan, *J. Sci. Food Agric.*, **28,** 571 (1977).
(*160*) C. G. Billiadersi, T. J. Maurice, and J. R. Vose, *J. Food Sci.*, **45,** 1669 (1980).
(*161*) K. Eberstein, R. Hoepcke, G. Konieczny-Janda, and R. Stute, *Starch/Staerke,* **32,** 397 (1980).
(*162*) M. Kugimiya and J. W. Donovan, *J. Food Sci.*, **46,** 765 (1981).
(*163*) K. D. Nishita and M. M. Bean, *Cereal Chem.*, **59,** 46 (1982).
(*164*) C. A. Knutson, V. Khoo, J. E. Cluskey, and G. E. Inglett, *Cereal Chem.*, **59,** 512 (1982).
(*165*) K. Takahashi, F. Shimizu, K. Shirai, and K. Wada, *Denpun Kagaku,* **29,** 34 (1982).
(*166*) J. Longton and G. A. LeGrys, *Starch/Staerke,* **33,** 410 (1981).
(*167*) R. G. McIver, D. W. E. Axford, K. H. Colwell, and G. A. H. Elton, *J. Sci. Food Agric.*, **19,** 560 (1968).
(*168*) K. H. D. Colwell, D. W. E. Axford, N. Chamberlain, and G. A. H. Elton, *J. Sci. Food Agric.*, **20,** 550 (1969).
(*169*) S. Y. Gerlsma, *Staerke,* **22,** 3 (1970).
(*170*) H. T. Lechert, *in* "Water Activity: Influences on Food Quality," L. B. Rockland and G. F. Franklin, eds., Academic Press, New York, 1981, p. 223.
(*171*) J. C. Doublier, *Starch/Staerke,* **33,** 415 (1981).
(*172*) A. Kawabata, E. Kataoka, S. Sawayama, and A. Murayama, *Tokyo Nosyo Daisaku Nosaku Shuho,* 195 (1981).
(*173*) R. D. Spies and R. C. Hoseney, *Cereal Chem.*, **59,** 128 (1982).
(*174*) I. D. Evans and D. R. Haisman, *Starch/Staerke,* **34,** 224 (1982).
(*175*) B. J. Oosten, *Starch/Staerke,* **34,** 233 (1982).
(*176*) R. B. K. Wong and J. Lelievre, *Starch/Staerke,* **34,** 231 (1982).
(*177*) A. C. Eliasson and L. Bohlin, *Starch/Staerke,* **34,** 267 (1982).
(*178*) J. W. Donovan, K. Lorenz, and K. Kulp, *Cereal Chem.*, in press.
(*179*) M. M. MacMasters and I. A. Wolff, *in* "Chemistry and Technology of Cereals as Food and Feed," (S. A. Matz, ed.) AVI Publishing Co., Westport, Connecticut, 1959, p. 573.

CHAPTER X

STARCH DERIVATIVES: PRODUCTION AND USES

By Morton W. Rutenberg and Daniel Solarek

Research Department, Industrial Starch and Food Products Division, National Starch and Chemical Corporation, Bridgewater, New Jersey

STARCH, 2nd ed.

ISBN 0-12-746270-8

I. Introduction

This chapter covers starch derivatives that are produced and marketed in quantities of sufficient size to be considered as commercial scale. However, commercial interest can arise whenever it becomes economically feasible to meet a particular need with a starch derivative which might heretofore have been interesting only from an academic or scientific viewpoint. This commercial interest could result from a unique property, a new, lower-cost method of manufacture, increased demand allowing economies of scale or, perhaps, a completely new and unforeseen application.

The term ''starch derivative'' includes those modifications which change the chemical structure of some of the D-glucopyranosyl units in the molecule. These modifications usually involve oxidation, esterification, or etherification. Other types of modifications such as hydrolysis (Chapter XVII) and dextrinization (Chapter XX) are the subjects of other chapters, although the products of these modifications may serve as bases for derivatization.

Derivatization of starch is conducted to modify the gelatinization and cooking characteristics of granular starch, to decrease the retrogradation and gelling tendencies of amylose-containing starches, to increase the water-holding capacity of starch dispersions at low temperature thereby minimizing syneresis, to enhance hydrophilic character, to impart hydrophobic properties, and/or to introduce ionic substituents. Modification of starch properties by derivatization is an important factor in the continued and increased use of starch to provide thickening, gelling, binding, adhesive, and film-forming functionality.

A starch derivative is fully defined by a number of factors: plant source (corn, waxy maize, potato); prior treatment (acid-catalyzed hydrolysis or dextrinization); amylose/amylopectin ratio or content; some measure of molecular weight distribution or degree of polymerization (DP; usually described commercially in terms of viscosity or fluidity); type of derivative (ester, ether, oxidized); nature of the substituent group (acetate, hydroxypropyl); degree of substitution (DS)[1] or molar substitution (MS)[1]; physical form (granular, pregelatinized); presence of associated components (proteins, fatty acids, fats, phosphorus compounds) or native substituents.

Starches commonly used for commercial derivatization in the United States are corn, waxy maize, tapioca, and potato. Derivatives of other less available starches such as sorghum, waxy sorghum, wheat, rice, and sago may be offered commercially. The properties, availability in large quantities, and economics are deciding factors in determining which starch is used for derivatization. In Australia, for example, wheat starch is commonly used.

Multiple treatments may be employed to obtain the desired combination of properties. Thus, an acid-converted starch or dextrin may be used for derivatization to obtain a lower viscosity product which is dispersable at higher solids than one made from the native starch and one whose dispersions are still able to be pumped and handled. Hypochlorite-oxidized starch may also be used for further dervatization. Sometimes a derivative made from undegraded starch may subsequently be subjected to the acid-conversion, dextrinization, or oxidation treatment to obtain the desired range of viscosity. Cross-linking is often used in combination with other derivatization treatments to maintain dispersion viscosity upon exposure to high temperature cooking, high shear, or acid. The sequence of the treatments is based on the stability of the substituent groups or the treated starch to subsequent reactions.

The amylose or amylopectin content can be varied by using a high-amylose starch or a starch which is essentially 100% amylopectin (waxy maize starch). Isolated amylose or amylopectin fractions may be used as such.

Most commercially produced derivatives have a DS, generally less than 0.2. Since starch is inherently water-soluble after disruption of the granular structure,

[1]The degree of substitution (DS) is a measure of the average number of hydroxyl groups on each D-glucopyranosyl unit which are derivatized by substituent groups. DS is expressed as moles of substituent per D-glucopyranosyl unit (commonly called an anhydroglucose unit and abbreviated as AGU). Since the majority of the AGUs in starch have 3 hydroxyl groups available for substitution, the maximum possible DS is 3. When the substituent group can react further with the reagent to form a polymeric substituent, molar substitution (MS) expresses the level of substitution in terms of moles of monomeric units (in the polymeric substituent) per mole of AGU. Thus, MS can be greater than 3.

$$\mathrm{MS} \geq \mathrm{DS}$$

$$\mathrm{DS} = \frac{162\,W}{100\,M - (M - 1)W}$$

Where W = % by weight of substituent and M = molecular weight of the substituent, considered as a whole, whether monomeric or polymeric. Where polymeric substituents are present, M and W refer only to the monomeric units of the polymeric substituent and the formula then yields the MS. For example, the DS is used for starch acetates, and MS is used for ethylene oxide-treated starch.

a relatively high DS is not required to impart solubility or dispersibility as is necessary with cellulose.

1. Manufacture of Derivatives

Low DS derivatives are generally manufactured by reacting starch in an aqueous suspension of 35–45% solids, usually at pH 7–12. Sodium hydroxide and calcium hydroxide are commonly used to produce the alkaline pH. Calcium hydroxide (lime), generally used in suspension because of its limited solubility in water, usually produces some noticeable degradation of starch molecules, presumably because of air oxidation. Sometimes pH is controlled by the metered addition of dilute aqueous alkali, such as a 3% sodium hydroxide solution. Reactions may be done at temperatures ranging up to 60°. Conditions are adjusted to prevent gelatinization of granular starch and to allow recovery of the starch derivative in granular form by filtration or centrifugation and drying. The derivative may be washed to remove unreacted reagent, by-products, salts, and other solubles before final recovery in dry form. To prevent the swelling of starch under strongly alkaline reaction conditions, sodium chloride or sodium sulfate may be added to a concentration of 10–30%. Because the granule gelatinization temperature is lowered as the DS increases, there is a limit to the level of substitution that can be made in aqueous slurry while retaining the starch in granular form.

If the level of hydrophilic substituent becomes high enough, the starch derivative gelatinizes, becomes dispersible at room temperature, and is said to be cold-water-dispersible. These starch granules swell on contact with water due to effects of the introduced hydrophilic groups. Thus, a nonswelling solvent such as isopropanol or acetone, neat or mixed with water, is used to prepare higher DS, cold-water-dispersable derivatives in granule form.

Derivatization is also done by treating the "dry" starch with the required reagents and then heating to temperatures up to 150° to yield granular products of DS up to 1 in granule form. Reagents may be mixed with the starch by dry blending, by spraying an aqueous or nonaqueous solution onto dry starch or a filter cake prior to drying, or by suspending the starch in a solution of the reagents and filtering and drying. Generally, the presence of 5–25% moisture is desirable for efficient reaction. There are some reactions which are more efficient at lower moisture. Derivatives made by this type of semidry reaction often contain salts and reaction by-products as well as unreacted reagent because the product is usually not washed.

Higher DS derivatives containing hydrophobic substituents can be made in water dispersion with recovery of the precipitated product by filtration. Derivatized starches can be recovered from aqueous, dispersed reactions by drum-drying or spray-drying either directly or after removal of salts and low-molecular-weight by-products by dialysis or ultrafiltration.

II. Hypochlorite-Oxidized Starches

1. Introduction

Although many reagents oxidize starch, alkaline hypochlorite is the most common commercial reagent. Starches oxidized with hypochlorite are termed "chlorinated starches," although no chlorine is introduced into the starch molecules by this treatment. Hypochlorite oxidation has been practiced since the early 1800s. Early work has been extensively reviewed, including investigations with other oxidizing agents such as periodate, dichromate, permanganate, persulfate, chlorite, and hydrogen peroxide (*1–3*). Oxidants, such as potassium permanganate, hydrogen peroxide, sodium chlorite, hydrogen peroxide, and peracetic acid that have been used to bleach starch are used in such small amounts that the starch is not changed significantly and the products are not considered as oxidized starches. Ammonium persulfate is used in paper mills with continuous thermal cookers to prepare *in situ* high-solids, low-viscosity, aqueous dispersions of degraded starch for coating (*4*) and sizing operations (*5, 6*). Hydrogen peroxide is also reported as useful for starch depolymerization in a continuous thermal cooking process (*7, 8*) as well as on the granular starch (*9, 10*) or starch derivative (*11, 12*).

Starch is oxidized to obtain low-viscosity, high-solids dispersions and resistance to viscosity increases or gelling in aqueous dispersion. The oxidation causes depolymerization, which results in a lower viscosity dispersion, and introduces carbonyl and carboxyl groups, which minimize retrogradation of amylose, thus giving viscosity stability.

2. Manufacture

The oxidizing agent is prepared on site by diffusing chlorine into a dilute solution of sodium hydroxide (caustic soda) cooled to about 4°. The concentrations of the caustic and hypochlorite in the reagent solution must be controlled and monitored because excess sodium hydroxide and available chlorine are determining factors in the type of starch modification obtained (*643*).

$$2NaOH + Cl_2 \rightarrow NaCl + NaOCl + H_2O + 24{,}650 \text{ calories}$$

Undesirable formation of chlorate is facilitated by temperatures above 30°.

Hypochlorite oxidation of starch is conducted in aqueous slurry with the alkaline sodium hypochlorite solution while controlling the pH, temperature, and concentrations of hypochlorite, alkali, and starch. Since the treatment facilities are usually located at the corn wet-milling plant, starch slurry from the refinery is added to the treatment tank at 18°–24° Bé (approx. 33–44% dry starch). 20,000–100,000 lb (9,000–45,000 kg) of starch may be reacted in a tank equipped with efficient impellers to keep the starch in suspension and to mix the added reagents rapidly and uniformly. The pH is adjusted to 8–10 with ~3% sodium hydroxide solution, and the hypochlorite solution, containing 5–10% available chlorine, is added during the allotted time. The pH is controlled by addition of dilute sodium hydroxide solution to neutralize the acidic substances produced. The temperature of the exothermic oxidation reaction is controlled to 21°–38° by the rate of addition of the hypochlorite solution or by cooling. A wide variety of products can be made by adjusting the variables of time, temperature, pH, starch, and hypochlorite concentrations and the rate of addition of the hypochlorite.

When oxidation has reached the required level, usually as determined by a viscosity measurement (*13, 14*), the reaction is stopped by lowering the pH to 5–7 and destroying excess chlorine with sodium bisulfite solution or sulfur dioxide gas. Starch is separated from the reaction mixture by filtration or centrifugation and washed to remove soluble reaction by-products, salts, and carbohydrate degradation products.

3. Oxidative Mechanisms

Investigations of the oxidation of starch by hypochlorite (*15–22*) and hypochlorite oxidation of amylose (*23, 24*) and amylopectin (*25*), as well as the action of chlorine on starch (*26–28*) and cellulose (*29*) and the hypobromite oxidation of amylopectin (*30*) and bromine oxidation of starch (*31*) have been reported. One investigation indicates that the oxidizing agent penetrates deeply into the granule, apparently acting mainly on the less crystalline areas of the granule. This was indicated by the lack of change in the birefringence (*14*) and x-ray patterns (*15*) in oxidized starches. There appears to be a drastic localized attack on some molecules that result in the formation of highly degraded, acidic fragments which become soluble in the alkaline reaction medium and are lost when the oxidized starch is washed (*15*). Cracks and fissures that develop in oxidized starch have been attributed to this localized overoxidation (*15, 32*).

The reaction rate of hypochlorite oxidation of amylopectin is markedly influenced by pH (*25*). The rate is most rapid at pH 7 and very slow at pH 11–13. Similar results were obtained with granular corn starch (*20, 21*), waxy corn starch, and wheat starch (*16*). The reaction rate decreases with increasing pH from 7.5 to 10 and remains constant from pH 10 to 11.7 (*16*). Similar results were noted with amylopectin (*25*) and amylose (*23*).

A mechanism for the course of the hypochlorite oxidation was proposed to explain the decrease in reaction rate under acidic and alkaline conditions (*20, 21*). In acidic medium, hypochlorite is rapidly converted to chlorine which reacts with the hydroxyl groups of starch molecules with the formation of an hypochlorite ester and hydrogen chloride as shown below. The ester then decomposes to a keto group and a molecule of hydrogen chloride. In both steps, hydrogen atoms are removed as protons from the oxygen and carbon atoms. Thus, in acid medium with excess protons, liberation of protons would be hindered and the rate would be expected to decrease with increasing acidity. Under alkaline conditions, formation of negatively charged starchate ions would occur, increasing with increasing pH. Since negatively charged hypochlorite ion predominates at higher pH, reaction between two negatively charged ions would be difficult because of repulsion. Hence, the oxidation rate would be hindered by increasing pH. Under neutral or slightly acidic or basic conditions, hypochlorite is primarily undissociated and the starch is neutral. Undissociated hypochlorite (hypochlorous acid) would produce starch hypochlorite ester and water, and the ester would decompose to give the oxidized product and hydrogen chloride. Any hypochlorite anion present would act on undissociated starch hydroxyl groups in a similar manner. This hypothesis would explain lower oxidation rates in acidic and basic media and the liberation of acid.

Acid:

$$\mathrm{H{-}\overset{|}{\underset{|}{C}}{-}OH + Cl{-}Cl \xrightarrow{\text{fast}} H{-}\overset{|}{\underset{|}{C}}{-}O{-}Cl + HCl}$$

$$\mathrm{H{-}\overset{|}{\underset{|}{C}}{-}O{-}Cl \xrightarrow{\text{slow}} \overset{|}{\underset{|}{C}}{=}O + HCl}$$

Alkaline:

$$\mathrm{H{-}\overset{|}{\underset{|}{C}}{-}OH + NaOH \longrightarrow H{-}\overset{|}{\underset{|}{C}}{-}O^{-}Na^{+} + H_2O}$$

$$\mathrm{2\,H{-}\overset{|}{\underset{|}{C}}{-}O^{-} + OCl^{-} \longrightarrow 2\,\overset{|}{\underset{|}{C}}{=}O + H_2O + Cl^{-}}$$

Neutral:

$$\mathrm{H{-}\overset{|}{\underset{|}{C}}{-}OH + HOCl \longrightarrow H{-}\overset{|}{\underset{|}{C}}{-}OCl + H_2O}$$

$$\mathrm{H{-}\overset{|}{\underset{|}{C}}{-}OCl \longrightarrow \overset{|}{\underset{|}{C}}{=}O + HCl}$$

$$\mathrm{H{-}\overset{|}{\underset{|}{C}}{-}OH + OCl^{-} \longrightarrow \overset{|}{\underset{|}{C}}{=}O + H_2O + Cl^{-}}$$

The reaction rate is considerably higher when oxidation is conducted in a starch solution as compared to a suspension of starch granules, indicating that only a part of the granule is available for oxidation (*16*). At 27° and pH 8.0–10.0, the ratio of reaction rates in gelatinized dispersion and in suspension is constant: for wheat starch, 1.45 at 10.2 g/L; for waxy maize starch, 3.02 at 13.4 g/L and 2.25 at 30 g/L.

Hypochlorite oxidation in an unbuffered starch suspension at pH 10 causes the pH to drop to pH 4. The initial rate at pH 10 is high, drops to a lower rate at pH 8.5, reaches a maximum at about pH 6.2, and then decreases as the pH decreases to about 4. The high initial reaction rate at pH 10 is attributed to soluble starch leaching from the partially swollen granules (about 7.5% of the starch). Correcting for this brings the rate of reaction of the insoluble starch to less than the rate at pH 8.5. About 1–2% of the starch is solubilized at neutral or acid pH (*20*).

The energy of activation for hypochlorite oxidation of wheat starch at pH 8–10 is 21–23 kcal/mole and for waxy maize 16–20 kcal/mole. The reaction rate increases 2–4 times with each 10° rise in temperature (*16*). The activation energy of corn starch is 17.6 kcal/mole at pH 7.0, 18.4 kcal/mole at pH 8.5, and 23.7 kcal/mole at pH 10.0 (*21*).

The rate of hypochlorite oxidation of corn starch in a borate buffer is slower than with a carbonate–bicarbonate buffer owing to deactivation of the reactive sites in the starch by complex formation (*21*).

In oxidation of amylopectin it was found that the conversion of hypochlorite to chlorate is greatest at pH 7 (28.8%). The conversion at pH 3, 5, 9, 10.5, and 12 was 1.5%, 7.1%, 4.6%, 0.7%, and 0.5%, respectively. At pH 3, 5, and 7, the chlorate formed is not an oxidant for amylopectin or hypochlorite-oxidized amylopectin (*25*). Other workers report no chlorate formation in the pH range 7.5–11.0 at 27° and 37° and less than 2% at 47° (*16*).

In the hypochlorite oxidation of potato starch, the presence of bromide and cobalt ions exert a catalytic effect at pH 9 and 10, increasing the rate of oxidation as measured by the decrease in active chlorine content, bringing it close to the rate at pH 7. As the active chlorine content of the hypochlorite reaction medium decreases from 30 g/L to 3 g/L, the reaction rate at pH 9–11 becomes more rapid than at pH 7 (*19*). Nickel sulfate is said to have a catalytic effect on hypochlorite oxidation (*33*).

4. Chemical Properties

Carboxyl and carbonyl groups are formed by hypochlorite oxidation of starch hydroxyl groups. Scission of some of the glucosidic linkages also occurs, resulting in a decrease in molecular weight. Some of the starch is solubilized and removed during the commercial separation and washing process. Degradative effects induced by the presence of alkali as well as oxidation by atmospheric oxygen can also result in structural changes.

Hypochlorite treatment solubilizes 70–80% of the nitrogen-containing impurities of starch and removes or decolorizes pigmented material (*34, 35*). Free fatty acid content is reduced by 15–20% on prolonged treatment, with most reduction taking place in the early stages of the reaction.

Whistler and co-workers (*23–25, 36*) studied the hypochlorite oxidation of amylose, amylopectin and methyl D-glucopyranosides and found the oxidation to be highly pH dependent. Analysis of potato amylose oxidized at pH 2–12 with 0.1 mole hypochlorite per mole of α-D-glucopyranosyl unit showed an increase in total carboxyl content and a decrease in aldehyde groups with increasing pH while the number of uronic acid carboxyl groups remained about the same. Increasing the concentration of hypochlorite at pH 7 in the oxidation of aqueous suspensions of potato and wheat amylose resulted in increasing numbers of aldehyde and carboxyl groups per AGU[1], with uronic acid carboxyl groups comprising one-half to one-third of the total carboxyl groups (*24*). Based on data obtained by hydrolysis of hypochlorite-oxidized corn amylopectin and amylose, it was concluded that some of the D-glucopyanosyl units (AGU[1]) are not oxidized, even at high levels of oxidant. Further, the isolation of glyoxylic and D-erythronic acids as major products on hydrolysis of hypochlorite-oxidized amylopectin, with a maximum of these products resulting from oxidation at pH 7, suggests initial oxidation at C-2 or C-3 with the formation of a carbonyl group. The formation of an enediol at C-2—C-3 with subsequent oxidation to a dicarboxylated unit in the amylopectin chain was postulated (*25*). Similar results were obtained in the hypochlorite oxidation of corn amylose in dilute aqueous dispersion at pH 9 and 11, again suggesting initial oxidation at C-2 and/or C-3 to give keto groups which are oxidized further. It was suggested that about 25% of the hypochlorite was consumed in the oxidative cleavage of the C-2—C-3 bonds. The presence of a carbonyl group at C-2 or C-3 leads to depolymerization via β-elimination. A large portion of the hypochlorite is apparently consumed in oxidizing fragments cleaved from the starch chains (*23*). Alkaline hypochlorite oxidation of methyl 4-*O*-methyl-α- and β-D-glucopyranosides also indicates preferential oxidation at the C2 and C3 positions, the α form being oxidized at a slower rate (*23, 36*).

Although the secondary hydroxyl groups are apparently oxidized at a faster rate than are primary hydroxyl groups, the reaction rate on modified starches suggests attack at other sites (*20*). McKillican and Purves (*37*) found carboxyl and carbonyl groups in gelatinized wheat starch oxidized by hypochlorite at pH 4.0–4.2. The carboxyl groups appeared to be those of D-glucuronic acid. The carbonyl groups were predominantly aldehyde groups in the C-6 position (65–80%); ~9% were found as keto groups in the C-2 position. Thus, under these conditions, oxidation occurred mainly at the primary hydroxyl group. Hypochlorite oxidation of dispersed wheat starch at about pH 12 gave evidence of attack at the C-2 position followed by cleavage at C-2—C-3, as postulated by Whistler, as well as some oxidation at C-6 (*38*). Prey and Siklossy (*39*) found that about 85% of the total carbonyl groups in potato starch oxidized at pH 7–12 with hypochlorite were aldehyde groups, the amount of carbonyl groups decreasing as pH increased.

Hypochlorite oxidation of suspensions of corn starch or high-amylose corn starch (50% amylose) at pHs above 8.5 for 8 h produced carboxyl and carbonyl groups, in a ratio of very roughly 2:1 (*40, 41*). An increase in hypochlorite concentration increases the carboxyl and carbonyl group content (*39–41*). In oxidized potato starch, the ratio of carboxyl to aldehyde groups increases with increasing hypochlorite treatment at pH 8–9, climbing from 0.17 at 5 mg active chlorine per gram of starch to 4.79 at 95 mg/g (*42*).

Schmorak and co-workers (*16–18*) investigated the hypochlorite oxidation of granular wheat and waxy corn starches at pH 7.5–11.0. In wheat starch, roughly equal amounts of carboxyl and carbonyl groups were produced at pH 7.5. With increasing pH, the number of carboxyl groups per 100 AGU[1] increased from 1.03 at pH 7.5 to 2.78 at pH 11, while the number of carbonyl groups decreased from 0.91 at pH 7.5 to almost none at pH 9. Waxy corn starch develops 1.5 carbonyl groups per 100 AGU at pH 8, decreasing to 1 per 100 AGU at pH 9 and to practically none at pH 10. The number of carboxyl groups per 100 AGU in hypochlorite-oxidized waxy corn starch increased from 1.0 at pH 8 to 1.4 at pH 9 and 1.8 at pH 10 (*16, 17*). With an initial hypochlorite concentration of 30 millimoles/liter and a starch concentration of 54 g/L, 5.45% of wheat starch dissolved on oxidation at pH 8, and 1.74% dissolved at pH 10. The apparent DP of the soluble products ranged from 3 at pH 8 to 17 at pH 10 (*16, 17*).

Schmorak and colleagues (*16–18*) found that the amylose from wheat starch oxidized with hypochlorite at pH 8 had a number average DP of 131 and a weight average DP of 128 and that glycosidic cleavage was random.

The consumption of about 0.0525 oxygen atoms/AGU during hypochlorite oxidation of wheat starch, as calculated by Schmorak and co-workers (*16–18*), accompanied by extensive degradation of both amylose and amylopectin components, indicates that scission of each D-glucosidic bond consumes 4–5 oxygen atoms and produces 1–2 carboxyl and 0–1 carbonyl groups. The corresponding figures for waxy corn starch are 2–3 oxygen atoms consumed and about 1.5 functional groups formed per scission.

Hypochlorite oxidation of granular potato starch with cobalt and bromide ion catalysis increases the carboxyl content and shifts the pH maximum for carboxyl group formation from pH 8 without catalyst to about pH 9 (*19*).

Prey and Fischer (*43*) investigating hypochlorite-oxidized potato starch found that a stepwise variation of pH from 8 to 10 resulted in a higher proportion of carboxyl groups. A two-step oxidation is claimed to give more rapid oxidation as well as to provide control of the carbonyl:carboxyl ratio (*44*).

Schmorak and co-workers (*16, 17*) found that, with wheat starch, the iodine binding capacity does not change much upon oxidation in the pH range 7.5–11.

On the other hand, Fischer and Piller (*42*) found that, with increasing levels of hypochlorite (5–95 mg Cl_2/g starch) at pH 8–9, there is a decrease in the iodine affinity of the potato starch, reaching a constant low value at about 55 mg Cl_2/g starch. Based on these results and β-amylase digestibility measurements, they concluded that the amylose is more rapidly degraded than the amylopectin in the initial stages of the oxidation. The lower β-amylase digestibility and iodine binding capacity could also result from the presence of carboxyl and carbonyl substituents in the amylose chain. Degradation of hypochlorite-oxidized starch with amyloglucosidase or a mixture of α- and β-amylase shows decreased digestibility with increasing levels of oxidation (*45*). It was concluded that introduction of a functional group on one AGU[1] protects neighboring units from enzymic attack.

Commercial hypochlorite oxidation of starch under alkaline conditions can also bring about degradative effects. The alkaline degradation of starch has been discussed by BeMiller (*46*) and Greenwood (*47*); Corbett (*48*) has treated the corresponding cellulose reactions.

The presence of carboxyl groups in oxidized starch imparts a negative charge and causes starch granules to absorb methylene blue, a cationic dye. The intensity of the staining is related to the level of electronegativity and hence roughly to the degree of oxidation. Other functional groups such as carboxymethyl, phosphate, or sulfonate will also impart a negative charge and therefore induce staining with methylene blue (*49*).

5. Physical Properties

Hypochlorite-oxidized starch is supplied in granule form. By virtue of the bleaching effect of the hypochlorite treatment, as well as the solubilization and washing out of protein and associated pigments from, for example, corn starches, the oxidized starch is whiter than is the base starch. Within limits, the degree of whiteness increases with the extent of treatment (*50*).

In general, oxidized starches are sensitive to heat, tending to yellow or brown when exposed to high temperature. This yellowing tendency during drying has been related to the aldehyde content (*51*). With increasing aldehyde content, the oxidized starch becomes increasingly yellow on storage (*39*). Yellowing of oxidized starch dispersed in water by cooking or by alkali is also related to aldehyde content (*19*).

The hypochlorite-oxidized starch granules exhibit polarization crosses as well as an unchanged x-ray diffraction pattern, indicating that the oxidation takes place mainly in the amorphous regions of the granule (*15, 17, 18*). Scanning electron microscopy (SEM) shows that the surface of the corn starch granule is unchanged by hypochlorite oxidation up to about 6% active chlorine, with some change apparent at the 8% level (*50*). Similar results were seen with potato starch (*32*). Light microscopy shows the presence of fissures developing at the hilum in potato starch granules, extending to the distal poles of the granule and widening with increased hypochlorite treatment. This indicates that the oxidation takes place in the interior of the granule as well as at the surface. SEM shows hollows in the potato granule after hypochlorite oxidation (*32, 52*, see also this Volume, Chap. VII).

Although waxy maize starch granules showed no change in size on oxidation, wheat starch granules oxidized at pH 7.5–11.0 with 1 mole of hypochlorite per 20 AGU[1] increased approximately 16% in diameter (*16*).

The molecular weight distribution of hypochlorite-oxidized wheat starch determined by gel chromatography (*22, 53*) shows a breakdown of molecules by a second order reaction as oxidation proceeds. Others have found indication of a first-order reaction (*16, 20, 21*). The intrinsic viscosity of wheat starch, amylose, and amylopectin decreases as the oxidant absorbed increases (*17*).

Oxidized starches gelatinize at a lower temperature than do native starches when measured by loss of birefringence or by Brabender Viscograph pasting viscosity (*13, 49, 54, 55*). The shape of the Brabender pasting curve is related to the oxidation conditions. Potato starch oxidized at pH 11 generally has lowest peak viscosity and least paste viscosity stability compared to products made at pH 7–8 (*19*). Prey and co-workers (*54*) found that the Brabender peak viscosity decreased and then increased with the amount of hypochlorite treatment of potato starch, although the final viscosity was low for moderate to high oxidant levels (*54*). Low peak viscosity was observed when granules fragmented without the maximum swelling that occurs with unmodified potato starch. The higher peak viscosity of highly oxidized starch was observed when the granules were greatly swollen although the interiors were solubilized and leached out. The empty hulls collapse rapidly with continued agitation, giving a low final viscosity (*15*). Changes in viscosity and retrogradation during aging of oxidized potato starch pastes are related to the extent of hypochlorite treatment; higher levels of oxidation produce less retrogradation and viscosity change (*56*).

Hypochlorite-oxidized starches produce aqueous dispersions of greater clarity and lower viscosity than those of native starch. Further, the dispersions have less tendency to set back or gel. Thus, the pastes are more fluid. Clear solutions are obtained if the degree of oxidation is high (*15, 57*). The higher the level of hypochlorite treatment, the lower the gelatinization temperature and paste viscosity, the less the setback, and the greater the clarity. Although these effects are found with normal corn starch, it is necessary to use much higher levels of hypochlorite on high-amylose corn starch to obtain increased dispersibility as well as increased clarity and reduced setback (*41*).

6. Uses

About 80–85% of the hypochlorite-oxidized starch produced is used in the paper industry. It is primarily a paper coating binder where its high fluidity and good binding and adhesive properties make it effective in high solids pigmented coating colors. Compatibility with the pigment, usually clay, is important and it should not adversely affect water-holding capacity and rheology of the coating color (*13, 58–63, 65, 67, 68,* see also this Volume, Chap. XVIII).

Until recently, large quantities of oxidized starches were used for paper and paperboard surface sizing to seal pores, tie down loose surface fibers, improve surface strength, and provide holdout of printing inks. Viscosity stability of oxidized starch dispersions as well as the range of viscosities available made them particularly suitable. However, the on-site conversion of native pearl starch by continuous enzyme conversion (*69*) or thermal-chemical conversion (*4, 5, 6, 59*) has led to a decrease in the use of oxidized starch in surface sizing and has had some effect on coating use. Thermochemical-converted starch does not have the viscosity stability of the oxidized starch and tends to retrograde on storage making it difficult to use in coating. Methods developed to overcome this retrogradation tendency (*70*) may lead to a decrease in the use of hypochlorite-oxidized starch in paper coating. Oxidized starch in the wet end of the paper machine, mainly introduced in repulped coated paper (broke), has a detrimental effect on pigment retention by acting as a dispersant (*71*).

Starch has a long history in the manufacture of textiles; it is used primarily as warp sizing but also in finishing and printing (*61, 66, 72, 73*). Ground has been lost to synthetic polymers in finishing and to new printing styles. Little starch is used in sizing synthetic filament yarns, and oxidized starches are not used as extensively as other starches in warp sizing. In some instances, the high fluidity, stable viscosity, and flow properties at high solids of oxidized starches allows for greater add-on to the yarn and provides good abrasion protection. Such starches are readily soluble and can be desized from the woven cloth. Oxidized starches may be used in back-filling where a mixture of starch and a filler such as clay is applied to the back of a fabric to fill the interstices of the weave and impart opacity and stiffness. The lower viscosity oxidized starch penetrates fabric to a greater extent than do higher viscosity starches. Finishes are applied to give weight, hand, and draping quality to the fabric. The hypochlorite-oxidized starch may be used, particularly with cotton, in printed fabrics where the less opaque film does not dull colors.

Oxidized starches have been used in laundry finishing, sometimes in aerosol cans for home use. Oxidized starches are also used in the fabrication of construction materials, such as insulation and wall boards and acoustical tile, to provide adhesive, binding, and sizing properties.

Slightly oxidized starches have been used in batters and breadings for foodstuffs such as fried fish where it is claimed to give good adhesion to the food (*74–78*).

Hypochlorite oxidation provides a convenient method for viscosity reduction in conjunction with other types of modification of starch.

III. Cross-Linked Starch

1. Introduction

When starch is treated with multifunctional reagents, cross-linking occurs. The reagent introduces intermolecular bridges or cross-links between molecules, thereby markedly increasing the average molecular weight. Because starch contains many hydroxyl groups, some intramolecular reaction which does not increase the molecular weight also takes place. Intramolecular reaction is not significant in the usual granular reactions because the close packing of starch molecules favors intermolecular cross-linking.

Reaction of starch with multifunctional agents may not only be used to interconnect starch molecules but may also be used to bind starch to a substrate such as cellulose (*79*). Thus, starch-containing wet-rub-resistant paper coatings (*65, 80*) and water-resistant adhesives (*79, 81–83*) are produced through formulation with cross-linking agents such as glyoxal and thermosetting resins of the urea-formaldehyde or resorcinol–formaldehyde types. Cross-linking may be effected through weak, temporary bonds such as those formed by the reaction of starch with borax (sodium tetraborate) (*644*).

Cross-linking granular starch reinforces hydrogen bonds holding the granule together. This produces considerable change in the gelatinization and swelling properties of the starch granule with reagent in amounts as low as 0.005% to 0.1% by weight of starch. This toughening of the granule leads to restriction in the swelling of the granule during gelatinization, the degree of which is related to the amount of cross-linking. Hence, cross-linking reagents are sometimes referred to as "inhibiting reagents."

2. Preparation

Cross-linking reagent is generally added to an aqueous alkaline suspension of starch at 20°–50°. After reaction for the required time, the starch is recovered by filtration, washed, and dried.

$$\text{starch—OH} + \underset{\substack{\text{phosphoryl}\\\text{chloride}\\\text{(phosphorus}\\\text{oxychloride)}}}{POCl_3} \xrightarrow{NaOH} \text{starch—O—}\overset{\overset{\displaystyle O}{\|}}{\underset{\underset{\displaystyle O^- Na^+}{|}}{P}}\text{—O—starch} + NaCl$$

$$\text{starch—OH} + \underset{\substack{\text{sodium}\\\text{trimetaphosphate}}}{(NaPO_3)_3} \xrightarrow{Na_2CO_3} \text{starch—O—}\overset{\overset{\displaystyle O}{\|}}{\underset{\underset{\displaystyle O^- Na^+}{|}}{P}}\text{—O—starch} + \underset{\substack{\text{sodium dihydrogen}\\\text{pyrophosphate}}}{Na_2H_2P_2O_7}$$

A number of multifunctional cross-linking reagents have been suggested. Maxwell (*84*) proposed treatment of gelatinized starch or starch derivatives with bifunctional etherifying and/or esterifying agents such as epichlorohydrin, β,β'-dichlorodiethyl ether, or dibasic organic acids reacted under conditions such that both carboxyl groups esterify starch hydroxyl groups. Treatment of granular starch in aqueous alkaline suspension at pH 8–12 with 0.005–0.25% phosphorus oxychloride (phosphoryl chloride) yields a product with high viscosity, even after long cooking or exposure to acid or shear (*85*). Granules become very resistant to gelatinization when treated with 1% or more of phosphorus oxychloride.

$$\text{starch}-\text{OH} + \underset{\text{epichlorohydrin}}{H_2C\overset{O}{-}CH-CH_2Cl} \xrightarrow{OH^-} \text{starch}-O-CH_2-\underset{OH}{CH}-CH_2Cl \longrightarrow$$

$$\text{starch}-O-CH_2-HC\overset{O}{-}CH_2 \xrightarrow[+\ \text{starch}]{OH^-} \text{starch}-O-CH_2-\underset{OH}{CH}-CH_2-O-\text{starch}$$

$$\downarrow OH^-$$

$$\text{starch}-O-CH_2-\underset{OH}{CH}-CH_2OH \xrightarrow[OH^-]{HC_2\overset{O}{-}CH-CH_2Cl}$$

$$\text{starch}\left[O-CH_2-\underset{OH}{CH}-CH_2-O\right]CH_2-CH\overset{O}{-}CH_2 \xrightarrow[+\ \text{starch}]{OH^-}$$

$$\text{starch}-O\left[CH_2-\underset{OH}{CH}-CH_2-O\right]_2\text{starch}$$

Other common cross-linking agents are epichlorohydrin (*86–90*), trimetaphosphate (*91–94*), and linear mixed anhydrides of acetic and di- or tribasic carboxylic acids (*88, 95*). Still other cross-linking agents include vinyl sulfone (*96*),

$$\text{starch}-\text{OH} + CH_3-\overset{O}{\overset{\|}{C}}-O-\overset{O}{\overset{\|}{C}}-(CH_2)_4-\overset{O}{\overset{\|}{C}}-O-\overset{O}{\overset{\|}{C}}-CH_3 \xrightarrow[\text{pH 8}]{\text{NaOH}}$$

$$\text{starch}-O-\overset{O}{\overset{\|}{C}}-(CH_2)_4-\overset{O}{\overset{\|}{C}}-O-\text{starch} + CH_3-\overset{O}{\overset{\|}{C}}-O^-Na^+ +$$

$$\text{starch}-O-\overset{O}{\overset{\|}{C}}-(CH_2)_4-\overset{O}{\overset{\|}{C}}-O^-Na^+ + \text{starch}-\overset{O}{\overset{\|}{O}}-C-CH_3$$

diepoxides (*97*), cyanuric chloride (*98*), hexahydro-1,3,5-trisacryloyl-*s*-triazine (*99, 100*), hexamethylene diisocyanate and toluene 2,4-diisocyanate (*101*), *N,N*-

methylenebisacrylamide (*102*), *N,N'*-bis(hydroxymethyl)ethyleneurea (*103*), phosgene (*104*), tripolyphosphate (*105*), mixed carbonic–carboxylic acid anhydrides (*106*), imidazolides of carbonic and polybasic carboxylic acids (*107*), imidazolium salts of polybasic carboxylic acids (*108*), guanidine derivatives of polycarboxylic acids (*109*), and esters of propynoic acid (*110*). Aldehydes, such as formaldehyde (*111–115*), acetaldehyde (*114*), and acrolein (*116*), react bifunctionally to cross-link starch (*79*), forming acetals. The use of 2,5-dimethoxytetrahydrofuran, as a donor of succinaldehyde via acid hydrolysis, for the acid-catalyzed cross-linking of starch pastes (*117*) to give a water-resistant adhesive might be applied to granule inhibition. Water-soluble, urea–formaldehyde and melamine–formaldehyde resins have also been claimed as starch cross-linking agents (*118, 119*). Claims have also been made for cross-linking granular starch with dichlorobutene (*120*) and less reactive dihalides (*86*).

The presence of small amounts (0.1–10% on starch) of neutral alkali or alkaline earth metal salts, such as sodium chloride or sulfate, during the cross-linking reaction with phosphorus oxychloride appears to impart better control and a more uniform and efficient reaction, perhaps by retarding hydrolysis of the reagent and thereby increasing penetration into the granule (*121*). It may be that the salts affect the aqueous environment inside the granule, altering the structure of the water and thereby modifying the juxtaposition of the starch molecules as well as the rate of reagent hydrolysis and penetration. Salts prevent leaching of starch molecules from granules (*122*).

Inhibited granular starch products may also be made by treatment with a combination of glycine or a glycine precursor and a chlorine-containing oxidizing agent (*123*). These cross-linkages are heat-labile.

Reactions are conducted mainly in starch slurries. Impregnation of starch with metaphosphate salts at pH 5–11.5, drying, and heating at 100°–160° will also produce a cross-linked distarch phosphate (*124*).

3. Physical Properties (see also Chapter IX)

Because very low degrees of cross-linking reaction are difficult to determine directly, characterization of cross-linked starches as well as manufacturing in-process and quality control are dependent upon measurement of physical properties such as viscosity, swelling power, solubility pattern, and resistance to shear (*125–128*). For these measurements, the Corn Industries Viscometer, the Brabender Viscometer, or the Brabender Visco-Amylo-Graph are particularly useful (*128–130*).

The viscosity of pastes, produced by cooking suspensions of starch, is primarily dependent on the size of swollen, hydrated starch granules. Swollen, hydrated granules, particularly those of potato, tapioca, and waxy corn starches are quite fragile and tend to be fragmented by continued heating or agitation.

Cooking at low pH also causes a rapid breakdown from the initially high paste viscosity (Fig. 1). A cross-linking agent provides covalent bonds which are not readily disrupted by cooking and hold the granule together. Thus, cross-linked granules are less fragile and are more resistant to fragmentation by shear, high temperature, and low pH. Cross-linked starches will maintain higher working viscosities and show less viscosity breakdown than do untreated starches (Fig. 1) (*131, 132*). At low DS values, the peak viscosity of a cross-linked starch will be higher than that of natural starch. As the amount of cross-linking increases, viscosity breakdown becomes less and the high peak viscosity tends to stabilize. With increasing DS, the peak viscosity disappears and the rate of gelatinization and swelling of the granules decreases, resulting in a continuing increase in viscosity with prolonged cooking. This property can be seen with phosphorus oxychloride-treated potato starch (*133*) and with tapioca starch reacted with epichlorohydrin, sodium trimetaphosphate, or phosphorus oxychloride (*134*). Epichlorohydrin is most efficient (Fig. 2).

At high cross-linking levels, granules no longer gelatinize in boiling water (*134*) nor even under autoclave conditions. Thus, highly cross-linked starch has been used as a dusting powder for surgeon's gloves (*90, 92, 112, 135*). With intermediate cross-linking, starch is absorbed by the body (*90*), although there have been some reports of postoperative peritonitis, presumably caused by starch (*136, 137*).

Cross-linking at low levels will minimize or eliminate the rubbery, cohesive, stringy nature of the aqueous dispersions of waxy corn, tapioca, and potato starch and the accompanying clinging mouth-feel texture that is unpalatable in foods (*86, 88, 91, 114, 131*). It has been suggested that this viscoelasticity of native root or root-type starches is the result of the interaction of highly swollen,

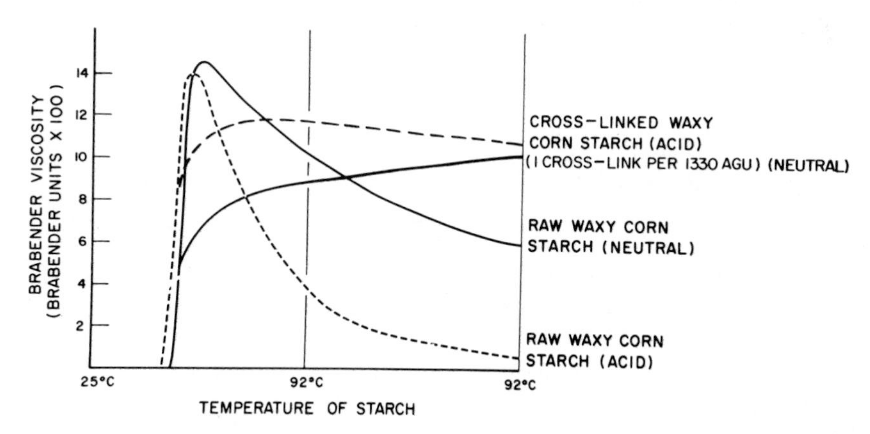

FIG. 1.—Effect of mild cross-linking on viscosity of waxy corn starch and raw waxy corn starch (*131*).

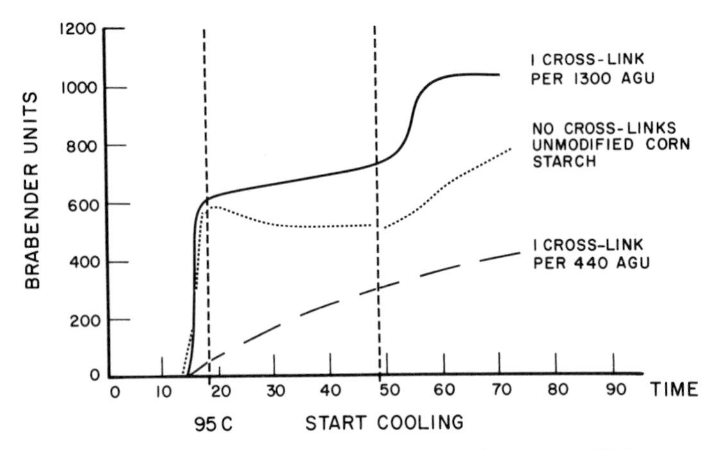

FIG. 2.—Effect of cross-linking on viscosity of corn starch (*131*).

fragile, hydrated granules so that they interpenetrate, become entangled, and hence generate resistance to flow. Cross-linked granules give little or no interpenetration and, therefore, dispersions can have a noncohesive, salve-like structure. The effect of cross-linking on viscoelastic behavior has been related to the fine structure of the starch (*138*).

Because cross-linking imparts commercially important viscosity-textural properties, it is employed in conjunction with other types of reactions such as oxidation (*87, 89*), phosphorylation (*91, 114*), hydroxyalkylation (*91*), and esterification (*88*). The possibility of making highly substituted granular starch derivatives by first cross-linking to produce acetal cross-links, reacting with a monofunctional esterifying or etherifying reagent under aqueous alkaline conditions, and removing the acetal cross-linkages under acid conditions has been claimed (*139*). The use of combinations of cross-linking bonds that are non-labile (ether linkages) and labile to heat, acid (acetal), or alkali (ester) could provide controlled cooking properties whereby a low viscosity on gelatinization could subsequently be changed to a high viscosity with a short texture. A product of this type would be useful as a thickener in retorted can foods (*140*). To obtain a particular pattern of viscosity development and high viscosity in a cold-water-dispersible, pregelatinized starch, a dual inhibition treatment is proposed (*141*). This involves adding 0.03–0.2% sodium trimetaphosphate and sodium chloride to a suspension of granular trimetaphosphate-cross-linked starch at pH 7.8–8.1 prior to drum-drying. Additional cross-linking occurs during the drum-drying.

4. Chemical Properties

Roberts (*142*) has suggested that starch reacts with epichlorohydrin in a sequence of three reactions catalyzed by alkali to yield a cross-linked network as shown on page 325. A monoglycerol ether of the starch may be produced by competitive hydrolysis of the starch epoxypropyl ether. Marchessault and co-workers (*143*) suggest the possibility that another molecule of epichlorohydrin (or more) might react with the starch monoglycerol ether leading to formation of a single cross-link (see page 325).

The amount of epichlorohydrin reacted is estimated by deducting the amount of glycerol formed and residual epichlorohydrin from the total reagent added (*144*). As little as one cross-link/1000 AGU[1] produced a marked reduction in hot-water solubles. The relationship between the amount of hot-water solubles and reacted epichlorohydrin is nearly linear. Approximately 78% of the epichlorohydrin added reacts with the starch at 25° in 18 h.

Gough and Pybus (*145, 146*) examined the products of the reaction of epichlorohydrin and alkali-gelatinized wheat starch by acid- and enzyme-catalyzed hydrolysis. Epichlorohydrin both produced cross-links and reacted monofunctionally at the 2-, 3-, 6-positions of the D-glucopyranosyl units. Higher starch-water ratios appeared more favorable to cross-linking.

Marchessault and co-workers (*143, 147–150*) examined the cross-linking reaction between epichlorohydrin and starch in homogeneous alkaline solutions and heterogeneous granular suspensions. The monoetherification reaction consumes 5–25% of the epichlorohydrin and is strongly dependent on the reaction conditions. Best cross-linking efficiency occurred at high starch concentration and a NaOH/starch mole ratio of 0.5–1.0. High temperature increases the cross-linking rate, but low temperature favors uniformity of product. The extent of the monoether side reaction in the heterogeneous granule reaction is comparable to that in the homogeneous solution reaction. The granule product has a polarization cross, and reaction presumably occurs in the noncrystalline areas. The vapor-phase epichlorohydrin cross-linking reaction is very efficient (*147, 148*).

Epichlorohydrin-cross-linked granular potato starch has been examined by differential scanning calorimetry and by swelling in solvents (glycerol, water, and dimethyl sulfoxide) (*143*). Degree of swelling varied regularly with the degree of cross-linking. Calorimetric and X-ray results indicate that the crystallinity of the granule did not change significantly with cross-linking. Water sorption and acid-catalyzed hydrolysis show increased accessibility of the cross-linked granules to small molecules. The percent water in fully hydrated, uncooked, epichlorohydrin-cross-linked corn starch granules, however, is the same as in normal corn starch granules (*122*). The greater thermal stability of the cross-linked starch as indicated by an increase in the gelatinization temperature measured in the calorimeter is attributed to a decrease in the entropy of melting.

Cross-linking of amylose by epichlorohydrin in a homogeneous reaction in aqueous alkaline solution shows that the reactivity of the C-2 hydroxyl group is much greater than that of the C-6 hydroxyl group, while that of the C-3 hydroxyl group is considerably lower (*149*). This suggests that the cross-linking reaction proceeds by a reaction similar to etherification of starch by ethylene and propylene oxides. It fosters the assumption that the extent of direct nucleophilic substitution of chlorine is insignificant, favoring the mechanism of the epoxy ring opening reaction.

In contrast to α-amylase digestion of epichlorohydrin-cross-linked, granular potato starch, which decreases as the degree of cross-linking increases, acid-catalyzed hydrolysis increases with increasing cross-linking. It is suggested that cross-linking prevents molecular and structural rearrangement on drying, providing a more open and internally strained structure more accessible to small molecules (*150*).

A method for determining the adipate content of acetylated starch adipate, produced by the reaction of the mixed linear acetic–adipic anhydride, has been developed (*151*). The method measures total adipate introduced and does not differentiate the cross-linking adipate di-ester from the mono-ester (page 325). An efficiency of 32.5% is reported for the reaction utilizing 0.1–0.3% adipic acid on the starch. Although methods for the determination of the phosphate content of starch phosphate esters have been available, they have not been suitable for the very low DS of cross-linked distarch phosphates. A method, based on titration of the free acid groups available on the monophosphate and diphosphate ester groups, developed by Mitchell (*152*) and refined by Koch and co-workers (*153*), appears useful. Data obtained shows that the cross-linking reaction predominates at a ratio of approximately 3.5:1 when starches are treated with 1% phosphorus oxychloride and 1.5% sodium hydroxide solution for 2 hours at 25° to give a DS of about 0.002. In waxy corn starch, maximum viscosity appears at a DS of approximately 0.00001 phosphate diester. At a DS of 0.005 phosphate di-ester, the products no longer completely disperse on cooking (*153*).

Potato starch treated with phosphorus oxychloride binds with iodine, but the cross-linked starch forms an iodine complex at a much slower rate than does untreated starch. At high levels of epichlorohydrin treatment of pasted wheat starch, the cross-linked product gives a purple to reddish-brown color indicative of a restricted chain length (*145*).

5. Uses

Cross-linked starches are used when a stable, high-viscosity starch paste is needed and particularly when the dispersion is to be subjected to high temperature, shear, or low pH. While cross-linking may be the only modification, it is usually employed in combination with other types of derivatization or modification. For resistance to rigorous conditions, high degrees of cross-linking are required. The trend toward continuous cookers generally requires higher cross-linking owing to increased shear and contact with hot surfaces. In extrusion, a higher level of cross-linking is required.

Food starches, especially those made from waxy corn, potato, and tapioca starches, are usually phosphates, acetates, or hydroxypropyl ethers that are cross-linked to provide appropriate gelatinization, viscosity, and textural properties, including a short, salve-like consistency. Cross-linked starches are also needed for salad dressings to provide the thickening without allowing viscosity breakdown by low pH and the high shear of the homogenization process. Storage stability at the low pH is also increased. In retort sterilization of canned foods, cross-linked starches with a slow gelatinization or swelling rate are used to provide low initial viscosity, high heat transfer, and rapid temperature increase for quick sterilization, with subsequent thickening to provide suspending and texture properties (*123, 140, 155–159*). Cross-linked starches are used in canned soups, gravies, sauces, baby foods, and cream style corn, as well as in fruit pie fillings, puddings (*88, 91, 114, 124, 132, 160–165*), and batter mixes for deep-fried foods (*166,* see also this Volume, Chap. XIX).

Cross-linking of granular starch before drum-drying preserves texture, viscosity, and water-holding power against the shear of the process as well as providing special effects. Drum-dried, cross-linked starches are used to impart a pulpy texture to food systems (*167, 168*). Drum-dried, cross-linked starches with low amylose content, as in waxy sorghum or corn, are claimed to improve cake volume, crumb softness, and keeping qualities of cakes (*169*). Acid-converted, cross-linked potato or tapioca starches are drum-dried to produce cold-water-dispersible starches that gel (*170*).

Cross-linked starches, particularly waxy types or starch derivatives, can be treated with β-amylase to improve low-temperature stability of their aqueous dispersions. Cross-linking is needed to provide optimum thickening and rheological properties in food systems (*171,* see also this Volume, Chap. XIX).

Cross-linked starches are used in anti-perspirants (*172, 173*). It has been claimed that cross-linked starch ethers containing carboxymethyl (*174*) and/or hydroxyalkyl groups (*175*) are suitable absorbents for personal sanitary applications.

Cross-linked starches are used in alkaline textile printing pastes, contributing high viscosity and the short, non-cohesive texture needed (*116*). They are used in corrugating adhesives to provide high viscosity under strongly alkaline conditions (*116;* see also Chapter XX). Other application areas are oil well drilling muds, printing inks, binders for coal and charcoal briquettes, electrolyte-holding media for dry cells (*176–178*), fiberglass sizing (*179–181*), and textile sizing (*182*).

Cross-linking is also used to obtain starches which are not gelatinized on exposure to high temperatures such as sterilization temperatures (*183*), strongly alkaline solutions, or other conditions where gelatinization would normally occur; such products are generally very high DS granular starch derivatives. Highly cross-linked insoluble granular starches are also used in the preparation of starch xanthates for water treatment (*184–186*) ion-exchangers (*187*), stilt material for micro-encapsulated coatings (*188*), and anti-blocking agents for films (*189*).

An interesting proposal is the reaction of epichlorohydrin with cross-linked starch under acid conditions to obtain the 3-chloro-2-hydroxypropyl starch ether that can be further reacted to make an insoluble starch containing anionic, cationic or chelating groups for removal of heavy metal ions from aqueous solutions (*190–192*).

IV. Starch Esters

1. Introduction

Extensive literature reviews on starch esters have been published (*62, 193–202*). Starch esters of commercial value are those which provide sol stability and functional properties such as hydrophobic, cationic, or anionic character at relatively low cost. Currently, these esters are the starch acetates and the half-esters of some dibasic carboxylic acids, particularly the alkenylsuccinic acids. As mentioned in the preceding section, the diesters of the dibasic carboxylic acids are valuable because of the cross-linking introduced.

Because the tendency of a starch dispersion to increase in viscosity on cooling and aging and finally to gel is related to the association of the amylose molecules, any treatment which retards or eliminates this crystallization or retrogradation phenomenon will affect stabilization of the starch sol. Acetylation prevents or minimizes association of outer branches of amylopectin molecules. This is of practical value in food applications where it prevents cloudiness and syneresis in aqueous dispersions of waxy starches stored at low temperature. Since a relatively low DS is adequate to achieve this sol-stabilization effect, commercial starch acetates are generally granular and less than 0.2 DS (5% acetyl) (*625*).

Although there has been considerable interest in the higher DS acetates, particularly amylose acetate, because of their organic solvent solubility, thermoplasticity, and film properties, they have not developed commercially.

2. Preparation of Starch Acetates

A number of reagent–catalyst–solvent systems have been reported, some mainly for laboratory esterification to DS 2–3. Reagents used include acetic acid (203–205), acetic anhydride (*88, 95, 200, 201, 206–232*), vinyl acetate (*233–239*), ketene (*211, 240, 241*), *N*-acetylimidazole (*107*), ethyl carbonic–acetic anhydride (*242*), acetyl guanidine (*109*), acetyl phosphate (*243*), and *N*-acetyl-*N′*-methylimidazolium chloride (*108*).

In many cases, some type of ''activation'' treatment is used to improve reactivity. All activation treatments involve disruption of the intermolecular hydrogen bonding of granules (*193–198, 200, 201, 217*). Activation of isolated amylose and amylopectin fractions also involves rupture of intermolecular hydrogen bonds. Reassociation (reorganization) is prevented by insuring that drying does not take place from a hydrated state; that is, water is removed by azeotropic distillation or by washing with organic solvents such as acetic acid, alcohol, or pyridine.

a. *Acetic Acid*

Aqueous acetic acid has little effect on starch. Heating starch with 25–100% glacial acetic acid at 100° for 5–13 h introduced 3–6% acetyl groups in the granular product (*205*). Refluxing dry starch in a large excess of glacial acetic acid dissolves the starch in about 18 h, yielding 44% acetyl content in 296 h (*203, 204*). Glacial acetic acid may be suitable for producing partially degraded, low-DS starch acetates, but is not efficient for the preparation of DS 2–3 acetates.

b. *Acetic Anhydride*

Acetic anhydride is generally used for acetylation of starch. It is used alone or with catalysts (*203, 210, 215, 216*) as well as in conjunction with acetic acid (*209, 211–213, 217*), pyridine (*200, 201, 206–208, 218–221*), and dimethyl sulfoxide (*223*) and in aqueous alkaline solution (*88, 95, 207, 231*).

Heating starch in acetic anhydride at 90°–140° results in acetylation accompanied by degradation. After 8 h at 140°, 1.8% acetyl is introduced; the acetyl content rises to 8.7% in 15 h and 34% in 74 h (*199, 203*). Use of acid catalysts, such as sulfuric acid and hydrogen halides, with acetic anhydride produces considerable degradation (*194, 199, 202*). Alcohol-precipitated, gelatinized potato starch is acetylated to 44.9% acetyl in 1 day in acetic anhydride containing 5% sulfuric acid at 25°, whereas untreated potato starch is 80% unchanged after 6 days (*215*); corn starch in refluxing acetic anhydride containing sodium acetate yielded a degraded product (43.5% acetyl) in 4 days (*216*). Ammonium acetate–acetamide catalysis is claimed to yield acetates of DS 2.5–2.8 (*244*). Potato, corn, and high-amylose corn starches can be fully acetylated with only minor degradation after a 5 h refluxing in fourfold quantities of acetic anhydride and 11% (w/w starch) of sodium hydroxide added as a 50% aqueous solution (*226*).

Rapid, uniform acetylation of starch by acetic anhydride in liquid sulfur dioxide at 12 bars and 10°–15° is claimed (*245*). Granular starch acetates having a DS below 0.1 are claimed from heating starch with 8–20% moisture in neat acetic anhydride or with added dimethylformamide (*210*).

High-amylose corn starch (70%), activated by 1.5% sodium hydroxide added as a 50% solution, can be acetylated to a DS of 0.1–0.31 by treatment with acetic anhydride under acid conditions in a heavy-duty mixer for 30 min at 90–95° (*224, 225, 227*).

Mixtures of acetic anhydride–acetic acid are slow to acetylate starch at 50° without a catalyst. Acetylation is more rapid in the presence of 1% sulfuric acid, reaching 40.9% acetyl in 6 h at 50°. Acetylation rate increases with temperature and acid concentration as does degradation (*211*). Rapid acetylation occurs at reflux without catalyst (*211*). Almost complete acetylation is claimed for esterification in 1:1 or 3:2 (v/v) acetic anhydride–acetic acid at 95° with 0.2% perchloric acid and 0.6% sulfuric acid (*212*) or at 105° after dispersing corn starch in 85% phosphoric acid (*213*). Preparation of granular starch acetates with DS up to 1.5 is claimed by incorporating an inert organic liquid, such as xylene, in the acetic anhydride–acetic acid reaction mixture with a sulfuric acid catalyst (*213*).

A process was proposed for making amylose and amylopectin triacetates using acetic anhydride–acetic acid at 28°–32° for 6 h (*217*). Potato amylose is acetylated to a DS of 2.8 with little degradation by reaction in an acetic anhydride–acetic acid mixture at 60°–90° for 3–4 h in presence of 30% poly(styrenesulfonic acid) resin (*209*).

Acetic anhydride–pyridine is the most common laboratory procedure for acetylation of starch, giving high DS with minimum degradation (*194, 207*). The system can be used to prepare starch acetates with a wide range of substitution by control of reaction time, temperature, and anhydride concentration (*200, 201, 208*).

Activation is necessary for reaction in acetic anhydride–pyridine. Refluxing in pyridine at 115° for 1 h activates starch without gelatinization (*218*). Gelatinization of starch by heating in 60% aqueous pyridine, removing the water as a pyridine azeotrope while adding pyridine, then adding acetic anhydride when the temperature reaches 115° yields a triacetate (*200, 201*). Another activation method involves disintegration of the starch granules by cooking at 95°–100°, rupture by high shear agitation, and recovery by alcohol precipitation, washing, and drying under reduced pressure (*206, 207, 246*). The preferred ratio of acetic anhydride:pyridine:starch is 3.2:3.7:1. Activation of starch in formamide has been recommended (*219–221, 247*). Dispersion in formamide followed by addition of acetic anhydride without the pyridine gives a corn starch diacetate (*222*). The influence of various activation treatments on high-amylose corn starch (71% apparent amylose) prior to the acetic anhydride–pyridine reaction at 100° to prepare triacetates was examined (*226*). Dimethyl sulfoxide or hot water cooking were most effective in disrupting the granules.

Low-DS starch acetates can be made by treatment of an aqueous starch suspension with acetic anhydride at pH 7–11 (*95, 207, 230–232*). Although sodium hydroxide is the preferred alkali, other alkali metal hydroxides, sodium carbonate, trisodium phosphate, and calcium hydroxide are claimed to be suitable (*95*). Magnesium oxide or hydroxide are claimed as superior alkaline agents for pH control without requiring continuous monitoring (*228*). At 25°–30°, the pH optimum is 8–8.4, but at 38°, the optimum pH is about 7 (*207*). Reaction efficiencies of about 70% are obtained.

A. Reaction with acetic anhydride

$$\text{starch—OH} + \left(CH_3-\overset{O}{\overset{\|}{C}}\right)_2O + NaOH \longrightarrow \text{starch—O—}\overset{O}{\overset{\|}{C}}-CH_3 + CH_3-\overset{O}{\overset{\|}{C}}-O^-Na^+ + H_2O$$

Side reactions

$$\left(CH_3-\overset{O}{\overset{\|}{C}}\right)_2O + 2\,NaOH \longrightarrow 2\,CH_3-\overset{O}{\overset{\|}{C}}-O^-Na^+ + H_2O$$

$$\text{starch—O—}\overset{O}{\overset{\|}{C}}-CH_3 + NaOH \longrightarrow \text{starch—OH} + CH_3-\overset{O}{\overset{\|}{C}}-O^-Na^+$$

B. Vinyl acetate

$$\text{starch—OH} + CH_2{=}CH-O-\overset{O}{\overset{\|}{C}}-CH_3 \xrightarrow{Na_2CO_3} \text{starch—O—}\overset{O}{\overset{\|}{C}}-CH_3 + CH_3-\overset{O}{\overset{\|}{C}}-H$$

The maximum DS level attainable without gelatinization varies with the particular starch, but the upper limit is about 0.5 DS. To reach a DS of 0.5, it is necessary to repeatedly increase reagent concentrations by filtering the starch from the reaction mixture, resuspending it in 1.25–1.5 parts of water per part of starch, and continuing the acetylation. When higher DS products are required, acetylation of gelatinized starch is preferred (*207*). A complex between hydroxyl groups and acetic anhydride may precede acetylation (*248–250*).

It has been suggested that sodium acetate formed as a by-product causes acetyl odors and off-flavors in puddings made from these starch acetates; therefore, the starch acetate must be thoroughly washed to remove by-products and maintain ash content below 0.2% (*229*).

Acetylation of starch in aqueous suspension by treatment with acetic anhydride in the presence of hydrogen peroxide and ferrous salts is claimed to lower the gelatinization temperature without significant degradation (*230*). The product may be more readily liquefied by α-amylases.

Acetylation of granular starch in aqueous suspension by acetic anhydride at alkaline pH is used commercially to produce starch acetates of low DS. The type of starch to be acetylated is determined mainly by the properties desired in the product, as well as by the cost of the starch. Potato, wheat, tapioca, corn, and sorghum starches, the last two in regular or waxy varieties, are commonly used. In the United States, regular or waxy corn or sorghum starches are normally chosen for starch acetates. The acetylation treatment may be combined with other modifications such as cross-linking, or a lower viscosity acid-converted or oxidized starch may be used as the base. The starch may come directly from the corn wet-milling process in aqueous suspension or dry starch, such as imported tapioca or potato starch, may be slurried in water and acetylated. Amylose or amylopectin may be required to provide an acetate with specific properties. The granular starch is the preferred raw material since it can be handled and recovered easily.

c. *Ketene*

Acetylation of starch with ketene in glacial acetic acid using catalytic amounts of sulfuric acid has been claimed (*240*). Ketene has also been reacted with starch suspended in acetone or ether with sulfuric acid as the catalyst to introduce 2.2–9.4% of acetyl groups in 0.5–2 h at 25° (*240*). Gaseous ketene may be used (*240*). Activation pretreatment of starch is recommended (*241*). Acetic acid is used for both the pretreatment agent and the reaction medium, although acetone, chloroform, methyl ethyl ketone or tetrachloroethane may also be used as the medium. Acid catalysts are required, the most effective being sulfuric acid; the next most effective is *p*-toluenesulfonic acid. The acid concentration must be regulated to obtain good efficiency without extensive degradation. Reaction at 90° in glacial acetic acid for about 5 h yielded a product containing 42.5% acetyl.

d. *Vinyl Acetate*

Starch has been acylated by alkaline-catalyzed transesterification with vinyl esters in aqueous medium (*233*). With vinyl acetate, starch acetate is formed with acetaldehyde as a by-product (*234*) as shown on page 335.

Acetylation of either granular or gelatinized starch requires the presence of water, since reaction efficiencies of only 2–5% are obtained with less than 10% moisture. Reaction efficiency rises with water content when vinyl acetate is added to starch blended with sodium carbonate and reacted at 24° for one hour. Although the reaction may be run at pH 7.5–12.5 using alkali metal hydroxides, quaternary ammonium hydroxides, ammonium hydroxide and aliphatic amines as catalysts, the preferred range is pH 9–10 where reaction efficiencies of about 65–70% can be expected. Reaction of granular starch suspension in water containing the carbonate buffer gives an acetyl content of 1–4%. Use of the liberated acetaldehyde to cross-link the granular starch at pH 2.5–3.5 is claimed (*235*). A linear relationship was found between the reaction rate and the amount of sodium hydroxide "adsorbed" by potato starch when it is reacted with vinyl acetate (*239*).

Vinyl acetate can be used to acetylate amylose in a continuous reactor (*237*). Amylose suspended in water containing sodium carbonate is mixed with vinyl acetate, and the reaction is done at 166°–177° and 100–140 psi at one gallon per minute throughput. Products have up to 5–8% acetyl content and give solutions with less tendency to gel. Low-DS acetylated potato starch can also be made by a continuous production method (*638*).

3. Properties of Low-DS Starch Acetates

Microscopic examination of the granules of DS up to 0.2 reveals no discernible difference from native granules. Anionic or cationic dyes will not differentiate starch acetate from the native starch granule (*49*). A granular starch acetate containing 1.85% acetyl can be completely deacetylated by suspension in water at pH 11 for 4 hours at 25°. The deacetylated starch granule is practically identical to the original. Brabender viscosity curves of corn starch before and after acetylation to 1.8% acetyl are shown in Figure 3. The acetylated starch has a 6° lower gelatinization temperature and a higher hot peak viscosity, reached at a temperature 10° lower than required for the untreated starch. This reduction in gelatinization temperature on acetylation is common to all starches. An aqueous slurry of slightly cross-linked waxy corn starch has a gelatinization temperature of 71°, whereas the same starch with 1.8% acetyl content gelatinizes at 64°, and with 3.5% acetyl, at 62°. Similar effects are reported for corn starch acetate and hydroxyethyl corn starch (*49, 55*). Acetylated starches are, therefore, more readily dispersed on cooking than are the corresponding native starches.

Corn starch acetates increase in viscosity more slowly than does untreated corn

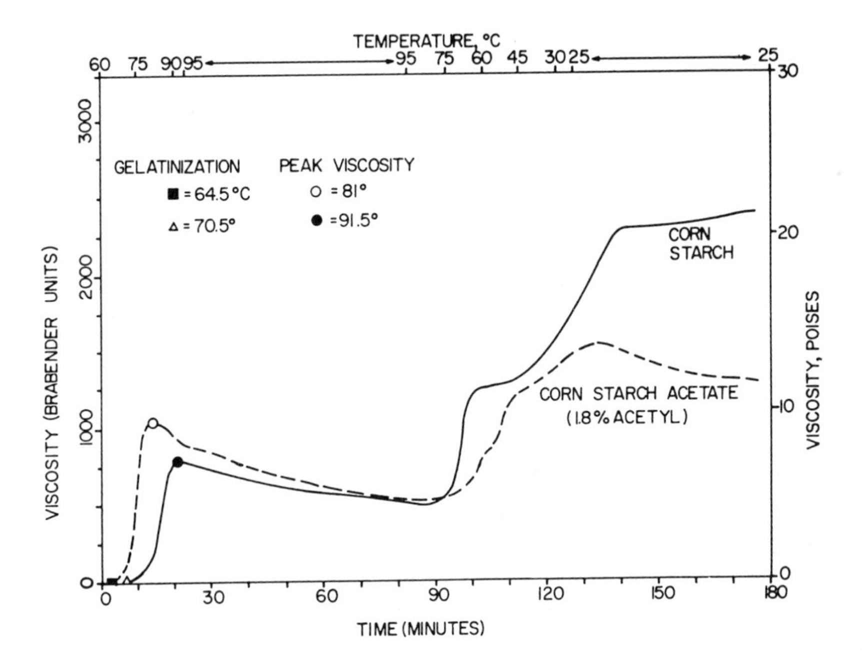

FIG. 3.—Brabender viscosity curves of corn starch and acetylated corn starch (1.8% acetyl). The concentration was 31.5 g of dry starch suspended in 418.5 g distilled water. The pH of the slurry was 5.9. The 350 cm·g Brabender cartridge was used.

starch on cooling and do not reach as high a cold viscosity. Potato starch acetates of 0.9–4.23% acetyl content also have lower gelatinization temperatures and hot and cold paste viscosities as the acetyl content increases (*251, 252*). The ratio of 30° to 80° viscosity decreases with increasing acetyl content up to 2.36% acetyl but does not change significantly at higher acetyl contents.

Use of acetylation to lower gelatinization temperature via interference with molecular association is of value when applied to high-amylose starches. Starches containing at least 50% amylose are so highly associated that they do not gelatinize on cooking in boiling water but require temperatures as high as 160°. The sols produced by high temperature cooking are unstable, tending to gel as the temperature decreases. Acetylation to a DS of 0.1–0.2 lowers the gelatinization temperature so that the product disperses in boiling water and produces relatively stable sols. Substituent groups other than acetates are also effective (*223*).

Acetylation increases the clarity of sols by increasing the degree of swelling and dispersion of the starch granule as well as reducing retrogradation. Clarity, viscosity and stability of acetylated starches are of value in food, paper, and textile applications (*88, 95, 236*).

4. Uses of Low-DS Starch Acetates

a. *Foods (see also Chapter XIX)*

The major use of starch in foods is for thickening. This requires the starch to have a bland taste and impart appealing texture. In food preparation and sterilization, starch may be exposed to a pH as low as 3, high shear in mixing and pumping, high temperatures, temperatures of 5° in the refrigerator or −18° in the freezer and moderately high temperatures in warehousing and transportation. Acetylation alone or in combination with other treatments such as cross-linking can produce starches with the required properties.

Starch acetates containing 0.5–2.5% acetyl groups are used primarily because of their viscosity stability and clarity. This stability is effective under low temperature conditions. Even tapioca, potato, and the waxy corn starches, which have a relatively high degree of stability, require acetylation for low temperature stability. Such stabilization prevents syneresis (weeping or watering) and the development of cloudiness at low temperature (*95, 253*). Cross-linking is needed to provide desired textural properties and viscosity breakdown resistance.

Cross-linked acetylated starches are used in canned, frozen, baked, and dry foods. They are also used in baby foods and fruit and cream pie fillings in cans and jars to meet the requirement of long shelf life under varying temperature conditions. Frozen fruit pies, pot pies, and gravies maintain their stability under low-temperature storage. In baked goods, the pies, tarts, and filled cakes have greater "weeping" resistance. Acetylated starches are pre-gelatinized for use in dry mixes and instant gravies and pie fillings. A can-filling starch which has a high viscosity to facilitate uniform can filling and then exhibits viscosity breakdown on heating to increase heat penetration for sterilization and a thin final viscosity is claimed for cross-linked acetylated tapioca, potato, and waxy corn starches (*238*).

FDA (Food and Drug Administration) regulations permit up to 2.5% acetyl content in starches used in foods. Acetylation in combination with certain other treatments is also permitted (*254*).

b. *Textile, Paper, and Other Uses (see also Chapter XVIII)*

Ready dispersibility, viscosity stability and the noncongealing character of pastes make low-DS starch acetates convenient for preparation, storage, handling, and application in textile and paper manufacture.

In textiles, the major market for starch acetates is in warp sizing where they have good yarn adhesion, tensile strength, and flexibility. In addition, the film solubility makes it easily removed (*255*). Poly(vinyl alcohol) is used in warp sizing because of its high film strength and flexibility, abrasion resistance, and good adhesion. Because of the expense of poly(vinyl alcohol), starch acetates are sometimes blended with it to lower costs and extend properties.

Low-DS starch acetates, with or without cross-linking, are claimed to be effective glass fiber forming sizes (*256*).

Starch acetates are also used in combination with thermoset resins to produce inexpensive finishes. The starch acetate in the formulation gives weight and "hand" to the fabrics. Starch acetate is used in finishes for inter-liners to give stiffness.

The major use of starch acetates in the paper industry is in surface sizing for improved printability and functional properties by imparting low and uniform porosity, surface strength, abrasion resistance, oil holdout, and solvent resistance, as well as by adhering stray fibers to the substrate (*255, 257*). Starch acetates compete with oxidized starches, hydroxyethylstarches and low-viscosity starches prepared via enzymes or thermochemical processes. Other competitive products are other film-formers such as carboxymethylcellulose, poly(vinyl alcohol), and alginates (*257*). When paper containing starch acetates is reintroduced into the paper-making system (by repulping), there is no adverse dispersant effect on the retention of fillers and pigments, as noted with oxidized starches.

In gummed tape formulations, the flexibility, high gloss, and remoistenability of starch acetate films are useful. A gummed tape containing hypochlorite-oxidized waxy corn starch acetate with 1.5–2% acetyl groups and 0.3–0.5% carboxyl groups is similar to a tape made with animal glue, the industry standard (*258*).

Acetylation has been used to direct amylase hydrolysis to produce maltodextrins (*259*). The higher the DS, the higher is the proportion of oligosaccharides with DP of 30 or greater.

5. Properties of High-DS Starch Acetates

Early work on high-DS starch acetates was done to find a substitute for cellulose acetate (*260, 261*). However, starch and amylopectin acetates give weak and brittle films, fibers, and molded products (*208, 262, 263*) and have no economic advantages over cellulose acetates (*208, 262–265*).

Properties of starch esters with DS of 2–3 depend upon the type of starch, chain length of the substitutent, and the conditions of activation and reaction (*206, 263, 266*). The specific gravity, specific rotation, and melting temperatures generally decrease with increasing acetyl content (*261*). The specific rotations of the triacetates of whole starch, amylose, and amylopectin from corn are 170–174.5° (in chloroform) (*206*). In pyridine at 22°, the specific rotations of corn, potato, tapioca, wheat, and rice starch triacetates are 151°–158° (*201*). The amylose esters of the C_2 to C_{16} alkanoic acids usually have a narrow but higher melting range than do the corresponding amylopectin esters, with whole starch esters intermediate. Triacetates of corn amylose, amylopectin, and whole starch melt at 300°–301°, 205°–265°, and 270°–292°, respectively (*263*). Melting or softening points of the amylose triesters decrease with increasing length of the alkanoic acid, the melting range of amylose palmitate being 90°–140° (*200, 201, 263, 266*).

Solubility characteristics of starch acetates are dependent upon DS and DP (*194, 260*). Whole starch acetates of up to 15% acetyl content are soluble in water at 50°–100° and are insoluble in organic solvents (*261*). Water solubility has been reported for degraded products containing as much as 25% acetyl groups (*203, 204*). Starch acetates with 40% acetyl content or higher are soluble in aromatic hydrocarbons, halogenated aliphatic hydrocarbons (except carbon tetrachloride), ketones, glycol ethers, and nitroparaffins (*261*). The aqueous paste disruption activation technique (*206*) yields more soluble products than either the liquid ammonia or pyridine pretreatment methods (*200, 201, 206*). Whole starch triacetates do not form true solutions when prepared by the pyridine activation method, but yield dispersions (*199, 208*). Similar solvent solubilities are reported for corn starch, amylose, and amylopectin triacetates (*267*).

The thermoplastic molding properties of starch esters, including acetates, have been reported (*208*). Investigation has been made of amylose triacetate fibers (*264, 265*), films (*262, 263, 266*), and plasticization of films (*262, 263, 268*). X-Ray data on stretched films are available (*263, 269*). Alkali amyloses with well-defined fiber x-ray diffraction patterns can be prepared by deacetylation of stretched amylose triacetate films (*270*). Mixed esters of amylose, such as acetate–propionate or acetate–butyrate esters, have been evaluated (*271*).

6. Acetyl Analysis

The acetyl content of both high- and low-DS starch acetates can be determined by measuring the amount of alkali used in saponification (*194, 207, 272*). Trans-esterification in anhydrous methanol with sodium methoxide as the catalyst and distillation of the methyl acetate formed is also a suitable analytical method (*194, 273*). It has the advantage that saponification of the collected methyl acetate is unaffected by traces of nonvolatile acidic materials that may be formed from reducing end-groups in the first method. Combining the latter procedure with gas chromatography of the methyl esters formed can give a qualitative and quantitative analysis for acyl groups in any starch ester. Infrared spectroscopy of the methyl esters can also be used for identification of substitutent ester groups. The infrared carbonyl peak at 1724 cm^{-1} has been used to determine the acetyl content in starch acetates (*274*); the method was also applied to carboxymethyl and cyanoethyl ethers of starch and to starch sulfate. ^{1}H-nmr can be used to detect and determine acetyl groups as well as other ester groups (*275*).

7. Miscellaneous Esters

a. *Half-Esters of Dicarboxylic Acids*

$$\text{starch—OH} + \text{R—HC(—C(=O)—O—C(=O)—CH}_2\text{, cyclic)} \xrightarrow[\text{pH 8}]{\text{OH}^-} \text{starch—O—}\overset{\text{O}}{\overset{\|}{\text{C}}}\text{—CH}_2\text{—}\underset{\text{R}}{\underset{|}{\text{CH}}}\text{—}\overset{\text{O}}{\overset{\|}{\text{C}}}\text{—O}^-\text{Na}^+$$

$$\text{starch—OH} + \text{HC(—C(=O)—O—C(=O)—)=CH, cyclic} \xrightarrow[\text{pH 8}]{\text{OH}^-} \text{starch—O—}\overset{\text{O}}{\overset{\|}{\text{C}}}\text{—CH=CH—}\overset{\text{O}}{\overset{\|}{\text{C}}}\text{—O}^-\text{Na}^+$$

$$\text{starch—O—}\overset{\text{O}}{\overset{\|}{\text{C}}}\text{—CH=CH—}\overset{\text{O}}{\overset{\|}{\text{C}}}\text{—O}^-\text{Na}^+ \xrightarrow{\text{NaHSO}_3} \text{starch—O—}\overset{\text{O}}{\overset{\|}{\text{C}}}\text{—CH}_2\text{—}\underset{\text{SO}_3^-\text{Na}^+}{\underset{|}{\text{CH}}}\text{—}\overset{\text{O}}{\overset{\|}{\text{C}}}\text{—O}^-\text{Na}^+$$

Utilizing the same controlled-pH, aqueous, alkaline starch suspension reaction used to prepare granular starch acetates, low-DS starch half esters of dicarboxylic acids can be prepared (*95*). Cyclic dibasic acid anhydrides such as succinic anhydride yield starch esters containing a free carboxylate group that increases the water-holding power of the product (*95*). If a maleate half-ester is made, a sulfonic acid group can be introduced by reaction with bisulfite to give a sulfosuccinate half-ester. This derivative has a lower gelatinization temperature compared to the maleate half-ester and gives clearer, more translucent, stable dispersions with high water-holding capacity (*276*). If a cross-linked starch is esterified with a dicarboxylic acid, viscosity stability is obtained. Reaction of cross-linked, thin-boiling, acid-converted or oxidized starch with a polycarboxylic acid anhydride yields an effective warp size (*277*). A combination corn starch acetate–succinate of DS 0.037 is useful as a corrugating adhesive (*278*).

Treatment of a starch suspension with a cyclic dicarboxylic acid anhydride containing an hydrophobic substitutent group yields products with emulsion stabilizing properties (*279*). Such an anhydride is 1-octenylsuccinic anhydride. Starches esterified by treatment with up to 3% of this reagent meet the requirements of a ''Food Starch-Modified'' designation and may be used in food (*254*). These derivatives are used for emulsion stabilization and encapsulations. If these hydrophobic half-esters are treated in granular form with compounds containing a polyvalent metal ion, such as chromic chloride or aluminum sulfate, a free-flowing water-repellent powder results (*280*). A product with not more than 2% 1-octenylsuccinic anhydride and a maximum of 2% aluminum sulfate meets the FDA requirements for ''Food Starch-Modified'' (*254*). Similar products are recommended as water-repellent encapsulating agents (*281*). A lipophilic starch derivative suitable for encapsulation of flavoring oils and perfumes and for forming stable emulsions is provided by a waxy corn starch half-ester made with a hydrophobic anhydride and subjected to controlled enzyme hydrolysis with α-amylase to adjust the viscosity before spray-drying (*282*).

A water-soluble half-ester containing hydrophobic groups, such as starch alkenylsuccinate, in combination with gelatin provides a protective overcoat for photographic film (*283*). This type of product based on an acid-converted starch may be a sizing for paper (*284*).

Low-molecular-weight, high-DS half-esters made from hydrolyzed starches with DE up to 40 may act as dispersants in aqueous coating formulations such as latex paints (*285*). Reaction is conducted for 4 h at 82° in an organic medium such as pyridine, formamide, and quinoline, which can act as a solvent for reactants and products. An acylating agent such as octenylsuccinic anhydride is used at ratios of 0.25–3.0 moles per AGU[1]. Products of a similar type consisting of mixed esters of mono- and dicarboxylic acids (half-esters) are recommended as dispersants (*275, 286*).

Products suitable for adhesives are made from thin-boiling, acid-converted or hypochlorite-oxidized starches and dextrins by dry-blending them with dibasic acid anhydrides, such as maleic or succinic anhydride, and heating the mixture at low moisture and 100°–150° to a DS of 0.01–0.1 (*287*). Mixed hydroxyalkyl ether–esters are made using the hydroxyalkyl dextrin as starting material (*288*). Dextrin esters of maleic acid copolymers are prepared using ethylene–maleic anhydride or methyl vinyl ether–maleic anhydride copolymers in the dry heat reaction (*289*).

b. *Starch Ether–Esters*

A water-insoluble, elastic chewing gum base is prepared from hydroxypropylstarch acetate with an MS of 3–6 hydroxypropyl groups and a DS of 1.0–2.5 acetate groups (*290*). The preparation involves suspending corn or waxy corn starch in toluene, adding a small amount of 25% aqueous sodium hydroxide solution, and then reacting the starch with propylene oxide at 110° for 1–6 h followed by reaction with acetic anhydride at 40°–80° for 1–3 h. After neutralization, the product is purified by extensive washing, steaming, and precipitation from ethanol with water. Hydroxypropyl- or hydroxyethylstarch of MS 0.66 or cellulose of MS 2–8 esterified with acetic or lauric acids to DS 1.2–3 are recommended as gelling agents for organic liquids including jet fuel (*291*).

Esterification and etherification of high-amylose corn starch can produce granular amphoteric starches with anionic, cationic, and non-ionic groups together with acetyl groups for the preparation of textile warp sizes (*292*).

c. *Esters for Slow Release*

There is interest in the use of polysaccharides as carriers for the slow release of herbicides and pesticides. Treatment of corn starch with (2,4-dichlorophenoxy)acetyl chloride in pyridine gives a 2,4-D-starch ester. Hydrolysis at pH 6–8 liberates varying amounts of 2,4-D and soluble 2,4-D esters, suggesting its use as a controlled release herbicide (*293*). However, field tests showed that 2,4-D-starch did not hydrolyze sufficiently to control weed growth. However, various herbicide acid chlorides have been reacted with pregelatinized starch in pyridine to yield esters containing 35–48% active ingredient. One pyridine-based herbicide, picloram, with low volatility and sensitivity to photochemical degradation, was converted to the acid chloride and reacted with starch derivatives and a dextrin. Subsequent work produced pregelatinized starch ester which hydrolyzed more readily than the earlier 2,4-D ester (*294*). Polysaccharides, including amylose and amylopectin, were dissolved in 5% lithium chloride or bromide solution in dimethylacetamide and treated with the acid chloride or isocyanate pesticide at 90° to obtain polymeric pesticide esters or carbamates (*295*).

There is interest in polymer–drug adducts in which drugs are linked to macromolecular matrices by covalent bonds which are unstable in the biological environment. Starch trisuccinate converted to a stable benzotriazolide appears to offer a good approach (*296*).

V. HYDROXYALKYLSTARCHES

1. Introduction

A limited selection of literature will be included in this section; reference should be made to other reviews (*62, 297, 646*).

Like the starch acetates, the dispersion stability and nonionic character of the hydroxyalkyl ethers of starch are properties of major commercial interest. Ready availability and low cost also provide commercial interest. Hydroxyethyl- and hydroxypropylstarches are produced in quantities of about 200 million pounds (91×10^6 kg) per year.

2. Preparation

Parents have been issued for the preparation of low-substituted hydroxyalkyl starch ethers (DS 0.05–0.1; 1.3–2.6% hydroxyethyl; 1.8–3.5% hydroxypropyl). (*155–158, 161, 298–314*). Reaction with alkylene oxide is usually run on a 40–45% starch suspension in water under strongly alkaline conditions at temperatures up to 50°. A blanket of nitrogen is recommended in closed pressure

$$\text{starch—OH} + H_2C\underset{O}{—}CHR \xrightarrow{OH^-} \text{starch—O—}CH_2\text{—}\underset{R}{\underset{|}{C}}HOH$$

Side reactions

$$\text{starch—O—}CH_2\text{—}\underset{R}{\underset{|}{C}}HOH + nH_2C\underset{O}{—}CHR \xrightarrow{OH^-} \text{starch—O}\left(CH_2\text{—}\underset{R}{\underset{|}{C}}H\text{—O}\right)_n CH_2\text{—}\underset{R}{\underset{|}{C}}HOH$$

$$H_2C\underset{O}{—}CHR + NaCl + H_2O \longrightarrow R\text{—}\underset{OH}{\underset{|}{C}}H\text{—}CH_2Cl + R\text{—}\underset{Cl}{\underset{|}{C}}H\text{—}CH_2OH + NaOH$$

vessels because of the explosibility of alkylene oxide–air mixtures. Since high alkalinity is needed for good reaction efficiency and since the gelatinization temperature of the granular starch is lowered by the introduction of the hydroxyalkyl groups, salts, such as sodium sulfate or chloride, are added to repress the swelling tendency (*313*). As the level of hydroxyalkylation increases, more salt is required and, at a DS greater than 0.1, the product becomes difficult to purify because the granules swell as the salt is washed out. To obtain more uniform catalyst distribution, a 30% aqueous sodium hydroxide solution containing 26% salt can be injected into a turbulent flowing starch suspension before charging a tank for the etherification reaction (*314*).

Hydroxyalkylation can be accomplished by treating starch at 10–13% moisture with gaseous alkylene oxide (*307–310*). The starch is first impregnated with alkali and a salt such as sodium chloride (*308*). Another approach employs catalysis by quaternary ammonium bases formed from reaction of tertiary amines with the added epoxide (*310*). High levels of substitution can be obtained in granular form by these "dry" reactions without concern for swelling (*308*). Ungelatinized, cold-water-soluble (CWS), hydroxypropylated starches can be prepared using phosphate salts (*315*) or carboxylic acid salts (*316*) as catalysts. By using concentrated solutions of strong bases, for example, 50% sodium hydroxide solution, sprayed onto dry starch or by dry-blending starch and an alkali followed by a water spray, the starch is activated sufficiently to react with etherifying agents, including alkylene oxides, to yield ungelatinized CWS starch ethers (*317*). Other similar processes with dry starch and cereal flours are reported (*318–322*).

To prepare a derivative with a DS greater than 0.1–0.3, steps must be taken to reduce swelling. Suspending potato starch in water at 55° and pH about 6.5 raises the gelatinization temperature sufficiently to prevent swelling in a propylene oxide reaction containing sodium sulfate and sodium hydroxide, that will produce 18% hydroxypropyl content (*323*). It is also claimed that a DS of 0.2–1.5 can be obtained by suspension of the starch in a 5–10% sodium sulfate solution containing 1.5% sodium hydroxide on a starch basis and reacting at 38°, while providing ethylene oxide in small increments over 24-hour periods. Products are ungelatinized and can be washed and filtered (*324*). Cross-linking of starch is effective in preventing swelling during hydroxyalkylation (*182, 325*). In one case (*182*), removal of vinyl sulfone cross-links by cooking in slightly alkaline pH gave a dispersion suitable for warp sizing.

Highly substituted, granular, hydroxyalkyl ethers of starch are readily prepared by suspending the moist, alkaline starch in organic solvents such as lower aliphatic alcohols or ketones (*326, 327*), higher alcohols (*328*), and mixtures of organic liquids (*329*). The alkylene oxide reacts preferentially with the starch, which has adsorbed alkali, rather than with the hydroxyl group of the alcohol. Hydroxyethylation in the range of DS 0.75–1.0 can be obtained without granule swelling. The solubility of the hydroxyalkylstarch in the lower aliphatic alcohols increases with DS. Alcohols can be used to prepare highly substituted products with thermoplasticity and solubility in water for use as adhesives and coatings (*330*).

Hydroxyalkyl waxy corn starch of MS 0.7–0.9 with water solubility can be made in alcohol solution (*326*) and used as a cryoprotective agent for human erythrocytes and as a blood plasma volume extender (*331, 332*).

Cold-water-soluble, hydroxyalkylstarches may be obtained by drum-drying low-DS granular ethers, by gelatinization during or after the hydroxyalkylation reaction and then drum-drying (*303, 333, 334*), or by reacting to a high DS in solvent to retain the granular form (*308*). Drum-drying with additives such as borax (sodium tetraborate), boric acid, gum arabic, and sulfate salts are claimed to give improved dispersibility (*311, 335–337*). Preparation of starch ethers by salting out the product has been recommended (*338*).

Preparation of di- and poly(hydroxyalkyl) starch ethers via reaction with 2,3-epoxy alcohols is claimed (*339*). The reaction of styrene oxide with starch to form a phenyl-hydroxyethyl ether is reported (*340*).

3. Structure of Hydroxyalkylstarches

Considerable work has been reported on the distribution of the substituents in hydroxyethyl- and hydroxypropylstarches (*341–350, 639*). Merkus and coworkers (*341*) have reviewed the literature and determined the distribution of hydroxyethyl groups in hydroxyethylstarches with an MS range of 0.03–1.9. In agreement with other findings, 0-2 substitution predominates. In mono-substituted derivatives, 70–80% of the hydroxyethyl groups are reported to be on O-2. Merkus and co-workers (*341*) suggest that part of the literature differences in substitutent location can be attributed to variations in degrees of substitution and/or in the preparation procedures used. O-3 substitution is favored over O-6 substitution by a factor of 2 in products made in water suspension and of MS in the range 0.028–0.041. At high MS (0.84–1.93) and for ethers made in solution reactions, the O-3/O-6 monosubstitution ratio decreases to 0.2. Part of this change can be attributed to more polysubstitution at O-3 as compared to O-6. At levels of substitution of MS about 2, there is a preponderance of O-2 hydroxyethoxyethyl substitution, indicating preference for reaction at the primary hydroxyl group of the O-2 hydroxyethyl group. Calculations of the nominal degree of substitution of derivatized D-glucopyranosyl units shows that the value never reaches 1, not even at high MS levels, for example, the nominal DS is 0.85 at MS 1.93. This indicates that most of the ethylene oxide in the last stage of the reaction goes to polysubstitution on hydroxyethyl substituents already present (*341, 351*). Even at low MS (0.028–0.041), there is some minimal diethylene oxide substitution (*341*). This confirms previous observations that there is little formation of poly(ethylene oxide) chains in low-substituted, hydroxyethyl starch ethers (*297, 345*).

Graft copolymers of poly(alkylene oxides) on starch have been prepared by reaction of the potassium starch alkoxide with ethylene or propylene oxide in dimethyl sulfoxide under anhydrous conditions (*354*).

Limited data on hydroxypropyl starch is conflicting (*348, 349*). Leegewater and co-workers (*349*) reported that over 80% of the hydroxypropyl groups in hydroxypropyl potato starch of DS 0.07 were found at O-2 and 7% at O-6. Similarly, at DS 0.66, the majority of the hydroxypropyl groups were found at O-2. In contrast, for a hydroxypropyl starch, DS 0.47, Gminder and Wagner (*348*) reported 34% substitution at O-2, 38% at O-3, and 28% at O-6, with 41% of the D-glucopyranosyl units unsubstituted.

The location of the hydroxypropyl groups in a granular hydroxypropyl tapioca starch phosphate of MS 0.045 was deduced from enzyme digestion to be concentrated near the branch points in the amorphous areas of the granule (*352*).

Infrared and x-ray analysis of hydroxyethyl starches of varying MS showed that substitution of MS 0.18 increases the volume of the unit cell by 4.3% and changes molecular conformation at the equatorial anomeric hydroxyl to axial to relieve ring strain (*353*).

4. Properties

Low- DS hydroxyethyl- and hydroxypropylstarches behave like low-substituted starch acetates (*297*). The effects generally increase with increasing DS

(*351*); thus, the gelatinization temperature decreases (*49, 55, 640*). The rate of granule swelling and dispersion on cooking increases, clarity (*57*) and cohesiveness of dispersions increase, and the tendency to increase in viscosity and gel on cooling and aging decreases (*640*). Hydroxyalkylation also improves low-temperature stability and the clarity, solubility (*355*), and flexibility of films.

The hydroxyalkyl ether linkage is stable, and substituents remain in place during acid-conversion, oxidation, or dextrinization of the starch derivative. Hydroxyalkyl starches can be used under alkaline conditions as occur in some adhesives or paper coating colors where starch acetates tend to undergo hydrolysis. These non-ionic starch ethers are less affected by electrolytes and pH than are ionic starches (*297*).

The gelatinization temperature drops as the level of substitution increases. Hydroxyethyl corn starch of MS 0.12 has a birefringence gelatinization temperature range of 45.5°–54.5° compared to corn starch which has a gelatinization temperature range of 62°–70° (*49*). As MS increases from 0.4 to 1.0, granular hydroxypropylstarch exhibits better dispersibility in cold water. As MS rises above 1.0, solvation by lower aliphatic alcohols increases until the products become soluble in methanol or ethanol (*297*).

Film prepared from hydroxypropyl high-amylose corn starch (71% apparent amylose; 1.11% hydroxypropyl, intrinsic viscosity 1.02) is water-soluble, transparent, and impermeable to oxygen at 25° over a wide range of humidities (*356*). Hydroxypropylation reduces tensile strength of the film but increases elongation and burst strength and maintains folding endurance. Plasticization with 5% glycerol is not beneficial.

With hydroxypropyl potato starch of MS 0.02–0.45, pancreatin digestibility decreased exponentially with increasing MS (*357, 358*). Pancreatic α-amylase digestibility is decreased to a greater extent by hydroxypropyl groups in wheat starch than by acetyl groups (*359*). Digestibility of gelatinized wheat starch ethers decreased to 55% at MS 0.17 while digestibility of the corresponding granular starch ethers increased from 31% to 61% (*360*). The difference in digestibility may be explained by retrogradation in the gelatinized sample. On partial digestion, there is a concentration of the hydroxypropyl groups in the low-DP oligosaccharide fraction. The relative amount of this fraction increases with MS. This indicates a greater degree of preferred attack by the enzyme with increasing MS (*360*). Pancreatic alpha-amylase hydrolysis of hydroxyethylstarch, MS 0.55–0.94, showed decreased hydrolysis rates with increasing MS with indication that substitution at O-2 conferred more resistance to the enzyme-catalyzed hydrolysis than did O-6 substitution (*347*).

5. Uses

Hydroxyethylstarches of low MS have been widely used in the paper industry, particularly for surface sizing and coating (*65, 255, 257, 361, 362*, see also this

Volume, Chap. XVIII). In surface sizing, they provide strength and stiffness to the paper as well ink hold-out.

Low-MS hydroxyethylstarches are used as a binder, alone or in combination with synthetic polymers, in pigmented paper coatings (*363–366*). They provide good leveling and viscosity stability in the coating color. The high water-holding properties control binder penetration into the base sheet (*366*). These starches provide high binder strength for the pigment and adhesion to the base stock. Printing quality of the resulting coating is good. Hydroxyethylated starches are claimed to be superior for the preparation of wet-rub-resistant coatings using glyoxal (*367, 368*). They are suitable for admixture with synthetic polymers in clay coating (*369*). Molecular weight distributions of acid-converted hydroxyethylstarches used in coating have been determined (*370, 371*).

The low gelatinization temperature of hydroxyethylstarch is of value when granular starch is added to the wet end of the paper-making machine and expected to gelatinize when the paper sheet passes over the drying rolls. The starch granules swell and spread through the sheet and increase internal strength (*361*).

The water-holding and filming properties of hydroxyethylated starches are useful in adhesives such as bag pastes and case-sealing, label, and envelop adhesives (*297*). Some is used for the carrier portion of corrugating adhesives (*372*).

Low-substituted hydroxyethylstarches are used in warp sizing, alone or in blends with poly(vinyl alcohol). The stable viscosity and filming properties give the hydroxyalkylstarches use in liquid laundry starches or, when pregelatinized, as instant laundry starches (*297*). They may be used in aerosol spray starch formulations (*373*).

Hydroxypropylstarches are of importance in food applications where they provide viscosity stability in the food thickener and insure water-holding under low-temperature storage conditions. These starches are usually cross-linked to obtain desired texture and resistance to the high temperatures, low pHs, and shear degradation often encountered in processing. They may be used in granular form or as pregelatinized, cold-water-soluble products. They may be used in conjunction with other thickeners, for example, with carrageenan in milk systems to be retorted (*374*) and with xanthan gum in salad dressing (*375*). They are used as thickeners in gravies, sauces, fruit pie fillings, and puddings where they must impart a smooth, thick, clear, non-granular texture that will hold up under various storage temperatures, including freezing, and also impart no taste (*376*). Hydroxypropyl tapioca starch is a good starch for frozen pudding (*377*). Hydrolysis of hydroxypropylstarches gives low viscosity, bland-tasting, low bulk-density maltodextrin-like materials (*378, 379*). Hydroxypropylated high-amylose starch produces an edible, water-soluble film coating for foods (*380*). Hydroxypropylstarches with specific levels and types of treatment have been designated as "Food Starch - Modified" under the F.D.A. regulations for use in foods (*254*, this Volume, Chap. XIX).

There has been considerable interest in the use of hydroxyethylstarch as a blood volume extender and as a cryoprotective agent for erythrocytes (*331, 351, 381*).

6. Analysis

To determine total hydroxyalkyl content, use is normally made of the standard Zeisel hydriodic acid digestion as modified by Morgan (*382*) and Lortz (*383, 384*). A number of attempts have been made to couple the Zeisel digestion with gas chromatography (*385, 386*). One procedure for cellulose ethers which uses xylene and adipic acid to catalyze the hydriodic acid cleavage of the ether substituent to form an alkyl iodide appears quite effective for starch ethers (*386*). Another method uses pyrolysis–gas chromatography (*387*). Concentrated sulfuric acid digestion of hydroxypropylstarch releases propionaldehyde, which is determined spectrophotometrically (*388*). ^{1}H-nmr may also be used to determine hydroxypropyl DS (*389*). O-Hydroxyalkyl-D-glucose obtained by hydrolysis with concentrated sulfuric acid can be determined by gas chromatography (*341, 342, 347, 349, 390*).

VI. Starch Phosphate Monoesters

1. Preparation

a. *Orthophosphates*

Sodium phosphate monoesters can be prepared by heating intimate blends of 10% moisture starch and orthophosphates (particularly a mixture of monohydrogen and dihydrogen phosphates) at pH 5–6.5 for 0.5 to 6 h at 120°–160° (*391–400*).

$$\text{starch—OH} + NaH_2PO_4 \longrightarrow \text{starch—O—}\overset{\overset{\displaystyle O}{\|}}{\underset{\underset{\displaystyle O^-Na^+}{|}}{P}}\text{—}O^-Na^+$$

In a typical laboratory reaction, starch is suspended in water containing dissolved phosphate salts, and the mixture is stirred for 10–30 min and filtered. The filter cake is air-dried or dried at 40°–45° to 5–10% moisture and then heat-reacted. Representative data is shown in Table I. Using these reaction conditions, products up to 0.2 DS can be made (*391, 397*).

It is not commercially feasible to dry the starch–phosphate salt filter cake at low temperature. Kerr (*124*) suggests that typical continuous-belt driers of the Proctor and Schwartz type are satisfactory. In this equipment, starch can be dried at a temperature of 48°–124° without gelatinization. The temperature of the starch–phosphate salt mixture should not exceed 60°–70° until the moisture content has been reduced to about 20%. This prevents gelatinization and minimizes undesirable side reactions such as hydrolysis of the phosphate reagents (*124*).

Table I

Phosphorylation with Orthophosphates

	NaH_2PO_4/Na_2HPO_4,[a] g	Starch/water, g/mL	Temp./time, °C/h	%P/DS[b]	Ref.
A	23.2/—	162/240	160°/0.5	0.45/—	*394*
B	34.5/96	180/190	150°/4.0	1.63/—	*393*
C	57.7/83.7	100/106	155°/3.0	2.50/0.15	*397*
D	7.5/11.2	50/65	145°/2.5	0.56/0.03	*398*

[a] A,B,C used $NaH_2PO_4 \cdot H_2O$; B used $Na_2HPO_4 \cdot 12H_2O$; C, $Na_2HPO_4 \cdot 7H_2O$. ABD: Filter cakes were air-dried. C: filter-cake dried overnight at 40–45° in a forced-draft air oven; then at 65° for 90 min. B,D: heat-reacted in vacuum oven. A,C: heated with continuous stirring (for example, in a stainless steel beaker placed in an oil bath). Slurry pH: A(5.5), B(—), C(6.1), D(6.5).

[b] %P = bound phosphorus; DS = degree of substitution.

Treating starch with a proteinase such as pepsin or papain appears to enhance sodium phosphate retention by the starch granule (*401*). Products containing 6–12% phosphorus can be prepared by slurrying starch in 45–55% orthophosphate solutions at 50°–60° followed by filtration, drying, and heat reaction at 140°–155° (*400*). Similarly, starch has been phosphorylated with tetrasodium pyrophosphate (TSPP) (*124*) or mixtures of TSPP with orthophosphates (*402*) or phosphoric acid (*403, 404*). Alkyl pyrophosphates such as dimethyl or bis(2-ethylhexyl) pyrophosphate can be used to phosphorylate starch in dry reactions (*406*). Monostarch phosphates can be prepared by heating mixtures of starch and ammonium metaphosphate or ammonium polyphosphate at 110°–140° for 1 to 4 h at a pH range of 5–9 (*405*). The use of sodium (or other alkali metals) metaphosphate (*124*) or polyphosphate (*405*) in the same pH range results in substantial cross-linking. For these alkali metal salts, the reaction pH should be less than 5 for mono ester formation (*124*); pHs lower than 4 promote hydrolysis of the starch (*407*).

b. *Sodium Tripolyphosphate*

Starch can be phosphorylated with sodium tripolyphosphate (STP) (*124, 391, 397, 408, 409*).

$$\text{starch—OH} + Na^{+}\,{}^{-}O\text{—}\overset{\overset{Na^{+}O^{-}}{|}}{\underset{\underset{O}{\|}}{P}}\text{—}O\text{—}\overset{\overset{O}{\|}}{\underset{\underset{Na^{+}}{\underset{O^{-}}{|}}}{P}}\text{—}O\text{—}\overset{\overset{O^{-}Na^{+}}{|}}{\underset{\underset{O}{\|}}{P}}\text{—}O^{-}Na^{+} \longrightarrow \text{starch—}O\text{—}\overset{\overset{O}{\|}}{\underset{\underset{O^{-}Na^{+}}{|}}{P}}\text{—}O^{-}Na^{+} + Na_3HP_2O_7$$

The initial pH of the STP–starch mixture is typically 8.5, decreasing to 7 or less after heat reaction. For example, a starch filter cake containing 5% STP is dried at 60° to about 12% moisture, then heated at 120°–130° for 1 h. A water-washed sample contains 0.37% phosphorus. Extending the heat treatment to 2 hours yields a product with 0.47% phosphorus (*124*). The amount of STP retained in the starch filter cake is adjusted by varying the amount of STP and/or the amount of water in the slurry.

In an alternative procedure, acidified (pH 4.2–4.8) 20–36% aqueous solutions of STP were used to impregnate starch filter cakes (*408, 409*). The starch-STP blends were dried in a flash drier to 5–8% moisture followed by heat reaction at 110°–130° or the filter cake-STP blends were dried directly at 130° to 6% moisture or less. Products containing 0.07–0.10% phosphorus were produced.

The distribution of the phosphate ester groups in a STP-derivatized starch (heat-reacted at 150°, 0.30% bound phosphorus) was determined (*410*).

Position	*% Distribution of phosphate groups*
O-2	28
O-3	9
O-6	63

c. *Urea–Phosphates*

Combinations of urea and phosphate salts have been used to modify starch (*411–422*). Urea appears to act as a catalyst promoting higher reaction efficiencies and more rapid reaction between the starch and phosphates (*412, 414*). The final product viscosities are higher when urea is present (*411–417*). Using 2–5% urea (on starch), varying amounts of orthophosphates, pH 4–8, and heating times of several hours at 140°–160° gave 1 to 5% bound phosphorus (*411*). Hjermstad (*412*) indicated that urea (or other amides) plus orthophoshate and starch produced nitrogen-substituted phosphate groups, but Alexander (*414*) found that 2–3% urea and 1–6% sodium tripolyphosphate produced derivatives containing phosphate (<0.3% bound phosphorus) and carbamate groups (1.3–1.7% nitrogen).

In other work, the ratio of nitrogen to phosphorus was controlled by heating the reaction mixture under reduced pressures. When the vacuum was decreased, nitrogen substitution, cold-water solubility, and the rate of attainment of maximum dispersion viscosity increased (*413*).

d. *Organic Phosphorylating Reagents*

Starch phosphate monoesters can be produced by aqueous reactions with *o*-carboxyaryl phosphates (**I,** *423*), *N'*-substituted, *N*-phosphorylimidazole salts (**II,** *424*), or *N*-acylphosphoramidic acid salts (**III,** *425*). Typically, reactions are run at 30°–50°. I and III react at pH 3–8 while II requires more alkaline reaction conditions, (pH 11–12).

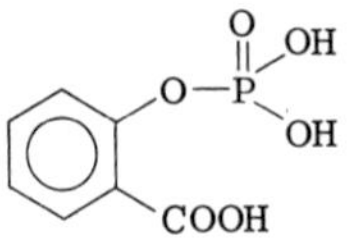

(I) 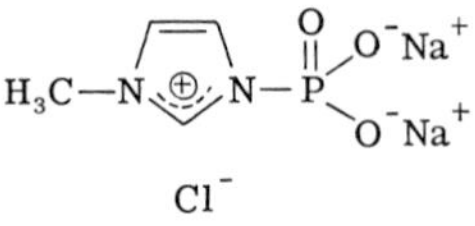 (II)

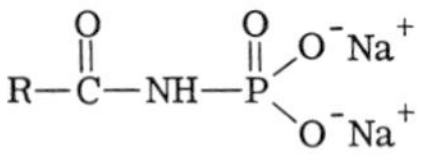

(III)

Starch phosphonate ether derivatives have been claimed by alkali-catalyzed slurry reactions with vinyl phosphonates, vicinal halohydroxyalkane phosphonates, or vicinal epoxyalkane phosphonates (*426*).

Derivatizing reagents have also been prepared by reacting 2-alkyl-2-oxazolines with phosphorus oxychloride (phosphoryl chloride) (*427*). Reactions with starch at pH 11.4 and at 40° for 16 hours produce derivatives containing anionic, phosphoramidic acid groups. Corn starch products yield dispersions that are clear and stable.

Starch monophosphate esters with degrees of substitution of 1.75 are obtained by heating starch at 120° with a mixture of trialkylamine and tetrapolyphosphoric acid in *N,N*-dimethylformamide. The starch is recovered by alcohol precipitation (*428*).

2. Properties

Dispersions of corn starch monophosphates have clarity, high viscosity, a long cohesive texture, and stability to retrogradation (*391*). Viscosity is controlled by manipulating the amount of phosphate salts, reaction temperature, time, and pH (*391, 393*). Starch phosphate derivatives are useful as emulsification agents (*124, 405, 406*), and dispersions of the phosphate esters have freeze-thaw stability (*391*).

3. Uses

a. *Paper (see also Chapter XVIII)*

Starch phosphates find use in the paper industry as wet-end additives providing improved strength and filler retention (*361, 391, 401, 407, 429*). Alum must be present in the system for optimum performance (*391, 407*). Starch phosphates have outstanding dispersing properties for clay-satin white coating colors (*430*). Low viscosity urea–phosphate starches are useful in paper coating, particularly when they are combined with satin white pigment (*391, 431, 432*). For example, tapioca phosphate of 34 cP at 25° and 15% concentration is prepared by heating a mixture of 1000 parts of tapioca starch, 75 parts of orthophosphoric acid, and 140 parts of urea at 126° for 45 min. Coating colors consisting of the urea–phosphate starches, satin white, and china clay SPS yield excellent wet-rub resistance of 90–91 in the Adams test. If satin white is eliminated, the wet-rub decreases to zero (*431*). Combinations of phosphoric acid and urea are approved for starch modification for internal sizing, surface sizing, and coating under the FDA indirect food additive regulations (*433*).

Amphoteric starches prepared by phosphorylating cationic derivatives are useful wet-end additives (*408, 409, 423–426, 434, 435*). Starch phosphates have also been used in emulsions with ketene dimers as internal sizing agents for paper (*436, 437*).

b. *Textile*

Compositions containing 50–90 parts of starch monophosphate and 10–50 parts of poly(vinyl alcohol) or polyacrylate are useful textile sizing agents of cotton, polyester–rayon, and polyester–cotton yarns (*438*). Phosphorylated starches are useful thickeners for textile printing, providing improved uniformity and penetration for printing cotton (*439*).

c. *Adesives (see also Chapter XX)*

Adhesives with improved storage stability for use in corrugated paperboard are prepared by blending starch monophosphate with native starch, borax, sodium hydroxide, and water (*440*). In another adhesive, Neoprene rubber latexes are combined with 0.01–10% starch phosphate to produce adhesives with good, rapid bond strength. In one example, a latex containing about 4.2% starch phosphate is used to bond wood pieces. Shear strengths of 108, 290, and 464 lb/in.^{2} (7.59, 20.4, and 32.6 10^4 kg/m^2) were observed after 10 minutes, 24 h, and 10 days, respectively. Without the starch, shear strengths were 9, 249, and 440 lb/in.^{2} (0.633, 17.5, and 30.9 10^4 kg/m^2) for the same time periods (*441*).

d. *Scale Inhibition*

Starch phosphates prevent or inhibit scale-forming deposits when added to water in amounts of 10 mg/L (*442, 443*). In a similar application, 10–50 ppm of starch phosphate was added to salt brine containing 2–5.5 g of calcium sulfate. The starch phosphate prevented precipitation of the calcium sulfate and enabled recovery of high purity salt (*444, 445*).

e. *Flocculation*

A flocculant for coal washery tailings has been prepared by drum-drying a starch slurry containing sodium dihydrogenphosphate and heat-treating the resulting product at 100° for 2 h and at 160° for 0.5 h. The product had properties of a flocculant in the 4–10 ppm range (*446*). Starch phosphates (0.62 lb, 0.28 kg) were also used in combination with 0.035 lb (0.016 kg) of water-soluble poly(acrylamide) per ton (907 kg) of tailings (*447*).

f. *Food (see also Chapter XIX)*

Starch phosphate monoesters have useful properties in food systems (*124, 391, 448*). They are good emulsifiers for vegetable oil in water (*124, 449, 450*) and are effective as pudding starches (*124, 391*). Starches spray-dried with 0.5–1.0% of phosphate salts such as tetrasodium pyrophosphate or sodium tripolyphosphate are claimed to have improved taste; some phosphorylation occurred during the spray-drying. Additional heat treatment of the spray-dried material increased phosphorylation (*124*).

The U.S. Food and Drug Administration has established regulations that permit only monosodium orthophosphate or sodium tripolyphosphate and sodium trimetaphosphate to be used for esterification of starch to be added to food (*254*). The residual phosphate in the starch must not exceed 0.4% calculated as phosphorus.

g. *Miscellaneous*

An interesting application of starch phosphates is in the production of films containing gelatin and glycerol (*451*) or 2:3 glycerol–sorbitol (*452*) useful for dressing skin wounds and burns (*451*). The films appeared to promote more rapid healing and lower infection (*451*).

A cornstarch phosphate–Sn(II) mixture, developed as a carrier for Tc radionuclide, is useful as a radiolabeled diagnostic agent (*453*).

Starch monoester phosphates have been added in concentrations of 0.01% to cement to improve workability and decrease bleeding (*454*).

4. Analysis

In the analysis of phosphorus-containing starches, all organic material is destroyed by combustion (*455, 456*). The orthophosphates remaining can be determined spectrophotometrically using a colorimetric reagent consisting of ammonium molybdate, ascorbic acid, sulfuric acid, and/or bismuth subcarbonate (*456, 457*).

VII. Cationic Starches

1. Introduction

Cationic starches are important commercial derivatives used in large quantities by the paper industry as wet-end additives, where they provide retention, drainage and strength improvements, as size press starches, and as binders in coatings (*361, 458–460*). Estimated annual usage by U.S. papermakers has grown from 42 million lb (19×10^6 kg) in 1966 to about 132 million lb (60×10^6 kg) in 1977 (*458*).

2. Preparation

a. *Tertiary Aminoalkyl Ethers of Starch*

Although cationic sulfonium (*461*) and phosphonium (*462*) starches are known, the most important cationic derivatives are those that contain tertiary amino or quaternary ammonium groups. An example of the former type is that produced with diethylaminoethyl chloride (*463–465*).

$$\text{starch—OH} + \text{Cl—CH}_2\text{CH}_2\text{N(CH}_2\text{CH}_3)_2 \xrightarrow{\text{OH}^-} \text{starch—O—CH}_2\text{CH}_2\text{N(CH}_2\text{CH}_3)_2$$

$$\downarrow \text{HCl}$$

$$\left[\text{starch—O—CH}_2\text{CH}_2\overset{\text{H}}{\overset{|}{\text{N}}}\text{(CH}_2\text{CH}_3)_2\right]^+ \text{Cl}^-$$

Under alkaline reaction conditions, the starch tertiary amino ether is obtained as the free base. Neutralization converts the free amine to the cationic tertiary ammonium salt. Presumably, the reaction proceeds through a highly reactive, cyclic ethyleneimmonium intermediate. Cationic starches have been made with

$$\left[(\text{C}_2\text{H}_5)_2\text{N}\underset{\text{CH}_2}{\overset{}{\diagdown\!\diagup}}\text{—CH}_2\right]^+ \text{Cl}^-$$

various dialkylaminoalkyl chloride reagents (*463, 464*). Starch tertiary amines can be converted to quaternary ammonium salts by reaction with methyl iodide in refluxing ethanol (*463*) or by treatment in aqueous solution with alkylene oxides or esters such as dimethyl sulfate (*466*). Glycidyl tertiary amines (*467, 468*) and the reaction products of 2,3-dihalopropionamides with secondary amines (*469*) have been used to prepare tertiary amino ethers of starch. The reaction efficiencies of glycidyl tertiary amines are lower than those of the β-haloalkylamines, owing to intramolecular and intermolecular diquaternary amine formation (*460*).

b. *Quaternary Ammonium Starch Ethers*

The facile reaction of epichlorohydrin with tertiary amines has been used to prepare reagents which add quaternary ammonium groups to the starch molecule (*470, 471*).

$$(CH_3)_3N + Cl{-}CH_2{-}\underset{\diagdown O \diagup}{CH{-}CH_2} \longrightarrow \left[H_2\overset{O}{C{-}C}HCH_2N(CH_3)_3 \right]^+ Cl^-$$

$$\text{starch}{-}OH + \left[H_2\overset{O}{C{-}C}HCH_2N(CH_3)_3 \right]^+ Cl^- \longrightarrow \left[\text{starch}{-}O{-}CH_2{-}\underset{OH}{CH}{-}CH_2{-}N(CH_3)_3 \right]^+ Cl^-$$

A wide variety of tertiary amines can be used, but the ones most reactive with epichlorohydrin contain at least two methyl groups. Care must be taken to remove or substantially reduce, by vacuum distillation or solvent extraction, any residual epichlorohydrin or by-products such as 1,3-dichloropropanol that can cross-link the starch and reduce dispersibility and effectiveness of the cationic starch. The use of tertiary amine salts of weak acids, such as acetic acid, in aqueous reactions with epichlorohydrin is claimed to result in lower by-product formation (*472*).

The chlorohydrin form of the reagents is stable in water and can be rapidly converted to the "active" epoxide form by the addition of sodium hydroxide.

$$\left[Cl{-}CH_2\overset{OH}{CH}{-}CH_2N(CH_3)_3 \right]^+ Cl^- + NaOH \longrightarrow \left[H_2\overset{O}{C{-}C}HCH_2N(CH_3)_3 \right]^+ Cl^- + NaCl$$

Recently, an investigation was made on the aqueous reaction of 3-chloro-2-(hydroxypropyl)trimethylammonium chloride with corn starch (*473*). Products ranging from 0.01 to 0.07 DS were made in the presence of sodium hydroxide and sodium sulfate to prevent swelling of starch. Using a sodium hydroxide to reagent mole ratio of 2.8:1, a reagent to starch mole ratio of 0.05:1, and a 35% starch concentration at 50°, the reaction efficiency was 84% after 4 h. Lower temperatures required longer reaction times. The concentration of both reagents and starch influence the reaction efficiency. A DS of 0.02–0.04 is satisfactory for most paper uses.

These non-volatile quaternary ammonium reagents are particularly useful for dry or "semi-dry" reactions with starch. The epoxide or glycidyl forms of the reagents can react with starch without alkaline catalyst, but at 120°–150° for 1 h (*474, 475*). Reaction efficiencies of 75–100% are obtained if an alkaline catalyst is used in combination with the quaternary ammonium reagents (*476, 477*). Alkaline-catalyzed reactions can be run at lower temperatures, typically 70°–80° for 1–2 h.

The reagents can also be prepared by quaternizing a tertiary amine with allyl chloride and reacting the resulting *N*-allyl quaternary ammonium chloride with hypochlorous acid to produce the cationizing reagent (*478–480*). Preparation of 3-chloro-2-(hydroxypropyl)triethylammonium chloride was claimed by the reaction of triethylamine with α-chloroglycerol followed by treatment with hydrogen chloride dissolved in acetic anhydride (*481*). Quaternary pyridinium salt–starch derivatives have also been described (*482*).

A similar class of cationizing derivatives is represented by the 4-halo-2-butenyltrialkylammonium halides (*483–486*).

$$\text{starch—OH} + \left[\text{Cl—CH}_2\text{—CH=CH—CH}_2\text{—N(CH}_3)_3\right]^+ \text{Cl}^-$$

$$\downarrow$$

$$\left[\text{starch—O—CH}_2\text{CH=CHCH}_2\text{N(CH}_3)_3\right]^+ \text{Cl}^-$$

The reaction products of trimethyl or triethylamine with dihalo substituted alkylene compounds have been used to prepare cationic starch ethers (*487*).

$$\left[\text{X—CH}_2\text{—(CH}_2\text{)}_y\text{—CH}_2\text{N(CH}_3)_3\right]^+ \text{Cl}^-$$

$$\text{X} = \text{Cl, Br};\ y = 1\text{-}3$$

c. *Aminoethylated Starches*

Although commercially useful cationic starches are primarily the tertiary amino and quaternary ammonium derivatives, other types have received some attention. Ethyleneimine reacts with starch to yield the 2-aminoethyl ether (*488*).

$$\text{starch—OH} + \underset{\text{N—H}}{\text{H}_2\text{C—CH}_2} \longrightarrow \text{starch—O—CH}_2\text{CH}_2\text{NH}_2$$

Typical processes involve mixing the dry, or semi-dry, starch with gaseous ethyleneimine at 75°–120° without catalysts (*489–491*). Poly(ethyleneimine) is a by-product in these reactions, but its formation can be reduced or eliminated by control of the temperature and pressure of the reaction. Aminoalkylation of starch in inert solvents such as carbon tetrachloride and benzene has been accomplished with the use of an aziridine–sulfur dioxide complex (*492*).

Cationic starches have been prepared by treating aqueous starch suspensions with ethyleneimine in the presence of ethylene oxide and an alkaline catalyst at 35° for 16 hours (*493*). Without ethylene oxide, almost no cationic groups are introduced. Organic halides or esters of strong acids may be used in place of an alkylene oxide.

The efficiency of the reaction of ethyleneimine with starch is increased if a tertiary amino starch (*494*) derivative is used. In this case, ethyleneimine side chains are grafted onto the substituent tertiary amino alkyl groups or onto starch hydroxyl groups.

Aziridinyl starch derivatives can be prepared by reacting starch containing keto or aldehyde substituents with ethyleneimine in water at temperatures below 20° (*495*).

Similar derivatives of acid-modified corn starch can be prepared by sodium hydroxide-catalyzed reactions with *N*-(4-chloro-2-butenyl)aziridine or *N*-(1-hydroxy-2,3-epoxypropyl)aziridine. The resulting derivatives can be reacted further and are useful as latent cross-linking agents (*495*).

$$\text{starch—O—CH}_2\text{CHO} + \text{H}_2\text{C—CH}_2 \text{ (aziridine, N—H)} \longrightarrow \text{starch—O—CH}_2\text{—C(OH)(H)—N}\langle\text{CH}_2\text{—CH}_2\rangle$$

The preparation of aminoalkylated starches has also been claimed by reduction of cyanoalkylated starch with sodium borohydride, hydrazine hydrate, sodium hydrosulfite, and 1,3-dihydroxypropanone in aqueous ammonium hydroxide (*496*).

d. *Cyanamide–Starch*

Reaction of starch with a disubstituted cyanamide in the presence of a strong base yields a starch iminodialkyl carbamate (*497*). Protonation of the imino

$$\text{starch—OH} + \text{R}_2\text{N—C}\equiv\text{N} \longrightarrow \text{starch—O—}\overset{\text{NH}}{\overset{\|}{\text{C}}}\text{—NR}_2$$

nitrogen atom forms the ionic salt which makes the substituent stable to hydrolysis. Similar derivatives have been prepared using cyanamide salts (R = H) (*498–502*). Iminoalkyl derivatives produced in granular form by normal slurry reactions tend to be cross-linked. Cyanamide–starch pastes tend to thicken with time. Phosphate salts and aluminum sulfate appear to stabilize the pastes (*501*).

The dry, starch–cyanamide reaction products become water-insoluble with storage, but can be redispersed in boiling water at pH <2 (*503*). Storage-stable cyanamide–starch has been produced by neutralizing the reaction mixture to pH 1.0, followed by gelatinization and drum-drying to 10% moisture (*504*). Similar products are prepared by drying the granular derivatives at 40°–50° after neutralization to pH 3.0 (*505, 506*).

e. *Starch Anthranilates*

Cationic starch esters have been prepared by reacting starch with isatoic anhydride and its derivatives (*507–511*). The starch slurry reaction is conducted at pH of 7.5–9.0. The product (when R_1 = H) can be diazotized by treatment with nitrous acid (*507, 508*) and subsequently coupled with aromatic type compounds to produce azo dyes (*507*) or further reacted with starch to form cross-linked dervatives (*508*).

$$R^2\text{-}C_6H_3\langle C(=O)\text{-}O\text{-}C(=O)\text{-}N(R^1)\rangle + \text{starch-OH} \longrightarrow R^2\text{-}C_6H_3(\text{-}C(=O)\text{-}O\text{-starch})(\text{-}NH\text{-}R^1)$$

$$\xrightarrow{\text{HONO} \mid \text{HCl } (R^1=H)} R^2\text{-}C_6H_3(\text{-}C(=O)\text{-}O\text{-starch})(\text{-}\overset{+}{N}\equiv N) \quad Cl^-$$

f. *Cationic Dialdehyde Starch*

Dialdehyde starch has served as a base for preparing cationic derivatives. The aldehyde groups are reacted with hydrazines or hydrazides (*512–514*). The reaction with betaine hydrazide hydrochloride is illustrated (*515*).

Dialdehyde starch at 15% concentration in water is reacted with 3–5% reagent. The initial pH of 4.5 decreases to 2.5–3.2 after reaction. The dispersion forms an initial gel which liquefies with continued heating at 90°–95°. To obtain a dispersion which remains fluid at 25°, holding times of at least 2.3–3.0 hours at 95° are required. Cationic dialdehyde starch has also been prepared by reacting unmodified starch with *N,N*-dimethylaminoethyl chloride or with the quaternary ammonium reagents described previously, followed by oxidation with periodic acid to form the dialdehyde starch (*515*).

$$\text{starch-CHO} + \left[H_2NNH\overset{O}{\overset{\|}{C}}CH_2N(CH_3)_3\right]^+ Cl^- \longrightarrow \left[\text{starch-}CH{=}NNH\overset{O}{\overset{\|}{C}}CH_2N(CH_3)_3\right]^+ Cl^-$$

g. *"In Situ" Cationization*

Generally, commercial cationic starches are prepared by slurry reactions with a cationizing reagent followed by recovery of the starch in granular form.

Another method involves the simultaneous cooking and reaction of the starch at pH 8 with a monofunctional reagent such as β-diethylaminoethyl chloride or 3-chloro-2-(hydroxypropyl)trimethylammonium chloride or another of the reagents previously described (*516, 517*). Typically, temperatures of 100°–160° are used.

Many derivatizing agents require the preparation of epichlorohydrin–ammonia (*518, 519*) and/or amine condensates (*520–523*). Similar polymeric, polyamine reagents have been prepared by the reaction of alkylene dihalides and amines

(*524*) and the reaction of poly(epichlorohydrin) with tertiary amines (*525*). All the reagents are multi-functional and complex in nature. A possible representation of the polyepichlorohydrin–trimethylamine polymer is shown.

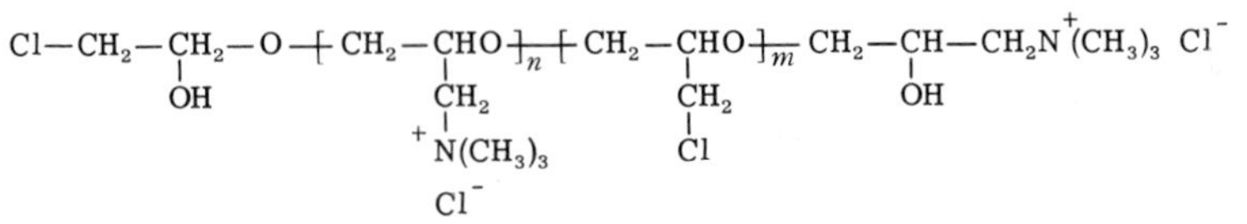

$$\mathrm{Cl{-}CH_2{-}\underset{\substack{|\\ OH}}{CH_2}{-}O{+}CH_2{-}\underset{\substack{|\\ CH_2\\ |\\ {}^{+}N(CH_3)_3\\ Cl^-}}{CHO}{+}_n{+}CH_2{-}\underset{\substack{|\\ CH_2\\ |\\ Cl}}{CHO}{+}_m\,CH_2{-}\underset{\substack{|\\ OH}}{CH}{-}CH_2\overset{+}{N}(CH_3)_3\ Cl^-}$$

Reactions with the polymeric reagents are alkali-catalyzed and can be done with gelatinized starch or simultaneously with gelatinization in a continuous cooker. Cross-linking can occur because of the multifunctional nature of the cationizing agents. It is substantially reduced or avoided by using dispersions with less than 5% solids, degraded starches, and/or cross-linking inhibitors such as amines, alkanolamines, or α-chlorocarboxylic esters and acids (*518, 519*).

Similar products were prepared by adding epichlorohydrin and amines directly to a dilute starch slurry containing sodium hydroxide, followed by immediate reaction/gelatinization in a continuous cooker (*526*). This type of reaction has been conducted in a granular starch suspension (*527, 528*). A depolymerization step involving acid conversion, oxidation, or dextrinization is required to overcome cross-linking and make the starch gelatinizable.

Additional polyfunctional derivatizing reagents that have been used are *N*-β-chloroethyl-4-(3-chloropropyl)piperidine hydrochloride, tris-β-chloroethylamine and formaldehyde–acetone–amine condensates (*529*). By using oxidized starches, it is claimed that cationic derivatives can be prepared with epichlorohydrin–amine condensates without the use of alkaline catalysts (*530, 531*). Poly(ethyleneimine) (PEI) has also been reacted with oxidized starches (*532*). Cationic starch compositions have also been prepared by heating dilute native starch dispersions to 70°–110° with poly(alkyleneimine) or polyalkylenepolyamine having molecular weights of at least 50,000 (*533*).

PEI (MW = 100,000) has been reacted with low-DS starch xanthates to form starch poly(ethyleneiminothiourethane) derivatives (*534*).

Similarly, starch–polyamide–polyamine interpolymers were prepared by treatment of starch xanthates with polyamide–polyamine–epichlorohydrin resins (*535, 536*).

h. *Cationization by Complex Formation*

Starch can be made cationic through complex formation with amine compounds containing, preferably, one long, unbranched, aliphatic chain of 12–22 carbon atoms (*535–542*). Usually the fatty acid amines are used as acetate or hydrochloride salts. Suitable compounds are carbethoxymethyldimethyloctadecylammonium chloride, hexadecyltrimethylammonium chloride, quaternary ammonium glycine hydrazide salts, and tallow-1,3-propylenediamine diacetate. The complexes are formed by cooking or dispersing the starch in the presence of the amine. In one example using a hydrazide salt–unmodified cornstarch complex, 81% of the starch was retained by cellulose pulp (*538*). Other work indicates that strength improvements are also obtained with starch–fatty amine complexes (*537, 539*). Likewise, cationic starch complexes have been prepared with polyamines (*543*). A starch with cationic characteristics is prepared by coating a native or anionic starch with a conventionally prepared cationic starch (*544*).

i. *Amphoteric Starches*

Cationic modification, particularly tertiary amino or quaternary ammonium etherification of starch, has been combined with other treatments to introduce anionic and/or nonionic groups into the same starch molecule. The objective was to develop products with enhanced performance, particularly in paper-making, textile sizing, and flocculation. For example, the introduction of anionic phosphate groups into cationic starches through reaction with phosphate salts (*434*) or phosphate etherifying reagents (*435*) results in products with improved pigment retention under varying paper-making conditions. An important aspect of this modification is the balance of cationic to anionic groups. A preferred range for a starch diethylaminoethyl ether containing phosphate ester substitutents appears to be 0.07–0.18 mole of anionic groups per mole of cationic groups (*434*).

Amphoteric starches have also been made by the introduction of sulfosuccinate groups into cationic starches (*545*). This is accomplished by adding maleic acid half-ester groups to a cationic starch and reacting the maleate double bond with sodium bisulfite. The resulting product contains sulfosuccinate ester groups (0.02 DS) and diethylaminoethyl ether groups (0.03 DS). A similar derivative can be prepared by etherifying cationic starch with 3-chloro-2-sulfopropionic acid (*546*).

Other types of reagents have been used to introduce anionic groups (*547–557*). For example, carboxyl groups were introduced by reacting starch with sodium chloroacetate or by hypochlorite oxidation. Hypochlorite oxidation of cationic starches resulted in products for use as paper coating binders (*548–551*). Amphoteric starches useful in textile warp sizing are produced by treating cationic starches with propane sultone (*552–555*). The modification can be extended by the introduction of nonionic or hydroxyalkyl groups by treatment with ethylene oxide or propylene oxide (*556, 557*). Similarly, cationic hydroxyalkyl derivatives have been prepared and found to be useful papermaking additives (*468, 558, 559*).

Amphoteric starches useful as wet and dry strength additives were prepared by xanthation of diethylaminoethyl- or 2-(hydroxypropyl)trimethylammonium starch ethers (*560*).

3. Properties

The gelatinization temperature of cationic starches decreases as the number of cationic substituent groups increases (*460*). The dispersions show improved stability and clarity (*463*). Zeta potential measurements of typical cationic starches show a net positive charge at pH 4–9 (*458*).

4. Uses

a. *Paper: Wet-End Additive (see also Chapter XVIII)*

Electrochemical affinity of cationic starch for negatively charged cellulose fibers results in nearly 100% irreversible adsorption of the starch derivative (*561*). The cationic starch acts as an ionic bridge between the cellulose fibers and mineral fillers and pigments. The preferential adsorption of cationic starch on pulp fines results in increased retention of fines and an improvement in strength by inclusion of fines with long fibers in a cohesive network (*458*). This can lead to a better drainage as well (*562*).

A typical paper-making furnish contains numerous additives used to affect sheet properties or control the runnability of the furnish on the paper machine (*562*). The most popular additive, alum (*563*), may be present in various amounts (*564*) and can adversely affect the pulp adsorption of cationic starch (*561, 565, 566*). To improve performance under varying conditions, numerous cationic starches were developed (*361*). Amphoteric starches (*434, 435, 545, 546*) are examples of products developed to yield consistent performance under varying paper mill conditions. Use of a starch quaternary ammonium ether (0.033 DS) and an anionic poly(acrylamide) (2% dry basis on starch) starch derivative improves pigment retention with alum at pH 4–5 (*567*).

In another process, filler retention and strength are increased by coating Kaolin particles with 5% of a starch quaternary ammonium ether and incorporating the product into a pulp system at a 30% concentration (*568*).

In addition to new products and experimentation with combinations of additives (*563, 569–571*) and changes in papermaking systems, better wet-end control methods have been devised to ensure optimum performance from cationic starches (*564*).

b. *Sizing*

Of increasing importance in paper making is the use of cationic starch to emulsify synthetic sizing agents such as alkyl ketene dimers or alkenyl succinic anhydrides (*458, 572–578*).

In addition to their wet end use, cationic starches are used as surface sizing agents. The irreversible, ionic attraction to cellulose fibers results in lower penetration as well as lower BOD in mill effluents (*257, 579, 580*). Cationization is often combined with degradative reactions such as oxidation, acid-catalyzed hydrolysis, or dextrinization because starch dispersions with high solids are required. An improved surface sizing starch is made by combining the film-forming properties of high-amylose starch with cationic modification (*581*).

c. *Coating*

As a coating binder, cationic starch offers increased strength due to the electrochemical binding of clay to fiber (*582, 583*). However, if the amine content is too high, agglomeration or even coagulation of the clay can occur. A starch product containing amine (0.015–0.025 DS) and carboxyl substituents (0.5–1.5% COOH groups) yields good binding and shock-free clay dispersion (*548–551*).

d. *Textile*

Both cationic and amphoteric starches find use as warp sizing agents (*552–557, 584*). The starch size provides lubrication and abrasion resistance. Cationic quaternary ammonium starches in combination with a cationic lubricant and an emulsified wax have been used to provide protective coatings for glass fibers during twisting and texturizing (*585*).

e. *Flocculation*

Cationic starches are effective agents for flocculating aqueous suspensions of negative organic or inorganic particles such as clay, titanium dioxide, coal, carbon, iron ore, silt, anionic starch, and cellulose (*564, 586, 587*). In an effort to develop improved products, high-DS types are prepared by reacting gelatinized starch with quaternary amine reagents (*588, 589*). Derivatives of 50–90 fluidity are used because they will form higher solids dispersions. Products with 0.4 to 0.5 DS (*588*) and >0.7 DS (*589*) have been prepared. They can be used directly or purified by dialysis, ultrafiltration, or ion exchange (*589*). These products, as well as lower DS (0.15–0.25 amine substituent DS) derivatives of hypochlorite-oxidized starch (*590, 591*) or acid-converted waxy maize starch (*592*), are effective in treating raw primary sludge.

f. *Miscellaneous*

Starch quaternary ammonium ethers of 0.1–0.45 DS are useful demulsifiers for breaking water-in-oil and oil-in-water emulsions (*593*). Both tertiary amino and quaternary ammonium starches act as reducing and suspension agents for preparing antihalation layers for photographic film from manganese dioxide (*594*). Insoluble (cross-linked) cationic starches are effective in removing heavy metal anions such as chromate, dichromate, ferrocyanide, ferricyanide, molybdate, and permanganate from industrial effluents (*595*). A tertiary aminoalkyl derivative of high-amylose corn starch is useful as the active ingredient in a hair-holding spray (*596*).

5. FDA Regulations

The Federal Food and Drug Administration has regulations (*433*) for modified starches that may be in any material that comes in contact with food. Cationic starches may be present as additives in the paper or paperboard food package. The cationizing reagents that may be used in the preparation of these starches are (4-chlorobutene-2)-trimethylammonium chloride, β-diethylaminoethyl chloride, or 2,3-(epoxypropyl)trimethylammonium chloride. The regulation specifies maximum treatment levels, and, in some cases, the use of the derivative and the addition levels in a particular application.

6. Analysis

Cationic starches are conveniently analyzed by determining the nitrogen content using the Kjeldahl method (*597, 598*).

When starch is analyzed, a dextrose blank is commonly used. Unmodified starches typically contain very small amounts of protein corresponding to nitrogen values of less than 0.1% (*455*). An examination of 17 commercial cationic starches indicated a nitrogen content ranging from 0.18% to 0.37% (*361*).

VIII. Other Starch Derivatives

1. Introduction

Some starch derivatives are commercially manufactured and sold on a small scale; others appear to have properties of commercial interest. Two derivatives in these categories are mentioned here.

2. Starch Xanthates

The reaction of polysaccharides with carbon disulfide under strongly alkaline conditions to form xanthates has long been known. Interest in starch xanthates

$$\text{starch—OH} + CS_2 + \text{NaOH} \longrightarrow \text{starch—O—}\overset{\overset{\displaystyle S}{\|}}{C}\text{—SNa} + H_2O$$

has been rekindled by work showing their effectiveness in removal of heavy metal ions from process waters (*184–186, 599–602*) and as reinforcing agents in rubber (*603–608*) and slow-release encapsulating agents for pesticides (*609–612*).

Early work on starch xanthation reported the distribution of xanthate groups (*613*) and the effect of various alkali metal hydroxides (*614*) on the distribution. A continuous process for xanthation of starch involving the metered addition of granular starch, sodium hydroxide solution and carbon disulfide to a continuous, high-shear mixer–reactor and discharging the viscous paste after roughly 2 minutes reaction time has been developed (*185, 615, 616*). Low-concentration pastes (10%) with DS 0.05–0.15 xanthate were prepared in a low-powered mixer using only small amounts of alkali (*185, 617–620*). For the water treatment, the product is prepared in granular form by using starch cross-linked sufficiently with epichlorohydrin to resist gelatinization in the strongly alkaline xanthation reaction.

The distribution of the xanthate groups on the starch xanthates (DS 0.12 or 0.33) prepared by the rapid continuous process has been shown to take place initially at O-2 and O-6 (*621*). Xanthation at O-3 is much slower. The product of 0.12 DS showed the presence of unsubstituted AGU[1] and their mono- and dixanthates in mole ratios of 34:4:0.1, while the corresponding ratios for the DS 0.33 product were 32:13:1. The monosubstituted component of the DS 0.33 product showed a 67:27:6 ratio for O-6:O-2:O-3 xanthate substitution while the corresponding ratio for the DS 0.12 product was 56:44:0 (*621*). Thus, the primary hydroxyl group on C-6 is most readily xanthated under the conditions of the continuous reaction (*613, 621*).

Aqueous solutions of the starch xanthates are unstable owing to both hydrolysis and oxidation. Dry xanthates usually have sufficient moisture to be unstable. Starch xanthate powder, stable for several months at 25° or indefinitely at 0°, has been made by decreasing the alkali content before spray-drying (*622*). The magnesium salt of starch xanthate has room temperature stability for several months if the xanthate is completely in the magnesium salt form. Storage at 0° increases stability. A moisture content below 2% is important for storage stability (*184, 185, 602*). This instability and the difficulty in handling, storage and transportation has been a major commercial drawback to an otherwise useful product.

$$2\ \text{starch}-O-\overset{\overset{S}{\|}}{C}-SNa \xrightarrow{[O]} \text{starch}-O-\overset{\overset{S}{\|}}{C}-S-S-\overset{\overset{S}{\|}}{C}-O-\text{starch}$$

Starch xanthates can be oxidized to cross-linked xanthides, which are stable and effective in heavy metal ion removal from water (*619, 620, 623*). Metal ions, such as zinc, copper, and iron react with starch xanthate to form insoluble metal dithiocarbonates; the preferred ion is zinc (*617, 619, 620*).

$$2\ \text{starch}-O-\overset{\overset{S}{\|}}{C}-SNa + ZnCl_2 \longrightarrow \left(\text{starch}-O-\overset{\overset{S}{\|}}{C}-S\right)_2 Zn + 2\ NaCl$$

Addition of the soluble starch xanthate (DS 0.05–0.75) to the wet end of the paper-making process with precipitation on the pulp through xanthide or metal starch dithiocarbonate formation results in increases in the dry and wet tensile, burst, and fold strength with a reduction in tear strength (*619, 620, 624*). Similarly, mixing starch xanthates with elastomer latexes, coprecipitating the two polymers by cross-linking the starch via oxidation or with zinc ions, and destabilizing the latex gives a curd which dries to a friable starch xanthide-encased rubber crumb. Milling gives coalescence of the rubber to form a continuous phase with starch xanthide reenforcement. The products are light in color (*603–607, 625*).

The starch xanthates are most effective in removing heavy metal ions from waste waters (*184, 599–602*).

3. Dialdehyde Starch

Large scale use of dialdehyde starch has not developed because of high cost. A comprehensive coverage has been given in various reviews (*63, 626–628, 647*).

Periodic acid and periodates oxidize 2,3-glycol structures specifically to aldehydes and this reaction has been used for structure determination of carbohydrates (*629–632*). The degree of oxidation is determined by the amount of periodate used and, hence, can be controlled (*628*). Even though the periodate-oxidized starch is called "dialdehyde starch" (DAS), very few free aldehyde groups are present; rather the principal structures are hydrated hemialdal and intra- and intermolecular hemiacetals (*626, 628, 631*). However, DAS reacts as an aldehyde-containing material; the latent aldehyde groups react with bisulfite ions, alcohols, amines, hydrazines, hydrazides, and other reagents that condense with aldehydes. Sodium bisulfite dialdehyde starch yields tetrahydroxybenzoquinone via glyoxal on alkaline degradation (*642*). Chlorous acid oxidizes the aldehyde groups to carboxylic acids, yielding a polycarboxylate polymer (*87, 626*).

For small scale production of DAS in a range of oxidation levels, a chemical process has been proposed in which the spent oxidant is oxidized with alkaline hypochlorite to sodium paraperiodate. This salt can be recovered by filtration for recycling (*635*). A commercial process renews the periodate ion by electrolytic oxidation (*627, 628, 633, 634*).

The proposed uses of DAS are based on its activity as a reactive polymeric polyaldehyde that acts as a cross-linking agent for substrates containing hydroxyl, amino, and imino groups. Thus, it is useful in imparting temporary and permanent wet strength to paper (*636*), in hardening gelatin, in tanning leather, and in making water-resistant adhesives (*626, 627*). Reactions to prepare cationic DAS have been reported (*637*).

IX. References

(*1*) J. Newton and G. Peckham, *in* "Chemistry and Industry of Starch," R. W. Kerr, ed., Academic Press, New York, 2nd Ed., 1950, pp. 325–343.

(*2*) E. F. Degering, *in* "Starch and Its Derivatives," J. A. Radley, ed., Wiley, New York, 3rd Ed., 1954, Vol. I, pp. 343–364. See also the 4th Ed., Chapman and Hall, London, 1968, pp. 306–353.

(3) R. P. Walton, "A Comprehensive Survey of Starch Chemistry," The Chemical Catalog Co., New York, Part 2, 1928, pp. 100–103.
(4) K. E. Craig, E. F. Oltmanns, and F. P. Loppnow, *TAPPI,* **51,** No. 11, 82A (1968).
(5) G. E. Lauterbach, U.S. Patent 3,211,564 (1965); *Chem. Abstr.,* **64,** 22681 (1966). (Kimberly-Clark Corp.)
(6) D. A. Brogly, *Tappi Paper Makers Conference,* **1978,** 87–92.
(7) H. W. Maurer, U.S. Patent 3,475,215 (1969); *Chem. Abstr.,* **72,** 68463 (1970). (West Virginia Pulp and Paper Co.)
(8) F. G. Ewing, U.S. Patent 3,539,366 (1970); *Chem. Abstr.,* **74,** 65923 (1971). (Standard Brands Inc.)
(9) E. L. Speakman, U.S. Patent 3,935,187 (1976); *Chem. Abstr.,* **84,** 123768 (1976) (Standard Brands Inc.)
(10) J. A. Lotzgesell, K. B. Moser, and T. L. Hurst, U.S. Patent 3,975,206 (1976); *Chem. Abstr.,* **85,** 162290 (1976). (A. E. Staley Mfg. Co.)
(11) H. W. Durand, U.S. Patent 3,655,644 (1972); *Chem. Abstr.,* **77,** 36744 (1972). (Grain Processing Corp.)
(12) C. L. Mehltretter and C. A. Wilham, U.S. Patent 3,515,718 (1970); *Chem. Abstr.,* **73,** 36821 (1970). (U.S. Department of Agriculture.)
(13) D. E. Lucas and C. H. Fletcher, *Paper Ind.,* **40,** 810 (1959).
(14) C. H. Hullinger, *Methods Carbohydr. Chem.* **4,** 313 (1964).
(15) F. F. Farley and R. M. Hixon, *Ind. Eng. Chem.,* **34,** 677 (1942).
(16) J. Schmorak, D. Mejzler, and M. Lewin, *Staerke,* **14,** 278 (1962); *J. Polymer Sci.,* **49,** 203 (1961).
(17) J. Schmorak and M. Lewin, *J. Polymer Sci., Part A,* **1,** 2601 (1963).
(18) J. Schmorak and M. Lewin, *Bull. Res. Counc. Isr., Sect. A,* **11,** 228 (1962).
(19) J. Potze and P. Hiemstra, *Staerke,* **15,** 217 (1963).
(20) K. F. Patel, H. U. Mehta, and H. C. Srivastava, *Staerke,* **25,** 266 (1973).
(21) K. F. Patel, H. U. Mehta, and H. C. Srivastava, *J. Appl. Polym. Sci.,* **18,** 389 (1974).
(22) H. Henriksnäs and H. Bruun, *Starch/Staerke,* **30,** 233 (1978).
(23) R. L. Whistler, E. G. Linke, and S. Kazeniac, *J. Amer. Chem. Soc.,* **78,** 4704 (1956).
(24) C. H. Hullinger and R. L. Whistler, *Cereal Chem.,* **28,** 153 (1951).
(25) R. L. Whistler and R. Schweiger, *J. Amer. Chem. Soc.,* **79,** 6460 (1957).
(26) T. R. Ingle and R. L. Whistler, *Cereal Chem.,* **41,** 474 (1964).
(27) R. L. Whistler, T. W. Mittag, and T. R. Ingle, *Cereal Chem.,* **43,** 362 (1966).
(28) N. Uchino and R. L. Whistler, *Cereal Chem.,* **39,** 477 (1962).
(29) R. L. Whistler, T. W. Mittag, T. R. Ingle, and G. Buffini, *TAPPI,* **49,** 310 (1966).
(30) W. M. Doane and R. L. Whistler, *Staerke,* **16,** 177 (1964).
(31) I. Ziderman and J. Bel-Ayche, *Carbohydr. Res.,* **27,** 341 (1973).
(32) S. K. Fischer, *Staerke,* **29,** 380 (1977).
(33) M. C. Patel, B. N. Mankad, and R. D. Patel, *J. Indian Chem. Soc., Ind. News Ed.,* **18,** 81 (1955); *Chem. Abst.,* **50,** 6823 (1956).
(34) A. D. Fuller, U.S. Patent 1,942,544 (1934); *Chem. Abstr.,* **28,** 1888 (1934). (National Adhesives Corp.)
(35) G. Pollack and C. Campbell, *Abst. Papers, Amer. Chem. Soc.,* **121,** 8P (1952).
(36) R. L. Whistler and S. J. Kazeniac, *J. Org. Chem.,* **21,** 468 (1956).
(37) M. E. McKillican and C. B. Purves, *Can. J. Chem.,* **32,** 312 (1954).
(38) A. A. Eisenbraun and C. B. Purves, *Can J. Chem.,* **39,** 61 (1961).
(39) V. Prey and St. Siklossy, *Staerke,* **23,** 235 (1971).
(40) R. L. Mellies, C. L. Mehltretter, and I. A. Wolff, *Ind. Eng. Chem.,* **50,** 1311 (1958).
(41) R. L. Mellies, C. L. Mehltretter, and F. R. Senti, *J. Chem. Eng. Data,* **5,** 169 (1960).
(42) S. K. Fischer, and F. Piller, *Starch/Staerke,* **30,** 4 (1978).

(*43*) V. Prey and S. K. Fischer, *Staerke,* **27,** 192 (1975).
(*44*) R. G. Hyldon, D. E. Fink, and H. G. Aboul, U.S. Patent 3,450,692 (1969); *Chem. Abstr.,* **71,** 40551 (1969). (Keever Co.)
(*45*) S. K. Fischer and F. Piller, *Staerke,* **29,** 262 (1977).
(*46*) J. N. BeMiller, *in* "Starch: Chemistry and Technology," R. L. Whistler and E. G. Paschal, eds., Academic Press, New York, 1965, Vol. I, pp. 521–532.
(*47*) C. T. Greenwood, *Advan. Carbohydr. Chem.,* **11,** 335 (1956).
(*48*) W. M. Corbett, *in* "Recent Advances in the Chemistry of Cellulose and Starch," J. Honeyman, ed., Interscience, New York, 1959, pp. 106–133.
(*49*) T. J. Schoch and E. C. Maywald, *Anal. Chem.,* **28,** 382 (1956).
(*50*) D. M. Hall, E. Van Patten, J. L. Brown, G. R. Harmon, and G. H. Nix, *Ind. Eng. Prod. Res. Develp.,* **10,** 171 (1971).
(*51*) V. Prey and S. K. Fischer, *Staerke,* **28,** 125 (1976).
(*52*) V. Prey, S. K. Fischer, and St. Klinger, *Staerke,* **28,** 166 (1976).
(*53*) H. Bruun and H. Henriksnäs, *Staerke,* **29,** 122 (1977).
(*54*) V. Prey, S. K. Fischer, and St. Klinger, *Staerke,* **28,** 259 (1976).
(*55*) T. J. Schoch and E. C. Maywald, *in* "Starch: Chemistry and Technology," R. L. Whistler and E. F. Paschall, eds., Academic Press, New York, 1967, Vol. II, pp. 637–647.
(*56*) S. K. Fischer and F. Piller, *Staerke,* **29,** 232 (1977).
(*57*) T. J. Schoch, *TAPPI,* **35,** No. 7, 22A (1952).
(*58*) E. J. Heiser, *Paper Trade J.,* 30 (May 1, 1978).
(*59*) C. W. Cairns, *TAPPI,* **57,** No. 5, 85 (1974).
(*60*) S. Roguls and R. L. High, *Paper Trade J.,* 44 (Dec. 13, 1965).
(*61*) J. A. Radley, "Starch Production Technology," Applied Science Publishers, London, 1976, pp. 457–469.
(*62*) Reference *61,* pp. 481–542.
(*63*) Reference *61,* pp. 423–448.
(*64*) J. A. Radley, "Industrial Uses of Starch and Its Derivatives," Applied Science Publishers, London, 1976.
(*65*) A. H. Zuderveld and P. G. Stoutjesdijk, *in* Reference *64,* pp. 199–228.
(*66*) Reference *64,* pp. 149–197.
(*67*) "Starch and Starch Products in Paper Coating," TAPPI Monograph Series, No. 17, Technical Association of the Pulp and Paper Industry, New York, 1957, p. 15.
(*68*) K. J. Huber, J. F. Johnston, E. K. Nissen, and D. R. Pourie, U.S. Patent 3,255,040 (1966); *Chem. Abstr.,* **65,** 4087 (1966). (Union Starch and Refining Co., Inc.)
(*69*) H. S. DeGroot and F. G. Ewing, *Paper Trade J.,* 32 (Sept. 5, 1966).
(*70*) R. D. Harvey and L. H. Welling, *TAPPI,* **59,** No. 12, 92 (1976).
(*71*) H. C. Brill, *TAPPI,* **38,** No. 9, 522 (1955).
(*72*) C. R. Russell, *in* "Industrial Uses of Cereals," Y. Pomerantz, ed., American Association of Cereal Chemists, St. Paul, Minnesota, 1973, pp. 262–284.
(*73*) C. A. Moore, "An Economic Evaluation of Starch Use in the Textile Industry," Agr. Econ. Rept. 109, U.S. Department of Agriculture, Washington, D.C., 1967.
(*74*) J. A. Antinori and M. W. Rutenberg, U.S. Patent 3,208.851 (1965); *Chem. Abstr.,* **63,** 17042 (1965). (National Starch and Chemical Corp.)
(*75*) N. G. Marotta, H. Bell, and K. S. Ronai, U.S. Patent 3,482,984 (1969); *Chem. Abstr.,* **71,** 90137 (1969). (National Starch and Chemical Corp.)
(*76*) R. R. Gabel, R. M. Hamilton, and W. E. Dudecek, U.S. Patent 3,607,393 (1971); *Chem. Abstr.,* **72,** 20720 (1970). (CPC International Inc.)
(*77*) C. S. Campbell, U.S. Patent 3,655,443 (1972); *Chem. Abstr.,* **77,** 18387 (1972). (American Maize Products Co.)

(*78*) J. C. Fruin, U.S. Patent 3,767,826 (1973); *Chem. Abstr.*, **80,** 36038 (1974). (Anheuser-Busch, Inc.)

(*79*) R. W. Kerr, "Chemistry and Industry of Starch," Academic Press, New York, 2nd Ed., 1950, pp. 466–472.

(*80*) E. D. Mazzarella and E. Dalton, U.S. Patent 3,320,080 (1967); *Chem. Abstr.*, **64,** 16121 (1966). (National Starch and Chemical Corp.)

(*81*) J. E. Schoenberg and D. K. Ray-Chaudhuri, U.S. Patents, 3,944,428 (1976); 4,009,311 (1977); *Chem. Abstr.*, **85,** 23016 (1976). (National Starch and Chemical Corp.)

(*82*) P. R. Demko, F. J. Washabaugh, and R. H. Williams, U.S. Patent 4,018,959 (1977); *Chem. Abstr.*, **87,** 7105 (1977). (National Starch and Chemical Corp.)

(*83*) A. Sadle and T. J. Pratt, U.S. Patent 4,157,318 (1979); *Chem. Abstr.*, **91,** 93337 (1979). (International Paper Co.)

(*84*) R. M. Maxwell, U.S. Patent 2,148,951 (1939); *Chem. Abstr.*, **33,** 4453 (1939). (E. I. Dupont de Nemours Co.)

(*85*) G. E. Felton and H. H. Schopmeyer, U.S. Patent 2,328,537 (1943); *Chem. Abstr.*, **38,** 889 (1944). (American Maize Products Co.)

(*86*) M. Konigsberg, U.S. Patent 2,500,950 (1950); *Chem. Abstr.*, **44,** 6666 (1950). (National Starch Products, Inc.)

(*87*) B. T. Hofreiter, C. L. Mehltretter, J. Bennie, and G. E. Hammerstrand, U.S. Patent 2,929,811 (1960); *Chem. Abstr.*, **54,** 13704 (1960). (U.S. Department of Agriculture.)

(*88*) O. B. Wurzburg, U.S. Patent 2,935,510 (1960); *Chem. Abstr.*, **54,** 16886 (1960). (National Starch and Chemical Corp.)

(*89*) F. R. Senti, R. L. Mellies, and C. L. Mehltretter, U.S. Patent 2,989,521 (1961); *Chem. Abstr.*, **59,** 3000 (1963). (U.S. Department of Agriculture.)

(*90*) C. G. Caldwell, T. A. White, W. L. George, and J. J. Eberle, U.S. Patent 2,626,257 (1953); *Chem. Abstr.*, **47,** 3528 (1953). (National Starch Products, Inc. and Johnson & Johnson.)

(*91*) R. W. Kerr, and F. C. Cleveland, Jr., U.S. Patent 2,801,242 (1957); *Chem. Abstr.*, **51,** 18666 (1957). (Corn Products Co.)

(*92*) R. W. Kerr and F. C. Cleveland, Jr., U.S. Patent 2,938,901 (1960); *Chem. Abstr.*, **54,** 16886 (1960). (Corn Products Co.)

(*93*) R. W. Kerr and F. C. Cleveland, Jr., U.S. Patent 2,852,393 (1958); *Chem. Abstr.*, **53,** 1797 (1959). (Corn Products Co.)

(*94*) S. M. Chang, C. J. Wang, and C. Y. Lii, *Proc. Natl. Sci. Counc. Repub. China*, **3,** 449 (1979); *Chem. Abstr.*, **92,** 148880 (1980).

(*95*) C. G. Caldwell, U.S. Patent 2,461,139 (1949); *Chem. Abstr.*, **43,** 3222 (1949). (National Starch and Chemical Corp.)

(*96*) D. L. Schoene and V. S. Chambers, U.S. Patent 2,524,400 (1950); *Chem. Abstr.*, **45,** 4474 (1951). (U.S. Rubber Co.)

(*97*) J. D. Commerford and I. Ehrenthal, U.S. Patent 2,977,356 (1961); *Chem. Abstr.*, **55,** 19289 (1961). (Anheuser-Busch, Inc.)

(*98*) T. S. W. Gerwitz, U.S. Patent 2,805,220 (1957); *Chem. Abstr.*, **52,** 767 (1958). (Anheuser-Busch, Inc.)

(*99*) E. L. Wimmer, U.S. Patent 2,910,467 (1959); *Chem. Abstr.*, **54,** 1906 (1960). (C. A. Krause Milling Co.)

(*100*) D. Trimnell, C. P. Patel, and J. F. Johnston, U.S. Patent 3,086,971 (1963); *Chem. Abstr.*, **59,** 3000 (1963). (Union Starch and Refining Co.)

(*101*) I. A. Wolf, P. R. Watson, and C. E. Rist, *J. Amer. Chem. Soc.*, **76,** 757 (1954).

(*102*) D. Trimnell, C. P. Patel, and J. F. Johnson, U.S. Patent 3,035,045 (1962); *Chem. Abstr.*, **57,** 6194 (1962). (Union Starch and Refining Co.)

(*103*) E. A. Sowell, J. F. Voight, and R. J. Horst, U.S. Patent 3,001,985 (1961); *Chem. Abstr.*, **56,** 3712 (1962). (Anheuser-Busch, Inc.)
(*104*) W. Jarowenko and M. W. Rutenberg, U.S. Patent 3,376,287 (1968); *Chem. Abstr.*, **68,** 115682 (1968). (National Starch and Chemical Corp.)
(*105*) W. Jarowenko, U.S. Patent 3,553,195 (1971); *Chem. Abstr.*, **74,** 65922 (1971). (National Starch and Chemical Corp.)
(*106*) M. M. Tessler and M. W. Rutenberg, U.S. Patent 3,699,095 (1970); *Chem. Abstr.*, **77,** 163247 (1972). (National Starch and Chemical Corp.)
(*107*) M. M. Tessler, U.S. Patent 3,720,663 (1973), reissued as Re 28,809 (1973); *Chem. Abstr.*, **78,** 99447 (1973). (National Starch and Chemical Corp.)
(*108*) M. Tessler, U.S. Patent 4,020,272 (1977); *Chem. Abstr.*, **86,** 191620 (1977). (National Starch and Chemical Corp.)
(*109*) M. M. Tessler and M. W. Rutenberg, U.S. Patent 3,728,332 (1973); *Chem. Abstr.*, **78,** 161154 (1973). (National Starch and Chemical Corp.)
(*110*) M. M. Tessler, U.S. Patent 4,098,997 (1978); *Chem. Abstr.*, **89,** 217103 (1978). (National Starch and Chemical Corp.)
(*111*) B. W. Rowland and J. V. Bauer, U.S. Patent 2,113,034 (1938); *Chem. Abstr.*, **32,** 3963 (1938). (Stein Hall Mfg. Co.)
(*112*) J. E. Fenn, U.S. Patent 2,469,957 (1949); *Chem. Abstr.*, **43,** 5910 (1949).
(*113*) G. G. Pierson, U.S. Patent 2,417,611 (1947); *Chem. Abstr.*, **41,** 4326 (1947). (Perkins Glue Co.)
(*114*) R. W. Kerr and N. F. Schink, U.S. Patent 2,438,855 (1948); *Chem. Abstr.*, **42,** 4380 (1948). (Corn Products Refining Co.)
(*115*) E. I. Speakman, U.S. Patents 3,549,618 (1970) and 3,705,046 (1972); *Chem. Abstr.*, **74,** 100819 (1971); **78,** 113035 (1973). (Standard Brands, Inc.)
(*116*) C. E. Smith and J. V. Tuschoff, U.S. Patent 3,069,410 (1962); *Chem. Abstr.*, **58,** 5874 (1963). (A. E. Staley Mfg. Co.)
(*117*) J. F. Walker and S. E. Kokowicz, U.S. Patent 2,548,455 (1951); *Chem. Abstr.*, **45,** 5927 (1951). (E. I. duPont de Nemours and Co.)
(*118*) L. O. Gill and J. W. McDonald, U.S. Patent 2,407,071 (1946); *Chem. Abstr.*, **41,** 612 (1947). (A. E. Staley Mfg. Co.)
(*119*) T. Porowski, U.S. Patent 2,838,465 (1958); *Chem. Abstr.*, **52,** 15106 (1958). (A. E. Staley Mfg. Co.)
(*120*) C. Patel and R. E. Pyle, U.S. Patent 3,152,925 (1964); *Chem. Abstr.*, **62,** 1835 (1965). (Union Starch and Refining Co.)
(*121*) H. W. Wetzstein and P. Lyon, U.S. Patent 2,754,232 (1956); *Chem. Abstr.*, **50,** 13490 (1956). (American Maize Products Co.)
(*122*) J. N. BeMiller and G. W. Pratt, *Cereal Chem.*, **58,** 517 (1981).
(*123*) R. B. Evans, L. H. Kruger, and C. D. Szymanski, U.S. Patent 3,463,668 (1969). (National Starch and Chemical Corp.)
(*124*) R. W. Kerr and F. C. Cleveland, Jr., U.S. Patent 2,884,413 (1959); *Chem. Abstr.*, **53,** 16569 (1959). (Corn Products Co.)
(*125*) T. J. Schoch, *Staerke*, **11,** 156 (1959).
(*126*) F. E. Kite, T. J. Schoch, and H. W. Leach, *Baker's Dig.*, **31,** No 4, 42 (1957).
(*127*) H. W. Leach, L. D. McCowen, and T. J. Schoch, *Cereal Chem.*, **36,** 534 (1959).
(*128*) T. J. Schoch, *Methods Carbohydr. Chem.* **4,** 106 (1964).
(*129*) R. J. Smith, *Methods Carbohydr. Chem.*, **4,** 114 (1964).
(*130*) E. G. Mazurs, T. J. Schoch, and F. E. Kite, *Cereal Chem.*, **34,** 141 (1957).
(*131*) O. B. Wurzburg, *in* "Seminar Proceedings, Products of the Corn Refining Industry in Food," Corn Refiners Association, Washington, D.C., May, 1978, pp. 23–34. See also "Sym-

posium Proceedings, Products of The Wet-Milling Industry in Food," Corn Refiners Association, Washington, D.C., 1970.

(*132*) O. B. Wurzburg and C. D. Szymanski, *Agr. Food Chem.*, **18,** 997 (1970).

(*133*) F. K. Gotlieb and P. Woldendorp, *Staerke,* **19,** 263 (1967).

(*134*) H. C. Srivastava and M. M. Patel, *Staerke,* **25,** 17 (1973).

(*135*) A. A. Stonehill, U.S. Patent 3,072,537 (1963); *Chem. Abstr.*, **58,** 6655 (1963).

(*136*) C. R. Blair and J. Blumenthal, *N.Y. State J. Med.*, 1202 (May 15, 1964).

(*137*) D. Pelling and K. R. Butterworth, *J. Pharm. Pharmacol.*, **32 (Suppl.),** 81P (1980); *Chem. Abstr.*, **94,** 197497 (1981).

(*138*) R. Murakami, *Kenkyu Hokoku-Kumamoto Kogyo Daigaku*, **5,** 127 (1980); *Chem. Abstr.*, **93,** 74285 (1980).

(*139*) M. W. Rutenberg, W. Jarowenko, and M. M. Tessler, U.S. Patent 4,048,435 (1977); *Chem. Abstr.*, **87,** 169527 (1977). (National Starch and Chemical Corp.)

(*140*) M. W. Rutenberg, M. M. Tessler, and L. Kruger, U.S. Patents 3,832,342 (1974); 3,899,602 (1975); *Chem. Abstr.*, **82,** 3026 (1975). (National Starch and Chemical Corp.)

(*141*) R. W. Rubens, U.S. Patents 4,183,969 (1980); 4,219,646 (1980); *Chem. Abstr.*, **92,** 145268 (1980). (National Starch and Chemical Corp.)

(*142*) H. J. Roberts, *in* "Starch, Chemistry and Technology," R. L. Whistler and E. F. Paschall, eds., Academic Press, New York, Vol. I, 1965, p. 482.

(*143*) G. Hollinger, L. Kuniak, and R. H. Marchessault, *Biopolymers,* **13,** 879 (1974).

(*144*) G. E. Hammerstrand, B. T. Hofreiter, and C. L. Mehltretter, *Cereal Chem.*, **37,** 519 (1969).

(*145*) B. M. Gough, *Staerke,* **19,** 240 (1967).

(*146*) B. M. Gough and J. W. Pybus, *Staerke,* **20,** 108 (1968).

(*147*) L. Kuniak and R. H. Marchessault, *Staerke,* **24,** 110 (1972).

(*148*) L. Kuniak and R. H. Marchessault, Can. Patent 949,965 (1974); *Chem. Abstr.*, **81,** 123350 (1974).

(*149*) P. Luby and L. Kuniak, *Makromol. Chem.*, **180,** 2213 (1979).

(*150*) G. Hollinger and R. H. Marchessault, *Biopolymers,* **14,** 265 (1975).

(*151*) G. A. Mitchell, M. J. Vanderbist and F. F. Meert, *J. Assoc. Off. Anal. Chem.*, **65,** 238 (1982).

(*152*) W. A. Mitchell, *Food Technol.*, 34–36, 38, 40, 42, 79 (March, 1972).

(*153*) H. Koch, H. D. Bommer and J. Koppers, *Starch/Staerke,* **34,** 16–21 (1982).

(*154*) R. Collison and J. O. Ogundiwin, *Staerke,* **24,** 258 (1972).

(*155*) F. del Valle, J. V. Tuschoff, and C. E. Streaty, U.S. Patent 4,000,128 (1976); *Chem. Abstr.*, **86** 70412 (1976). (A. E. Staley Mfg. Co.)

(*156*) C. D. Szymanski, M. M. Tessler, and H. Bell, U.S. Patents 3,804,828 (1974) and 3,857,976 (1974); *Chem. Abstr.*, **81** 118820 (1974); **82,** 138020 (1975). (National Starch and Chemical Corp.)

(*157*) M. M. Tessler, W. Jarowenko, and R. A. Amitrano, U.S. Patent 3,904,601 (1975); *Chem. Abstr.*, **81,** 154960 (1974). (National Starch and Chemical Corp.)

(*158*) J. V. Tuschoff, G. L. Kessinger, and C. E. Hanson, U.S. Patent 3,422,088 (1969); *Chem. Abstr.*, **67,** 63097 (1967). (A. E. Staley Mfg. Co.)

(*159*) M. M. Tessler and W. Jarowenko, U.S. Patents 3,969,340 (1976) and 3,970,767 (1976); *Chem. Abstr.*, **85,** 122115 and 122140 (1976). (National Starch and Chemical Corp.)

(*160*) J. W. Robinson, G. N. Bookwalter, and J. V. Tuschhoff, U.S. Patents 3,437,493 (1969) and 3,719,661 (1973); *Chem Abstr.*, **71,** 11930 (1969) and **78,** 122967 (1973); (A. E. Staley Mfg. Co.)

(*161*) J. R. Caracci, F. J. Germino, and T. D. Yoshida, U.S. Patent 3,751,410 (1973); *Chem. Abstr.*, **79,** 103863 (1973). (CPC International)

(*162*) V. J. Kelly and W. G. Fry, U.S. Patent 3,685,999 (1972). (Gerber Products Co.)

(*163*) W. A. Mitchell, H. D. Stahl, and R. A. Williams, Can. Patent 991,908 (1976); *Chem. Abstr.*, **85,** 122141 (1976). (General Foods Corp.)

(*164*) R. Van Schanefelt, J. E. Eastman, and M. F. Campbell, Ger. Patent 2,541,513 (1976); *Chem. Abstr.*, **84,** 166584 (1976). (A. E. Staley Mfg. Co.)

(*165*) M. Jonason, Brit. Patent 1,409,769 (1975); *Chem. Abstr.*, **84,** 15908 (1976). (Slimcea, Ltd.)

(*166*) J. J. Ducharme, H. S. Black, Jr., and S. J. Leith, U.S. Patent 3,052,545 (1962). (National Starch and Chemical Corp.)

(*167*) N. G. Marotta, P. C. Trubiano, and K. S. Ronai, U.S. Patent 3,443,964 (1969).

(*168*) N. G. Marotta and P. C. Trubiano, U.S. Patent 3,579,341 (1971); *Chem. Abstr.*, **75,** 34226 (1971). (National Starch and Chemical Corp.)

(*169*) J. W. Evans and C. S. McWilliams, U.S. Patent 3,346,387 (1967); *Chem. Abstr.*, **68,** 38376 (1968). (American Maize-Products Co.)

(*170*) C. W. Chiu and M. W. Rutenberg, U.S. Patents 4,207,355 (1980) and 4,228,199 (1980); 4,229,489 (1980); *Chem. Abstr.*, **94,** 45859 and 29079 (1981). (National Starch and Chemical Corp.)

(*171*) O. B. Wurzburg and C. D. Szymanski, U.S. Patent 3,525,672 (1970); *Chem. Abstr.*, **72,** 68454 (1970). (National Starch and Chemical Corp.)

(*172*) J. Pomot and J. P. Chalaye, Ger. Patent 2,405,216 (1978); *Chem. Abstr.*, **82,** 21727 (1975). (Oreal S.A.)

(*173*) D. Chaudhuri and M. R. Stebles, Ger. Patent 2,837,088 (1979); *Chem. Abstr.*, **91,** 27184 (1979). (Unilever N.V.)

(*174*) W. Kelly and A. A. McKinnon, Brit. Patent 1,550,614 (1979); *Chem. Abstr.*, **92,** 153189 (1980). (Unilever, Ltd.)

(*175*) A. Holst, M. Kostrzewa, and G. Buchberger, U.S. Patent 4,117,222 (1978); *Chem. Abstr.*, **88,** 122902 (1978). (Hoechst, A.G.)

(*176*) Societe des Accumulateurs Fixes et de Traction, Japanese Patent 78 132,739 (1978); *Chem. Abstr.*, **90,** 94413 (1979).

(*177*) Y. Uetani, S. Matsushima, and Y. Taniguchi, Japanese Patents 78/129,824, 78/129,825, and 78/129,826 (1978); *Chem. Abstr.*, **90,** 78442, 78443, and 74413 (1979). (Hitachi Maxell, Ltd.)

(*178*) K. Iwamaru and K. Tsuchiyama, Japanese Patents 79/45,745, 79/45,746, 79/45,747, 79/45,748, 79/45,749, and 79/45,750 (1979); *Chem. Abstr.*, **91,** 41976, 23986, 23987, 41977, and 23992 (1979). (Hitachi Maxell, Ltd.)

(*179*) J. C. Hedden, U.S. Patent 3,887,389 (1975); *Chem. Abstr.*, **83,** 194901 (1975). (PPG Industries, Inc.)

(*180*) R. R. Graham, U.S. Patent 4,002,445 (1970); *Chem. Abstr.*, **86,** 74356 (1977). (PPG Industries, Inc.)

(*181*) Y. Mukai and M. Hosaka, Jap. Patent 74/87,893 (1974); *Chem. Abstr.*, **82,** 87603 (1975). (Nitto Boseki Co., Ltd.)

(*182*) E. T. Hjermstad, U.S. Patent 3,438,913 (1969); *Chem. Abstr.*, **71,** 4729 (1969). (Penick and Ford, Ltd.)

(*183*) J. F. Fox, T. W. Roylance, and A. C. Mair, U.S. Patent 4,053,379 (1977); *Chem. Abstr.*, **85,** 182439 (1976). (Arbrook, Inc.)

(*184*) R. E. Wing, W. M. Doane, and C. R. Russell, *J. Appl. Polym. Sci.*, **19,** 847 (1975).

(*185*) R. E. Wing, B. K. Jasberg, and L. L. Navickis, *Starch/Staerke,* **30,** 163 (1978).

(*186*) R. E. Wing, U.S. Patents 3,979,286 (1976), 4,051,316 (1977); *Chem. Abstr.*, **86,** 47109 (1977). (U.S. Department of Agriculture.)

(*187*) L. Kuniak, Czech. Patent 160,814 (1975); *Chem. Abstr.*, **84,** 166585 (1976).

(*188*) S. Rogols and J. W. Solter, U.S. Patent 4,139,505 (1979); *Chem. Abstr.*, **90,** 139328 (1979). (General Mills Chemicals, Inc.)

(*189*) S. Saito, Jap. Patent 73/95,471 (1973); *Chem. Abstr.*, **80,** 97653 (1974). (Nikka Co., Ltd.)
(*190*) W. E. Rayford, and R. E. Wing, *Starch/Staerke,* **31,** 361 (1979).
(*191*) W. E. Rayford and R. E. Wing, U.S. Patent 4,237,271 (1980); *Chem. Abstr.*, **94,** 67600 (1981). (U.S. Department of Agriculture.)
(*192*) R. E. Wing and W. M. Doane, U.S. Patent 3,795,671 (1974); *Chem. Abstr.*, **81,** 27522 (1974). (U.S. Department of Agriculture.)
(*193*) E. F. Degering, *in* "Chemistry and Industry of Starch," R. W. Kerr, ed., Academic Press, New York, 2nd Ed., 1950, pp. 259–323.
(*194*) R. L. Whistler, *Adv. Carbohydr. Chem.*, **1,** 279 (1945).
(*195*) E. F. Degering, *in* "Starch and Its Derivatives," J. A. Radley, ed., Wiley, New York, 3rd Ed., 1954, Vol. 1.
(*196*) Reference *195,* pp. 298–325.
(*197*) Reference *195,* pp. 326–331.
(*198*) Reference *195,* 4th Ed., Chapman and Hall, London, 1968, pp. 354–419.
(*199*) J. Seiberlich, *Rayon Textile Monthly,* **22,** 605, 686 (1941).
(*200*) J. W. Mullen, II, and E. Pacsu, *Ind. Eng. Chem.*, **34,** 1209 (1942).
(*201*) J. W. Mullen and E. Pacsu, U.S. Patent 2,372,337 (1945); *Chem. Abstr.*, **39,** 4250 (1945). (Research Corp.)
(*202*) C. A. Burkhard and E. F. Degering, *Proc. Indiana Acad. Sci.*, **51,** 173 (1941).
(*203*) J. Traquair, *J. Soc. Chem. Ind.*, **28,** 288 (1909).
(*204*) H. T. Clark and H. B. Gillespie, *J. Amer. Chem. Soc.*, **54,** 2083 (1932).
(*205*) C. F. Cross, E. J. Bevan, and J. Traquair, *Chem. Ztg.*, **29,** 527 (1905).
(*206*) I. A. Wolff, D. W. Olds, and G. E. Hilbert, *J. Amer. Chem. Soc.*, **73,** 346 (1951).
(*207*) O. B. Wurzburg, *Methods Carbohydr. Chem.*, **4,** 286 (1964).
(*208*) J. W. Mullen, II, and E. Pacsu, *Ind. Eng. Chem.*, **35,** 381 (1943).
(*209*) H. P. Panzer, *Abstr. Papers, Amer. Chem. Soc.*, **144,** 16C (1963).
(*210*) E. F. Paschall, U.S. Patent 2,914,526 (1959); *Chem. Abstr.*, **55,** 25306 (1961). (Corn Products Co.)
(*211*) C. A. Burkhard and E. F. Degering, *Rayon Textile Monthly,* **23,** 340 (1942).
(*212*) R. H. Treadway, U.S. Patent 2,399,455 (1946); *Chem. Abstr.*, **40,** 4238 (1946). (U.S. Department of Agriculture.)
(*213*) G. V. Caesar, U.S. Patent 2,365,173 (1944); *Chem. Abstr.*, **39,** 3956 (1945). (Stein Hall and Co.)
(*214*) J. T. Lemmerling, U.S. Patent 3,281,411 (1966); *Chem. Abstr.*, **66,** 12115 (1967). (Gevaert Photo-Producten, N.V.)
(*215*) G. K. Hughes, A. K. MacBeth, and F. L. Winzor, *J. Chem. Soc.*, 2026 (1932).
(*216*) R. W. Kerr, O. R. Trubell, and G. M. Severson, *Cereal Chem.*, **19,** 64 (1942).
(*217*) J. Mutgeert, P. Hiemstra, and W. C. Bus, *Staerke,* **10,** 303 (1958).
(*218*) R. Lohmer and C. E. Rist, *J. Amer. Chem. Soc.*, **72,** 4298 (1950).
(*219*) J. F. Carson and W. D. Maclay, *J. Amer. Chem. Soc.*, **68,** 1015 (1946).
(*220*) Upjohn Co., Brit. Patent 810,306 (1959); *Chem. Abstr.*, **53,** 18401 (1959).
(*221*) E. Husemann and H. Bartl, *Makromol. Chem.*, **18/19,** 342 (1956).
(*222*) Societa Nationale Industria Applicazioni Viscosa, Belg. Patent 610,875 (1962); *Chem. Abstr.*, **57,** 11440 (1942).
(*223*) M. W. Rutenberg, W. Jarowenko, and L. J. Ross, U.S. Patent 3,038,895 (1962); *Chem. Abstr.*, **57,** 10086 (1962). (National Starch and Chemical Corp.)
(*224*) A. M. Mark and C. L. Mehltretter, *Staerke,* **21,** 92 (1969).
(*225*) A. M. Mark and C. L. Mehltretter, U.S. Patent 3,553,196 (1971); *Chem. Abstr.*, **74,** 75380 (1971). (U.S. Department of Agriculture.)
(*226*) A. M. Mark and C. L. Mehltretter, *Staerke,* **22,** 108 (1970).

(227) A. M. Mark and C. L. Mehltretter, U.S. Patent 3,549,619 (1970); *Chem. Abstr.*, **74,** 100814 (1971). (U.S. Department of Agriculture.)
(228) C. D. Bauer, U.S. Patent 3,839,320 (1974); *Chem. Abstr.*, **82,** 172931 (1975). (Anheuser-Busch Inc.)
(229) J. H. Katcher and J. A. Ackilli, U.S. Patent 4,238,604 (1980); *Chem. Abstr.*, **94,** 123497 (1981). (General Foods Corp.)
(230) L. C. Martin and R. D. Harvey, U.S. Patent 3,557,091 (1971); *Chem. Abstr.*, **74,** 113549 (1971). (Penick and Ford, Ltd.)
(231) L. Mezynski, *Przem. Chem.*, **51** 289 (1972); *Chem. Abstr.*, **77,** 63755 (1972).
(232) S. Zelenka, *Staerke*, **26,** 81 (1974).
(233) C. E. Smith and J. V. Tuschhoff, U.S. Patent 2,928,828 (1960); *Chem. Abstr.*, **54,** 13703 (1960). (A. E. Staley Mfg. Co.)
(234) J. V. Tuschhoff, U.S. Patent 3,022,289 (1962); *Chem. Abstr.*, **56,** 14520 (1962). (A. E. Staley Mfg. Co.)
(235) C. W. Smith and J. V. Tuschhoff, U.S. Patent 3,081,296 (1963); *Chem. Abstr.*, **58,** 8128 (1963). (A. E. Staley Mfg. Co.)
(236) J. V. Tuschhoff and C. E. Smith, U.S. Patent 3,238,193 (1966); *Chem. Abstr.*, **58,** 8128 (1963). (A. E. Staley Mfg. Co.)
(237) R. B. Evans and W. G. Kunze, U.S. Patent 3,318,868 (1967); *Chem. Abstr.*, **67,** 55327 (1967). (National Starch and Chemical Corp.)
(238) J. E. Eastman, U.S. Patents 3,959,514 (1976) and 4,038,482 (1977); *Chem. Abstr.*, **85,** 45179 (1976) and **87,** 119625 (1977). (A. E. Staley Mfg. Co.)
(239) W. Jetten, E. J. Stamhuis, and G. E. H. Joosten, *Starch/Staerke*, **32,** 364 (1980).
(240) E. B. Middleton, U.S. Patent 1,685,220 (1928); *Chem. Abstr.*, **22,** 4536 (1928). (E. I. duPont de Nemours & Co.)
(241) E. A. Talley and L. T. Smith, *J. Org. Chem.*, **10,** 101 (1945).
(242) M. M. Tessler and M. W. Rutenberg, U.S. Patent 3,720,662 (1973); *Chem. Abstr.*, **79,** 20608 (1973). (National Starch and Chemical Corp.)
(243) M. M. Tessler, U.S. Patent 3,928,321 (1975); *Chem. Abstr.*, **84,** 76098 (1976). (National Starch and Chemical Corp.)
(244) M. G. Groen, U.S. 2,412,213 (1946); *Chem. Abstr.*, **41,** 2265 (1947). (Hoogezand, Netherlands.)
(245) A. Schmidt, G. Balle, and A. Lange, U.S. Patent 1,928,269 (1933); *Chem. Abstr.*, **27,** 5976 (1933). (I. G. Farbenindustrie.)
(246) R. L. Whistler, A. Jeanes, and G. E. Hilbert, *Abstr. Papers, Amer. Chem. Soc.*, **104,** 3R (1942). See also reference *193*, p. 270.
(247) A. L. Potter and W. Z. Hassid, *J. Amer. Chem. Soc.*, **70,** 3774 (1948).
(248) V. Prey and A. Aszalos, *Montsh. Chem.*, **91,** 729 (1960); *Chem. Abstr.*, **55,** 12306 (1961).
(249) A. Aszalos and V. Prey, *Staerke*, **14,** 50 (1962).
(250) F. Grundschober and V. Prey, *Staerke*, **15,** 225 (1963).
(251) K. Ramaszeder, *Staerke*, **23,** 176 (1971).
(252) I. Koubek, *Staerke*, **26,** 81 (1974).
(253) R. W. Kerr and W. J. Katzbeck, U.S. Patent 3,061,604 (1962); *Chem. Abstr.*, **58,** 3593 (1963). (Corn Products Co.)
(254) Code of Federal Regulations, Title 21, Chapter I, Part 172, Food Additives Permitted in Food For Human Consumption, Section 172.892, Food Starch-Modified, U.S. Govt. Printing Office, Washington, D.C., 1981.
(255) A. Harsveldt, *Chem. Ind. (London)*, 2062 (1961).
(256) A. P. Doering, U.S. Patent 3,481,771 (1969); *Chem. Abstr.*, **72,** 46966 (1970). (National Starch and Chemical Corp.)

(*257*) M. L. Cushing, *in* "Pulp and Paper Chemistry and Technology," J. P. Casey, ed., Wiley, New York, 3rd Ed., 1981, pp. 1667–1709.
(*258*) E. M. Bovier and J. A. Carter, U.S. Patent 4,231,803 (1980); *Chem. Abstr.*, **94,** 17392 (1981). (Anheuser-Busch, Inc.)
(*259*) F. Verbanac, U.S. Patent 4,052,226 (1977); *Chem. Abstr.*, **88,** 36131 (1978). (A. E. Staley Mfg. Co.)
(*260*) J. A. Radley, *Paint Manuf.*, **17,** 83 (1947).
(*261*) C. A. Burkhard and E. F. Degering, *Rayon Text. Monthly*, **23,** 416 (1942).
(*262*) R. L. Whistler and G. E. Hilbert, *Ind. Eng. Chem.*, **36,** 796 (1944).
(*263*) I. A. Wolff, D. W. Olds, and G. E. Hilbert, *Ind. Eng. Chem.*, **43,** 911 (1951).
(*264*) R. W. Whistler and G. N. Richards, *Ind. Eng. Chem.*, **58,** 1551 (1958).
(*265*) I. A. Wolff, *Ind. Eng. Chem.*, **58,** 1552 (1958).
(*266*) A. T. Gros and R. O. Feuge, *J. Amer. Oil Chem. Soc.*, **39,** 19 (1962).
(*267*) A. Jeanes and R. W. Johns, *J. Amer. Chem. Soc.*, **74,** 6116 (1952).
(*268*) C. A. Burkhard and E. F. Degering, *Rayon Text. Monthly*, **23,** 676 (1942).
(*269*) R. L. Whistler and N. C. Schieltz, *J. Amer. Chem. Soc.*, **65,** 1436 (1943).
(*270*) F. R. Senti and L. P. Witnauer, *J. Amer. Chem. Soc.*, **68,** 2407 (1968); **70,** 1438 (1948).
(*271*) I. A. Wolff, D. W. Olds, and G. E. Hilbert, *Ind. Eng. Chem.*, **49,** 1247 (1957).
(*272*) L. B. Genung and R. D. Mallot, *Ind. Engl. Chem., Anal. Ed.*, **13,** 369 (1941).
(*273*) R. L. Whistler and A. Jeanes, *Ind. Eng. Chem., Anal. Ed.*, **15,** 317 (1943).
(*274*) J. R. van der Bij and W. F. Vogel, *Staerke*, **14,** 113 (1962).
(*275*) S. E. Rudolph and R. C. Glowaky, *J. Polymer Sci.*, **16,** 2129 (1978).
(*276*) C. G. Caldwell, U.S. Patent 2,825,727 (1958); *Chem. Abstr.*, **52,** 8601 (1958). (National Starch Products, Inc.)
(*277*) F. J. Germino, J. F. Stejskal, E. H. Christensen, and F. E. Kite, U.S. Patent 3,661,895 (1972); *Chem. Abstr.*, **75,** 78103 (1972). (CPC, International.)
(*278*) G. H. Klein, H. L. Arons, J. F. Stejskal, D. G. Stevens, and H. F. Zobel, Ger. Patent 2,633,048 (1977); *Chem. Abstr.*, **86,** 173394 (1977). (CPC, International.)
(*279*) C. G. Caldwell and O. B. Wurzburg, U.S. Patent 2,661,349 (1953); *Chem. Abstr.*, **48,** 1720 (1954).
(*280*) C. G. Caldwell, U.S. Patent 2,613,206 (1952); *Chem. Abstr.*, **47,** 899 (1953). (National Starch Products, Inc.)
(*281*) O. B. Wurzburg, W. Herbst, and H. M. Cole, U.S. Patent 3,091,567 (1963); *Chem. Abstr.*, **59,** 5357 (1963). (National Starch and Chemical Corp., and Firmenich, Inc.)
(*282*) C. Richards and C. D. Bauer, U.S. Patent 4,035,235 (1977); *Chem. Abstr.*, **87,** 137597 (1977). (Anheuser-Busch, Inc.)
(*283*) E. S. Mackey, K. Pechmann, and D. E. Tritten, U.S. Patent 3,870,521 (1975); *Chem. Abstr.*, **82,** 17188 (1975). (GAF Corp.)
(*284*) R. W. Best and R. J. Sherwin, Eur. Patent 14,520 (1980); *Chem. Abstr.*, **94,** 5126 (1981). (A. E. Staley Mfg. Co.)
(*285*) R. C. Glowaky, S. E. Rudolph, and G. P. Bierwagen, U.S. Patent 4,061,610 (1977); *Chem. Abstr.*, **89,** 131161 (1978). (Sherwin-Williams Co.)
(*286*) R. C. Glowaky, S. E. Rudolph, and G. P. Bierwagen, U.S. Patent 4,061,611 (1977); *Chem. Abstr.*, **88,** 171957 (1978). (Sherwin-Williams Co.)
(*287*) L. P. Kovats, U.S. Patent 3,732,207 (1973); *Chem. Abstr.*, **79,** 20611 (1973). (Anheuser-Busch, Inc.)
(*288*) L. P. Kovats, U.S. Patent 3,732,206 (1973); *Chem. Abstr.*, **79,** 20612 (1973). (Anheuser-Busch, Inc.)
(*289*) L. P. Kovats, U.S. Patent 3,730,925 (1973); *Chem. Abstr.*, **79,** 7135 (1973). (Anheuser-Busch, Inc.)

(*290*) M. C. Stubits and J. Teng, U.S. Patents 3,883,666 (1975) and 4,035,572 (1977); *Chem. Abstr.*, **83,** 95207 (1975); **87,** 137598 (1977). (Anheuser-Busch, Inc.)
(*291*) J. Teng, M. C. Stubits, R. E. Pyler, and J. M. Lucas, U.S. Patent 3,824,085 (1974); *Chem. Abstr.*, **82,** 60301 (1975). (Anheuser-Busch, Inc.)
(*292*) L. H. Elizer, U.S. Patents 3,793,310 (1974) and 3,887,752 (1975). *Chem. Abstr.*, **81,** 65509 (1971); **83,** 116828 (1975). (Hubinger Co.)
(*293*) C. L. Mehltretter, W. B. Roth, F. B. Weakley, T. A. McGuire, and C. R. Russell, *Weed Sci.*, **22,** 415 (1974); *Chem. Abstr.*, **82,** 12137 (1975).
(*294*) T. A. McGuire, R. E. Wing, and W. M. Doane, *Starch/Staerke*, **33,** 138 (1981).
(*295*) C. L. McCormick and D. K. Lichatowich, *J. Polymer Sci., Polymer Lett.*, **17,** 479 (1979).
(*296*) P. Ferruti, M. C. Tanzi, and F. Vaccaroni, *Makromol. Chem.*, **180,** 375 (1979).
(*297*) E. T. Hjermstad, *in* "Industrial Gums," R. L. Whistler and J. N. BeMiller, eds., Academic Press, New York, 1973, pp. 601–615.
(*298*) C. C. Kesler and E. T. Hjermstad, U.S. Patent 2,516,633 (1950); *Chem. Abstr.*, **44,** 11142 (1950). (Penick and Ford, Ltd.)
(*299*) C. C. Kesler and E. T. Hjermstad, Can. Patent 520,865 (1956). (Penick and Ford, Ltd.)
(*300*) K. C. Hobbs, U.S. Patent 2,801,241 (1957); *Chem. Abstr.*, **51,** 18666 (1957). (Corn Products Co.)
(*301*) K. C. Hobbs, R. W. Kerr, and F. E. Kite, U.S. Patent 2,833,759 (1958); *Chem. Abstr.*, **52,** 14205 (1958). (Corn Products Co.)
(*302*) K. C. Hobbs, U.S. Patent 2,999,090 (1961); *Chem. Abstr.*, **56,** 1658 (1962). (Corn Products Co.)
(*303*) R. A. Brobst, U.S. Patent 3,049,538 (1962); *Chem. Abstr.*, **58,** 651 (1963). (Hercules Powder Co.)
(*304*) E. E. Fischer and R. R. Estes, U.S. Patent 3,127,392 (1964); *Chem. Abstr.*, **57,** 10086 (1962). (A. E. Staley Mfg. Co.)
(*305*) J. V. Tuschhoff, U.S. Patent 3,176,007 (1965); *Chem. Abstr.*, **62,** 14918 (1965). (A. E. Staley Mfg. Co.)
(*306*) C. C. Kesler and E. T. Hjermstad, *Methods Carbohydr. Chem.*, **4,** 304 (1964).
(*307*) C. C. Kesler and E. T. Hjermstad, U.S. Patent 2,516,632 (1950); *Chem. Abstr.*, **44,** 11141 (1950). (Penick and Ford, Ltd.)
(*308*) C. C. Kesler and E. T. Hjermstad, U.S. Patent 2,516,634 (1950); *Chem. Abstr.*, **44,** 11142 (1950). (Penick and Ford, Ltd.)
(*309*) C. C. Kesler and E. T. Hjermstad, Can. Patent 520,866 (1956). (Penick and Ford, Ltd.)
(*310*) R. W. Kerr and W. A. Faucett, U.S. Patent 2,733,238 (1956); *Chem. Abstr.*, **50,** 6824 (1956). (Corn Products Refining Co.)
(*311*) R. W. Kerr, U.S. Patent 2,732,309 (1956); *Chem. Abstr.*, **50,** 6824 (1956). (Corn Products Refining Co.)
(*312*) E. L. Speakman, U.S. Patent 4,048,434 (1977); *Chem. Abstr.*, **87,** 169528 (1977). (Standard Brands, Inc.)
(*313*) T. Tsuzuki, U.S. Patent 3,378,546 (1968); *Chem. Abstr.*, **69,** 3883 (1968). (American Maize Products Co.)
(*314*) E. T. Hjermstad and J. Rajtora, U.S. Patent 3,632,803 (1972); *Chem. Abstr.*, **76,** 10152 (1972). (Penick and Ford, Inc.)
(*315*) J. V. Tuschhoff and C. E. Hanson, U.S. Patents 3,705,891 (1972); 3,725,386 (1973); *Chem. Abstr.*, **71,** 100666 (1972); **78,** 161153 (1973). (A. E. Staley Mfg. Co.)
(*316*) W. Jarowenko, U.S. Patent 4,112,222 (1978); *Chem. Abstr.*, **90,** 89091 (1979). (National Starch and Chemical Corp.)
(*317*) C. G. Caldwell and I. Martin, U.S. Patent 2,802,000 (1957); *Chem. Abstr.*, **51,** 3168 (1957). (National Starch Products, Inc.)

(*318*) M. A. Staerkle and E. Meier, U.S. Patent 2,698,936 (1955); *Chem. Abstr.*, **49,** 4314 (1955).
(*319*) L. T. Monson and W. J. Dickson, U.S. Patent 2,854,449 (1958); *Chem. Abstr.*, **53,** 12209 (1959).
(*320*) J. C. Rankin, C. L. Mehltretter, and F. R. Senti, *Cereal Chem.*, **36,** 215 (1959).
(*321*) J. C. Rankin, J. G. Rall, C. R. Russell, and C. E. Rist, *Cereal Chem.*, **41,** 111 (1964).
(*322*) K. J. Gardenier and W. Heimbuerger, Ger. Patent 2,900,073 (1980); *Chem. Abstr.*, **94,** 17391 (1981). (Henkel K-G.a.A.)
(*323*) E. T. Hjermstad, U.S. Patent 3,577,407 (1971); *Chem. Abstr.*, **75,** 65461 (1971). (Penick and Ford, Ltd.)
(*324*) E. T. Hjermstad, U.S. Patent 3,706,731 (1972); *Chem. Abstr.*, **78,** 73904 (1972). (Penick and Ford, Ltd.)
(*325*) L. O. Gill and J. A. Wagoner, U.S. Patent 3,014,901 (1961); *Chem. Abstr.*, **56,** 7564 (1962). (A. E. Staley Mfg. Co.)
(*326*) C. C. Kesler and E. T. Hjermstad, U.S. Patent 2,845,417 (1958); *Chem. Abstr.*, **53,** 2657 (1959). (Penick and Ford, Ltd.)
(*327*) E. T. Hjermstad and L. C. Martin, U.S. Patent 3,135,739 (1964); *Chem. Abstr.*, **61,** 4580 (1964). (Penick and Ford, Ltd.)
(*328*) A. E. Broderick, U.S. Patent 2,682,535 (1954); *Chem. Abstr.*, **48,** 11100 (1954). (Union Carbide and Carbon Corp.)
(*329*) D. B. Benedict and A. E. Broderick, U.S. Patent 2,744,894 (1956); *Chem. Abstr.*, **50,** 16865 (1956). (Union Carbide and Carbon Corp.)
(*330*) W. Jarowenko, U.S. Patent 2,996,498 (1961); *Chem. Abstr.*, **56,** 1658 (1962). (National Starch and Chemical Corp.)
(*331*) C. T. Greenwood, D. D. Muir, and H. W. Whitcher, *Staerke*, **27,** 109 (1975).
(*332*) C. T. Greenwood, U.S. Patent 4,016,354 (1977); *Chem. Abstr.*, **86,** 191619 (1977). (U.S. Navy.)
(*333*) E. D. Klug, U.S. Patent 3,033,853 (1962); *Chem. Abstr.*, **57,** 2484 (1962). (Hercules Powder Co.)
(*334*) J. Lolkema, U.S. Patent Reissue 23,443 (1951); *Chem. Abstr.*, **46,** 2830 (1952). (U.S. Attorney General.)
(*335*) R. W. Kerr, U.S. Patent 2,903,391 (1959); *Chem. Abstr.*, **54,** 1850 (1960). (Corn Products Co.)
(*336*) N.V.W. A. Scholten's Chemische Fabrieken, Brit. Patent 816,049 (1959); *Chem. Abstr.*, **53,** 20868 (1959).
(*337*) G. Moes, Ger. Patent 1,117,510 (1960); *Chem. Abstr.*, **56,** 8984 (1962). (W. A. Scholten's Chemische Fabrieken.)
(*338*) L. Balassa, U.S. Patent 2,588,463 (1952); *Chem. Abstr.*, **46,** 4830 (1952).
(*339*) H. J. Roberts and J. T. Saatkamp, U.S. Patent 3,313,803 (1967); *Chem. Abstr.*, **67,** 65699 (1967). (Corn Products Co.)
(*340*) G. Ezra and A. Zilkha, *Eur. Polym. J.*, **6,** 1313 (1970).
(*341*) H. G. Merkus, J. W. Mourits, L. Galan, and W. A. deJong, *Staerke*, **29,** 406 (1977).
(*342*) C. E. Lott and K. M. Brobst, *Anal Chem.*, **38,** 1767 (1966).
(*343*) H. C. Srivastava and K. V. Ramalingam, *Staerke*, **19,** 295 (1967).
(*344*) H. C. Srivastava, K. V. Ramalingam, and N. M. Doshi, *Staerke*, **21,** 181 (1969).
(*345*) G. N. Bollenback, R. S. Golick, and F. W. Parrish, *Cereal Chem.*, **46,** 304 (1969).
(*346*) T. Ozaki, M. Tada, and T.Irikura, *Yakugaku Zasshi*, **92,** 1500 (1972); *Chem. Abstr.*, **78,** 72481 (1973).
(*347*) M. Yoshida, T. Yamashita, J. Matsuo, and T. Kishikawa, *Staerke*, **25,** 373 (1973).
(*348*) L. E. Gminder and D. Wagner, *Text.-Prax.*, **23,** 392, 479 (1968).
(*349*) D. C. Leegwater, J. W. Marsman, and A. Mackor, *Staerke*, **25,** 142 (1973).

(*350*) A. W. Belder and B. Norman, *Carbohydr. Res.*, **10,** 391 (1969).
(*351*) W. Banks, C. T. Greenwood, and D. D. Muir, *Br. J. Pharmacol.*, **47,** 172 (1973).
(*352*) L. F. Hood and C. Mercier, *Carbohyd. Res.*, **61,** 53 (1978).
(*353*) S. J. El-Hinnawy, H. M. El-Saied, A. Fahmy, A. E. El-Shirbeeny, and K. M. El-Sahy, *Starch/Staerke,* **34,** 92 (1982).
(*354*) A. Zilkha, M. Tahan, and G. Ezra, U.S. Patent 3,414,530 (1968); *Chem. Abstr.*, **69,** 53051 (1968). (U.S. Department of Agriculture.)
(*355*) G. A. Hull and T. J. Schoch, *TAPPI,* **42,** 438 (1959).
(*356*) W. B. Roth and C. L. Mehltretter, *Food Technol.*, **21,** 72 (1967).
(*357*) D. C. Leegwater and J. B. Luten, *Staerke,* **23,** 430 (1971).
(*358*) D. C. Leegwater, *Staerke,* **24,** 11 (1973).
(*359*) M. Wooton and M. A. Chaudry, *Starch/Staerke,* **31,** 224 (1979).
(*360*) M. Wooton and M. A. Chaudry, *Starch/Staerke,* **33,** 135, 168, 200 (1981).
(*361*) B. T. Hofreiter, *in* "Pulp and Paper Chemistry and Chemical Technology," J. R. Casey, ed., Wiley, New York, 3rd Ed., 1981, pp. 1475–1514.
(*362*) C. T. Beals, *in* "Dry Strength Additives," W. F. Reynolds, ed., TAPPI Press, Atlanta, GA, 1980, pp. 33–65.
(*363*) W. C. Black, *TAPPI,* **39,** 1991A (1956).
(*364*) E. D. Klug, U.S. Patent 3,117,021 (1964); *Chem. Abstr.*, **60,** 8234 (1964).
(*365*) R. E. Weber, U.S. Patent 3,372,050 (1968); *Chem. Abstr.*, **69,** 3858 (1968).
(*366*) A. Harsveldt, TAPPI, **45,** 85 (1962).
(*367*) G. W. Buttrick and N. R. Eldred, *TAPPI,* **45,** 890 (1962).
(*368*) J. D. Lohnas, R. D. Bourdeau, and K. K. Kalia, U.S. Patent 3,442,685 (1969); *Chem. Abstr.*, **71,** 31595 (1969). (West Virginia Pulp and Paper Co.)
(*369*) R. V. Hershey and G. M. Hein, U.S. Patent 4,154,899 (1979); *Chem. Abstr.*, **91,** 59110 (1979). (Potlatch Forests, Inc.)
(*370*) M. Papantonakis, *TAPPI,* **63,** 65 (1980).
(*371*) R. G. Stone and J. A. Krasowski, *Anal. Chem.*, **53,** 736 (1981).
(*372*) "Preparation of Corrugating Adhesives," W. O. Kroeschell, ed., TAPPI Press, Atlanta, Georgia, 1977, pp. 26–28.
(*373*) R. P. Messina, U.S. Patent 4,023,978 (1977); *Chem. Abstr.*, **87,** 7845 (1977). (Colgate Palmolive Co.)
(*374*) R. P. Vilim and H. Bell, U.S. Patent 3.628,969 (1971); *Chem. Abstr.*, **76,** 84678 (1972). (National Starch and Chemical Corp.)
(*375*) J. S. Racciato, U.S. Patent 4,105,461 (1978); *Chem. Abstr.*, **88,** 188488 (1978). (Merck and Co.)
(*376*) A. J. Ganz and G. C. Harris, U.S. Patent 3,369,910 (1968); *Chem. Abstr.*, **68,** 104007 (1968). (Hercules, Inc.)
(*377*) A. D. D'Ercole, U.S. Patent 3,669,687 (1972). (General Foods Corp.)
(*378*) M. Huchette and G. Fleche, U.S. Patent 3,890,300 (1975); *Chem. Abstr.*, **81,** 27518 (1974). (Roquette Freres.)
(*379*) W. A. Mitchell and W. C. Seidel, U.S. Patent 4,009,291 (1977); *Chem. Abstr.*, **86,** 123284 (1977). (General Foods Corp.)
(*380*) F. J. Mitan and L. Jokay, U.S. Patent, 3,427,951 (1969); *Chem. Abstr.*, **70,** 105300 (1969). (American Maize-Products Co.)
(*381*) C. T. Greenwood, D. D. Muir, and H. W. Whitcher, *Staerke,* **29,** 343 (1977).
(*382*) P. W. Morgan, *Ind. Eng. Chem., Anal. Ed.*, **18,** 500 (1946).
(*383*) H. J. Lortz, *Anal. Chem.*, **28,** 892 (1952).
(*384*) A. N. DeBelder, A. Persson, and S. Markstrom, *Staerke,* **24,** 361 (1972).
(*385*) J. G. Cobler, E. P. Samsel, and G. H. Beaver, *Talanta,* **9,** 473 (1962).

(*386*) K. L. Hodges, W. E. Kester, D. L. Widerrich, and J. A. Grover, *Anal. Chem.*, **51,** 2172 (1979).
(*387*) H. Tai, R. M. Powers, and T. F. Protzman, *Anal. Chem.*, **36,** 108 (1964).
(*388*) D. P. Johnson, *Anal. Chem.*, **41,** 859 (1969).
(*389*) H. Stahl and R. P. McNaught, *Cereal Chem.*, **47,** 345 (1970).
(*390*) J. W. Maurits, H. G. Merkus, and L. de Galan, *Anal. Chem.*, **48,** 1157 (1976).
(*391*) R. M. Hamilton and E. F. Paschall, *in* "Starch Chemistry and Technology," R. L. Whistler and E. F. Paschall, eds., Academic Press, New York, 1967, Vol. II, pp. 351–368.
(*392*) H. Neukom, U.S. Patent 2,865,762 (1958); *Chem. Abstr.*, **53,** 5538 (1959). (International Minerals and Chemical Corp.)
(*393*) H. Neukom, U.S. Patent 2,884,412 (1958); *Chem. Abstr.*, **53,** 15612 (1959). (International Minerals and Chemical Corp.)
(*394*) R. W. Kerr and F. C. Cleveland, U.S. Patent 2,961,440 (1960); *Chem. Abstr.*, **57,** 1138 (1962). (Corn Products Co.)
(*395*) F. Schierbaum and O. Borner, East Ger. Patent 36,806 (1965); *Chem. Abstr.*, **63,** 3150 (1965).
(*396*) W. Traud and G. Schlick, Ger. Offen. 2,308,886 (1974); *Chem. Abstr.*, **82,** 5545 (1975). (Hoffmann's Staerkefabriken A.-G.)
(*397*) E. F. Paschall, *in* "Methods in Carbohydrate Chemistry," R. L. Whistler, ed., Academic Press, New York, 1964, Vol. 4, pp. 294–296.
(*398*) M. E. Carr and B. T. Hofreiter, *Staerke,* **31,** 115 (1979).
(*399*) W. Bergthaller, *Staerke,* **23,** 73 (1971).
(*400*) D. S. Greidinger and B. M. Cohen, U.S. Patent 3,320,237 (1967); *Chem. Abstr.*, **67,** 91923 (1967). (Chemicals and Phosphates, Ltd.)
(*401*) S. Rogols, R. L. High, and J. F. Green, U.S. Patent 3,753,857 (1973); *Chem. Abstr.*, **79,** 116474 (1973) (A. E. Staley Mfg. Co)
(*402*) F. Krueger, German Patent 1,443,522 (1972); *Chem. Abstr.*, **77,** 7645 (1972). (Chemische Fabrik.)
(*403*) U. Schobinger, K. Berner, and C. Christoffel, Ger. Patent 1,925,322 (1970); *Chem. Abstr.*, **72,** 68464 (1970). (Blattmann and Co.)
(*404*) U. Schobinger, C. Christoffel, and K. Berner, Swiss Patent 544,779 (1974); *Chem. Abstr.*, **80,** 122663 (1974). (Blattmann and Co.)
(*405*) N. E. Lloyd, U.S. Patent 3,539,553 (1970); *Chem. Abstr.*, **74,** 32885 (1971). (Standard Brands, Inc.)
(*406*) N. E. Lloyd, U.S. Patent 3,539,551 (1970); *Chem. Abstr.*, **74,** 23799 (1971). (Standard Brands, Inc.)
(*407*) R. W. Kerr and F. C. Cleveland, U.S. Patent 3,132,066 (1964); *Chem. Abstr.*, **61,** 2055 (1964). (Corn Products Co.)
(*408*) O. B. Wurzburg, W. Jarowenko, R. W. Reubens, and J. K. Patel, U.S. Patent 4,166,173 (1979); *Chem. Abstr.*, **91,** 212879 (1979). (National Starch and Chemical Corp.)
(*409*) O. B. Wurzburg, W. Jarowenko, R. W. Reubens, and J. K. Patel, U.S. Patent 4,216,310 (1980); *Chem. Abstr.*, **93,** 188098 (1980). (National Starch and Chemical Corp.)
(*410*) R. E. Gramera, J. Heerema, and F. W. Parrish, *Cereal Chem.*, **43,** 104 (1966).
(*411*) H. Neukom, U.S. Patent 2,824,870 (1958); *Chem. Abstr.*, **52,** 8601 (1958). (International Minerals and Chemicals)
(*412*) E. T. Hjermstad, U.S. Patent 3,069,411 (1962); *Chem. Abstr.*, **58,** 5874 (1963). (Penick and Ford, Ltd.)
(*413*) C. Christoffel, E. A. Borel, A. Blumenthal, and K. Mueller, U.S. Patent 3,352,248 (1967); *Chem Abstr.*, **64,** 2267 (1966). (Blattmar and Co.)

(*414*) R. J. Alexander, U.S. Patent 3,843,377 (1974); *Chem. Abstr.*, **82,** 74776 (1975). (Krause Milling Co.)
(*415*) Hoffmann's Staerkefabriken A.-G., Ger. Patent 2,114,305 (1972); *Chem. Abstr.*, **78,** 18008 (1973).
(*416*) F. Swiderski, Polish Patent 74,513 (1975); *Chem. Abstr.*, **84,** 6836 (1976). (Academia Rolnicza)
(*417*) F. Swiderski, *Acta Aliment. Pol.*, **3,** 115 (1977); *Chem. Abstr.*, **87,** 182880 (1977).
(*418*) N. Mochizuki, K. Katsuura, and K. Inagaki, Jpn. Patent 77/51,481 (1977); *Chem. Abstr.*, **87,** 54848 (1977). (Japan Maize Products Co., Ltd.)
(*419*) Y. Tsunematsu, M. Okane, and I. Matsubara, Jpn. Patent 70/20, 512 (1970); *Chem. Abstr.*, **73,** 89420 (1970). (Matsutani Chemical Industry Co., Ltd.)
(*420*) W. Traud and G. Schlick, Ger. Patent 2,152,276 (1973); *Chem. Abstr.*, **79,** 20607 (1973). (Benckiser-Knapsack G.m.b.H.)
(*421*) M. Takase, et al. Jpn. Patent 77/28,833 (1977); *Chem. Abstr.*, **88,** 75568 (1978). (Japan Maize Products Co. Ltd.)
(*422*) A. I. Zhushman, V. A. Kovalenok, and N. N. Naumova, *Sakh. Prom-st*, **9,** 51 (1979); *Chem. Abstr.*, **91,** 194924 (1979).
(*423*) M. M. Tessler, U.S. Patent 3,719,662 (1973); *Chem. Abstr.*, **79,** 7134 (1973). (National Starch and Chemical Corp.)
(*424*) M. M. Tessler, U.S. Patent 3,838,149 (1974); *Chem. Abstr.*, **82,** 74773 (1975). (National Starch and Chemical Corp.)
(*425*) M. M. Tessler, U.S. Patent 3,842,071 (1974); *Chem. Abstr.*, **82,** 74774 (1974). (National Starch and Chemical Corp.)
(*426*) F. Verbanac and K. B. Moser, U.S. Patent 3,553,194 (1971); *Chem. Abstr.*, **74,** 100820 (1971). (A. E. Staley Mfg. Co.)
(*427*) M. M. Tessler, U.S. Patent 3,969,341 (1976); *Chem. Abstr.*, **85,** 145150 (1976). (National Starch and Chemical Co.)
(*428*) G. A. Towle and R. L. Whistler, *Methods Carbohydr. Chem.*, **6,** 408 (1972).
(*429*) P. J. Malden, Ger. Patent 2,728,111 (1978); *Chem. Abstr.*, **88,** 107066 (1978). (English Clays Lovering Pochin and Co. Ltd.)
(*430*) H. Benninga, A. Harsveldt, and A. A. DeSturler, *TAPPI*, **50,** 577 (1967).
(*431*) G. Smit, U.S. Patent 3,682,733 (1972); *Chem. Abstr.*, **69,** 44675 (1968). (Scholten Research N.V.)
(*432*) G. Smit, U.S. Patent 3,591,412 (1971); *Chem. Abstr.*, **69,** 44675 (1968). (Scholten Research N.V.)
(*433*) Code of Federal Regulations, Title 21, Part 178, Indirect Food Additives, Section 178.3520, Industrial Starch-Modified, U.S. Government Printing Office, Washington, D.C., 1981, p. 283.
(*434*) C. G. Caldwell, W. Jarowenko, and I. D. Hodgkin, U.S. Patent 3,459,632 (1969); *Chem. Abstr.*, **69,** 60169 (1968). (National Starch and Chemical Co.)
(*435*) K. B. Moser and F. Verbanac, U.S. Patent 3,562,103 (1971); *Chem. Abstr.*, **74,** 143578 (1971). (A. E. Staley Mfg. Co.)
(*436*) V. Schobinger, K. Berner, and C. Christoffel, German Patent 1,966,452 (1973); *Chem. Abstr.*, **79,** 106300 (1973). (Blattmann and Co.)
(*437*) N. H. Yui and L. R. Cohen, U.S. Patent 3,524,796 (1970); *Chem. Abstr.*, **74,** 23784 (1971). (American Maize Co.)
(*438*) A. Kling, W. Traud, W. Hansi, and H. Jalke, German Patent 2,426,404 (1975); *Chem. Abstr.*, **84,** 91582 (1976). (Benckiser-Knapsack G.m.b.H; Hoechst A.-G.)
(*439*) K. Schneider and V. Specht, French Patent 2,005,952 (1969); *Chem. Abstr.*, **72,** 134056 (1970). (Benckiser-Knapsack)

(*440*) C. H. Hoepke, A. Mueller, and H. J. Ruge, German Patent 2,345,350 (1975); *Chem. Abstr.*, **83,** 195589 (1975). (Hoffmann's Staerkefabriken A.-G.)
(*441*) H. Nakazawa, T. Shinitini, and S. Okamoto, Jap. Patent 77/74,648 (1977); *Chem. Abstr.*, **87,** 137111 (1977). (Toyo Soda Mfg. Co.)
(*442*) (J. A. Benckiser, G.m.b.H.; Chemische Fabrik.) Brit. Patent 1,233,637 (1971); *Chem. Abstr.*, **75,** 112794 (1971).
(*443*) F. Kruger and L. Baver, French Patent 2,017,582 (1970); *Chem. Abstr.*, **74,** 102949 (1971). (J. A. Benckiser, G.m.b.H.; Chemische Fabrik.)
(*444*) H. W. Fiedelman and R. L. Lintvedt, U.S. Patent 3,155,458; *Chem. Abstr.*, **62,** 257 (1965). (Morton Salt Co.)
(*445*) T. Goto and H. Murotani, *Kogyo Kagaku Zasshi,* **71,** 1833 (1968); *Chem. Abstr.*, **70,** 60717 (1969).
(*446*) M. Jonason, British Patent 1,309,473 (1973); *Chem. Abstr.*, **79,** 116473 (1973). (Slimcea, Ltd.)
(*447*) M. Jonason, British Patent 1,314,431 (1973); *Chem. Abstr.*, **79,** 80657 (1973). (Slimcea, Ltd.)
(*448*) D. L. Lin, *Shih P'in Kung Yeh,* **7,** 17 (1975); *Chem. Abstr.*, **83,** 112441 (1975).
(*449*) R. H. Klostermann, U.S. Patent 3,108,004 (1963); *Chem. Abstr.*, **60,** 1041 (1964). (Vita-Zyme Laboratories, Inc.)
(*450*) National Dairy Products Corporation, British Patent 938,717 (1963).
(*451*) H. C. A. Meyer, R. L. Milloch, V. Shreeram, T. Tsuzuki, U.S. Patent 3,238,100 (1966); *Chem. Abstr.*, **64,** 14041 (1966). (American-Maize Co.)
(*452*) W. Traud and G. Schlick Ger. Patent 2,428,133 (1975); *Chem. Abstr.*, **84,** 91928 (1976). (Benckiser-Knapsack Gm.b.H.)
(*453*) F. A. Hartman, H. C. Kretschmar, and A. J. Tofe, Ger. Patent 2,600,539 (1976); *Chem. Abstr.*, **85,** 130522 (1976). (Procter and Gamble Co.)
(*454*) K. Hamabe and N. Oku, Japan. Patent 78/126,030 (1977); *Chem. Abstr.*, **90,** 108910 (1979).
(*455*) R. J. Smith, *in* "Starch Chemistry and Technology," R. L. Whistler and E. F. Paschall, eds., Academic Press, New York, 1967, Vol. II, pp. 569–655.
(*456*) K. D. Fleischer, B. C. Southworth, J. H. Hodecker, and M. M. Tuckerman, *Anal. Chem.*, **30,** 153 (1958).
(*457*) D. N. Fogg and N. T. Wilkinson, *Analyst,* **83,** 406 (1958).
(*458*) D. S. Greif and L. A. Gaspar, *in* "Dry Strength Additives," W. F. Reynolds, ed., TAPPI Press, Atlanta, Georgia, 1980, pp. 95–117.
(*459*) R. D. Harvey, R. E. Klem, M. Bale, and E. D. Hubbard, *in* "Cationic Starches in Papermaking Applications," TAPPI Retention and Drainage Seminar Notes, TAPPI Press, Atlanta, Georgia, 1980, pp. 41–51.
(*460*) E. F. Paschall, *in* "Starch Chemistry and Technology," R. L. Whistler and E. F. Paschall, eds., Academic Press, New York, 1967, Vol. II, pp. 403–422.
(*461*) M. W. Rutenberg and J. L. Volpe, U.S. Patent 2,989,520 (1961); *Chem. Abstr.*, **55,** 22878 (1961). (National Starch and Chemical Corp.)
(*462*) A. Aszalos, U.S. Patent 3,077.469(1963); *Chem. Abstr.*, **59,** 3000 (1963). (National Starch and Chemical Corp.)
(*463*) C. G. Caldwell and O. B. Wurzburg, U.S. Patent 2,813,093 (1957); *Chem. Abstr.*, **52,** 2438 (1958). (National Starch and Chemical Corp.)
(*464*) C. H. Hullinger and N. H. Yui, U.S. Patent 2,970,140 (1961); *Chem. Abstr.*, **57,** 2483 (1962). (American Maize Prod. Co.)
(*465*) C. P. Iovine and D. K. Ray-Chaudhuri, U.S. Patent 4,104,307 (1978); *Chem. Abstr.*, **90,** 71740 (1979). (National Starch and Chemical Corp.)

(466) W. A. Scholten's Chemische Fabrieken, N.V., French Patent 1,493,421 (1967); *Chem. Abstr.*, **69,** 107797 (1968).
(467) Y. Merle, *Compt. Rend.*, **246,** 1425 (1958); *Chem. Abstr.*, **52,** 14533 (1959).
(468) C. G. Harris and H. A. Leonard, U.S. Patent 3,070,594 (1962); *Chem. Abstr.*, **58,** 11460 (1963). (Hercules Powder Co.)
(469) M. M. Tessler, U.S. Patent 4,060,683 (1977); *Chem. Abstr.*, **38,** 63495 (1978). (National Starch and Chemical Co.)
(470) E. F. Paschall, U.S. Patent 2,876,217 (1959); *Chem. Abstr.*, **53,** 12720 (1959). (Corn Products Co.)
(471) J. B. Doughty and R. E. Klem, U.S. Patent 4,066,673 (1978); *Chem. Abstr.*, **88,** 89091 (1978). (Westvaco Corp.)
(472) A. M. Goldstein, E. M. Heckman, and J. H. Katcher, U.S. Patent 3,649,616 (1972); *Chem. Abstr.*, **77,** 7644 (1972). (Stein Hall and Co.)
(473) M. E. Carr and M. D. Bagby, *Starch/Staerke*, **33,** 310 (1981).
(474) G. V. Caesar, U.S. Patent 3,422,067 (1969); *Chem. Abstr.*, **70,** 58564 (1969).
(475) J. M. Billy and J. A. Seguin, U.S. Patent 3,448,101 (1969); *Chem. Abstr.*, **71,** 28270 (1969). (Ogilvie Flour Mills Co., Ltd.)
(476) J. C. Rankin and B. S. Phillips, U.S. Patent 4,127,563 (1978); *Chem. Abstr.*, **88,** 107093 (1978). (U.S. Department of Agriculture.)
(477) W. Jarowenko and D. B. Solarek, U.S. Patent 4,281,109 (1981); *Chem. Abstr.*, **95,** 171419 (1981). (National Starch and Chemical Corp.)
(478) P. R. Shildneck and R. J. Hathway, U.S. Patent 3,346,563 (1967); *Chem. Abstr.*, **65,** 14279 (1968). (A. E. Staley Mfg. Co.)
(479) R. R. Langher, J. C. Walling, and R. T. McFadden, U.S. Patent 3,532,751 (1970); *Chem. Abstr.*, **73,** 120064 (1970). (Dow Chemical Co.)
(480) T. A. McGuire and C. L. Mehltretter, U.S. Patent 3,558,501 (1971); *Chem. Abstr.*, **74,** 87349 (1971). (U.S. Department of Agriculture.)
(481) D. Burmeister and H. K. Klein, German Patent 2,055,046 (1972); *Chem. Abstr.*, **77,** 77030 (1972). (Hoffmann's Staerkefabriken-A.-G.)
(482) G. A. Hull, U.S. Patent 3,669,955 (1972); *Chem. Abstr.*, **77,** 77032 (1972). (CPC International, Inc.)
(483) C. P. Patel, M. A. Jaeger, and R. E. Pyle, U.S. Patent 3,378,547 (1968); *Chem. Abstr.*, **69,** 3882 (1968). (Union Starch and Refining Co., Inc.)
(484) C. P. Patel and R. E. Pyle, U.S. Patent 3,417,078 (1968); *Chem. Abstr.*, **70,** 59126 (1969). (Union Starch and Refining Co., Inc.)
(485) W. G. Hunt, U.S. Patent 3,624,070 (1971); *Chem. Abstr.*, **76,** 87508 (1972). (Anheuser-Busch, Inc.)
(486) W. G. Hunt and L. P. Kovats, U.S. Patent 3,959,169 (1976); *Chem. Abstr.*, **85,** 77632 (1976). (Anheuser-Busch, Inc.)
(487) K. W. Kirby, U.S. Patent 3,336,292 (1967); *Chem. Abstr.*, **67,** 91920 (1967). (Penick and Ford, Ltd., Inc.)
(488) R. W. Kerr and H. Neukom, *Staerke*, **4,** 255 (1952).
(489) J. C. Rankin and C. R. Russell, U.S. Patent 3,522,238 (1970); *Chem. Abstr.*, **73,** 86758 (1970). (U.S. Department of Agriculture.)
(490) J. C. McClendon and E. L. Berry, U.S. Patent 3,725,387 (1973); *Chem. Abstr.*, **78** 31774 (1973). (Dow Chemical Co.)
(491) J. C. McClendon, U.S. Patent 3,846,405 (1974). (Dow Chemical Co.)
(492) D. A. Tomalla, J. L. Brewbaker, and N. D. Ojho, U.S. Patent 3,824,269 (1974); *Chem. Abstr.*, **81,** 171804 (1974). (Dow Chemical Co.)
(493) AVEBE GA, Neth. Appl. 73/13,113; *Chem. Abstr.*, **81,** 93373 (1974).

(545) W. Jarowenko and H. R. Hernandez, U.S. Patent 4,029,544 (1977); *Chem. Abstr.*, **85,** 48590 (1976). (National Starch and Chemical Corp.)
(546) M. M. Tessler, U.S. Patent 4,119,487 (1978); *Chem. Abstr.*, **90,** 56660 (1979). (National Starch and Chemical Corp.)
(547) H. Benninga, U.S. Patent 3,467,647 (1969); *Chem. Abstr.*, **68,** 14281 (1968). (W. A. Scholten's Chemische Fabrieken.)
(548) R. M. Powers and R. W. Best, U.S. Patent 3,598,623 (1971); *Chem. Abstr.*, **75,** 153138 (1971). (A. E. Staley Mfg. Co.)
(549) R. M. Powers and R. W. Best, U.S. Patent 3,649,624 (1972); *Chem. Abstr.*, **77,** 7596 (1972). (A. E. Staley Mfg. Co.)
(550) R. W. Cescato, U.S. Patent 3,654,263 (1972); *Chem. Abstr.*, **77,** 21891 (1972). (CPC International, Inc.)
(551) R. W. Cescato, U.S. Patent 3,706,584 (1972); *Chem. Abstr.*, **78,** 73864 (1973). (CPC International, Inc.)
(552) L. H. Elizer, U.S. Patent 3,622,563 (1971); *Chem. Abstr.*, **76,** *76,* 101129 (1972). (Hubinger Co.)
(553) L. H. Elizer, U.S. Patent 3,650,787 (1972); *Chem. Abstr.*, **77,** 7646 (1972). (Hubinger Co.)
(554) L. H. Elizer, U.S. Patent 3,676,205 (1972); *Chem. Abstr.*, **77,** 103229 (1972). (Hubinger Co.)
(555) L. H. Elizer, U.S. Patent 3,793,310 (1974); *Chem. Abstr.*, **81,** 65509 (1974). (Hubinger Co.)
(556) L. H. Elizer, U.S. Patent 3,673,171 (1972); *Chem. Abstr.*, **77,** 90370 (1972). (Hubinger Co.)
(557) L. H. Elizer, U.S. Patent 3,751,411 (1973); *Chem. Abstr.*, **79,** 106370 (1973). (Hubinger Co.)
(558) E. J. Barber and C. E. Maag, U.S. Patent 3,219,578 (1965); *Chem. Abstr.*, **64,** 6895 (1966). (Hercules Powder Co.)
(559) E. J. Barber, R. H. Earle, and G. C. Harris, U.S. Patent 3,219,519 (1965); *Chem. Abstr.*, **64,** 6895 (1966). (Hercules Powder Co.)
(560) M. E. Carr, U.S. Patent 4,093,510 (1978). (U.S. Department of Agriculture.)
(561) H. W. Moeller, *TAPPI,* **49,** 211 (1966).
(562) J. E. Unbehend and K. W. Britt, *in* "Pulp and Paper Chemistry and Technology," J. P. Casey, ed., Wiley, New York, 1981, Vol. III, pp. 1593–1607.
(563) J. Marton, *TAPPI,* **63,** 87 (1980).
(564) D. D. Halabisky, *TAPPI,* **60,** 125 (1977).
(565) H. R. Hernandez, *TAPPI,* **53,** 2101 (1970).
(566) L. P. Avery, *TAPPI,* **62,** 43 (1979).
(567) J. E. Voigt and H. Pender, Jr., U.S. Patent 4,066,495 (1978); *Chem. Abstr.*, **88,** 91301 (1978). (Anheuser-Busch, Inc.)
(568) Kloninklijke Scholten-Honig N.V., French Patent 2,396,831 (1979); *Chem. Abstr.*, **91,** 176937 (1979).
(569) K. W. Britt and J. E. Unbehend, *TAPPI,* **59,** 60 (1964).
(570) E. E. Moore, *Papermakers Conf., (Prepr.),* TAPPI Press, Atlanta, Georgia 1976, pp. 36–36.
(571) C. P. Vallete and J. F. Lafaye, *ATIP,* **24,** 75 (1970).
(572) O. B. Wurzburg, U.S. Patent 3,968,005 (1976); *Chem. Abstr.*, **85,** 12676 (1976). (National Starch and Chemical Corp.)
(573) O. B. Wurzburg and E. D. Mazzarella, U.S. Patent 3,102,064 (1963); *Chem. Abstr.*, **59,** 13014 (1963). (National Starch and Chemical Corp.)
(574) G. C. Harris and C. A. Weisgerber, U.S. Patent 3,070,452 (1962); *Chem. Abstr.*, **58,** 10398 (1963). (Hercules Powder Co.)
(575) A. R. Savina, U.S. Patent 3,223,543 (1965); *Chem. Abstr.*, **64,** 6895 (1966). (American Cyanamid Co.)

(*576*) A. R. Savina, U.S. Patent 3,223,544 (1965); *Chem. Abstr.*, **64,** 6895 (1966). (American Cyanamid Co.)
(*577*) O. B. Wurzburg, U.S. Patent 3,821,069 (1974); *Chem. Abstr.*, **81,** 171803 (1974). (National Starch and Chemical Corp.)
(*578*) J. J. Keavney and R. J. Kulick, *in* "Pulp and Paper Chemistry and Technology," J. P. Casey, ed., Wiley, New York, 1981, Vol. III, pp. 1547–1592.
(*579*) G. E. Hamerstrand, H. D. Heath, B. S. Phillips, J. C. Rankin, and M. I. Schulte, *TAPPI*, **62,** 35 (1979).
(*580*) G. Heuten, W. H. Thomin, *Wochenbl. Papierfab.*, **103,** 329 (1975); *Chem. Abstr.*, **83,** 81652 (1975).
(*581*) G. H. Brown and E. D. Mazzarella, U.S. Patent 3,671,310 (1972); *Chem. Abstr.*, **72,** 45240 (1970). (National Starch and Chemical Corp.)
(*582*) J. Kronfeld, U.S. Patent 3,052,561 (1962). (National Starch and Chemical Corp.)
(*583*) E. D. Mazzarella and L. J. Hickey, *TAPPI*, **49,** 526 (1966).
(*584*) H. C. Olsen, U.S. Patent 2,946,705 (1960); *Chem. Abstr.*, **54,** 21780 (1960). (National Starch and Chemical Corp.)
(*585*) H. L. Haynes and M. J. Harvey, U.S. Patent 3,971,871 (1976); *Chem. Abstr.*, **85,** 194045 (1976). (Owens-Corning Fiberglas Corp.)
(*586*) E. F. Paschall, U.S. Patent 2,876,217 (1959); *Chem. Abstr.*, **54,** 12720 (1959). (Corn Products Co.)
(*587*) C. G. Caldwell and O. B. Wurzburg, U.S. Patent 2,975,124 (1961); *Chem. Abstr.*, **55,** 16043 (1961). (National Starch and Chemical Corp.)
(*588*) E. F. Paschall and W. H. Minkema, U.S. Patent 2,995,513 (1961); *Chem. Abstr.*, **76,** 155916 (1972). (Corn Products Co.)
(*589*) K. B. Moser and F. Verbanac, U.S. Patent 3,842,005 (1974); *Chem. Abstr.*, **76,** 155916 (1972). (A. E. Staley Mfg. Co.)
(*590*) W. G. Hunt and R. J. Belz, U.S. Patent 3,835,114 (1974); *Chem. Abstr.*, **81,** 171857 (1974). (Anheuser-Busch, Inc.)
(*591*) W. G. Hunt and R. J. Belz, U.S. Patent 3,875,054 (1975); *Chem. Abstr.*, **83,** 65164 (1975). (Anheuser-Busch, Inc.)
(*592*) W. G. Hunt and R. J. Belz, U.S. Patent 3,962,079 (1976); *Chem. Abstr.*, **85,** 181971 (1976). (Anheuser-Busch, Inc.)
(*593*) T. R. Tutein, A. E. Harrington, and J. T. Jacob, U.S. Patent 4,088,600 (1978); *Chem. Abstr.*, **89,** 91425 (1978). (Chemed Corp.)
(*594*) E. S. Mackey, U.S. Patent 3,627,694 (1971); *Chem. Abstr.*, **77,** 41346 (1972). (General Aniline and Film Corp.)
(*595*) R. E. Wing, W. E. Rayford, W. M. Doane, and C. R. Russell, *J. Appl. Polymer Sci.*, **22,** 1405 (1978).
(*596*) C. D. Szymanski, A. L. Micchelli, and P. C. Trubiano, German Patent 2,038,986 (1971); *Chem. Abstr.*, **74,** 130285 (1971). (National Starch and Chemical Corp.)
(*597*) R. B. Bradstreet, "The Kjeldahl Method for Organic Nitrogen," Academic Press, New York, 1965.
(*598*) D. A. Skoog and D. M. West, "Fundamentals of Analytical Chemistry," Holt, Rinehart and Winston, New York, 1969, pp. 313–318.
(*599*) J. F. Zievers, U.S. Patent 4,238,329 (1980); *Chem. Abstr.*, **94,** 126868 (1981). (Industrial Filter Corp.)
(*600*) C. L. Swanson, R. E. Wing, and W. M. Doane, U.S. Patent 3,947,354 (1976); *Chem. Abstr.*, **85,** 98782 (1976). (U.S. Department of Agriculture.)
(*601*) C. L. Swanson, R. E. Wing, W. M. Doane, and C. R. Russell, *Environ. Sci. Technol.*, **7,** 614 (1973).

(*602*) R. E. Wing, C. L. Swanson, W. M. Doane, and C. R. Russell, *J. Water Pollut. Control Fed.*, **46,** 2043 (1974).
(*603*) R. A. Buchanan, W. F. Kwalek, H. C. Katz, and C. R. Russell, *Staerke*, **23,** 350 (1971).
(*604*) R. A. Buchanan, H. C. Katz, C. R. Russell, and C. E. Rist, *Rubber J.*, **153,** No. 10, 23, 30, 32, 35, 88–91 (1971).
(*605*) T. P. Abbott, W. M. Doane, and C. R. Russell, *Rubber Age*, **105,** No. 8, 43 (1973).
(*606*) T. P. Abbott, C. James, W. M. Doane, and C. E. Rist, *Rubber World*, **169,** No. 4, 40 (1974).
(*607*) H. F. Conway and V. E. Sohns, *J. Elast. Plastics*, **7,** 365 (1975).
(*608*) R. A. Buchanan, *Staerke*, **26,** 165 (1974).
(*609*) E. I. Stout, B. S. Shasha, and W. M. Doane, *J. Appl. Polym. Sci.*, **24,** 153 (1979).
(*610*) M. E. Foley and L. M. Wax, *Weed Sci.*, **28,** 626 (1980).
(*611*) W. M. Doane, B. S. Shasha, and C. R. Russell, *in* "Controlled Release Pesticides," Amer. Chem. Soc. Symp. Series, No. 53, American Chemical Society, Washington, D.C., 1977, pp. 74–83.
(*612*) B. S. Shasha, W. Doane, W. McKee, and C. R. Russell, German Patent 2,658,221 (1977); *Chem. Abstr.*, **87,** 179047 (1977). (Stauffer Chemical Co.)
(*613*) E. G. Adamek and C. B. Purves, *Can. J. Chem.* **38,** 2425 (1960).
(*614*) E. G. Adamek and C. B. Purves, *Can. J. Chem.*, **35,** 960 (1957).
(*615*) C. L. Swanson, T. R. Noffziger, C. R. Rusell, B. T. Hofreiter, and C. E. Rist, *Ind. Eng. Chem. Prod. Res. Dev.*, **3,** 22 (1964).
(*616*) W. M. Doane, C. R. Russell, and C. E. Rist, *Staerke*, **17,** 77 (1965).
(*617*) E. B. Lancaster, L. T. Black, H. F. Conway, and E. L. Griffin, Jr., *Ind. Eng. Chem. Prod. Res. Dev.*, **5,** 354 (1966).
(*618*) E. B. Lancaster, H. F. Conway, L. A. Winecke, and E. L. Griffin, Jr. U.S. Patent 3,385,719 (1968); *Chem. Abstr.*, **69,** 29762 (1968). (U.S. Department of Agriculture.)
(*619*) C. R. Russell, R. A. Buchanan, C. E. Rist, B. T. Hofreiter, and A. J. Ernst, *TAPPI*, **45,** 557 (1962).
(*620*) E. I. Stout, D. Trimnell, W. M. Doane, and C. R. Russell, *Staerke*, **29,** 299 (1977).
(*621*) W. M. Doane, C. R. Russell, and C. E. Rist, *Staerke*, **17,** 176 (1965).
(*622*) D. J. Bridgeford, U.S. Patents, 3,291,789 (1966); 3,339,069 (1968); 3,484,433 (1969); *Chem. Abstr.*, **63,** 3162 (1965); **71** 40552 (1969). (Teepack, Inc.)
(*623*) R. E. Wing and W. E. Rayford, *Starch/Staerke*, **32,** 129 (1980).
(*624*) G. E. Lauterbach, E. J. Jones, J. W. Swanson, B. T. Hofreiter, and C. E. Rist, *Staerke*, **26,** 58 (1974).
(*625*) R. A. Buchanan, O. E. Weislogel, C. R. Russell, and C. E. Rist, *Ind. Eng. Chem. Prod. Res. Dev.*, **7,** 155 (1968).
(*626*) C. L. Mehltretter, *Staerke*, **15,** 313 (1963).
(*627*) C. L. Mehltretter, *Staerke*, **18,** 208 (1966).
(*628*) C. L. Mehltretter, *Methods Carbohydr. Chem.*, **4,** 316 (1964).
(*629*) E. L. Jackson, *Org. Reactions*, **2,** 341 (1944).
(*630*) J. M. Bobbit, *Adv. Carbohydr. Chem.*, **11,** 1 (1956).
(*631*) R. D. Guthrie, *Adv. Carbohydr. Chem.*, **16,** 105 (1961).
(*632*) E. L. Jackson and C. S. Hudson, *J. Amer. Chem. Soc.*, **59,** 2049 (1937); **60,** 989 (1938).
(*633*) V. F. Pfeifer, V. E. Sohns, H. F. Conway, E. B. Lancaster, S. Dabic, and E. L. Griffin, Jr., *Ind. Eng. Chem.*, **52,** 201 (1960).
(*634*) J. Slager, U.S. Patent 3,086,969 (1963); *Chem. Abstr.*, **59,** 11644 (1963).
(*635*) T. A. McGuire and C. L. Mehltretter, *Staerke*, **23,** 42 (1971).
(*636*) B. T. Hofreiter, H. D. Heath, A. J. Ernst, and C. R. Russell, *TAPPI*, **57,** No. 8, 81 (1974).
(*637*) C. L. Mehltretter, T. E. Yeates, G. E. Hammerstrand, B. T. Hofreiter, and C. E. Rist, *TAPPI*, **45,** 750 (1962).

(*638*) G. E. H. Joosten, E. J. Stamhuis, and W. A. Roelfsema, *Starch/Staerke,* **34,** 402 (1982).
(*639*) S. I. El-Hinnaway, A. Fahmy, H. M. El-Saied, A. F. El-Shirbeeny, and K. M. EL-Sahy, *Starch/Staerke,* **34,** 65 (1982).
(*640*) S. I. El-Hinnaway, H. M. El-Saied, A. Fahmy, A. E. El-Shirbeeny, and K. M. El-Sahy, *Starch/Staerke,* **34,** 112 (1982).
(*641*) T. Shiroza, Furihata, T. Endo, H. Seto, and N. Otake, *Agr. Biol. Chem.,* **46,** 1425 (1982).
(*642*) M. Abdel-Akher, *Starch/Staerke,* **34,** 169 (1982).
(*643*) B. L. Scallet and E. A. Sowell, *in* ''Starch: Chemistry and Technology,'' R. L. Whistler and E. F. Paschall, eds., Academic Press, New York, Vol. II, 1967, pp. 237–251.
(*644*) C. H. Hullinger, *in* ''Starch: Chemistry and Technology,'' R. L. Whistler and E. F. Paschall, eds., Academic Press, New York, Vol. II, 1967, pp. 445–450.
(*645*) L. H. Kruger and M. W. Rutenberg, *in* ''Starch: Chemistry and Technology,'' R. L. Whistler and E. F. Paschall, eds., Academic Press, New York, Vol. II, 1967, pp. 369–401.
(*646*) E. T. Hjermstad, *in* ''Starch: Chemistry and Technology,'' R. L. Whistler and E. F. Paschall, eds., Academic Press, New York, Vol. II, 1967, pp. 423–432.
(*647*) C. L. Mehltretter, *in* ''Starch: Chemistry and Technology,'' R. L. Whistler and E. F. Paschall, eds., Academic Press, New York, Vol. II, 1967, pp. 433–444.

CHAPTER XI

CHEMICALS FROM STARCH

By Felix H. Otey and William M. Doane

Northern Regional Research Center, Agricultural Research, Science and Education Administration, U.S. Department of Agriculture, Peoria, Illinois 61604*

I. Introduction

Rapidly increasing prices and dwindling supplies of petroleum have intensified interest in natural products as alternative sources of energy and raw materials for the chemical industry. Starch, a polysaccharide produced in great abundance in nature, is a prime candidate for use as a raw material because it is available at a low cost and can be converted readily into a variety of useful monomeric and polymeric products by chemical and biochemical means.

While there is social and economic controversy over using such natural products as starch for chemicals instead of food, the fact remains that the United States produces more carbohydrate than is needed for a balanced diet. The six major cereal grain crops produced in 1977 contained 400 billion pounds of starch. Most of this starch was fed to animals, despite tests showing better weight gain with higher protein feeds (*1*). Hence, the nutritional value of grain products is enhanced by removing part of the starch or by fermenting the starch to alcohol, leaving a high quality feed known as distillers dried grains. However, these processes are economically feasible only when a market exists for the starch or starch-derived products. Hundreds of products from and applications of starch

*The mention of firm names or trade products does not imply that they are endorsed or recommended by the U.S. Department of Agriculture over other firms or similar products not mentioned.

STARCH, 2nd ed.
ISBN 0-12-746270-8

have been developed over the years. Here we have chosen to emphasize only the more recent and highly promising applications of starch-derived products.

II. Chemicals from Starch via Biosynthesis

Although techniques for fermenting starch to various chemicals have been known for many years, a few processes are discussed here because of current high interest in developing alternative sources of raw materials. The production of some industrial-grade ethanol from natural products is one example of the rising trend toward fermentation biosynthesis (*2*).

Ethanol is produced in the United States by two major procedures, fermentation and chemical synthesis. Practically all synthetic ethanol is now produced from ethylene. Table I shows the effects of increasing raw material cost from 1973 to 1980 on the cost of producing alcohol from ethylene. These data were based on fixed conversion costs reported in 1976 by Miller (*3*).

$$\underset{\text{ethylene}}{CH_2{=}CH_2 + H_2O} \xrightarrow{\text{Catalyst}} \underset{\text{ethanol}}{C_2H_5OH}$$

All beverage alcohol is produced in the United States by the fermentation of starch and sugar or from grains and molasses which are high in starch and sugar content, respectively.

Table I

Effect of Ethylene Cost on Ethanol Cost[a]

	Ethylene, cents/pound[b]							
	3.3	*7.5*	*8.8*	*11.2*	*12.0*	*13.0*	*19.0*	*23.0*
Alcohol	*Cost/gallon, cents*							
190° Proof								
Ethylene	13.2	30.0	35.2	44.8	48.0	52.0	76.0	92.0
Conversion	26.9	26.9	26.9	26.9	26.9	26.9	26.9	26.9
Total manufacturing cost/gallon (exclusive of profit, packaging, and sales expenses)	40.1	56.0	62.1	71.7	74.9	78.9	102.9	118.9
200° Proof								
Ethylene	13.8	31.3	36.8	46.4	50.1	54.3	79.4	96.1
Conversion	31.3	31.3	31.3	31.3	31.3	31.3	31.3	31.3
Total manufacturing cost/gallon (exclusive of profit, packaging, and sales expenses)	45.1	62.6	68.1	77.7	81.4	85.6	110.7	127.4

[a] Based on conversion costs reported by D. L. Miller in 1976 (*3*).
[b] Approximate prices of ethylene from 1973 to 1980.

Table II

Effect of Corn Cost on Ethanol Cost
(Basis: 2.7 gal. of 200° proof alcohol/bushel)

Corn price/bushel, dollars	Alcohol cost/gallon, cents		
	Corn	Conversion[a]	Total base cost[b]
2.00	74.0	25.4	99.4
2.25	83.3	25.4	108.7
2.50	92.6	25.4	118.0
3.00	111.0	25.4	136.4
3.50	129.5	25.4	154.9
4.00	148.4	25.4	173.8

[a] By-product grains credited at $100/ton in conversion cost.
[b] These costs do not include profits, packaging, and sales expenses.

$$\underset{\text{starch}}{(C_6H_{10}O_5)_n} + nH_2O \rightarrow \underset{\text{glucose}}{nC_6H_{12}O_6}$$

$$\underset{\text{glucose}}{C_6H_{12}O_6} \rightarrow \underset{\text{ethanol}}{2\ C_2H_5OH} + 2\ CO_2$$

One pound of starch should theoretically yield 0.568 lb of ethanol. In actual practice, yields generally are about 90–95% of theory. By allowing a 34 cents per gallon of alcohol of by-product feed credit ($100/ton), D. L. Miller estimated in 1976 that the net cost of converting corn to absolute ethanol was 25.4 cents/gal. The effects of corn prices on the cost of producing fermentation alcohol are shown in Table II (*3, 4*). While this conversion cost has increased since 1976, the by-product feed credit increased so that the cost per gallon reported in Table II remains fairly accurate. The price of fermentation alcohol may be less than these production figures reflect because industrial grain alcohol comes largely from integrated grain milling plants where potable and industrial ethanol are produced along with other corn products.

Interest in fermentation alcohol is further intensified by the prospects of fermenting surplus and low quality grains, sugar, and cellulosic residues into fuel alcohol. Many companies now sell gasohol (90% gasoline, 10% ethanol) and other gasoline–ethanol blends for motor fuel. However, many experts believe that several persistent problems remain to be solved before fermentation alcohol can be expected to replace a significant amount of gasoline fuel. Yet proponents of fuel alcohol maintain that the fermentation of grain to fuel alcohol is already feasible. Regardless of the controversy, considerable state and federal funds have been appropriated for fermentation alcohol research. Undoubtedly, much of these research efforts will be directed toward reducing the energy needed to produce and recover pure ethanol.

Tong (*2*) reviewed the bioconversion of carbohydrate raw materials into isopropanol, *n*-butanol, acetone, ethanol, 2,3-butylene glycol, glycerol, and fumaric acid and the subsequent catalytic conversion of butylene glycol to methyl ethyl ketone, fumaric acid to maleic anhydride, and ethanol to butadiene. Of these three catalytic conversions, only the butadiene process from ethanol attained commercial-scale success. Prior to 1950, the bulk of our acetone and *n*-butanol was obtained by the anaerobic fermentation of carbohydrates. Several anaerobic glycerol fermentation processes were developed prior to 1960, but have since been displaced by synthetic routes. From this list of 10 chemicals, Tong reports that, of the total U.S. production in 1976, 5% of the acetone, 10% of the *n*-butanol, 15% of the fumaric acid, and 30% of the ethanol were produced by fermentation.

III. Polyhydroxy Compounds from Starch

Although starch is being considered for the production of alcohol and hydrocarbons as a direct replacement for petroleum, any such conversions are accompanied by drastic mass losses due to the elimination of CO_2 and H_2O. For example, the maximum possible mass yield of ethylene from starch is 52% owing to the loss of oxygen. The only process under consideration for producing ethylene from starch is based on fermenting starch to ethanol followed by dehydration of the ethanol. Based on this process, the loss of both CO_2 and H_2O reduces the maximum mass yield of ethylene from starch to only 34.6%. A more feasible approach to utilizing starch may depend on developing substitutes, rather than direct replacements, for petroleum such that the hydroxyl groups are retained to perform a useful function. Several polyhydroxy compounds have been developed from starch for industrial applications that are less expensive than comparable products made from petroleum.

Glucose, the most common polyhydroxy compound made from starch, is obtained in 90–95% yields by acid or enzymic depolymerization of starch, and because a molecule of water is added, the mass yield of glucose is slightly more than 100%. Glucose can be converted to a variety of cyclic and acyclic polyols, aldehydes, ketones, acids, esters, and ethers, some of which are now used industrially and others which could increase in importance if the price for petroleum and petroleum products continues to rise. Fermentation of simple sugars is now employed as the preferred route to commercial organic acids such as D-gluconic and itaconic. Sorbitol (D-glucitol) is probably the most widely used polyol made from starch or glucose. It is used extensively for making surfactants and emulsifiers, especially for food applications. An estimated 1 to 2 million pounds (900,000 kg) of sorbitol go into making specialty polyethers for urethane foam production. Sorbitol, now priced at 60 cents/lb, is too expensive for many industrial applications.

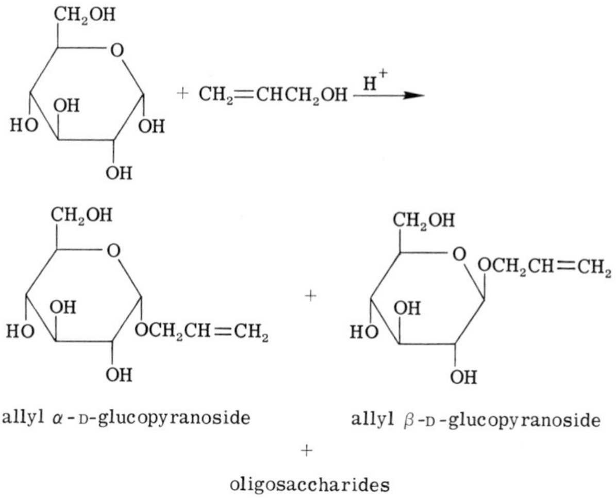
CH2OH
O
OH
HO
OH
OH
+ CH2=CHCH2OH
H+
CH2OH
O
OH
HO
OCH2CH=CH2
OH
+
CH2OH
O
OCH2CH=CH2
OH
HO
OH
allyl α-D-glucopyranoside
allyl β-D-glucopyranoside
+
oligosaccharides

FIG. 1.—Conversion of D-glucose to allyl glucosides.

Methyl α-D-glucopyranoside (methyl glucoside), obtained by the reaction of glucose or starch with methanol, has been produced and evaluated for a number of industrial applications. The most promising of these was for making polyethers for rigid polyurethane foam production (*5*); but unfavorable economics forced its removal from this market for several years, except for minor specialty uses. In 1982, a corn wet-milling company announced plans to start producing this polyol. Higher alkyl glucosides are made from methyl glucoside by a double alcohol interchange, first with *n*-butanol then with a C_8 to C_{18} alcohol (*6*) or by reacting glucose with higher alcohols (*7*). These higher alkyl glucosides are sold commercially as nonionic detergents, in part because of their high alkali resistance and biodegradability. Methyl glucoside has been extensively evaluated in other surfactant applications (*8*) as well as in oil-modified urethane coatings (*9*) and alkyds. Most of the work on methyl glucoside was conducted by J. P. Gibbons and co-workers and published in technical literature by CPC International.

Allyl glucoside was prepared and evaluated in the laboratory as a polyol initiator for making polyethers for flame retardant urethane foam application. One approach involves the stepwise reaction of glucose with allyl alcohol, propylene oxide, and bromine, illustrated in Figures 1 and 2 (*10*). Alternatively, the halogen was incorporated through free radical addition of CCl_4 and $CBrCl_3$ to the double bond of propoxylated allyl glucoside (*11*). Similar products were obtained by an alcohol interchange of a commercial-grade methyl glucoside polyether with allyl alcohol followed by halogenation (*12*). All these polyethers produced good quality, flame-resistant foams.

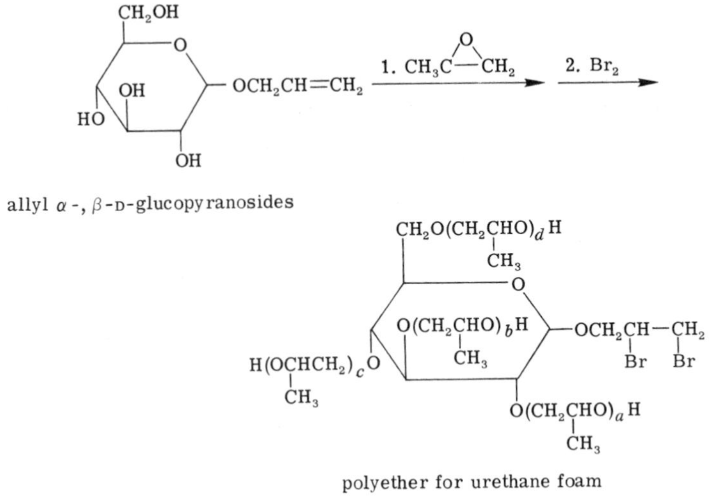

CH_2OH / OH / HO / OH — $OCH_2CH{=}CH_2$

1. $CH_3C(O)CH_2$ (propylene oxide) → 2. Br_2 →

allyl α-, β-D-glucopyranosides

$CH_2O(CH_2CHO)_dH$ (CH_3)

$O(CH_2CHO)_bH$ (CH_3)

$H(OCHCH_2)_cO$ (CH_3)

$-OCH_2CH-CH_2$ (Br, Br)

$O(CH_2CHO)_aH$ (CH_3)

polyether for urethane foam

FIG. 2.—Conversion of allyl glucosides to brominated polyethers.

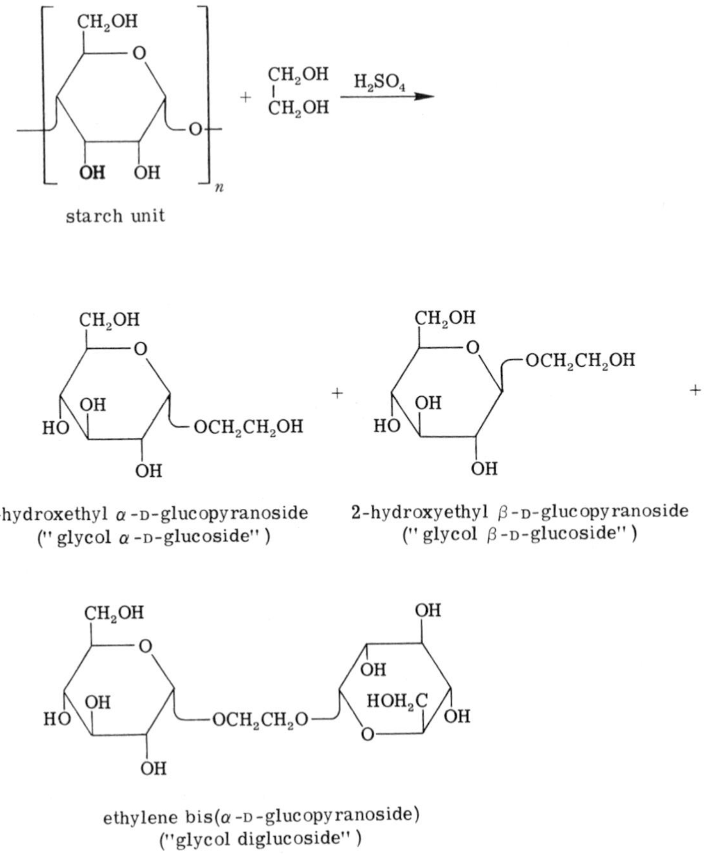

starch unit (CH_2OH, OH, OH, $-O-$)$_n$ + CH_2OH–CH_2OH $\xrightarrow{H_2SO_4}$

2-hydroxethyl α-D-glucopyranoside
("glycol α-D-glucoside")

+

2-hydroxyethyl β-D-glucopyranoside
("glycol β-D-glucoside")

+

ethylene bis(α-D-glucopyranoside)
("glycol diglucoside")

FIG. 3.—Conversion of starch to glycol glucosides.

An experimental polyol, obtained from the reaction of starch with ethylene glycol, may find large-scale application in rigid urethane foams, surfactants, and alkyds. The polyol is made by heating a mixture of 1 mole of starch, 4 moles of glycol, and an acid catalyst at 130° for 45 min. About 0.8 mole of glycol reacts per anhydroglucose unit and a part or all of the remaining is removed by vacuum distillation, depending upon the intended application for the polyol. The product is a mixture of glycol glucoside polyols shown in Figure 3 (*13*). Similar products were obtained by reacting starch with glycerol. In 1969, Ashland Chemical Company, under a USDA contract, estimated a raw material cost of 8.81 cents/lb ($0.194/kg) and an operating cost of 2.03 cents/lb ($0.045/kg) for producing glycol glycosides in a plant designed to make 7 million pounds (3 million kg) per year (*14*). Updating these figures to reflect the 1980 cost of dry starch at 10 cents/lb and ethylene glycol at 34 cents/lb, the approximate raw material cost to produce the glycol glucoside is now 18.5 cents/lb. Conventional polyols such as pentaerythritol and glycerol now sell for 60–65 cents/lb.

The most promising application for these glycol or glycerol glucosides is as polyol initiators for making polyethers for rigid urethane foam production. If glycerol glucosides are used for this application, all the unreacted glycerol need not be removed because it will help lower polyether viscosity and will not greatly affect foam properties. Polyethers are obtained by reacting the crude glucosides with 6 moles of propylene oxide (Fig. 4). Treatment of the polyethers with polyisocyanates in the presence of blowing agent, surfactant, and catalysts produces good quality foams (*15*). The projected 1 billion pounds (450 million kg) per year foam market could utilize about 100 million pounds (45 million kg) of starch. Several companies have evaluated the production of glucoside-based polyethers and one has made large-scale runs.

The glycol and glycerol glucosides were extensively evaluated in the laboratory (*16*) and pilot plant (*17*) as raw materials for biodegradable surfactants. First, the glucosides were reacted with ethylene or propylene oxide or a mixture of the oxides to form alkoxylated glucosides. These products were not fully characterized but apparently the alkoxides were randomly substituted at the hydroxyl sites. The alkoxylated products were then esterified with long-chain fatty acyl groups or etherified with long-chain α-olefin oxides. To illustrate the general class of products, a conceptual structure of an alkyl ether obtained by reacting 1 mole of dodecane 1-oxide with an ethoxylated glycol glucoside is shown. Ob-

$CH_2O(CH_2CH_2O)_eH$

$H(OCH_2CH_2)_dO$ — $O(CH_2CH_2O)_cH$ — $OCH_2CH_2O(CH_2CH_2O)_{\bar{a}}{-}CH_2CH(OH)(CH_2)_9CH_3$

$O(CH_2CH_2O)_bH$

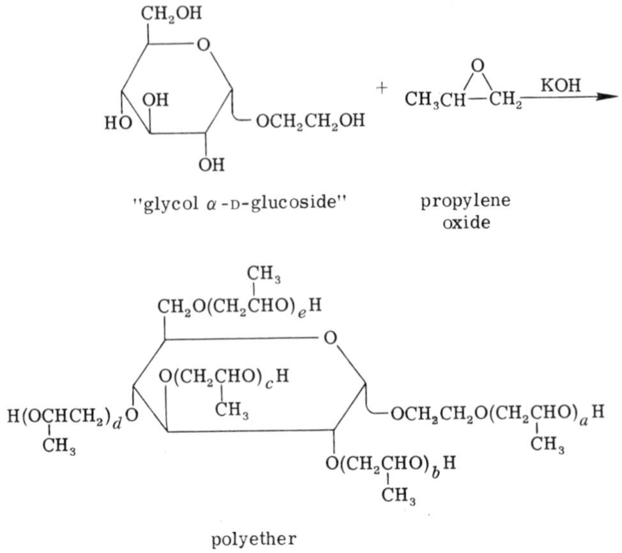

FIG. 4.—Starch-derived polyethers.

viously, this is only one of five possible isomers since the fatty ether could be attached at any of the hydroxyl sites. The surfactancy and detergency performance of the new products were rated as equivalent to those of ethoxylated octyl phenol. However, they show a significant improvement in biodegradability. Langdon suggested the application of fatty ether glycosides as alkali-stable surfactants (*18*). Feuge and co-workers (*19, 20*) have extensively evaluated glycerol glucoside esters in emulsifier applications. Others have made and evaluated food emulsifiers from the glycerol glucosides (*21*).

Glycol glucosides were studied for use in both conventional and urethane alkyds (*14*). These alkyds were evaluated in both clear and pigmented films and were found to have superior drying and hardness, and equivalent flexibility, but slightly inferior gloss and yellowing characteristics when compared with industrial controls based on pentaerythritol–glycerol polyols. Preferably, the alcoholysis or esterification reactions were conducted with a mixture of up to 55% glycol glucoside and at least 45% glycerol, which suggests that glycerol glucosides made without removing all excess glycerol would be the more economically feasible route to making these alkyds.

Other polyols have been made from starch, including erythritol and ethylene glycol which were prepared in 90% of theory by reductive hydrolysis of dialdehyde starch (*22*). This process is not expected to become industrially feasible unless dialdehyde starch is made available at the lowest possible cost. Numerous studies have been conducted on the hydrogenolysis of saccharidic materials, such as dextrin, starch, and glucose; a variety of polyols are produced (*23*). The low cost of starch and increasing costs of petroleum-based polyols suggest that additional research could lead to economically feasible routes for enzymically or chemically depolymerizing starch into low-molecular-weight polyols for such applications as antifreeze and synthetic fabric production.

$$[\text{dialdehyde starch}]_n + 3H_2 + H_2O \longrightarrow n\,HOCH_2\text{-}CHOH\text{-}CHOH\text{-}CH_2OH + n\,HOCH_2\text{-}CH_2OH$$

dialdehyde starch erythritol ethylene glycol

IV. Starch in Plastics

A large portion of the chemicals produced from petroleum is utilized to manufacture the nearly 40 billion pounds (18 billion kg) of synthetic polymers now used annually in the United States. Plastics, which now account for about 75% of the total synthetic polymer production, will double in usage volume during the next decade if raw materials are available. In recognition of this growth potential and the uncertainties in availability of sufficient petroleum feed stocks, interest has increased in the use of natural polymers as extenders for plastics and as total replacements for certain types of plastics. Not only is it the renewable aspect of raw materials such as starch that has piqued the interest of industry, but also the potential of such natural polymers to impart biodegradability to fabricated materials. Because of concern over buildup in the environment of discarded plastic goods due to their resistance to microorganisms, the plastics industry is giving considerable attention to this area. If plastics can be made readily biodegradable, new markets for such materials would materialize, and the growth for plastics would likely exceed even the most liberal estimates.

Starch was evaluated as an inert filler in poly(vinyl chloride) (PVC) plastics, as a reactive filler in rigid urethane foams, and as a component in poly(vinyl alcohol) and ethylene–acrylic acid copolymer films. Three techniques were investigated for incorporating large amounts of starch as a filler in PVC plastics (*24*). In one, a starch derivative was coprecipitated with a PVC latex and the coprecipitate was filtered off, dried, hammer-milled to a fine powder, blended with dioctyl phthalate (DOP), and molded in an aluminum cavity. In a second, starch was gelatinized and mixed with PVC latex, and water was removed from the mixture in an oven. The dry product was milled, mixed with DOP, and molded as before. The third technique involved dry blending starch, PVC, and

Table III

Properties of Plastics Made with Starch and 25 Parts of DOP per 100 Parts of PVC[a]

Starch, %	Tensile strength, psi	Elongation, %	Specific gravity	Fungi resistance[b]	Clarity[c]
	Coprecipitate: Starch Xanthate–PVC (Geon 151)				
0	2,560	133	1.27	0	12
12.3	2,720	66	1.23	1	12
32.9	2,240	8	1.36	4	16
51.0	1,920	5	1.35	3	20
	Coconcentrate: Starch–PVC (Geon 151)				
0	3,650	140	—	0	13
12.9	3,220	44	1.24	1	16
34.0	3,360	13	1.31	4	15
	Dry Mix: Starch–PVC (Geon 126)				
0	2,600	150	1.25	0	6
13.4	2,170	110	1.27	1	43
38.2	1,370	31	1.30	4	85
	Dry Mix: Starch–PVC (Geon 102)				
0	3,600	140	—	0	—
30	2,930	104	—	—	70
40	3,040	35	—	—	78

[a] DOP = dioctyl phthalate; PVC = poly(vinyl chloride).
[b] ASTM D-1924-70, fungus growth; 0 = none; 1 = 10%; 2 = 10–30%; 3 = 30–60%; 4 = 60–100%.
[c] Relative values: 0 is completely clear and 100 is opaque.

DOP on a rubber mill and then molding. Tensile strength remained good even with as much as 50% starch in the plastic (Table III). Clarity of the plastics also was good except for those made by dry blending, but elongation decreased rapidly as the starch level increased.

The three formulations were also blended on a rubber mill until films could be removed from the rolls. Properties of the film were measured and their longevity in Weather-Ometer and outdoor exposure tests was evaluated (Table IV). By varying the composition, films were obtained that lasted from 40 to 900 h in the Weather-Ometer and from 30 to more than 120 days in the soil. All samples tested under standard conditions with common soil microorganisms showed mold growth, with the greatest amount of growth recorded for samples containing the highest amount of starch.

One area of application for biodegradable plastic film is agricultural mulch. An estimated 60 million pounds (27 million kg) of plastic mulch is used in the

United States to improve the yield and quality of vegetable and fruit crops such as tomatoes, peppers, melons, and sweet corn. Mulch helps to control soil moisture and temperature, reduce nutrient leaching, prevent weed growth, and increase crop yields by 50–350%. Polyethylene film, commonly used for mulch, does not degrade between growing seasons and must be removed from the field and buried or burned at an estimated cost of $100 per acre ($250/hectare). Considerable interest has been expressed in using a degradable mulch to obviate the need for removal and disposal (*25*).

Griffin reported that starch can be incorporated into low-density polyethylene (LDPE) film to impart biodegradable properties (*26*). Based on this work, Coloroll Ltd., England, is now producing a polyethylene bag that contains 7–10% starch and is reportedly biodegradable.

Starch–poly(vinyl alcohol) (PVA) films may have application as a degradable agricultural mulch. In preliminary studies, a composition containing 60–65% starch, 16% PVA, 16–22% glycerol, 1–3% formaldehyde, and 2% ammonium chloride was combined with water to give 13% solids and heated at 95° for 1 h. The hot mixture was then cast and dried at 130° to a clear film. Films, on removal from the hot surface, were passed through a solution of PVC or Saran to coat the film with a water-repellent coating. Uncoated films are insoluble in water, but their wet strength is low. Coated films retain good strength even after water soaking. Weather-Ometer tests suggest that film with 15–20% coating might last 3–4 months in outdoor exposure (*27*).

Since 1977, one company has been using the starch–PVA technology to produce a water-soluble laundry bag for use by hospitals in which to keep soiled or contaminated clothing prior to washing. The bags and their contents are placed directly into washing machines, where the bag dissolves. In order to provide enhanced solubility, a slightly derivatized starch is used for this application.

Table IV

Properties of Starch–PVC–DOP Films

Starch		*PVC resin*					
Type	*%*	*Type*	*%*	*DOP, PHR*[a]	*Tensile strength, psi*	*Outdoor exposure,*[b] *days*	*Weather-Ometer,*[b] *h*
Whole	25.1	Geon 126	57.5	25	1,720	120	458
Xanthate	40.5	Geon 151	39.7	50	1,860	>150	120
Gelatinized	44.0	Opalon[c]	37.2	50	740	52	170
Gelatinized	36.4	Geon 151	36.4	75	420	66	—

[a] Parts of DOP per 100 parts of PVC.

[b] Films showed major deterioration after given time.

[c] Monsanto Company.

Such water-soluble bags are also being suggested for packaging agricultural chemical pesticides to improve safety during handling.

$$-CH_2-CH_2-CH_2-\underset{\displaystyle COOH}{\underset{|}{CH}}-CH_2-CH_2-CH_2-CH_2-CH_2-\underset{\displaystyle COOH}{\underset{|}{CH}}-$$

EAA

Films that have potential application in biodegradable mulch, packaging, and other products were obtained from various combinations of starch and ethylene–acrylic acid copolymer (EAA). Three forms of EAA obtained from Union Carbide Corp. were evaluated: PCX-300, dry pellets; PCX-140, a mechanically produced latex containing 40% EAA; and PCX-100, a 25% ammonium solution of EAA. The EAA has a carbon backbone with occasional pendant carboxyl groups that can react with ammonium hydroxide to solubilize the resin. The EAA evaluated contained about 20% reacted acrylic acid. Films were prepared by casting heated aqueous dispersions of starch and the copolymer followed by oven drying or by fluxing dry mixtures of starch and EAA on a rubber mill. Properties of these films are listed in Table V. The starch–EAA system can also be ex-

Table V

Properties of Starch–Ethylene Acrylic Acid Copolymer Films from Aqueous Dispersions and Dry Mixing on a Rubber Mill

Composition, %			*Tensile strength, psi*	*Elongation, %*	*Folding endurance, no. double folds*	*Burst factor*	*Fungi[a] resistance*
Starch	*EAA*	*Stearic acid*					
			Starch and PCX-100				
0	100	0	3288	295	No break	21.8	0
20	76	4	2483	116	2972	20.7	2
40	56	4	2528	44	531	21.7	4
60	37	3	3186	21	349	20.6	4
80	19	1	5345	10	61	15.0	4
			Starch and PCX-140				
0	100	0	3336	349	—	—	0
20	80	0	3898	226	—	—	4
40	60	0	4077	87	—	—	4
60	40	0	4993	44	—	—	4
			Dry Mixtures of Starch and PCX-300				
0	100	0	6699	48	—	—	0
20	80	0	4519	48	—	—	0–1
40	60	0	1897	53	—	—	4
60	40	0	2685	21	—	—	4

[a] ASTM D 1924-63; mold coverage: 0 = none; 2 = 10–30%; 3 = 30–60%; 4 = 60–100%.

truded. Duplicate samples of the cast films containing 30, 40, 50, 70, and 90% starch were exposed to outdoor soil contact, with the ends buried in soil, to observe their resistance to sunshine, rain, and soil microorganisms. Films with more than 40% starch deteriorated within 7 days while those containing 30 and 40% starch remained flexible and provided mulch protection for at least 30 days. When the films containing up to 50% starch were formulated with 2% paraformaldehyde, they provided good soil protection for at least 70 days. These preliminary soil exposure tests demonstrate that starch–EAA films are biodegradable. Furthermore, some control in the rate of biodegradation can be built into the films by varying the amount of starch used and by adding a fungicide such as paraformaldehyde. Other tests revealed that the starch–EAA films have sufficient strength, flexibility, water resistance, and heat sealability for a variety of mulch and packaging applications (*28*).

In a subsequent study, films were prepared and evaluated from various combinations of starch and pelletized EAA, using the less expensive extrusion blowing technique (*29*). Extrusion blowing is a common, economical method for producing film in which a tubular extrudate is expanded and shaped by internal air pressure to form a bubble several times the size of the die opening. The process also allowed the incorporation of LDPE as a partial replacement of the EAA, which further reduced the film cost and in some instances improved properties. Properties of the films are shown in Table VI.

Not only can starch be mixed with synthetic polymers and exhibit utility as a filler, extender, or reinforcing agent, it can also become an integral part of such polymers through chemical bonds. Urethane is an example of a system where starch has been chemically bonded to a resin. To produce relatively low-cost, rigid urethane plastics that might have application in solvent-resistant floor tile, a system was developed based on 10–60% starch, castor oil, the reactive products of castor oil and starch-derived glycol glycosides, and polymeric diisocyanates (*30*). The addition of starch to the isocyanate resins substantially reduced chemical costs and improved solvent resistance and strength properties. Evidence showed that the starch chemically combined with the resin molecules.

The degree of reactivity between starch and isocyanates can be greatly enhanced by modifying starch with non-polar groups, such as fatty acid esters, before the isocyanate reaction (*31*). Maximum reactivity of the modified starches was achieved when the degree of substitution was about 0.7.

Elastomers have been prepared with starch as a filler and crosslinking agent for diisocyanate-modified polyesters (*32*). Dosmann and Steel (*33*) reported that starch can be incorporated into urethane systems to yield shock-resistant foams. Bennett and co-workers (*34*) found that up to 40% starch or dextrin can be incorporated into rigid urethane foam and that such foams are more flame resistant and more readily attacked by microorganisms.

An alternative approach to making plastics from starch may depend upon the

Table VI

Effect on Film Properties of Starch and LDPE Levels

Run no.	Formulation, %[a,b]			Tensile strength, psi	Elongation, %	Ammonia[c]	MIT fold, no. folds	Burst factor	Weather-Ometer, h	Fungi susceptibility, weeks[d]			
	Starch	EAA	LDPE							1	2	3	4
1	10	90	0	3470	260	4.9	—	—	402	0	0	0	0
2	20	80	0	4140	120	4.3	—	—	212	0	0	0	0
3	30	70	0	3225	150	3.8	—	—	168	0	0	0	0
4	40	60	0	3870	92	3.3	—	—	90	1	1	1	1
5	50	50	0	3940	61	2.7	—	—	90	1	2	3	3
6	40	50	10	3570	80	3.6	3800	24	111	2	3	4	4
7	40	40	20	3477	66	2.2	7000	24	134	1	2	3	4
8	40	30	30	3150	85	1.7	2700	21	151	1	2	3	4
9	40	20	40	2920	34	2.8	4800	19	199	—	—	—	4
10	40	10	50	1840	10	2.8	470	9	559	—	—	—	4
11[e]	30	32.5	32.5	2000	62	1.6	—	—	710	—	—	—	—

[a] Based on combintion dry weight of starch, EAA, and PE, exclusive of water and NH_3.
[b] Formulations of examples 4, 6, 9, and 11 additionally contained about 1% antioxidant (''Irganox 1035,'' Ciba Geigy Corp.)
[c] Parts of ammonia per 100 parts of formulation dry weight.
[d] ASTM D 1924-70, fungus growth, 0 = none, 1 = 10%, 2 = 10–30%, 3 = 30–60%, 4 = 60–100%.
[e] Formulation contained 5% carbon black (Industrial Reference Black, No. 3).

use of microbial polysaccharides. One such example includes pullulan, a linear chain of maltotriose units, obtained by the action of a yeast-like fungus on starch. In a review of this subject, Shipman and Fan (*35*) report that pullulan-derived plastics resemble styrene in gloss, hardness, and transparency, but are much more elastic.

V. Starch Graft Copolymers

1. Starch Graft Copolymers in Plastics

Another approach to chemically bonded natural polymer–synthetic polymer compositions is through graft polymerization. This technique has received considerable attention of scientists at the Northern Regional Research Center, especially for those systems where the natural polymer is starch. Basically, the procedure used for synthesizing starch graft polymers is to initiate a free radical on the starch backbone and then allow the radical to react with polymerizable vinyl or acrylic monomers. A number of free-radical initiating methods have been used to prepare graft copolymers and these may be divided into two broad categories: initiation chemically and by irradiation. The choice depends in part on the particular monomer or combination of monomers to be polymerized. A conceptual structure of a starch graft polymer is shown in Figure 5. Both chemical and irradiation initiating systems have been employed to graft polymerize onto starch a wide variety of monomers, both alone and in selected combinations. Fanta and Bagley (*36*) recently reviewed these systems.

For plastic or elastomeric copolymer compositions which can be extruded or milled, monomers such as styrene, isoprene, acrylonitrile, and various alkyl acrylates and methacrylates were employed.

Starch-graft-polystyrene, -poly(methyl methacrylate), -poly(methyl acrylate), and -poly(methyl acrylate-*co*-butyl acrylate) polymers have been prepared with approximately 50% add-on and evaluated for extrusion processing characteristics (*37*). A 20:80 (by weight) mixture of starch-graft-polystyrene and commercial polystyrene produced an extrudate which was filled with particles of unfluxed graft copolymer after two passes through the extruder at 150°. Addition of glycerol to the mixture as a plasticizer for starch did not greatly improve extrudate properties. However, when the graft copolymer was extruded at 175° in the absence of additives, a continuous, well-formed extrudate was produced. Tensile strengths for specimens milled from the extrudate were in the range 7500–9100 psi (527–640 kg/cm^2) (Table VII).

Two starch-graft-poly(methyl acrylate) products, one prepared from granular starch and the other from gelatinized starch, were extrusion processed under various conditions. The graft copolymer prepared from granular starch was extruded (three passes) through a 1 × 0.020-in. (25 × 0.5-mm) slit die at 160° and

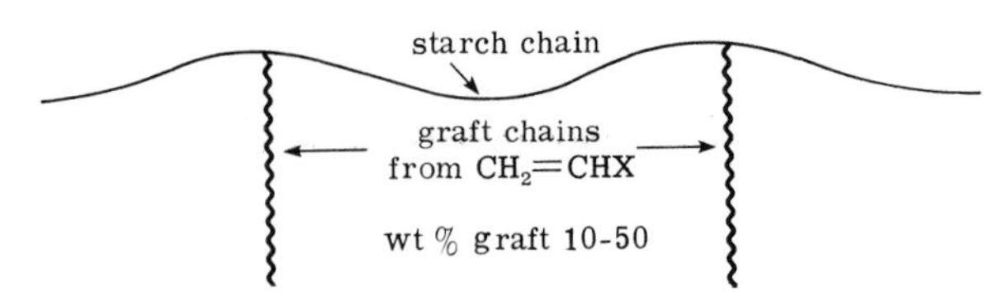

FIG. 5.—Conceptual structure of starch graft polymers. When X = —CO_2H, —$CONH_2$, —$CO_2(CH_2)_n^+$—NR_3Cl^- products are water soluble and useful as thickeners, absorbents, sizes, adhesives, and flocculants. When X = —CN, —CO_2R, ⬡— products are water insoluble and potentially useful as resins and plastics.

a barrel temperature of 150°. The extruded ribbon was smooth and translucent and exhibited good tensile strength (3000 psi, 211 kg/cm^2) and little die swell. Lower extrusion temperatures produced an extrudate which contained unfluxed polymer, and only a crumbly mass was obtained at 125°.

The graft copolymer prepared from gelatinized starch was extruded (one pass) at 125° with a die temperature of 140° to give an extrudate which resembled that obtained from granular starch, but which was less brittle. When extruder and die temperatures were lowered to 100°, a continuous plastic was still produced, although there was an appreciable amount of unfluxed polymer. Tensile strength of the higher temperature extrudate was on the order of 2500 psi (176 kg/cm^2) and die swell was minimal. Prolonged soaking of the extrudate in water at room temperature produced a material that was white, soft, and pliable and that showed appreciable increases in both weight and thickness (Table VIII). The specimen surface, however, was not sticky, and the plastic remained continuous and showed no tendency to disintegrate. To estimate biodegradability, a portion of the extrudate was incubated for 5 days at 25° with three different cultures in a nutrient solution as suggested by ASTM DI924-70. *Aspergillus niger* and *Trichoderma viride* gave good growth and sporulation. *Penicillium funiculosum* produced little sporulation but gave good growth.

Ceric ion-initiated graft polymerization of acrylonitrile, methyl acrylate, and chloroprene onto gelatinized cationic starch yielded copolymers with up to 60% polyvinyl side chains (*38, 39*). When starch-graft reaction mixtures of any of the copolymers were sonified at 20 kHz for 1–3 min, latexes that dried at room temperature or 70° to clear adhesive films resulted. Viscosity of the sonified dispersions of about 8% concentration was in the range 10–40 cP. Molecular weights of synthetic side chains were influenced by the type of cationic charge on the starch and on the type of stirring action employed during the polymerizations (Table IX). Number-average molecular weights in the range of 100,000 to 1,000,000 were readily obtainable. This work was recently extended to the preparation of starch-graft-poly(styrene-*co*-acrylonitrile) latexes (*40*). Interest in the preparation of latexes with higher solids content at low viscosities and in the evaluation of the latexes in various end-use applications continues.

Table VII

Polysaccharide Graft Copolymers

Monomer, g	Polysaccharide, g	Water, mL	Initiator	% Add-on[a]		Graft MW	UTS,[b] psi
				Weight gain	Weight loss		
Styrene (40)	Granular starch (40)	10	^{60}Co, 1.0 mrad	48[c]	—	—	7,500–9,100
Styrene (40)	Granular starch (40)	10	^{60}Co, 1.0 mrad	40	41	710,000	—
Methyl methacrylate (160)	Gelatinized starch (100)	2,500	Ce^{4+}	47	50	1,360,000	—
Methyl acrylate (300)	Granular starch (250)	2,000	Ce^{4+}	39	41	845,000	3,000
Methyl acrylate (120)	Gelatinized starch (100)	2,500	Ce^{4+}	44	42	861,000	2,500
Methyl acrylate (63.7)	Cellulose (53.2)	2,000	Ce^{4+}	49	—	—	—
Methyl acrylate (75) and butyl acrylate (75)	Gelatinized starch (100)	2,500	Ce^{4+}	53	—	—	—

[a] Weight gain: % add-on determined by gain in weight of starch after graft polymerization. Weight loss: % add-on determined by loss in weight of graft copolymer after depolymerization and removal of starch.

[b] UTS: Ultimate tensile strength of the extruded plastic.

[c] Crude graft copolymer; homopolymer not extracted.

Table VIII

Effect of Water on Starch-Graft-Poly(methyl acrylate) Extrudate[a]

Immersion time, h	% Increase		
	In weight	In thickness	In width
0	—	—	—
22.5	54	41	12.5
476.0	53	41	12.5

[a] 1 × 0.022 in. extrudate from polymer 5, Table VI. Extruded at 125° with die temperature of 140°. Soaked in water at room temperature.

Young and co-workers (*41*) report the development and evaluation of a modified starch containing acrylamidomethyl groups, starch-$OCH_2NHCOCH{=}CH_2$, capable of polymerization to produce cross-linking within the starch or copolymerization with water-soluble acrylate monomers to yield an acrylic network. Among a variety of potential applications for the starch product, plastics have been formulated for ultraviolet and thermal cure.

2. Starch Graft Copolymers as Water Absorbers

A wide variety of other monomers have been graft polymerized onto granular and gelatinized starch, and several of the graft polymers show promise as thick-

Table IX

Influence of Cationic Functionality (CF) on Starch on Number-Average Molecular Weight ($\overline{M}_n$) of Grafted Polyacrylonitrile [Poly(AN)]

CF[a]	N, %	Agitator	AGU/Ce(IV)	Starch graft product, N, %	Poly(AN), $\overline{M}_n$
TA	0.42	ST[b]	40	12.25	1,094,000
TA	0.42	SH[c]	40	13.50	506,000
QA	0.39	ST	45	13.07	232,000
QA	0.39	SH	45	12.81	178,000

[a] Cationic starch (24 g), gelatinized in 600 mL of water by heating under nitrogen for 30 min, was cooled to 25°, and reacted with AN and cerium(IV) reagent. The AN:AGU molar ratio was 3.1 and the reaction time, 3 h. TA = tertiary amine; QA = quaternary amine.

[b] Reaction was conducted in a round-bottomed flask using a rotating blade stirrer at 240 rpm.

[c] Reaction was conducted in a stoppered Erlenmeyer flask clamped to a Burrell wrist-action shaker.

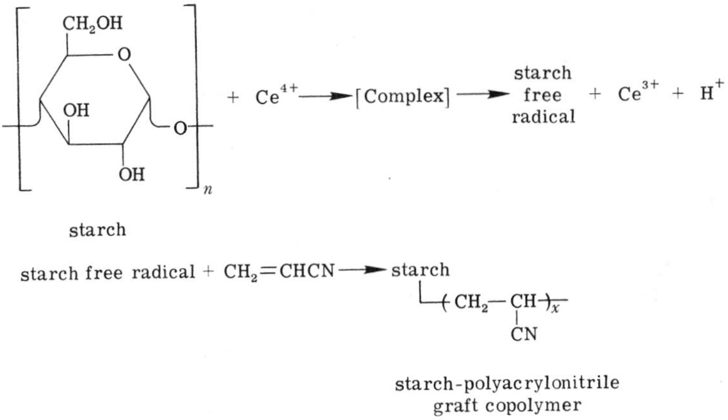

$$\left[\text{starch anhydroglucose unit}\right]_n + Ce^{4+} \longrightarrow [\text{Complex}] \longrightarrow \text{starch free radical} + Ce^{3+} + H^+$$

starch

$$\text{starch free radical} + CH_2{=}CHCN \longrightarrow \text{starch}{-}\!\left(CH_2{-}\underset{\displaystyle CN}{CH}\right)_x$$

starch-polyacrylonitrile graft copolymer

FIG. 6.—Preparation of starch–polyacrylonitrile graft copolymer.

eners for aqueous systems, flocculants, clarification aids for wastewaters, retention aids in papermaking, and many other uses. The polymer that has received the most attention and is now being marketed by four U.S. companies is made by graft polymerizing acrylonitrile onto gelatinized starch and subjecting the resulting starch-graft-polyacrylonitrile copolymer to alkaline saponification to convert the nitrile groups to a mixture of carbamoyl and alkali metal carboxylate groups (Fig. 6). Removing the water from this polymer provides a solid which absorbs many hundreds of times its weight of water but which does not dissolve (Fig. 7). Because of its ability to rapidly absorb such large amounts of water, it has been named Super Slurper. As of this writing the U.S. Department of Agriculture has granted more than 50 non-exclusive licenses to parties interested in practicing the technology covered in the four patents issued in 1976 on work with this starch graft polymer (*42*).

Aqueous dispersions of saponified gelatinized starch-graft-polyacrylonitrile can be cast to yield films on drying. These films are brittle but can be plasticized to improve flexibility. Flexible films can be obtained on casting aqueous dispersions of the graft copolymer containing an anionic or nonionic latex (*43*). The films will absorb several hundred times their weight of distilled water to give

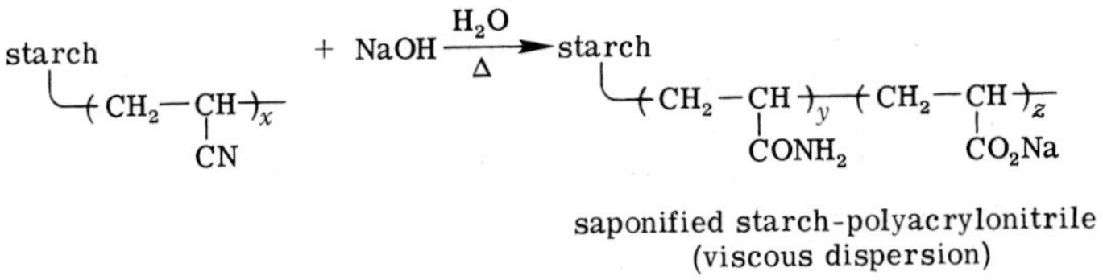

$$\text{starch}{-}\!\left(CH_2{-}\underset{\displaystyle CN}{CH}\right)_x + NaOH \xrightarrow[\Delta]{H_2O} \text{starch}{-}\!\left(CH_2{-}\underset{\displaystyle CONH_2}{CH}\right)_y\!\left(CH_2{-}\underset{\displaystyle CO_2Na}{CH}\right)_z$$

saponified starch-polyacrylonitrile (viscous dispersion)

FIG. 7.—Saponification of starch–polyacrylonitrile graft copolymer.

FIG. 8.—Absorbency of Super Slurper in film form.

clear sheets of gel, which remain strong enough to maintain their integrity. Also, despite a thirtyfold increase in surface area upon imbibing water, the exact shape of the dry film is retained (*44*) (Fig. 8). Reducing the pH of the water to about 3 causes the gel sheet to retract to its original size, and raising the pH back to about 7 or higher causes the film to return to a highly swollen gel sheet.

Other forms of the absorbent polymer are obtained by alternative methods of drying. Alcohol precipitation yields a granular or powdery product, whereas drum drying affords flakes and freeze-drying gives spongy mats. Selected end-use applications may dictate the form of the product most desirable.

For application as an additive to absorbent soft goods, such as disposable diapers, incontinent pads, bandages, hospital bed pads, and catamenials, interest has been expressed in the powder, film, and mat forms and the first two forms are being used commercially in incontinent pads and hospital bed pads. The ability of the absorbent polymer to retain most of its absorbed fluid under pressure is a desirable property for such applications (*45*) (Table X). Under a pressure of $45 \times g$, a Super Slurper product which had absorbed 648 times its weight of water still retained 409 times its weight. Cellulose fibers, on the other hand initially absorbed 40 times their weight of water and retained only 2.1 times their weight under this pressure.

Partial hydration of the starch graft copolymer provides a hydrogel especially effective in treating skin wounds of animals (*46*). The hydrogel absorbs large quantities of fluids secreted by the wounds, provides relief from pain, and prevents drying of subcutaneous tissue. Clinical trials with the hydrogel conducted with human patients suffering from decubitus ulcers or stasis ulcers gave excellent results (*47*). All skin ulcers responded favorably to the treatment, and the wounds either healed completely or developed a cleaner bed of granulation tissue. In every case, the hydrogel dressing resulted in less eschar formation, fewer infections, and less odor than when ulcers were treated by other methods.

Agricultural applications, such as for seed and root coating and as an additive to fast-draining soils to retain water, appear most promising for Super Slurper. Large-scale field trials with corn, soybean, and cotton seed coated with the polymer have shown increased germination and seedling emergence, and in most trials, increased yield. The dipping of bare-rooted seedlings in hydrated polymer before transplanting overcomes transplant shock and greatly increases survival. For such applications, the powder or granular form is being used.

Grafting of acrylonitrile onto granular (rather than gelatinized) starch followed by saponification in an alcoholic medium also produces an absorbent product (*48*). The polymer produced by this route is under commercial development by Grain Processing Corporation.

Salts of the hydrolyzed starch-graft-polyacrylonitrile copolymer have been evaluated as thickening agents for aqueous systems (*49*). These studies showed that the saponified product was readily recoverable from the saponification mass by lowering the pH to about 3. At this pH, the carboxylate groups are converted to the acid form which causes the polymer to precipitate for easy recovery by filtration or centrifugation. Resuspension of the dry product in water and adjustment of the pH to about 7 produces a low solids viscous dispersion with visible,

Table X

Absorbency of Hydrolyzed Starch-Graft-Polyacrylonitrile (H-SPAN) vs. Cellulose Fibers

		g Fluid/g absorbent		
Absorbent	*Fluid*	*Free draining*	*45 × g*	*180 × g*
Cellulose fibers	Water	40	2.1	1.05
	Simulated urine	32	1.8	1.0
H-SPAN (insoluble) air dried	Water	648	409	—
	Simulated urine	54	40	37
H-SPAN (insoluble) drum dried	Water	896	—	—
	Simulated urine	60	—	—

highly swollen, gel particles. Taylor and Bagley (*50*) confirmed the presence of a substantial gel fraction by isoionic dilution experiments, which failed to yield linear reduced viscosity-concentration plots, and also by an ultracentrifugation study, which showed that water dispersions of the polymer contained only about 20% solubles. They propose that the tremendous thickening action of the polymer in water is due to the nearly complete absorption of solvent by gel to give a system consisting of highly swollen, deformable gel particles closely packed and in intimate contact. Neither the minor amounts of graft copolymer in solution nor the size of the gel particles exerts a large influence on rheological properties. Under high-dilution or high-ionic strength conditions, solvent is in excess; the gel particles are no longer tightly packed, and the viscosity therefore drops sharply.

VI. Starch Xanthide

1. Starch Xanthide in Rubber

The use of starch as a replacement for carbon black has been studied in detail at the Northern Regional Research Center. Cross-linked starch xanthate was incorporated into rubber to provide reinforcement to the same extent as medium grades of carbon black (*51*) (Table XI). Domestic use of carbon black in rubber is about 3 billion pounds, virtually all of which is derived from petroleum. By modification of the process for incorporating the starch derivative into rubber, a

Table XI

Comparative Vulcanizate Properties of Extrusion Processed Starch-SBR and Black Reinforced SBR Rubbers

Vulcanizate[a]	*Hardness shore A*	*Tensile strength, psi*	*Elongation, %*
50 phr starch xanthide-SBR 1,500	70	2,730	310
50.4 phr FEF black-SBR 1,500	63	2,900	360
50.4 phr SRF black-SBR 1,500	58	2,380	480
50.4 phr FT black-SBR 1,500	51	1,880	690
50.4 phr MT black-SBR 1,500	51	1,340	510
Premium grade passenger tread	58	2,990	600
100 level passenger tread	56	2,550	620
First line competitive tread	58	2,290	590

[a] Blacks used were fine extrusion furnace black (FEF), semi-reinforcing black (SRF), fine thermal black (FT), and medium thermal black (MT).

Table XII

Formulation Designation

Compound type[a]	Form of polymer base		
	Slab	Powdered rubber with 5 or 6 phr SX[b]	Powdered rubber with 20 phr SX
I. Shoe heel	I.S	I.5	I.20
II. Shoe sole	II.S	II.5	II.20
III. Tire tread	III.S	III.5	III.20
IV. Mechanical compound (mixed polymer base)	IV.S	IV.5	IV.20
V. Mechanical compound (SBR polymer base)	V.S	V.5	V.20
VI. White NBR compound	VI.S	VI.5	VI.20

[a] Roman numerals correspond to the formula type and S, 5, and 20 correspond to the nature of the raw polymer used. S corresponds to slab rubber; 5, to powdered rubber with 5 or 6 phr SX; and 20, to powdered rubber with 20 phr SX.
[b] SX = starch xanthide.

simple and economically feasible process was developed for making powdered rubber, a long-sought goal of the rubber industry. A crosslinked starch xanthate–rubber coprecipitate containing 3–5% starch and 95–97% rubber can be readily blended with various rubber additives (Table XII) and injection molded to finished rubber goods with good properties (*52, 53*) (Table XIII). Preliminary calculations, based on an estimated 50% market penetration, suggest that the United States could save 2.5 billion kilowatt hours of electricity annually by using the new process (Fig. 9).

Table XIII

Tensile Properties of Milled and Compression-Molded Shoe Sole and Tire Tread Compounds

Compound No.[a]	At break			Modulus	
	Tensile strength, psi	Elongation, %	10 min set, %	100%	300%
II.S	1,500	420	35	140	600
II.5	1,500	340	20	360	1,200
II.20	1,490	340	30	320	1,150
III.S	2,200	420	25	240	1,560
III.5	2,400	350	30	440	2,000
III.20	2,880	470	35	400	1,700

[a] Designations refer to products listed in Table XI.

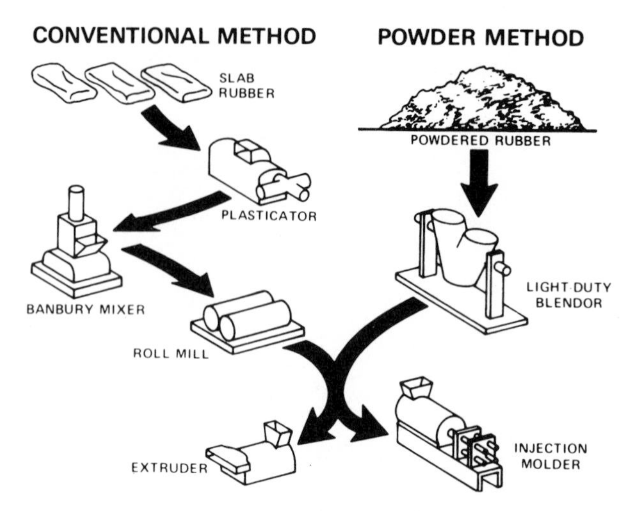

FIG. 9.—Comparison of processing rubber by conventional vs. powder methods.

2. Starch Xanthide Encapsulation

A new technology has been developed for encapsulating a broad range of chemical pesticides within a starch matrix to improve safety in handling and reduce losses of the pesticide in the environment due to volatility, leaching, and decomposition by light (*54, 55*). The procedure, based on starch xanthate, consists of dispersing the pesticide in an aqueous starch xanthate solution and subsequently cross-linking the starch xanthate either oxidatively, or with multi-valent metal ions, or with difunctional reagents such as epichlorohydrin (Fig. 10). Cereal flours, which contain about 10% protein along with starch, also can be xanthated and used as an encapsulating matrix. Upon crosslinking, which is effected within a few seconds under ambient conditions, the entire mass becomes gel-like and, on continued mixing for an additional few seconds, becomes a particulate solid which can be dried to low moisture content with only minimal or no loss of the entrapped chemical. That only a single phase is produced on cross-linking with no supernatant is important in assuring essentially complete entrapment of both water-soluble and water-insoluble pesticidal chemicals.

Both solid and liquid pesticides have been encapsulated by this procedure. Where the pesticide is a liquid or finely divided solid, it is added as such to the aqueous solution. For pesticides provided as granular solids, they are first dissolved in any appropriate solvent and then added to the solution. Formulations were made containing up to 55% by weight of liquid pesticides and even higher amounts of solid ones.

Starch-encapsulated formulations have excellent shelf life: no loss of pesticide was recorded on storage in closed containers for up to a year. During storage in

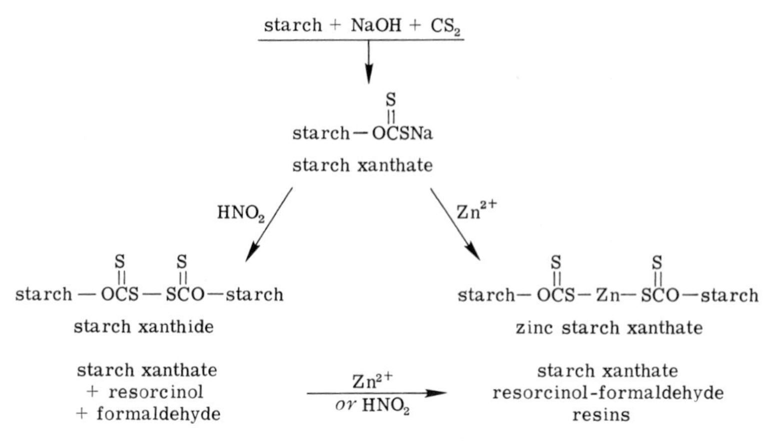

FIG. 10.—Xanthation and cross-linking of starch.

open containers for several weeks, loss of even volatile pesticides was negligible. However, when products are placed in water or soil, active agent is released from the matrix.

Other polymers can be incorporated readily into the products to modify release properties. Polymers like polystyrene, polyethylene, and poly(vinyl chloride) are just dissolved in a small amount of an appropriate solvent such as benzene or acetone then added to the xanthate solution. Poly(styrene–butadiene), commercially provided as a latex, is conveniently added in this form. Upon cross-linking the xanthate, the other polymers are entrapped along with active agents.

Another modification easily made provides products which are doubly encapsulated. This is achieved on addition of more starch xanthate, either alone or containing another polymer, after the initial cross-linking reaction has been effected, and then adding additional cross-linking agent. Schreiber (*56*) recently reported results of greenhouse and field tests of two cross-linked starch xanthate–EPTC (*S*-ethyl dipropylthiocarbamate) products. One formulation con-

Table XIV

Starch–EPTC Formulations in Weed Control

	Total weed	
Treatment[a]	*Count*	*Weight, g*
Untreated	95.3	924.8
EC–EPTC	48.0	247.8
Starch–EPTC (single)	33.7	186.0
Starch–EPTC (double)	33.0	31.0

[a] Applied at a rate of EPTC of 3.36 kg/ha.

tained 14% EPTC and the other was a double-encapsulated product containing 20% latex polymer and 22% EPTC. In greenhouse and field tests, both starch formulations gave better control of weeds than the commercial formulation applied at an equivalent weight of active agent (Table XIV). The double encapsulated product gave excellent control of weeds for 120 days, whereas the commercial product controlled weeds for only about 45 days. Four companies have licensed this technology and several others are now evaluating it with their own particular pesticides. Two alternative methods to the xanthate procedure for encapsulating pesticides within a starch matrix recently have been reported. In both, the pesticide is dispersed in aqueous alkali-gelatinized starch, and the encapsulating matrix is formed by adding either calcium chloride (*57*) or boric acid (*58*).

VII. References

(*1*) E. V. Anderson, *Chem. Eng. News,* **56,** No. 31, 8 (1978).
(*2*) G. E. Tong, *Chem. Eng. Prog.,* **74,** 70 (1978).
(*3*) D. L. Miller, *Biotechnol. Bioeng. Symp.,* No. **6,** 307 (1976).
(*4*) D. L. Miller, *Biotechnol. Bioeng. Symp.,* No. **5,** 345 (1975).
(*5*) D. W. Kaiser and S. Fuzesi, U.S. Patent 3,167,538 (1965).
(*6*) F. E. Boettner, U.S. Patent 3,219,656 (1965).
(*7*) R. C. Mansfield, U.S. Patent 3,839,318 (1974).
(*8*) F. H. Otey, T. E. Yeates, C. L. Mehltretter, and C. E. Rist, *J. Am. Oil Chem. Soc.,* **38,** 517 (1961).
(*9*) H. M. Kennedy and J. P. Gibbons, *Paint Varn. Prod.,* **55,** No. 4, 47 (1965).
(*10*) F. H. Otey, R. P. Westhoff, and C. L. Mehltretter, *J. Cell. Plast.,* **8,** 156 (1972).
(*11*) C. A. Wilham, F. H. Otey, C. L. Mehltretter, and C. R. Russell, *Ind. Eng. Chem., Prod. Res. Dev.,* **14,** 189 (1975).
(*12*) F. H. Otey, C. A. Wilham, and C. R. Russell, *Ind. Eng. Chem. Prod. Res. Dev.,* **17,** 162 (1978).
(*13*) F. H. Otey, B. L. Zagoren, and C. L. Mehltretter, *Ind. Eng. Chem. Prod. Res. Dev.,* **2,** 256 (1963).
(*14*) W. J. McKillip, J. N. Kellen, C. N. Ipola, R. W. Buckney, and F. H. Otey, *J. Paint Technol.,* **42,** 312 (1970).
(*15*) R. H. Leitheiser, C. N. Impola, R. J. Reid, and F. H. Otey, *Ind. Eng. Chem., Prod. Res. Dev.,* **5,** 276 (1966).
(*16*) F. H. Otey, C. L. Mehltretter, and C. E. Rist, *J. Am. Oil Chem. Soc.,* **40,** 76 (1963).
(*17*) P. E. Throckmorton, R. R. Egan, D. Aelony, G. K. Mulberry, and F. H. Otey, *J. Am. Oil Chem. Soc.,* **51,** 486 (1974).
(*18*) W. K. Langdon, U.S. Patent 4,011,389 (1977).
(*19*) R. O. Feuge, M. Brown, and J. L. White, *J. Am. Oil Chem. Soc.,* **49,** 672 (1972).
(*20*) K. M. Decossas, R. O. Feuge, J. L. White, and L. W. Mazzeno, *J. Am. Oil Chem. Soc.,* **55,** 567 (1978).
(*21*) M. M. Bean, C. L. Mehltretter, C. A. Wilham, and T. A. McGuire, *Food Prod. Dev.,* **7,** 30 (1973).
(*22*) F. H. Otey, J. W. Sloan, C. A. Wilham, and C. L. Mehltretter, *Ind. Eng. Chem.,* **53,** 267 (1961).

(*23*) J. A. Finneran, D. H. Martin, and S. M. Frank, U.S. Patent 3,278,398 (1966).
(*24*) R. P. Westhoff, F. H. Otey, C. L. Mehltretter, and C. R. Russell, *Ind. Eng. Chem., Prod. Res. Dev.,* **13,** 123 (1974).
(*25*) D. Carnell, presented at National Agricultural Plastics Association, 14th Agricultural Plastics Congress, Miami Beach, Florida, November 10–13, 1978.
(*26*) G. J. L. Griffin, *Am. Chem. Soc., Div. Org. Coat. Plast. Chem.,* **33,** 88 (1973).
(*27*) F. H. Otey, A. M. Mark, C. L. Mehltretter, and C. R. Russell, *Ind. Eng. Chem., Prod. Res. Dev.,* **13,** 90 (1974).
(*28*) F. H. Otey, R. P. Westhoff, and C. R. Russell, *Ind. Eng. Chem. Prod. Res. Dev.,* **16,** 305 (1977).
(*29*) F. H. Otey, R. P. Westhoff, and W. M. Doane, *Ind. Eng. Chem. Prod. Res. Dev.,* **19,** 592 (1980).
(*30*) F. H. Otey, R. P. Westhoff, W. F. Kwolek, C. L. Mehltretter, and C. E. Rist, *Ind. Eng. Chem., Prod. Res. Dev.,* **8,** 267 (1969).
(*31*) F. H. Otey, R. P. Westhoff, and C. L. Mehltretter, *Staerke,* **24,** 107 (1972).
(*32*) F. W. Boggs, U.S. Patent 2,908,657 (1959).
(*33*) L. P. Dosmann and R. N. Steel, U.S. Patent 3,004,934 (1961).
(*34*) F. L. Bennett, F. H. Otey, and C. L. Mehltretter, *J. Cell. Plast.* **3,** 369 (1967).
(*35*) R. H. Shipman and L. T. Fan, *Process Biochemistry,* **13,** No. 3, 19 (1978).
(*36*) G. F. Fanta and E. B. Bagley, *Encycl. Polym. Sci. Technol., Suppl.* **No. 2,** 665 (1977).
(*37*) E. B. Bagley, G. F. Fanta, R. C. Burr, W. M. Doane, and C. R. Russell, *Polym. Eng. Sci.,* **17,** 311 (1977).
(*38*) L. A. Gugliemelli, C. L. Swanson, F. L. Baker, W. M. Doane, and C. R. Russell, *J. Polym. Sci.,* **12,** 2683 (1974).
(*39*) L. A. Gugliemelli, C. L. Swanson, W. M. Doane, and C. R. Russell, *J. Polym. Sci., Polym. Lett. Ed.,* **14,** 215 (1976).
(*40*) L. A. Gugliemelli, C. L. Swanson, W. M. Doane, and C. R. Russell, *J. Polym. Sci., Polym. Lett. Ed.,* **15,** 739 (1977).
(*41*) A. H. Young, F. Verbanac, and T. F. Protzman, *Coat. Plast. Preprints* **36,** 411 (1976).
(*42*) M. O. Weaver, E. B. Bagley, G. F. Fanta, and W. M. Doane, U.S. Patents 3,935,099; 3,981,100; 3,985,616; and 3,997,484 (1976).
(*43*) E. L. Skinner and L. F. Elmquist, U.S. Patent 4,156,664 (1979).
(*44*) M. O. Weaver, E. B. Bagley, G. F. Fanta, and W. M. Doane, *Appl. Polym. Symp.,* No. 25, 97 (1974).
(*45*) M. O. Weaver, G. F. Fanta, and W. M. Doane, *Proc. Tech. Symp., Nonwoven Product Technol.,* Intern. Nonwovens Disposables Assoc., Washington, D.C., March 5–6, 1974, pp. 169–177.
(*46*) H. Valdez, *J. Equine Practice,* **2,** No. 3, 33 (1980).
(*47*) W. R. Spence, U.S. Patent 4,226,232 (1980).
(*48*) T. Smith, U.S. Patent 3,661,815 (1972).
(*49*) L. A. Gugliemelli, M. O. Weaver, C. R. Russell, and C. E. Rist, *J. Appl. Polym. Sci.,* **13,** 2007 (1969).
(*50*) N. W. Taylor and E. B. Bagley, *J. Appl. Polym. Sci.,* **18,** 2747 (1974).
(*51*) R. A. Buchanan, W. F. Kwolek, H. C. Katz, and C. R. Russell, *Staerke,* **23,** 350 (1971).
(*52*) T. P. Abbott, W. M. Doane, and C. R. Russell, *Rubber Age,* **105,** No. 8, 43 (1973).
(*53*) T. P. Abbott, C. James, W. M. Doane, and C. R. Russell, *J. Elast. Plast.,* **7,** No. 2, 114 (1975).
(*54*) B. S. Shasha, W. M. Doane, and C. R. Russell, *J. Polym. Sci., Polym. Lett. Ed.,* **14,** 417 (1976).
(*55*) W. M. Doane, B. S. Shasha, and C. R. Russell, *Am. Chem. Soc., Symp. Ser.,* **53,** 74 (1977).

(56) M. Schreiber, Efficacy and Persistence of Starch Encapsulated EPTC, presented at the North Central Weed Control Conference, Omaha, Nebraska, December 7–9, 1976.

(57) B. S. Shasha, D. Trimnell, and F. H. Otey, *J. Polym. Sci., Polym. Chem. Ed.*, **19,** 1891 (1981).

(58) D. Trimnell, B. S. Shasha, R. E. Wing, and F. H. Otey, *J. Appl. Polym. Sci.*, **27,** 3919 (1982).

CHAPTER XII

CORN AND SORGHUM STARCHES: PRODUCTION

By Stanley A. Watson

Ohio Agricultural Research and Development Center, The Ohio State University, Wooster, Ohio

I. Introduction

The word "corn" is used in the United States as the common name for the cultivated member of the grass family (Gramineae) known to botanists as *Zea mays* L. More specifically, "corn" here means the seed produced by this plant. Outside of the United States, this crop is commonly known as maize. Corn is believed to be a product of domestication in Central Mexico beginning 5000–7000 years ago. The evidence is now quite strong that domestic corn was derived by human selection from mutants of the grass, teosinte (*Zea mays,* ssp. *Mexicana*) which grows wild in Mexican central highlands (*1*). Recent discovery of a distinct new species, *Zea diploperennis* Iltis (*2*), a wild perennial teosinte with the same number of chromosomes as corn has given new validity to the teosinte

STARCH, 2nd ed.

ISBN 0-12-746270-8

origin. Corn has reached its present state of development through continual mutations, hybridizations, segregations, and selections by random, natural processes and conscious selection. By this process, a number of types have developed which differ primarily in structure of the seed. Examples are popcorn, sweet corn, dent corn, flint corn, and flour corn.

Corn has been the staple food for countless generations of Indians of North and South America. The diverse seed types were probably selected and cultivated by these primitive farmers in response to food preferences. ''Mahys'' specimens taken to Spain by Columbus on an early voyage, introduced corn to Europe. The first recorded planting of corn was near Seville in 1494 (*3*) from whence it subsequently spread to all of Europe, Asia Minor, and eventually to the Far East. It is interesting that a mutant of considerable industrial importance, waxy maize, was first discovered in China in 1909 (*4*). Flint types of corn gradually became an important crop across Southern Europe and Turkey. However, it was the development of American dent types and their eventual hybridization that propelled corn to the position of agricultural dominance as the most cost effective feed grain so important in development of today's highly specialized animal agriculture.

The ready availability of corn at relatively low and steady prices, its storability from season to season, and its high starch content led naturally to development of commercial processes for recovery of corn starch. From the early nineteenth century, when Yankee inventors first discovered that corn starch was fairly easily recovered by grinding the soaked grain, the process has gradually evolved into today's highly sophisticated automated process, which produces a multitude of useful food and industrial products. Mechanical innovations developed by trial and error during full-scale operations stimulated much of the early growth of the corn starch industry (*5*). Today, process and product improvements more commonly follow research and engineering studies and thorough pilot plant evaluations.

Process innovations have resulted from pressures to improve worker efficiency and work place environment, to reduce energy consumption, to reduce air and water pollution, to improve end-product quality, and to introduce new products. In the last 20 years, older facilities have been completely redesigned and expanded, and ten completely new wet milling facilities have been built in the United States since 1970. Such installations now require much less space than the units they replaced; working conditions and plant sanitation have vastly improved; continuous operation and product monitoring have improved product quality and uniformity.

Grain sorghum (*Sorghum bicolor Moench*) is a cereal grain also known in some localities as milo, milo maize, or kaffir corn. Sorghum culture probably began in Eastern Africa 5000 to 7000 years ago and spread to all of Africa, Europe, and Asia and eventually to the United States in the mid-nineteenth century (*6*). The sorghum plant resembles corn, but the seeds are borne on a terminal, bisexual rachis (head) and are about the size and shape of No. 5 shot (*7*). Some sorghum varieties have juicy stalks of high sugar content and are grown for syrup production on very limited acreage. They bear small, inedible seeds and are not to be confused with the grain sorghums.

Although grain sorghums are a major world crop, they generally are considered to be inferior to corn for food, feed, or industrial uses. They require less water for growth than corn and, therefore, are grown in more arid regions. In this chapter, some space will be devoted to grain sorghum starch production, but it is of marginal interest, because the only known manufacturer of sorghum starch in the United States stopped production in 1975. The wet milling of grain sorghum is similar to that of corn and has been thoroughly described elsewhere (*8*).

II. Structure, Composition, and Quality of Grain

Wet milling, as a process to recover starch, is essentially a method of disrupting the corn or sorghum kernel in such a way that the component parts can be separated in an aqueous medium into relatively pure fractions. Dry milling using screening and air classification for component separation of corn and sorghum achieves less efficient fractionation (*9*). Satisfactory starch recovery cannot be practically attained by dry milling methods even when fine grinding and air classification processing are used (*10*). Only wet milling can achieve a commercially satisfactory yield and quality of starch from corn or grain sorghum. Attempts to combine the wet and dry milling processes have been reported (*163, 164*). The energy saving obtained by dry separation of germ and fiber and wet processing of endosperm only would be great. However, no practical application has been achieved because germ (oil) yield by dry milling is significantly lower than the yield from the wet milling process.

1. Structure

Mature corn or sorghum kernels (seeds) are unique, well-organized entities that exist for the purpose of reproducing the species. A description of their structure and composition is helpful in understanding the process of disruption that is achieved in the wet milling process.

a. *Corn*

The corn kernel is a caryopsis or berry, a one-seeded fruit, borne on a female inflorescence commonly known as the ear. Each ear is comprised of a central stem, the cob, on which up to 1000 seeds develop. The seeds (kernels) mature in about 60 days after pollination and are harvested in late summer or early fall in

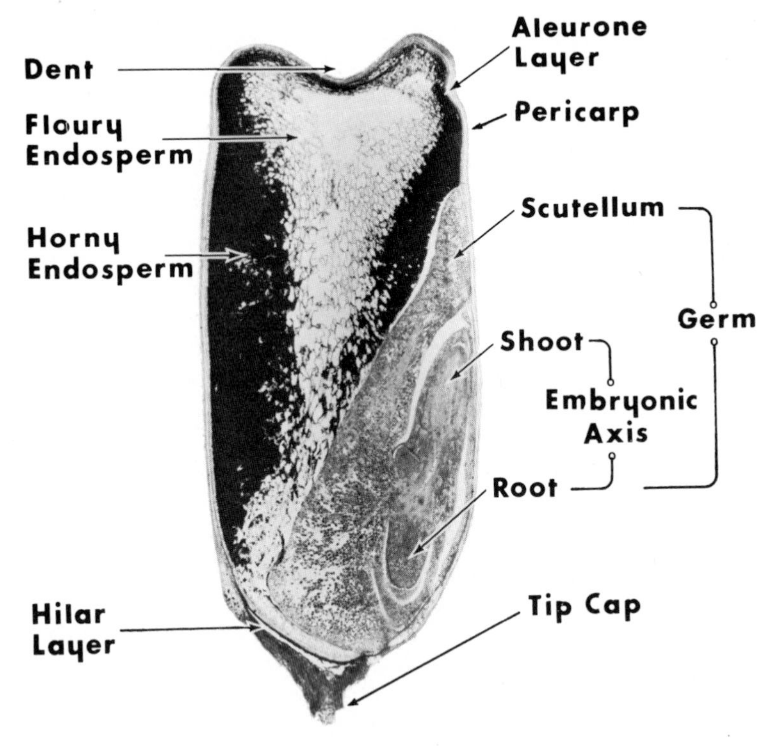

FIG. 1.—Longitudinal 10 μm bisection of steeped dent corn kernel; magnified 6×; iodine stained. Note how starch has been lost from floury endosperm cells as a result of sulfur dioxide action.

the Northern hemisphere when kernel moisture content has dropped below 30% (wet weight basis).

The mature corn kernel is composed of four principle parts: Tip caps, 0.8%; pericarp, 5%; germ, 11%; and endosperm, 82% (Fig. 1) (*11*).

The tip cap is the remnant of the pedicle which attaches the seed to the cob. It is composed of insoluble fibrous elements arranged in a spongy structure well adapted for rapid water adsorption (*12*). When the tip cap is removed, a black tissue known as the hilar layer, which apparently functions as a sealing layer across the base of the germ upon kernel maturation, is revealed (*12, 13*).

The pericarp (hull) is the smooth, dense, outer covering of the kernel (*12*). Its outer layer has a thin coating of wax-like cutin, beneath which are several layers of dead, hollow, elongate cells compressed into a dense tissue. Beneath this layer is a spongy tissue known as the tube cell and cross cell layer which acts as a natural channel for absorption of water. Directly beneath the spongy layer is a very thin suberized membrane known as the seed coat or testa. This layer is thought to act as a semi-permeable membrane, limiting passage of large molecules into or out of the endosperm and germ.

Immediately beneath the testa is the first layer of the endosperm, the aleurone, a single layer of cells having thick, tough cell walls. Aleurone comprises about 3% of the weight of the kernel (*14, 15*). Although structurally part of the endosperm, the aleurone, along with the pericarp and tip cap, is included in the "fiber" fraction in wet milling and in the bran fraction in dry milling.

The mature endosperm is comprised of a large number of cells, each packed with starch granules embedded in a continuous matrix of protein surrounding each granule and the whole cell contents surrounded by a cellulosic cell wall (Fig. 2) (*14, 16, 17*). Mature endosperm of dent corns contains a central core of soft or floury endosperm extending to the crown which shrinks on drying, causing a "dent." The floury endosperm is surrounded by a glassy appearing region known as horny or hard endosperm. The average ratio of these two regions is about 1:2, floury:horny regions, but varies considerably depending on the protein content of the kernel (*11, 18*). Flour corns contain virtually no horny endosperm, while flint corn and popcorn endosperms are comprised of a small core of floury endosperm entirely surrounded by horny endosperm which resists denting when the kernel dries. Although the dividing line between horny and floury regions of the endosperm is morphologically indistinct, the floury region is characterized by larger cells, large round starch granules, and a relatively thin protein matrix (*14*). The thin strands of protein matrix rupture during drying of the kernel, causing air pockets which give the floury region a white opaque appearance (*16*) and a porous texture which makes for easy starch recovery.

In the horny endosperm, the thicker protein matrix shrinks during drying but does not rupture. The resulting pressure produces a dense glassy structure in which the starch granules are forced into an angular, close-packed conformation. The dense nature of horny endosperm requires adequate steeping to insure recovery of starch. The horny endosperm contains only about 1.5–2.0% higher protein content (*15, 18*) and more of the yellow carotenoid pigments (*19*) than the floury region; but just under the aleurone layer there is a dense row of cells known as subaleurone or dense peripheral endosperm (*20*) containing as much as 28% protein (*15*). These small cells comprise probably less than 5% of the endosperm. They contain very small starch granules and a thick protein matrix and can cause difficulty in starch purification (*20*).

The germ is comprised of two major parts: the scutellum and the embryonic axis (Fig. 1). The embryonic axis is the structure that grows into the seedling upon germination and makes up only 10% of the weight of the germ (*21*). The scutellum functions as a storage organ from which nutrients can be quickly mobilized during initial seedling growth. The surface of the scutellum adjacent to the endosperm is covered by secretory epithelium, a layer deeply furrowed by canals or glands lined with elongated secretory cells. The function of these cells is to secrete enzymes (*22*) which diffuse into the endosperm where they digest starch and other constituents to provide nourishment for the embryo.

The scutellar epithelium adheres to the endosperm by an insoluble cementing

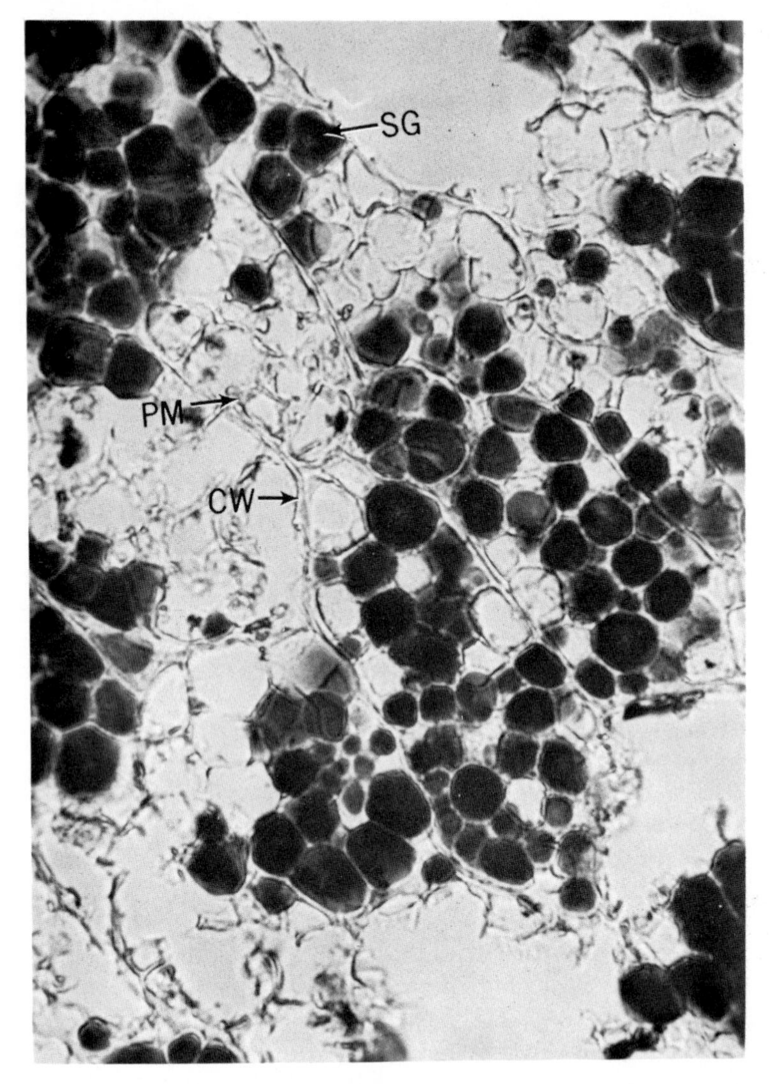

FIG. 2.—Section of steeped corn horny endosperm; 10 μm thick; iodine stained; magnified 612×; SG, starch granules; PM, protein matrix; CW, cell wall.

substance, apparently consisting of degradation products of crushed cells and composed largely of pentoglycans and protein (*23*). This layer provides a strong bond that resists many chemical and physical means of separating the germ and endosperm; hence, prolonged steeping is required for an effective separation of whole intact germ from the endosperm. The major portion of the scutellum is composed of thick-walled isodiametric cells densely packed with cytoplasm. Oil droplets may be seen in sections of such cells and have been shown to be nearly identical to the spherosome bodies (L. Yatsu, Southern Regional Research Center, USDA, New Orleans, Louisiana, personal communication) shown to be the repository of oil in cottonseed and peanut cotyledons (*24*).

b. *Grain Sorghum*

The structure and composition of the sorghum kernel is quite similar to that of corn. The kernels are flattened spheres measuring about 4.0 mm long, 3.5 mm wide, and 2.5 mm thick. The weight of individual kernels ranges from 8 to 50 mg, with an average of 28 mg. Colors of the prehybrid varieties range from white through pale orange, tan and red, to dark red-brown (*6*). Most grain now in commercial channels is a brownish red color because one or both parents of the best hybrids have that color.

A longitudinal section of a typical sorghum grain (Fig. 3) is nearly identical to

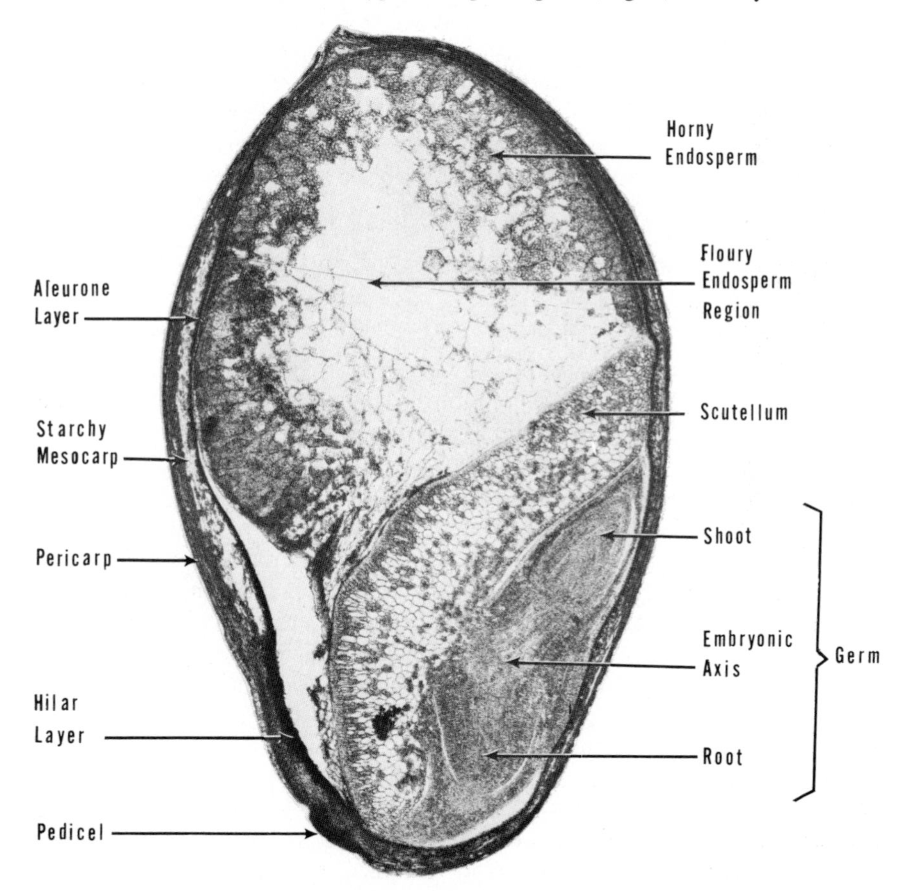

FIG. 3.—Longitudinal 10 μm bisection of steeped grain sorghum kernel; magnified 22×; iodine stained.

the corn kernel in gross morphology. The structure of the germ is identical to that of corn germ. The endosperm differs from that of corn only in the relative proportions of floury and horny endosperms; all sorghums are like flint corn in that the horny endosperm entirely surrounds the floury region. The dense peripheral endosperm layer comprises a larger proportion of the kernel than in corn and results in much greater problems during starch purification (*20*). Common sorghums have white endosperm. However, yellow (xanthophyll pigmented) endosperm types were introduced from Africa for breeding purposes. Several hybrids have been developed which have a high percentage of horny endosperm where the pigment is concentrated, but they have not been successful because the carotenoid pigments bleach out rapidily in the field before harvest.

The pericarp of grain sorghums show the most differences from corn. It is covered with a thick layer of wax (*25*); most varieties have a thick mesocarp layer in the pericarp containing very small unrecoverable starch granules; some varieties have a thick orange pigmented testa layer (*26*) which shatters during wet milling and adds colored particles to the starch. The epidermis layer of pericarp in some varieties contains water-soluble, flavone-type pigments ranging in color from red to orange. In some varieties, a purple flavonoid contained in the glume which surrounds each seed leaches into the seed and gives a gray cast to isolated starch.

2. Composition

a. *Corn*

The chemical composition of commercial corn based on proximate analyses is shown in Table I. The average values for main components are averages for corn purchased for several large wet milling plants in 1970–74. The range of values cover analyses from numerous sources (*27*). Commercial corn, which is a thorough mixture from many sources, has a much narrower range. Two fiber values are given in Table I. The crude fiber value is a crude measure of cellulose, but is in disrepute because of non-specificity. A newer preferred method measures neutral detergent residue (NDR) which includes all insoluble, cell-wall polysaccharides (*28*). Acid detergent residue (ADR) measures the sum of cellulose and lignin. About 90% of the kernel NDR fiber content is in the pericarp and tip cap. Because lignin content is negligible, the ADR value is essentially a measure of the cellulose content. The difference between the two values (NDR − ADR) is equal to the pentoglycan (hemicellulose) content.

The chemical composition of the kernel parts is given in Table II. Data were obtained by hand dissecting kernels of nine typical corn belt hybrids (*29*). These data show that endosperm contains over 98% of the total starch. The endosperm also contains 74% of the kernel protein and 16% of the kernel fat. Endosperm lipids, mostly bound to cell contents, are only 18% triglycerides, whereas germ

Table I

Proximate Analysis of Corn Grain

	Range[a]	*Average*[b]
Moisture (%, wet basis)	7–23	16.7
Starch (%, dry basis)	64–78	71.3
Protein[c] (%, dry basis)	8–14	9.91
Fat (%, dry basis)	3.1–5.7	4.45
Ash (oxide) (%, dry basis)	1.1–3.9	1.42
Pentoglycans (as xylose) (%, dry basis)	5.8–6.6	6.2
Crude fiber (%, dry basis)	1.8–3.5	2.66
Fiber (neutral detergent residue) (%, dry basis)	8.3–11.9	9.5
Cellulose + lignin (acid detergent residue) (%, dry basis[d])	3.3–4.3	3.3
Sugars, total (as glucose) (%, dry basis)	1.0–3.0	2.58
Total carotenoids (mg/kg)	5–40	30

[a] Numerous sources, including reference *27* and private data.
[b] Moisture, starch, protein, and fat values average of corn purchased on open market 1970–1974.
[c] $N \times 6.25$.
[d] Lignin content less than 0.1% (*158*).

lipids are nearly all free triglycerides (*30*). Thus, the 84% of the kernel fat that is in germ is all that can be economically recovered. Germ also contains 22% of the kernel protein, 82% of the total ash, and 65% of the sugar. Germ contains all the phytin phosphorus in corn (*37*), probably as a Mg^{2+} salt.

There are three classes of corn recognized in the U.S. Grain Standards:

Table II

Weight and Composition of Component Parts of Dent Corn Kernels from Twelve Midwest Hybrids (29)

Part		*Percent dry weight of whole kernel*	*Composition of kernel parts (% d.b.)*				
			Starch	*Fat*	*Protein*	*Ash*	*Sugar*
Germ	Range	10.5–13.1	5.1–10.0	31.1–38.9	17.3–20.0	9.38–11.3	10.0–12.5
	Mean	11.5	8.3	34.4	18.5	10.3	11.0
Endosperm	Range	80.3–83.5	83.9–88.9	0.7–1.1	6.7–11.1	0.22–0.46	0.47–0.82
	Mean	82.3	86.6	0.86	8.6	0.31	0.61
Tip-cap	Range	0.8–1.1	—	3.7–3.9	9.1–10.7	1.4–2.0	—
	Mean	0.8	5.3[a]	3.8	9.7	1.7	1.5
Pericarp	Range	4.4–6.2	3.5–10.4	0.7–1.2	2.9–3.9	0.29–1.0	0.19–0.52
	Mean	5.3	7.3	0.98	3.5	0.67	0.34
Whole kernels	Range	—	67.8–74.0	3.9–5.8	8.1–11.5	1.27–1.52	1.61–2.22
	Mean	100	72.4	4.7	9.6	1.43	1.94

[a] Composite.

yellow, white, and mixed. White corn and mixed white and yellow corn classes are not purchased by the wet milling industry because yellow corn is what farmers produce and therefore is cheaper. Furthermore, the yellow pigment, which is valued in poultry feed, is concentrated in the gluten.

Composition of corn can be altered by plant breeding methods. Corn selection studies at the University of Illinois (*18, 31*) have shown that the oil or protein content can be increased or decreased over a wide range. Breeding corn for higher oil content has been seriously pursued for many years and has resulted in development of good performing hybrids containing 6–9% of oil (dry basis) (*32*). Higher oil corn has been strongly desired by the wet milling industry (*34*) because oil is a product of high value. However, adapted hybrids have not been developed by the seed corn industry because of low economic incentive (*33*).

Average oil content of commercial corn hybrids planted in the corn belt has declined over the last 25 years from 4.8–5.0% to 4.4% (*34*). The apparent cause of this decline has been the widespread use of one or two high yielding, low-oil-content corn inbreds. At the same time, the iodine value of corn oil has increased from 122–124 to 128–130, equivalent to a linoleic acid content increase of from 55% to 60%. This change is apparently associated with the decline in oil content (*35*).

Increasing the total protein content of corn by breeding has never resulted in development of high-protein, commercial hybrids, partly because use of low or high levels of nitrogen fertilizers can also cause corn to vary from 8% to 12% protein. Furthermore, the increased protein is predominantly zein (*36, 37*) which is nearly devoid of two essential amino acids, tryptophan and lysine.

Search for corn of better protein nutritional value led to the discovery (*38*) that the floury type of endosperm mutants, termed opaque-2 and floury 1, are significantly higher in lysine and tryptophan than is normal dent corn. A great deal of breeding work has since been conducted, but hybrids having acceptable yields and kernal hardness have not yet been developed. However, if high-lysine hybrids are eventually developed, they probably will not be of much significance to the wet milling industry because of poor economics and little product advantage (*39*). Laboratory studies indicate a higher steepwater yield, a lower yield of gluten meal, and a lower starch yield might be expected. The higher lysine content of the gluten meal would be of little economic significance.

The most important genetic change in corn for starch production was development of waxy corn hybrids (*40*). The starch from waxy corn, which contains 100% of the branched starch fraction, amylopectin, has become a major food starch (Chaps. III and XIX). This corn must be grown in fields isolated from dent corn to prevent cross pollination, and it must be grown under contract to assure delivery of uncontaminated shipments of waxy corn to the wet milling establishment. The wet milling properties of waxy corn are very similar to those of regular corn except that the yield of solubles is a little higher. Also, starch and gluten slurries filter and dry a little slower than those from normal starch. This

Table III

Proximate Analysis of Grain Sorghum

	Range[a], *% dry basis*	*Average*[b], *% dry basis*
Water (% wet basis)	8–20	15.5
Starch	60–77	74.1
Protein ($N \times 6.25$)	6.6–16	11.1
Fat (CCl_4 extract)[c]	1.4–6.1	3.7
Ash	1.2–7.1	1.5
Crude fiber	0.4–13.4	2.6
Pentoglycans (as xylose)	1.8–4.9	2.5
Sugars, total (as glucose)	0.5–2.5	1.8
Tannin	0.003–0.17	0.1
Wax	0.2–0.5	0.3

[a] Numerous sources including reference *27* and private data.

[b] Average of grain purchased in South and West Texas, 1959–1970, for use in wet milling.

[c] Includes wax.

characteristic has been shown to be caused by the presence of a small amount of a phytoglycogen. Addition of a small amount of an amylolytic enzyme to the slurries gives them normal filtering and drying characteristics (*41*).

High-amylose corn having starch which contains 55–80% of the linear starch fraction, amylose, have also been developed (*42*) (Chapter III). High-amylose corn must also be grown under contract, but the acreage grown is much smaller than for waxy hybrids because of more limited uses for the starch. Starch yields of high-amylose corn are lower than those of dent corn (*43*).

b. *Grain Sorghum*

The chemical composition of commercial grain sorghum based on proximate analyses is shown in Table III. The average values for the major components were obtained on grain purchases from South and West Texas for wet milling at Corpus Christi, Texas, 1959–70. Grain sorghum differs from corn in minor ways. Moisture content is generally lower because the grain dries more in the field before harvest. Compared with corn, starch and protein contents are 1–2% higher, while total fat is always lower because the germ constitutes a smaller proportion of the kernel than it does in corn (Table IV). Furthermore, the pericarp wax amounts to about 8% of the total fat (*44*) and must be removed in refining the crude oil. Analysis of the wax indicates that it is comprised of saturated fatty acids, aldehydes, and alcohols of 27–30 carbon atoms (*45*). All sorghum varieties contain tannins, but the dark-brown, bird-resistant varieties contain tannin levels so high as to reduce digestibility when fed (*46*).

Table IV

Weight and Composition of Component Parts of Kernels from Five Grain Sorghum Varieties (44)

		Percent dry weight of whole kernel	Composition of kernel parts (% dry basis)			
			Starch	*Protein*	*Fat*	*Ash*
Germ	Range	7.8–12.1	—	18.0–19.2	26.9–30.6	—
	Mean	9.8	13.4[a]	18.9	28.1	10.36[a]
Endosperm	Range	80.0–84.6	81.3–83.0	11.2–13.0	0.4–0.8	0.30–0.44
	Mean	82.3	82.5	12.3	0.6	0.37
Bran	Range	7.3–9.3	—	5.2–7.6	3.7–6.0[b]	—
	Mean	7.9	34.6[a]	6.7	4.9	2.02[a]
Whole kernel	Range	—	72.3–75.1	11.5–13.2	3.2–3.9[b]	1.57–1.68
	Mean	—	73.8	12.3	3.6	1.65

[a] Composite.
[b] Includes wax.

As with corn, the only recoverable crude oil is in the germ; it amounts to 76% of the total fat (Table IV). Total fat in sorghum is 80–85% that of corn and includes 8% wax (Table III). Germ contains 62% of the ash in the kernel. About 92% of the total kernel starch is in the endosperm. Unlike corn, another 3.7% of the total starch is in the mesocarp layer (Fig. 3) and is not recoverable because of the small size of the granules.

Plant breeding methods has also been used to modify grain sorghum kernel properties. Waxy varieties and hybrids which contain starch composed of 100% amylopectin have been developed (*6, 47*). White-pericarp, waxy hybrids were processed for production of waxy starch in the United States between 1955 and 1970. High-amylose sorghum has not been found, but a high-lysine mutant has (*48*).

3. Quality of Grain

a. *Corn*

Kernels with the optimum quality desired are those that are fully matured, unblemished, carefully harvested, and carefully dried. However, commercial corn is seldom up to this peak of perfection. Problems with quality of corn relate to the degree of negative deviations from optimum quality. Reductions in quality are caused by harvesting immature kernels, insects, molds, mechanical damage, and drying damage. Freeman (*49*) has thoroughly described all quality parameters that have a bearing on wet-milling performance.

One method of measuring grain quality is provided by the U.S. Grade Standards (*50*). The official standards (Table V) are based on five criteria which have been designated official methods of determination. The grade factor which is the lowest of the five criteria determines the grade of a particular lot of corn. Grade is determined at the marketing point by licensed inspectors who issue official inspection certificates. As corn moves from farm to end user, the lots become more uniform owing to extensive blending, both conscious and circumstantial by each handler. As a result, corn moving toward the end user tends to average about U.S. No. 2 grade. Nevertheless, differences in grade do exist, and buyers apply discounts for each grade factor lower than U.S. No. 2. The discount is related to the negative effect that factor may have on the wet milling and yield and quality of products (*49*).

Discounts for high moisture content is a matter of dilution of the dry weight of the grain substance; however, presence of moisture above 14.5% may result in the development of mold if grain temperature exceeds 22°–24° over a long period of time. Corn containing over 18% moisture rapidly becomes moldy at favorable temperatures. Molds, usually species of *Aspergillus* or *Penicillium*, grow on the moist kernels if temperature and moisture conditions are right for germination of their ubiquitous spores (*51*). Mycelium growth is especially vigorous in the nutrient-rich germ. The resulting decomposition causes reduction in germ recovery and increased oil refining losses owing to fragmentation of damaged germs during milling and hydrolysis of glycerides to free fatty acids (*52*). The damage factor in the U.S. Grade is a visual estimation of the percent of kernels having mold-invaded germs.

Table V

United States, Grading Standards for Corn[a] *(50)*

Grade	Minimum test weight per bushel (lb.)	Maximum limits: Moisture (%)	Broken corn and foreign material (%)	Damaged kernels: Total (%)	Heat-damaged kernels (%)
1	56	14.0	2.0	3.0	0.1
2	54	15.5	3.0	5.0	0.2
3	52	17.5	4.0	7.0	0.5
4	49	20.0	5.0	10.0	1.0
5	46	23.0	7.0	15.0	3.0
Sample grade[b]	—	—	—	—	—

[a] Grades and grade requirements for the classes Yellow Corn, White Corn, and Mixed Corn.

[b] Sample grade shall be corn which does not meet the requirements for any of the grades from No. 1 to No. 5, inclusive; or which contains stones; or which is musty, or sour, or heating, or which has any commercially objectionable foreign odor; or which is otherwise of distinctly low quality.

Another undesirable result of mold growth on corn is the production of mycotoxins, especially the carcinogenic aflatoxins caused by growth of *Aspergillus flavus* (*53*). Frequency of occurrence in the U.S. Midwest is low (*54, 55*), but corn grown in Southern climates is especially vulnerable (*56*), mainly because of damage and inoculation of kernels in the fields (*57, 58*). Laboratory wet milling tests have shown that fractionation of corn containing aflatoxin results in starch that is free of the mycotoxin, but aflatoxin is concentrated in the byproduct streams (*59*). Corn infected with *A. flavus* shows a blue-green-yellow (BGY) fluorescence under ultraviolet light (*60*). Wet-milling companies now monitor incoming corn for presence of aflatoxin using an ultraviolet fluorescence test for presence of aflatoxin (*61*) and avoid purchasing obviously moldy corn.

Other mycotoxins of concern in corn in some years is zearalenone and deoxynivalenol (vomitoxin). These compounds are produced by ear-rotting molds of the *Fusarium* (*Giberella*) genus (*53*) which frequently invade corn in the field but may become epidemic during years of cold, wet fall weather (*62*). Laboratory wet-milling tests have shown that zearalenone in corn is concentrated in feed and is not present in starch (*63, 64*).

The grade factor of "Test Weight" is an indirect measure of grain maturity. An early frost, such as happened in Northern Illinois and Indiana in 1974, usually stops maturation of corn. When the corn was finally harvested and dried, test weights as low as 35 lb/bu were recorded. Needless to say, starch yield from this corn was quite low and discounts were large.

The grade category of "broken corn, foreign material, and other grains" (BCFM) is predominantly broken corn which is removed at the wet milling plant before steeping because it interferes with water flow in the steeps. It is usually added to gluten feed. If the foreign material is removed, the broken corn can be steeped in a separate slurry tank of SO_2 solution for 8–12 h, then added to the millhouse stream for starch recovery.

High levels of broken corn in commercial channels result from the way it is harvested and dried. Practically all marketed corn is now harvested with field shellers or combines at a moisture content of 22–28% (*65*). Corn at this high moisture level is quickly attacked by spoilage organisms and, therefore, must be dried quickly. Many kinds of corn dryers are in use on farms and in elevators (*66*). The preferred dryers are continuous-flow, heated-air dryers. Corn that has been heated at temperatures greater than 60° during drying produces starch of lowered yield and paste viscosity on wet milling (*67–69*). Corn dried from highest moisture (28–30%) at temperatures above 82° also exhibits reduced oil yield and lower protein in steepwater. The latter observation is an indication that the adverse wet milling separations obtained with high-temperature-dried corn results from protein denaturation (*70, 105*). Unfortunately for the wet miller, dryer damage is difficult to detect and does not adversely affect the feeding value of corn. Therefore, it cannot be penalized by applying price discounts. However, extensive publicity by the wet-milling industry of the adverse effects of use of excessive temperatures in drying corn and higher energy costs resulting from poor thermal efficiency have reduced the severity of the problem. The heat damage category in the U.S. Grade standard (Table V) is a measure of severe microbial heating that causes black germs.

Another problem that has resulted from artificial drying of corn is the formation of stress cracks in the horny endosperm (*71*) caused by a high rate of drying and rapid cooling. The degree of stress cracking is directly related to the breakage during subsequent handling (*72*). Stress cracks themselves do not adversely affect wet-milling results, but are the major cause of high levels of BCFM. Brown and co-workers (*73*) developed a steeping index which predicts factory performance of dryer-damaged corn.

b. *Grain Sorghum*

The quality parameters of grain sorghum for wet milling are similar to those of corn. The U.S. Grades for sorghum (Table VI) are similar to those of corn, except that test weight levels are a little higher, moisture levels a little lower, total damage a little lower, heat damage a little higher, and "broken kernels and foreign material" (BKFM) twice as high.

The lower level of total damage in commercial grain sorghum is related to the fact that most grain sorghum is seldom harvested at moisture contents greater than 18%. This low moisture content results because the kernels are individually

Table VI

U.S. Grading Standards for Grain Sorghum[a] (50)

		Maximum limits of			
			Damaged kernels		
Grade	Minimum test weight per bushel (lb.)	Moisture (%)	Total (%)	Heat-damaged kernels (%)	Broken kernels, foreign material, and other grains (%)
1	57	13.0	2.0	0.2	4.0
2	55	14.0	5.0	0.5	8.0
3[b]	53	15.0	10.0	1.0	12.0
4	51	18.0	15.0	3.0	15.0
Sample grade[c]	—	—	—	—	—

[a] Grades and grade requirements for the classes Yellow Grain Sorghum, White Grain Sorghum, Brown Grain Sorghum, and Mixed Grain Sorghum.

[b] Grain sorghum which is distinctly discolored shall not be graded higher than No. 3.

[c] Sample grade shall be grain sorghum which does not meet the requirements of any of the grades from No. 1 to No. 4, inclusive; or which contains stones; or which is musty, or sour, or heating; or which is badly weathered; or which has any commercially objectionable foreign odor except of smut; or which is otherwise of distinctly low quality.

exposed to wind and sun; thus in the more arid climates, where major production is located, the kernels dry rapidly. Since little artificial drying of sorghum is practiced, there is little dryer damage and less stress cracking. However, some stress cracking probably takes place in the field due to wetting and drying as was shown to be the case with wheat (*74*).

The harvesting of sorghum grain at low moisture levels also makes it less vulnerable to mold damage. However, grain that is to be stored must be dried to 13–14% moisture to prevent mold growth. The same mold species that attack corn will invade grain sorghum and produce mycotoxins. Germ damage and mycotoxins in grain sorghum are just as detrimental in starch production as they are in corn, but frequency of occurrence is very low (*75*). A much greater problem with commercial sorghum in the warmer climates where much sorghum is grown is infestation by weevils and other storage insect pests.

The major quality problem with grain sorghum for wet milling relates to pericarp pigmentation. The U.S. Grades divide grain sorghum into four classes: Yellow Grain Sorghum, White Grain Sorghum, Brown Grain Sorghum, and Mixed Grain Sorghum. The Yellow Grain Sorghum is defined as those varieties which have yellow, salmon-pink, or red seedcoats, and which contain less than 10% of grain sorghum of other colors. It does not refer to yellow (carotenoid containing) endosperm, as is the case with corn. Yellow endosperm types have never achieved a large enough place in commercial channels to be recognized as a grade.

The colors of finished starches are related to the intensity of pericarp colors (*76*). Steepwater color, and probably it's tannin content, varies with the intensity of the pericarp pigments. Some varieties contain pigments which are colorless at acid pH but are bright yellow at pH 9–10. These pigments cause starch to have a gray cast which can be improved by bleaching (*77*).

As mentioned previously, some sorghum varieties have a highly pigmented testa or subcoat. Some varieties have a white pericarp masking a brown subcoat. Such varieties are not classed as White Grain Sorghum, but must be classed as Brown Grain Sorghum. Any variety with a brown subcoat or dark brown pericarp, or both, is undesirable for starch production.

III. Wet Milling

The basic principles of steeping and milling corn for separation of starch are universal because they are dictated by the nature of the corn kernel and the properties of its components, but the machinery assembled to accomplish starch recovery has changed greatly over the years. Individual facilities around the world differ somewhat in the kinds of machinery employed and the configurations of product flow because of differing product outputs, machinery preferences, degree of automation, and age. Most plants in the United States have been modernized to optimize worker output and the process described will be that which is now in most general use. The generalized flow diagram of the starch recovery process is given in Figure 4.

1. Cleaning the Grain

Shelled grain is received in bulk at the wet-milling plants. It is prepared for milling by screening to remove all large and small pieces of cob, chaff, sand, and other undesirable foreign material. Dust and light chaff are removed by aspiration. Cleaning the corn is an important first step in the wet milling process because the presence of small pieces of broken kernels can alter the normal flow of steepwater through the grain mass, resulting in non-uniform steeping. Furthermore, starch granules are washed into the steepwater and are gelatinized during evaporation, resulting in viscous steepwater.

2. Steeping

a. *Principles*

Prior to wet milling, the corn must be softened by a steeping process developed specifically to produce optimum milling and separation of corn components. Steeping is more than simple water soaking of corn. It involves maintaining the correct balance of water flow, temperature, sulfur dioxide concentration, and pH. Corn is normally steeped 30–40 hours at a temperature of 48°–52°. By the end of the steeping period, the kernels should have: (a) absorbed water to about 45% (wet basis); (b) released about 6.0–6.5% of their dry substance as solubles into the steepwater; (c) absorbed about 0.2–0.4 g of sulfur dioxide per kg; and (d) become sufficiently soft to yield when squeezed between the fingers. At this point, the kernel can be easily pulled apart with the fingers. The germ is easily liberated intact and free of adhering endosperm or hull. When the endosperm is mascerated under water, the starch easily separates as a white floc and gluten is obtained as a yellow floc.

b. *Mechanics of Commerical Steeping*

Steeps are large tanks that may be constructed of any material resistant to the corrosive action of solutions of sulfur dioxide and lactic acid at pH 3–4. Stainless steel construction is now most common. Steep tanks usually hold 2000–3500 bushels (50–90 metric tons) of grain.

The tanks are filled with raw grain from an overhead conveyor and are emptied through an orifice at the apex of the conical bottom. The inside surface of the cone bottom is covered with a strainer (slatted wood or stainless steel) for

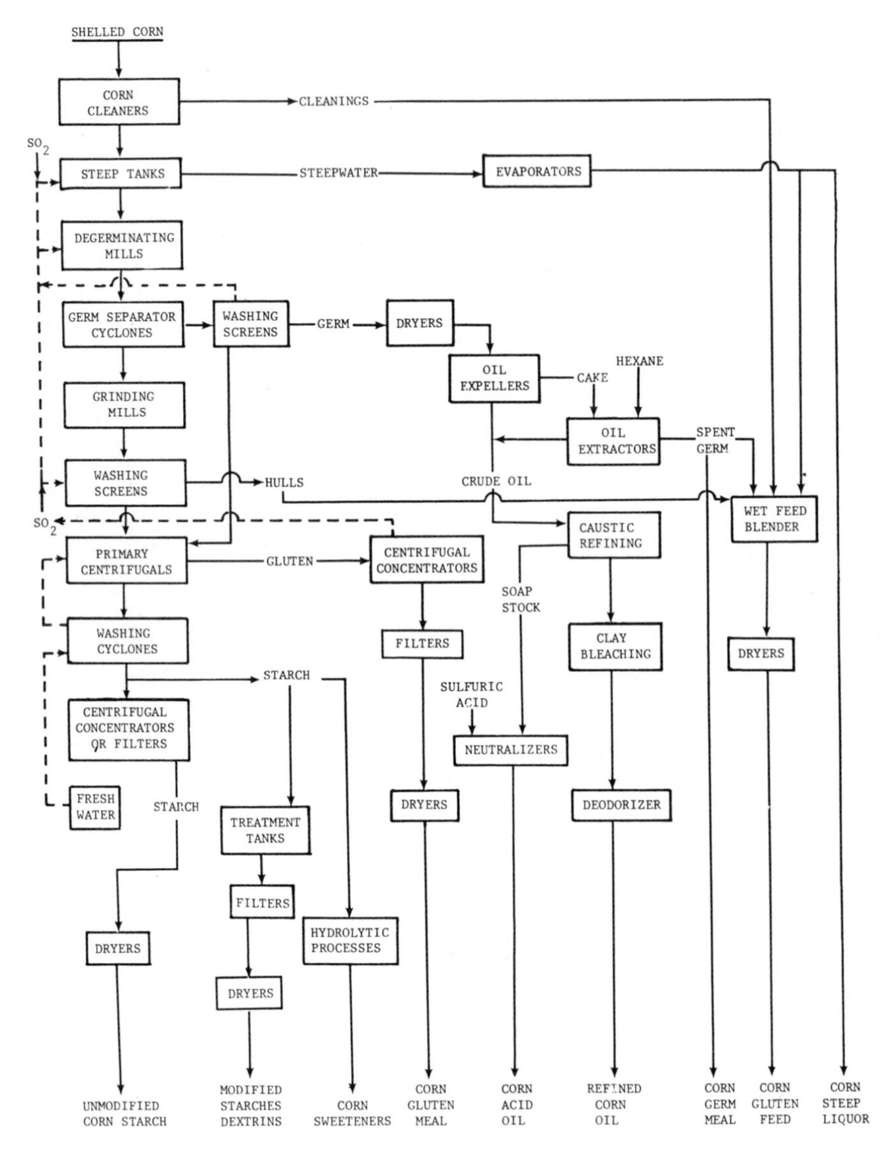

FIG. 4.—Flow diagram of the corn starch manufacturing process.

drainage. Each steep is equipped with piping and a pump to move steepwater from one steep to another, to pass it through a heat exchanger, or to withdraw it from the system.

To prevent the loss of valuable organic matter, and to avoid sewage disposal problems, careful management in the reuse of water is essential. For this reason the entire milling and steeping installation is operated as a countercurrent washing unit. Fresh water, usually a mixture of steam condensate and demineralized tap water, is introduced at the final starch washing step just prior to drying or conversion. Water of low hardness must be used to prevent the formation of calcium sulfate haze in starch hydrolyzates. The filtrate from the final starch washing is used to slurry the starch for the next to last washing operation. Water works its way from starch washing to steeps (Fig. 4), gradually increasing in solubles content during passage through a series of dilution and reconcentration steps, and finally attaining a solubles level of 0.5–1.5%. It is then ready to be used in steeping the corn.

Steeps are normally operated as countercurrent batteries of 7–12 tanks, although much longer batteries are sometimes used. Process wash water containing 0.1–0.2% sulfur dioxide is placed on corn that has been in the steeps longest and, therefore, has the lowest residual solubles content. Figure 5 shows schematically the movement of steepwater from steep to steep. The high-SO_2 process wash water (steep acid) first enters the steep containing corn that has been

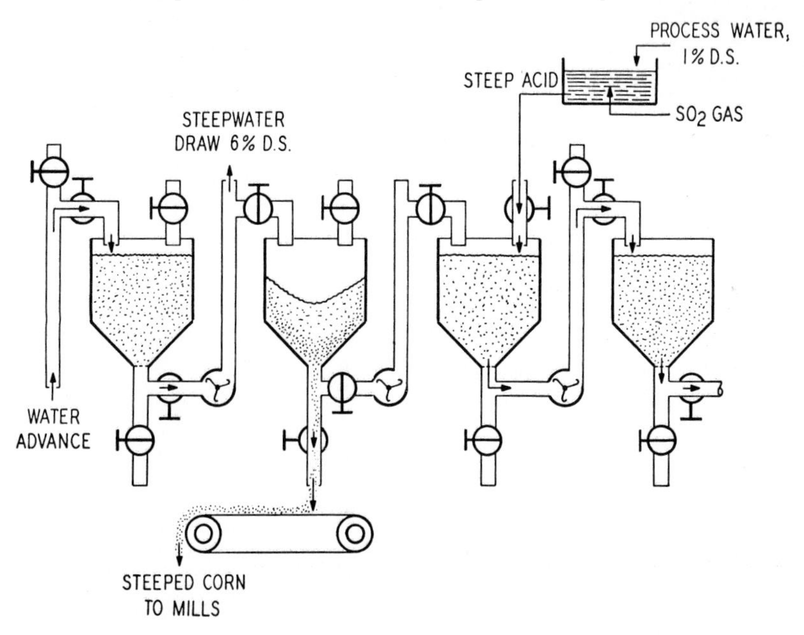

FIG. 5.—Detail of countercurrent continuous advance steep battery operation.

steeped for 40 h and passes down through the corn. It then is pumped to the top of the next steep and moves down through this next-to-longest steeped corn. The process is repeated successively for each steep in the battery. Finally, when it emerges from the bottom of the steep containing corn that has been steeped least long, it is removed from the system and sent to the evaporators.

Most commonly, the water is moved by a small pump attached to each steep tank. In older "batch advance" systems, one large pump serves an entire battery. At least one system moves water by gravity through sealed steeps. Batteries using gravity or individual pumps are termed "continuous advance systems" (Fig. 5) because the steepwater must be moved continuously during grinding operations. Oldest light steepwater, now containing 5–7% solubles content, is pumped continuously out of the steep containing the newest corn into the steepwater "draw" tank and thence to the evaporator at a rate of about 4–6 gallons per bushel (596–894 L/metric ton). Instead of being withdrawn from the system, it may be by-passed through a heater and passed into the adjacent steep when necessary to cover and heat new corn just added to the system. The system should be maintained at a temperature of 48°–52° (118°–125°F). Addition of steep acid to the oldest corn is stopped in time to be able to pump out (advance) all the water to uncover the oldest corn for grinding. About 2–3 volumes of water pass through the battery for each volume of corn. The corn is not moved except for loading and unloading the steeps. Although several authors (*78, 79*) have claimed faster steeping if corn and water are both moved, the evidence is not convincing and the cost in energy and equipment would be prohibitive.

Figure 6 is a plot of steepwater composition taken from successive steeps across a ten-steep, continuous-flow battery in 1950 when process water used for steep acid provided a rich bacterial inoculum. Today's steep battery has approximately the same types of changes in every property except Relative Bacterial Activity, which today will be at a much lower level. Therefore, a lower level of lactic acid is observed in steeps 6 to 10, but the steepwater has a fairly constant pH of 4.0–4.1.

In some factories, grind increase is accomplished by adding additional steeps to a continuous-advance system, eventually achieving from 15 to 50 steeps in a single battery. To accomplish operation of a long battery, several steeps are ground out simultaneously. The sulfur dioxide is added in liquid form by injection into the water entering 2–4 oldest corn steeps, but may be preceeded by a "washing phase," in which process water of low sulfur dioxide content is passed over the oldest corn in preparation for grinding. The sulfur dioxide concentration in the steepwater declines as the water moves over newer corn, dropping to a level that is tolerated by lactic acid bacteria in the three or four steeps in the new corn end of the battery, termed the "fermentation phase."

A steep battery having only one large pump is termed a "batch-advance system." In this sytem steeping is quiescent except during periods of water movement which may occur once, commonly twice, or possibly thrice during the

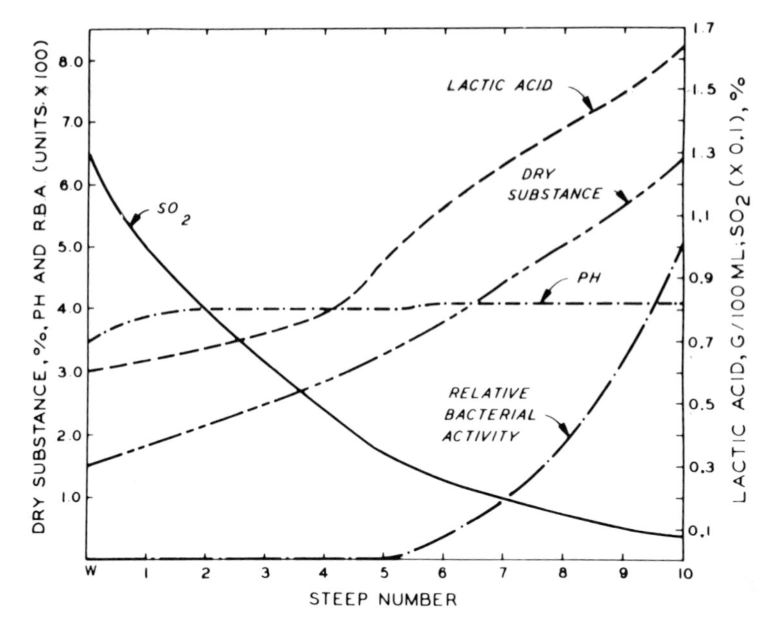

FIG. 6.—Composition of steepwater in individual steeps of a continuous-flow countercurrent steep battery. Ten steeps sampled simultaneously. "W" under steep number is composition of steep acid (input water). Water flow is from left to right. Steep number 10 contains newest corn.

period when a steep is being ground out. To prepare a steep for grinding, a "draw" of the volume scheduled (4–6 gal/bu) is made by pumping all the water off the newest corn and enough from the second oldest to make up the required draw volume. Water is then advanced on to newer corn in "leap-frog" fashion to cover the corn uncovered by the draw pumping. Water is advanced in this manner until older corn steeps are emptied in succession. The empty steeps, except for the one next in line to be ground out, are then covered with steep acid containing fresh sulfur dioxide. If a second or third advance is made during grinding of each steep, then the draw volume is proportioned accordingly and all old steeps are covered with steep acid. The advance pump is usually attached to a heat exchanger that maintains the desired temperature each time water is moved. This pump must deliver a high volume so that water movement is rapid and considerable aeration occurs when water is advanced into the next steep. No marked differences in type of lactic acid bacteria or milling efficiency have been observed when compared with a continuous-advance system.

c. *Water Absorption and Solubles Removal*

Raw corn entering a steeping battery will start at 15–20% moisture. Higher moisture levels are encountered for several months after the harvest season when good quality, high moisture corn is available at a discount.

Water enters the kernel mostly at the tip cap and moves quickly through voids in the pericarp by capillary action (*12, 80*). Diffusion of water into the endosperm and germ follows standard diffusivity laws that apply equally to corn, grain sorghum, or wheat (*81*). Fan and co-workers (*82*) showed that the effect of SO_2 dissolved in the water is to increase the rate of water diffusion into dent corn kernels. Increasing the temperature up to 60° increased the rate of water diffusion. After 1 7-h steep at 49°, the moisture increase was 25% higher in the SO_2 solution than in water.

Water moves quickly up the sides of the kernel and enters through the dent region into the porous floury endosperm region as it does in wheat (*83*). Water penetrates quickly into the germ because the aleurone layer is thin and testa is absent. The germ becomes wet in about 4 h at 49° and the endosperm in about 8 h, but the kernel is not appreciably softened (*11, 80*). Work with wheat (*83*) indicates that a dense, proteinaceous subaleurone layer appreciably slows the penetration of water into the endosperm, which may explain why grain sorghum requires longer steeping than does corn.

Although mass movement of water into the kernel parts is relatively rapid, the hydration of cellular components required for thorough softening is slower and takes longer, possible 12–18 h. For this reason, the addition of wetting agents or scratching the pericarp (*84*) in efforts to increase the rate of water penetration do not appear to decrease steeping time appreciably. Corn artificially dried at elevated temperatures attains lower final water content during steeping than does unheated corn (*67*). This probably results from the reduction of water binding sites by protein denaturation. Another factor which might slow water adsorption is the formation of gas bubbles which can be observed evolving from the tip cap for many hours after corn has been immersed in water. A first small volume of gas evolved may result from replacement of air as water enters the voids, but formation of carbon dioxide by glutamic acid decarboxylase is probably the major source (*86*).

Light steepwater, into which the raw corn is initially immersed, is a complex solution of organic and inorganic molecules (refer to discussion under feed products). Major components are peptides and amino acids, lactic acid, and cations, mainly K^+, Mg^{2+}, and H^+. Lactate salts generated by bacteria growing in the steepwater buffer it at pH 3.8–4.1. Lactic acid is absorbed into the kernel with the water and reduces internal pH in about 12 h, which in concert with the high temperature soon results in death of all living cells in the germ. As a result, cell membranes become porous and soluble sugars, amino acids, proteins, minerals and the large assortment of organic molecules required for growth of living cells leach into the steepwater. Total dry substance extraction is most rapid in the first 12–20 h (*87, 88*) but may continue at a slower rate as the solubles, formed by reaction of kernel proteins with SO_2, diffuse. Calculations show that about one-half of the steepwater solubles come from the germ, which represents only 11–12% of the kernel weight (Table II). Verification of this is seen in the increase of fat content of the raw germ from 30–38% to 55–60% in steeped germ. The other half of the solubles comes from the endosperm. Calculations indicate that the naturally soluble substances in the germ account for 95% of its contribution of solubles. About 60% of the germ protein is soluble in salt solutions, but only about 9% of the endosperm protein is soluble (*36, 37*). Since sugars, ash, and soluble proteins can account for less than half of the solubles contribution of the endosperm, the remainder must be made during steeping, as will be discussed later.

Water absorption by both corn and sorghum kernels in water or sulfur dioxide solutions causes a volume expansion of 55–65% (*89, 90*). However, the volume expansion of the two grains in a large bulk is different because of differences in shape of the kernels. Corn in a large steep does not appear to increase in volume because the weight of grain forces the swelling kernels to reorient to fill the irregular voids between kernels. The spherical milo kernals, being initially closely packed, exhibit a volume increase during the water absorption phase of steeping and create a force great enough to break steeps. Grain sorghum must be introduced into a steep in several increments to prevent rupture of the tank. Intact corn kernels steeped in an excess volume of sulfur dioxide solution, such as the laboratory semisynthetic steeping method (*20*), are quite turgid, but shrink when placed in a salt solution. This indicates that the kernel has a semi-permeable membrane which retains soluble molecules. Unpublished experiments indicate that these molecules are large peptides or soluble proteins.

d. *Effect of Sulfur Dioxide*

Sulfur dioxide was first used in corn steeping probably to prevent growth of putrefactive organisms. Eventually this chemical was found indispensible for obtainment of maximum starch yield. Cox and co-workers (*80*) were the first to study the effects of steeping agents by microscopic observations of thin sections of corn kernels. They demonstrated that, during steeping with sulfur dioxide over a 24-h period at 50°, the protein matrix gradually swells, becomes globular, and finally disperses. The degree of protein globularity was shown to be directly related to the ease of starch recovery on grinding or on agitation of the thin sections of the endosperm from the steeped kernels. Protein dispersion increased as the sulfur dioxide concentration was increased to 0.4%. The presence of lactic acid produced some apparent softening action, but other acids had little effect. Other reducing agents can replace sulfur dioxide (*80, 91*) but have no practical value in commercial practice. Wagoner (*92*) found that rate of protein swelling is faster and dispersion more complete with freshly harvested corn than with old dry corn.

Sulfur dioxide dissolved in water forms an equilibrium mixture:

$$H_2SO_3 \rightleftharpoons H^+ + HSO_3^- \qquad K_a = 1.54 \times 10^{-2}$$
$$HSO_3^- \rightleftharpoons H^+ + SO_3^{2-} \qquad K_a = 1.02 \times 10^{-7}$$

The reaction of bisulfite ion with endosperm protein is completed in whole kernels in 6–10 h (*87, 88*). Both SO_3^{2-} and HSO_3^- ions are capable of reducing disulfide bonds, but pH conditions inside the kernel probably determine which reaction predominates. The kernel pH of 3.6–4.0 reported by Wahl (*93*) would favor HSO_3^- as the reactant, but the pH levels are normally closer to 4.0–4.5 in most commercial steeps with a solubles level of 1–2% in the steep acid (unpublished data).

Steeping sections of horny endosperm (10 μm thick) bathed in a solution of constant bisulfite ion concentration provides a means of studying the rate of the sulfur dioxide reaction under conditions where diffusion is not limiting (*94*). These data showed that loss of starch granules from the protein matrix on gentle agitation was 50–70% complete in one hour and was complete in 4 h at 52° in 0.2% SO_2 solution. Under these conditions, the rate and extent of starch release from dent corn endosperm sections increased with increasing bisulfite concentration (0.05 to 0.2%) and with increasing temperature (52° to 60°). Further experiments showed that at pH levels below 9, the pH of the medium did not have a significant effect on starch release. Other acids, including lactic, did not affect starch release from the thin endosperm sections even on agitation for 24 h at 52°–60°. This research emphasized the unique role of sulfur dioxide in starch release. Figure 2 shows horny endosperm cells from which most starch granules have been released by action of sulfur dioxide and clearly shows the globular nature of protein strands which surrounded the granules.

Similar effects have been observed on steeping isolated corn endosperm (degerminated corn meal) in bisulfite solutions buffered at pH 4 with lactic acid–potassium lactate (Y. Hirata and S. Watson, unpublished data). During the first 2 h of contact with the bisulfite solution at 50°, the endosperm loses more nitrogenous material than control endosperm steeped in lactate buffer alone. The bulk of the dissolved nitrogen is precipitated when the steepwater is adjusted to pH 7 and heated to boiling, indicating that a relatively high-molecular-weight protein is dissolved from the endosperm by the action of bisulfite. The appearance of this protein in solution coincides with the release of starch from the endosperm cells.

Swan (*95*) demonstrated that sulfite or bisulfite ion reacts with the sulfhydryl groups of cystine in wool protein, reducing the disulfide linkage and giving a protein fraction (P′) having one cysteine SH and a second protein (P″) on which a *S*-sulfo derivative of the cysteine moiety has been formed:

$$P'S{-}SP'' + HSO_3^- \rightarrow P'SH + P''SSO_3^-$$

This reaction has been confirmed for corn (*96, 97*) and grain sorghum (*98*). The *S*-sulfo reaction product permanently increases the protein solubility by preventing reformation of the disulfide bond and because of the ionic nature of the P″ protein. The reaction rate is greatest at pH 5.0 (*99*), but the pH inside the corn kernel in a countercurrent battery is 4.0–4.5 in the old corn end of the battery where the sulfur dioxide concentration is highest. Nierle (*100*) estimates that only 5.7% of the sulfur dioxide used in steeping is absorbed by the corn, with 45% in the endosperm protein, 2% in the starch, and 40% in the germ. Only 12% of the absorbed sulfur dioxide reacts with protein.

Knowledge of the protein components of corn has been clarified considerably in recent years and is thoroughly reviewed by Wall (*101*) and Paulis (*85*). Normal corn endosperm contains a mixture of proteins which can be extracted with a succession of solvents (*36, 102, 103*) with the following representative results:

Solvent	*Protein class*	*Percent of total nitrogen*
Water	Albumin plus NPN	4–5
0.5 *M* Na_2SO_4	Globulin	3–4
71% Ethanol	Zein	40–50
0.2 *M* NaOH	Glutelin	30–40
Insoluble	Glutelin	10–20

The albumins and globulins are probably cellular enzymes and are quickly dissolved in the steeping process (*103*). Zein occurs in spherical bodies (*16, 17, 104*) which are embedded in a glutelin matrix. Treatment of corn endosperm with 71% ethanol to remove zein does not improve starch separation over untreated endosperm, but steeping in an alkaline solution does liberate starch. This indicates that the matrix surrounding starch granules is composed of glutelin. Wall (*101*) has characterized glutelin as a large and complex protein molecule composed of about 20 different protein subunits ranging in molecular weight from 11,000 to 127,000. These subunits are united through disulfide bridges in such a complex way that the native glutelin molecular weight may not be definable. Hydrogen bonds also take part in the binding, because only a combination of reducing agents and hydrogen bond breaking chemicals (NaOH, urea, detergent) will completely dissolve the glutelin. The glutelin of grain sorghum is similar to that of corn (*101*).

The action of sulfur dioxide in steeping, then, is to weaken the glutelin matrix by breaking inter- and intramolecular disulfide bonds (Fig. 2). Some zein is solubilized during steeping owing to disulfide crossbond disruption by SO_2, but this solubilization probably doesn't contribute much to matrix weakening. The solubilized polypeptides resulting from these reactions are nondialyzable (*96*) and too large to pass through cellular and seed coat membranes. They are retained inside intact corn kernels until liberated in first break milling. Corn that has been dried at excessively high temperatures yields less soluble protein on steeping (*70, 105*) than undried corn, presumably because of increasing the degree of intramolecular hydrogen and non-polar bonding. However, temperatures ordinarily used in commercial corn drying have little effect on protein solubility.

e. *Role of Lactic Acid Bacteria*

Although corn can be adequately steeped in sterile, aqueous solutions of sulfur dioxide or sodium or potassium bisulfite in the laboratory (*106, 160*), commercial steeping involves microorganisms. Raw corn carries small natural populations of bacteria, yeasts, and molds which are capable of rapid multiplication in aqueous systems. Practical men learned early that corn steeped at temperatures of 45°–55° was "sweet," but that putrefaction and butyric acid or alcohol production occurred at lower temperatures. Steeping at 45°–55° is now known to favor development of lactic acid bacteria. The lactic acid produced lowers the pH of the medium and restricts growth of most other organisms. Substrate for the bacteria is the sugar that quickly leaches from the corn. Sucrose is quickly hydrolyzed to D-glucose and D-fructose, each of which is converted almost completely to two moles of lactic acid with release of energy needed for bacterial development. The amino acids required by these bacteria are also supplied by corn, partly as free amino acids, but largely as solubilized protein. Although lactic acid bacteria generally do not show much proteolytic activity, the organisms endemic to the steeps do affect significant protein degradation (*88*). Only about one-tenth of the nitrogen in corn is in the form of non-protein nitrogen (NPN) and, therefore, account for only about one-fourth of the soluble nitrogen obtained during steeping. Only enzymic degradation of dissolved protein can account for the fact that 85% of the nitrogen in incubated steepwater is dialyzable NPN.

The microbial species existing in commercial steeps have not been positively identified nor seriously studied. Under steeping conditions that produce optimum conversion of sugar to lactic acid, elongate rods that have been identified as homofermentative lactobacilli are observed as the dominant organism. They appear to closely resemble *Lactobacillus bulgaricus*. Also, steepwater does appear to contain nutrients which stimulate growth of lactic streptococci (*107*). The grain is obviously the primary source of the initial microbial inoculum. However, because specific steeping conditions are maintained on a continuous basis in a commercial steephouse, natural selection or adaptation is probably quickly established for each particular environment; resulting biota may differ significantly from traditional species descriptions.

Examination of fermentation characteristics in commercial steeping in 1952–55 showed that raw corn incubated at 50° (122°F) with fresh milling process water develops a vigorous culture of lactic acid bacteria in about 32 h (*88*). Lactobacilli normally do not function well at pH 4 which exists in a steep battery, so it was not surprising when Wahl (*93*) found 70% of the bacterial activity to be inside the corn kernels (in the new grain end of a battery) where the pH is closer to 5 and sugar concentration is highest (probably in the spaces inside the tip cap).

In today's automated process, the process water is a poor source of inoculum and light steepwater has a low population of bacteria. Conversion of sugar to lactic acid is not complete at the time steepwater is drawn off for evaporation, but is more complete than expected from observed populations of organisms. A higher population probably would be found inside the kernel. The cause of the low steep population can be found in the millhouse. In 1952, starch was still separated from gluten on large surface, open, wood tables; and gluten was recovered from the dilute slurries by allowing it to settle by gravity in large tanks for 5 days at ambient temperature. Sulfur dioxide was added to prevent development of putrefactive organisms. Today, starch is separated on a continuous basis and gluten is concentrated in centrifugal machines. The entire process requires a transit time of 2–3 h in stainless steel equipment maintained at 43°–45° (110°–113°F) in the presence of sulfur dioxide. All tanks are small and act as surge reservoirs for process streams. Thus, there is little time for an inoculum to develop in the millhouse process water that is used in steeping.

Although every wet-milling plant has different operational parameters, the principles of bacterial development and their influence on steeping, as expounded above, are still important for optimum performance. A process water pH of 4.0–5.0 generally is most favorable for optimum separation of components of the system, especially the starch/gluten separation; the effectiveness of sulfur dioxide in preventing growth of undesirable microorganisms is most operative in that pH range. Maintaining a pH of 3.8–4.2 in light steepwater for evaporation is important because, at higher pH levels, mineral scales may be deposited on th heat-exchange surfaces, reducing the rate of water evaporation. The generation of acidity from sugars in the corn is a low cost source of acidity. Steepwater is so highly buffered that a large volume of mineral acid addition would be required to achieve adequate acidity. In addition to producing acids by fermentation, lactic acid bacteria also effect hydrolysis of high-molecular-weight, soluble proteins. If these proteins are not degraded, they produce stable foams that can interfere with steepwater evaporation and may be deposited as gelatinous precipitates on the heat-exchange surfaces of the evaporator. Reduced evaporation rate and shutdowns for evaporator cleanout result.

Several remedies for the problems caused by low natural bacterial inoculum have been tried. The simplest solution is to ensure that the sulfur dioxide concentration in the fermentation zone of the battery is not so high as to inhibit bacterial growth and to maintain the steeping temperature between 48° and 51° (118–124°F) for optimum growth of the desired bacterial species. Additional incubation after light steepwater is withdrawn from the steeps may also be necessary. A temperature of 46°–48° and a pH maintained at 4.5 by addition of ammonia will give most rapid fermentation. Time required to achieve the desired degree of lactic acid development will depend on local conditions, but probably will not be less than 4 h. The problem of steepwater foaming and gel formation may be solved by adding a proteolytic enzyme to the steeps. A low dosage of a crude enzyme preparation is quite adequate and apparently has no effect on the corn (R. E. Heady, CPC International, Inc., unpublished data). A more complete solution has been demonstrated by Balana and Caixes (*108*) who prepared active cultures of one of several *Lactobacillus* species, such as *L. delbrukii* or *L. leichmanii*. Inocula prepared with synthetic culture media were added to steeps at a rate of 40 L per 68 tons of corn and allowed to pass through the battery with the advancing water. As expected, the concentration of reducing sugars was lower, the pH was lower, the concentration of lactic acid was higher, and the concentrated steepwater gave no evaporation problems; in addition, the concentrated steepwater gave improved penicillin yield when used in commercial media (R. Balana, CPC International, Inc., personal communication). Claims that inoculation permitted reduction of steeping time from 40–60 to 25 h and that the sulfur dioxide concentration in steep acid could be reduced from 0.15% to 0.75% have not been confirmed on subsequent evaluation (C. D. Gillece, CPC International, Inc., unpublished data).

3. Milling and Fraction Separation

a. *Component Yields*

The objective of the milling process is to provide for as complete a separation of component parts of the corn kernel as is possible and practical. While commercial yield figures are not available, laboratory steeping and milling procedures (*106, 109–111*) give results close to those obtained in commercial wet milling. Data are given in Table VII for regular dent hybrid corn, a high-oil corn variety, and regular red grain sorghum. All three grains were steeped 48 h at 52° in a 0.1% sulfur dioxide solution at pH 4 and processed as described in reference *106*. The starch separation was conducted by tabling. In tabling, the denser starch settles on the table; the lighter gluten containing many of the small starch granules flows off the end of the table. The surface of the starch is then gently hosed or "squeeged" to remove traces of gluten. The material washed off the table is termed the "squeegee" fraction; it has a high starch content.

The data in Table VII show that product yields are closely related to initial grain composition. For example, high-oil corn gives a greater germ and oil yield than dent corn, while red milo gives a lower germ and oil yield. Likewise, high-oil corn gives lower starch and gluten yield because the proportion of endosperm is less than that in dent corn. For the same reason, red milo gives a higher gluten yield, but a lower starch yield is obtained because starch is more difficult to separate from other components.

Table VII

Yield and Composition of Laboratory Wet Milled Fractions of Regular and High-Oil Dent Corn and Red Grain Sorghum (Milo)[a]

Fraction	Regular dent corn (%)	High-oil dent corn (%)	Red milo (%)
Whole grain analysis			
Moisture	14.3	13.6	14.9
Starch	71.5	67.0	73.1
Protein	10.5	10.4	13.0
Fat	5.10	7.96	3.6
Wax	Trace	Trace	0.32
Solubles			
Yield[b]	7.6	10.8	7.20
Protein[c]	46.1	46.9	41.5
Starch			
Yield[b]	63.7	59.7	60.17
Protein	0.30	0.26	0.32
Fat	0.02	0.03	0.03
Germ			
Yield[b]	7.3	10.9	6.17
Starch	7.6	7.2	19.1
Protein	10.7	7.2	11.9
Fat	58.9	65.5	39.6
Oil yield[b]	4.30	7.14	2.44
Fiber			
Yield[b]	9.5	9.8	9.30
Starch	11.4	12.3	36.7
Protein	11.3	11.0	19.7
Fat	1.8	2.7	3.8
Gluten			
Yield[b]	7.4	6.3	9.57
Starch	25.8	32.0	39.9
Protein	50.7	42.3	47.2
Fat	3.7	4.4	5.4
Squeegee			
Yield[b]	3.9	3.6	5.57
Starch	91.7	93.8	74.8
Protein	6.1	3.6	20.7
Fat	0.3	0.4	1.6
Total dry substance[b]	99.4	101.0	98.0
Recalculation to Expected Centrifugal Results			
Starch			
Yield	68.5	65.7	67.2
Gluten			
Yield	5.76	4.0	8.1
Protein	70.0	70.0	70.0

[a] All percentages other than moisture are expressed on a dry basis. [b] Percent of original grain. [c] Analytical values expressed as percent of the fraction.

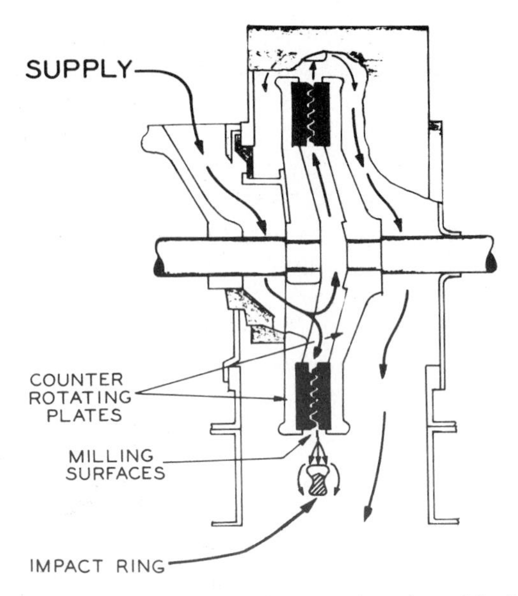

FIG. 7.—Diagram of Bauer attrition mill. Configuration shown is used for fine milling degerminated residue. With wider spacing of interlocking teeth and no impact ring, it is used for degermination (first break) milling.

Gluten obtained by tabling seldom contains more than 50% protein. To approximate the results obtained by continuous commercial centrifugation, the gluten and squeegee data were recalculated to 70% protein, which increases starch yield. Similar experiments with Argentine flint corn have shown that component separation is about equal to that of dent corn even though flint corn steeped kernels appear to be much less soft (S. A. Watson and Y. Hirata, CPC International, Inc., unpublished data).

Many simplified laboratory procedures have been developed (*20, 49, 92, 94*) to determine effectiveness of steeping or to predict the performance of a corn sample in a commercial process. The writer agrees with Kempf (*112*) that, although the complete laboratory fractionation method is slow and cumbersome, there is no good substitute for adequate evaluation of wet milling properties of grain.

b. *Germ Separation*

After steeping in the commercial process, the grain is coarsely ground or "pulped" with water in an attrition mill in preparation for degermination. The most commonly used mill has one stationary and one rotating milling surface as shown in Figure 7. When used for degermination, the plates are covered with pyramidal knobs and the impact ring is absent. The bulk of the germ is freed in the first pass, but a second pass is usually provided after free germ has been removed. The mill gap is adjusted to give the most free germ with minimum germ breakage. Any germ cells that are cut or disrupted in any way during any step in the process will lose oil which is mostly absorbed by gluten and cannot be recovered. Over half of the starch and gluten is also freed in this first milling step. The large density difference between the oil-rich germ and the heavier kernel components provides a basis for easy separation. This is accomplished by continuous flow through liquid cyclones or hydroclones. The type used for germ separation is a conical tube 6 in. in diameter at the top of its 3-ft length (Fig. 8). Pulped corn adjusted to 7°–8° Baumé with suspended starch is forced into the tube under pressure. The orifice angle, aperture, and pump pressure are chosen to produce a rotational velocity sufficient to cause a separation of the particles of differing density (*113, 114, 143*). The heavier endosperm and fiber particles pass out the bottom of the tube at a dry solids concentration of 20–24%, while the

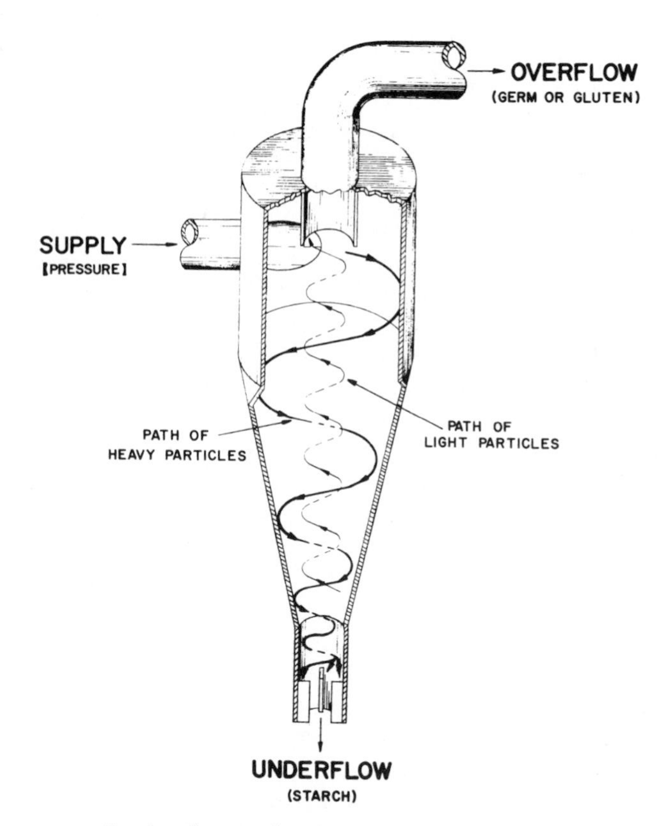

FIG. 8.—Cutaway diagram of hydroclone separator.

Fig. 9.—Battery of hydroclones used for separation of germ from milled, steeped corn. (Courtesy of Dorr-Oliver, Inc., Stamford, Connecticut.)

lighter germ is drawn off the top of the vortex. A bank of germ hydroclones is shown in Figure 9.

The cyclone type of germ recovery equipment occupies less floor space, is easier to maintain and clean, and allows greater response to changes in operating conditions than do the earlier germ flotation methods (*5*). As later shown in Figure 15, recovered germ is washed two or three times countercurrently with process water on screen bend devices to remove occluded starch. (Screen bends will be described in Section III.3.c.) The washed germ is dewatered to 50–55% water content (wet weight basis) by passing it through mechanical squeezers prior to drying in preparation for oil recovery.

The free starch and protein, which comprise about one-half the dry substance in the germ cyclone underflow, are separated from the unmilled endosperm and fiber by screening over a screen bend device to reduce the load of solids and water to the subsequent milling operations. The slurry of starch and gluten obtained by screening at this point is called "prime mill starch" because it derives mainly from the floury endosperm. Prime mill starch is lower in protein than whole starch. Laboratory work has shown that it is easier to separate starch from gluten using this product as opposed to the final mill starch combined from both milling steps (unpublished data).

One company in the United States has utilized the prime starch phenomena in a unique way. In this patented process (*115*), prime starch is recovered and marketed while the starch retained in endosperm pieces and fiber is converted to ethanol. The process is identical to the process being described here except that the endosperm milling step is omitted and "starch yield" is maximized by enzymically digesting the horny endosperm starch that is the most difficult to recover by milling. The resulting glucose is fermented to ethanol.

c. *Second Milling and Fiber Separation*

In processes where all the starch is to be recovered, the underflow from the germ cyclones, containing fiber and pieces of horny endosperm, is more thoroughly milled ("grinding mills" in Fig. 4) to recover the maximum yield of starch. One type, the Bauer mill (Fig. 7), employs a combination of attrition and impact milling (*116*). The corn material must pass between counterrotating grooved plates made of a hardened steel alloy and on discharge strikes an outer impact ring (*117*).

An impact-type mill known as the Entoleter mill is preferred by many operators (*118*) (Fig. 10). Endosperm slurry dropped onto the rotating horizontal disk is flung with great force against both rotating and stationary pins. Rapid and complete starch release is obtained with minimum fiber attrition. The larger pieces of fiber permit occluded starch to be more efficiently recovered by washing. These two types of mills produce roughly equivalent results. They have advantages of high throughput rates, low maintenance costs, uniform operation, and improved quality and yield of starch over formerly used Buhr stone mills (*5*). The Entoleter has an added advantage that there is no mill gap adjustment and, therefore, requires less operator attention.

Starch and gluten released by milling must be separated from fiber. This is

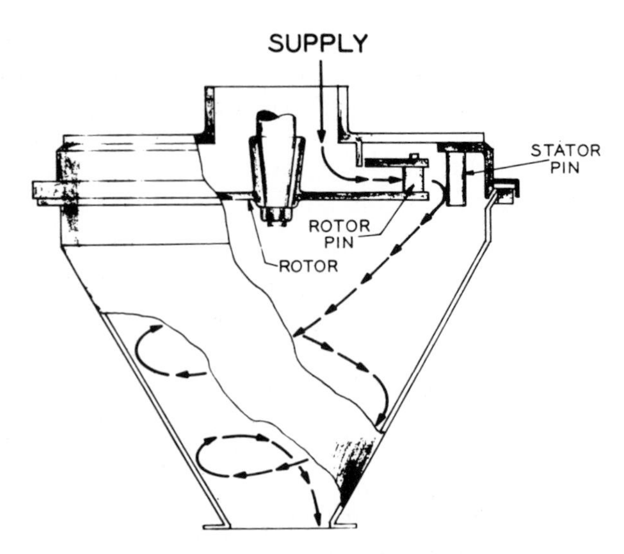

FIG. 10.—Cutaway diagram of Entoleter mill used for fine milling degerminated residue.

best accomplished by taking advantage of the difference between the fine particle sizes of starch granules and gluten particles, and the larger endosperm and pericarp particles. (There always remain a few dense pieces of horny endosperm that are not disrupted in milling.) For many years this separation was accomplished over fine nylon screens attached to a horizontally agitated frame. These devices (shakers, Rotex, etc.) required a large area, continual maintenance, and sanitation attention. But now all modern wet milling establishments use a fixed concave screen arranged in a vertical position and over which the slurry to be screened is pumped with considerable force (Fig. 11). The degree of concavity of the screen is varied to suit each screening problem and is designated in angular degrees as 50°, 120°, etc. This device has been made possible by the development of wedge-bar screening material of uniform slit width (*119, 120*) (Fig. 12) which has a unique slicing action that affords high capacity and eliminates clogging. These devices are known as screen bends (proprietarily DSM Screens) and have the advantage of effecting very sharp separations. They also have the advantages of low maintenance, minimal operator attention, and high capacity, thus requiring much less floor space and fewer employees than shaker stations. They are manufactured with screens having slit widths of 50, 75, 100, and 150 μm. A 50° DSM Screen with a large bar spacing is used to wash germ discharging from the germ recovery cyclones as described above (*121*).

A 120° DSM Screen having bar screen slit width of 50 μm is used for washing fiber free of occluded starch and gluten particles. The fiber mat discharging from the screen bend surface is reslurried in process water and passed over a second screen bend. As shown in Figure 15, the process of slurrying and screening is repeated 2–3 times. Units are connected in a manner that will give a countercurrent passage of water with respect to fiber in order to achieve high water use efficiency. Several different operating configurations for using these devices in a starch factory have been described (*121–123*). The number of successive units may vary with the manufacturer's goal for completeness of starch recovery. Typically, finished fiber contains 15–20% starch, about half free and half bound.

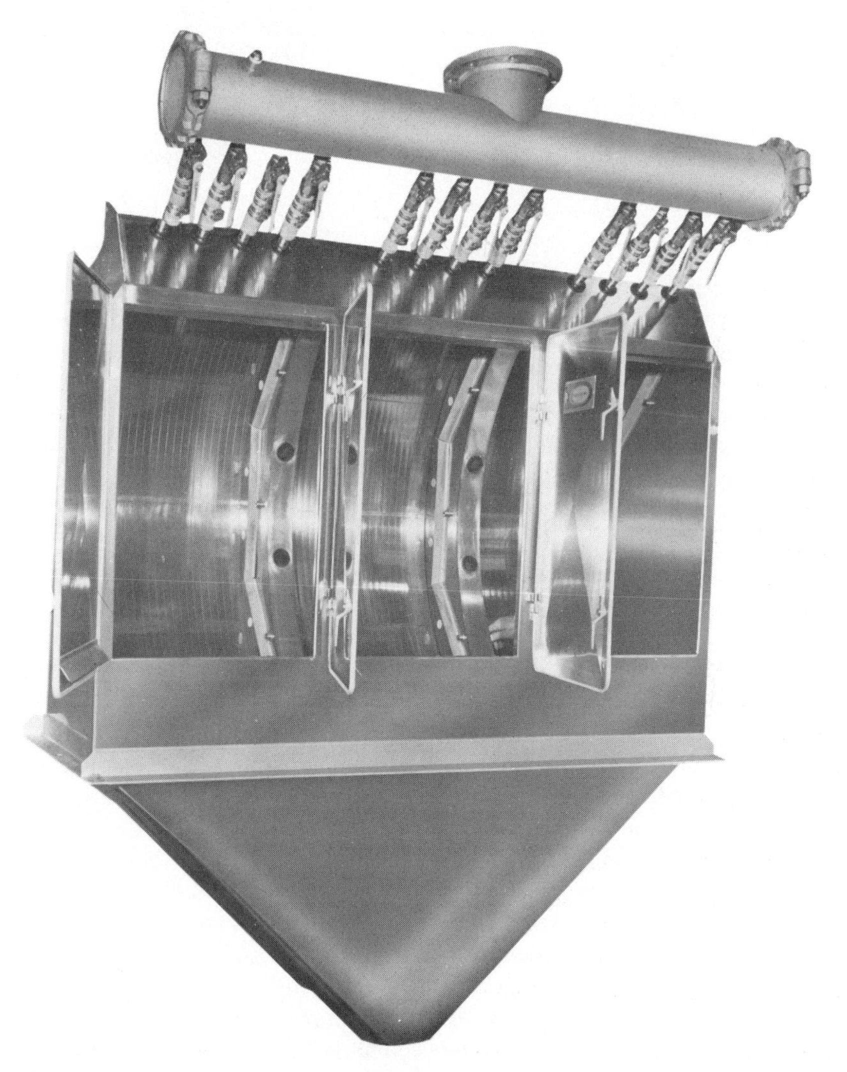

FIG. 11.—DSM fiber washing unit (120° model) employing wedge-bar screening surface. (Courtesy Dorr-Oliver, Inc., Stamford, Connecticut.)

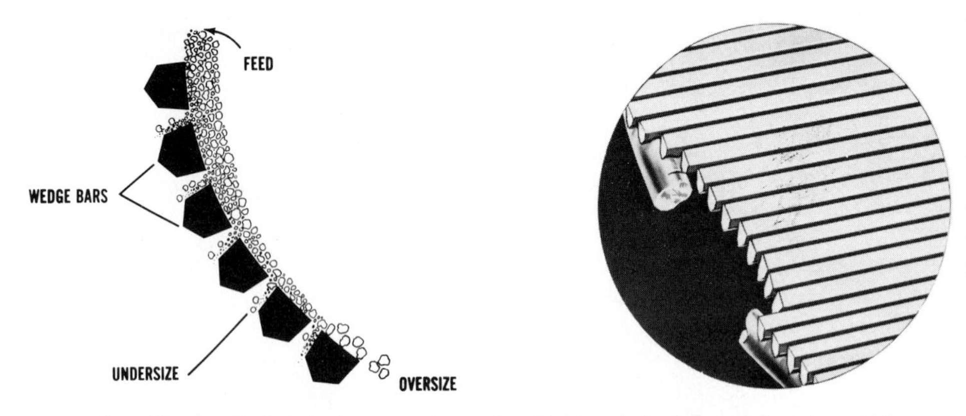

FIG. 12.—Detail of wedge-bar screening surface (right) and simulation of the unique slicing action of this screen.

Following the final screening, the fiber is further dewatered in a centrifugal screen, mechanical squeezers, and/or a horizontal solid-bowl, continuous-discharge centrifuge (*121*). Typically, final moisture content is 65–75% (wetweight basis). The fiber is then blended with concentrated steep liquor either before or after partial drying, then is further blended with corn cleanings, spent (deoiled) germ flakes and starch hydrolyzate residue, when available, and dried to make 21% protein Corn Gluten Feed (Fig. 4).

d. *Starch–Gluten Separation*

The defibered mixture of starch and protein, known as "mill starch," carries 5–8% insoluble protein content, depending on the protein content in the original corn or sorghum endosperm. The mill starch streams from both the degermination and fiber washing steps are combined and centrifugally concentrated to reduce solubles and to adjust the concentration of solids in preparation for the final step of starch separation. Any one of several bowl type, nozzle discharge centrifuges are used for this purpose, especially the Westfalia, the deLaval, and the Merco centrifuges (Fig. 13). For this use, the centrifuge is equipped with a clarifier assembly as a decanter to permit discharge of water in the overflow and all solids through the nozzles (*124, 129*).

The low density of hydrated gluten particles (1.1 g/cm^3) as compared with starch (1.5 g/cm^3) permits their ready separation by settling or centrifugation. A container of mill starch obtained from adequately steeped corn will deposit in about 10 minutes a layer of white starch with a layer of yellow gluten on top. In about one hour, the gluten layer will have further settled to leave a layer of supernatant liquid. This separation is accomplished commercially with the same type of centrifuges used for clarification but equipped for particle classification. These machines contain a stack of conical discs (Fig. 13), each separated by a narrow space to accentuate separation of discreet particles having distinctly different specific gravities (*124, 125*). The heavier starch granules are thrown to the periphery of the centrifuge bowl and are ejected through the nozzles. The lighter gluten particles are carried up between the discs by a stream of water and are ejected at a low solids concentration and a protein content of 68–75% (dry basis). The starch discharge contains 1–2% protein depending on the mode of operation. The mill starch is preferably supplied to the centrifuge at 10°–12° Baumé (18–21% dry solids) while the starch slurry discharges at a density of 20°–24° Baumé (35–42% dry solids).

The light gluten discharges from the centrifuge at 1–2% dry solids concentration and must be concentrated in centrifuges to 12–15% for efficient solids recovery. One means of gluten recovery is by filtration. Formerly, manually discharged filter presses were used. Now it may be accomplished with rotary vacuum filters arranged for continuous discharge of cake from the filter cloth belt and equipped with a mechanism to continuously wash the filter cloth to prevent

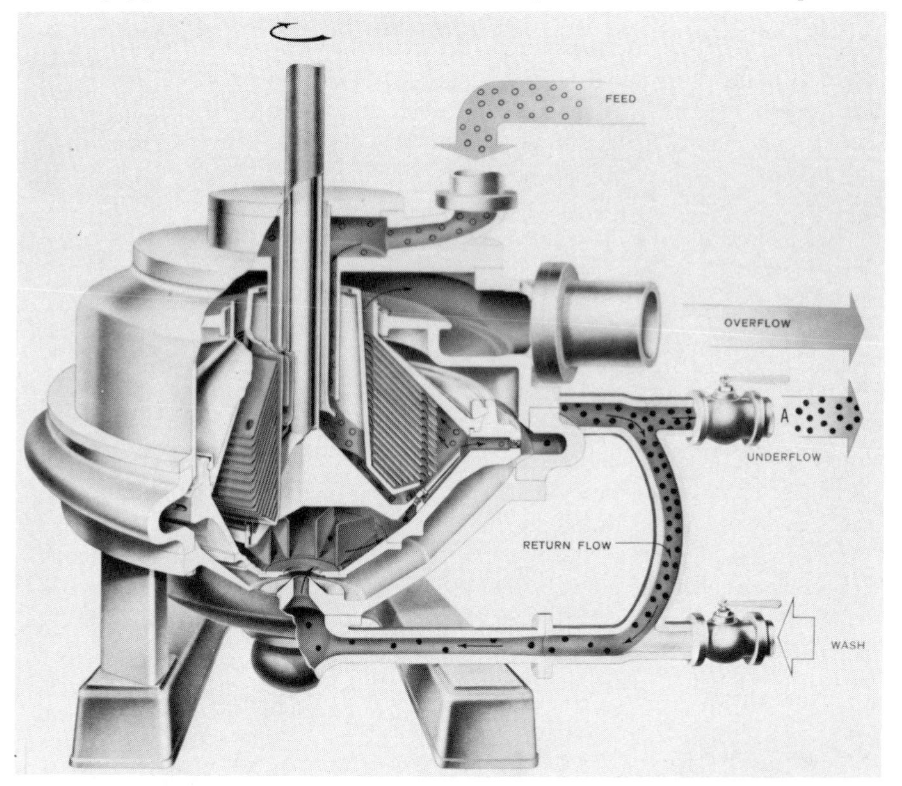

FIG. 13.—Cutaway diagram of Merco starch separation centrifugal. (Courtesy of Dorr-Oliver, Inc., Stamford, Connecticut.)

blinding by the very fine gluten particles. An alternative gluten recovery system dewaters the concentrated gluten in a horizontal decanter solids-discharge centrifuge (*126*). The rate of gluten sedimentation has a sharp peak between pH 4.5 and 5.5 requiring addition of sodium hydroxide solution to the gluten slurry to maximize operation of the decanter. In order to prevent microbial growth, the process temperature must be maintained above 45°, which temperature also coincides with minimum viscosity.

The next step in the process is removal of remaining soluble and insoluble protein. The starch discharging from the primary centrifugal separator must be diluted with process water to slurry density of 10°–12° Bé. The starch then may be further purified in a second centrifugal separator to a final insoluble protein level of under 0.38% dry basis (preferable 0.27–0.32% d.b.). However, since solubles content of the slurry must be reduced next by filtration or centrifugal decantation, the second step currently preferred is to utilize 8 to 14 stages of liquid cyclones which simultaneously remove residual gluten and wash the starch (*127, 128*).

The individual hydroclones are molded plastic devices with a tangential port for entry of the slurry into the top of a cone-shaped separation chamber which has two outlets. One outlet is an overflow vortex finder port for discharging water and the low density fraction from the top of the chamber. The other opening at the apex of the cone is a discharge port for the high density fraction. The inside diameter is 10 mm at the top of the cone and the cone length is usually 16 mm.

The hydroclones are assembled into manifolds holding numerous tubes of several different designs. The standard in the past has been the "clamshell" unit in which the manifold containing several hundred tubes is covered top and bottom, forming a three-partitioned unit. In a newer design (Fig. 14), 24 individual tubes are radially oriented into compact modules which may be "stacked" into compact units having any desired number of stages (*162*). In either type of unit, the starch–gluten slurry enters the central chamber and is forced by means of pump pressure (normally 80–100 psi, 5.4–6.8 atm) simultaneously into each of the individual tubes. The pressure energy creates rotational motion to the liquid entering the conical chamber, producing a centrifugal force throwing the heavier starch granules out the underflow port into a common collection chamber; the gluten particles discharge through the vortex finder into an overflow side collection chamber (*121, 127*). The concentrated starch stream is rediluted with water overflow from a succeeding hydroclone and passes into a second set of hydroclones. Six to ten sets of hydroclones arranged in countercurrent sequence provide for final removal of insoluble protein and give a complete washing operation.

As previously described, the only fresh water entering the milling operation contacts the starch in the last hydroclone stage. The overflow from each cyclone stage is used to dilute underflow entering the next to last stage. The water

FIG. 14.—Interior (foreground) and exterior views of hydroclone starch washing-separation units. The individual 10-cm hydroclone tubes appear as ''spokes'' in the circular modules (foreground). (Courtesy of Dorr-Oliver, Inc., Stamford, Connecticut.)

progresses back through the hydroclone station from stage to stage and, when it emerges from the first stage, it carries away solubles, gluten, very fine particles of fiber, small and damaged starch granules and some larger occluded starch granules. This stream is termed "middlings" and analyzes about 2% insoluble and 0.5% soluble dry substance. The dry substance is composed of about 85% starch and 10% protein. This starch is typically recovered by recycling it into the dilute mill starch stream for another pass through the separation system. Bonnyay (*129*) recommends that the middlings stream be concentrated in a clarifier centrifuge, thus employing the high-density, four-stage centrifugation system depicted in Figure 15. This system gives the purest form of a countercurrent system possible. Gluten concentrator overflow water is used for germ washing and steeping and middlings concentrator overflow is used for injection into the primary separation centrifuge and for fiber washing.

Centrifuges do have disadvantages of high capital cost, significant cleaning time, and frequent operational adjustment. A method has now been disclosed by which the entire starch–protein separation system can be completed in a hydroclone station, thus eliminating the primary separator centrifuge (*130*). Furthermore, the hydroclones will operate at normal mill starch concentrations of 7.5°–8.5° Bé (13–15% dry solids), thus eliminating the need for centrifugal concentrators ahead of the primary separation step. The system described utilizes two or three primary hydroclone stages for gluten and solubles recovery followed by 9–10 starch washing stages. Higher pump pressure differentials of 120–180 psi must be used for the primary stages, and pressure differentials of 100–150 psi for the washing stages. Operation of this system will normally produce gluten of 70% protein (dry weight basis) and starch of 0.33% protein (dry weight basis) at a lower capital and operating cost than the presently used centrifuge–hydroclone system.

e. *Grain Sorghum Processing Innovations*

In the wet milling of grain sorghum, the starch and gluten in mill starch usually does not separate sharply into two components (Table VII). Extensive laboratory studies (*131*) have indicated that the sorghum pericarp is the cause of this difficulty because it is much more fragile than that of corn. Small particles of fiber apparently interfere with starch and gluten separation and also may give the starch a pink color. Mold infection of the pericarp in field "weathering" accentuates this tendency. A solution to this problem is to dehull the kernels before steeping, preferably by an aqueous slurry technique (*131*, *132*). Dry dehulling techniques can be used, but starch losses are much higher (*133*). When the dehulled grain is steeped in conventional manner, the starch gluten separation is much sharper and gives a starch of lower color and lower protein content. Gluten has higher protein content. Germ separation is also improved, indicating a possible 20% increase in oil recovery (*131*).

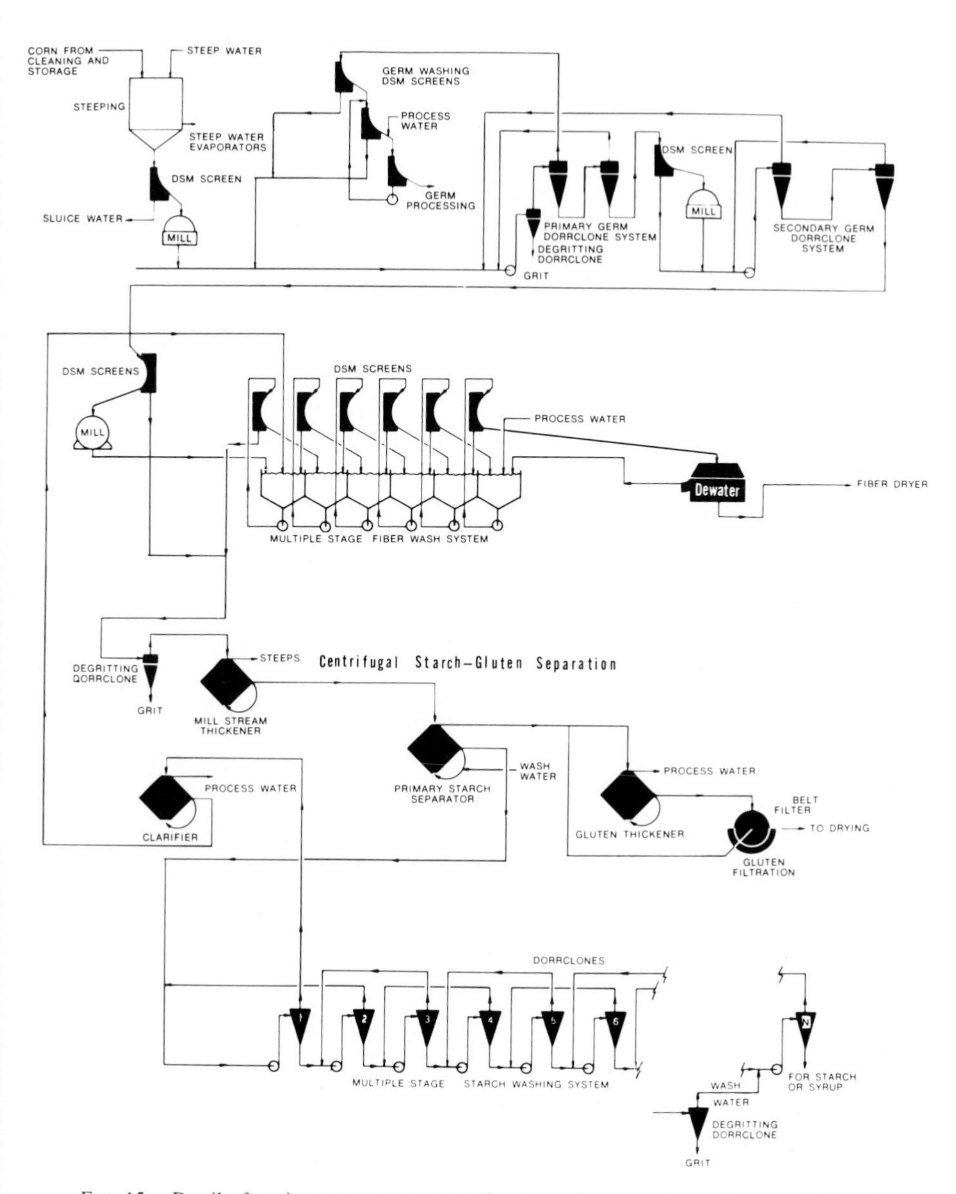

Fig. 15.—Detail of equipment arrangement and water and process stream flow employing the high-density four-stage starch separation centrifugation system. (Courtesy of Dorr-Oliver, Inc., Stamford, Connecticut.)

Alternatively, the dehulled kernels can be degerminated prior to steeping by subjecting them to a pressure just sufficient to rupture the endosperm, for example, by passing the kernels through a roller mill gapped at 0.114–0.165 mm (*132*). Gentle milling of the cracked kernels allows the germs to be readily separated by flotation. The isolated germs may be used for oil recovery or as a human food item. Steeping of the endosperm pieces for easy recovery of high quality starch can then be accomplished in sulfur dioxide solution for 8–10 h.

4. Starch Processing

The finished starch slurry from the final hydroclone stage may be further processed in alternative ways: (a) dried directly and sold as unmodified corn starch; (b) Modified by chemical or physical treatment in a way that preserves the granule structure, then rewashed to remove residual reactants and dried (Chaps. X, XVII, and XX); (c) gelatinized and dried; (d) chemically or physically modified and then dried (Chaps. X, XVII, and XX); (e) hydrolyzed either completely to the D-glucose monomer or partially hydrolyzed to give mixtures of soluble oligosaccharides and sugars (Chap. XXI), which can be fermented to yield ethenol and other products (Chap. XI).

5. Product Drying, Energy Use, and Pollution Control

Proper operation of a wet-milling system requires the use of 17–22 gallons of fresh water per 100 pounds of corn (1420–1830 L per metric ton), all added at the final hydroclone stage. This water must eventually be removed from the products by evaporation since most wet milled products are marketed at 10–12% moisture content.

The flash drying method is now most universally used for drying unmodified starch, gluten, and gluten feed. In the case of starch or gluten, finished slurry is dewatered in a centrifuge or filter, and the filter cake is injected into the bottom of a stream of rapidly moving air heated to temperatures of 200°–260°F (93°–127°C) with natural gas or steam. The particles are dried instantly and are collected in cyclones. Baunack (*134*) has described several flash dryer arrangements. He reports that drying conditions in flash dryers can be varied to control particle size and bulk density of starch, gluten, and fiber products. Rotary steam tube dryers are used for drying germ; rotary direct fired dryers are used in some plants for drying gluten and gluten feed. Fluidized bed dryers, such as the device described by Idaszak (*135*), are coming into use for drying starch, especially chemically modified types. Steepwater for use as a feed ingredient is generally concentrated to a viscous fluid containing 45–50% moisture in triple-effect evaporators with forced feed to reduce the problem of fouling of the tubes and to improve energy efficiency. A recent trend is use of vapor recompression evaporators to save energy during the initial evaporation of dilute solutions.

Evaporation and drying require large inputs of electrical and heat energy. The corn wet-milling industry has been characterized as the second most energy intensive of all the food industries in the United States (*136*). In 1972, the average energy consumption was 433,214 Btu per 100 pounds of corn (1×10^6 kilojoules per metric ton) of which about 11% was used for feed drying, 4% for starch drying, 21% for evaporation, 33% for process steam and heating, 15% for mechanical power and lighting, 17% for boiler losses, and 9% miscellaneous (*137*). One of the major activities for recent years in the corn wet-milling industry has been development of methods for energy use reduction.

Control of air and water pollution has been a major concern of the wet-milling industry for many years. Both have been reduced gradually by development of new by-products and by treatment of wastewater streams so that they may be returned to the process, put into community sewers, or sent to dedicated waste treatment facilities. In the 1930s, the practice of disposing of steepwater and unwanted process water in waterways was stopped, resulting in the development of the "bottled up" process (*5, 138*).

Although the "bottled up" process increased plant dry substance yields to 99%, there are still many dilute streams containing organic and inorganic solutes that cannot be returned to the process. Stavenger (*139*) has estimated that the average wet mill discharges about 7.0 g BOD_5 per metric ton of corn processed. Standard municipal sewer treatment methods have not always proved adequate because of large surges in pH and volume of effluents and variations in the type of contained organic matter. One corn wet-milling plant, using an aerobic digestion system (*140*), found it necessary to receive plant effluents in a large basin equipped for agitation, pH, and temperature control. Biomass develops in surface aerated tanks and is recovered by gravity and flotation collectors to reduce BOD and suspended solids in the water to the limits established by the Environmental Protection Agency. A recently announced biological waste treatment device for wet-milling plants achieves a ten-fold greater biomass conversion rate than standard air-activated sludge systems (*139*). This procedure utilizes a sand bed fluidized with pure oxygen.

Other methods of processing waste streams or dilute process water streams, which has good possibilities for wet milling, make use of membrane systems to effect reverse osmosis, ultrafiltration, or electrodialysis. Reverse osmosis (*141, 142*) appears to be a low energy alternative to evaporation of waters containing low concentrations of solids, such as those from the first step in steepwater evaporation and from sugar washing systems. Energy use of this method is significally lower than that of evaporative methods up to a concentration of 12–15% dissolved solids. Reverse osmosis for purifying starch wash water should be preferable to use of ion-exchangers.

Ultrafiltration membranes are manufactured with different porosities which have high selectivity in separation of molecules of different sizes and may eventually be found useful in processing soluble streams (*144*). For example, a concept of using ultrafiltration to separate sugars in starch hydrolyzates from residual fat and protein which has the potential of improving hydrolyzate refining with energy and equipment savings has been described (*145*).

Exhaust gases from feed by-product dryers contain particulates, aerosols, and gases. The gases contain low-molecular-weight compounds that are volatilized from the steepwater. These compounds have characteristic odors that may be offensive to some people. Air pollution problems can be ameliorated by reducing drying temperatures, by passing the exhaust gas through scrubber columns containing potassium permanganate solution (*146*) or by incineration (*147*) with recovery of heat.

6. Automation

The changes in the wet-milling process during the past twenty years have been in the direction of drastically reducing the number of operating personnel per 1000 bushels ground. This change has required the development of automated computer process control stations as reviewed by Simms (*148*). The sweetener processes have been more easily computerized than the milling processes because of the availability of liquid analysis systems such as polarimetry, refractive index measurements, and pump-automated analysis systems (Technicon, etc.). However, the technology of analysis of solid material by infrared reflectance analysis (*122*) now makes possible the computerized automation of the entire milling and product drying process. All plants are operated with continuous flow on a 24 h per day basis.

IV. The Products

1. Starch

Dry, unmodified corn starch is a white powder with a pale yellow cast; grain sorghum starch has a faint pink tint. Absolute whiteness of either starch, such as is required for aspirin tablets, must be achieved by bleaching. Table VIII gives general properties of regular corn and sorghum starch, and of waxy corn starch (*149–153*). The starch produced by different manufacturers is quite uniform with respect to most of these properties. The most important properties with respect to utilization of starch are largely related to properties of the gelatinized pastes as described in other Chapters IX, XVII, XVIII, XIX, and XX.

2. Sweeteners (see Chapter XXI)

Many different types of starch hydrolyzate products are produced for a multitude of uses, mostly in food products. Table IX provides definitive data on

Table VIII

Analysis and Properties of Powdered Corn and Sorghum Starches[a]

	Corn		Sorghum
	Waxy	Regular	Regular
Starch (%)	88	88	88
Moisture (%)	11	11	11
Protein ($N \times 6.25$) (%)	0.28	0.35	0.37
Ash (%)	0.1	0.1	0.1
Fat (by ether extraction) (%)	0.04	0.04	0.06
Lipids, total[b] (%)	0.23	0.87	0.72
SO_2 (mg/kg[c])	—	49	—
Crude fiber (%)	0.1	0.1	0.2
pH	5	5	5
Linear starch fraction (amylose) (%[d])	0	28	28
Branched starch fraction (amylopectin) (%[d])	100	72	72
Granule size (microns[e])	—	5–30	4–25
Average granule size (microns[e])	—	9.2	15
Granule gelatinization temperature range[f] (°)	63–72	62–72	68–75
Swelling power at 95°[g]	64	24	22
Solubility at 95°[g]	23	25	22
Specific gravity	1.5	1.5	1.5
Weight per cubic foot (pounds)	44–45	44–45	44–45

[a] Reference *149*, except as noted. Values for waxy corn and sorghum from unpublished data except as noted.

[b] Reference *151*.

[c] Reference *153*.

[d] % of carbohydrate.

[e] Reference *152*.

[f] Initial and end temperatures for loss of microscopic birefringence (*150*, *152*).

[g] Reference *150*.

properties of the most popular of the liquid sweetener products (*154*). Three other products are normally sold in dry form. They are D-glucose (dextrose) in the monohydrate and anhydrous crystal forms, very low DE (22–30) corn syrup, and maltodextrins (5–20 DE). The latter two products are sold as amorphous powders.

3. Ethanol (see Chapter XI)

Ethanol produced by fermentation of starch hydrolyzates is regarded legally as equivalent to grain alcohol and may be used in beverages. It also qualifies for tax-free status when blended with gasoline at a level of 10% for use as a motor fuel.

Table IX

Properties of Commercial Corn Syrups (154)

	Acid conversion, DE level			*Acid–enzyme*			*Enzyme–enzyme*		
Typical analysis	*Low*	*Regular*	*Intermediate*	*High maltose*	*Regular A–E*	*High DE*	*Glucose syrup*	*High fructose*	
Commercial Baume	43°	43°	43°	43°	43°	43°	—	—	
Solids (%)	80	80.3	81	80.3	82	82.2	71	71	
Moisture (%)	20	19.7	19	19.7	18	17.8	29	29	
Dry basis									
Dextrose equivalent	37	42	52	42	62	69	96	(95)	
Ash (sulfated) (%)	0.4	0.4	0.4	0.4	0.4	0.4	0.03	0.03	
Carbohydrate composition									
Monosaccharides (%)									
D-Glucose (%)	15	19	28	6	39	50	93	52	46
Fructose (%)	0	0	0	0	0	0	0	42	55
Disaccharides (%)	12	14	17	45	28	27	4	3	2
Trisaccharides (%)	11	12	13	15	14	8			
Tetrasaccharides (%)	10	10	10	2	4	5			
Pentasaccharides (%)	8	8	8	1	5	3	3	3	2
Hexasaccharides (%)	6	6	6	1	2	2			
Higher saccharides (%)	38	31	18	30	8	5			
Viscosity, centipoises at									
24°	150,000	56,000	31,500	56,000	22,000	—	—		—
37.7°	30,000	14,500	8,500	14,500	6,000	—	—		—
44°	8,000	4,900	2,900	4,900	2,050	—	—		—

4. Corn Oil

About 70 kg of crude corn oil is recovered from the germ isolated from a metric ton of corn (1260 bushels). The crude oil is refined by standard methods to reduce content of free fatty acids, waxes, phospholipids, color, and odor to acceptable food levels (*34*). Refining produces a bland, pale-yellow triglyceride oil. Its low solidifying point, low smoke point, and slightly "corny" flavor make it a favorite oil for household uses where 50–60% of the production is utilized. Nearly all the remainder is used in manufacture of oleomargarine. The high level of linoleic acid is claimed to be a dietary advantage. The low level of linolenic acid and an adequate level of tocopherols contribute to corn oil's good oxidative stability (*34*). Grain sorghum oil is similar in fatty acid composition to corn oil; the crude oil has a higher wax content and is more difficult to refine.

5. Feed Products

Total by-product volume amounts to about one-third of the total mill output. Most of the volume is sold as animal feed ingredients, except for the corn oil and a small amount of steep liquor used in antibiotic fermentation media (*156*). The composition, nutritional values, and animal feeding uses have been thoroughly described by Shroder and Heiman (*155*). Listed in the approximate order of volume of sales, the major feed by-products are briefly described as follows:

Corn gluten feed, 21% protein, is composed of fiber (bran) and steep liquor, plus corn germ residue at locations where germ is processed. The product is dried to 11% moisture. It usually contains no gluten. Primary use is as a dairy ration ingredient. The average corn gluten feed contains 25–30% steep liquor solids, but several companies market feed products containing up to 50% steep liquor solids.

Corn gluten meal is the 60% protein, gluten overflow stream from the first starch separation step dried to 11% moisture content. It is bright yellow as result of the 200–400 mg/kg of yellow xanthophyll pigments it contains. A small amount of 41% corn gluten meal is made by blending with fine ground gluten feed. Primary use of gluten meal is in broiler rations as a protein concentrate and to supply yellow pigmentation. It is also a good pet food ingredient.

Corn germ meal, 21% protein, is the dried residue left from oil recovery from germ. It has high absorbancy for liquids such as molasses and tallow, but the primary use is as an ingredient in corn gluten feeds.

Corn starch molasses is the concentrated mother liquor remaining from dextrose crystallization. It is used with cane molasses in cattle feeding regimes.

Concentrated steep liquor is officially identified for feeding purposes as "Condensed corn fermentation extractives." This product carries a 25% protein content at a 48% moisture content. Its only feeding use is as an ingredient in liquid feeds for cattle, but it is also an excellent nitrogen source in nutrient media for antibiotic production (*156*).

Corn bran, dehydrated, is the dried fiber fraction. Protein content is about 10%. Its only use is in beef or dairy cattle feeds. Recently, a purified form of corn fiber has been offered to the food industry for use in high-fiber food products (*159*).

Hydrolyzed vegetable oil (HVO) is a by-product of the alkali refining of crude corn oil and is obtained by acidulating alkaline soapstock. HVO must contain at least 92% total fatty acids. It is used to control dust and as an energy source in beef and poultry rations.

Feed products produced during grain sorghum starch production have names and protein contents similar to the corn starch feed by-products (*8, 157*). Milo (sorghum) steep liquor (condensed milo fermentation extractives) has been used as an ingredient in liquid feed supplements for beef cattle feeding. Because grain sorghums are brown in color and contain condensed tannins, all these feed products are dark brown in color. Therefore, they are used almost exclusively in ruminant feeds (*157*).

V. References

(*1*) G. W. Beadle, *in* "Maize Breeding and Genetics," D. B. Walden, ed., Wiley, New York, 1978, pp. 93–112.

(*2*) H. H. Iltis, J. F. Doebley, R. Guzmán M., and B. Pazy, *Science*, **203,** 186 (1979).

(*3*) Trifunovic, *in* reference *1*, pp. 41–58.

(*4*) G. N. Collins, *U.S. Bur. Plant Inds.*, No. 161 (1909), 31 pp.

(*5*) R. W. Kerr, "Chemistry and Industry of Starch," Academic Press, New York, 1942.

(*6*) J. H. Martin, *in* "Sorghum Production and Utilization," J. S. Wall and W. M. Ross, eds., AVI Publishing Co., Westport, Connecticut, 1970, pp. 1–27.

(*7*) J. E. Freeman, reference *6*, 28–72.

(*8*) S. A. Watson, reference *6*, pp. 602–626.

(*9*) O. L. Brekke, *in* "Corn: Culture, Processing, Products," G. E. Inglett, ed., AVI Publishing Co., Westport, Connecticut, 1970, pp. 262–291.

(*10*) V. E. Headley, J. Spanheimer, J. E. Freeman, and R. E. Heady, *Cereal Chem.*, **49,** 142 (1972).

(*11*) M. J. Wolf, C. L. Buzan, M. M. MacMasters, and C. E. Rist, *Cereal Chem.*, **29,** 321 (1952).

(*12*) M. J. Wolf, C. L. Buzan, M. M. MacMasters, and C. E. Rist, *Cereal Chem.*, **29,** 334 (1952).

(*13*) T. A. Kieselbach and E. R. Walker, *Am. J. Bot.*, **39,** 561 (1952).

(*14*) M. J. Wolf, C. L. Buzan, M. M. MacMasters, and C. E. Rist, *Cereal Chem.*, **29,** 349 (1952).

(*15*) J. J. C. Hinton, *Cereal Chem.*, **30,** 441 (1953).

(*16*) D. Duvick, *Cereal Chem.*, **38,** 374 (1961).

(*17*) M. J. Wolf and V. Khoo, *Cereal Chem.*, **52,** 771 (1975).

(*18*) C. B. Hopkins, L. H. Smith, and E. M. East, *University of Illinois Agr. Exp. Sta. Bull.*, No. 87 (1903).

(*19*) C. W. Blessin, J. D. Beicher, and R. J. Dimler, *Cereal Chem.*, **40,** 582 (1963).

(*20*) S. A. Watson, E. H. Sanders, R. D. Wakely, and C. B. Williams, *Cereal Chem.*, **32,** 165 (1955).

(*21*) M. J. Wolf, C. L. Buzan, M. M. MacMasters, and C. E. Rist, *Cereal Chem.*, **29,** 362 (1952).

(22) L. S. Dure, *Plant Physiol.* **35,** 925 (1963).
(23) H. L. Seckinger, M. J. Wolf, and M. M. MacMasters, *Cereal Chem.,* **37,** 121 (1960).
(24) T. J. Jacks, L. Y. Yatsu, and A. M. Altschul, *Plant Physiol.,* **42,** 585 (1967).
(25) F. A. Kummerow, *Oil Soap,* **23,** 167, 273 (1946).
(26) E. H. Sanders, *Cereal Chem.,* **32,** 12 (1955).
(27) D. F. Miller, "Composition of Cereal Grains and Forages," Publ. No. 585, National Academy of Science, National Research Council, Washington, D.C., 1958.
(28) P. J. Van Soest and R. H. Wine, *J. Off. Anal. Chem.* **51,** 780 (1968).
(29) F. R. Earle, J. J. Curtis, and J. E. Hubbard, *Cereal Chem.,* **23,** 504 (1946).
(30) S. L. Tan and W. R. Morrison, *J. Amer. Oil Chemists Soc.,* **56,** 759 (1979).
(31) J. W. Dudley, R. J. Lambert, and D. E. Alexander, *in,* "Seventy Generations of Selection for Oil and Protein Concentration in the Maize Kernel," J. W. Dudley, ed., Crop Sci. Soc. of Amer., Special Publication, 1974, pp. 181–212.
(32) D. E. Alexander and J. R. Creech, *in* "Corn and Corn Improvement," G. F. Sprague, ed., Amer. Soc. Agron., Madison, Wisconsin, 2nd Ed., 1977, pp. 336–390.
(33) S. A. Watson and J. E. Freeman, *in* "Proceedings of the Thirtieth Annual Corn and Sorghum Research Conference," Amer. Seed Trade Assn., Washington, D.C., 1975, pp. 251–275.
(34) R. A. Reiners, *in,* "Corn: Culture Processing, Products," G. L. Inglett, ed., AVI Publishing Co., Westport, Connecticut, 1970, p. 243.
(35) E. J. Weber and D. E. Alexander, *J. Amer. Oil Chemists Soc.,* **52,** 370 (1975).
(36) E. O. Schneider, E. B. Early, and E. E. DeTurk, *Agron. J.,* **41,** 30 (1949).
(37) T. H. Hamilton, B. C. Hamilton, B. C. Johnson, and H. H. Mitchell, *Cereal Chem.,* **28,** 163 (1951).
(38) E. T. Mertz, L. S. Bates, and O. E. Nelson, *Science,* **145,** 279 (1964).
(39) S. A. Watson and K. R. Yahl, *Cereal Chem.,* **44,** 488 (1967).
(40) R. M. Hixon and G. F. Sprague, *Ind. Eng. Chem.,* **34,** 959 (1944).
(41) J. E. Freeman, M. Abdullah and B. J. Bocan, U.S. Patent 3,928,631 (1975); *Chem. Abstr.,* **84,** 88301 (1976).
(42) R. P. Bear, M. M. Vineyard, M. M. MacMasters, and W. L. Deatherage, *Agron. J.* **50,** 598 (1958).
(43) R. A. Anderson, C. Vojonovich, and E. L. Griffin, Jr., *Cereal Chem.,* **38,** 84 (1961).
(44) J. E. Hubbard, H. H. Hall, and F. R. Earle, *Cereal Chem.,* **27,** 415 (1950).
(45) G. Bianchi, P. Avato, and G. Mariani, *Cereal Chem.,* **56,** 491 (1979).
(46) E. D. Maxon and L. W. Rooney, *Cereal Chem.,* **49,** 719 (1972).
(47) R. D. Karper, *J. Heredity,* 24, 257 (1933).
(48) R. Singh and J. D. Axtell, *Crop Sci.,* **13,** 535 (1973).
(49) J. E. Freeman, *Trans. Amer. Soc. Agr. Eng.,* **16,** 671 (1973).
(50) Official Grain Standards of the United States, USDA, Consumer Marketing Service, Grain Division, Washington, D.C., 1970.
(51) C. M. Christensen and H. H. Kaufmann, "Grain Storage: The Role of Fungi in Quality Loss," The University of Minnesota Press, Minneapolis, Minnesota, 1969, p. 153.
(52) J. E. Freeman, H. J. Heatherwick, and S. A. Watson, *in* "Proc. Conf. Res. Corn Quality," University of Illinois, Apr. 28–29, 1970, AE-4251.
(53) "Interactions of Mycotoxins in Animal Production," Nat. Acad. Sci. Washington, D. C., Proc. Symp., July 13, 1978 (Publ. 1979).
(54) O. L. Shotwell, M. L. Goulden, and C. M. Hesseltine, *Cereal Chem.,* **51,** 492 (1973).
(55) S. A. Watson and K. R. Yahl, *Cereal Sci. Today,* **16,** 153 (1971).
(56) E. B. Lillehoj, W. F. K. Wolek, E. S. Horner, N. W. Widstrom, L. M. Josephson, A. O. Franz, and E. A. Catalano, *Cereal Chem.,* **57,** 255 (1980).
(57) M. S. Zuber and E. B. Lillehoj, *J. Environ. Qual.,* **8,** 1 (1979).

(*58*) H. W. Anderson, E. W. Nehring, and W. R. Wichser, *J. Agr. Food Chem.*, **23,** 775 (1975).
(*59*) K. R. Yahl, S. A. Watson, R. J. Smith, and R. Barabolak, *Cereal Chem.*, **48,** 385 (1971).
(*60*) O. L. Shotwell and C. W. Hesseltine, *Cereal Chem.*, **58,** 124 (1981).
(*61*) R. Barabolak, C. R. Colburn, D. E. Just, and E. A. Schleichert, *Cereal Chem.*, **55,** 1065 (1978).
(*62*) J. Tuite, G. Shaner, G. Rambo, J. Foster, and R. W. Caldwell, *Cereal Sci. Today*, **19,** 238 (1974).
(*63*) G. A. Bennett, E. E. Vandergraft, O. L. Shotwell, S. A. Watson, and B. J. Bocan, *Cereal Chem.*, **55,** 455 (1978).
(*64*) G. A. Bennett and R. A. Anderson, *J. Agr. Food Chem.*, **26,** 1055 (1978).
(*65*) G. C. Shove, *in* "Corn: Culture, Processing, Products," G. E. Inglett, ed., AVI Publishing Co., Westport, Connecticut, 1970, pp. 60–72.
(*66*) F. W. Bakker-Arkema, R. C. Brook, and L. E. Lerew, *Adv. Cereal Sci. Technol.*, 1 (1977).
(*67*) S. A. Watson and Y. Hirata, *Cereal Chem.*, **39,** 35 (1962).
(*68*) G. H. Foster, *in*, "Twentieth Hybrid Corn Industry-Research Conference," Amer. Seed Trade Association, Washington, D.C., 1965, pp. 75–85.
(*69*) C. Vojnovich, R. A. Anderson, and E. L. Griffin, Jr., *Cereal Foods World*, **20,** 333 (1975).
(*70*) J. S. Wall, C. James, and G. L. Donaldson, *Cereal Chem.*, **52,** 779 (1975).
(*71*) R. A. Thompson and G. H. Foster, "Stress Cracks and Breakage in Artificially Dried Corn," *U.S.D.A. Marketing Research Report*, No. **631,** Agr. Marketing Service, 1963.
(*72*) G. H. Foster and L. E. Holman, "Grain Breakage Caused By Commerical Handling Methods," *U.S.D.A. Agr. Marketing Report*, No. **968,** p. 13 (1976).
(*73*) R. B. Brown, G. N. Fulford, T. B. Daynard, A. G. Meiering, and L. Otten, *Cereal Chem.*, **57,** 529 1979.
(*74*) G. M. Grosh and M. Milner, *Cereal Chem.*, **36,** 260 (1959).
(*75*) O. L. Shotwell, C. W. Hesseltine, H. R. Burmeister, F. W. Kwolek, G. M. Shannon, and H. H. Hall, *Cereal Chem.* **46,** 446 (1969).
(*76*) S. A. Watson and Y. Hirata, *Agron. J.*, **47,** 11 (1955).
(*77*) J. E. Freeman and S. A. Watson, *Cereal Sci. Today*, **16,** 378 (1971).
(*78*) W. Kempf, *Staerke*, **23,** 89 (1971).
(*79*) H. J. Vegter, British Patent 1,238,725 (1971).
(*80*) M. J. Cox, M. M. MacMasters, and G. E. Hilbert, *Cereal Chem.*, **21,** 447 (1964).
(*81*) L. T. Fan, P. S. Chu, and J. A. Shellenberger, *Cereal Chem.*, **40,** 303 (1963).
(*82*) L. T. Fan, H. C. Chen, J. A. Shellenberger, and D. S. Chung, *Cereal Chem.*, **42,** 385 (1965).
(*83*) N. L. Stenvert and K. Kingswood, *Cereal Chem.*, **53,** 141 (1976); **54,** 627 (1977).
(*84*) M. Roushdi, Y. Ghali, and A. Hassanean, *Starch/Staerke*, **31,** 78 (1979).
(*85*) J. W. Paulis, *J. Agr. Food Chem.*, **30,** 14 (1982).
(*86*) G. M. Bautista and P. Linko, *Cereal Chem.*, **39,** 455 (1962).
(*87*) C. Franzke and G. Wahl, *Staerke*, **22,** 64 (1970).
(*88*) S. A. Watson, Y. Hirata, and C. B. Williams, *Cereal Chem.*, **32,** 382 (1955).
(*89*) L. T. Fan, P. S. Chu, and J. A. Shellenberger, *Biotechnol. Bioeng.*, **4,** 311 (1962).
(*90*) R. A. Anderson, *Cereal Chem.*, **39,** 406 (1962).
(*91*) E. M. Montgomery, K. R. Sexson, and R. J. Dimler, *Staerke*, **16,** 314 (1964).
(*92*) J. A. Wagoner, *Cereal Chem.*, **25,** 354 (1948).
(*93*) G. Wahl, *Staerke* **21,** 77 (1969).
(*94*) S. A. Watson and E. H. Sanders, *Cereal Chem.*, **38,** 22 (1961).
(*95*) J. N. Swan, *Nature*, **180,** 643 (1957).
(*96*) J. A. Boundy, J. E. Turner, J. S. Wall, and R. J. Dimler, *Cereal Chem.*, **44,** 281 (1967).
(*97*) J. S. Wall, *J. Agr. Food Chem.*, **19,** 619 (1971).
(*98*) A. C. Beckwith, *J. Agr. Food Chem.*, **20,** 761 (1972).

(*99*) J. H. Kolthoff, A. Anastasi, and B. H. Tan, *J. Amer. Chem. Soc.*, **82,** 4147 (1960).
(*100*) W. Nierle, *Staerke*, **24,** 345 (1972).
(*101*) J. S. Wall, *Adv. Cereal Sci. Technol.*, **2,** 135 (1978).
(*102*) J. Landry and T. Moureaux, *Bull. Soc. Chim. Biol.*, **52,** 1021 (1970).
(*103*) J. W. Paulis and J. S. Wall, *Cereal Chem.*, **46,** 263 (1969).
(*104*) B. Burr and F. A. Burr, *Proc. Natl. Acad. Sci. USA*, **73,** 515 (1976).
(*105*) A. W. Wight, *Starch/Staerke*, **33,** 165 (1981).
(*106*) S. A. Watson, *Methods Carbohydr. Chem.*, **4,** 3 (1964).
(*107*) I. C. Johnson, S. E. Gilliland, and M. L. Speck, *Appl. Microbiol.*, **21,** 316 (1971).
(*108*) R. C. Balana and A. M. Caixes, U.S. Patent 4,086,135 (1978); *Chem. Abstr.* **84,** 57, 393 (1976).
(*109*) S. A. Watson, C. B. Williams, and R. D. Wakely, *Cereal Chem.*, **28,** 105 (1951).
(*110*) E. Lindemann, *Staerke*, **6,** 274 (1954).
(*111*) L. Saint-Lebe, M. Joussoud, and C. Andre, *Staerke*, **17,** 341 (1965).
(*112*) W. Kempf, *Staerke*, **25,** 376 (1973).
(*113*) Stamicarbon, B. V., Heerlen, Netherlands, British Patent 701,613 (1953).
(*114*) P. L. Stavenger and D. E. Wuth, U.S. Patent 2,913,112 (1959); *Chem. Abstr.*, **54,** 2844 (1960).
(*115*) N. B. Smith, H. A. McFate, and E. E. Eubanks, U.S. Patent 3,236,740 (1966); *Chem. Abstr.*, **64,** 14401 (1966).
(*116*) M. E. Ginaven, U.S. Patent 3,040,996 (1962).
(*117*) R. R. Dill and M. E. Ginaven, U.S. Patent 3,118,624 (1964).
(*118*) D. W. Dowie and D. Martin, U.S. Patent 3,029,169 (1962); *Chem. Abstr.*, **57,** 3684 (1962).
(*119*) F. J. Fontein, U.S. Patent 2,916,142 (1959).
(*120*) F. J. Fontein, U.S. Patent 2,975,068 (1961); *Chem. Abstr.*, **55,** 12898 (1961).
(*121*) H. Bier, J. C. Eisken, and R. W. Honeychurch, *Staerke*, **26,** 23 (1974).
(*122*) W. L. Bell, *Cereal Foods World*, **28,** 249 (1983).
(*123*) V. P. Chwalek, U.S. Patent 3,813,298 (1974).
(*124*) H. Huster, "The Use of Separators and Decanters in the Starch Industries," Westfalia Separator AG, Oelde, West Germany, Undated.
(*125*) A. Peltzer, Sr., U.S. Patent 2,973,896 (1961).
(*126*) C. H. Hoepke and H. Huster, *Staerke*, **28,** 14 (1976).
(*127*) H. J. Vegter, U.S. Patent 2,689,810 (1954); *Chem. Abstr.*, **48,** 14624 (1954).
(*128*) H. J. Vegter, U.S. Patent 2,778,752 (1957); *Chem. Abstr.*, **51,** 4745 (1957).
(*129*) L. Bonnyay, *Starch/Staerke*, **30,** 61 (1978).
(*130*) V. P. Chwalek and C. W. Schwartz, U.S. Patent 4,144,087 (1979); *Chem. Abstr.*, **89,** 7960 (1978).
(*131*) J. E. Freeman and S. A. Watson, *Cereal Sci. Today*, **14,** 10 (1969).
(*132*) J. E. Freeman, U.S. Patent 3,477,855 (1969).
(*133*) R. L. Zipf, R. A. Anderson, and R. L. Slotter, *Cereal Chem.*, **27,** 463 (1950).
(*134*) F. Baunack, *Staerke*, **15,** 299 (1963).
(*135*) L. R. Idaszak, U.S. Patent 4,021,927 (1977); *Chem. Abstr.*, **86,** 92222 (1977).
(*136*) "Industrial Energy Study of Selected Food Industries for the Federal Energy Office," U.S. Department of Commerce, Final Report, 1974.
(*137*) M. E. Casper, "Energy Saving Techniques for the Food Industry," Noyes Data Corp., Park Ridge, New Jersy, 1977, pp. 83–94.
(*138*) A. L. Pulfrey, R. W. Kerr, and H. R. Reintjes, *Ind. Eng. Chem.*, **21,** 205 (1929).
(*139*) P. L. Stavenger, *Starch/Staerke*, **31,** 81 (1979).
(*140*) H. O. Bensing, D. R. Brown, and S. A. Watson, *Cereal Sci. Today*, **17,** 304 (1972).
(*141*) S. Sourirajan, "Reverse Osmosis," Academic Press, New York, 1970.

(*142*) O. D. Spatz, presentation at meeting of AIChE, St. Louis, Missouri, May 24, 1972.
(*143*) D. Bradley, "The Hydroclone," Permagon Press, Oxford, 1965.
(*144*) T. D. Brock, "Membrane Filtration," Sci. Technol., Madison, Wisconsin, 1983.
(*145*) H. Muller, German Patent 2,618,131 (1976); *Chem. Abstr.* **86,** 31230 (1977).
(*146*) V. E. Healey, *Trans. Amer. Soc. Agr. Eng.*, **25,** 788 (1982).
(*147*) A. Grant, T. K. Tailor, and J. Powers, *Chem. Process.* (February 1975).
(*148*) R. L. Simms, *Sugar y Azucar,* p. 50 (March 1978).
(*149*) "Corn Starch," Corn Industries Research Foundation, Washington, D.C., 3rd Ed., 1964.
(*150*) H. W. Leach, *in* "Starch: Chemistry and Technology," R. L. Whistler and E. F. Paschall, eds., Academic Press, New York, 1965, Vol. I.
(*151*) W. R. Morrison, *Adv. Cereal Sci. Technol.*, **1,** 221 (1976).
(*152*) T. J. Schoch and E. C. Maywald, *Anal. Chem.*, **28,** 382 (1956).
(*153*) W. Bergthaller and G. Tegge, *Staerke,* **24,** 348 (1972).
(*154*) J. M. Newton, *Proc. 16th Ann. Symp., Central States Section, Amer. Assoc. Cereal Chemists,* St. Louis, Missouri, 1975.
(*155*) J. D. Shroder and V. Heiman, *in* "Corn: Culture and Processing," G. E. Inglett, ed., AVI Publishing Co., Westport, Connecticut, 1970, Chapter 12.
(*156*) J. P. Bowden and W. H. Peterson, *Arch. Biochem.*, **9,** 387 (1946).
(*157*) "Grain Sorghum By-Product Feeds for Farm Animals," *Texas Agr. Expt. Stn., Bull.* **743,** 1–32 (1951).
(*158*) P. J. VanSoest, J. Fadel, and C. J. Sniffen, *Proc. Cornell Nutr. Conf.*, Cornell University, Ithaca, New York, 63–75 (1979).
(*159*) H. L. Kickle, W. J. Ball, and R. V. Schanefelt, U.S. Patent 4,181,747 (1980); *Chem. Abstr.* **92,** 109437 (1980).
(*160*) T. Suzuki and M. Sugimoto, *Denpun Kogyo Gakkaishi,* **20,** 161 (1973).
(*161*) V. P. Chwalek and C. W. Schwartz, U.S. Patent 4,244,748 (1981).
(*162*) K. D. Lewis, A. P. Charlton, P. Nyrops, and T. O. Merediz, U.S. Patent 4,260,480. (1981).
(*163*) E. F. Powell, U.S. Patent 3,909,288 (1975); *Chem. Abstr.*, **83,** 195629 (1975).
(*164*) V. P. Chwalek and R. M. Olson, U.S. Patent 4,181,748 (1980); *Chem. Abstr.*, **92,** 109,438 (1980).

CHAPTER XIII

TAPIOCA, ARROWROOT, AND SAGO STARCHES: PRODUCTION

BY DOUGLAS A. CORBISHLEY

Research Department, Industrial Starch and Food Products Division, National Starch and Chemical Corporation, Bridgewater, New Jersey

AND

WILLIAM MILLER

Tapioca Associates, Inc., Wilton, Connecticut

I. MANUFACTURE OF TAPIOCA STARCH

1. Historical Background and Botanical Nomenclature

Starch is available from large tuberous roots of a plant which thrives in most equatorial regions between the Tropic of Cancer and the Tropic of Capricorn. This plant is known by many names such as Ubi kettella or kaspe in Indonesia, manioca, rumu, or yucca in Spanish America, mandioca or Aipim in Brazil, manioc in Madagascar and French-speaking Africa, tapioca in India and Malaysia, and cassava and sometimes cassada in English-speaking regions of Africa, Thailand, and Ceylon.

The cassava plant has been classified as *Manihot utilissima* Pohl of the family Euphorbiaceae. The name *Manihot esculenta* is being increasingly adopted.

Cassava is the term usually applied in Europe and in the United States to the roots of the plant, whereas, tapioca is the name given the processed products of cassava. The word tapioca is derived from tipioca, the Tupi Indian name for the meal which settles out of the liquid expressed from rasped tubers and made into pellets and then called triiocet (*1*). The cassava plant was first believed to have been cultivated in the tropical regions of North and South America. The roots

STARCH, 2nd ed.

ISBN 0-12-746270-8

from the plant are an important carbohydrate food source in many regions of the tropics.

2. Cassava Root Production and Composition

There are hundreds of commercial varieties of cassava in various equatorial regions. These varieties fall into two main categories: *Manihot palmata* and *Manihot aipi,* or bitter and sweet cassava. The distinction between the varieties is dependent upon the content of cyanohydrin leading to hydrogen cyanide. Sweet or nontoxic roots contain less than 50 mg of evolvable HCN per kilogram of fresh root matter, whereas the bitter varieties can contain up to 250 mg, or more, of evolvable HCN per kilogram of fresh root. ''Sweet'' cassava root varieties are cultivated for human consumption, whereas the ''bitter'' cassava root varieties are cultivated for industrial purposes. The latter roots tend to have higher starch contents. During the starch manufacturing process, the hydrogen cyanide content is reduced to acceptable levels.

In many regions cassava is grown in poor soil that is no longer suitable for other crops. Fertile sandy-clay or sandy-loam soils are preferred, but cassava will not tolerate wet, marshy soil or standing water. Cassava, like all plants yielding carbohydrates, has high nutrient requirements. Hence, soil is exhausted rapidly unless fertilized or crops are rotated on it (*1*). Cassava can be cultivated as a single crop or in conjunction with crops such as corn and other vegetables.

The cassava plant is a semi-shrubbery perennial that grows under cultivation to a height of 2–4 m. The leaves are large and palmate, ordinarily with 5–7 lobes, borne on a long slender petiole. The roots or tubers radiate from the stem just below the surface of the ground. Feeder roots growing vertically from the stem and from the storage roots penetrate the soil to a depth of 50–100 cm, and this capacity of the cassava plant to obtain nourishment at some distance below the surface helps in understanding its growth on inferior soils (*1*).

The cassava plant is propogated vegetatively by cuttings taken from the lower stems of plants which are 9–10 months old. The cuttings should be at least 20 cm in length and contain a minimum of four nodules. The usual method is flat planting, by hand, 5–10 cm below the soil surface in rows about 120 cm apart (4,000 plants/acre, 10,000 plants/hectare). Time of planting is affected by weather conditions, and in general, the best root yields occur if planting takes place at the beginning of a rainy, humid weather period. The plant produces best with abundant rainfall; however, it can be grown at annual rainfall levels of 20–200 in.

Young plants do not require much cultivation. During the first few months, weeds should be removed once or twice and hoeing may be necessary to preserve the subsoil moisture. Some herbicides have been used, but in most countries the cost of chemicals cannot be justified. The plant responds well to fertilizer; but again, in most countries, fertilizers are not used except on some commercial plantations.

Particularly, in some regions, the cassava plant may be subject to attack by various diseases. A major effort has been put forward to select varieties that are disease resistant. Virus diseases, such as mosaic, the brown streak and leaf curl of tobacco, may attack the leaves, stems, and branches. Common bacterial diseases, such as *Phytomonas manihotis* in Brazil, *Bacterium cassava* in Africa, and *Bacterium solanacearum* in Java, may attack the roots, stems, or leaves. Mycoses attack the roots, stems, or leaves. Some insects, such as locusts, beetles and ants, affect the plant directly; and some others, such as aphids, affect the plant indirectly by the transfer of virus. Rats, goats, and wild pigs are most troublesome in areas adjacent to forests (*1*).

Crop yields vary depending on agricultural practices and soil conditions. The length of time for manioc roots to reach full maturity varies with climate. In Brazil, the growing period can vary from 10 to 18 months. The yield of roots usually varies from 5 to 20 tons/acre (12 to 48 tons/hectare), with some yields as high as 60 tons/hectare.

Harvesting of roots is a manual operation, although many mechanical devices have been tried over the years. The mechanical devices, such as plows and chisels, have not been successful due to high root damage and the number of roots left in the ground. The day before harvesting, the plant tops are cut 40–60 cm above the ground. Selected top material can be used for next year's planting. The roots are usually pulled slowly from the ground by hand. Depending on the soil type, a simple lifting device or a hoe may be used to assist root removal. Tubers are cut from the stems and must be used within 48 h of harvest to prevent loss of starch owing to enzymic changes and rot. Ideally, roots are used within 24 h of harvest. The chemical composition of cassava roots differs depending on variety, soil type, climate, and age of the root. A typical root analysis would indicate 70% moisture, 24% starch, 2% fiber, 1% protein, and 3% fats, minerals, and sugar. Young roots, under 10 months, have a low starch content; but roots over 24 months are woody and difficult to handle.

Starch, extracted from the cassava root and dried, is known by many names depending on geographical location. Names such as Cassava, Mandioca, Manioc and Tapioca, followed by the word flour or starch, are common. Sliced roots (chips) dried in the sun for 2–3 days are called gaplek or raspa de mandioca. The milled product (gaplek meal) is used as an animal fodder and in some starch consuming industries such as alcohol production.

An important product, farinha de mandioca, in the diet of many persons living in Latin America comes from peeled roots which are grated or ground to a pulp, toasted slowly, and dried. Gari, a normal dietary component of many Africans, is made by fermenting the grated tubers then toasting and drying the product (*1*).

3. Tapioca Starch Manufacture

Manufacturing plants for cassava starch are located close to root growing areas to minimize root transport costs and, more importantly, to enable the processing of tubers in the shortest time.

The roots are delivered to the factory and stored in wooden or concrete bunkers. Bunker filling and emptying must be closely supervised to insure that the roots that are harvested first are consumed first. The roots are usually delivered to a washing station by a belt conveyor. After washing, the outer skin, or corky portion is removed. The inner part of the peel, or cortex, is not removed because it has starch recoverable in modern processes. The washer is usually a U-shaped trough with paddles that carry the roots to the peeler. The peeler can be an integral part of the washer or a separate unit. The roots are peeled by the abrasion of one against another and against the walls and paddles of the washing and peeling device.

To recover the starch, all cell walls must be ruptured. This has been accomplished, at times, by mild fermentation of the roots, then grinding them to a pulp with starch recovered by screening and washing, or by centrifugation. The fermentation process does not give good starch yields, and the quality of the starch is generally inferior.

For high quality starch production, the washed roots are chopped first into slices of 30–50 mm and conveyed to a disintegrating (rasping) device. A variable-speed conveyor is used to control the supply of raw material. Efficient disintegration is necessary to achieve high starch yield. This function can be accomplished in one or two stages depending on the efficiency of the machinery. The rasping device is an impacting machine with high peripheral speed. After rasping, the hydrogen cyanide in the roots is set free and dissolved in wash water. Reaction of the acid with iron may lead to the formation of ferrocyanide, which gives a bluish color. Therefore, the rasp and other machinery and piping coming in contact with the starch are made of stainless steel or other resistant materials.

After disintegration of the pulp, it is washed on screens where fiber remains on the screen and starch passes through. Screens are often in the form of rotating cones, angled or trough shaped (Figs. 1 and 2). In all cases, counter current washing is needed (*2, 3*). Washed fiber can be used as a fertilizer or pressed and dried for cattle feed. A general starch process is shown in Figure 3.

Crude starch milk leaves the fiber washing stage at a concentration of about 3° Be (54 kg starch/m^3) and passes through a degritting screen, where small foreign matter is removed, and then on to continuous centrifuges where starch is separated from remaining fine fiber and solubles. The starch is discharged through nozzles in the bowl periphery, while the light fraction (fruit water) containing fine fiber and soluble materials is discharged through conical discs with the help

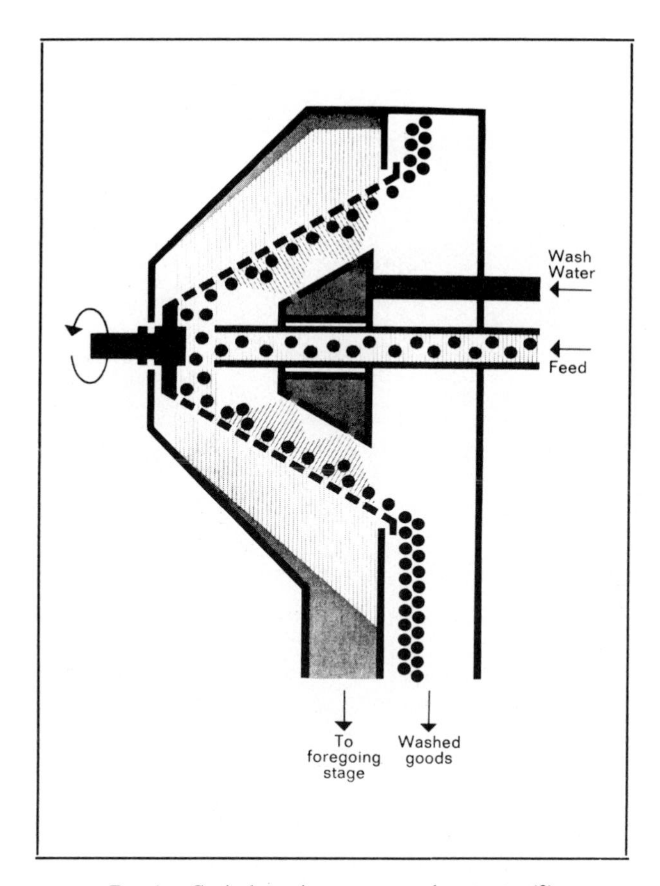

FIG. 1.—Conical rotating screen starch extractor (*2*).

of a centripetal pump. Clean wash water is introduced by nozzles near the collected starch. The fresh water displaces the impure water which is moved to the fiber washer and root washing sections (*2*). The starch may be reslurried and recentrifuged for further purification (Fig. 4).

In each centrifuge operation, starch is washed countercurrently with water containing 0.05% sulfur dioxide. Use of sulfur dioxide is essential to control microbial action in the starch water separation process (*2*). The starch-containing stream leaves the purification process as a slurry at 38–42% solids and is dewatered in a continuous vacuum filter or batch-operated basket-type centrifuge. The moisture content of the starch cake from a vacuum filter is 40–45% and 32–37% from a basket centrifuge. The dewatered starch cake is screw conveyed to the dryers which may be drum, belt, tunnel, or flash types. The most common is flash (pneumatic) drying. Air is drawn through a heater (steam coils or direct

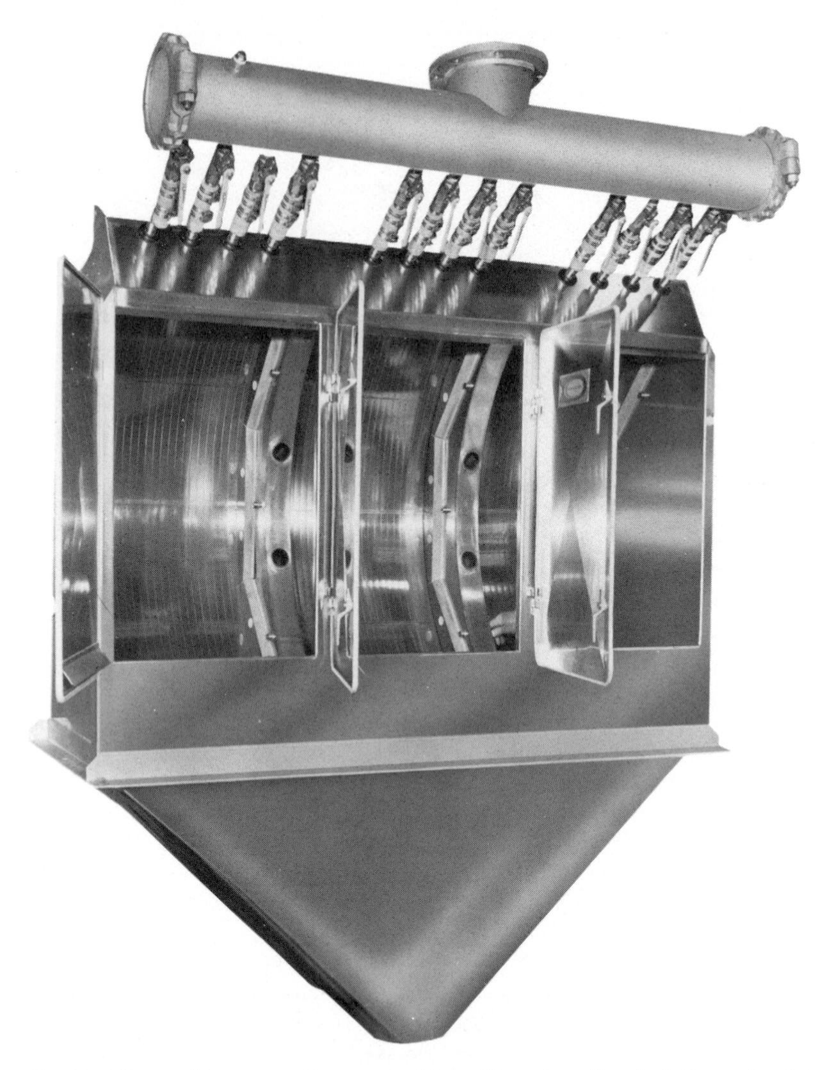

FIG. 2.—Fiber washing screen (*3*, courtesy of Dorr-Oliver, Inc., Stamford, Connecticut.)

gas or oil fired burner), where the temperature is raised to about 150°. The dried starch (12–14% moisture) is separated from the moist air in cyclones and then ground and sifted.

The equilibrium moisture for tapioca starch at 60% relative humidity and 21° is 12.5%. It will have a slurry pH of 5–6 and an ash of less than 0.15%. A high ash value would be indicative of poor washing of the starch.

Retained fiber is measured by the amount held on a 200-mesh screen. Its volume settling from screen washing is indicative of the amount passed. Various levels can be tolerated depending on the intended use of the starch. For starch destined for the food industry and certain dextrin applications, 0.2% fiber is a maximum tolerable level. High pulp content can indicate improper disintegration, poor fiber washing, poor starch purification, and poor starch sifting. At times, the contents of the material in the settling cylinder may contain gelatinized starch. Acid digestion of the fiber should then be used in its analytical determination.

The viscosity of tapioca starch depends on plant variety, growth area, time of harvest, age of roots, soil fertility, and rainfall during the growing period. Manufacturing practices used in making the starch are also important.

"Acid factor," a measured characteristic of tapioca starch, not often needed with other starches, is the number of milliliters of 0.1 *M* hydrochloric acid solution required to raise the pH of a 25 g slurry in 50 mL of distilled water to 3.0.

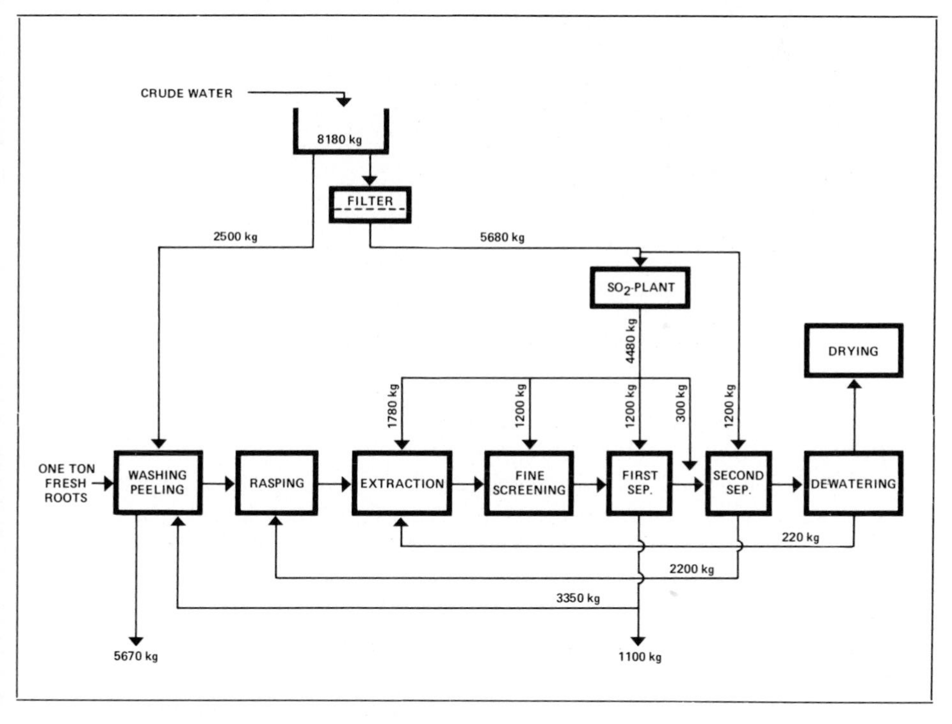

FIG. 3.—Water useage and basic manufacturing operations for tapioca starch production (*2*, courtesy of Alfa-Laval, Tumba, Sweden).

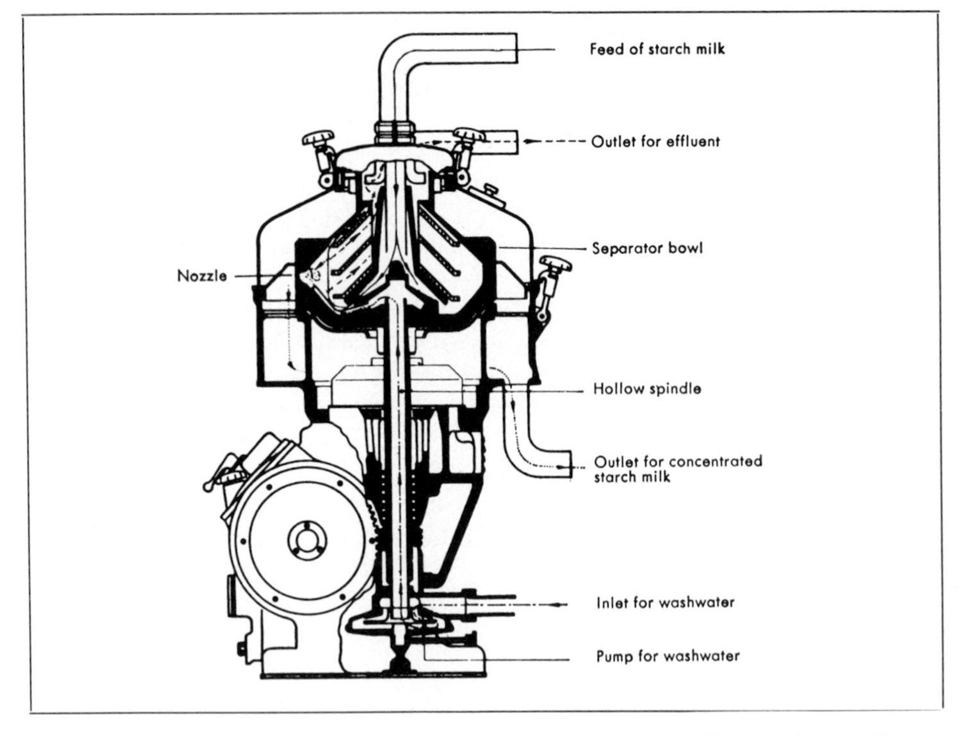

Fig. 4.—Continuous centrifugal starch separator and concentrator used in separating, washing, and concentrating the feed from the fiber washing stage (*2*, courtesy of Alfa-Laval, Tumba, Sweden.

II. Arrowroot Starch

Arrowroot starch is obtained from the root of the tropical plant *Maranta arundinacea,* a perennial that grows 2–5 ft (60–150 cm) high with oval lanceolate leaves and white oval flowers arranged in clusters. The plant has cylindrical elongated roots about 1-in.(2.5 cm) thick and 8–18 in.(20–45 cm) long with pale brown scales at each joint. The plant is propagated from root cuttings or seeds. In some areas, the plant grows as a weed and is harvested as an alternate crop.

Roots can be harvested after 6–12 months depending on the growing area and country. The roots can contain more than 20% starch, or which 17–18% is extractable in equipment of the type used to extract tapioca starch. Arrowroot requires more washing than does manioc root. The outer skin of the root must be thoroughly removed or the starch retains a bitter flavor and a yellow color. Arrowroot starch granules are somewhat larger (15–70 μm) than tapioca starch granules (5–15 μm) (*5*).

Arrowroot starch is produced mainly in Brazil, China, and on the Island of St.

Vincent in the West Indies. St. Vincent currently exports about 800 tons of arrowroot starch per year.

III. Sago Starch

Starch from the stems of palms has been mainly a product of local importance throughout the mainland and islands of Southeast Asia, in parts of Melanesia, certain islands of Micronesia, and various areas of tropical South America. Its production is confined to the humid tropics and swampy areas of tropical rain forests subject to frequent drenching downpours of rain (*4*). Although quite a number of tropical areas of the world can grow sago palms and some fourteen species belonging to eight genera have been used and to various degrees exploited, only *Metroxylon* and *Arenga* in the Eastern Hemisphere and *Mauritia* in the Western Hemisphere are of major importance as palm starch sources. The principal growing areas where there exists also moderate commercial sago starch production are Sarawak (where there are modern refineries at Sibu, Sabah, and Brunei) and New Guinea. In Papua, New Guinea, particularly along the Sepik River and its branches, where swamplands extend over a thousand square miles, great growths of *Metroxylon sagu* occur. This area commercially surveyed by the Australians and later the Japanese, represents a large source of sago starch.

The *Metroxylon* palms have large pinnate leaves, a stout erect trunk and reach a height of 30–50 feet (9–15 m) at maturity. Suckers grow from the base of the main stem and are used for propagation. In the wild, trees grow in clumps in fresh water swamps. At maturity, in about 15 years, each trunk produces a large terminal inflorescence after which the tree dies. For commercial starch production, trees of 8 years or older are usually cut.

Native production of sago starch consists of simple felling of the tree and a long strip of ''bark'' pried off to expose the pith containing the starch. The pith (*5*) is scrapped out and kneaded with water by hand, or trampled by foot, to loosen the starch which is washed away to settle out of the water as a white precipitate. Much of the sago starch produced locally in Southeast Asia is consumed for subsistence, usually in the form of small baked cakes.

Ethnobotanical information suggests a gradual replacement of ancient starch technology by higher levels of food production based on more developed agricultural systems; and most everywhere that sago starch is produced, the production of rice, much preferred as a food, is encroaching.

Commercial production of sago starch follows the same basic production line as used by natives. Thus, cut palm trunks are transported to mills where they are split into sections; the pith is rasped out, and the starch is kneaded out with water mechanically, but often by trampling, for 30 min. Frequently this first starch removal is done at local village sites, and the crude starch brought to the commercial plant for purification. Crude starch at the factory is slurried with water and sieved to remove coarse fiber; and the starch is removed and washed on a rotary filter, before drying with hot air. One trunk will have 600–800 lb (270–360 kg) of pith yielding 200–400 lb (90–180 kg) of sago starch, with an average of 40% starch in the pith. The extracted starch is composed of large granules 20–60 μm in diameter (*6*).

The historical and principal industrial uses of sago starch are as textile sizings and adhesives. In sago-producing areas of Sarawak, Subah, and Indonesia where cash economics operate, sago represents a cheap, locally available foodstuff. However, the preference for rice as a food and its adaptability to fresh water swamp lands has resulted in its encroachment, and rice now dominates as a crop for both food and industrial purposes.

IV. References

(*1*) M. Grace, "Processing of Cassava," Food and Agriculture Organization of the United Nations, Rome, Italy, 1970.

(*2*) Technical Bulletin, "The Alfa-Laval Manioc Starch Process," Bulletin No. TB 40324E2, Alfa Laval, Tumba, Sweden.

(*3*) "Processing of Cassava Roots into Cassava Starch," Technical Bulletin, Dorr-Oliver, Inc., Stamford, Connecticut.

(*4*) For an excellent review see: K. Ruddle, D. Johnson, P. K. Townsend and J. D. Rees, "Palm Sago, A Tropical Starch from Marginal Lands," East-West Center, University Press of Hawaii, Honolulu, 1978.

(*5*) R. M. Johnson and W. D. Raymond, *Colonial Plant Animal Prod.* (Gt. Brit.) **6,** No. 1, 20 (1957); *Econ. Bot.*, **11,** 326 (1957).

(*6*) O. B. Wurzburg, *Econ. Bot.*, **6,** 211 (1952).

CHAPTER XIV

POTATO STARCH: PRODUCTION AND USES

By Eugene L. Mitch

*Boise Cascade Corp., Portland, Oregon**

I. Introduction

In the late nineteenth century, potato starch was an important commodity in the United States. More than 150 factories were in operation throughout New England, the central and western states (*1*). Special varieties of potatoes were grown for their high starch content, as is still practiced in Europe. With the advent of modern corn wet-milling plants, corn starch became available at lower cost, and it eventually became uneconomical to grow potatoes just for starch.

There came also a demand for better cooking and tasting potatoes than those grown expressly for starch. This led to the cultivation of mostly food varieties for

*Present address: Process Chemical Development, Chemax, Inc., Portland, Oregon 97210.

STARCH, 2nd ed.

ISBN 0-12-746270-8

the fresh pack market. Anywhere from 10% to 20% of these potatoes are culled out because they are too small, too large, misshapen, or damaged. The culls and surplus potatoes provided raw material for continued starch production, but the industry declined due to corn starch competition. By 1950 there were less than thirty potato starch plants in the United States (*2*), twenty-two of which were in Aroostook County, Maine (*3*).

Up through the late 1950s, the availability of cull potatoes for starch was a function of the size of the potato crop and the demand for culls in the cattle feed market. There were no other markets. However, the development of instant potatoes, first marketed in 1946, created an alternative, higher valued use for culls. The rapid growth of snack foods and institutional and household markets for dehydrated potato flakes and granules in the late 1950s and early 1960s, plus fluctuating cattle feed markets, caused uncertainty in the supply and price of culls for starch production. As a consequence, there are now just five potato starch plants in the United States using significant quantities of cull potatoes and only one, Colby Starch Co. in 1982, has recently added major capacity.

There is, however, a resurgence in potato starch production based on starch recovery from plants that make french fries, hash browns, potato chips, and other specialty potato products. There are now seven major plants operating primarily on reclaimed starch, the most recent being built by Boise Cascade in 1977.

II. Manufacturing Locations

1. United States

Potato processing and starch production are confined to climates where late crop potatoes can be stored through the winter. Without a sufficient inventory of raw material to operate through a majority of the year, plant operation would be uneconomical. The main operating season runs from about October to about June of the following year. Some operators will begin earlier or run later depending on the market and availability of potatoes. Modern storage sheds have allowed an increase in operating days from about 200 to as high as 250 in some cases.

The potato starch manufacturers now operating in the United States are listed in Table I. This listing does not include several potato processing plants that have installed starch recovery and drying systems on site. Many of the latter, who have chosen starch recovery as a means of effluent control, are remote from manufacturers that can process and dry the wet starch cake collected. They have found it economically sound to install flash dryers and to supply an industrial grade starch for other starch manufacturers to convert into modified products (*4*).

2. Imports

The major potato growing and processing area outside of the United States is in northern Europe. Special varieties of potatoes are cultivated there for their

Table I

Potato Starch Production

Manufacturer	*Location*	*Production, tons*	*Raw material*
Domestic			
Boise Cascade Corp.	Stanfield, Oregon	6,000	Reclaimed[a]
Colby Starch Co.	Caribou, Maine	17,000	Culls
Non-Pareil Processing Corp.	Blackfoot, Idaho	1,000	Reclaimed
	Monte Vista, Colorado	3,000	Culls
J. R. Simplot Co.	Heyburn, Idaho	6,000	Culls and reclaimed
Western Polymer	Tulelake, California	5,000[b]	Culls
	Moses Lake, Washington	4,000	Reclaimed
Penick & Ford, Ltd.	Idaho Falls, Idaho	2,000	Reclaimed
A. E. Staley Manufacturing Co.	Monte Vista, Colorado	13,000	Culls
	Murtaugh, Idaho	2,000	Reclaimed
Worldwide[c]			
Avebe	Netherlands	550,000	
—	Poland	220,000	
Henkel	West Germany	77,000	

[a] Reclaimed starch is recovered from process waters and waste at processing plants as a slurry or wet cake and shipped to starch manufacturers.

[b] Maximum.

[c] Partial listing.

high starch content. The Netherlands is most prominent in potato starch manufacture and accounts for the majority of shipments to the United States. Somewhat over 40% of the potato starch consumed in the United States is imported. About two-thirds of the imports are dextrins and soluble or chemically treated starches; the remainder is mostly industrial-grade starch.

III. Starch From Cull Potatoes

1. Raw Material

The composition of the potato is influenced by the variety, area of growth, cultural practices, maturity at harvest, subsequent storage history, and other factors. Reported values are also a function of the different analytical methods employed. A proximate analysis (5), subject to all of these variables, is given in Table II.

Starch comprises 65 to 80%, typically about 75%, of the dry weight of the potato. Eastern varieties generally contain 18–20% solids which corresponds to 13.5–15.0% starch. The western Russet has about 22% solids or about 16.5% starch. A recently introduced variety, the Bell Rus, has been gaining favor among Maine potato growers. It is an excellent, high-solids potato (22%) that is well suited for food use, processing, or starch manufacture.

Table II

Proximate Analysis of White Potatoes

	Average, %	*Range, %*
Water	77.5	63.2–86.9
Total solids	22.5	13.1–36.8
Protein	2.0	0.7–4.6
Fat	0.1	0.02–0.96
Carbohydrate		
Total	19.4	13.3–30.53
Crude fiber	0.6	0.17–3.48
Ash	1.0	0.44–1.9

The starch content of potatoes may decrease during winter storage. Storage houses are normally kept at 38°–40°F (3°–4°C) or above. Lower temperatures result in conversion of up to one-fourth of the starch to sugars in about three months.

The percentage of culls available will vary with the potato variety, growing conditions, storage conditions, grading practices and the demand for cattle feed and dehydrated potato products. Typically about 10% of the crop is available as culls. In Aroostook County, Maine, it is as much as 15% and is expected to reach 20% as growers are tending to upgrade quality for the food market. This has justified the expansion at Colby Starch Co. because the culls in Main are mostly available for starch. Western cull availability for starch is limited by demands of cattle feeding and dehydration plants. With all factors considered, there are certainly over one million tons of cull potatoes available in the United States annually, of which about 150,000 tons are used for potato starch.

2. Manufacturing Process

When processing cull potatoes, the raw material is flumed into the plant. During this process, stones and much of the dirt are removed. A further washing step, employing either a trough with paddles for tumbling the potatoes or a barrel washer, is usually included.

After washing, the potatoes are disintegrated. European practice is to use a rotary saw blade rasp. In the United States, hammermills have largely displaced the rasps for breaking down cell walls. The most commonly used equipment is a vertical hammermill in a screen enclosure. The macerated potatoes are swirled against the screen and the pulp is forced through the holes. In the first stage of grinding, the outlet screen will have 3/16–1/4-in. (4.8–6.4 mm) holes. Sulfur dioxide is normally added at the time of disintegration to inhibit the action of oxidative enzymes that discolor the starch. It may be added once or twice more at other steps in processing.

In the next step, the pulp is screened or passed through a rotary seive to separate free starch from the pulp. When screening, screens of 80- to 120-mesh are used to remove much of the coarse fiber followed by 120- to 150-mesh screens to remove finer materials. The rotary seive consists essentially of an impeller that drives the disintegrated material through a slot screen with typical slot dimensions of 0.014 × 0.157 in. (0.35 × 4.00 mm).

For efficient starch recovery, the pulp is reground in a hammermill with screens of 1/16 in. (1.5 mm) or less. With two stage grinding, the starch yield will be at least 12% based on raw potatoes fed to the plant.

The starch from the screen or rotary sieves is diluted with fresh water to wash out soluble impurities, then concentrated in a continuous centrifugal separator. There may be two or more separators operating in series with further screening in between to remove traces of fiber. The starch slurry enters the separators at 3°–4° Baume and comes out at about 18° Baume. Attempting higher concentration at this stage results in loss of starch.

The final washing is sometimes accomplished by dilution with fresh water and reconcentration in a hydrocyclone. The hydrocyclone is a compact conical chamber with an opening centered in the top for water escapement and an opening at the bottom of the cone for starch slurry. The diluted starch enters at high pressure near the top and at a tangent to the chamber. The swirling action creates centrifugal force which concentrates the starch at the bottom and releases ''clean'' water at the top (see Chap. XII, Sections III.3b and c).

From the separators or hydrocyclones, the starch is fed to a batch centrifuge for further purification. In this operation, any remaining fiber or dirt is concentrated at the surface of the cake formed in the centrifuge. This is scraped off with a knife blade, and the residual clean starch is washed out to a holding tank where sulfur dioxide is usually added again. In some older plants, tabling is used instead of a batch centrifuge. In this process, the starch is allowed to settle in a tank and the impurities are flushed off the surface. The clean starch is then transferred to a holding tank.

The newest technology for starch processing involves the use of small hydrocyclones for both pulp separation and concentration (*12*). The hydrocyclones employed are about 15 mm wide at the top of the cone and about 100 mm in length. While the capacity of a single unit is only 300 liters per hour, a multiunit with 240 hydrocyclones operating in parallel which can handle 72,000 L/h has been developed. The hydrocyclones (or multiunits) are arranged in nineteen stages. Macerated potatoes are introduced at stage four without any prescreening or dilution, but the stream is joined by the underflow from stage three and the overflow from stage five. Pulp is rejected back through the first three stages, while the underflow from each carries starch forward. The starch moves on, with that from the fourth stage, through stage nineteen. The starch moves countercurrent to the wash water which is injected at stage nineteen. This system requires only about 5% of the water needed in a conventional system and achieves about

99% recovery of starch at about 22.5° Baumé. A variation of this technology is used in the new facilities at Colby Starch Co.

From the holding tank or last stage of the hydrocyclones, the starch slurry is dewatered over a vacuum filter to about 60–65% solids. The dewatered starch is then fed directly to a flash dryer. The most common drier is direct fired with natural gas or propane and has an induction fan that pulls the starch and hot air stream through a cyclone that separates the dry product. Temperature is controlled so that the outlet temperature is 115°–120°F (46°–49°C); an equilibrium moisture content of about 17–18% results. Care must be taken to avoid excessive inlet temperatures (over 350°F, 175°C) which may result in pre-gelled or "pasted" starch agglomerations. Screening over an 80–90-mesh shaker screen removes oversize or agglomerated material before the product is packaged in 50- or 100-lb bags.

3. Final Product

Industrial-grade potato starch is generally dried to 17–18% moisture. Typical levels of trace components will be 0.35% ash, 0.1% water solubles, traces of nitrogen and sugars, and nil fat. The product is pure white with a reflectance of at least 90 (relative to magnesium oxide = 100) at 450 nm. Total acidity, expressed as mL of 0.1 *M* sodium hydroxide required for neutralization of 100 g of dried starch, is of the order of 10 units.

Of the commercial starches, potato starch develops the highest consistency on pasting. Its viscosity breaks down on continued heating and agitation to give a paste of considerably lower viscosity. Because its pastes are electroviscous, they are sensitive to small concentrations of added electrolytes. Thus, potato starch exhibits a higher viscosity in distilled water than in hard water containing appreciable amounts of calcium salts or in saline solution.

IV. Starch Reclaimed from Potato Processing

1. Raw Material

In the processing of potatoes to make french fries, potato chips, dehydrated potatoes and other specialty potato products, starch granules are released into the plant processing and conveying water. The amount will vary depending on the extent of cutting required for each product. Typically, there will be about 7 lb (3 kg) of starch released per ton of potatoes processed in a frozen french fry plant.

For many years, the starch released from the potato processing plants was an undesirable effluent. It is high in biological oxygen demand (BOD) and requires large irrigation or flood fields for disposal because of its tendency to blind soil. Alternatively, high costs are incurred for disposal through municipal sewage plants or industrial treatment systems.

While the starch in potato processing waters is at a low concentration, perhaps 0.5%, it can be readily concentrated to 35% (about 18° Baume) by the use of hydrocyclones. It is desirable to screen the starch water over 140–150-mesh screens before concentrating. This removes the majority of the potato pulp from the starch before it is placed in agitated holding tanks.

The concentrated starch may be kept in slurry form in a holding tank equipped with an agitator. The slurry is conveniently picked up in a tank truck for transport to a central starch processing plant. An alternative is to allow the starch slurry to settle in tote bins to a wet cake at about 50% solids. The tote bins can then be similarly transported to a central starch plant. On balance, the labor cost of handling tote bins must be weighed against the cost of carrying added water in the slurry. In either case, it is desirable to add sulfur dioxide to the slurry or tote bins to retard starch degradation.

Most potato processing plants have large amounts of trim waste and scraps plus some cull potatoes. Often when certain specialty products are made, there is trim in excess of the product requirement. Frequently, this waste material is combined with peeler waste and sold as cattle feed. By installing a small grinder and pulp removal screens, 150–200 lb (70–90 kg) of additional starch per ton of potatoes processed can be obtained. While higher yield is possible, starch collection efficiency is not usually a high priority item in a plant making other potato products. The starch invariably has higher value, however, than cattle feed, and there is a rapid payback of installation costs.

2. Processing

Reclaimed starch from a large potato processing plant is sometimes cleaned, dried, and packaged at the same location. More often it is transported to a centrally located starch manufacturing plant. Slurried starch is pumped directly from the tank truck into holding tanks. If delivered in tote bins, the starch cake must be washed from the bins. Because reclaimed starch generally contains a fair amount of potato pulp, it must be cleaned up. This may be accomplished by tabling or by fine-mesh screening. The subsequent steps of dewatering, drying, and packaging are the same as when starting from culls. The final product from reclaimed starch is comparable to that obtained directly from grinding culls for most applications. However, unless care has been taken to prevent microbial action, the extended time from collection to drying may lead to some breakdown of the starch and a slight loss in viscosity.

V. By-Products and Effluent Control

When manufacturing potato starch from culls, there is obtained about 400–450 lb (180–200 kg) of pulp solids per ton of starch produced. This material has a high BOD and cannot be discharged into public waterways. Distribution of

Table III

Soluble Solids from a Typical Maine Starch Factory (8)

Component	*Relative amount, %*[a]
Total "protein" (N × 6.25)	33–41
Total sugars	35
Reducing sugars	28–32
Sucrose	1–7
Organic acids	4
Minerals	20
Other (probably carbohydrate)	6–9

[a] Moisture-free basis.

the pulp over a flood or irrigation field is unsatisfactory. The pulp soon blinds the soil, builds up on the surface, and decomposes with accompanying odor. Run off from the blinded field can pollute surrounding waterways.

In most locations, it is practical to dewater the pulp via a screen or vacuum filter and use it directly as cattle feed. It is also practical to dry the pulp for packaging and the shipping to livestock feeders. A sample of dried waste pulp had the following percentage composition (*7*): moisture, 4.5; starch, 54.6; uronic acid anhydride, 16; pectin, 12; "pentosans" (pentoglycans), 9.5; crude fiber, 15.6; ash, 1.0; fat, 0.4; protein (6.25 × % nitrogen), 5.9; sugars, trace. (The total is more than 100%, because of overlapping in the uronic acid anhydride, pectin, and pentosan determinations.)

The soluble constituents of the potato occurring in the protein water or juice after starch and pulp removal present a more difficult recovery problem. The quantity produced amounts to 450–460 lb of solids per ton of starch (*6*). The typical composition of the soluble solids is given in Table III.

While several methods of utilizing these solids have been proposed, the most common disposal is via flood or irrigation fields. Under proper management, this is a satisfactory means of handling this type of effluent. With sufficient land area under cultivation, only care to avoid excessive starch or pulp carryover is needed.

A more elegant and practical system for handling the soluble solids is based on the growth of single cell protein (SCP) (*9, 10*). This process employs innoculation with *Candida utilis* (torula yeast) followed by aerobic fermentation in a segmented reactor. The SCP is harvested by means of centrifugation and later dried to a high-protein animal feed. Automation keeps labor to a minimum; capital cost is less than conventional water treatment systems, and the value of the SCP largely covers operating costs. BOD is reduced by about 80%.

Other approaches to recovering starch plant wastes rely mainly on concentration and evaporation to obtain a dried product (*8, 11*). However, increasing energy costs are causing operators to look to other systems. Among systems under development are those which employ fermentation to generate methane from the soluble solids.

VI. Utilization of Potato Starch

1. General

Potato starch is preferred over corn starch and other starches in applications for which its properties are particularly suited. Its most important characteristics are (*a*) high consistency on pasting followed by a decrease in viscosity on further heating and agitation, (*b*) excellent flexible film formation, (*c*) binding power, and (*d*) low gelatinization temperature.

2. Modifications and Derivatives

The most important modification of potato starch is pregelatinization to render it cold water dispersible. Pre-gelled starch can be used directly in many applications without the need to cook it at the point of use. Some potato starch derivatives are also made in a pregelatinized form.

Cationic potato starch is next in importance. This derivative is made by the reaction of the starch under alkaline conditions with a reagent containing both a quaternary ammonium and a glycidyl or chlorohydrin group. The incorporation of the quaternary ammonium group gives cationic character to the starch. There is minimal loss of viscosity and film forming characteristics so that cationic potato starch is generally preferred over cationic corn starch. The electropositive charges on cationic starch causes it to bond to negatively charged substrates and to attract and hold other negatively charged additives to that substrate (Chap. X).

Dextrinized potato starch is prepared by dry heating, sometimes in the presence of an acid catalyst. Degradation in chain length occurs to give gums that are readily soluble in water (Chaps. XVII and XX).

The preparation of hydroxyethylated or carboxymethylated potato starch is accomplished by treatment with caustic and ethylene oxide or chloroacetic acid. The products are extremely stable in viscosity and give films of improved flexibility (Chap. X).

While some oxidized potato starch is used, it is less important in the United States than in Europe. It is prepared by the action of hypochlorite which degrades the starch to so called "thin boiling starch." In the United States, most thin boiling starches are derived from corn (Chap. X).

3. Paper Applications

Papermaking accounts for about 33% of the potato starch usage in the United States. A majority of the product used is the cationic derivative. Cationic potato starch improves retention of filler and fine fibers as well as the mechanical properties of the paper. It is mostly added on the wet end.

The addition of small quantities of cationic starch at the size press is also valuable in relation to broke recovery. The coating materials, including unmodified starches, are retained with the fiber for recycling. This significantly reduces BOD in the white water effluent.

Unmodified potato starch is preferred over other starches in coated paper for its film-forming properties, excellent binding power, and lower moisture retention. This leads to better coatings and reduces load on the drying section.

4. Food Applications (see also Chapter XIX)

Food use accounts for about 30% of the market for potato starch in the United States. Significant quantities are used in soups where its high initial viscosity effectively disperses ingredients during mixing and can filling. During subsequent autoclaving, the viscosity breaks down to the desired consistency for the final product.

Pregelatinized potato starch is effective in instant puddings. The dry formulation consists principally of soluble starch, sugar, and flavoring. Upon addition of cold milk, the starch quickly dissolves and sets to a gel. Other food applications include its use as a thickening agent for pie fillings, a gel body for cast candies such as jelly beans, a bodying agent for caramels and marshmallow, and a dusting agent, perhaps mixed with powdered sugar, for candy gums, chewing gums, and so on. There is also a use for unmodified food-grade starch as a filtering media in breweries. It is applied as a pre-coat to filters removing yeast cells from the wort.

5. Adhesive Applications (see also Chapter XX)

Dextrinized potato starch is the preferred form for adhesive use. About 19% of the potato starch consumed in the United States is for adhesive use. As binders in sandpaper and abrasive cloth, in bookbinding and in ring sizing, potato dextrins provide high paste tackiness and a flexible residual film. Potato dextrin films are also outstanding for their ease of remoistening, a property desired in mucilages used for gumming stamps, labels, envelopes, and paper tape.

6. Oil Field Applications

The fastest growing use for potato starch is in oil field applications which account for about 15% of its consumption in the United States. All the potato starch used in this application is pregelatinized and often stabilized against microbial attack. The starch functions as a viscosity and fluid loss control agent. The viscosity characteristics of potato starch are preferred over other starches in this use.

7. Other Applications

There are a variety of other uses for potato starch that do not fit neatly into the above categories. These include (a) cationic potato starch as a flocculant in water treatment, (b) tablet binders, (c) textile warp sizing, and (d) foundry binder and others.

VII. Outlook for Potato Starch

Because of its unique properties, potato starch has maintained a position in certain applications in the face of lower priced corn starch. In 1979, potato starch accounted for 4.4% of the 5.1 billion lbs (2.3 billion kg) of starch consumed in the United States. Corn starch held 91% of the market, and the remainder was mostly wheat and tapioca starches. If potato starch were available in sufficient quantity at corn starch prices, it would be preferred in most applications.

As pointed out earlier, of all the commercial starches, potato starch gives the highest consistency on pasting, and it excels in film forming and binding characteristics. These properties carry through in its derivatives. Thus, in specialty applications, where performance is needed, potato starch justifies its premium over corn starch.

Two factors are expected to bolster the position of potato starch in the near future. The first is the recent acceptance of high-fructose corn syrups in soft drinks. The demand thus created will divert significant amounts of corn starch to syrups. The second factor is the growing support for gasohol which may utilize huge quantities of corn. The combination of these events is expected to limit corn starch expansion and put upward pressure on corn starch prices. Potato starch, as a result, will become more competitive and some further expansion is likely as prices improve. There will undoubtedly be some diversion of cull potatoes to alcohol manufacture, but in the long run, potato starch will represent a higher valued use.

The future for potato starch thus seems secure and likely to improve. It will never again be the dominant starch in the United States, but it will remain an important product for both food and industrial use.

VIII. References

(*1*) *U.S. Tariff Comm. Report,* Ser. 2, No. 138 (1940).

(*2*) R. W. Kerr, "Chemistry and Industry of Starch," Academic Press, Inc., New York, 2nd Ed., 1950, p. 104.

(*3*) C. A. Brautlecht, "Starch, Its Sources, Production and Uses," Reinhold Publishing, New York, 1953, p. 60.

(*4*) B. Pettay, *Chipper,* 50 (January, 1975).

(*5*) W. F. Talburt and O. Smith "Potato Processing," Avi Publishing, Westport, Connecticut, 1967, p. 13.

(*6*) W. W. Howerton and R. H. Treadway, *Ind. Eng. Chem.*, **40,** 1402 (1948).
(*7*) R. H. Treadway and W. W. Howerton, *in* "Crops in Peace and War: 1950–51 Yearbook of Agriculture," A. Stefferud, ed., U.S. Department of Agriculture, U.S. Government Printing Office, Washington, D.C., 1951, p. 171.
(*8*) E. O. Strolle, N. C. Aceto, R. L. Stabile, and V. A. Turkot, *Food Technol.* **34,** 90 (1980).
(*9*) Anonymous, *Food Eng.*, 118 (January, 1977).
(*10*) I. Dambois, R. Reeves, and J. Forwalter, *Food Process.*, 156 (May, 1978).
(*11*) J. R. Rosenau, L. F. Whitney, and J. R. Haight, *Food Technol.* (Chicago), **32,** 37 (1978).
(*12*) P. Verberne, *Staerke*, **29,** 303 (1977).

CHAPTER XV

WHEAT STARCH: PRODUCTION, MODIFICATION, AND USES

BY J. W. KNIGHT

Fielder Gillespie Ltd., Sydney, New South Wales, Australia

AND

R. M. OLSON

Moffett Technical Center, Corn Products, Summit-Argo, Illinois

I. INTRODUCTION

Cultivation of wheat is believed to have originated in the fertile crescent of the Middle East. A radiocarbon date of 6700 B.C. for archaeological wheat starch is evidence that wheat grains existed in the neolithic site of Jarno, Northern Iraq (*1*). Starch produced from wheat was used to size or stiffen linens for wrapping mummies in ancient Egypt. Pliny the Elder reported that the natives of Chios, a Greek island, were pioneers in the manufacture of wheat starch (about 130 B.C.). Commercial scale manufacture of wheat starch is claimed to have started in England during the reign of Elizabeth in the 1500s (*2*).

II. PRODUCTION

Wheat and corn are the leading cereal grains grown worldwide, with more wheat grown than corn (Table I). The main use for wheat is food, while corn is used mainly as an animal feed. Wheat starch is produced in areas where wheat is

STARCH, 2nd ed.

ISBN 0-12-746270-8

Table I

World Production of Principal Cereal Crops 1978 (3)

Crop	Production, million metric tons
Wheat	425.2
Rice	377.0
Corn	359.8

Table II

World Production of Wheat (4)

Year	Production, million metric tons	Yield, tons/hectare
1981[a]	444	—
1980	429	—
1979	451	—
1978	425	1.88
1975	350	1.56
1970	315	1.52
1965	264	1.22
1960	239	1.18

[a] Estimated.

Table III

U.S. Wheat Gluten Consumption (5)

	Imported		
Year	Million pounds	Av. price, $/lb	Total U.S. consumption, million pounds
1981[a]	55.3	0.63[b]	110
1980[a]	53.1	0.59[b]	105.6
1979[a]	50.9	0.55	101.2
1978[a]	49.8	0.50	99.0
1977	48.3	0.44	96
1976	45.4	0.40	87
1975	33.1	0.33	69

[a] Estimated.
[b] Personal communication.

Table IV

Wheat Gluten Plant Capacity (5)

Country	Annual capacity,[a] millions of pounds
United States	67–75
Canada	24
Australia	70

[a] Based on 24 h day, 327 days/yr.

a better economic choice for wheat components or because of agropolitical considerations.

Wheat production has increased steadily as has the yield per hectare (Table II). Use of higher yielding varieties of wheat and rice has spread rapidly, particularly in Asia and developing areas. This "Green Revolution" has greatly reduced the threat of famine. In Europe, high yielding wheat varieties, called "Mass" wheats, have been developed. The combination of improved agricultural methods and higher yielding varieties has resulted in rapid growth of the world wheat crop. Wheat starch manufacture is competitive with corn starch manufacture because of the high value of the co-product, gluten. This is evident in Australia and New Zealand where virtually all starch is produced from wheat. The consumption and price of wheat gluten in the United States (*5*) is shown in Table III. The estimated capacity for producing wheat gluten in several countries is shown in Table IV (*5*).

III. Basic Processing Methods

An early method for producing wheat starch, known as the Halle fermentation process, consists of steeping wheat with water to soften the grain, followed by crushing and fermenting of the mash. After the gluten is degraded to release starch, large particles are removed by screening. Refining is accomplished by repeated dilutions of the starch with fresh water, and decantations of the water and light solids which collect above the settled starch. The refined starch is then dried in cakes. (*6*).

In the Alsatian process, non-fermentative steeping is conducted by replacing the water before fermentation starts. The softened grain is placed in mesh bags, and the starch is squeezed out by passage of the bags through pairs of rollers having progressively smaller gaps. Starch refining is accomplished by settling on tables. A low gluten yield is obtained by removing it from the bags and washing to remove the hulls (*6*).

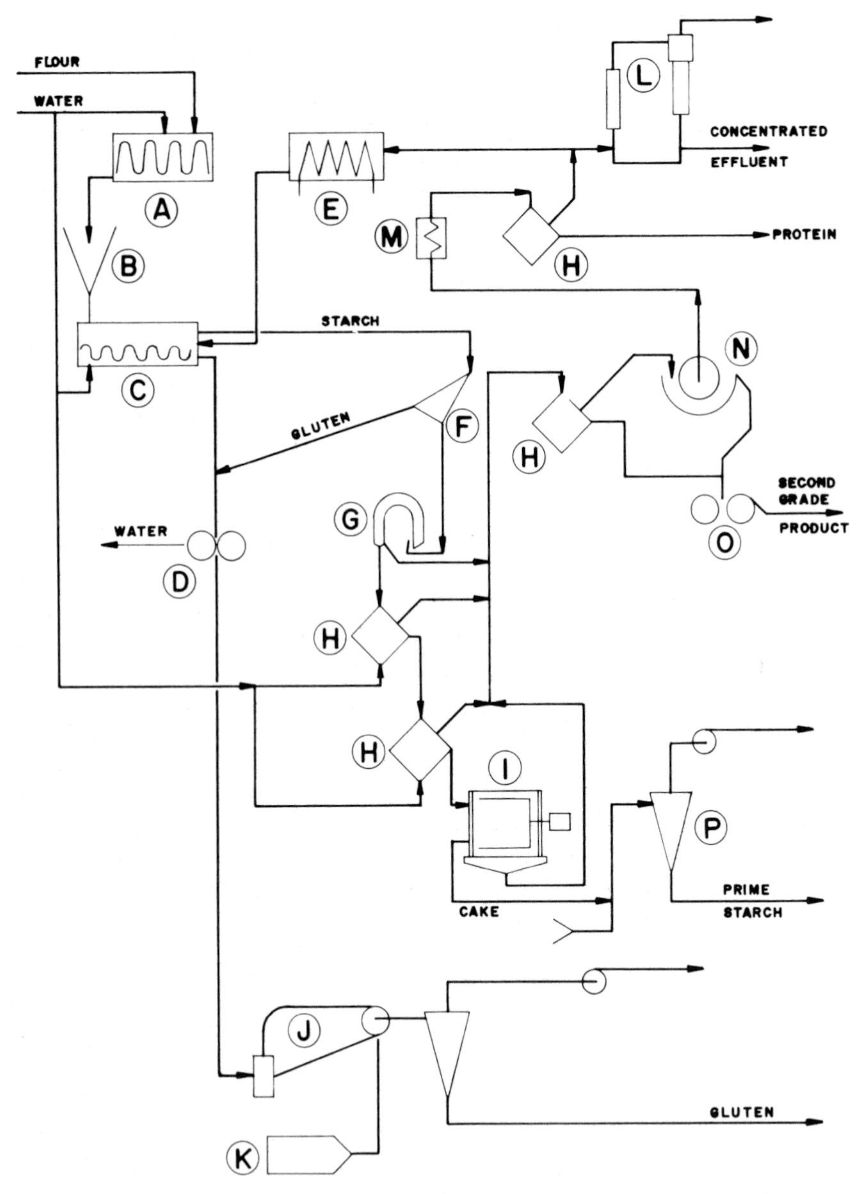

FIG. 1.—Martin process: A, Continuous dough mixer; B, rest hopper; C, extractor; D, dewater rolls; E, cooler; F, sieve; G, D.S.M. screen; H, continuous centrifuge; I, basket centrifuge; J, ring drier; K, heater; L, evaporator; M, heat exchanger; N, filter; O, roll drier; P, flash drier.

1. Martin Process

The Martin, or dough ball, process (*6*) was proposed in Paris about 1835, and, until recently, has been the most popular process (*5*). Wheat flour rather than wheat grain is used as the raw material. The process consists of five basic steps: mixing flour and water to a dough; washing out the starch; drying the remaining gluten; refining the starch; and drying the starch.

A flow sheet of the Martin process is shown in Figure 1, but variations of this process are widely practiced. Flour and water in a ratio of about 2:1 are metered into a blender designed to give a smooth, uniform, rather stiff dough, free of lumps. The flour to water ratio varies depending on the type of wheat flour used. Hard wheat flours form a strong elastic dough and require more water than soft wheat flours which produce a dough that crumbles and is easily torn apart. The water used should be about 20° and contain some mineral salts. Use of soft water, with low salt content, can cause the gluten to become slimy (*8*). The dough is allowed to stand to fully hydrate and strengthen the gluten before transfer to the washing stage.

The dough washing stage is designed to release the starch from the gluten without dispersing or breaking the gluten into small pieces. Sufficient water is used to wash the starch from the dough while it is kneaded or rolled. Many devices, such as ribbon blenders (*8*), rotating drums, (*8*), twin screw troughs (*9*), and agitated vessels (*10*), have been designed for this purpose. The ribbon blender (Fig. 2) is deep, narrow, and boat shaped, with twin open-paddle rotors extending the full length of the vessel. Grooves in the sides of the rotor beds assist the action of the paddles which rotate in opposite directions at different speeds. Dough is supplied continuously to the blender at a rate that keeps the paddles covered with dough. Fresh or treated process water is injected into the vessel along the bottom and sides. Wash-water and suspended starch overflow from the vessel, while gluten, at about 70% water and 70–80% protein (N × 5.7), discharges continuously through a take-off pipe in essentially plug flow with very little starch slurry being carried along.

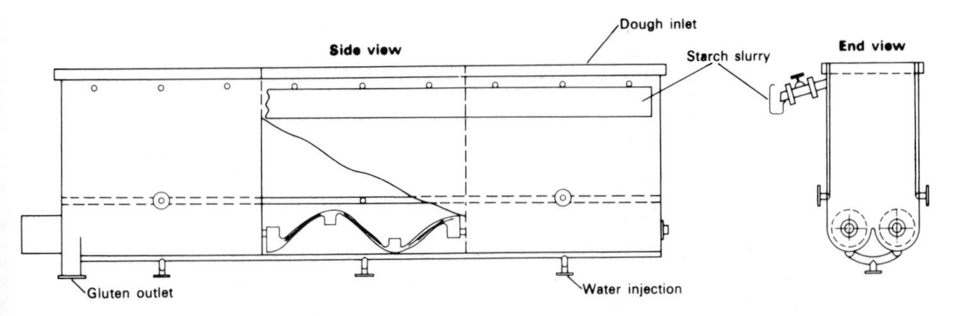

FIG. 2.—Dough washer.

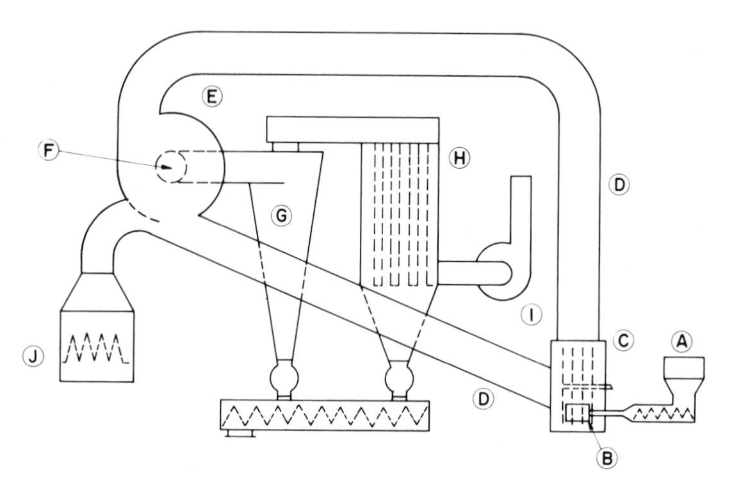

FIG. 3.—Ring drier: A, supply pump; B, extruder; C, disintegrator mill; D, ring duct; E, manifold; F, manifold exit; G, cyclone; H, bag dust collector; I, fan; J, air heater.

After additional dewatering by roller compression, the gluten may be dried to about 8% moisture in vacuum, spray, flash, or drum dryers. Usually it is dried in specially designed flash dryers. Because wet gluten vitality is lost by prolonged exposure to high temperatures, rapid drying at controlled temperature is needed. Wet gluten is mixed with dry material, and the mixture is dispersed as fine particles into the hot air stream. In the ring-type, flash dryer (*8*) (Fig. 3), the dry return, mixing, and dispersion are done within the dryer, thus producing a very fine, light-brown powder having satisfactory vitality when reconstituted with water.

Starch slurry from the dough washer at about 10% solids is passed over a vibratory sieve to recover any agglomerated gluten particles which are then combined with the gluten from the dough washer for further processing. Large particulates, such as bran, are removed from the starch slurry by fine screening with equipment such as the Dorr-Oliver D.S.M. screen using wedge wire media having about 100 μm openings (Chap. XII, Section III.3c) vibratory sieves, or centrifugal conical filters. Refining of the starch slurry is made by classification and washing in staged nozzle discharge centrifuges (*11*). The refined starch slurry is batch dewatered in basket type centrifuges to a cake of about 40% moisture. Continuous solid-bowl, scroll-discharge, decanter centrifuges are also used for dewatering, but these produce a cake of higher moisture content. Sufficient dry product is mixed with the cake to supply the flash dryer with starch at about 36% moisture. This is done to reduce the tendency of the starch to gelatinize when dried rapidly with hot air. Minimum moisture for gelatinization of wheat starch is 31%. Finished starch contains about 0.3% protein and about 10–12% moisture.

A secondary starch product is produced by concentrating and drying the small starch granules and other particulates removed during refining of the prime starch. Concentration in a nozzle-type centrifuge also reduces insoluble particulates in the waste water stream to a minimum. Because of large amounts of water required to wash the starch from the gluten (15 parts by weight of water per part of flour), effluent streams containing about 10–13% of flour dry substance are only 0.85–1.2% solids (*12*). These effluents are often sent to waste, although evaporation and recovery of the solids can be done (Fig. 1). A typical material balance for the process is given in Figure 4.

2. Batter Process

The Batter process, a variant of the Martin process developed during World War II at the Northern Regional Research Laboratory, USDA (*13*), is used in the United States and Australia. It differs from the Martin process in both the formation and treatment of the dough (Fig. 5). Flour and water in roughly equal amounts are mixed to a smooth paste-like dough, allowed to rest for about 30 min, and then vigorously mixed with additional water equal to two times the weight of flour. Gluten forms a curd-like suspension which can be separated and washed on 80-mesh screen before drying.

Screen filtrates contain starch and particulate contaminants of bran and gel-like complexes of pentosans (pentoglycans), proteins, starch, and solubles. Par-

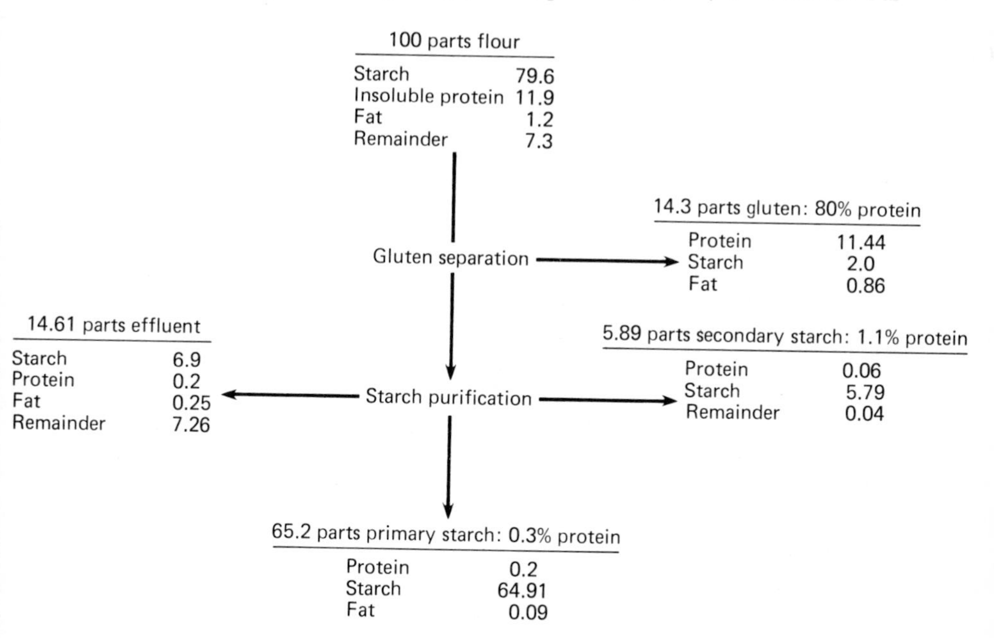

FIG. 4.—Martin process, materials balance.

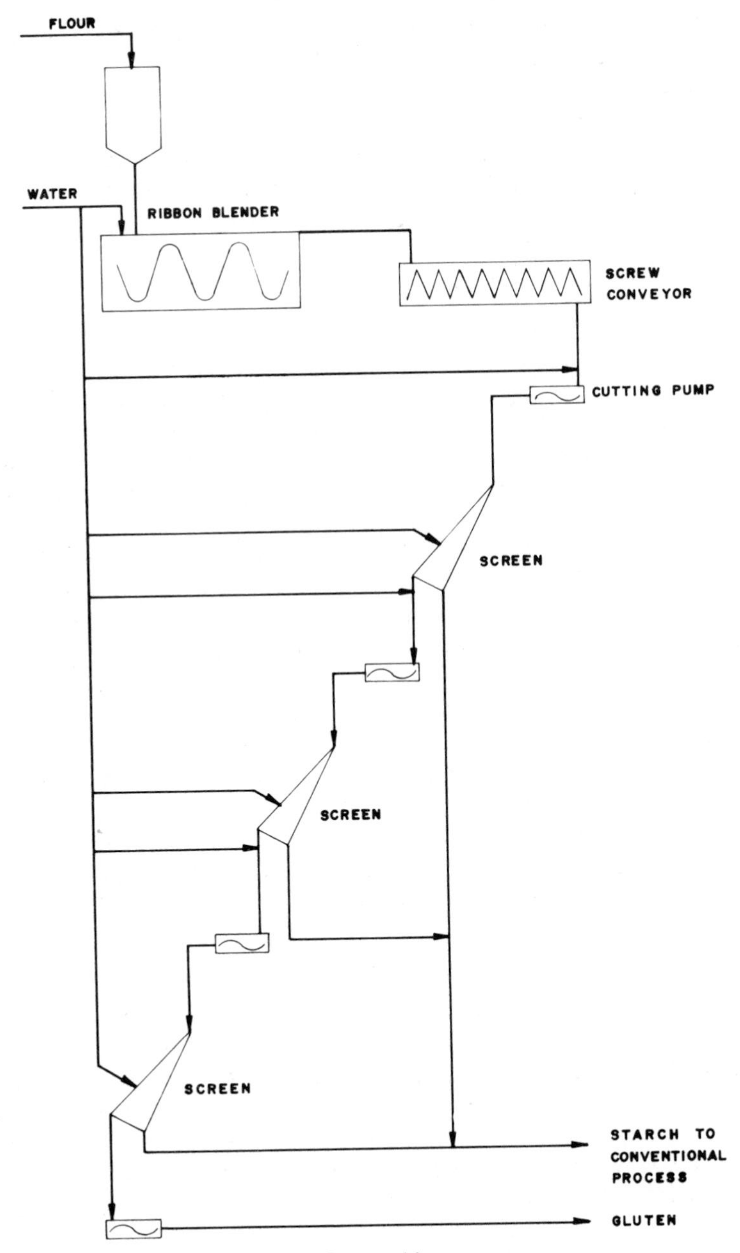

FIG. 5.—Improved batter process.

ticulate contaminants are removed by further screening on 150- and 300-mesh sieves. Starch contained in the filtrate has the bimodal granule size distribution characteristic of wheat starch (*14*). Essentially pure starch, known as A starch, is centrifugally separated from the impure B starch containing small granules, swollen damaged granules, and pentosan (pentoglycan) complexes. Washed A starch is concentrated, dewatered, and flash dried. Likewise B starch is dried to a low-purity starch. Clarified effluent streams obtained from the B starch concentrator approximate the concentration obtained from the Martin process and are usually sent to waste.

3. Fesca Process

The products of the Fesca process are starch and non-vital protein. Flour is rapidly mixed with water to form a thin batter which is dispersed by shearing to prevent gluten development. Centrifugation in a solid-bowl centrifuge settles the starch leaving a supernatent of protein concentrate. Starch is reslurried and refined. The protein concentrate can be evaporated to yield a product containing 22% protein and 67% starch (*15*).

The USDA Laboratories at Albany have developed a continuous process (*16*) in which the thin dispersed batter is separated into starch and a liquid protein concentrate by a solid-bowl, scroll-discharge, decanter centrifuge. Starch is further refined and dried. Protein concentrate can be spray-dried to give a 30% protein product containing starch, bran, and the solubles, a product that would normally be lost as effluent. Water usage is minimized, and solids recovery is almost 100%.

The Martin and Batter processes use large volumes of water and the resulting effluent poses a major disposal problem owing to low concentration and high biological oxygen demand. High capital and energy costs are required for evaporation to obtain a saleable product, and pollution controls restrict waste water disposal. As a result, several new processes have been developed.

4. Ammonia Process

Separation of starch from gluten using ammonium hydroxide was proposed by the National Research Council of Canada in 1966 (*17*). Flour is dispersed with intense mechanical shear in 5 parts of ammonium hydroxide solution, followed by centrifugation and spray-drying of the supernatant. A dry product of 75% protein that gives good baking performance is obtained. The centrifuged starch requires additional washing with ammonium hydroxide solution to reduce the protein content to acceptable levels. A flow sheet of the proposed process is shown in Figure 6.

Use of ammonium hydroxide in the Fesca-type process was proposed in 1969 (*16*). Flour is mixed with water in a ratio of 4:6–4:7 and at a pH of 7.8. Low

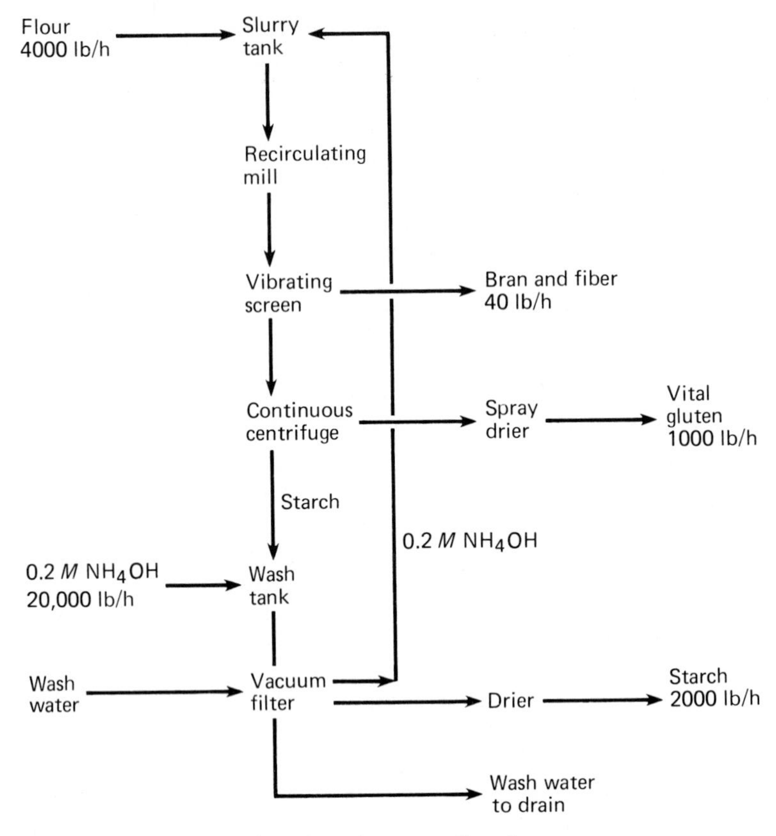

FIG. 6.—Ammonia process, flow diagram.

water usage and improved centrifugal separation are claimed, and all solubles are recovered with the solubilized gluten as a spray-dried product.

Treatment of wheat gluten at 65° with sodium hydroxide can destroy serine, threonine, cystine, lysine, arginine, and tyrosine residues and can lead to the formation of lysinoalanine and unidentified ninhydrin-positive substances (*18*). Observations of histological changes in the proximal tubules of rat kidneys caused by lysinoalanine (*19*) has caused concern about the nutritional quality and safety of alkali-treated food production.

5. Acid Processes (Pillsbury Hydromilling)

A hydroprocess for wet milling whole wheat was developed to obtain maximum endosperm recovery (*20*). Wheat is steeped in aqueous acid at pH 0.8–1.7 at 37°–40° for 12–24 h. After steeping, a Moyno pump can be used to crush or macerate the wheat, followed by dispersion in water maintained at pH 2.4–3.4.

Shear from agitation is sufficient to disengage the endosperm from the bran and germ. Separation on 700–1200 μm screens with water washing of the bran and germ at low pH results in endosperm recovery in the filtrates significantly higher than is obtained by dry milling. A recovery of about 80% wheat solids is obtained compared to a 73% yield of straight white flour by dry milling to comparable endosperm purities.

Purified starch and gluten can be prepared by separating the insoluble particulate fractions of the endosperm slurry from the soluble fraction at pH 5.5–6.5 to a concentration in excess of 55% solids. It is claimed that a dough can be obtained by mixing. After maturing the dough, a conventional dough (Martin) or Batter process can be used to recover and refine the starch.

6. Wheat Wet Milling

A wet-milling process for refining whole wheat was disclosed in 1979 (*21*). Wheat is steeped in 2000–4000 ppm sulfurous acid solution with agitation for 15 min–2 h, followed by screening and impact milling of the oversized solids. Screen filtrate and mill discharge are combined. Large endosperm particles are recovered in the underflow of a hydroclone and recycled to the first steep. The overflow, containing endosperm particles of about 53 μm, is steeped a second time to free starch from gluten. Bran and germ are separated on a 37- to 53-μm screen. The starch filtrate combined with filtrate from bran washing contains mill starch which can be refined in hydroclones, centrifuges or a combination of them similar to that used in corn starch refining systems (*22*). Because of sulfur dioxide steeping, the gluten is not vital. Gluten is concentrated and combined with the concentrated solubles and bran before drying to make animal feed. Prime starch yields (based on uncleaned wheat) of 53–59% and dry animal feed yields of 41–47% (25–32% by weight protein) are obtained by this process.

A variation of the sulfur dioxide steeping process was disclosed in which the wheat is dry milled prior to wet milling of the endosperm (*23*). A prime starch yield of 52% by weight and an animal feed yield of 48% are claimed for this process. Reduced energy requirement for animal feed drying can be expected when the bran and germ bypass the wet system.

7. Farmarco (Whole Wheat Fractionation)

The Farmarco process is similar to the Martin process, except that wheat endosperm replaces dry milled flour for making the dough. Endosperm is produced by pin milling tempered wheat, followed by sifting to remove bran and germ (*24*). Wheat is tempered to 14–22% moisture before milling in a pin mill, such as the Kolloplex made by the Alpine American Corporation, operating at about 17,000 rpm. Tempering reduces starch granule damage. Bran and germ, which are more plastic than the endosperm, are not ground as fine as the endosperm under these conditions, allowing them to be removed by sifting.

8. Rasio

Technicians at the Oy Vehna AB company in Finland (*25*), together with Alpha-Laval (*26, 27*), have further modified the Fesca process (*16*) to produce

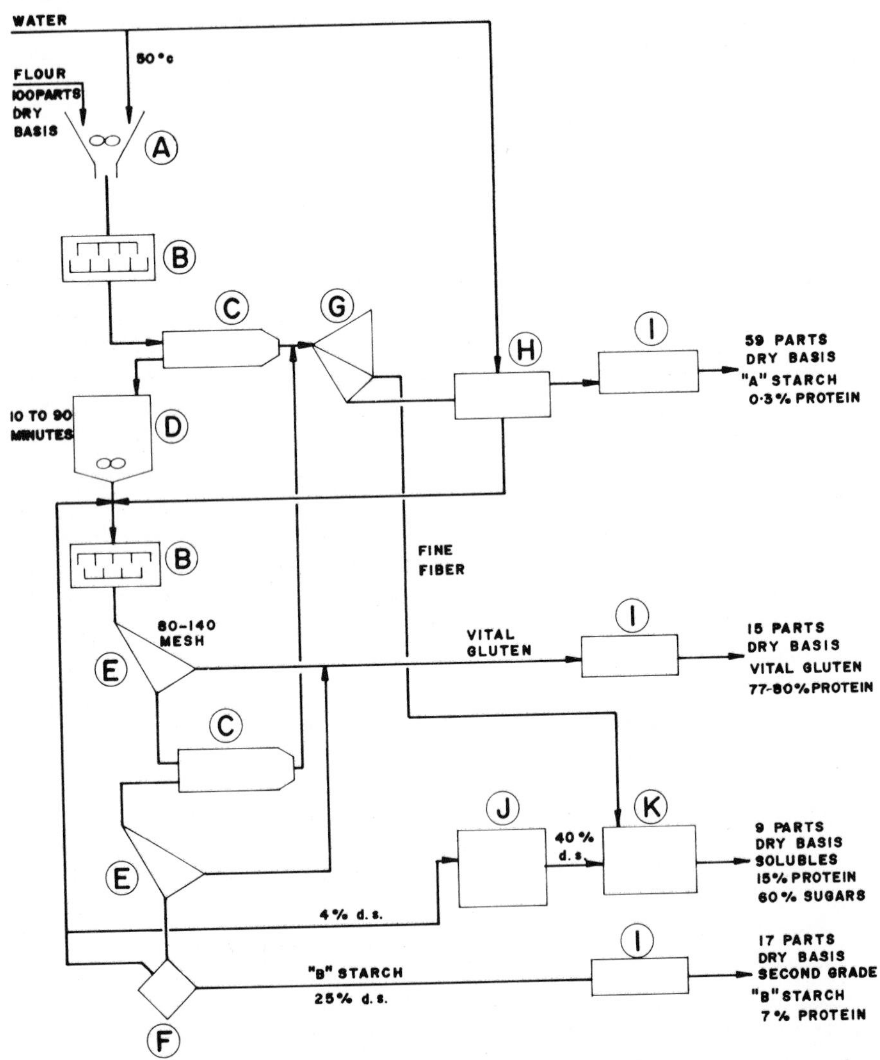

FIG. 7.—Raisio wheat starch and gluten process: A, mixer; B, pin mill; C, decanter; D, maturing tank; E, screen; F, solids-ejecting centrifuge; G, conical screen; H, two-stage decanters; I, drier; J, triple-effect evaporator; K, drum drier.

vital gluten (Fig. 7). Flour and water are mixed at a ratio of 5:6–5:10, depending upon the type of flour used. This mixture is pin milled to form a homogeneous suspension without gluten clots. The suspension is separated into a starch fraction and a gluten fraction in a decanter centrifuge. Starch is refined and washed by traditional techniques to produce a high quality A starch. Gluten is matured to form clots and threads, then agglomerated in a disc disintegrator with additional process water. Vital gluten is screened from the B starch, further dewatered, and finally dried. B starch filtrate is centrifuged in a decanter to remove residual A starch for recycling, and then concentrated in a solids-ejecting centrifuge prior to drying. Half of the process water obtained from the B starch concentrator is recycled, and the other half is evaporated to 40% dry substance before drum drying.

9. Hydroclone Process (KSH)

Koninklijke Scholten Honig Company of Holland (*28, 29*) has described a hydroclone process for producing starch and gluten from wheat flour (Fig. 8). A batter of flour and recycled process water is introduced into a multi-stage battery of 10-mm hydroclones. Washed A starch and bran fiber are discharged in the underflow of the last stage. Fiber is removed in a multi-stage screenbend installation (Chap. XII, Section III.3c), and the A starch is further concentrated to 21° Baumé in a three-stage hydroclone station prior to dewatering and drying. Overflow from the multi-stage hydroclone system is transferred to a three-stage hydroclone unit. Agglomerated gluten, in the form of clods and threads up to 10-cm long, B starch, and solubles are contained in the overflow. Recoverable A starch in the underflow is returned to the multi-stage washing system. Gluten is collected on a rotating gluten filter of 0.5 mm mesh and pneumatically flash dried. Part of the filtrate from the gluten washer is recycled for making up the flour batter; the remainder is evaporated after recovery of second grade B starch.

IV. Uses of Wheat Starch, Unmodified and Modified

Wheat starch produced in the United States and Canada may be considered a by-product in the manufacture of wheat gluten. Most of the starch is sold in unmodified form to industrial rather than food outlets. The highest proportion is consumed by the paper industry, where it is used as a wet-end adhesive, in surface coating, and as an adhesive for manufacture of corrugated board (Chaps. XVIII and XX). Other uses are in laundry sizing and cotton finishing, where wheat starch is considered to produce a superior finish.

In Europe, the main sources of starch are corn and potato, while wheat starch is produced on a small scale. Wheat starch production could increase if the European Economic Community decides to substitute local German and French wheat for imported American corn.

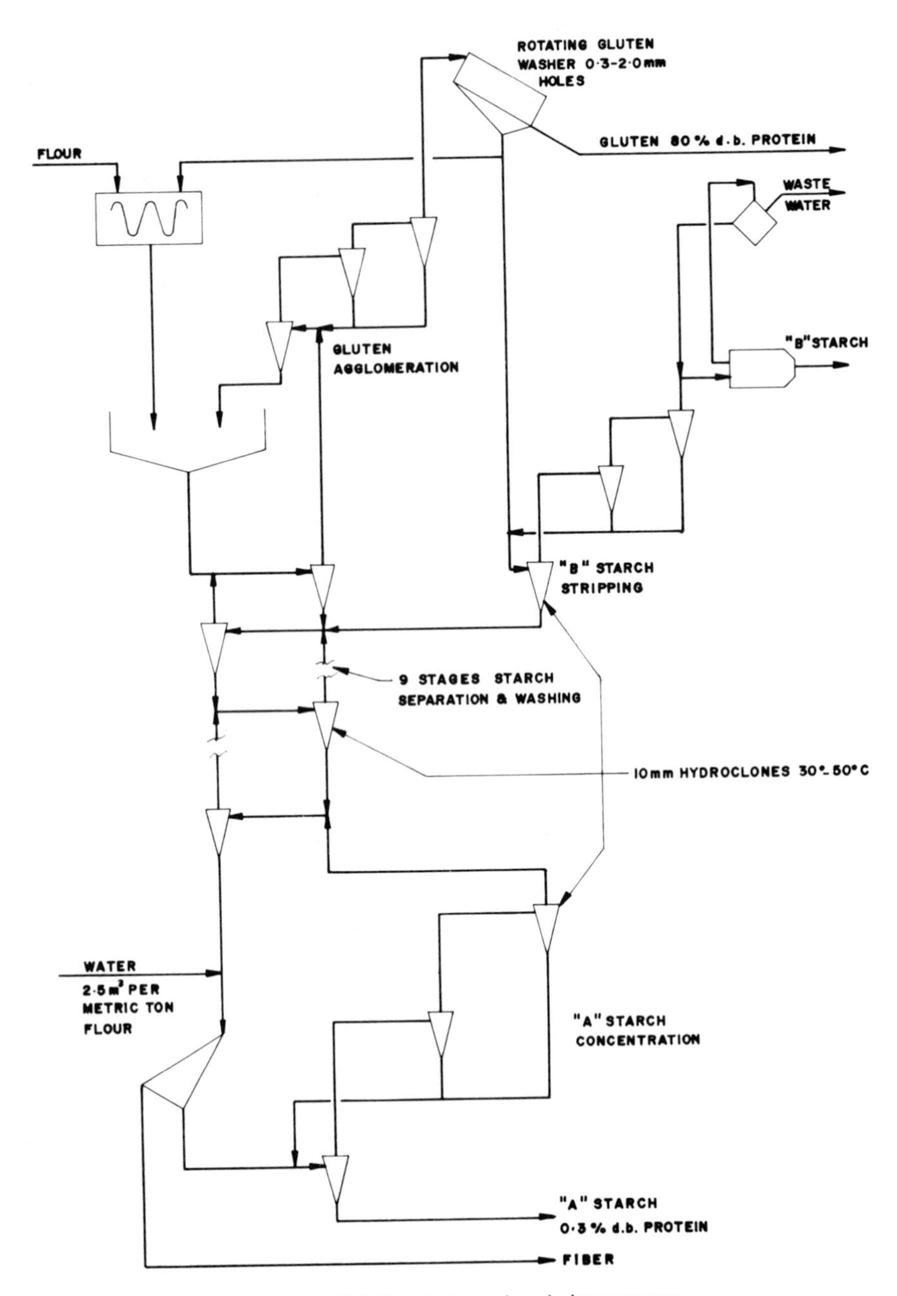

FIG. 8.—K.S.H. wheat starch and gluten process.

In Australia and New Zealand, starch is produced mainly from wheat, although there is some production from locally grown corn. The thickening power of wheat starch is less than that of corn starch; but paste texture, clarity, and strength are about the same. The lower gelatinization temperature of wheat starch gives it an advantage over corn starch for use in corrugating adhesives. Wheat starch is also preferred for laundry sizing, as it produces a stiffer finish at a lower temperature than does corn starch. Wheat starch is preferred for baked goods because no chemicals are used in its production. Modified wheat starches can have superior emulsifying property over other starches when used in some food products. This may owe to the high lipid content. High lipid and pentoglycan contents cause wheat starch to be more difficult to process than corn starch.

V. References

(*1*) G. E. Ingeltt, *in* "Wheat: Production and Utilization," G. E. Inglett, ed., AVI Publishing, Westport, Connecticut, 1974, Chap. 1.

(*2*) R. A. Anderson, *in* "Starch: Chemistry and Technology," R. L. Whistler and E. F. Paschall, eds., Academic Press, New York, 1967, Vol. II, Chap. 2.

(*3*) "Agricultural Statistics," U.S. Department of Agriculture, 1979.

(*4*) "Foreign Agriculture Circular—Grains," U.S. Department of Agriculture, FG-13-80, April 14, 1979.

(*5*) J. F. Mittleider, D. E. Anderson, C. E. McDonald, and N. Fisher, "An Analysis of the Economic Feasibility of Establishing Wheat Gluten Processing Plants in North Dakota," North Dakota Agricultural Experiment Station, North Dakota State University and the U.S. Department of Commerce, E.D.A., Bulletin No. 508, 1978.

(*6*) F. Rehwald, "Starch Making," Scott Greenwood, London, 1926.

(*7*) L. Eynon and J. H. Lane, "Starch: Its Chemistry, Technology and Uses," W. Heffer & Sons, Cambridge, 1928.

(*8*) J. W. Knight, *in* "Wheat Starch and Gluten," Leonard Hill, London, 1965.

(*9*) J. J. Van Edeskuty, U.S. Patent 2,555,908 (1951).

(*10*) K. O. Hyppola, U.S. Patent 3,651,768 (1972); *Chem. Abstr.*, **77,** 18386e (1972).

(*11*) H. Ferrin, U.S. Patent 2,543,281 (1951); *Chem. Abstr.*, **45,** 4953d (1951).

(*12*) A. G. Fane, *AIChE Symp. Ser.*, 73 (1963); 198 (1977).

(*13*) R. A. Anderson, J. F. Pfeifer, and E. D. Lancaster, *Cereal Chem.* **35,** 449 (1958); **37,** 180 (1960).

(*14*) A. Jackson, *in* "Starch Production Technology," J. A. Radley, ed., Applied Science Publishers, London, 1976, Chap. 9.

(*15*) P. H. Johnston and D. A. Fellers, *J. Food Sci.*, **36,** 649 (1971).

(*16*) D. A. Fellers, P. H. Johnston, S. Smith, A. P. Mossman, and A. D. Sheperd, *Food Technol.*, **23,** 560 (1969).

(*17*) K. Phillips and H. R. Sallans, *Cereal Sci. Today*, **11,** 61 (1966).

(*18*) M. Friedman, *Natl. Conf. Wheat Util. Res., 10th*, U.S. Agric. Res. Serv. West. Reg. Rept. 1977.

(*19*) D. H. Gould and J. T. MacGregor, *Adv. Exp. Med. Biol.*, **86B,** 29 (1977).

(*20*) N. E. Rodgers and R. G. Gidlow, U.S. Patent 3,851,085 (1974).

(*21*) V. P. Chwalek and R. M. Olson, U.S. Patent 4,171,383 (1979).

(*22*) V. P. Chwalek and C. W. Schwartz, U.S. Patent 4,144,087 (1976); *Chem. Abstr.*, **89**, 7960; (1978).
(*23*) V. P. Chwalek and R. M. Olson, U.S. Patent 4,171,384 (1979).
(*24*) G. V. Rao, W. E. Henry, and D. L. Hammond, U.S. Patent 3,979,375 (1976).
(*25*) H. K. Kerkkomen, K. M. J. Laine, M. A. Alanen, and H. V. Renner, U.S. Patent 3,951,938 (1976).
(*26*) M. Maijala, *Food Eng.*, **48,** 73 (1976).
(*27*) B. I. Dahlberg, *Starch/Staerke*, **30,** 8 (1978).
(*28*) P. Verbene and W. R. M. Zwitserloot, *Starch/Staerke*, **30,** 337 (1978).
(*29*) P. Verbene, W. R. M. Zwitserloot, and R. R. Nauta, U.S. Patent 4,132,566 (1979); *Chem. Abstr.*, **88,** 168706a (1978).

CHAPTER XVI

RICE STARCH: PRODUCTION, PROPERTIES, AND USES

By Bienvenido O. Juliano

Chemical Department, International Rice Research Institute, Los Baños, Laguna, Philippines

I. Rice Production and Processing

1. The Rice Grain and Its Production

Rice (*Oryza sativa* L.) is the staple food of East, Southeast, and South Asia, where 90% of the world rice crop is produced and consumed.

Rice probably originated in Asia (*1*). World rice production in 1980 is estimated at 397 million metric tons of rough rice grown on 144 million hectares. Mean yield is 2.7 tons/ha (*2,3*). The major rice-producing Asian countries are the Peoples Republic of China, India, Indonesia, Bangladesh, Thailand, Burma, Japan, Korea, Vietnam, and the Philippines (*2, 3*). The major areas of production other than Asia are Brazil, the United States, the Malagasy Republic, Egypt, Colombia, Nigeria, and Italy.

In the United States, 94% of the total rice crop is produced in Arkansas,

STARCH, 2nd ed.

ISBN 0-12-746270-8

Louisiana, Texas, and California, and the balance of the production is in Mississippi and Missouri (*4*). The average U.S. yield is 4.9 tons/ha. In the United States, the rice field is prepared by conventional means and leveled. It is flooded before pre-germinated seeds are broadcast by airplane. Fields are flooded throughout the growing season to facilitate weed control, as the rice plant can tolerate water because of air channels to the roots. Grain is harvested at about 20% moisture by combines and dried to less than 14% moisture in three drying cycles to prevent fissuring of the grain. In tropical Asia, mechanization in the form of hand tractors, hand pushed weeders, and power threshers (*5*) is slowly replacing the traditional water buffalo. Rice fields are flooded before plowing and harrowing, and the soil hardpan is broken or "puddled."

There are more than 20,000 rice varieties in the world. However, in the United States, because of the close cooperation between industry and breeders, there are only a few varieties, and each grain type—short, medium, and long—represents a specific grain quality (*6*).

The United States was once the largest exporter of milled rice, but was the second largest in 1981, accounting for 2.95 million out of 13.1 million metric tons total world export in 1981 (*2, 3*). In 1981 Thailand was the largest exporter at 3.05 million metric tons; third was the People's Republic of China at 1.0 million metric tons, then Pakistan, Burma, Brazil, and Australia. In 1980, Indonesia was the major rice importer at 2.0 million metric tons (*3*).

Traditional rices of tropical Asia are tall and slender with droopy leaves, photoperiod sensitivity, and high grain dormancy. The introduction in the late 1960s of the "Green Revolution" varieties, which are semi-dwarf and heavy tillering, have erect leaves, and are nonsensitive to changes in the photoperiod, has resulted in widespread replacement of traditional varieties by the higher yielding, lodging-resistant plant type (*7*). Modern varieties respond to added nitrogen fertilizer without lodging, but are comparable to traditional varieties at low levels of inputs. It is estimated that in 1974–1975, 26% of tropical Asia grew these new varieties, ranging from 62% in the Philippines to 6.4% in Burma (*2*). These new varieties were bred mainly for tropical lowland rice fields, which represent only 28% of the total rice area in tropical Asia (*2*). The current effort both at the International Rice Research Institute (I.R.R.I.) and the International Rice Testing Program is to develop specific varieties adapted to environmental stresses such as drought, flooding or deep water, high or low temperatures, and saline, alkaline, and acid sulfate soils. In addition, resistance to major disease and insect pests is also being sought and incorporated into the new varieties (*8*).

The rice grain, rough rice or paddy, consists of the hull (husk) and the edible portion (caryopsis) or brown rice (Fig. 1) (*9*). The outer layers of brown rice are the pericarp, seed coat, nucellus, and aleurone layer, which cover the starchy endosperm (*10*). The embryo or germ is adjacent to the starchy endosperm at the

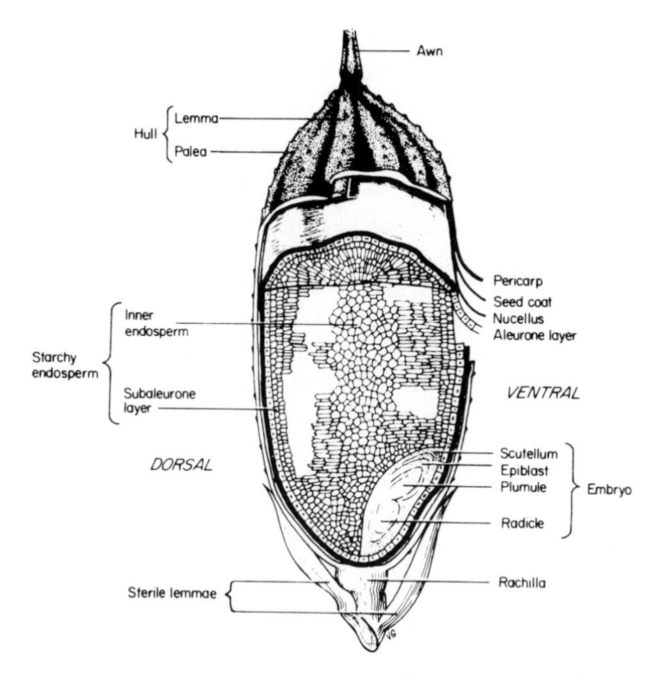

FIG. 1.—Longitudinal section of the rice grain (*9*).

basal ventral portion of the caryopsis. The embryo is 1–2% by weight of brown rice.

The lemma and palea of the hull interlock to form a tight protective cover over the brown rice or caryopsis. The pericarp has a relatively water-impervious outer layer which retards water adsorption during cooking of brown rice but which is removed by milling 1% by weight of brown rice (*11*). The aleurone layer completely covers the embryo and is thickest at the dorsal portion and thinnest at the lateral portion of brown rice, with the ventral portion being intermediate in thickness.

The starchy endosperm consists of thin-walled cells completely filled with compound starch granules (3–10 μm) and spherical protein bodies (0.5–4 μm) (*9*). Very little matrix protein is present. Crystalline-type protein bodies are observed only in the subaleurone layer where the starch granules are smaller (*12, 13, 120*). In his exhaustive study of the development of rice endosperm, Hoshikawa (*14*) reported the final size of the amyloplast to be as small as 6 μm in major diameter in the subaleurone layer and up to 39 μm in diameter in the center of the endosperm, with a mean size of 23 μm.

Non-waxy rices have a translucent endosperm, whereas waxy rices have an opaque endosperm (*9*). However, portions of non-waxy endosperm with loose arrangement of cell contents are opaque.

2. Parboiling and Rice Processing

Parboiling is a process whereby rough rice is steeped in cold or warm water, heated with steam under pressure or in boiling water to gelatinize starch with minimum grain swelling, followed by slow drying (*15, 16*). The process accelerates aging of the grain and loosens the hull. Opaque grains even of waxy rice grain are made translucent (*17*). Water-soluble vitamins are partially heat-degraded, but they diffuse into the endosperm. Thus, although parboiled brown rice has a lower thiamine content, its endosperm has a higher level of thiamine than the endosperm of raw rice (*18*).

The first step in rice processing, after precleaning, is removal of the hull, preferably with minimum grain breakage, using either an under-runner disc huller or a rubber roll huller (*19, 20*). Immature grains and brokens are separated from the hull–brown rice mixture prior to the hull separator or aspirator, which takes advantage of the relative lightness of the hull, and the paddy separator. The paddy separators are of the compartment, tray, and screen types, and utilize the differences in properties of rough rice and brown or dehulled rice: specific gravity, buoyancy, coefficient of surface friction, and length and width and thickness. Because of variation in grain thickness, the undehulled rough rice from the paddy separator may be hulled again once or twice but at a closer clearance.

The brown rice then undergoes abrasive milling in a vertical abrasive whitening cone, a horizontal abrasive whitening machine, or a horizontal jet pearler, to remove the outer layers and the embryo (*19, 20*). The powdery by-product is called bran. Multi-pass systems reduce grain breakage. The jet pearler is partly a milling machine and partly a polishing machine.

To remove traces of bran, the milled rice is then polished in a vertical cone-type or horizontal polisher (*19, 20*). The by-product collected is called polish. Bran and polish are richer in protein, free sugars, vitamins, and other nonstarch constituents than is milled rice (Table I) (*21*).

Milled rice is passed through a series of machines or classifiers which separate the grains according to size (*19, 20*). Whole and three-quarter rice grains are screened into a fraction called "head rice"; the one-third to three-quarter size grains are classed as "second-heads"; the one-quarter to one-third size grains are known as "screenings"; the smallest fragments are called "brewers," since they form a useful brewing adjunct.

In many countries, the dehulling and milling operations are done in a single step in an Engelberg-type mill (*20*). In addition, the total milled rice is not

Table I

Typical Mill Yields and Composition of Products and By-Products from Rough Rice (9)

Fraction	Yield from rough rice[a]	Protein, N × 5.95	Crude fat	Crude fiber, % dry basis	Crude ash	Nitrogen-free extract	Free sugars
Hull	18–28	2–4	0.4–0.8	48–53	15–20	26–34	—
Brown rice	72–82	7–15	2–4	1	1–2	79–90	1.3–1.5
Bran	4–5	12–17	15–22	9–16	9–16	40–49	6.4–6.5
Polish	3	13–16	9–15	2–5	5–9	54–71	—
Milled rice	64–74	6–13	0.3–0.6	0.1–0.6	0.3–0.7	84–93	0.2–0.5

[a] Based on 6% bran and 4% polish from brown rice at 90% milling.

classified, but is sold without removing the brokens. The bran produced is a mixture of hull, bran, and polish.

Milled rice may be conveyed to machines known as "trumbles," in which the grain is coated with talc and glucose (*19*). This inert coating is used to give the rice a gloss and is claimed to protect lipids in the surface of milled rice from oxidation (*22*).

Milled rice may be enriched either by addition of vitamin powder or by addition of a vitamin pre-mix (*23*). The pre-mix thiamine (vitamin B_1) is rinse-resistant (milled rice is usually rinsed with water before cooking), whereas the vitamin powder is leached out in the wash water.

A comparative proximate analysis of milled rice and its by-products is given in Table I.

II. Uses of Milled Rice and Rice By-Products

1. Uses of Milled Rice

Milled raw or parboiled rice is consumed mainly as boiled rice. Cooking methods differ widely in the amount of cooking water used. Use of excess water results in loss of vitamins, minerals, protein, and some starch, but results in a more flaky cooked rice.

Different rices are used for different purposes, with the amylose–amylopectin ratio as the major determinant of the texture of boiled rice (*24, 25*). Amylose content of milled rice is classified as waxy 0–2% dry basis, low 9–20%, intermediate 20–25%, and high >25% (*25*). Waxy rices are used for sweets, desserts, and salad dressing (*25, 26*). It is the staple food in Laos and northern and northeastern Thailand and is prepared by steaming pre-soaked milled rice (*27*). Low-amylose rices are used in baby foods, breakfast cereals, and yeast-leavened rice bread. (*28*). Intermediate amylose rices are used for Philippine fermented rice cakes (*29*) and canned soups (*30*). High-amylose rices are ideal for rice noodles.

In the United States, a waxy rice variety is Mochi gome. Examples of low-amylose, medium-grain varieties are Calrose, Nato, and Vista. Low-amylose, short-grain varieties are Caloro, Colusa, and Nortai. Intermediate-amylose, long-grain varieties are Belle Patna, Bluebelle, Labelle, and Starbonnet (*6*). There are no commercial high-amylose varieties grown in the United States.

Parboiled or converted rice is the staple in parts of India, Bangladesh, Sri Lanka, and Nepal (*31*). Because of the severity of steaming, the traditional milled, parboiled rice has a cream to tan color and is quite undermilled. Red-pericarped, parboiled rices are preferred over non-pigmented rices in some countries such as Sri Lanka (*32*).

Precooked rices are consumed by the military and are sold as ready-to-eat products like Minute Rice. Milled rice, usually parboiled, is cooked and subsequently dehydrated and packaged for consumer use as a quick-cooking product (*33*). Drying may be by freeze-drying, resulting in an opaque, porous product or by heat dehydration, resulting in a glossy product. For reconstitution, the product, in the proper amount of water, is brought to a boil, and the mixture is allowed to stand several minutes so that the rice completely absorbs the water.

Parboiled, milled rice is also processed as a canned, cooked product and as an ingredient in soups. The rice is boiled for 5 minutes, cooled, packed in the can, and passed through a retort to complete cooking and sterilization (*34*).

In addition, rice is used for noodles, often mixed with cassava starch or corn grits to reduce the cost. The rice is wet milled and drained; the cake is dropped in a boiling water bath until it floats and is then extruded under hydraulic pressure and dropped into a bath of cold water and sun-dried (*35*). The so-called "long rice noodle" is not rice at all, but is made from mung bean starch (*36*).

White-core, non-waxy Japanese rices are overmilled to remove 20–50% by weight of brown rice. The residue is used for making "sake" or Japanese rice wine (*37*). The core has a loose arrangement of cell contents and has a lower protein content than do the outer layers (*38*).

Rice grain undergoes some changes during storage or aging above 15° (*39, 40*). The grain gradually becomes harder, resulting in higher total and head rice yields. Aged rice absorbs more water, expands more during cooking, and leaves less dissolved solids in the cooking water. Its cooked rice is more flaky and harder than that of freshly harvested rice. The exact mechanism of this aging process is not fully understood (*39–41*).

Old, milled rice has loose, powdery starch granules on its surface and has a distinctive odor, indicating fat rancidity, both hydrolytic and oxidative. The volatiles of cooked stale rice come from carbonyl compounds formed from the oxidation of unsaturated fatty acids (*41*).

2. Uses of Rice By-Products

Hull or husk is used as fuel, particularly in the parboiling and drying of rough rice, and as a soil conditioner. Possible uses of hulls, including the production of silicon carbide, have been reviewed recently (*42, 43*), but disposal of hull waste is a problem.

Bran and polish are used mainly as animal feed. Processes utilizing alkali extraction of bran protein have been reported (*44, 45*), and pilot plant extraction is being studied in the Philippines. Extraction rates are close to 40%, and the protein extract has good nutritional value.

A major problem in the extraction of oil from bran is the susceptibility of the oil to hydrolysis via lipase activity after milling of the grain (*46*). Immediate solvent extraction or heat treatment of bran to inactivate the lipase is required. Otherwise, the extracted oil cannot be used for human consumption, but only for soap and industrial use as fatty acids. Fat hydrolysis is also a problem in stone-dehulled brown rice and in under-milled rice, wherein the fat at the surface becomes exposed to lipase action.

Brewer's rice is used as an ingredient in beer making and as starting material for starch manufacture (*47*).

III. Preparation of Rice Starch

1. Commercial Method

Commercial preparation of starch from rice is limited owing to the high cost of brewer's rice relative to other cereals and tubers. In the European Economic Community, only 7000 tons of rice starch (from 10,000 tons of broken rice) are produced annually in factories in Belgium, Germany, The Netherlands, and Italy (*48*). Factories in Egypt and Syria are also producing rice starch, but it has not been produced in the United States since 1943 (*49*).

Sodium hydroxide is used for purifying milled rice starch because at least 80% of the protein of milled rice is alkali-soluble protein (glutelin) (*50*). Commercial preparation of rice starch has been described by Hogan (*49*), and there has been little change in the process since the 1960s. The process consists of steeping broken rice in 0.3–0.5% sodium hydroxide solution, wet milling the steeped grain, removal of cell walls, extraction of the protein with sodium hydroxide solution, washing, and drying. Trash and dirt are removed by preliminary cleaning. Steeping softens the grain and aids in extracting the protein. Steeping temperature varies from room temperature to 50°, and the steeping period is usually 24 h. Wet grinding is preferred to dry grinding, as starch damage is more extensive in dry milling, resulting in greater starch losses through dissolution in alkali.

The steeped rice is usually ground or wet milled in pin mills, hammermills, or stone-mill disintegrators in the presence of sodium hydroxide solution. The starch suspension is then stored for 10–24 h, after which fiber is removed by passing the suspension through screens, and the starch is collected by centrifugation, washed with water, and dried in a dryer. Usually only one extraction is made, as 97% of the protein of milled rice flour is extracted in 3–6 h with 0.4% sodium hydroxide (*50, 51*).

Protein may be recovered from the sodium hydroxide effluent from steeping, dewatering, and centrifugation operations by neutralization with acid or sulfur dioxide to the isoelectric pH of the protein (pH 6.4) (*9*). The precipitate is allowed to settle and is recovered in filter presses or centrifuges. It is used mainly as an animal feed constituent.

2. Laboratory Methods

The most common laboratory method of preparing rice starch is by alkali extraction of the protein using 0.2–0.3% sodium hydroxide solution (*52, 53*). Milled rice or brokens is steeped in 5–6 volumes of 0.2–0.3% sodium hydroxide solution at 25° for 24 h to soften the endosperms. The steep liquor is drained off, and the endosperms are pressed and ground lightly in successive small fractions with a mortar and pestle. The slurry is then diluted to the original volume with 0.2–0.3% sodium hydroxide. The mixture is stirred for 10 min and allowed to settle overnight. The cloudy supernatant is drained off, and the sediment is diluted to the original volume with sodium hydroxide solution. The process is repeated until the supernatant becomes clear and gives a negative reaction to the biuret test for protein (10–20 days). Starch is suspended in distilled water, passed through a 100–200-mesh nylon cloth, and repeatedly washed with water until the supernatant no longer shows any pink color with phenolphthalein. The starch is collected by sedimentation or centrifugation, and the white middle portion is collected (thin dark surface and bottom layers are discarded) and air-dried. The whole process takes 20–30 days.

Improvements in the method include homogenization of steeped grain in a Waring blender and passing the resulting mixture through a 200-mesh sieve before the sodium hydroxide treatment. Continuous shaking for 3–6 h is sufficient for alkali extraction of protein and hastens the preparation of starch (*50*). Lower alkali concentrations are ineffective for complete extraction of proteins and starch lipids (Table II) (*51, 121*). Starch prepared by alkaline treatment has less x-ray crystallinity than starch prepared by the dodecyl benzene sulfonate (DoBS) method (*54*), which follows.

Another commonly used method involves treatment with a detergent solution (1–2% DoBS to which 0.2% sodium sulfite is added just before use) (*53*). Milled rice is steeped in 3–4 volumes of the detergent solution for 24–48 h. The

Table II

Effect of Method of Starch Preparation on Protein and Phosphorus Content and Non-Starch[a] and Starch[b] Lipids of IR480-5-9 Milled Rice (121)

Sample and method of preparation	Nitrogen content, % dry basis	Phosphorus content, ppm dry basis	Lipid fraction, % of starch		Ratio of lipid fractions of starch lipids[c]
			Non-starch	Starch	
Milled rice (control)	2.10	—	1.02	0.59	20/18/62
Starch, 1.2% sodium DoBS–0.12% Na_2SO_3	0.03	314	0.16	0.44	30/36/34
Starch, 0.5% SDS–0.6% β-mercaptoethanol	0.05	199	0	0.20	7/38/55
Starch, 0.2% $NaHSO_3$	1.00	158	0.08	0.84	32/14/54
Starch, 0.2% NaOH (0.05 *M*)	0.04	16	0.01	0.09	79/ 9/12
Starch, 0.1% NaOH (0.025 *M*)	0.04	58	0.02	0.28	80/ 9/11
Starch, 0.03% NaOH (0.0075 *M*), pH 10	0.71	397	0.02	0.80	32/ 6/62
Starch, 0.1 *M* Na_2CO_3–$NaHCO_3$, ph 10.2	0.38	357	0.02	0.62	41/ 4/55
Starch, pronase treated	0.08	230	0.01	0.82	44/ 7/49

[a] Extracted twice for 4 h each with 2:1 v/v $CHCl_3$:MeOH and twice for 30 min each with water-saturated 1-BuOH.
[b] Extracted three times for a total of 24 h with water-saturated 1-BuOH.
[c] Neutral lipids/glycolipids/phospholipids uncorrected for contaminants in glycolipid fraction.

supernatant is decanted off, and the residue is crushed in a mortar and pestle to pulverize the milled rice, adding detergent solution as needed. The slurry is passed through a 200-mesh cloth or sieve, and the large particles are ground further in a mortar and pestle. Low-speed grinding of the steeped grain in water may also be employed. The sieved starch is then shaken with 5 volumes of fresh detergent solution for 6 h, and the supernatant is removed by centrifugation. This treatment is repeated 3–4 times to remove any trace of protein and fat, and the purified starch is then washed repeatedly with distilled water until the washing is negative for sulfate. The starch is then again washed two to three more times with water. The purified starch is collected after discarding the thin layer of darker starch at the surface and bottom and air-dried. Yields are 33–50% of milled rice, and Kjeldahl nitrogen content is about 0.02% for japonica rice and higher for higher-amylose indica rices.

Sodium dodecyl sulfate (SDS) (0.5%)–β-mercaptoethanol (0.6%), pH 7, may also be employed for the purification of rice starch (*55*). This process results in lower bound lipids in the purified starch as compared to the DoBS method (*56*) (Table II). A lowering of gelatinization temperature is also observed.

Fujii (*57*), however, reported that detergents, DoBS in particular, may replace the bound lipids in rice starch. Although the decrease in bound lipids had little effect on the pasting properties of rice starch, detergent-treated starch is not the best sample for the study of starch-bound lipids.

Ultrasonication has been employed for purifying rice starch (*58*). About 5 g of milled rice powder suspended in 45 mL of distilled water in a test tube is subjected to 10 KHz for 10–20 min. The homogenate is filtered through a 200-mesh sieve, and the filtrate is allowed to settle. After scraping off the dark upper layer, the starch is collected, washed, and air-dried. The Kjeldahl nitrogen content of the starch preparation is 0.04% for japonica rice and 0.15–0.20% for indica rice.

The 0.1 *M* sodium chloride–toluene method of starch preparation (*59*) does not work well for rice starch (*56*), as only 15% of its protein is albumin–globulin (*43, 44*). Addition of sodium hydroxide to the system improves protein extraction 2.5-fold, but protein extraction efficiency is still poor at 52% (*56*).

IV. Physicochemical Properties of Rice Starch

1. Physical, Gelatinization, and Pasting Properties of Rice Starch Granules

Rice has one of the smallest starch granules of the cereal starches, varying in size from 3 to 10 μm in the mature grain. Mean granule size varies from 4 to 6 μm (*9, 60*). Only one set of starch granules is produced in the developing rice grain, and starch accumulation is mainly an increase in granule size (*14, 61*).

Table III

Range of Physicochemical Properties of Non-Waxy and Waxy Rice Starch[a]

Property	*Non-waxy starch*	*Waxy starch*
Final BEPT (°C)	58–79	58–78.5
Granule size (μm)	1.6–8.7	1.9–8.1
Density (xylene displacement)	1.49–1.51	1.48–1.50
Residual Kjeldahl N (% dry basis)	0.02–0.12	~0.02
Residual P (mg/g)	0.12–0.45	0.02–0.03
Bound lipid (% dry basis)	0.2–0.4	~0.03
6% gel viscosity (cP)	140–1200	64–1890
$[\eta]_{0.1\ M\ KOH}$ (mL/g)	160–194	46–164
Iodine binding capacity (% dry basis)	2.36–6.96	0.15–0.86

[a] Prepared by DoBS method.

Starch granules of rice are compound, and they are polyhedral or pentagonal dodecahedron (Chap. XXIII). As with other starches, waxy starch granules tend to have lower density than non-waxy granules (Table III), owing to the presence of micropores inside and on the surface (*38*).

Final gelatinization or birefringence end-point temperature (BEPT) (Chapter IX) of rice starch varies widely among rice varieties (Table III) and is classified in rice as low (58°–69.5°), intermediate (70°–74°), and high (74.5°–79°) (*25*). The variation probably reflects the compactness of the rice endosperm, as BEPT is determined in the breeding program by the resistance of the whole-grain milled rice to disintegration on soaking for 23 h in 1.7% potassium hydroxide solution at 30° (*62*). BEPT range varies from 2° to 12° and tends to be narrower for waxy rices (*60, 63*). The temperature of rapid increase in Brabender Amylograph viscosity of a 12.5% rice starch suspension (*64*) and the temperature at initial increase in Amylograph viscosity of 20% milled rice suspension (*65*) agreed with BEPT. In one commercial rice starch, a final BEPT of 79°–80° corresponded to 72% gelatinization of the granules in the hot stage, 76% gelatinization at initial Amylograph pasting temperature, and 79% gelatinization at the temperature of an Amylograph viscosity increase of >80 Brabender units (BU)/min (*64*).

Ambient temperature during grain development can also affect the BEPT and amylose content of the rice starch (*66*). High temperature decreases amylose content, but increases BEPT, while cool temperature has the opposite effect. Warmer ripening temperatures are also reported to give a qualitative change in x-ray diffraction pattern from C_A to A (*67, 68*).

Although BEPT and amylose content are independent properties of rice starch and are affected by temperature during grain development, only the following BEPT types are common among the various amylose types: waxy, low and high

BEPT; low-amylose, low and high BEPT; intermediate-amylose, low and intermediate BEPT; and high-amylose, low and intermediate BEPT. The intermediate BEPT type is rare in waxy and low-amylose rices, whereas the high BEPT type is rare for intermediate- and high-amylose rices.

Four-day Lintnerization loss of rice starch granules correlates negatively with final BEPT of starch regardless of amylose content (*69, 70*). A recent study showed that the first stage of Lintnerization in 2.2 *M* hydrochloric acid at 35° (involving the amorphous part of the granule) occurs in about 7–9 days, after which the corrosion rate decreases (*71*). The recovery of Lintnerized starch after 15 days of treatment ranged from 7 to 21% and was greater for starches with high BEPT and, to a lesser extent, high amylose content. Lintnerized starch had sharper x-ray diffraction peaks than did native starch. Their $\overline{DP}$ ranged from 19 to 30 D-glucose units and their β-amylolysis limit ranged from 74 to 92%. BioGel P-2 chromatography showed two peaks corresponding to $\overline{DP}$ 29–34 and 14–17 D-glucose units. Varietal differences were noted in the ratio of two amylodextrin peaks. Free sugars of Lintnerized starch were low (0.1–0.3%) except for the two derived from low-amylose starch granules that had 1.5 and 1.8% sugars.

Amylography of 10% rice starch pastes provides an insight into the textural changes of rice during cooking (*72, 122*). Rice flour has slightly lower viscosity than rice starch for non-waxy samples, but twice as much waxy rice flour is needed to obtain the same Amylograph viscosity as given by waxy rice starch (*73, 74*). The difference has been explained on the basis of higher α-amylase activity of waxy rice flour than in non-waxy rice flour; this α-amylase is readily extracted with water (*73*). However, α-amylase activity of mature rice grain is low (*75*). Boiling the water extract, cooling it, and adding it again to the waxy rice flour does not increase its Amylograph viscosity, which would be expected if α-amylase were present and inactivated by boiling (*56*). The non-starch fraction of waxy milled rice, particularly lipid, probably suppresses granule swelling of waxy granules more than that of non-waxy granules. Aside from waxy rices, high-amylose milled rice and starch also show greater varietal differences in Amylograph peak viscosity than do low-amylose and intermediate-amylose rice starches (*56*). Samples with lower peak viscosity also have lower setback viscosity and consistency. Amylograph setback viscosity (viscosity cooled to 50° minus peak viscosity) correlates positively with the amylose content of milled rice (*25*). Amylograph consistency (viscosity at 50° minus final viscosity at 94°) also correlates positively with the amylose content of milled rice (*25*).

Carboxymethylcellulose (0.9%)–starch (6.25%) amylography (*76*) of non-waxy rice starch showed the typical two-stage gelatinization of cereal starches (Fig. 2) (*77*). The pasting temperature obtained was within 2° of the gelatinization temperature. The second stage of pasting was less pronounced than the first stage and ranged from 81° to 88°. Its peak viscosity was in the same order of

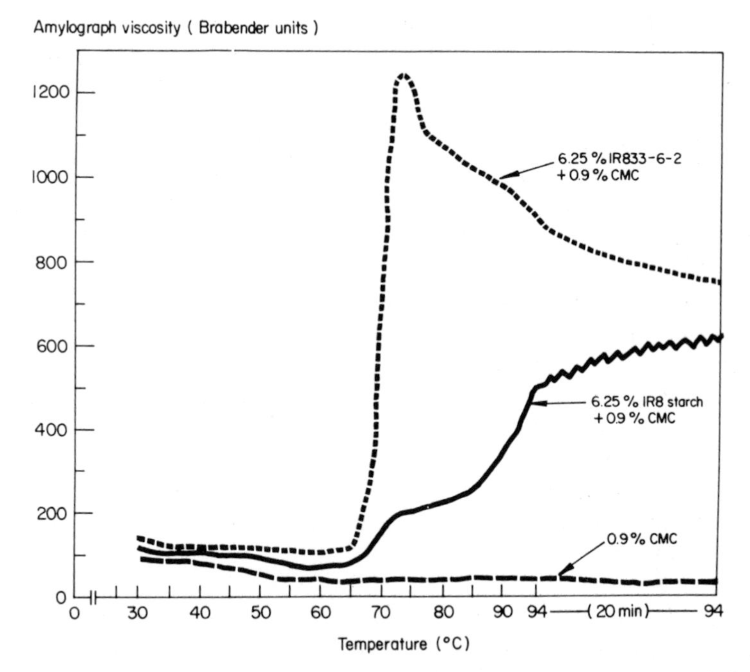

FIG. 2.—Typical 0.9% carboxymethylcellulose–6.25% (wet basis) starch amylograms of non-waxy (IR8) (*77*) and waxy (IR833-6-2) rice starch (*56*).

magnitude as the peak viscosity of 10% flours, but tended to be lowest for high-amylose, soft-gel-consistency samples. Similar trends are noted for peak viscosity in alkali viscography. Final viscosity at 94° is close to peak viscosity for most high-amylose samples, but lower for low- and intermediate-amylose starches. Waxy rice starch showed only the first stage of gelatinization and a higher peak viscosity than non-waxy rice starch (Fig. 2) in conformity with results on corn starches (*76*). The higher peak viscosity of starch from waxy varieties may be due to its single-stage gelatinization in the absence of amylose, in contrast to that of starch from non-waxy varieties.

Alkali viscograms of 2% non-waxy rice starch in potassium hydroxide solution have been used by Japanese chemists for characterizing the starch (*122*). Gelatinization molarity (0.29–0.49 *M*) was shown to closely follow final BEPT (Fig. 3) (*78, 79*). Among non-waxy starches, those with low BEPT correspond generally to gelatinization molarity (above 22.5 cP) of 0.29–0.36 *M;* intermediate BEPT, 0.40–0.43 *M;* and high BEPT, 0.46–0.49 *M* (*78*). Waxy rice starch (1.8% suspension) tends to have higher gelatinization normality, higher peak viscosity, and narrower ΔM range from initial viscosity increase to peak viscosity than non-waxy starch. For five waxy rice starches, gelatinization molarity

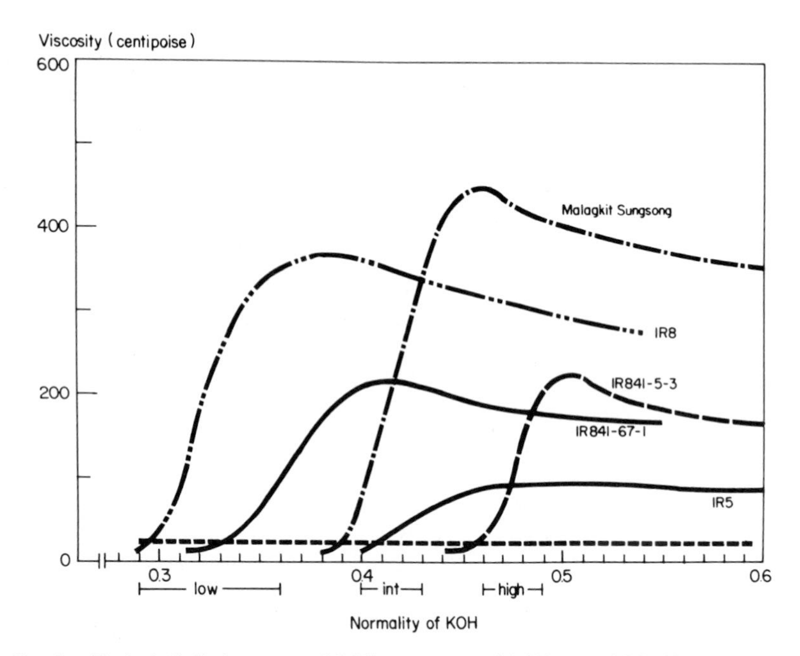

FIG. 3.—Typical alkali viscograms of 2.0% non-waxy and 1.8% waxy (Malagkit Sungsong) rice starches. (*78*).

was 0.39–0.49 *M*, Δ*M* from gelatinization to peak viscosity was 0.06–0.07 *M*, and peak viscosity was 435–540 cP. By contrast, 2% suspensions of 27 non-waxy rice starches had Δ*M* of 0.05–0.21 *M* and peak viscosities of 89–386 cP. Among high-amylose rices, samples with high Amylograph setback viscosity (>400 BU) (IR8) gave higher alkali viscogram peak viscosities than those with low setback viscosity (<300 BU) (IR5, Fig. 3).

Defatting of DoBS-prepared starches with cold, water-saturated butanol had little effect on the alkali viscograms except for a tendency for the gelatinization molarity to decrease (*56*). A Wells-Brookfield cone-plate micro viscometer may be used as a micro alkali viscograph using 2 mL of 2% starch dispersions at various potassium hydroxide molarities (*56*).

2. Chemical and Molecular Properties of Rice Starch and Its Fractions

The chemical and molecular properties of rice starch are probably similar to those of other cereal starches used for human consumption, except for the wide range of values among varieties (Table III) (*60, 63, 69*). High-amylose (40–80%) starches have not been reported in rice.

Dosage effects of the waxy gene *WxWxWx*, *WxWxwx*, *wxwxWx*, and *wxwxwx* (see Chapter III) in the rice endosperm are as follows in F_2 seeds: 19.6, 19.0,

17.1, and 0% amylose (*80*); 18.2, 15.7, 14.9, and 0% (*80*); 25.4, 21.4, 19.2, and 0% (*81*); 26.6, 24.2, 20.8, and 0% (*81*); 23.2, 21.0, 18.6, and 0% (*81*); and 12.2, 5.1, 2.2, and 0% (*82*). Similar results were obtained in Korea with normal and male sterile stocks (*83, 84*). Grain weights tended to decrease with an increase in the wx factor (*80, 85*). Reported dosage effects of the low-amylose gene (*L*) on high-amylose rices (*H*) *HHH, HHL, LLH,* and *LLL* are 25.7, 25.0, 16.1, and 13.1% (*81*); 26.4, 21.2, 18.2, and 6.1% (*81*); 25.4, 23.3, 16.9, and 12.9% (*81*); and 27.4, 25.8, 24.1, and 17.4% (*81*). Similar results were also obtained in Korea with normal and male sterile stocks (*83, 84*).

Both BEPT and amylose content are simply inherited properties (*86*). However, high BEPT usually results from crosses involving indica (intermediate BEPT) × japonica (low BEPT) parents, such as Century Patna 231. The three Taiwanese rices, which are the sources of the dwarfing gene utilized in deriving the modern semi-dwarf rices, all have high amylose content, low BEPT, and hard gel consistency (*78*). Although the initial modern rices, such as IR8 and IR22, have the same combination of starch properties as the dwarf parent (*78*), some of the newer varieties have intermediate BEPT and a soft gel consistency (*87*).

Even after repeated protein extraction, all rice starch preparations contain residual Kjeldahl nitrogen (protein N plus phospholipid N). A waxy rice sample on exhaustive DoBS extraction gave a residual nitrogen content of 0.007% as against 0.07% for a 30% amylose rice starch (*56*). Residual nitrogen closely follows amylose content (Table III) (*69, 71*) and probably involves amylose–protein binding, as in the case of adenosine diphosphoglucose starch synthetase (*88, 89*).

In DoBS-prepared rice starch, waxy starch contained much less bound lipid (0.03%) than non-waxy starch (0.2–0.4%), but the bound lipid content was similar (0.3–0.4%) in the intermediate- and high-amylose rices (*56*). Intermediate-amylose rice starch tended to have more lipids than high-amylose rice starch (*56*). Its bound lipids extracted with water-saturated butanol at 25° were a 1:1:1 mixture of neutral lipids (mainly fatty acids), glycolipids, and phospholipids (Table II). Starch lipids extracted from brown or milled rice directly with water-saturated butanol had much less glycolipids and more phospholipids. Pronase or *Streptomyces griseus* protease digestion produced starch granules containing 0.82% starch lipids, which is higher than that in the DoBS-prepared starch (Table II). The high glycolipid content of DoBS-prepared starch may be due to contaminant detergent. Glycolipids are also contaminated with at least three unknown sugar-containing spots as revealed by thin-layer chromatography (*56*).

Residual phosphorus of rice starch defatted with 80% dioxane was shown to be low (16–17 ppm) and mainly (92–94%) from D-glucose 6-phosphate residues in the amylopectin in waxy rice, but higher (89–201 ppm), mainly from phospholipids and only 8–10% D-glucose 6-phosphate residues in non-waxy rice starch (*90, 91*). Phospholipids identified in non-waxy rice starch are phosphatidylcholine (lysolecithin) (*92*) and lysophosphatidylethanolamine (*56, 92–94*). Defatting with water-saturated 1-butanol had little effect on the phosphorus content (12–22 ppm) of waxy starch, whereas about 40–50% of residual phosphorus (130–280 ppm) is extracted from non-waxy starch by water-saturated butanol (*56*); only a minor fraction of this phosphorus of defatted non-waxy rice can be attributed to D-glucose 6-phosphate residues (*90*). Part of the non-extracted phosphorus is in phosphatidylcholine (*95*). The residual Kjeldahl nitrogen of rice starch granules is not extracted with water-saturated butanol. Neutral lipids are mainly free fatty acids; mono-, di-, and triglycerides, phytosterol, and phytosterol esters are also present (*93, 94*).

Fractionation of starch to amylose and amylopectin usually employs starch pregelatinized either by autoclaving or ammonia or dimethyl sulfoxide treatment (*96, 97*). The standard method of Schoch (*98*) is employed to crystallize the amylose from solution of hot butanol-saturated water upon slow cooling in Dewar flasks. Greenwood and Thomson (*96*) and Taki (*99*) showed that thymol and secondary pentanols (isoamyl alcohol) result in purer amylopectin during the first crystallization. The relatively impure amylose–alcohol complex is readily purified by recrystallization from butanol-saturated water.

Amylose and amylopectin can be separated from each other by ultracentrifugation in dimethyl sulfoxide (*69*) or by gel filtration in 2–4% cross-linked agarose (*100*).

Amylose obtained by one recrystallization from butanol-saturated water has high (19%) iodine binding capacity (IBC) characteristic of amyloses from other starches (Table IV) (*9, 63, 101, 102*). Very little further increase in IBC is obtained on further recrystallization (*63*). Although higher β-amylolysis limits have been reported for rice amylose (*63*), this may be due mainly to the presence of maltase (α-amylase), even in crystalline sweet potato β-amylase (*103*). In the presence of glycerol (40% by volume), which inhibits maltase and α-amylase activity (*103*), β-amylolysis limits were 73–76% instead of 83–90% (*71*). Rice amylose from 1-butanol–isoamyl alcohol fractionation of gelatinized rice starch gave IBC by amperometry of 13.5–18.1% as against 18.3% for potato amylose (*104*). Corresponding amylopectin IBC ranged from 0.03–1.14%.

Rice amylopectins of both nonwaxy and waxy rice starch have mean chain lengths ($\overline{CL}$) by periodate oxidation similar to those of other amylopectins, that is, 20–28 D-glucose units (Table IV) (*60, 63, 105, 106*). Their β-amylolysis limits are also similar, indicating that the amylopectin of waxy starch is similar to that of other starches.

The ratios of A- and B-chains of amylopectins of two pairs of isogenic lines differing in the *waxy* gene have been determined from the amount of maltotriose released by subjecting the β-limit dextrins to the action of pullulanase and to the combined action of β-amylase and pullulanase (*107*). The fine structures of amylopectins of waxy and non-waxy rices are almost identical, differing in

Table IV

Range of Physicochemical Properties of Amylose and Amylopectin from Non-Waxy and Waxy Rices

Property	*Amylose*	*Non-waxy amylopectin*	*Waxy amylopectin*
$S_{20,\ DMSO}$ (Svedberg)	2.0–6.6	73–237	27–242
$[\eta]_{0.15\ or\ 1\ M\ KOH}$ (mL/g)	55–242	85–221	46–164
$\overline{CL}$ (anhydroglucose)	n.d.	20–28	20–27
β-Amylolysis limit (%)	73–76	49–58	49–56
6% gel viscosity (cP)	13–160	290–740	19–330
Iodine binding capacity (%)	17.4–20.2	0.37–2.74	0.07–0.09

BEPT, and the ratio of A-chains:B-chains is of the order 1.1–1.5. The two pairs differ in BEPT.

Although some correlation between final BEPT and [η] of amylose and amylopectin in sister lines differing in either amylose content or final BEPT has been found (*60, 102, 108*), the relationship was not consistent when studied on sister lines grown in two seasons and differing both in amylose content and BEPT (*69*).

Although [η] of amylose and amylopectin are similar (Table IV), gel viscosity (135 mg/2 mL 0.2 *M* KOH) was shown to be 26–73 cP for amylose and 354–604 cP for amylopectin for seven non-waxy starches, indicating that, at high concentrations, amylopectin contributes more to viscosity and consistency than does amylose (*109*). Gel viscosity is also higher in potassium hydroxide solution than in potassium acetate solution.

Schoch (*74*) reported on the superior freeze-thawing stability of waxy rice starch. A 5% aqueous paste of waxy rice starch shows no syneresis until after twenty freeze-thaw cycles, but waxy corn and sorghum starch is stable for only three freeze-thaw cycles. The stability of cooked pastes of waxy milled rice flours is almost as good as that of the starch. A comparison of stability of near-isogenic waxy and non-waxy rice starch amylopectins to freeze-thawing indicated varietal differences in stability, with three non-waxy amylopectins showing no syneresis only up to 3–5 freeze-thaw cycles and the corresponding waxy amylopectin showing no syneresis up to 8–14 cycles (*56*). Mean chain lengths were 24–25 and 23–24 glucose units, respectively (*108*).

V. Starch Properties in Relation to Quality of Milled Rice

The cooking and eating quality of milled rice is mainly determined by the properties of its starch. The major determinant of water absorption, volume expansion, and dissolved solids during cooking and of color, gloss, stickiness, and softness of cooked rice is its amylose content (*24, 25, 110, 111*).

However, differences in eating quality also exist among varieties of similar amylose content which were related to other quality factors, such as final BEPT and gel consistency (*112*). Waxy rices, preferred samples for rice cakes and related products because of softer and sticky boiled rice, have soft neutral gel consistencies and BEPT close to 65°–68° (*29, 106, 113, 114*). Neutral gel consistency is determined by dispersing 200 mg of milled rice flour in 1 mL of 0.3 *M* potassium hydroxide solution in a boiling water bath for 8 min, neutralizing with 1 mL of 0.3 *M* acetic acid, and measuring the length of the neutral gel after 1 h on a horizontal position in a 13 × 100 mm culture tube (*106, 113, 115*).

Among high-amylose rices, differences in hardness of cooked rice are related to differences in gel consistency of 100 mg of rice flour in 2 mL of 0.2 *M* potassium hydroxide solution (*116*). Samples with hard gel consistency give a harder cooked rice and have a very high Amylograph setback (*112*).

Among intermediate-amylose rices, differences in hardness of cooked rice were related to either differences in gel consistency or in final BEPT (*112*). Among low-amylose rices, hardness of cooked rice was also related to differences in modified gel (120 mg flour/2 mL of 0.2 *M* KOH) and Amylograph (48 g flour/400 g paste) consistencies (*112*).

Although grain size and shape of U.S. rices are useful indicators of their quality characteristics (*6*), rices from Asia are not distinguishable by the same characteristics. In addition, indica or tropical rices overlap in properties with japonica or temperate rices, but japonica rices have a narrower range of properties, including starch properties. Japonica rices have coarse, short grain, low BEPT starch, and the following milled rice amylose classes: waxy, low, and intermediate.

Flakiness and canning stability of cooked rice may be improved by crosslinking of starch molecules of raw milled rice with ethylene chlorohydrin (*117*).

Properties of rice starch related to varietal differences in processing characteristics of rice grain was recently reviewed (*118*).

VI. Uses of Rice Starch

The widespread use of rice starch is currently limited by its higher price relative to corn, wheat, and potato starches (*119*). Schoch (*74*) summarizes the principal uses of rice starch as (a) a cosmetic dusting powder, (b) a laundry stiffening agent in the cold-starching of fabrics, and (c) a "custard" or pudding starch. In the European Economic Community, rice starch (low-amylose) is used in baby foods, in specific paper and photographic paper powder, and in the laundry industry (*48*). The non-food applications take advantage of the small size of the rice starch granules.

VII. References

(*1*) T. T. Chang, *Phil. Trans. Roy. Soc. London, Ser. B.*, **275,** 143 (1976).

(*2*) A. C. Palacpac, "World Rice Statistics," Intern. Rice Res. Inst., Los Baños, Laguna, Philippines, 1982.

(*3*) Food Agr. Organ. U.N. FAO, *Rept. 24th Session Intergovt. Group on Rice to Common Commodity Problems,* **CCP 8114,** 1978, p. 3.

(*4*) C. R. Adair, *in* "Rice: Chemistry and Technology," D. F. Houston, ed., Amer. Assoc. Cereal Chemists, Inc., St. Paul, Minnesota, 1972, p. 1.

(*5*) H. Shimoda, *in* "Rice in Asia," Assoc. Japan. Agr. Sci. Soc., University Tokyo Press, Tokyo, 1975, p. 470.

(*6*) B. D. Webb, *Texas Agr. Expt. Sta. Res. Monograph,* **4,** Beaumont, Texas, 1975, p. 97.

(*7*) R. F. Chandler, *in* "Rice Breeding," Intern. Rice Res. Inst., Los Baños, Laguna, Philippines, 1972, p. 77.

(*8*) B. O. Juliano, *Cereal Foods World,* **22,** 284 (1977).

(*9*) B. O. Juliano, *in* "Rice: Chemistry and Technology," D. F. Houston, ed., Amer. Assoc. Cereal Chemists, Inc., St. Paul, Minnesota, 1972, p. 16.

(*10*) D. B. Bechtel and Y. Pomeranz, *Amer. J. Botany,* **64,** 966 (1977).

(*11*) H. S. R. Desikachar, S. N. Raghavendra Rao, and T. K. Anantachar, *J. Food Sci. Technol.* **2,** 110 (1965).

(*12*) N. Harris and B. O. Juliano, *Ann. Botany (London),* **41,** 1 (1977).

(*13*) D. B. Bechtel and Y. Pomeranz, *Amer. J. Botany,* **65,** 684 (1978).

(*14*) K. Hoshikawa, *Proc. Crop Sci. Soc. Japan* **37,** 207 (1968).

(*15*) F. Gariboldi, *in* "Rice: Chemistry and Technology," D. F. Houston, ed., Amer. Assoc. Cereal Chemists, Inc., St. Paul, Minnesota, 1972, p. 358.

(*16*) N. Ali and T. P. Ojha, *in* "Rice Postharvest Technology," E. V. Araullo, D. B. de Padua, and M. Graham, eds., Intern. Develop. Res. Centre, Ottawa, Canada, 1976, p. 161.

(*17*) S. N. Raghavendra Rao and B. O. Juliano, *J. Agr. Food Chem.*, **18,** 289 (1970).

(*18*) A. B. Padua and B. O. Juliano, *J. Sci. Food Agr.*, **25,** 697 (1974).

(*19*) G. C. Witte, Jr., *in* "Rice: Chemistry and Technology," D. F. Houston, ed., Amer. Assoc. Cereal Chemists, Inc., St. Paul, Minnesota, 1972, p. 188.

(*20*) H. Van Reuten, *in* "Rice Postharvest Technology," E. V. Araullo, D. B. de Padua, and M. Graham, eds., Intern. Develop. Res. Centre, Ottawa, Canada, 1976, p. 205.

(*21*) A. P. Resurreccion, B. O. Juliano, and Y. Tanaka, *J. Sci. Food Agr.*, **30,** 475 (1979).

(*22*) G. N. Stemmermann and L. N. Kolonel, *Amer. J. Clin. Nutr.*, **31,** 2017 (1978).

(*23*) C. L. Brooke, *in* "Rice: Chemistry and Technology," D. F. Houston, ed., Amer. Assoc. Chemists, Inc., St. Paul, Minnesota, 1972, p. 353.

(*24*) B. O. Juliano, L. U. Oñate, and A. M. del Mundo, *Food Technol.*, **19,** 1006 (1965).

(*25*) B. O. Juliano, *in* "Chemical Aspects of Rice Grain Quality," Intern. Rice Res. Inst., Los Baños, Laguna, Philippines, 1979, p. 69.

(*26*) J. G. Davis, J. H. Anderson, and H. L. Hanson, *Food Technol.*, **9,** 13 (1955).

(*27*) N. Kongseree, *in* "Chemical Aspects of Rice Grain Quality," Intern. Rice Res. Inst., Los Baños, Laguna, Philippines, 1979, p. 303.

(*28*) K. D. Nishita, R. L. Roberts, M. M. Bean and B. M. Kennedy, *Cereal Chem.*, **53,** 626 (1976).

(*29*) A. A. Perdon and B. O. Juliano, *Staerke,* **27,** 196 (1975).

(*30*) E. C. Hagberg, *in* "Proceedings National Utilization Conference 1966," U.S. Department of Agriculture ARS, **7253,** 1967, p. 58.

(*31*) K. R. Bhattacharya, *in* "Chemical Aspects of Rice Grain Quality," Intern. Rice Res. Inst., Los Baños, Laguna, Philippines, 1979, p. 363.
(*32*) C. Breckenridge, *in* "Chemical Aspects of Rice Grain Quality," Intern. Rice Res. Inst., Los Baños, Laguna, Philippines, 1979, p. 175.
(*33*) R. L. Roberts, *in* "Rice: Chemistry and Technology," D. F. Houston, ed., Amer. Assoc. Cereal Chemists, Inc., St. Paul, Minnesota, 1972, p. 381.
(*34*) E. E. Burns, *in* "Rice: Chemistry and Technology," D. F. Houston, ed., Amer. Assoc. Cereal Chemists, Inc., St. Paul, Minnesota, 1972, p. 417.
(*35*) International Rice Research Institute, "Annual Report for 1972," Los Baños, Laguna, Philippines, 1973, p. 16.
(*36*) D. M. Hilker, M. Fujikawa and K.-C. Chan, *J. Amer. Dietet. Ass.*, **60,** 315 (1972).
(*37*) Y. Nunokawa, *in* "Rice: Chemistry and Technology," D. F. Houston, ed., Amer. Assoc. Cereal Chemists, Inc., St. Paul, Minnesota, 1972, p. 449.
(*38*) T. Tashiro and M. Ebata, *Proc. Crop Sci. Soc. Japan,* **44,** 205 (1974).
(*39*) R. M. Villareal, A. P. Resurreccion, L. B. Suzuki, and B. O. Juliano, *Staerke,* **28,** 88 (1976).
(*40*) S. Barber, *in* "Rice: Chemistry and Technology," D. F. Houston, ed., Amer. Assoc. Cereal Chemists, Inc., St. Paul, Minnesota, 1972, p. 215.
(*41*) K. Yasumatsu, S. Moritaka, and S. Wada, *Agr. Biol. Chem.*, **30,** 483 (1966).
(*42*) D. F. Houston, *in* "Rice: Chemistry and Technology," D. F. Houston, ed., Amer. Assoc. Cereal Chemists, Inc., St. Paul, Minnesota, 1972, p. 301.
(*43*) E. C. Beagle, *in* "Proceedings Rice By-products Utilization International Conference 1974," Inst. Agroquím. Tecnol. Aliment., Valencia, Spain, 1977, Vol. 1, p. 1.
(*44*) D. F. Houston, *in* "Rice: Chemistry and Technology," D. F. Houston, ed., Amer. Assoc. Cereal Chemists, Inc., St. Paul, Minnesota, 1972, p. 272.
(*45*) S. Barber and C. Benedito de Barber, *in* "Proceedings Rice By-products Utilization International Conference 1974," Inst. Agroquím. Tecnol. Aliment., Valencia, Spain, 1977, Vol. 4, p. 1.
(*46*) H. S. R. Desikachar, *in* "Proceedings Rice By-products Utilization International Conference 1974," Inst. Agroquím. Tecnol. Aliment., Valencia, Spain, 1977, Vol. 2, p. 1.
(*47*) Y. Pomeranz, *in* "Rice: Chemistry and Technology" D. F. Houston, ed., Amer. Assoc. Cereal Chemists, Inc., St. Paul, Minnesota, 1972, p. 433.
(*48*) W. Kempf, personal communication, 1978.
(*49*) J. T. Hogan, *in* "Starch: Chemistry and Technology," R. L. Whistler and E. F. Paschall, eds., Academic Press, New York, 1967, Vol. II, p. 65.
(*50*) G. B. Cagampang, L. J. Cruz, S. G. Espiritu, R. G. Santiago, and B. O. Juliano, *Cereal Chem.*, **43,** 145 (1966).
(*51*) E. M. S. Tecson, B. V. Esmama, L. P. Lontok, and B. O. Juliano, *Cereal Chem.*, **48,** 168 (1971).
(*52*) S. Sato, "Studies on Rice Starch," Taigado, Tokyo, 1944, p. 12.
(*53*) Z. Nikuni and S. Hizukuri, *in* "Purification and Reaction of High Molecular Substances," Y. Ono, ed., Kyoritsu Publishing Co., Tokyo, Vol. 12, 1958, p. 46.
(*54*) J. C. Lugay and B. O. Juliano, *J. Appl. Polymer Sci.*, **9,** 3775 (1965).
(*55*) B. O. Juliano and D. Boulter, *Phytochem.*, **15,** 1601 (1976).
(*56*) International Rice Research Institute, unpublished data, 1978.
(*57*) T. Fujii, *J. Japan. Soc. Starch Sci.*, **19,** 159 (1972); *Chem. Abstr.*, **81,** 154906j (1974).
(*58*) H. Horiuchi and T. Tani, *J. Agr. Chem. Soc. Japan,* **38,** 23 (1964); *Chem. Abstr.*, **62,** 13362g (1965).
(*59*) G. K. Adkins and C. T. Greenwood, *Staerke,* **18,** 213 (1966).
(*60*) B. O. Juliano, M. B. Nazareno, and N. B. Ramos, *J. Agr. Food Chem.*, **17,** 1364 (1969).
(*61*) V. P. Briones, L. G. Magbanua, and B. O. Juliano, *Cereal Chem.*, **45,** 351 (1968).

(*62*) R. R. Little, G. B. Hilder, and E. H. Dawson, *Cereal Chem.*, **35,** 111 (1958).
(*63*) A. C. Reyes, E. L. Albano, V. P. Briones, and B. O. Juliano, *J. Agr. Food Chem.*, **13,** 438 (1965).
(*64*) K. J. Goering, D. H. Fritts, and K. G. D. Allen, *Cereal Chem.*, **51,** 764 (1974).
(*65*) J. V. Halick, H. M. Beachell, J. W. Stansel, and H. H. Kramer, *Cereal Chem.*, **37,** 670 (1960).
(*66*) A. P. Resurreccion, T. Hara, B. O. Juliano, and S. Yoshida, *Soil Sci. Plant Nutr.*, **23,** 109 (1977).
(*67*) Z. Nikuni, S. Hizukuri, K. Kumagai, H. Hasegawa, T. Moriwaki, T. Fukui, K. Doi, S. Nara, and I. Maeda, *Mem. Inst. Sci. Ind. Res. Osaka Univ.*, **26,** 1 (1969).
(*68*) K. Sato, S. Sitoji, and M. Ebata, *Proc. Crop Sci. Soc. Japan*, **40,** 439 (1971); *Chem. Abstr.*, **77,** 2906p (1972).
(*69*) N. Kongseree and B. O. Juliano, *J. Agr. Food Chem.*, **20,** 714 (1972).
(*70*) A. D. Evers and B. O. Juliano, *Staerke*, **28,** 160 (1976).
(*71*) C. C. Maniñgat and B. O. Juliano, *Starch/Staerke*, **31,** 5 (1979).
(*72*) J. V. Halick and V. J. Kelly, *Cereal Chem.*, **36,** 91 (1959).
(*73*) H. Horiuchi, *Agr. Biol. Chem.*, **31,** 1003 (1967).
(*74*) T. J. Schoch, *in* "Starch: Chemistry and Technology," R. L. Whistler and E. F. Paschall, eds., Academic Press, New York, 1967, Vol. II, p. 65.
(*75*) E. P. Palmiano and B. O. Juliano, *Plant Physiol.*, **49,** 751 (1972).
(*76*) R. M. Sandstedt and R. C. Abbott, *Cereal Sci. Today*, **9,** 13 (1964).
(*77*) B. O. Juliano and A. A. Perdon, *Staerke*, **27,** 115 (1975).
(*78*) H. Suzuki and B. O. Juliano, *Agr. Biol. Chem.*, **39,** 811 (1975).
(*79*) H. Suzuki and B. O. Juliano, *Tech. Bull. Fac. Agr. Kagawa Univ.*, **28,** 75 (1977).
(*80*) T. Sugawara, *Bull. Coll. Agr. Utsunomiya Univ.*, **7,** 97 (1968); *Chem. Abstr.*, **71,** 21037y (1969).
(*81*) International Rice Research Institute, "Annual Report for 1975," Los Baños, Laguna, Philippines, 1976, p. 85.
(*82*) K. Okuno, *Jap. J. Genet.*, **53,** 219 (1978).
(*83*) M. H. Heu and S. Z. Park, *Korean J. Breeding*, **8,** 48 (1976).
(*84*) M. H. Heu and S. Z. Park, *Seoul Natl. Univ. Coll. Agr. Bull.*, **1,** 39 (1976).
(*85*) K. Takeda, K. Nakajima, and K. Saito, *Japan J. Breeding*, **28,** 225 (1978).
(*86*) T. T. Chang and B. Somrith, *in* "Chemical Aspects of Rice Grain Quality," Intern. Rice Res. Inst., Los Baños, Laguna, Philippines, 1979, p. 49.
(*87*) G. S. Khush, C. M. Paule, and N. M. de la Cruz, *in* "Chemical Aspects of Rice Grain Quality," Intern. Rice Res. Inst., Los Baños, Laguna, Philippines, 1979, p. 21.
(*88*) L. C. Baun, E. P. Palmiano, C. M. Perez, and B. O. Juliano, *Plant Physiol.*, **46,** 429 (1970).
(*89*) A. A. Perdon, E. J. del Rosario, and B. O. Juliano, *Phytochem.*, **14,** 949 (1975).
(*90*) S. Tabata, K. Nagata, and S. Hizukuri, *Staerke*, **27,** 333 (1975).
(*91*) S. Tabata and S. Hizukuri, *Denpun Kagaku*, **22,** 27 (1975).
(*92*) A. Nakamura, T. Kono, and S. Funahashi, *Bull. Agr. Chem. Soc. Jap.* **22,** 320 (1958).
(*93*) L. Acker and H. J. Schmitz, *Staerke*, **19,** 275 (1967); *Chem. Abstr.*, **67,** 101255a (1967).
(*94*) H. Bolling and A. W. El Baya, *Chem. Microbiol. Technol. Lebensm.*, **3,** 161 (1975); *Chem. Abstr.*, **82,** 168923v (1975).
(*95*) S. Hizukuri, Y. Tanaka, and T. Matsubayashi, *Denpun Kagaku*, **26,** 112 (1979).
(*96*) C. T. Greenwood and J. Thomson, *J. Chem. Soc.*, **222** (1962).
(*97*) W. Banks and C. T. Greenwood, *Staerke*, **19,** 394 (1967).
(*98*) E. J. Wilson, Jr., T. J. Schoch, and C. S. Hudson, *J. Amer. Chem. Soc.*, **65,** 1380 (1943).
(*99*) M. Taki, *J. Agr. Chem. Soc. Jap.* **33,** 781 (1959); *Chem. Abstr.*, **58,** 2561a (1963).
(*100*) C. D. Boyer, D. L. Garwood, and J. C. Shannon, *Staerke*, **28,** 405 (1976).

(*101*) B. O. Juliano, *Intern. Rice Comm. Newsletter* (Special Issue), **93** (1967).
(*102*) B. O. Juliano, *in* "Rice Production and Utilization," B. S. Luh, ed., AVI Publishing, Westport, Connecticut, 1980, p. 403.
(*103*) W. Banks and C. T. Greenwood, *Carbohydr. Res.*, **6,** 177 (1968).
(*104*) H. Horiuchi and T. Tani, *Agr. Biol. Chem.*, **30,** 457 (1966).
(*105*) E. P. Palmiano and B. O. Juliano, *Agr. Biol. Chem.*, **36,** 157 (1972).
(*106*) A. A. Antonio and B. O. Juliano, *Philippine Agriculturist*, **58,** 17 (1974).
(*107*) B. S. Enevoldsen "Xth International Symposium on Carbohydrate Chemistry, Abstracts" Sydney, Australia, 1980.
(*108*) A. J. Vidal and B. O. Juliano, *Cereal Chem.*, **44,** 86 (1967).
(*109*) B. O. Juliano and A. A. Perdon, *Staerke*, **27,** 115 (1975).
(*110*) B. O. Juliano, L. U. Oñate, and A. M. del Mundo, *Philippine Agriculturist*, **56,** 44 (1972).
(*111*) B. O. Juliano, *in* "Rice Breeding," Intern. Rice Res. Inst., Los Baños, Laguna, Philippines, 1972, p. 389.
(*112*) C. M. Perez and B. O. Juliano, *Food Chem.*, **4,** 185 (1979).
(*113*) A. A. Antonio, B. O. Juliano, and A. M. del Mundo, *Philippine Agriculturist*, **58,** 351 (1975).
(*114*) C. M. Perez, C. G. Pascual, and B. O. Juliano, *Food Chem.*, **4,** 179 (1979).
(*115*) B. O. Juliano, A. A. Perdon, C. M. Perez, and G. B. Cagampang, *Proc., 4th Intern. Congr. Food Sci. Technol., Madrid, 1974,* **1,** 120 (1974).
(*116*) G. B. Cagampang, C. M. Perez, and B. O. Juliano, *J. Sci. Food Agr.*, **24,** 1589 (1973).
(*117*) J. E. Rutledge, M. N. Islam, and W. H. James, *Cereal Chem.*, **49,** 430 (1972).
(*118*) B. O. Juliano, *Denpun Kogyo Gakkaishi*, **29,** 305 (1982).
(*119*) S. Barber, *in* "Interregional Seminar for the Industrial Processing of Rice," Madras, India, U.N. Ind. Develop. Organ. Paper ID/WG.89/11, 1971 p. 52.
(*120*) D. B. Bechtel and B. O. Juliano, *Ann. Botany* (London), **45,** 503 (1980).
(*121*) C. C. Maniñgat and B. O. Juliano, *Starch/Staerke*, **32,** 76 (1980).
(*122*) H. Suzuki, *in* "Chemical Aspects of Rice Grain Quality," Intern. Rice Res. Inst., Los Baños, Laguna, Philippines, 1979, p. 261.

CHAPTER XVII

ACID-MODIFIED STARCH: PRODUCTION AND USES

By Robert G. Rohwer and Robert E. Klem

Grain Processing Corporation, Muscatine, Iowa

I. Introduction

Treatment of starch with acid, without substantially changing the granular form, results in modified starch with properties that have commercial value. The methods of production and the uses of acid-modified starch have not changed substantially since 1967, when this subject was reviewed by Shildneck and Smith (*1*). The usefulness of acid modification as a step in manufacture of other types of modified starch and the acid treatment of granular starches in slurries other than water will be briefly discussed.

This chapter excludes dextrins (Chap. XX) and acid treatments done above the gelatinization temperature of the parent starch.

STARCH, 2nd ed.

ISBN 0-12-746270-8

II. Properties

When starch in water slurry is treated with acid below the gelatinization temperature, the product can be expected to have the same granular appearance (*2*), similar birefringence (*3, 4*), and essentially the same insolubility in cold water (*2*) as the parent starch. Using the same comparison, an acid-treated starch can be expected to exhibit lower hot-paste viscosity (*5–13*), higher ratio of cold-to hot-paste viscosity (*5, 6, 11, 12, 14*), higher alkali number (*15, 16*), higher critical absorption of sodium hydroxide (*17*), lower intrinsic viscosity (*15, 16*), lower iodine affinity (*15, 16*), higher osmotic pressure (*18*), increased solubility in water at temperatures just below the gelatinization temperature (*7, 17, 19*), higher gelatinization temperature (*7, 17*), less granular swelling during gelatinization (*8, 20*), higher content of pentasol-precipitable fraction at the lower modification levels but not at high levels of modification (*21*), and a slower rate of solution in cold anhydrous dimethyl sulfoxide (*17*).

The above holds for yellow dent corn starch. Although experimental work is limited, waxy maize, wheat, rye, potato, sweet potato, tapioca, sago, and canna starches treated similarly can exhibit these same properties.

III. Theory (see also Chapter VII, Section V.8)

Acid modification of starch generally occurs by preferential attack of the acid upon the amorphous section of the granule (*22–24*), a conclusion based on electron microscopic work which indicates that electron-thin rings (amorphous section) of starch granules are completely disintegrated by acid modification, with only limited disintegration of electron-dense rings (crystalline section) (*24*, this Volume, Chap. VII). This means that the amylopectin fraction in corn starch is initially depolymerized to much greater extent than the amylose fraction (but see Chap. VII). The conclusions help explain some of the observed properties of acid-modified starch.

1. Lower Hot-Paste Viscosity

Work on dilute starch pastes in hot water has indicated that viscosity is primarily related to the suspension of swollen granules and granule fragments (discontinuous gel phase) in a water solution of starch substance (continuous liquid phase) (*6, 8, 19, 25–28*). Shildneck and Smith (*29*) proposed, based on greater hot-water solubility of acid-modified starch versus parent starch, that the lower hot-paste viscosity of acid-modified starch is due to a greater ratio of continuous to discontinuous phase.

Shildneck and Smith (*1*) question the effect of molecular size distribution upon viscosity at a constant ratio of continuous to discontinuous phase. Molecular size distribution of acid-modified hydroxyethylated starch has been measured in dimethyl sulfoxide–water solutions (*30*). Application of this technique to the different phases may resolve this question. The "discontinuous gel phase" accounts for significant viscosity and is probably composed of the high-molecular-weight components, which in turn are the major contributors to the hot-paste viscosity.

2. Ratio of Cold- to Hot-Paste Viscosity

The effect of differences in types of acid, concentration of acid, and reaction time upon modification of starch is outlined in Table I.

Gelling power and ratio of cold- to hot-paste viscosity, is increased by increasing the concentration of acid and by reducing the time of acid treatment (*11, 31*). This may be due to preferential destruction of the amorphous structure and/or relatively higher quantities of neutralization salts.

As shown in Table II (*11*), the ratio of gel strength to hot-paste viscosity is greater for acid-modified starch than for the parent starch.

Acid-catalyzed hydrolysis of starch, as opposed to enzymic (α-amylase) hydrolysis, can take place at branch points as well as in linear segments. Thus, the

Table I

Influence of Process Conditions on Fluidity of Starches Acid-Modified at 50°

Starch and acid	*Acid conc., % by wt.*	*Time, h*	*Fluidity*
Corn starch, sulfuric acid (*8*)	0.06	24	13.0
	0.13	24	32.0
	0.22	24	53.0
	0.29	24	64.0
	0.44	24	72.0
	0.61	24	74.0
Corn starch, hydrochloric acid (*6*)	2.05	0.25	10.0
		0.47	20.0
		0.67	30.0
		0.87	40.0
		1.13	50.0
		1.50	60.0
		2.25	70.0
Potato starch, hydrochloric acid (*7*)	2.05	0.67	3.0
		1.33	8.0
		2.0	15.5
		2.67	25.0
		3.33	37.0
		4.0	52.8

Table II

Hot-Paste Viscosity and Gel Properties of Corn Starch and Acid-Modified Corn Starches (11)

Starch fluidity	*Hot-paste viscosity, poises*	*Gel rigidity, dynes/cm²*	*Gel breaking strength, g/cm²*	*GP[a]/HPV[b]*	*GBS[c]/HPV[b]*
Unmodified	34.0	1850	194	54.4	5.71
10	15.1	1140	118	75.5	7.81
20	8.5	810	71.5	95	8.4
30	6.0	738	61.1	120	10
40	3.6	510	40.3	140	11.5
50	3.0	422	32.6	140	11
60	1.1	318	23.6	290	21
70	0.2	156	13.4	300	70

[a] Gel rigidity.
[b] Hot-paste viscosity.
[c] Gel breaking strength.

relatively greater gelling power or retrogradation of acid-modified starch can be attributed to reducing the degree of branching and increasing the percentage of linear segments.

3. Alkali Number, Intrinsic Viscosity, and Reducing Values

Alkali number (*32*) of acid-modified starch increases and intrinsic viscosity decreases with degree of acid treatment (*15, 16, 32*). Alkali number is the meq of alkali consumed per 10 g of calculated dry starch during a standard 1-h digestion in 0.1 *M* sodium hydroxide solution at the temperature of a boiling water bath. The alkali is consumed by acids produced from the reducing ends of the polymer chains (*33*) and is believed to be a measure of chain lengths (direct index of hydrolysis) rather than a quantitative determination of aldehyde content.

Alkali number is related to intrinsic viscosity and fluidities in (Table III) (*1*).

Below 75 fluidity, the reducing value, as measured by ferricyanide number, the meq of ferricyanide reduced per 10 g of starch (*34*), is substantially the same as that of its parent starch (*35*). Above 75 fluidity, a measurably large reducing value results (Table IV).

4. Osmotic Pressure, Iodine Affinity, and Film Strength

Kerr (*21*) made an extensive investigation of number-average degree of polymerization ($\overline{DP}_n$) by osmotic pressure for acid-modified starch. Kerr related ($\overline{DP}_n$) to several properties of amylose and amylopectin fractions.

Iodine affinity (*36*) of acid-modified corn starch shows little change up to 70 fluidity. Preferential attack of acid upon amylopectin fraction can be speculated. However, as shown in Table IV, the amylose fraction of acid-modified starches exhibits little change in iodine affinity as the fluidity of the parent starch is changed from 40 to 90. Over this same range, the $\overline{DP}_n$ changed from 470 to 190. Thus, it would seem to the authors that iodine affinity alone cannot be used to conclude that acid preferentially attacks the amorphous section of the granule first.

Film strength (tensile and elongation) data for acid-modified starch are presented in Table V.

Films were cast from hot pastes of various fluidities. Tensile strength was not significantly affected by acid modification of the starch. The authors hypothesize that hydrogen bonding compensates for the shorter chain lengths at higher fluidities (*37*).

Table III

Intrinsic Viscosities, Alkali Numbers, and Fluidities of Acid-Modified Corn Starches

Acid and temperature	*% by weight*	*Time h*	*Starch fluidity*	*Alkali number*	*Intrinsic viscosity*
Hydrochloric acid 50° (*16*)	0.274	5		13	1.06
		16		18	0.66
		26		21.6	0.47
		40		27.2	0.37
	1.095	2		13.9	0.89
		7		22.6	0.44
		16		34.8	0.24
Acid not reported (*15*)	—	—	20	11.5	1.05
			40	13.0	0.90
			60	16.5	0.60
			75	21.0	0.50
			90	26.5	0.35
Acid not reported (*32*)	—	—	20	14.5	
			40	15.0	
			60	15.7	
			75	20.7	
			90	41.5	
Hydrochloric acid 51.7° (unpublished)	0.72	8	10		
	0.85	8	20		
	0.97	8	30		
	1.04	8	40		
	1.23	8	50		
	1.86	8	60		
	1.93	8	70		
	2.03	8	80		

Table IV

Properties of Fractions from Acid-Modified Corn Starches (21)

Parent starch fluidity	Amylose fraction					Amylopectin fraction			
	$\overline{DP}_n$	Ferricyanide number	Alkali number	Iodine affinity	Yield, wt. % of parent starch	$\overline{DP}_n$	Ferricyanide number	Alkali number	Intrinsic viscosity
Unmodified	480	1.43	19.7	19.2	21.0	1450	0.46	4.8	1.25
10	—	—	—	11.9	34.9	920	0.59	7.05	1.07
20	525	1.59	20.4	16.6	37.0	625	0.85	9.7	0.70
40	470	1.80	22.8	17.1	28.8	565	0.91	10.8	0.65
60	425	2.01	27.9	18.0	25.2	525	1.00	11.1	0.58
80	245	3.72	43.0	18.1	23.1	260	3.31	25.9	0.26
90	190	6.90	—	16.3	12.0	210	4.27	27.6	0.29

Table V

Effects of Acid-Modification on Corn Starch Film Properties (37)

Starch fluidity	*Intrinsic viscosity dL/g*	*Film tensile strength kg/mm²*	*Film elongation %*
Unmodified	1.73	4.67	3.2
15	1.21	4.47	2.7
34	1.06	4.45	2.6
50	0.88	4.94	2.7
71	0.67	4.57	2.9
89	0.32	4.58	2.2

5. Additional Characteristics

Properties listed previously can be explained to some extent, especially for acid-modified corn starch, by preferential acidic degradation of the amorphous region of the granule leaving the crystalline fraction more intact as is suggested by examinations with the electron microscope (*24*). For example, the relatively intact amylose fraction is more difficult to hydrate and swell, thus the higher gelatinization temperature. The pentasol (amyl alcohols) precipitate (*21*) is generally amylose (Chapters II and VIII). Amylose, due to its crystalline nature, is more difficult to dissolve in dimethyl sulfoxide (*17*).

IV. History

Naegeli (*38*), in 1874, reported the treatment of starch with acid until the granules fragmented. Lintner (*39*) was the first to prepare an acid-treated starch that remained in granular form. He treated potato starch slurries with either aqueous 7.5% hydrochloric or 15% sulfuric acid for several days. He then filtered and washed the starch to obtain an acid-modified starch that would form a clear dilute solution in hot water, a solution that was stable for many days. Bellmas (*40*), in 1897, found that the same starch could be made using a lower concentration of acid at higher temperatures.

Duryea (*41*) patented processes for making acid-modified starches commercially in 1901–1902. Since then a number of workers have reported the making of acid-modified starches with either hydrochloric or sulfuric acid at various concentrations, at various temperatures, and for various times (*6–9, 16, 31, 42–46*).

V. Industrial Production

1. General

The commercial manufacture of acid-modified starches is usually done by treating approximately 40% starch in a slurry of dilute hydrochloric or sulfuric acid at 25°–55° for the length of time required to obtain the thinned starch that is sought. Corn and waxy maize starch are the principal commercial parent starches. Small amounts of tapioca, wheat, and potato starch are also manufactured.

The conditions for treatment are varied, depending upon the ratio of cold- to hot-paste viscosity and gel characteristics that are sought (*16, 31*). If two samples of the same starch are treated with different amounts of acid to the same fluidity, the starch treated with the higher concentration of acid will exhibit the greater gel strength (*47*).

The choice of acid can also strongly affect the gel and viscosity characteristics. Ferrara (*37*) combined hydrofluoric and hydrochloric acid and found that the starch so treated gelled much slower than starch treated with hydrochloric acid alone.

2. Process Control

Although Duryea (1902) manufactured acid-modified starches without analytical control (*41*), this is not done today. To modify a starch consistently, it is necessary to use the same starch concentration, the same acid, the same acid concentration, and the same temperature conditions. However, even with careful control, the reaction time to produce the same modified starch varies. Fluidity is the primary control to obtain the desired starch and a number of methods can be used successfully (*48, 49*). As the starch undergoes reaction, its fluidity is regularly checked and plotted against time. At the time predicted, by extrapolating the plot, the reaction is stopped by neutralizing to approximately pH 6 with soda ash (sodium carbonate) or a dilute caustic (sodium hydroxide) solution. The starch may then be filtered and dried.

3. Kiln or Dryer Process

Granular, acid-modified starches can be made by treating the starch at room temperature with acid, filtering the slurry, and then drying the acid-containing starch under controlled conditions (*35*). By this process, the solubles are not lost; this can be a major economic factor when making 75, and higher, fluidity starches. The process, however, is not easily controlled. Ferrara (*50*) absorbed acid onto a dry inert finely divided carrier and then mixed the acidified carrier into the dry starch to produce acid-thinned starches.

4. Effect of Nonaqueous Solvents

Smithies (*51*) prepared acid-modified starch gels for electrophoresis by treating potato starch slurried in acetone containing hydrochloric acid. The granular product, after neutralization, is thoroughly washed with water, acetone, and dried. The final product contains no cold-water-extractable matter. The gels are reasonably flexible and exhibit a clean break.

Schoch (*52*) demonstrated that solvent washing a starch either before or after acid modification markedly shortens the time required for the paste to gel. This has major value in the confectionary industry.

It seems apparent that nonaqueous solvents remove substances that inhibit gelation. These substances are probably lipids that inhibit some of the hydrogen bonding required for gelation.

If the cost of recovering the solvents used can be overcome, this process could become important for obtaining new useful starches.

VI. Acid Modification Combined with Other Starch Modifications

Acid modification plays an important roll in the manufacture of other types of modified starches. Acid modification may be used as a premodification step in some cases or a postmodification step in others. In both cases, the acid treatment is used to make available the modified starch with a range of fluidities. In other reactions, the medium is acidic owing to the reagent or conditions used, and the starch is simultaneously acid-thinned while undergoing other reactions.

A prime example of the usefulness of pre- or post-treatment is in the manufacture of starch ethers (*53, 54*). The products without thinning are inherently low-fluidity starches. Their utility is expanded by acid modification to obtain starch ethers having useful higher fluidities. It has been shown in unpublished work that high-fluidity starch ethers that appear to be the same can be manufactured either by acid-modifying the parent corn starch before etherification or after the parent starch has been etherified. Acid-modified starches can be used as the parent starch for the production of cationic and cross-linked starch. When starch is reacted with acid anhydrides, acid degradation or modification take place simultaneously unless special precautions are taken (*55*).

VII. Industrial Uses

1. Textile Manufacture

Usage of acid-modified starch, still the predominant type in the textile industry, has declined significantly over the last ten years. This has been due, in part, to inroads of poly(vinyl alcohol) (PVA). The use of synthetic fabrics has also contributed to this decline. Less PVA than starch can be used on yarn and less shedding is observed. Combinations of PVA and starch are being used.

Acid-modified starch functions as a warp sizing agent to increase yarn strength and abrasion resistance in the weaving operation. The starch is also used in fabric finishing, mostly on all cotton fabrics, to increase stiffness of the finished product.

Published reports as early as 1917 (*56*) and 1921 (*57*) indicated that acid-modified starch was useful in warp sizing. In addition, acid-modified starch can be produced economically and can be rather easily tailored to different fluidities and, consequently, end-use viscosities and solids. Oxidized starches may not perform as well as acid-modified starches, although oxidized starches make softer films and have less tendency to retrograde. Acid-modified hydroxyethylated starches form excellent films and, although more expensive, have been used.

Different fluidities allow selection of the optimum starch for the type of yarn encountered. Coarse materials such as two-ply yarn fabrics require little sizing, if any; but when starch is used, it is usually of 20–30 fluidity. Lighter yarns need a warp sizing, and a 40–60 fluidity product is utilized. Higher fluidities (e.g., 80 fluidity) are occasionally used, for example, in fabric finishing.

Starch concentration in the sizing solution is generally 10–12%, with softeners, lubricants, and preservatives as other essential ingredients. Pickup can be changed by varying the viscosity and the concentration. Higher pickup from the same viscosity sizing solution can be obtained by using a higher fluidity starch at greater concentration.

Most starch paste preparation systems are atmospheric cookers. Jet cookers or super atmospheric cookers are utilized for less modified products (*58*).

2. Building Products Manufacture

Acid-modified starch is used in the manufacture of gypsum board for dry wall construction. In this application, a thick plaster paste containing acid-modified starch is spread between two sheets of paper. Presumably, the starch (80–90 fluidity) migrates to the paper interface and helps form a bond between the plaster and the paper. Acid-modified corn flour may control the rate of hydrated calcium sulfate crystal formation (*59*). The hydrated calcium sulfate crystallizes into needles forming a bond between the paper and inner layer.

3. Starch Gum Candy Manufacture

Historically, in starch gum candy production, acid-modified starch is boiled with sugar, corn syrup, and water. The water is then evaporated to form a jelly, which upon cooling and aging in the molds, produces a gum confection of desired texture. Unmodified starch requires lower starch solids and generally resuslts in unsatisfactory confections of short texture and weak body. A food acid is generally required with the unmodified starch to depolymerize the starch partially during the candy "boil."

Non-waxy cereal starches with acid modification are advantageous owing to their ability to produce highly concentrated fluid pastes which form firm gels upon cooling and aging. Most manufacturers now use pressurized cooking to prepare starch gum candy.

Cooking temperature and retention time are adjusted to produce slightly less than complete granule disintegration. Otherwise, over cooking is likely to result, forming gels subject to syneresis. A typical composition of starch gum candy is 100 lb of granulated sugar, 150 lb of 63 DE corn syrup (44 Baumé), 40 lb of 70-fluidity acid-modified starch, 7 gal of water, and appropriate color and flavoring. The corn syrup, water, and all the starch is placed into a hemisphere double agitator kettle; the mixer is started, and the steam is turned on. The starch is stirred until all is dispersed, but the mixture is not allowed to boil. When the starch is dispersed, the sugar is added. The mass is brought to a rolling boil (approximately 103°, 218°F), and the steam is turned off to prevent moisture loss. Moisture is adjusted to 22–23%, and the batch is run through a suitable injection cooker at approximately 140° (285°F). Flavor, color, and mix are then added. The gel is deposited in starch molds, which are then placed in curing rooms for drying and setting. The molded gum candy is screened from the starch and coated with sugar by passage through a steamer and sugar sander.

4. Paper and Paperboard Manufacture (see also Chapter XVIII)

Use of acid-modified starch in all paper applications is approaching the amount used in the textile industry. Acid-modified starch is used for some fine paper size press applications. Maintenance of 66°–71° (150°–160°F) application temperature and pH 7.5–8.0 is essential for satisfactory performance. It has been used to increase strength of Kraft linerboard and to improve printability of bleached board. Calender stack applications generally improve surface strength and ink holdout for printing grades.

Acid-modified waxy starches have found usefulness in corrugated board manufacture (see also Chap. XVIII). A relatively high solids carrier starch can be produced to increase production rates (*36*).

VIII. References

(*1*) P. Shildneck and C. E. Smith, *in* "Starch: Chemistry and Technology," R. L. Whistler and E. F. Paschall, eds., Academic Press, New York, 1967, Vol. II, pp. 217–235.

(*2*) R. P. Walton, "Comprehensive Survey of Starch Chemistry," Chemical Catalog Co., New York, Vol. 1, 1928, p. 170.

(*3*) H. T. Brown and G. H. Morris, *J. Chem. Soc.*, **55,** 449 (1889).

(*4*) D. French, *in* "Chemistry and Industry of Starch," R. W. Kerr, ed., Academic Press, New York, 2nd Ed., 1950, p. 158.
(*5*) G. V. Caesar and E. E. Moore, *Ind. Eng. Chem.*, **27,** 1447 (1935).
(*6*) W. Gallay and A. C. Bell, *Can. J. Res., Sect. B,* **14,** 360 (1936).
(*7*) W. Gallay and A. C. Bell, *Can. J. Res., Sect. B,* **14,** 381 (1936).
(*8*) J. R. Katz, *Textile Res. J,* **9,** 146 (1939).
(*9*) H. H. Schopmeyer and G. E. Felton, U.S. Patent 2,319,637 (1943); *Chem. Abstr.*, **37,** 6488 (1943).
(*10*) W. G. Bechtel, *Cereal Chem.*, **24,** 200 (1947).
(*11*) W. G. Bechtel, *J. Colloid Sci.*, **5,** 260 (1950).
(*12*) E. G. Mazurs, T. J. Schoch, and F. E. Kite, *Cereal Chem.*, **34,** 141 (1957).
(*13*) Deutsche Forschungsgemeinschaft, "Report of the Commission for Examination of the Bleaching of Foods," No. 3, Jan. 18, 1961.
(*14*) E. G. Mayers, T. J. Schoch, and F. E. Kite, *Cereal Chem.*, **34,** 141 (1937).
(*15*) R. W. Kerr, *in* "Chemistry and Industry of Starch," R. W. Kerr, ed., Academic Press, New York, 2nd Ed., 1950, p. 682.
(*16*) S. Lansky, M. Kooi, and T. J. Schoch, *J. Amer. Chem. Soc.*, **71,** 4066 (1949).
(*17*) H. W. Leach and T. J. Schoch, *Cereal Chem.*, **39,** 318 (1962).
(*18*) M. Samec and S. Jencic, *Kolloidchem. Beih.*, **7,** 137 (1915).
(*19*) H. W. Leach, L. D. McCowen, and T. J. Schoch, *Cereal Chem.*, **36,** 534 (1959).
(*20*) O. A. Sjostrom, *Ind. Eng. Chem.*, **28,** 63 (1936).
(*21*) R. W. Kerr, *Staerke,* **4,** 39 (1952).
(*22*) R. W. Kerr, *in* "Chemistry and Industry of Starch," R. W. Kerr, ed., Academic Press, New York, 2nd Ed., 1950, p. 158.
(*23*) K. H. Meyer and P. Bernfeld, *Helv. Chim. Acta,* **23,** 890 (1940).
(*24*) W. C. Mussulman and J. A. Wagoner, *Cereal Chem.*, **45,** 162 (1968).
(*25*) W. Harrison, *J. Soc. Dyers Colour.*, **27,** 84 (1911).
(*26*) C. L. Alsberg, *Ind. Eng. Chem.*, **18,** 190 (1926).
(*27*) J. R. Katz, M. C. Desai, and J. Seiberlich, *Trans. Faraday Soc.*, **34,** 1258 (1938).
(*28*) K. H. Meyer and M. Fuld, *Helv. Chim. Acta,* **25,** 391 (1942).
(*29*) P. Shildneck and C. E. Smith, *in* "Starch: Chemistry and Technology," R. L. Whistler and E. F. Paschall, eds., Academic Press, New York, 1967, Vol. II, pp. 223–224.
(*30*) R. Stone and J. Krasowski, *Anal. Chem.*, **53,** 736 (1981).
(*31*) H. Meisel, U.S. Patent 2,231,476 (1941).
(*32*) W. Gallay, *Can. J. Res., B,* **14,** 391 (1936).
(*33*) R. J. Smith, *in* "Starch: Chemistry and Technology," R. L. Whistler and E. F. Paschall, eds., Academic Press, New York, 1967, Vol. II pp. 587–588.
(*34*) R. J. Smith, *in* "Starch: Chemistry and Technology," R. L. Whistler and E. F. Paschall, eds., Academic Press, New York, Vol. 2, 1967, p. 589.
(*35*) R. W. Kerr, *in* "Chemistry and Industry of Starch," R. W. Kerr, ed., Academic Press, New York, 2nd Ed. 1950, p. 77.
(*36*) C. B. Musselman and E. M. Bovier, U.S. Patent 4,014,727 (1977).
(*37*) P. J. Ferrara, U.S. Patent 3,692,581 (1972).
(*38*) W. Nägeli, *Ann. Chem.*, **173,** 218 (1874).
(*39*) C. J. Lintner, *J. Prakt. Chem.*, **34,** 378 (1886).
(*40*) B. Bellmas, German Patent 110,957 (1897).
(*41*) C. B. Duryea, U.S. Patents 675,822 (1901) and 696,949 (1902).
(*42*) R. W. Kerr, *in* "Chemistry and Industry of Starch," R. W. Kerr, ed., Academic Press, New York, 2nd Ed., 1950, p. 76.
(*43*) M. Witlich, *Kunstöffe,* **2,** 61 (1912).

(*44*) A. Oelker, *Kunststöffe,* **6,** 189 (1916).
(*45*) E. Parow, *Z. Spiritusind.,* **45,** 169 (1922).
(*46*) M. Samec, "Kolloidchemie der Staerke," Theodor Steinkopff, Dresden and Leipzig, 1927, pp. 221–226.
(*47*) J. A. Radley, *in* "Starch Production Technology," J. A. Radley, ed., Applied Science Publishers, Barking, Essex, England, 1976, p. 453.
(*48*) H. Bail, *Intern. Cong. Appl. Chem., 8th, Orig. Comm.,* **13,** 63 (1912).
(*49*) W. R. Fetzer and L. C. Kirst, *Cereal Chem.,* **36,** 108 (1959).
(*50*) P. J. Ferrara, U.S. Patent 3,479,220 (1969).
(*51*) O. Smithies, *Biochem. J.,* **61,** 629 (1955); **71,** 585 (1959).
(*52*) T. J. Schoch, D. F. Stella, and H. J. Wolfmeyer, U.S. Patent 3,446,628 (1969).
(*53*) E. T. Hjermstad, *in* "Starch: Chemistry and Technology," R. L. Whistler and E. F. Paschall, eds., Academic Press, New York, 1967, Vol. II, p. 425.
(*54*) J. A. Radley, *in* "Starch Production Technology," J. A. Radley, ed., Applied Science Publishers, Ltd., Barkwig, Essex, England, 1976, p. 498.
(*55*) H. J. Roberts, *in* "Starch: Chemistry and Technology," R. L. Whistler and E. F. Paschall, eds., Academic Press, New York, 1967, Vol. II, pp. 298–299.
(*56*) Anon., *Posselt's Textile J.,* **21,** 5 (1917).
(*57*) W. R. Cathcart, *Textile World,* **59,** 2896 (1921).
(*58*) R. B. Pressley, *Textile World,* **112,** No. 8, 33 (1962).
(*59*) G. H. Wells, *Cereal Foods World,* **24,** 333 (1979).

CHAPTER XVIII

STARCH IN THE PAPER INDUSTRY

By Merle J. Mentzer

CPC-Latin America, Division of CPC International Inc., Buenos Aires, Argentina

I. Introduction to the Paper Industry

Per capita consumption of paper and paperboard products in 1978 was more than 640 pounds and is steadily increasing. During 1978, U.S. production of paper and paperboard reached 62,156,000 tons, nearly double the production of 20 years earlier (Fig. 1) (*1*). The top 10 companies in the paper industry account for 26% of the total tonnage of paper produced, whereas the top 10 manufactur-

STARCH, 2nd ed.

ISBN 0-12-746270-8

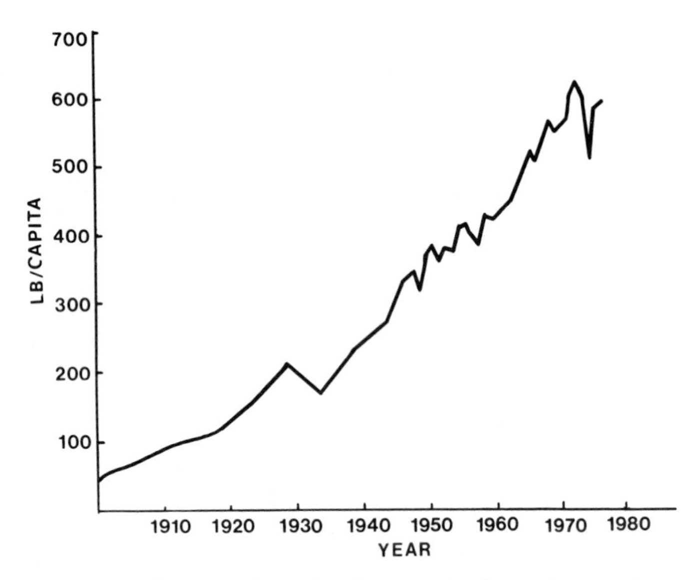

FIG. 1.—U.S. paper and paperboard consumption in pounds per capita.

ing companies in the United States account for 40% of the total goods produced (*2*).

While the basic component of paper products is cellulose fiber, large quantities of other materials are also used to process and modify the finished product so it can meet a wide variety of end-uses. In 1977, 7,050,000 tons of chemicals were used in the preparation and bleaching of the 49,080,000 tons of wood pulp, while an additional 5,030,000 tons of pigments and chemicals were used to produce 61,871,000 tons of finished paper products. A committee from The Technical Association of the Pulp and Paper Industry (TAPPI) reported (*3*) that the 1976 estimate of natural binders used by the U.S. paper industry exceeded

Table I

Natural Binder Consumption for Paper and Paperboard (1976)

Type	*MM pounds*	*Metric tons*
Corn starch	1730	786
Tapioca starch	105	48
Potato starch	81	37
Wheat starch	72	33
Soy protein	45	20
Casein	17	8

two billion pounds, with starch accounting for 97% of the usage (Table I). 87% of the starch was corn starch, of which two-thirds of the starch used was unmodified, while the remainder was modified by oxidation, hydroxyethylation, acid-catalyzed hydrolysis, or by other derivatizations.

II. Paper and Paperboard Manufacture

1. Pulping

Wood, the basic material used in papermaking, consists of approximately 50% cellulose fibers cemented together by about 25% lignin and 20% hemicellulose, along with small amounts of protein, resin, fat, and ash. Paper may be made directly from raw wood fiber or from wood fiber that has been chemically refined to remove the lignin and other soluble components.

When made from raw wood fibers, whole wood in either roundwood or chip form is mechanically ground into small fibrous bundles and groups of fibers. High fiber yields are obtained, and the paper produced is bulky and opaque. Strength is usually low, due to the low ratio of fiber length to diameter. Processes for producing these wood pulps are known as groundwood or mechanical pulping.

In another method for producing pulp, various chemicals are used to soften and remove the lignins and hemicelluloses that bind the fibers together, permitting the fibers to be freed. Pulp is produced, in which the ratio of fiber length to diameter is substantially higher than in groundwood pulp, thereby resulting in stronger paper. However, yields of less than 50% of the starting material are common. The main chemical pulping processes in use today are Kraft (sulfate) and sulfite.

By combining heat, chemical, and mechanical treatment, various other hybrid pulping methods have been developed to maximize yield without severe loss in strength. These pulping methods are known as semi-mechanical, semi-chemical, thermo-mechanical, and refiner groundwood pulping. A more detailed discussion of current pulping processes is found in TAPPI Monographs Nos. 6 and 21 (*4, 5*).

2. Papermaking

Papermaking is a process of first forming a water suspension of individual pulp fibers into an interlocking mat, and then removing the water from the mat by draining through a wire mesh, squeezing the mat between rollers, and drying it on heated cylinders.

Most paper is made on a Fourdrinier machine (Fig. 2) where a dilute pulp suspension, less than 1% fiber, is delivered through a narrow slot in the headbox

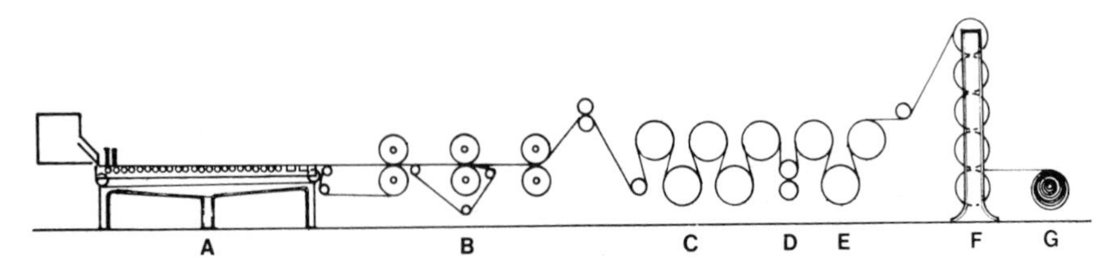

Fig. 2.—Fourdrinier machine: A, Fourdrinier or wire section; B, press section; C, first drier section; D, size press; E, second drier section; F, calender stack; G, winder.

onto a moving endless wire screen. The velocity of the pulp suspension is matched to the velocity of the moving screen, and a large amount of the water in the suspension drains through the screen, assisted at certain places by rolls, foils, and suction boxes fitted beneath the screen. At the end of the screen, the wet sheet has 85% moisture and is sufficiently strong for self-support. Next, the sheet is passed through the press section, where additional water is squeezed out and the fibers are brought into intimate contact for optimum bonding. Upon leaving the press section, moisture in the wet web is 70%. Remaining water is removed by drying the sheet against steam-heated drying cylinders until the sheet is more than 90% dry substance.

A surface treatment may be applied to the sheet at the size press. Starch is generally used to produce specific surface properties required in the finished sheet. If a size press is utilized, the moisture added must be removed by drying. After the sheet has been redried, it is usually passed through a vertical series of hard rolls, called a calender stack, to smooth and polish the surface. A light surface treatment, such as a starch size, may also be applied at the calender stack to control curl or to obtain additional surface properties. The web is finally wound onto reels for further processing if desired.

While the Fourdrinier is the most common paper machine, there are a number of other wet formers in operation. Most of these differ from the Fourdrinier only in the wet-forming section. One of these is a cylinder machine, where dilute pulp suspensions are placed in a series of vats. Each vat contains a submerged rotating wire mesh-covered cylinder that picks up thin layers of fiber. Fiber layers are then transferred continuously, one on top of another, onto a moving felt. As many as eleven cylinders may be used to make a heavy paperboard. Different pulps may be utilized in the various vats to produce paperboard that contains high-quality white paper on the outside and lower-grade paper in the center.

Various twin-wire machines, vertical-wire machines, slant-wire machines, combination cylinder/fourdrinier, and other machines are available under such trade names as Vertiformer, Ultraformer, Inverformer, Twinverformer, Stevensformer, and Rotoformer. A short history of their development has been presented by Attwood (*6*).

III. Wet-End Application of Starch

1. Fiber Refining

Use of unrefined cellulose fiber for papermaking results in a relatively low strength, poorly formed sheet. Therefore, the fiber is mechanically refined to promote fiber-to-fiber bonding. A fiber slurry is passed through disc- or plug-type refiners or circulated in a beater where the individual fibers are subjected to high mechanical stress. Primary cell walls of the fibers are ruptured to expose inner layers which are then frayed into minute fibrils. This increases the surface area available for fiber contact and promotes increased fiber-to-fiber bonding.

Strength development in the paper was originally attributed to physical interlocking of fibrils. It is now recognized that strength results from covalent bonding, polar bond attractions, or hydrogen bonding (*7*). Investigative work by Jayme and Hunger (*8*) revealed, through electron photomicrographs, that separate portions of the wood fiber cell become hydrogen-bonded during sheet formation. Swanson (*9*) supports the hydrogen bonding theory, while acknowledging the need for more investigative work. Cushing (*10*) points out that the cellulose molecule has a multitude of hydroxyl groups potentially capable of interacting through hydrogen bonding. These hydroxyl groups are also strongly polar, but since attractive forces decrease with the inverse seventh power of the distance, close association is necessary for maximum bonding. A practical limit exists in the amount of refining that can be used for strength development. Excessive refining causes the sheet to lose desirable characteristics, such as porosity, flexibility, brightness, and opacity. Starch may, therefore, be added at the wet-end to obtain the desired paper characteristics without the need for excessive refining.

2. Wet-End Starch Addition

Starches used as wet-end additives today are primarily corn and potato starches, and to a lesser extent wheat and tapioca starches. These may be unmodified or specially modified. They may be added in uncooked form, cooked directly before use, or pregelatinized and dried. They may be added as far back in the system as the beaters or refiners or directly onto the moving Fourdrinier wire.

Many factors determine which particular starch is most effective in a given mill. These factors include the grades of paper which are made, refining conditions and capacity, fiber types, individual mill preference, and economics. An ideal wet-end adhesive is one that gives maximum product strength and appearance with maximum starch retention without deleterious side effects, such as pigment fallout. No starch in use today meets all these requirements exactly, but many closely fit the needs of individual mills.

Pregelatinized starches are used in mills where starch cooking facilities are unavailable. Pregelatinized starches are added late in the beating–refining cycle to prevent breakdown under shear, but are added early enough to achieve good dispersion. Addition levels range from 20 to 40 lb per ton of fiber.

For heavyweight papers, mills with no cooking facilities may add uncooked starch early in the beating–refining cycle. The starch must be gelatinized in the wet sheet on the paper machine dryers because ungelatinized starch does not contribute to sheet strength. Sheet temperature must reach or exceed the starch gelatinization temperature (above 74° for unmodified corn starch) while there is still sufficient water available for gelatinization. This requirement tends to limit the use of uncooked starch to heavier weight papers where high temperature and moist conditions prevail for longer time periods.

Somewhat improved properties are attained by using high-viscosity starches that have been chemically modified to lower the gelatinization temperature. Hydroxyethylated corn starch (Chap. X), with a gelatinization temperature range of 55–60°, is preferred over oxidized starch because of stronger bonding strength. Also, oxidized starch has the disadvantage of dispersing pigments to prevent their retention in the paper. Addition levels, as in the case of pregelatinized starch, are in the range of 20 to 40 lb per ton of fiber.

The large size of the unmodified starch molecule contributes significantly to paper strength. Houtz (*11*) has shown that thoroughly cooked starches with decreasing molecular weights produce corresponding strength decreases. Mill experience has shown that, except for a very few special cases, degraded starches are not as useful for wet-end addition as are unmodified starches.

Unmodified starch for wet-end addition is typically batch-cooked at 95° to achieve good molecular dispersion. Alternatively, continuous starch cookers may be used to minimize the storage of cooked pastes. Most fine paper mills add cooked starch at the fan pump, delivering furnish to the headbox, to avoid high turbulence and shear, as is found in the beater–refiner sections.

Starch pastes containing swollen but unruptured starch granules are retained by the paper more effectively than is completely dispersed starch. These pastes are particularly effective for coarse papers or for papers containing unrefined fiber. They can be prepared by careful control of the cooking cycle or by the use of starch modified for such an application. Large swollen granules are mechanically entrapped by the fibers while the sheet is being formed. Such granules act as spot welds to bridge the gap between fibers, producing increased strength in the finished sheet. Large granule root starches, such as potato and tapioca, are especially useful for improving strength of low-density sheets. Corn starch is effective if modified to produce maximum granule swelling prior to addition to the furnish.

It is difficult to designate the exact quantity of unmodified starch needed. Depending upon mill conditions and the type of paper desired, effective quantities may range from 0.5% to 7.0%, based on the dry weight of fiber. Too much fully cooked starch impairs drainage on the wire and may adversely affect the properties of the finished sheet. According to Cobb and co-workers (*12*) maximum strength is attained at a level of 0.5% of starch, based on fiber. Initial rapid sorption of starch (*13*) on fiber is about 0.35–0.40%. However, in practice the addition rate is 2–5%.

Significant differences sometime occur between the quantity of starch added to the sheet and the amount retained. Simulation of starch retention in the laboratory is difficult because it is not easy to duplicate continuous refining, wire suction, or white water recirculation conditions found in the mill. Waters (*14*) evaluated starch as a wet-end adhesive by measuring starch concentrations in the headbox, first white water tray, and a finished 20-lb bond sheet. His data showed retention figures of 26.7% for potato starch and 9.45% for corn starch under comparable conditions.

Such low retention values may be attributed to coulombic repulsion resulting from the anionic nature of both starch and cellulose. To improve starch retention, retention aids are often used. Colloidal alumina floc can improve starch retention (*15*). The complex formed by alumina floc, starch, and fiber may be disrupted by mechanical shear. For best effect of alumina, the cooked starch must be added near the end of the beating cycle.

Since early 1960, cationic starches have been preferred wet-end starch additives. They are not only attracted to the negatively charged cellulose fiber, but also to negatively charged fillers where they increase fiber-to-fiber and fiber-to-filler bonding. Thus, they promote a high degree of filler retention as well as strength increases at low application levels compared with native corn starch. Work by Nissen (*16*) on unbleached sulfite pulp has shown comparable strength increases in finished sheets with either 1% cationic starch or 3% corn starch. Mill experience has shown cationic starches particularly effective in low-density, long-fibered sheets.

Tanaka and co-workers (*17*) examined the interaction of cationic polymers, such as aminoethylated and diethylaminoethylated starch, with bleached kraft pulp and concluded that the primary factor causing adsorption was charge interaction. Moeller (*18*) compared the retention of cationic starch to that of unmodified corn starch on a wide variety of pulps. He found consistently high retention and strength values when small amounts of cationic starch (less than 2%) were added. Similar strengths could be obtained with fully cooked unmodified starch only at addition levels of 6% or more. Hammerstrand and co-workers (*19*) described low-cost, dry-processed cationic starch, and Carr and co-workers (*20*) examined starch polyampholytes that improved both dry and wet strength of unbleached kraft paper. Jarowenko and Hernandez (*21*) described the use of cationic starch containing sulfosuccinate groups to improve pigment retention and to increase paper strength. Techniques developed to react commercial cationic reagents and/or polymers with starch by special cooking procedures produce results similar to those obtained with commercial cationic starch, but at lower cost (*22*).

Interest has periodically been shown in the use of dialdehyde starches as wet-end additives for wet strength improvement (*23*). Wet strength is generally obtained by the addition of urea–formaldehyde or melamine–formaldehyde resins. Urea–formaldehyde-treated sheets require a substantial curing time, and melamine–formaldehyde-treated paper is extremely difficult to reprocess.

3. Alternative Points for Starch Addition

Although starch is usually added at the wet end of the paper machine as a liquid feed directly to the furnish, other systems which place the starch directly on the formed sheet while it is still on the wire of the Fourdrinier machine or on the felt of the cylinder machine may be used. Advantages claimed are improved retention and better distribution of starch throughout the sheet, while permitting the use of low-cost unmodified starch.

In one system, a solution of cooked starch or a dispersion of starch granules is sprayed from nozzles directly onto the wet-web of fibers. By varying concentration, spray pressure, and spray location, a variety of effects can be achieved (*24, 25*). Three types of spray systems are in use: high-pressure air atomization, high-pressure airless atomization, and low-pressure airless atomization. With high-pressure systems, an electrostatic assist is used to prevent loss owing to misting (*26*).

In another system, low-density starch foam is applied directly on the wet-web immediately before it enters the wet press. The foam is mechanically broken at the press nip, and the starch is dispersed through the sheet. By controlling foam density, bubble size, and starch concentration, a wide variety of results can be achieved (*27*). As in the spraying system, very high retentions are possible, and low-cost unmodified starch may be used.

In another system, a thin curtain of liquid is applied to the wet-web (*28*) for high retention of chemicals, including starches. This system is claimed to be suitable for addition of starch to multi-ply paperboard where it increases ply bond strength.

IV. Surface Sizing with Starch

Application of non-pigmented starch pastes to the nearly dry sheet at the size press is referred to as surface sizing. Applications may also be made at the calender stack. In surface sizing, a continuous film is deposited on the sheet. All material added at the size press stays in the sheet, while wet-end additives are not always completely retained.

Surface sizing improves the finish, produces a better printing surface, minimizes scuffing and linting, prevents unwanted absorption of subsequent coatings, and improves strength characteristics of the paper. By proper control of viscosity, starch sizes can either be deposited mainly on the surface of the sheet or permitted to penetrate deeply into the sheet as required for control of physical properties.

1. Size Press Principles

The size press is located in the dryer section where the entering sheet has 5–15% moisture. After the size press, another dryer section is needed to dry the rewetted paper. Nearly one-third of the steam used in the total drying operation is consumed to remove water added at the size press (*29*).

A size press is simply a pair of squeeze rolls. The oldest type is the vertical press, where the size solution is showered on each roll to form a pond at the ingoing nip. One of the problems with this configuration is the tendency of the weight of the upper pond to deform or break the web. This is eliminated by use of a horizontal size press. Here, an equalized pond forms on each side of the sheet. To avoid the physically awkward vertical run required for the horizontal press, the inclined size press was developed (*30*). The size press nip has three regions, as described in Figure 3. As the paper enters the starch pond, the main parameter affecting size solution pick-up is absorption. Factors controlling absorption are porosity or void volume, degree of internal sizing, moisture content, starch viscosity, and machine parameters, such as pond size and machine speed. In the shear compression zone, additional parameters, such as roll hardness, roll diameter, and roll nip pressure, become important. As the sheet exits the nip, surface characteristics of the paper and of the size press roll, as well as the lead-out

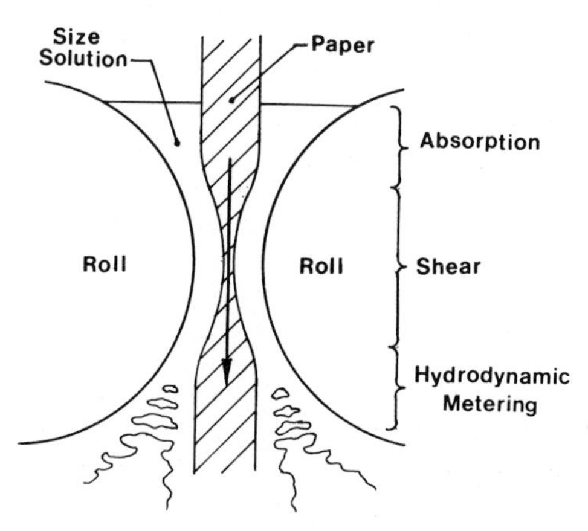

FIG. 3.—Size press wet pickup controlling factors.

angle, affect sheet properties (*30–32*). When the machine is operated at high speeds, size pick-up increases with increased starch viscosity. At low speeds, pick-up decreases with increasing viscosity.

By varying size press parameters and by selecting the proper starch product and concentration, the papermaker can control both the amount of size pick-up and its location in the sheet. Size forced deeply into the sheet, for example, can contribute 50% to the ultimate strength of the sheet (*33*), while linting in fine papers is substantially reduced if the starch is kept at the surface of the sheet (*34*). Solheim (*35*) showed a direct relationship between starch pick-up, internal strength, and surface strength, while factors such as ink receptivity, opacity, and water resistance were dependent upon the type of starch.

2. Starch Selection Parameters

In the paper industry, most starch is used for surface sizing. The current trend is to convert unmodified starch in the mill rather than to purchase more costly preconverted starches.

Parameters affecting starch pick-up are viscosity and size concentration (*31*). A papermaker can vary these parameters to obtain sizing solutions that range from high solids and low viscosity to low solids and high viscosity. Typical solids range from 2–12%, with viscosities usually less than 50 cP at 54°–77°. As a rule, the higher the inherent viscosity of the starch, the stronger is the adhesive quality. For this reason, starch with the highest possible inherent viscosity, consistent with pick-up and operating limitations, should be chosen (*30*).

The major starch used for surface-sizing applications is corn starch. Tapioca, potato, and wheat starches are sometimes used in areas where these starches are available at a reasonable cost. Unmodified starches are thermally or enzyme-converted to low viscosity before use. Pre-converted starches, such as oxidized, hydroxyethylated, acid-modified, and acetylated starches, are also widely used. More recently, pre-converted, low-viscosity, cationic starches have been recommended for size press application.

3. Preconverted Starches

Because at least a portion of the finished paper is returned to the wet-end of the machine as broke, the influence of the starch used for surface sizing on wet-end chemistry is important. Wilhelm (*36*) determined that a major reason for poor pigment and fines retention when using oxidized starch for sizing or coating is its negative charge. As little as 10 ppm of oxidized starch causes substantial loss of pigment. Hammerstrand and co-workers (*37, 38*) compared anionic, nonionic, and cationic starches for residual effects on wet-end activity and showed that retention is related to charge, with anionic starches being severely detrimental, cationic starches having beneficial effects, and nonionic starches showing little effect.

Hydroxyethylstarch is recommended as a surface size to improve grease and oil holdout. Kane (*39*) recommends the use of up to 25% poly(vinyl alcohol) in combination with hydroxyethylstarch to allow a 12–20% increase in filler loading of the sheet, thereby reducing costs. Reid (*40*) described the use of combinations of poly(vinyl alcohol) and starch at the size press for improving oil holdout. Fineman (*34*) added a hydrophobic agent to the starch size to improve printing properties. Powers (*41*) patented a process to size paper with cyanoethylstarch to reduce ink feathering. Fairchild (*42*) used a combination of starch, sodium alginate, and sodium aluminate to size food packaging materials to retard absorption of hot and cold liquids and to eliminate cereal flavor transfer to the food. Bristol and Brown (*43*) described the advantages of cationic surface-sizing starch in terms of improved physical properties of the paper and improved wet-end retention of the surface-sized broke.

4. Mill Conversion (Enzyme/Thermal) of Unmodified Starch

Unmodified starch is too high in viscosity for all but a few size press applications. Viscosity is usually reduced for effective use. Low-viscosity starches may be prepared by the starch manufacturer, or starches may be thinned at the paper mill. Today, thinning at the paper mill is widely practiced. Several low-cost methods may be used. Most important are batch or continuous enzyme conversion, continuous thermal conversion, and continuous thermochemical conversion.

a. *Enzyme (α-Amylase) Conversion*

Batch enzyme conversion is well-established in the paper industry. The operation is flexible, in that minor changes in either enzyme dose or other conversion parameters provide a wide range of finished starch paste viscosities. Continuous enzyme conversion is of recent origin and has the added capability of producing converted starch pastes as required.

To obtain a desired conversion, the starch dispersion must be buffered to the optimum pH for the enzyme. Also, the proper concentration of calcium ion is needed to provide heat stability for the enzymes. Starch manufacturers provide starches with the proper buffers and calcium levels for a wide range of water hardnesses.

Starches are converted during a fixed time period or to a fixed viscosity. In either case, the mixture of starch, water, and enzyme is heated to a specified conversion temperature, usually 71°–82°. For a fixed-time cycle, the mixture is held for the required period, usually 15–45 min at conversion temperature. For a

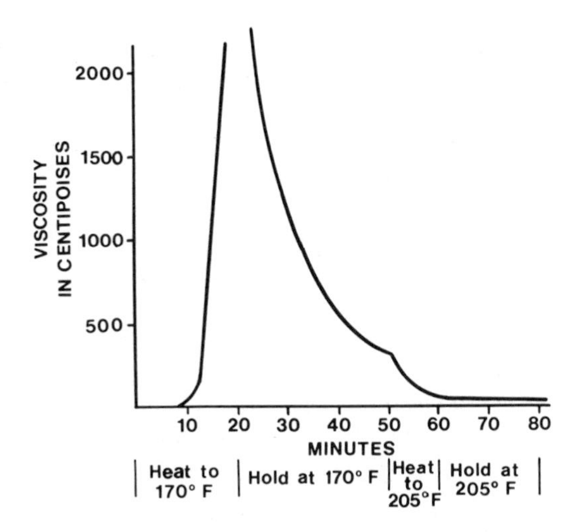

FIG. 4.—Time–temperature cycle for enzyme conversion (25% solids).

viscosity-controlled batch, the mixture is held at conversion temperature until a preset viscosity is reached. Finally, the conversions are raised quickly to 96° and held at temperature until the enzyme is inactivated. Inactivation requires 15–45 min, depending upon the concentration of conversion solids. In some cases, especially with very high solids conversions, two conversion hold-time periods are used. For example, one may heat to 70°, hold 10 min, then heat to 80° for 20 min and achieve a considerably lower peak viscosity with less strain on converting equipment. Viscosities reached during a typical batch enzyme conversion are shown in Figure 4.

Peak viscosities in excess of 80,000 cP are reached at 25% solids conversion, but a 200,000 cP peak will result from a 30% solids conversion (*44*). Extended action of α-amylase or contamination with β-amylase results in excessive reducing sugars, thereby decreasing the adhesive strength of the starch film. Generally, 3% reducing sugars is the maximum allowable level. Use of pure enzyme preparations and careful heat inactivation will ensure low levels of reducing sugars.

Some large paper mills employ continuous enzyme conversion. For this operation, an enzyme with high temperature stability is used. Operation of a typical continuous enzyme converter (*45*) is illustrated diagramatically in Figure 5.

A starch slurry containing the proper enzyme dosage is continuously pasted in a conversion column designed for plug flow in the main section. Residence time or conversion time is controlled by the level maintained in the tank. Converted paste is picked up at the agitated lower section of the conversion column by a positive displacement pump and heated to 115°–143° under high pressure for several seconds to inactivate the enzyme. After flashing to atmospheric pressure, excess steam is separated from the product flow. Starch pastes made by continuous conversion form stronger and more glossy films than are obtained by batch conversion (*46*).

Another type of continuous enzyme converter is a water-jacketed column, in which starch slurry enters the bottom and steam is injected at the top to inactivate the enzyme (*47*).

Hofreiter and co-workers (*48*) suggest a continuous conversion procedure in which a starch paste is passed through a column containing α-amylase bound to resin particles.Such a system eliminates the need for inactivation of the enzyme. In addition, data indicate that amylose is selectively hydrolyzed, producing a paste with less setback.

b. *Thermomechanical/Thermochemical Conversion*

By cooking starch at high pressure with an excess of steam and allowing the system to flash to atmospheric pressure, sufficient mechanical shear is generated to thin the starch. The system is called thermomechanical conversion. If acids or oxidants are also used, even further decrease in viscosity occurs. This latter process is called thermochemical conversion.

During the past several years, improvements in continuous starch cookers have enabled many paper mills to effect considerable raw material savings through the replacement of batch-cooked, highly converted, premodified starch with thermomechanical- or thermochemical-converted starch. Advantages, such as improved film formation, surface strength, and solution properties, compared to enzyme-converted starches and certain types of premodified starches, have been reported (*24*). Any one of a number of types of continuous, high-temperature cookers can be employed for viscosity reduction. Most corn starch manufac-

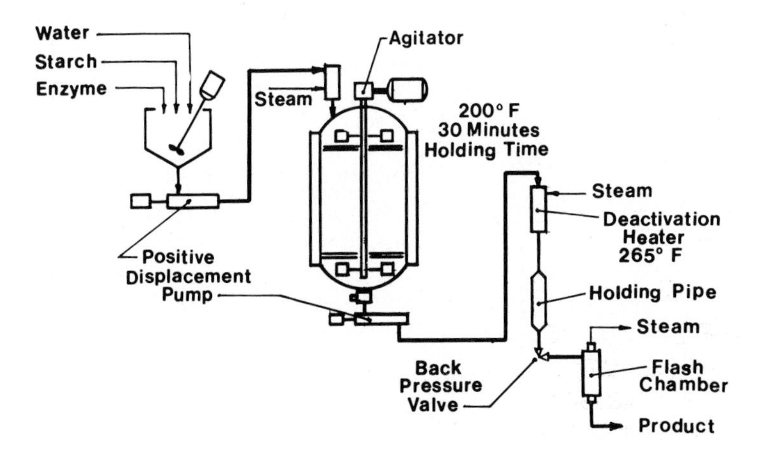

FIG. 5.—Flow diagram of enzyme thinning process.

turers can supply specially designed cookers (*49–56*). Equipment manufacturers and some paper mills can also provide such cookers (*57, 58*). Figure 6 shows the operation of a typical cooker and defines the variables that affect finished paste properties (*59*).

Operation of a cooker is relatively straightforward. A slurry of unmodified starch, usually at high solids, is pumped through a heater where high-pressure steam is injected to raise the temperature and impart mechanical shear to the paste. While under pressure, the hot paste enters a retention chamber or coil to maintain the high temperature for several minutes. By means of a back-pressure valve, the paste is flashed to atmospheric pressure, producing extremely high

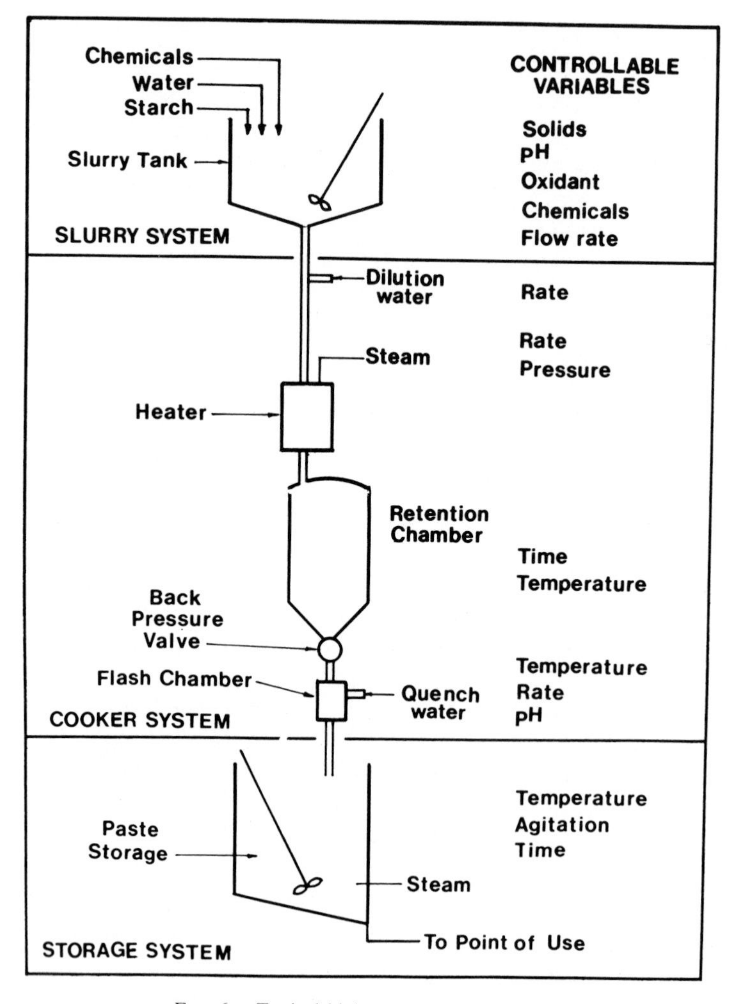

FIG. 6.—Typical high-temperature converter.

Table II

Methods to Change Converted Paste Properties

To Increase Solids	*To Decrease Color*
Decrease pre-cook dilution rate	Decrease slurry pH slightly
Decrease quench water rate	Decrease steam pressure
Increase slurry solids	Decrease steam rate
To Increase Temperature	Decrease retention time
Decrease quench water rate	Decrease quench water temperature
Increase quench water temperature	Decrease finished hold paste temperature
Increase steam pressure	Decrease oxidant level
Increase steam rate	Add chemical stabilizers
To Decrease Viscosity	Add chemical bleaches
Increase steam pressure	*To Decrease Sludge*
Increase steam rate	Increase steam pressure
Increase retention time	Increase steam rate
Decrease slurry solids	Increase retention time
Increase finished paste hold temperature	Increase quench water temperature
Increase oxidant level	Increase slurry pH slightly
Decrease slurry pH slightly	Decrease oxidant level
To Increase Stability	Increase finished paste hold temperature
Increase steam pressure	Add chemical bleaches
Increase steam rate	
Increase quench water temperature	
Increase finished paste hold temperature	
Add chemical stabilizers	

turbulence and shear. Excess steam is separated, and product is collected. By use of both excess steam and oxidizing agents, such as ammonium persulfate or hydrogen peroxide (*51, 59, 60*), required thinning for both paper sizings and coatings may be obtained.

Controllable variables described in Figure 6 have multiple effects on the finished paste properties. As shown in Table II, lowest viscosity and maximum paste stability are obtained at the highest oxidant level and steam pressure and at longest retention times (*59*). However, these conditions also promote undesirable color and sludge formation. Process variables, therefore, must be controlled to give low-viscosity stable pastes consistent with acceptable color and sludge. Paper sized with properly converted unmodified starch will have properties at least equivalent to those of enzyme-converted starch and usually as good as many of the more expensive premodified starches.

5. Calender Sizing

A starch size may be applied also at the calender stack, the last point on the machine where starch can be applied. At the calender, paper passes around a series of steel rolls, one or more of which transfers starch film to the paper surface. Next, the sheet is wound directly on reels. Further drying can only be accomplished by a combination of absorption of water into the sheet and evaporation of moisture by the latent heat of the sheet. Because of these restrictions, calender sizing is usually limited to heavy papers, such as liner board, food board, and box boards.

A calender-box is used to apply surface treatment on nearly all paperboard machines (*61*). Some of the major functions of calender-box treatment are curl control, surface strength, laying of surface fuzz, clay coating holdout, printing ink receptivity and gloss ink holdout, and grease or oil resistance.

For uniform surface treatment and best operating efficiencies, paste viscosity, paste solids, and operating temperature need to be controlled. The amount of solution metered at the calender stack is a function of size viscosity and calender loading pressure (*62*). Starch pastes of 20–75 cP can be applied at solids of 2–20% and at temperatures of 57°–73°. At higher viscosities, picking and calender wrapping occur frequently (*61*).

For heavyweight kraft liners, thin-boiling starch at solids of 2–5% is used with a lubricant to improve scuff resistance and printability. Gloss ink holdout and plybonding of cylinder board can be improved with a 10–20% solids low-viscosity, oxidized or hydroxyethylated starch containing an emulsifiable wax to reduce tackiness of the paste. Food board mills use a 6–12% solids, medium-viscosity, oxidized or hydroxyethylated starch, sometimes in combination with small amounts of pigment to prepare the board for subsequent functional coatings. Low-viscosity, oxidized starch at 18–24% solids is sometimes used for curl control. Nitz (*61*) suggests that filmformers used for surface strength and coating holdout have a maximum efficiency at about 7% solids.

Recently, the use of enzyme-converted or thermochemical-converted starches has replaced a large portion of the expensive highly modified starches for calender stack application. Improved formulating techniques have resulted in consistent, high-performance sizings.

V. Starch as Binder for Pigmented Coating

Use of starch as an adhesive for pigmented coatings for both paper and paperboard is described in numerous reviews (*63–68*).

1. Purposes of Coating

Paper coating refers to the application of a layer of pigment, adhesive, and other supplementary materials to the surface of dry paper or paperboard. The coating mixture, commonly called a coating color, is applied to the paper surface in the form of an aqueous suspension. Important coating systems contain pigment as the primary coating material and a starch adhesive to bond the pigment particles to each other and to the paper. The most commonly used pigments are clay, calcium carbonate, titanium dioxide, or their combinations. Satin white (calcium sulfoaluminate), zinc sulfide, barium sulfate, calcium sulfite, calcium sulfate, and diatomaceous silica pigments are used also, but to a much lesser extent. A coating may be applied to the paper during papermaking. More often, coating is applied as a separate step. In either case, the basic process differs only in the auxiliary equipment needed to perform the coating operation.

Paper or paperboard is coated to produce a surface adequate for printing processes. Coatings provide whiteness, brightness, gloss, and opacity to the paper along with a smoother, more uniform surface. Each printing process requires slightly different sheet properties and surface properties. In letter-press printing, low-viscosity inks that dry by absorption are used and strong surface strength is not required. However, compressibility and good ink receptivity are needed. Gravure printing requires an extremely smooth, flat surface. In lithography, or offset printing, tacky inks separated by water are employed. Paper for offset printing requires a degree of water resistance and a high surface strength, even when moist. Detailed discussions on the various printing processes are available (*69*). Also, basic information on pigments used in coatings can be found in a TAPPI Monograph (*70*).

Adhesive in pigmented coatings binds the pigment to the fiber and to itself to form a strong, smooth, continuous surface for subsequent printing. While starch is the major adhesive used, synthetic and protein-based adhesives are used to achieve specific effects. Casein or isolated soy protein is used in coatings requiring high water resistance. Latices, such as acrylics and styrene–butadiene, are used alone or in combination with other binders for improved strength and water resistance. More details on synthetic and protein-based coating binders may be found elsewhere (*71*).

2. Adhesive Requirements

While the primary purpose of the adhesive is to act as a binder for the pigment, it must also act as a carrier for the pigment, impart desirable flow characteristics to the coating color for proper application and leveling, regulate the degree of water retention of the coating color, and produce the desired strength, ink receptivity, and ink holdout for optimum printing characteristics. The adhesive should be easy to prepare, have a high adhesive strength, be inexpensive and readily available, have a stable viscosity during storage, and be unaffected by external conditions, such as wetting, after being applied to the sheet. No single adhesive meets all these requirements with complete satisfaction. Starch, however, adequately meets enough of them to make it the largest volume adhesive used in paper coating.

3. Rheology

Rheology of starch-based coating colors is an important factor in the selection of a starch for any given coating application. (Rheology is the science of liquid flow and gel deformation. Viscosity is the resistance to flow and is a measure of internal fluid friction.) Prior to the development of the concept of plastic flow by Bingham (*72*), coating colors were generally accepted as exhibiting straight-line flow patterns regardless of shear rate. Yet colors with identical single-point viscosity values might perform differently on the coaters. Following the work of Bingham, the concept of thixotropy was developed and explained the behavior of coating colors on different machines. Many detailed papers on thixotropy have been published; examples are those by Cobb and co-workers (*12, 73*) and Pryce-Jones (*74, 75*).

Newtonian flow is characterized by an immediate response to shear stress. Flow shear stress (shear force) and rate of flow (shear rate) are directly proportional at all rates of shear and viscosity remains the same at all shear rates. A pseudoplastic solution is shear thinning. Pseudoplastic flow is characterized by immediate response to shear; but the rate of flow increases less rapidly than the shearing force, so the viscosity will be lower as more shear is applied. Dispersions of high-molecular-weight, linear polymers exhibit this type of flow. Plastic flow is characterized by a lack of immediate response to shear; but as soon as sufficient shear has been applied to overcome initial resistance, the system exhibits Newtonian flow. In other words, a plastic solution has a yield value. After the yield value has been exceeded, a plastic solution may be Newtonian, pseudoplastic, or thixotropic. A clay-water suspension is a good example of plastic flow. Dilatant flow is characterized by an immediate response to shear; but as shear is increased, the resistance to flow becomes greater and the material increases in viscosity as shear force increases. A typical example is quicksand (*76*). Thixotropy is a time-dependent function. Thixotropic solutions thin with shear, but a time interval is required before the viscosity returns to its original value. Thixotropic flow is associated with an internal structure that is destroyed temporarily by the application of force, but reforms on standing. A force-flow curve of a thixotropic color shows a hysteresis loop (*78*). The shape and area of the hysteresis loop will vary with the time interval required for viscosity recovery and the rate of shear change. The area of the loop is a measure of thixotropic breakdown.

Except for very-low-solids coating colors sometimes used in size press coating, coating colors used for paper coating are thixotropic. It is apparent that no single point viscosity determination will adequately characterize such colors. Consequently, rotational-type viscometers are used to measure the viscosities of the colors produced. The instruments used may be generally divided in two types: low-shear viscometers, such as the Stormer, MacMichael, and Brookfield; and high-shear viscometers, such as the Hercules Hi-Shear and Hagan. Low-shear viscometers are generally used for routine control purposes with established formulations, while the high-shear instruments are useful for obtaining fundamental rheology data (*79*).

Factors in a coating system that affect flow properties are the pigment species, particle size of pigment, base exchange properties of pigment, absorptive qualities of raw stock, solids content of coating mixture, amount and type of adhesive, amount and type of mineral dispersing agent, pressure of coating rolls, and speed of machine. Thus, it can be seen that an understanding of pigment rheology as well as starch rheology is necessary in the evaluation of coatings. Much work has been done in this area by Albert (*80*) and Asdell (*81, 82*), particularly with respect to clay, the common coating pigment.

A first step in the preparation of a coating color is to produce a well-dispersed pigment. Pigment particles may be strongly bonded initially, but must be disassociated as completely as possible. Particle dispersion or disaggregation is generally accomplished by mechanical force, and the dispersion is stabilized by dispersing agents to prevent flocculation. A method of measuring the degree of dispersion based on relative sediment volume has been devised by Robinson (*83*).

Starch properties that effect dispersion of pigment will have an effect on color rheology. The effect of starch on flow characteristics of clay slurry has been described by Rowland (*84*). More recently, Kline (*85*) investigated adhesive–pigment interaction and concluded from rheological evidence that starch acts as a pigment dispersant. He theorized that, in the absence of other dispersants, starch stabilizes the dispersion by means of its protective colloid action. When pigment is already dispersed, the starch, through mass action effect, causes increased absorption of the dispersant. In this case, an overdispersed, slightly flocculated system is formed. The pseudoplastic–thixotropic rheology of such systems has been considered desirable for leveling in high-speed coating systems.

In addition to rheology, other important properties of starch in coating colors are its water-holding capacity and viscosity. Both factors may be altered by changing the type of starch or its concentration. In general, if the water-holding capacity is too low, water will leave the color rapidly, resulting in high viscosity on the sheet surface. This high viscosity prevents natural leveling of the coating surface with concomitant patterning on the finished sheet. If the water-holding capacity is too high, the color will remain wet on the paper surface.

Frost (*86*) has shown the relationship between water retention and wax pick values for starch in the coating color. In general, the higher the water retention, the greater the adhesive retention, and the stronger the finished coating.

Hemstock and Swanson (*87*) used a roll-inclined plane test, previously described by Arnold (*88*), for examining the water-holding capacity of coating colors. An instrumental method measuring water retention has been described by Stinchfield and co-workers (*89*). This equipment utilizes conductivity of the system to determine the rate of water loss from coating colors.

Migration of water and adhesive occurs during coating application and drying. Just as part of the binder penetrates into the base paper, a portion will also migrate to the surface of the coating during the drying operation. Binder that migrates to the surface occupies pore volume between pigment particles and decreases ink receptivity, but produces a smoother, more dense surface (*90*).

Because coating color formulations vary widely, no specific relation between water loss and binder migration have been derived. At comparable viscosities, however, the water-holding capacity of hydroxyethylstarch is generally higher than that of oxidized starch which, in turn, is superior to that of converted pearl starch.

Choice of a starch for a coating binder depends upon rheology requirements of the coating machine and water-holding capacity desired in the coating color. The adhesive has a great influence on the characteristics and printability of the finished paper (*77*). Optimum printability is achieved by selecting the proper starch and using it at the correct level. Selection of the coating formulation, therefore, must be done with care to ensure a proper balance among paper properties, coater operation, and economics.

4. Mechanics of Coater Operation

Coaters are designed to apply the desired amount of coating uniformly on the paper and to level the coating. Coating color requirements for optimum operation are different for each coater. A partial list of coaters includes the size press, contra-roll coater, reverse roll coater, Levelon coater, smoothing bar coater, Massey coater, knife-over-roll coater, several kinds of trailing blade coaters, air knife or air doctor coaters, brush coaters and cast coaters. Each has a special advantage and provides certain coated paper characteristics. Crane and Majani (*76*) provide more information on coaters. Van Derveer (*91*) provides a description of modern blade coating operations, and the machinery used for paper coating is described in a TAPPI Monograph (*63*).

All coaters, except the air knife or air doctor, tend to cover the base sheet with a film of varying thickness. Metering devices on the various coaters, whether they be knives, blades, or rolls, tend to fill the hills and valleys of the sheet to produce a smooth surface. On the other hand, the air knife produces a uniform film thickness and requires a uniform base sheet for optimum operation. Coater operation depends largely upon hydrodynamic pressure and film splitting. Coating color that is squeezed between a roll and a paper surface generates high hydrodynamic pressure. This pressure is proportional to the viscosity of the

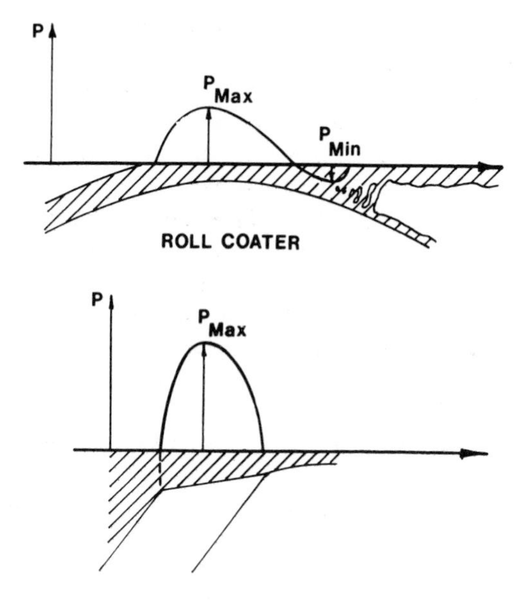

FIG. 7.—Pressure profiles.

coating color, the sum of the speeds of the paper and roll, and the roll diameter. It is inversely proportional to the thickness of the coating color film. The amount of coating applied is primarily determined by the pressure between the applicator roll and the sheet (*92*).

When roll coaters or roll metering devices are used, the diverging nip creates a negative pressure within the film of coating color as shown in Figure 7. When this exceeds the cohesive forces within the coating color, the film splits, leaving a pattern on both the roll and the paper. Proper thixotropy of the coating color will permit this pattern to level out before the viscosity increases from the combination of shear force reduction and water loss. Smith, Trelfa, and Ware (*93*) have reported some empirical limits for the leveling index of adhesives that exhibit thixotropic and plastic flow.

With blade coaters, the amount of coating applied is regulated by pressure of the blade edge. Action of a blade coater is somewhat different since no diverging rolls are used. High pressures in the nip produce a dewatering action on the coating color to set partially the coating color when it leaves the blade.Some of the smoothest coatings are applied with blade coaters (*92*).

5. Starches as Binders

High solids coatings require starch pastes with low viscosities. The largest volume of starch used in paper coatings is unmodified corn starch that has been enzyme-converted at high solids. De Groot and Ewing (*94*) found no differences between continuous enzyme conversion and batch-enzyme conversion with respect to paste properties, coating color rheology, or physical properties of the finished paper. Cave and Adams (*46*), while agreeing with these conclusions, pointed out that continuously enzyme-converted starch produces clear glossy films as opposed to the more opaque films from batch conversions. Continuously converted starch pastes were considerably lower in viscosity for the same coating color viscosity, and the stability of pastes from a continuous converter was better than paste made by the batch process.

Use of high-pressure, thermochemical conversion of starch for making high solids coating adhesives is of recent origin. With modern high temperature converters, unmodified starch can be converted at solids levels ranging up to 40% by the use of oxidants and excess steam to produce suitable coating binders at low cost.

Pre-converted starch products continue to hold a substantial, but decreasing, share of the coating binder market. Oxidized starch, once considered the ultimate coating binder because of its good pigment dispersion properties in coating make-up and film-forming properties, has lost a considerable share of the coating binder market because of pigment dispersion. When coated broke (waste paper) is recycled to the wet-end of the paper machine, pigment fillers are dispersed and not retained in the paper. Dispersed pigments are undesirable in the paper mill effluents because of waste treatment problems (*95*).

Hydroxyethylstarch is considered one of the best for use in coating binders. Higher cost tends to limit its use in conventional coatings. However, higher reactivity with insolubilizing resins makes it particularly useful for water-resistant or wet-rub-resistant coatings.

Thin-boiling cationic starch has been used in certain areas as a coating binder. Because it causes pigment flocculation at the wet-end of the paper machine when coated broke is recycled, it is recommended where control of waste effluent solids cannot be more economically controlled by sewage treatment plants. Mazzarella and Hickey (*96*) reported that one-third as much cationic starch as preconverted starch produces an equal level of binding strength along with increased opacity. Special, but simple, formulating techniques are necessary when using cationic starch to eliminate dilatancy under shear.

Yeates and co-workers (*97*) suggested the use of cyanoethylstarch and dextrins as coating adhesives. Use of oxidized, acid-modified and various derivatized starches in typical coating color formulations is described in a TAPPI Monograph (*98*).

Adhesive strength of starch is lower than that of other major coating binders. Thorndyke (*99*) reported that 18 parts of starch are equivalent to 15 parts of casein or α-protein or 12 parts of latex. However, with latex having a cost about four times that of starch, economics strongly favor starch wherever it can provide the necessary properties. Combinations of starch and latex or protein binders are often used to attain a compromise between properties and economics. Excellent descriptions of coating color preparation and specific coating formulations are available to the interested reader (*77, 97, 100, 101*).

6. Size Press Coating

While the size press is used primarily for applying a non-pigmented starch film to paper, it may also be used to apply a pigmented coating. Grant (*102*) described the use of a size press to improve sheet uniformity and as a precoater to lay down a base for subsequent higher quality coatings. Enzyme-converted, thermal-converted, hydroxyethylated, and oxidized starches are suggested as suitable adhesives. Coating formulations must be carefully developed for application at the size press. Oakleaf and Janes (*103*) found a close relationship between coating color rheology and patterning at the size press. Most patterning is attributable to the leveling index of the coating. Patterning is reduced at higher solids and adhesive levels. Flow modifiers, such as calcium stearate, also reduce surface pattern.

Typical size press coating colors are higher in adhesive and lower in solids and viscosity than those used for off-machine coatings. Size press coating colors may range from 10–50% total solids, but are usually in the 30–40% solids range. Viscosities are generally less than 300 cP and adhesive to pigment ratios as high as 1:1 are common. With the proper formulations, printing properties can be dramatically improved. Margotta and Gaither (*104*) outlined a computer cost analysis system as a useful tool in selecting size press coating formulas.

VI. Starch Adhesive for Corrugated Board

Corrugated board manufacture is an integral part of the paper industry and maintains an important position in U.S. economy. The Fiber Box Industry's Annual Report (*105*) shows that more than 250 billion ft^2 (23 billion m^2) of corrugated board were produced in 1978 with nearly 70% of the corrugated boxes used to ship non-durable goods. Consumption in the U.S. has risen to over 1100 ft^2 (102 m^2) per person (*106*). The value of corrugated board shipped in 1978 was $7.6 billion. Nearly all the corrugated board produced today is bonded with a starch-based adhesive.

Corrugators spend more for starch than for any other raw material except paper. In 1978, nearly $100 million were spent on adhesives to produce 256 billion ft^2 (23.8 billion m^2) of corrugated board (*107*). At an average of 2.5 lb (1.1 kg) of starch per thousand ft^2 (92.9 m^2) of board, the corrugating industry consumed about 650 million lb (295 million kg) of starch in 1978.

1. Principles of Corrugator Operation

Corrugated board is composed of liners and the fluted medium. Various weights, thicknesses, and combinations of liners and medium are used for different applications. Edge views of a selection of these combinations are shown in Figure 8. Double-backer or single-wall board is the workhorse of the corrugated board industry, representing more than 90% of 1977 production. It is used primarily for standard-type shipping containers. Typically, double-wall and triple-wall constructions, just over 8% of 1977 production, are used for packaging large heavy objects such as refrigerators and for uses where great stacking strength is required. Generally, single facer board is used as a cushioning material. A schematic of machinery used for manufacture of these board combinations is shown in Figure 9.

Principles of manufacture are simple. Starting at the single facer, flat corrugating medium is softened with heat and moisture and passed between a set of corrugating rolls to form it into flutes. Adhesive is applied to the flute tips on one side of the medium, and a single-facer linerboard is brought into contact with the fluted medium under heat and pressure to produce a single-facer web. This web is continuously conveyed by a bridge conveyor to the double-backer station where adhesive is applied to the exposed flute tips and the double-backer liner is applied. This combination is then conveyed over a series of hot plates to set the adhesive. The finished product is single-wall board. Additional single-facer piles may be added during the process to produce double- or triple-wall board.

2. Corrugating Adhesive Theory

Modern corrugating machines are run at high speeds ranging from 100–600 ft/min (30.5–183 m/min) depending on the type of board made. High speeds are made possible by the *in situ* gelatinization of corn starch to produce extremely high viscosities in the glue line. This process was developed in about 1935 by Jordon Bauer of the Stein Hall Company. A mixture of raw starch, borax (sodium tetraborate), caustic soda (sodium hydroxide), and water is suspended in a

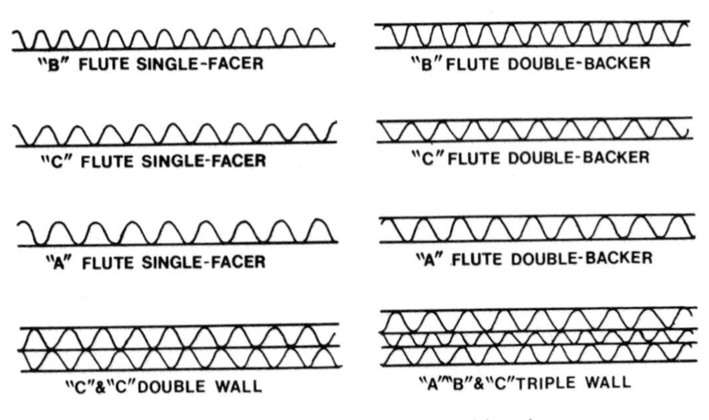

FIG. 8.—Major types of corrugated board.

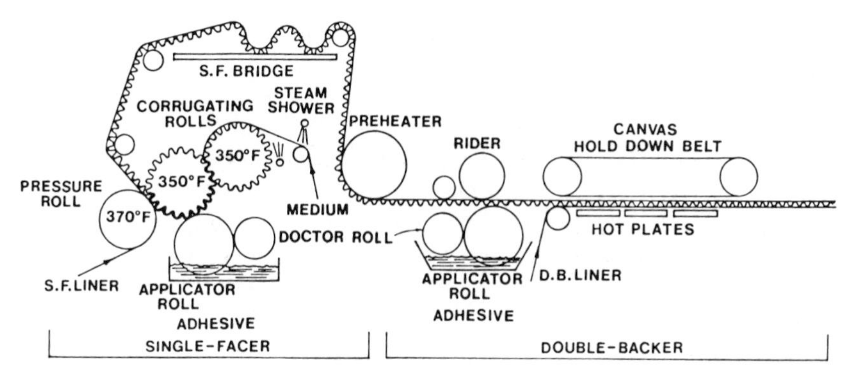

FIG. 9.—Corrugated board manufacture.

paste of cooked starch. Usually 15–20% of the starch is fully cooked in part of the water to provide a carrier for the raw starch. Total starch solids are typically 18–21%, although for special purposes, starch solids may approach 30%. Viscosity of the finished adhesive is 1000 cP or less. After the adhesive is applied to the flute tips and subsequently heated, the raw starch gelatinizes in place.

Caustic soda is added to increase the rate of wetting of the cellulosic fibers and to reduce the gelatinization temperature of the raw starch to about 63°–66° for easier, more complete cooking of the starch in the glue line. Borax crosslinks the cooked carrier starch to improve water-holding capacity. This prevents water in the glue line from penetrating the paper before the raw starch has gelatinized. If insufficient gelatinization occurs, poor bonding and a white glue line will result. Borax also reacts with the raw starch during *in situ* gelatinization to further increase viscosity (*108*).

Immediately after the adhesive is applied to the flute tips of the hot medium at the single facer, starch begins to penetrate the medium. The medium is then brought into contact with linerboard with a pressure roll for only 0.03 sec. During this brief period, starch swells sufficiently to form a permanent bond. It is unlikely that the starch is completely gelatinized at this point, but the residual heat in the paper will complete the gelatinization process on the bridge. Thayer and Thomas (*109*) suggested that the initial single facer bond is formed by amylose migrating out of the starch granules, causing their surface to become tacky.

Adhesive applied at the double-backer is under lower pressure for a longer period of time. This can cause excessive penetration of the adhesive, which sometimes can be a problem. Heat applied from hot plates must pass through the double-backer liner before reaching the glue line. The liner is a poor heat conductor, and machine speed sometimes must be reduced to permit adequate heat transfer. This is especially true with heavy paper or double- and triple-wall combinations. Because of the lower pressure and longer times used at the double-

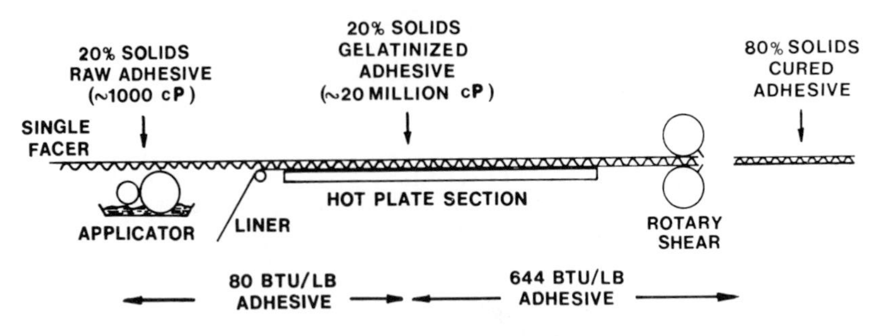

FIG. 10.—Adhesive energy requirements.

backer, a stronger bond is usually formed. As the board is heated, the temperature gradually increases until the gelatinization temperature of the starch is reached. At this point, the starch granules rapidly absorb water, swell, and react with the borax to form a highly viscous gel of at least 20 million cP. Even this high viscosity, however, will not hold the components together firmly. Additional water must be driven out of the glue line by evaporation. Dehydration must be sufficient to form at least the minimum green bond strength necessary for the board to resist delamination during the mechanical operations of slitting, scoring, cut-off, and take-away. Most of the heat energy applied at the double-backer is used for evaporation of water after starch gelatinization rather than for gelatinization of the starch (Fig. 10).

Adhesive viscosity is extremely important. Fortunately, it is one of the easiest variables to control by formulation changes. A higher adhesive viscosity will tend to inhibit wetting of the paper and will hold water in the glue line where it is needed for full starch gelatinization. A higher borax level will also inhibit the release of water from the gel structure of the carrier starch. Too much borax, however, will produce only a shallow surface bond with insufficient adhesive penetration for good strength. White glue lines, loose edges, and the need for slowing the corrugator speed have usually been attributed to a gelatinization temperature that was too high. However, too low a gelatinization temperature may cause premature gelatinization and thickening of the starch to prevent proper transfer to the flute tips.

3. Adhesive Formulating

Specific formulations for adhesive preparation are given in the literature (*110*). A standard Stein Hall adhesive is given in Table III. Adhesive viscosities are measured with a brass cup viscometer about 6.5 inches (16.5 cm) long with a 2-in. (5-cm) inside diameter. Inside are two pins with a volume of 100 mL between them. The cup is filled with adhesive, and as it flows through a calibrated orifice, the number of seconds for the surface of the adhesive to pass between the pins is recorded as the Stein Hall viscosity. Water, at room temperature, should have a viscosity of 15 sec. A single-facer adhesive will usually have a viscosity of about 25–35 sec, and the double-backer adhesive will have a slightly higher viscosity of 40–45 sec.

Basic equipment for adhesive preparation consists of a cylindrical primary mixer with a low-speed, high-torque mixer situated above a secondary tank with a high-speed, lower-torque agitator. A drop valve is provided to allow the contents of the primary mixer to flow into the secondary tank at a controlled rate.

There are a number of variations of the basic Stein Hall preparation, but they are all essentially two-component mixing systems with cooked starch in the primary mixer and raw starch in the secondary mixer. Borax may be added to the primary instead of the secondary mixer, and the carrier may be dropped quickly without cooling water addition to speed preparation time if a heavy duty primary mixer is available to handle the higher viscosities encountered during the cook. In other formulations, only the secondary mixer is used. The carrier starch may be cooked with caustic soda followed by addition of borax and additional water to bring down the mixture to the proper temperature. Raw starch is then dispersed in the mixture to complete adhesive make-up. In another variation, pregelatinized starch is used as the carrier portion. There are also many formulations which employ a steam jet to cook the carrier starch continuously.

Perhaps the only starch adhesives that differ significantly in preparation and properties are the no-carrier types. Rather than a cooked and raw starch portion, essentially all the starch granules in a no-carrier adhesive are preswollen. Swelling is accomplished by precise, automatic addition of dissolved caustic soda to the starch slurry (*111*). Viscosity development is monitored to measure the degree of granule swelling. When the desired viscosity is reached, boric acid is

Table III

Basic Stein Hall Adhesive

Primary Mixer	
1. Add water	100 gal (378.5 L)
2. Add unmodified corn starch	200 lb (90.7 kg)
3. Add caustic soda [dissolved in about 10 gal (37.8 L) of water]	30 lb (13.6 kg)
4. Heat to	71°
5. Hold under agitation	15 min
6. Add cooling water	60 gal (227 L)
Secondary Mixer	
7. Add water	400 gal (1.5×10^3 L)
8. Heat to	29°
9. Add unmodified corn starch	1000 lb (454 kg)
10. Add borax (10 mol)	30 lb (13.6 kg)
11. Transfer primary to secondary in about	30 min
12. Mix until smooth	

added to neutralize some of the caustic soda and thereby stabilize the adhesive. This adhesive is more stable to shear and has better flow properties than adhesives made with a two-component system. Another method utilizes controlled steam injection to achieve controlled swelling (*112*). In a third concept, unmodified starch is milled at 10–40% moisture to damage the starch granules so that they will partially swell when added to water. This adhesive is similar to the above-mentioned no-carrier type (*113*).

4. Special Adhesives

For some types of corrugated board applications, special adhesives are used to fulfill special end-use requirements. One special requirement is water resistance. Because a starch bond is normally not water-resistant, it is necessary to insolubilize it for special packaging requirements, such as refrigerated boxes, or boxes that must be stored outdoors. This corrugated board is produced by the addition of an alkaline curing resin to the adhesive. These resins are usually urea–formaldehyde, ketone–formaldehyde, or resorcinol-based resins. Each has specific advantages and disadvantages, and a resin supplier should be consulted before one is chosen. Water resistance increases as resin addition is increased to a maximum of about 15%. Higher resin levels usually require higher starch solids. When used as a supplementary reactant with a formaldehyde compound, cyanamide, or one of its alkaline salts, produces increased water resistance (*114*).

Musselman (*115*) discloses the use of a thin-boiling, acid-modified waxy starch to produce a higher solids carrier portion of the corrugating adhesive to increase production rates on heavyweight board. Desmarais (*116*) described the use of hydroxyethylstarch as a carrier in combination with calcium hydroxide to allow corrugator operation over a wider than normal speed range. Klein and coworkers (*117*) suggested the use of starch esters as corrugating adhesives to provide increased corrugating machine speed. Casey and Lehman (*118*) formulated a mixture of raw starch and sodium carboxymethylcellulose as the carrier to eliminate the need for cooking and cooling a starch carrier. High-amylose starches are also used as carrier starches to produce stable viscosity adhesives.

A cold process for manufacture of corrugated board is claimed by The Institute of Paper Chemistry (*119*). No heat is used either for forming the flutes or for combining liner board with the fluted medium. An adhesive is prepared by cooking unmodified corn starch with oxidizing chemicals at 138°–143° for about 90 sec. The paste at 88°–93° is applied to the cold flute tips. Cooling of the paste causes a rapid increase in viscosity to form a glue line having good strength. This combination is referred to as a setback adhesive (*120, 121*). Because the cold corrugating process is expected to reduce process heat consumption by 95% and drive-energy consumption by 30–40%, its introduction could save as much as $70 million per year in U.S. energy costs (*122*).

Other types of special adhesives represent only a very small percentage of total consumption. These include moldproof and fireproof formulations, as well as low pH water-resistant formulas. Synthetic adhesives, mostly poly(vinyl acetate)-based, are sometimes used on double or triple wallboard because of their ability to form a bond with little or no heat. High cost has prevented their widespread use, however.

VII. References

(*1*) Statistics of Paper and Paperboard, American Paper Institute, Inc., New York, 1978, p. 67.

(*2*) J. Rauch, ed. ''The Kline Guide to the Paper and Pulp Industry,'' Charles H. Kline & Co., Fairfield, New Jersey, 3rd Ed., 1976, p. 10.

(*3*) J. E. Maryanski, ''Natural Binder Outlook,'' Presented at TAPPI Annual Meeting, 14–16 Feb., CPC International Reprint #ST–27–36, 1977, pp. 5, 6.

(*4*) Anon., ''Nature of the Chemical Components of Wood,'' TAPPI Monograph No. 6, 1948.

(*5*) Anon., ''Mechanical Pulping Manual,'' TAPPI Monograph No. 21, 1960.

(*6*) B. W. Attwood, ''How We Got Here,'' Introductory paper from EUCEPA '79, Connaught Rooms, London, 21–24 May, 1979.

(*7*) J. P. Casey, ''Pulp and Paper,'' Interscience Publishers, New York, Vol. 1, 1952, p. 552.

(*8*) G. Jayme and J. Hunger, *Zellstoff Papier,* **6,** 341 (1957).

(*9*) J. W. Swanson, *TAPPI,* **43,** No. 3, 176A (1960).

(*10*) M. L. Cushing, *TAPPI,* **44,** No. 3, 191A (1961).

(*11*) H. H. Houtz, *Tech. Assoc. Papers,* **24,** 131 (1941).

(*12*) R. M. Cobb, D. V. Lowe, E. Pohl, and W. Weiss, *Paper Trade J.,* **105,** No. 7, 33 (1937); *Tech. Assoc. Papers,* **20,** 299 (1937).

(*13*) B. T. Hofreiter, G. E. Hamerstrand, D. J. Kay, and C. E. Rist, *TAPPI,* **45,** 177 (1962).

(*14*) J. R. Waters, *TAPPI,* **44,** No. 7, 185A (1961).

(*15*) B. W. Rowland, *Paper Trade J.,* **97,** No. 21, 249 (1933).

(*16*) E. K. Nissen, unpublished data, 1962.

(*17*) H. Tanaka, K. Tachiki, and M. Sumimoto, *TAPPI,* **62,** No. 1, 41 (1979).

(*18*) H. W. Moeller, *TAPPI,* **49,** No. 5, 211 (1966).

(*19*) G. E. Hamerstrand, B. S. Phillips, J. C. Rankin, and B. T. Hofreiter, *TAPPI,* **61,** No. 9, 59 (1978).

(*20*) M. E. Carr, B. T. Hofreiter, M. I. Schulte, and C.R. Russell, ''A Starch Polyampholyte for Paper,'' presented at 1977 TAPPI Annual Meeting, 14–16 February, 1977.

(*21*) W. Jarowenko and H. R. Hernandez, U.S. Patent 4,029,544 (1977); *Chem. Abstr.,* **87,** 54768w (1977); Ger. Offen. 2,547,700 (1976); *Chem. Abstr.,* **185,** 48590z (1976).

(*22*) P. D. Buikema, U.S. Patent 4,029,885 (1977); *Chem. Abstr.,* **87,** 54779a (1977).

(*23*) C. L. Mehltretter, T. E. Yeates, G. E. Hamerstrand, B. T. Hofreiter, and C. E. Rist, *TAPPI,* **45,** 750 (1962).

(*24*) J. G. Reid, *Paper Technol. Ind.,* 39 (February, 1975).

(*25*) A. A. Procter and J. G. Reid, *Paper Technol.,* 283 (October 1974).

(*26*) R. F. Bryn Davies, *Paper Technol Ind.,* 186 (July, 1977).

(*27*) M. C. Riddell and B. Jenkins, *Paper Technol. Ind.,* 176 (July, 1977).

(*28*) Australian Patent No. 69818/74 (June, 1974).

(*29*) G. Herwig, *TAPPI,* **62,** No. 4, 49 (1979).

(*30*) C. T. Beals, *Pulp Paper,* 131 (April, 1978).

(*31*) R. W. Hoyland, P. Howarth, C. J. Whitaker and C. J. H. Pycraft, ''Mechanism of the Size-Press Treatment of Paper,'' Paper presented at Spring Conference on the Application of Chemicals and Coatings to Paper, Swiss Cottage, London, 16 March, 1977).

(*32*) R. W. Hoyland and P. Howarth, *Paper Technol.*, 38 (February, 1972).
(*33*) H. R. Bryson, *Paper Trade J.*, 30 (October, 1978).
(*34*) I. Fineman and M. Hoc, *TAPPI*, **43,** No. 5, 43 (1978).
(*35*) P. F. Solheim, *TAPPI*, **49,** No. 4, 82A (1966).
(*36*) L. K. Wilhelm, 19th TAPPI Coating Conference, Preprint, 13–16 May, 1968, p. 459. Miami Beach, Florida.
(*37*) G. E. Hamerstrand, B. S. Phillips, J. C. Rankin, H. D. Heath, M. I. Schulte, and B. T. Hofreiter, *TAPPI*, **61,** No. 1, 81 (1978).
(*38*) G. E. Hamerstrand, H. D. Heath, B. S. Phillips, J. C. Rankin, and M. I. Schulte, *TAPPI*, **62,** No. 7, 35 (1979).
(*39*) T. G. Kane, *Pulp Paper*, **52,** No. 2, 125 (1978).
(*40*) J. G. Reid and R. F. Bryn Davis, *Paper Technol.*, **14,** No. 5, 265 (October, 1973).
(*41*) R. M. Powers, U.S. Patent 3,387,998 (1968).
(*42*) W. P. Fairchild, U.S. Patent 3,255,028 (1966); *Chem. Abstr.*, **65,** 4086h (1966).
(*43*) K. J. Bristol and G. H. Brown, *Paper Trade J.*, **153,** No. 27, 42 (1969).
(*44*) J. E. Maryanski, "Enzyme Conversion of Starch in the Paper Industry," Presented at Joint TAPPI/PIMA Meeting, Gearhart, Oregon, 3–5 June, 1965.
(*45*) L. E. Cave and F. R. Adams, *TAPPI*, **51,** No. 11, 109A (1968).
(*46*) L. E. Cave and F. R. Adams, "Mill Runs Validate Continuous Enzyme Converting of Starch for Paper Surface Treatment," Presented at TAPPI Coating Conference, Miami Beach, Florida, 12–16 May, 1968.
(*47*) J. L. Bidrawn, F. G. Ewing, B. H. Landis, and H. R. Wheeler, "Continuous Enzyme Conversion of Corn Starch," Presented at the 16th Coating Conference of TAPPI, Portland, Oregon, 10–13 May, 1965.
(*48*) B. T. Hofreiter, K. L. Smiley, J. A. Boundy, C. L. Swanson, and R. J. Fecht, *Cereal Chem.*, **55,** 995 (1978).
(*49*) O. R. Etheridge, U.S. Patent 2,919,214 (1959).
(*50*) O. R. Etheridge, U.S. Patent 3,101,284 (1959).
(*51*) V. L. Winfrey and William C. Black, U.S. Patent 3,133,836 (1964).
(*52*) N. F. Schink, U.S. Patent 3,197,337 (1965).
(*53*) K. J. Huber, J. F. Johnston, F. K. Nissen, and D. R. Pourie, U.S. Patent 3,220,884 (1965).
(*54*) K. J. Huber, J. F. Johnston, E. K. Nissen, and D. R. Pourie, U.S. Patent 3,276,907 (1966); *Chem. Abstr.*, **65,** 18816d (1966).
(*55*) K. J. Huber, J. F. Johnston, E. K. Nissen, and D. R. Pourie, Canadian Patent 759, 405 (1967).
(*56*) E. Frank, N. H. Yui, and A. A. Silvasi, U.S. Patent 3,374,115 (1968); *Chem. Abstr.*, **68,** 106245c (1968).
(*57*) H. Goos, H. W. Mauer, and A. W. Kurz, U.S. Patent 3,219,483 (1962).
(*58*) H. W. Maurer, U.S. Patent 3,485,667 (1969); *Chem. Abstr.*, **72,** 68456v (1970).
(*59*) Anon., "Thermal and Thermal-Chemical Conversion of Unmodified Mill Starch for Paper Sizing and Coating," CPC International Publication ST-27-38, 1976.
(*60*) D. A. Brogly, *TAPPI*, **61,** No. 4, 43 (1978).
(*61*) R. N. Nitz, *Pulp Paper*, **51,** No. 14, 120 (1977).
(*62*) Anon., *Pulp Paper*, **51,** No. 5, 67 (1977).
(*63*) Anon., "Machinery for Paper Coating," TAPPI Monograph No. 8, 1950.
(*64*) J. P. Casey, "Pulp and Paper," Interscience Publishers, New York, Vol. 3, 1961, pp. 1007–1130.
(*65*) Anon., "Preparation of Paper Coating Colors," TAPPI Monograph No. 11, 1954.
(*66*) Anon., "Starch and Starch Products in Paper Coating," TAPPI Monograph No. 17, 1957.
(*67*) Anon., "Paper Coating Pigments," TAPPI Monograph No. 29, 1958.

(*68*) Anon., "Pigmented Coating Processes for Paper and Board," TAPPI Monograph No. 28, 1964.
(*69*) A. H. Nadelman and G. H. Baldauf, "Coating Formulations—Principles and Practices," Lockwood Trade Journal Co., New York, 1966, Chap. I, p. 3.
(*70*) H. M. Murray, "Paper Coating Pigments," TAPPI Monograph Series No. 30, 1966.
(*71*) L. H. Silvernail and W. M. Bain, "Synthetic and Protein Adhesives for Paper Coating," TAPPI Monograph Series No. 22, 1961.
(*72*) E. C. Bingham, Natl. Bur. Stds. Sci. Paper No. 278 (1916).
(*73*) R. M. Cobb and D. V. Lowe, *J. Rheology*, **1,** 158 (1930).
(*74*) J. Pryce-Jones, *J. Oil Colour Chem. Ass.*, **19,** 295 (1936).
(*75*) J. Pryce-Jones, *J. Sci. Inst.*, **18,** 39 (1941).
(*76*) L. P. Crane and B. E. Majani, *TAPPI*, **41,** No. 8, 196A (1958).
(*77*) A. H. Nadelman and G. H. Baldauf, "Coating Formulations—Principles and Practices," Lockwood Trade Journal Co., New York, 1966, Chap. 3, p. 22.
(*78*) J. B. Batdorf, *in* "Industrial Gums," R. L. Whistler, ed., Academic Press, New York, 1959, p. 659.
(*79*) Anon., "Testing of Adhesives," TAPPI Monograph No. 26, 1965.
(*80*) C. G. Albert, *TAPPI*, **34,** No. 10, 453 (1951).
(*81*) B. K. Asdell, *Paper Mill News* (31 May, 1947).
(*82*) B. K. Asdell, *Paper Mill News* (26 June, 1948).
(*83*) J. V. Robinson, *TAPPI*, **42,** No. 6, 432 (1959).
(*84*) B. W. Rowland, *Paper Trade J.*, **112,** No. 26, 311 (1941).
(*85*) J. E. Kline, An Investigation of Adhesive-Pigment Interactions in Coating Mixtures, *TAPPI*, **55,** No. 4, 556 (1972).
(*86*) F. H. Frost, *TAPPI*, **35,** No. 7, 16A (1952).
(*87*) G. A. Hemstock and J. W. Swanson, *TAPPI*, **40,** No. 10, 833 (1957).
(*88*) K. A. Arnold, *Paper Trade J.*, **117,** No. 9, 28 (1943).
(*89*) J. C. Stinchfield, R. A. Clift, and J. J. Thomas, *TAPPI*, **41,** No. 2, 77 (1958).
(*90*) P. Hiemstra and J. Vandermeeren, TAPPI 19th Coating Conference, Miami Beach, Florida, 13–16 May, 1968, p. 491.
(*91*) P. D. Van Derveer, *Paper Trade J.*, **147,** No. 13, 33 (1963).
(92) J. P. Escarfail and J. E. Penkala, *Pulp Paper*, **38,** No. 43, 52 (1964).
(*93*) J. W. Smith, R. T. Trelfa, and H. O. Ware, *TAPPI*, **33,** No. 5, 212 (1950).
(*94*) H. S. DeGroot and F. G. Ewing, *Paper Trade J.*, **150,** No. 36, 32 (1966).
(*95*) J. Kline, *Pulp Paper*, **60,** No. 12, 154 (1976).
(*96*) E. D. Mazzarella and L. J. Hickey, *TAPPI*, **49,** No. 12, 526 (1966).
(*97*) T. E. Yeates, M. E. Carr, C. L. Mehltretter, and B. T. Hofreiter, "Cyanoethylated Starches and Dextrins as Coating Adhesives," 16th Coating Conf. TAPPI, Portland, Oregon, 10–13 May, 1965.
(*98*) Anon., "Starch and Starch Products in Paper Coating," TAPPI Monograph No. 17, 1957, pp. 26–38.
(*99*) L. W. Zabel, ed., "Lectures on Papermaking," 1972, Vol. 2, p. 392.
(*100*) Anon., "Pigmented Coating Processes," TAPPI Monograph Series No. 28, 1964.
(*101*) Anon., "Paper Coating Additives," TAPPI Monograph Series No. 25, 1963.
(*102*) R. Grant, *TAPPI*, **53,** No. 2, 261 (1970).
(*103*) S. L. Oakleaf and R. L. Janes, *TAPPI*, **60,** No. 11, 95 (1977).
(*104*) D. G. Margotta and H. E. Gaither, *TAPPI*, **53,** No. 3, 401 (1970).
(*105*) Anon., Annual Report, Fiber Box Industry, Chicago, Illinois, 1978, p. 35.
(*106*) Anon., *Paperboard Packaging*, **63,** No. 8 (August, 1978).
(*107*) Anon., *Paperboard Packaging*, **64,** No. 1, 51 (January, 1979).

(*108*) J. A. Radley, "Industrial Uses of Starch and Its Derivatives," Applied Science Publishers, London, 1976, pp. 36–38.
(*109*) W. S. Thayer and C. E. Thomas, "Analysis of the Glue Lines in Corrugated Board," *TAPPI*, **54,** No. 11, 1853 (1971).
(*110*) W. O. Kroeschell, ed., "Preparation of Corrugating Adhesives," TAPPI Press Books, Atlanta, Georgia, 1977.
(*111*) J. J. Schoenberger and R. P. Citko, U.S. Patent 3,355,307 (1967).
(*112*) J. M. Billy, U.S. Patent 3,524,750 (1970); *Chem. Abstr.*, **73,** 132230s (1970).
(*113*) J. P. Casey, L. O. Gill, and R. A. Sherman, U.S. Patent 2,610,136 (1946); *Chem. Abstr.*, **47,** 864e (1953).
(*114*) J. V. Bauer and L. H. Elizer, U.S. Patent 3,019,120 (1962); *Chem. Abstr.*, **63,** 10160a (1965).
(*115*) C. B. Musselman and E. M. Bovier, U.S. Patent 3,912,531 (1975); *Chem. Abstr.*, **83,** 207825g (1975).
(*116*) A. J. Desmarais, U.S. Patent 3,151,996 (1964); *Chem. Abstr.*, **61,** 16292d (1964).
(*117*) G. H. Klein, H. L. Arons, J. F. Stejskal, D. G. Stevens, H. F. Zobel, and L. Wondolowski, Ger. Offen. 2,633,048 (1977); *Chem. Abstr.*, **86,** 173394k (1977).
(*118*) J. P. Casey and E. R. Lehman, U.S. Patent 3,015,572 (1962); *Chem. Abstr.*, **56,** 7565d (1962).
(*119*) Anon., *Pulp Paper,* **51,** No. 8, 107 (1977).
(*120*) C. H. Sprague, *TAPPI,* **62,** No. 6, 45 (1979).
(*121*) R. H. Williams and L. J. Hickey, U.S. Patent 3,300,360 (1967); *Chem. Abstr.*, **66,** 66965n (1967).
(*122*) S. E. Boie, *Boxboard Containers,* **86,** No. 1, 90 (August, 1978).

CHAPTER XIX

APPLICATIONS OF STARCHES IN FOODS

By C. O. Moore, J. V. Tuschhoff, C. W. Hastings, and R. V. Schanefelt

A. E. Staley Manufacturing Company, Decatur, Illinois

I. Introduction

The earliest utilization of starch in food was as a nutritional material obtained from fruits, vegetables, and roots or seeds, the latter having the advantage of stability for storage. At some point, it was discovered that the seed could be crushed and cooked in water to become more palatable and effective as a binder in cereal or dough products, and the refining of starch for its functional properties was in motion.

Today, starch and its subsequent refinements is a nutritive, abundant, and economical food source in a rapidly overpopulating world (*1*, *2*). Consumption of starch occurs via two routes. The preparation of fruit and vegetables by the consumer for table use provides starch in essentially its native state. However, a large amount of food must be mass produced and must be shelf stable to ensure a supply at affordable price. This food, termed processed food, is produced from what are basically home recipes modified to produce convenient, shelf-stable,

STARCH, 2nd ed.

ISBN 0-12-746270-8

and available products (*3, 4*). In the processed food industry, basic ingredients such as starches, sweeteners, proteins, and oils are separated from the source materials and re-incorporated into food products as needed. During the course of separation, the functionality of basic ingredients can be altered and improved to meet requirements of food processing.

Food starch manufacturing has developed during the past 40 years from a basic separation process for obtaining powdered starch into the production of diversified lines of sweetener and starch products (*5, 6*). Food starches perform two basic roles. As a nutritive stabilizer, starches provide the characteristic viscosity, texture, mouth-feel, and consistency of many food products (*7, 8*). Starch can serve this purpose either gelatinized or in uncooked granular form. Examples of food products in which starch serves this purpose are sauces, puddings, gumdrops, and tableted products. The second primary use of starch is as a processing aid to facilitate manufacturing, the most obvious example being the classic use of powdered corn starch to dust work surfaces or inprocess material to prevent sticking. Molding starch is an integral component of the confectionery operation, where it serves both to form the candy piece and to dry it. Selected starches can be included in a formula to improve processability. Examples include the use of modified thick-thin starches to provide uniformity in the filling operations for liquid and particulate formulations, thin-thick starches that can increase production capacity and quality in continuous retorts, and the use of pregelatinized starches to control the spread or cold flow of products in extrusions. Hundreds of food starches are used in the canning, dairy, confectionery, baking, packaged mix, brewing, frozen food, and pharmaceutical industries (*3, 5, 7, 9–11*).

II. Food Starch Sources

Starches differ from plant to plant and, hence, from one commercial source to another. With proper refining, many starches can be prepared so that they perform almost equivantly. On the other hand, the uniqueness of some starches can be brought out by refining so that they excel in certain applications. However, locale, growing conditions, and crop availability often dictate the choice of starch for specific uses. Unique properties command a premium price for starches such as tapioca, waxy maize and sorghum, and high-amylose corn starches.

A large volume of starch is available as a food ingredient in the form of various flours and ground meals. These products are not pure starches, but contain quantities of protein, oils, fiber, and other plant components. Large volumes of flour and meal are used in the baking and snack industries, but in many other foods, starch alone is required (*4*). The wet-milling process provides a more complete separation, and wet-milled starch can be obtained with only trace amounts of other plant components. The principal food starches used in the

United States are corn, tapioca, potato, and wheat starches. In certain regional or geographical areas, sorghum, rice, arrowroot, and sago starches are used (*3, 12*).

1. Corn Starch

Corn starch accounts for the major share of U.S. food starch usage which is about 4 billion pounds (1.8×10^9 kg) per year. Dent corn, which is the common commodity used for animal feed, sweetener, and starch production, is available in nearly unlimited quantity for food starch manufacturing. Corn starch is the most inexpensive food starch available and is used in food products both in its native powdered form and in modified forms. For many years, powdered corn starch was the mainstay of the food starch industry. With growth in the industry, a succession of simple modifications evolved which improved specific properties. Simple treatments such as pH adjustment, bleaching, oiling, agglomerating, and re-drying have resulted in a line of products commonly referred to as "unmodifieds," corn starch, or powdered starches. Powdered corn starch is cheap and is the practical choice for use as dusting powder, molding starch, fillers, bulking agents, and as a cooked stabilizer in a number of products in both cook and serve foods and shelf-stable canned goods (*5, 13*).

When used as a stabilizer, corn starch can be cooked at atmospheric pressure in most formulations if the moisture content is 50% or greater. Cooking develops a thick, heavy viscosity and translucent appearance. On cooling, the starch sets to an opaque, resilient, short-textured gel. Corn starch is an acceptable thickener for retail or institutional uses involving cook and serve products such as gravies, sauces, puddings, or pie fillings where shelf life expectancy is a matter of hours or a few days.

A large volume of corn starch is consumed in spreadable salad dressings, where its gel property helps to develop the stiff, short consistency[1] of these products. However, some dressings, on lengthy storage, may undergo severe textural changes due to aging of unmodified starch gel. The most common change is that the gel shrinks until cracks appear and free liquid is released from the matrix. Freezing and cold storage accelerate this problem more than canning and room temperature storage. Generally, unmodified corn starch is avoided in frozen foods, but in certain canned goods like soups, stews and chili, unmodified starch gives marginal performance. Canned goods of this type are generally disturbed, diluted, and reheated by the consumer, and the product recovers enough consistency to be an acceptable serving. However, such products generally have a poor appearance in the can and suffer accordingly when compared with products made from high-performance modified starch.

[1]A "short" paste is one that breaks abruptly when allowed to flow from a stirring rod, as opposed to a "long" (cohesive) paste which forms long strings under the same conditions.

Many food products require a stabilizer that thickens to a stable viscosity without the subsequent gel formation of unmodified corn starch. Clarity, flavor, and durability to withstand rigorous processing conditions are also common requirements.

Starch owes much of its functionality as a stabilizer to two structural characteristics. One, is that it is composed of two distinct polymer fractions, amylose and amylopectin. A second, is that the organization of these large molecules into spherocrystalline granules allows swelling in heated water while maintaining a degree of granule integrity and identity (*12*, see also Chaps. VII and IX).

Amylose and amylopectin differ markedly in their rheological and viscolastic properties (see also Chap. VIII). Amylose, when released into solution, readily hydrogen bonds to form rigid, opaque gels. Amylopectin, because of the branched structure, has a limited ability to hydrogen bond and solutions remain relatively clear and fluid. Both polymers are hydrophilic, and when heated in a food system, are readily hydrated by the uncombined water to form a paste. Starches generally contain from 17% to 24% amylose and from 76% to 83% amylopectin. However, waxy maize starch contains 99% or more of amylopectin and high-amylose corn starch can contain 70% or more of amylose (see also Chap. III). The interaction of these polymers is an important aspect of the rheological properties exhibited by cooked starch paste.

Intact swollen starch granules are major contributors to the rheological properties of a starch paste. Native starch granules are insoluble in cold water. However, when heated in water, the granule begins to swell when sufficient heat energy is present to overcome the bonding forces of the starch molecules (see also Chap. IX). With continued heating, the granule swells to many times its original volume. It is the physical force or friction between these highly swollen granules that is a major factor contributing to starch paste viscosity. With continued heating or cooking, the starch granule ruptures and disintegrates to give a dispersion of amylose, amylopectin, and granule fragments. In some food processing, it is necessary to cook the product until all granules are gelatinized and mainly dispersed free molecules are present. On cooling, these dispersions set to special types of gels. However, optimum paste properties for many food applications generally are obtained when the starch granule is only highly swollen. Overcooking usually results in rubbery or gummy gels, while an undercooked starch paste is opaque and watery. Optimum degrees of swelling are not always easy to obtain because of the swollen granule's susceptibility to fragmentation resulting from acidity, pressure, and shear in processing. The range of tolerance between undercooking and overprocessing is often very narrow for unmodified starch.

Problems associated with the use of unmodified starches in foods occur because granules are not always easy to gelatinize due to lack of water or low temperatures and because, once gelatinized, the granules are easily disrupted and changes in rheology and paste characteristics result. Furthermore, when amylose leaches into solution, it produces undesirable set back (retrogradation), gel hardening, or precipitate formation (see Chaps. VI and VIII).

2. Waxy Maize and Waxy Sorghum (Milo) Starch

Waxy maize starch has an amylopectin content of approximately 99%. The outward appearance of the grain and the starch separated from it is similar to that of common dent corn, but the rheological properties of the starch are quite different. Paste properties of waxy starches are more like those of the root starches than those of other cereal starches. Cooked pastes of waxy starches develop heavy viscosity and good clarity. Being essentially free of amylose, pastes have above average resistance to syneresis and exhibit a somewhat elastic, stringy consistency when cooled, in contrast to the short, resilient corn starch gel. Many users of waxy maize starch are of the opinion that it imparts less cereal-type flavor to foods than common corn starch. It is an excellent material for modification to produce a versatile line of food stabilizers. Waxy starches command a major share of the market for all-purpose modified thickeners.

3. High-Amylose Corn Starch

High-amylose corn, also called amylomaize starch, has an amylose content of from 50 to 70% or higher and is a relatively newcomer to the field of food starches. High-amylose corn starch is resistant to swelling and cannot be cooked effectively in typical open-kettle processing. The elevated temperatures of pressure cooking are required, and most processing is accomplished with continuous cookers utilizing steam injection or swept surface heat exchangers. Retorting, high-temperature extrusion, and deep fat frying can be used also. The increase in amylose content makes high-amylose corn starch gel rapidly and form high-strength gels. These properties are useful in the confectionery industry where candy pieces require a stabilizer to supply individual piece shape and integrity. Modification of high-amylose starch produces starches for new applications (*6, 14*).

4. Potato Starch (see also Chapter XIV)

Potato starch is the primary starch of commerce in Europe, but in the United States it is a distant second to corn starch. Rheological properties of potato starch are typical of root starches. Root and tuber starches that develop in high moisture plant tissues generally have large granules and cook and solubilize more readily than do the grain starches. When cooked, the large swollen granules impart extra high viscosity and a subtle grainy effect in the thickened paste. Easily over-processed, the starch forms a smooth pituitous paste that retrogrades on cooling to a gummy, but pliable, gel. The paste has good clarity, but synereses like corn starch, especially in cold storage or in freeze-thaw applications. Food technologists describe cooked potato starch as having a slightly earthy flavor.

Commercial food applications for potato starch in the United States are primarily in the canning industry where it is used in retorted products. It is also used in pregelatinized products and in blends with other starches where its exceptionally high thickening ability provides advantages (*3*).

5. Tapioca Starch (see also Chapter XIII)

Tapioca products are imported primarily from Asia and South America in either powdered form or as tapioca pearls. The native starch forms a clear but stringy and cohesive paste when cooked and, like unmodified waxy starches, finds limited usage in this form since the texture is not desirable. Some food manufacturers prize tapioca for its bland flavor, and it commands a premium price when modified to eliminate the textural problem. Delicately flavored puddings, pastry fillings, and baby food products have traditionally been prepared with tapioca starch because of its flavor advantage. Some manufacturers also prefer modified tapioca starches over waxy starch products, as tapioca produces a slight gel property in puddings and pie fillings, while duplicating the other desirable attributes of waxy starches. Pregelled tapioca starch is widely used because of its excellent flavor characteristics. However, use of modified tapioca starch as a general thickener has given way to use of waxy maize starch and mixtures of waxy and common corn starches, largely because of cost differences.

Tapioca pearls are dense agglomerates of uncooked and partially gelatinized starch granules that are commonly made by stirring or rolling damp starch in a container to the desired particle size and dried. Some success has been had in modernizing this procedure through extrusion forming (*12*).

6. Wheat Starch (see also Chapter XV)

Wheat starch is a by-product in the manufacture of gluten. Residual protein in the starch gives it a flour-like odor, flavor, and appearance. Rheological properties of the starch are similar to those of corn starch, although viscosity and gel strength are not as high. Its principal use is in baking where it replaces portions of wheat flour. Functional advantages over other starches are in the increased volume and tenderization of cakes and in the reduction of fat absorption in doughnuts. Limited usage has occurred in the confectionery and canning industry (*3*).

7. Miscellaneous

Other starches that have limited specialty use in the food industry are arrowroot, sago, and rice starches. Arrowroot starch, which is reported to be easily digested, is used in biscuits for children (see Chap. XIII). Sago starch has a high gel strength which is desirable in gum candy (see Chap. XIII). Rice starch has application in the brewing industry (see Chapter XVI). Waxy rice starch has minor uses as a thickener in white sauces and gravies because of its good freeze-thaw stability (*3*, *12*).

III. Modification of Food Starches (see also Chapter X)

Improved performance of starches may be realized by chemical modifications; and the rapid growth of new food technology, especially in the area of convenience foods, has resulted from this fact. In applications requiring specific and narrow process and product requirements, chemically modified or ''tailor-made'' food starches have been most successful. Without the special rheological properties obtained by chemical modification, frozen, instant, dehydrated, encapsulated, and heat and serve foods and cold water swelling products would not be economically competitive.

Several basic properties of starches of concern to food processors are improved by starch modification; among these are improved heat, shear, and acid stability. Either one or a combination of these characteristics is required in most food processes where the properties of unmodified starch are insufficient. Improved gel characteristics provide the smooth spoonable or flowable consistency required in certain food products. Native starches may have suitable properties when freshly cooked, but they do not have shelf stability. Likewise, the temperature at which the starch thickens when subjected to heat can be adjusted to improve processing and finished product quality. Cold storage, temperature cycling, and freezing accelerate the breakdown of systems stabilized by native starch, especially in foods of high moisture content. Viscosity, moisture binding, stabilization, and specific textural properties can be controlled by modification.

Most modified or derivatized starches are made in aqueous slurry by reactions between starch and small percentages of FDA approved chemicals. The quality and purity of chemically modified starches meet performance standards and comply with manufacturing practices specified (*15*) in FDA regulation 21 CFR 172 892.

The most important reaction in the modification of food starches is the introduction of substituent chemical groups. These chemical modifiers are of two types, monofunctional and di- or polyfunctional. The number of reactive groups determines the manner in which a chemical modifier will alter rheological properties (*16–19*).

Monofunctional reagents react with one or more hydroxyl groups per sugar unit to alter the polarity of the unit, sometimes making it ionic, and markedly influence the rheological properties of the starch. In general, monofunctional substitution lowers the swelling (pasting) temperature, increases paste clarity, reduces gel formation, and improves freeze-thaw and water-holding properties.

Monofunctional reagents most often used for food starch are acetic anhydride and propylene oxide (*3, 5, 7, 9, 20, 21,* Chap. X). Acetic anhydride reacts to produce starch acetate (ester linkage), while propylene oxide forms hydroxypropylstarch (ether linkage). These chemical reagents have been in use for some 30 years to produce modified food starches. Conditions governing their use are listed in FDA Regulations.

Reaction of starch with di- or polyfunctional reagents is possibly the most dramatic and useful reaction in modification of food starch. Here, reagents with two or more reactive groups react with the hydroxyl groups of starch to form a crossbond or bridge between polymer chains. Reaction with as little as 0.0005% (based on the weight of starch) of a cross-linking reagent can affect rheological properties. Cross-linking alters starch properties by increasing the pasting (swelling) temperature, by stabilizing swollen granules during cooking to obtain desirable texture and mouth-feel, by increasing and/or stabilizing viscosity in severe cooking or in conditions of high shear and low pH, by giving improvement in freeze-thaw and water-holding properties, and by reducing clarity of starch pastes (*22*).

Reagents most often used for cross-linking of starch are phosphoryl chloride (phosphorus oxychloride) and epichlorohydrin, which form respectively ester and ether linkages (see Chap. X). These reagents have also been used for more than 30 years.

Sometimes double derivatization is employed, usually by first reacting the starch with a monofunctional reagent and then by carefully adjusting rheological properties to the desired endpoint by use of a polyfunctional reagent. Careful control is required to obtain optimum balance of mono- to polyfunctional reagent for desired properties. Thus, the food starch is "tailor-made" for specific, and usually narrow, processing conditions and product functionality (*23–25*).

Other food starch modifications of lesser importance are bleaching (oxidation) and acid-catalyzed hydrolysis. Bleaching is accomplished by treating starch with low levels of sodium hypochlorite, potassium permanganate, or hydrogen peroxide to improve dry starch color and/or to destroy microorganisms. Reagent levels of about 0.5% are usually employed, inducing some loss of starch viscosity due to breakage of starch chains (*21, 14*).

Acid-catalyzed hydrolysis, which shortens the molecular chains in the granule, is used to reduce hot viscosity and to improve adhesive properties (*26*). Acid treatment is conducted in wet (slurry), dry, or semi-dry conditions (see Chaps. XVII and XX). Slurry converted products are known in the food industry as thin-boiling starches and are generally made from corn starch. These products have a low hot-paste viscosity when cooked and yet develop good gel characteristics when cooled. Thin-boiling starch is used most extensively in candy production where a hot fluid material is deposited into molds and must then gel to a fixed consistency and shape (*20*).

Acid modification of starches by the dry process results in dextrins with characteristic thin viscosity, increased solubility, and a loss of gel properties. As adhesives, they are used in pan coating to prevent separation of the sugar shell from the base center material. Another use of special dextrins is to replace gum arabic in the encapsulation of flavor oils (*20*).

Pregelatinization of starch is a process of precooking starch to produce material that hydrates and swells in cold water. Products are called pregelatinized starches or "pregels." Drum drying is the most common method of preparation, but spray drying and extrusion cooking are alternative methods. In drum drying, a starch slurry or paste is spread on hot rolls as a thin film and cooked or dried before being scraped from the rolls. Water, time, and temperature can be varied to produce different products. The feed starch can be a chemically modified product to further extend the range of finished properties. Pregels, generally, have slightly less viscosity than their counterpart cooked starches and are less glossy and less smooth when dispersed in water.

IV. Food Starch Processing

1. Starch Properties and Food Processing Technology

Selection and use of starches in food products for consistent production quality and for innovative new products requires an understanding of basic food processing technology and an understanding of starch properties. Certain factors should be kept in mind. One is temperature. Although most starch granules begin to swell and lose their birefringency at ~60° when heated in water, different starches have different properties. For the most part, viscosity develops through the 65°–95° range; fragmentation and thinning occurs thereafter, if the material is held at an elevated temperature. This phenomenon is well demonstrated by the Brabender Visco-Amylograph which traces a curve of changing viscosity through a cooking cycle (Chap. IX). A typical procedure with this instrument begins with a 50° starch slurry being heated at the rate of 1.5°/min, up to 95°. A cooling cycle, which can indicate cold paste viscosity, may follow. The curves in Figure 1 demonstrate initial pasting temperature, rate of viscosity development, peak viscosity, and breakdown or thinning. Figures 2 and 3 show how modification treatments described earlier can alter a starch's cooking performance.

The cooking time of a starch is related to the cooking temperature. An example is high-amylose starch, which will not cook at atmospheric pressure in boiling water, requires ~145° for processing in a continuous pressure cooker, and yet will hydrate in 30 min at ~115°. In practice, hot filled containers of pie filling starch that are allowed to cool slowly, will overcook in the container, giving a thin paste at a temperature considerably below that of the original batch.

Another factor to be considered is moisture and dissolved solids. Dissolved

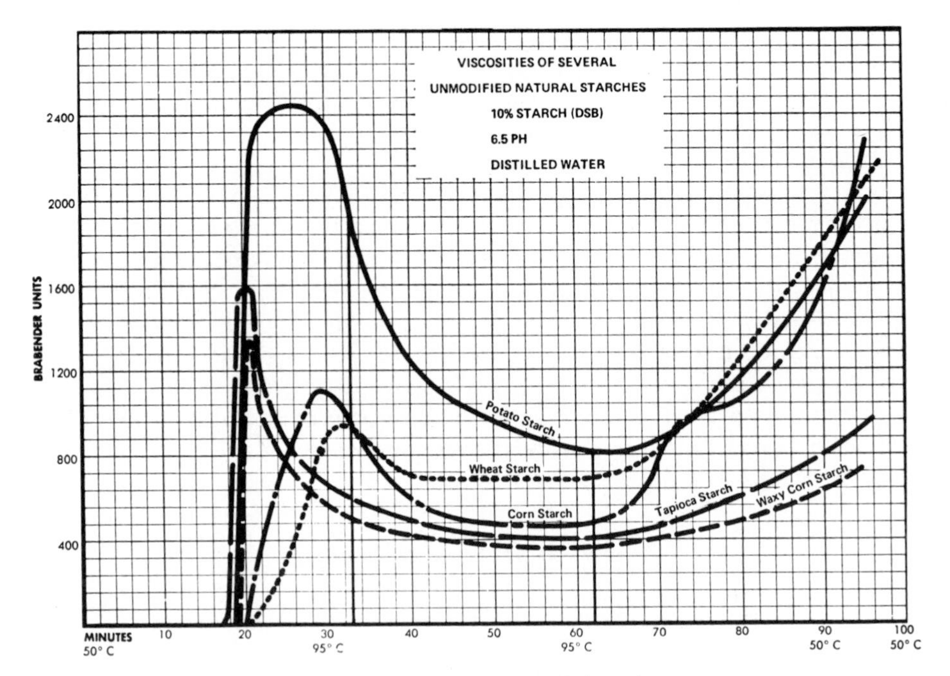

FIG. 1.—Viscosities of unmodified starches.

solids present in most formulations that include starch can very effectively inhibit hydration of granules. Dissolved solids may reduce the viscosity or gelling properties of the finished product. Such water-soluble substances include sucrose, corn sweeteners, sorbitol, milk solids, glycerol, and various salts. Sweeteners are used in the highest concentrations and have the most pronounced effects. Salts can produce changes in the rheology of cooked starch pastes, but sodium chloride, the salt most commonly present, is used at insignificant levels and does not ordinarily interfere with starch functionality. Phosphate setting salts are used to precipitate milk protein gels in combination with pregelatinized starch in convenience pudding mixes; here some type of complexing may occur (5).

Most food products have soluble solids contents in the range of 10–45%, and most starches are easily cooked in these formulations at about 100°. Some food items have an intermediate to high solid content of 50–80%; in these products, available moisture is insufficient to hydrate the starch in atmospheric cooking. In such products, the ingredients are subjected to continuous pressure cooking, for an increased cooking temperature partially compensates for lack of water. Current commercial practice of this type includes systems where live steam is mixed into the product and then flashed off upon discharge from the pressure unit, or swept surface heat exchangers are used in which the product receives heat through a thin metal wall but does not come into contact with the steam. Typical

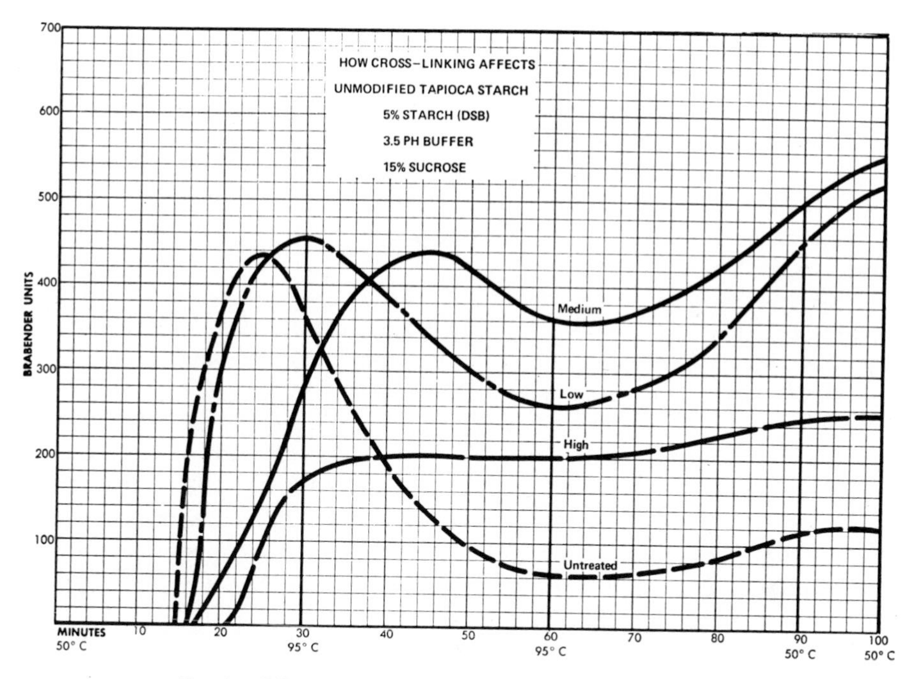

FIG. 2.—Effect of cross-linking on the viscosity of tapioca starch.

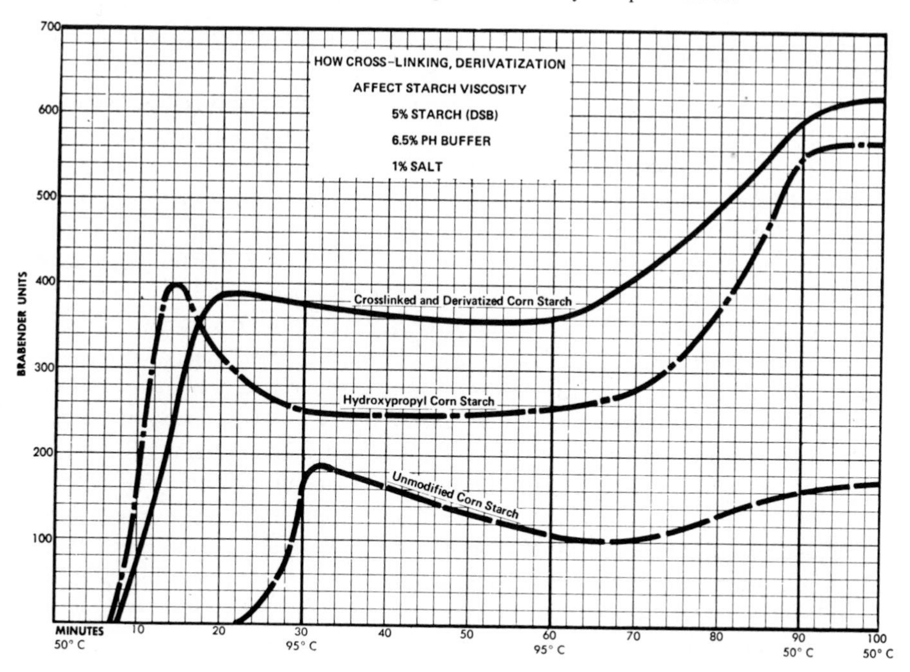

FIG. 3.—Effect of cross-linking and hydroxypropylation on the viscosity of corn starch.

pressure cooking temperatures run from 105° to 168°, depending on the product involved. In the absence of pressure cooking facilities, there are some procedural techniques that may be used in an open kettle: (a) A common baker's practice is to add dry sugar after the starch is cooked. (b) In formulations using thin-boiling starch, an excess of water is used to cook the starch, and the cooked product is evaporated to the desired finished solids content. (c) A third procedure is to cook a concentrated sweetener solution to a rolling boil and then add the starch as a water slurry.

Pregelatinized starches, although instant thickeners in water, are also affected by dissolved sugars. The hydration rate of a pregel can be delayed significantly by the density of the sugar solution. Although only a matter of minutes may be involved in the preparation of a convenience pudding mix, a pregel dispersed in a high solids corn syrup may not completely thicken for several hours. This property can be used advantageously for smoothly incorporating a pregel into water.

A third factor that can moderate cooking requirements is the use of a starch that has had its pasting temperature lowered by chemical modification. Some modified starches will swell even in cold water, without previous pregelatinization. Products of this type are currently expensive but may become important in our energy conscious future.

A fourth factor to be considered is pH. The presence of acids in many foods serves a number of key functions. Preservation is most important. Acids also have an important place as flavoring components. Interactions of acids with other ingredients occur. The effect of acids on starch can be either useful or detrimental.

Acids catalyze the hydrolysis of starch molecules. Hydrolysis is used to produce thin-boiling starches; carried further, it converts starch to a syrup. Although food acids are weak acids, the starch granule is weakened when cooked in an acidic system, and an undesirable loss of viscosity can occur if the treatment is severe. The effect is most noticeable in hot aqueous systems, but dry mix products can also be affected. Starch viscosity is lowered if acidified dry mixes of 5–10% moisture content are submitted to long storage.

FDA regulations give pH 4.6 as the dividing line between acid and low-acid foods for the purpose of securing proper sterilization in hermetically sealed food containers. However, in the pH range of 3.0–4.0, substantial thinning of cooked starch is of concern. Therefore, the starch-refining industry generally uses a pH of 3.5 to test the acid cooking performance of food starches. A pH value of 6.0 is used to evaluate starches intended for use in neutral or mildly acid food products.

Cross-linked starches have more resistance to effects of acid than other starches, the most highly cross-linked, being the most resistant to thinning. By using various levels of cross-linking, starches are produced that give proper performance under a particular processing condition (22).

Buffering agents are sometimes used in foods, if acidity is not needed to prevent microbial problems. High-solids, starch-gelled candy formulas, subjected to long heating, can be buffered with sodium citrate to control acid activity while allowing acid flavor to complement the fruit flavors.

Physical agitation is another factor that must be considered. In most manufacturing operations where starches are used as thickeners or stabilizers, a portion of the process involves pumping, stirring, or scraping to ensure uniform mixing and thorough cooking. Where swollen starch granules are used to develop viscosity in food products, the granule is susceptible to disintegration if subjected to physical impact or pressure. In open kettles, it is important to replace high-speed propeller or turbine agitators with slowly turning scrapers that sweep against the kettle wall or cooking surface. Raw starch granules are generally not damaged by vigorous motion in the uncooked slurry and can be well dispersed with high-speed mixing prior to cooking. Inadequate stirring can prevent thorough cooking, as thickened starch will form an insulating layer on the cooking surface if scraping is not efficient. Large kettles should have baffles that generate currents to work the center slurry outward to the cooking surface, since thickened starch paste has poor heat transfer properties (*5*).

Continuous pressure heat exchangers generally have swift scraping that subjects the product to more physical manipulation than occurs in open kettles, but modified starches are handled well. Steam injection at near 100° provides a gentle treatment; however, steam injection at 165° produces the high turbulence needed in the processing of high-amylose starch, which is otherwise difficult to cook, especially if it is in high-sugar candy formulas.

Process equipment, such as gear or centrifugal transfer pumps, can cause viscosity reductions in certain cooked starch formulations, especially if recycling is involved. Homogenizers, colloid mills, and high shear mixers will disintegrate cooked starch granules under vigorous operating conditions. For products such as salad dressing, some starch granule destruction in the colloid mill is an acceptable trade-off to obtain desired finished product consistency (*27*). In some pudding products, homogenization is a beneficial treatment if it occurs prior to cooking, but is detrimental afterward.

It is important to remember that a promising laboratory formulation can change significantly when scaled up, owing to the effect of equipment on food starch functionality.

2. Complexing Agents and Interactions

When cooked starch becomes part of a matrix with other common food ingredients, the character of the starch–water gel changes. For example, the addition of dry sugar to a hot starch paste lowers viscosity, although the percent of water available is reduced. If moisture is removed from a high-solids, sugar–starch gel, the optical transparency increases significantly. When starch and milk protein are cooked together, the resultant viscosity may be greater than that of the individual ingredients used alone at the same total concentration. Fats and emulsifiers modify mouth-feel by smoothing and shortening the consistency of starch pastes. The interactions are, for the most part, simply physical blending of characteristics; in some cases starch molecules are complexed with the ingredients.

Ingredients that produce changes in starch formulations are surfactants, emulsifiers, and complexing agents such as sodium 2-steroyl-lactate. A good example of a food starch formulation utilizing a complexing agent is the processing of aseptic puddings. These formulations are thickened with modified food starch and milk protein cooked at high temperature for a short time. Granules break up, and free starch molecules impart stringiness and textural graininess. Use of sodium 2-stearoyl-lactate in these formulas helps prevent these textural deficiencies.

Distilled monoglyceride products are also effective in the prevention of starch retrogradation in food products. Addition of 0.1% of monoglyceride to a starch jelly candy prior to high-temperature cooking reduces final gel strength. The effect of addition of monoglyceride after cooking is less pronounced.

V. Food Starch Applications

Starches are generally considered in three categories: (a) powdered food starches, (b) modified food starches, and (c) pregelatinized starches. This arbitrary classification system is a convenience of the industry for discussing functionality or marketing and should not be confused with FDA labeling requirements for food starch modified. FDA regulation 21 CRF 172.892 describes how products must be labeled, and products from all three categories may qualify.

Powdered food starches in various forms represent the earliest diversification of the food starch product line, but are still used in large volumes. Their applications are well understood throughout the industry. Products of this type include basic native starches, and variations of it with simple physical or chemical treatments that do not restructure the granule. The use of powdered corn starch for dusting and molding purposes has been mentioned earlier along with its use as a cooked thickener. Other products of this type include redried starches, where the equilibrium moisture of 12% has been dried to 3–5% for use in packaged dry mixes. Starch grits are produced by adding a small portion of gelatinized starch to a slurry of raw starch prior to drying, producing a coarse agglomerate that is much less dusty for use as a grinding aid or for fermentation. Bleaching produces whiteness desired for tableting purposes. The adjustment of powdered starch to an alkaline pH (pH 8–10) for use in acid foods is disappearing, but products are still available. Small amounts of oil dry blended into powdered starch lowers mobility for use in molding starch. Products made by blending powdered corn starch with pregelatinized starch or waxy maize starch for products such as batter coatings, candy jellies, and salad dressing complete the major use areas (*3, 20*).

In more recent years, greater emphasis has been placed on the development of specialized modified starches and applications for the relatively new genetic varieties. A broad spectrum of these specialty starch products has been commercialized, ranging from all purpose thickeners to specific starches tailored for one product application. Some food manufacturers have found that a good general purpose modified starch will furnish desired results in a number of items, but there still is a place for a special situation starch in many products (*18, 19, 28*).

The canning industry requires both retort temperatures of 120° for 20–60 min and processing at 145° for a matter of seconds. In the retort, the canned product must be heated to sterilization temperature and held for a prescribed length of time. If the food product contains a native starch, a general purpose modified starch, or flour-thickened sauce, thickening occurs quickly and slows heat penetration, thus, lengthening the heating treatment needed. If cans are rolled through the heating process, agitation of the ingredients occurs. If a modified starch is used that will gelatinize at elevated temperatures, the canned material remains fluid in the early heating stages and quick heat penetration can occur. Sterilization then occurs in a significantly shortened time, and the quality of the final product and nutrients suffer less.

Starches that have been modified to cook at a lower temperature can be useful. They can be used beneficially in cream-style canned corn, for example. The heat of retorting denatures the soluble protein early in the process. This curdling would result in an unsightly product unless starch modified to thicken at a low temperature is added to surround the protein with a protective colloid and eliminate the problem.

Vegetable soup can be thickened initially with low temperature gelatinizing starches, thus maintaining complete suspension and dispersal of particulate ingredients during the filling operation. When retorted, the starch breaks down to produce a broth consistency (*5*).

Not all canned foods are preserved by pressure cooking treatment. A large volume of products are acidic in nature, and the container can be hot filled and sealed to destroy the organisms that could survive this pH. The hot canned material slowly returns to ambient temperature which insures destruction of the organisms. Many pie fillings, pastry centers, and prepared sauces fit into this category. Native starches generally fail when exposed to the combination of heat and acid over long periods of time, but products of this type can be made with modified starches and hot filled at adequate temperatures and acidity with good safety margins against spoilage. Yet they are free from stabilizer breakdown.

The frozen food industry incorporates thickeners in many of their formulations. A homemade pie or pudding dessert made from modified corn starch will not crack or release free moisture into the container when left in the refrigerator (*8*).

Unmodified waxy corn starch has been used in canned pet foods and marshmallow candies.

The exceptional gelling properties of high-amylose starch has led to improvements in the manufacture of jelly candy. Through its use, the traditional method of drying thin-boiling starch gum drops for two days in molding starch has been reduced to a curing time of 8 h. Thus, the starch load and floor space requirements are reduced by one-half, and the daily production can be doubled on existing equipment.

Pregelatinized or cold-water swelling starches have enjoyed a fast rate of growth in recent years due to the increased public demand for convenience package mixes. The first generation of food products made with pregels included puddings, gravy mixes, pie fillings, glazes, and dressing preparations, where the purpose of the starch is primarily that of a thickener. The cake mix industry has rapidly converted to the extra-moist cake formulas, where pregelatinized starch softens the cake crumb and retains moisture in the baked product. Another growth area is the manufacturing process that extrudes and cuts doughs for baking or enrobing. Much of this material is prepared in cold mix processing. Pregel starches are used to control cold flow and slumping of the extruded shapes. Some of these starches are designed for snack formulations, where they contribute to the puffing properties of baked snack concepts (*5*, *29*).

Crop shortages and price spirals of natural ingredients such as tomatoes, cocoa, coconut, and fruit products have encouraged the investigation and development of suitable alternatives when needed. A line of starch products that impart textural characteristics similar to fruit pulp, tomato paste, and shredded coconut is now available, and their use has become established in non-standardized products such as pizza sauces, coconut candies, and baker's jellies.

VI. References

(*1*) Anon., "Products of the Corn Refining Industry in Food," Corn Refiners Assn., Inc., Washington, D.C., 1978.

(*2*) Anon., "Products of the Wet Milling Industry in Food," Corn Refiners Assn., Inc., Washington, D.C., 1979.

(*3*) C. A. Brautlecht, "Starch—Its Sources, Production, and Uses," Reinhold, New York, 1953, p. 19.

(*4*) L. F. Hood, *in* "Cereals for Food and Beverages," G. E. Inglett and L. Munck, eds., Academic Press, New York, 1980, p. 41.

(*5*) E. M. Osman, *in* "Starch: Chemistry and Technology," R. L. Whistler and E. F. Paschall, eds., Academic Press, New York, 1967, Vol. II, p. 163.

(*6*) J. O'Dell, *in* "Polysaccharides in Food," J. M. V. Blanshard and J. R. Mitchell, eds., Butterworths, London, 1979, p. 171.

(*7*) A. L. Elder and T. J. Schoch, *Cereal Sci. Today*, **4**, 202 (1959).

(*8*) C. T. Greenwood, *in* "Advances in Cereal Science and Technology," Y. Pomeranz, ed., American Association of Cereal Chemists, Inc., St. Paul, Minnesota, 1976, Vol. 1, p. 119.
(*9*) M. Glicksman, "Gum Technology in the Food Industry," Academic Press, New York, 1969, p. 274.
(*10*) A. A. Lawrence, "Edible Gums and Related Substances," Noyes Data Corp., Park Ridge, New Jersey, 1973, p. 326.
(*11*) Anon., "The Story of Starches," National Starch Co., Plainsfield, New Jersey, 1953.
(*12*) F. E. Kite, E. C. Maywald, and T. J. Schoch, *Staerke,* **15,** 131 (1963).
(*13*) Anon., "Corn Starch," Corn Refiners Association, Inc., Washington, D.C., 1979.
(*14*) Anon., "Review of Safety and Suitability of Modified Food Starches in Infant Food," Committee on Nutrition, American Academy of Pediatrics, Washington, D.C., 1978.
(*15*) O. B. Wurzburg, *in* "Handbook of Food Additives," T. E. Furia, ed., The Chemical Rubber Co., Cleveland, Ohio, 1968, p. 378.
(*16*) H. H. Schopmeyer, U.S. Patent 2,431,512 (1947); *Chem. Abstr.,* **42,** 1442 (1948).
(*17*) O. B. Wurzburg and C. D. Szymanski, *J. Agr. Food Chem.,* **18,** 997 (1970).
(*18*) T. A. White, *Cereal Sci. Today,* **8,** 48 (1963).
(*19*) R. W. Kerr, "Chemistry and Industry of Starch," Academic Press, New York, 2nd Ed., 1950, p. 535.
(*20*) L. H. Kruger and M. W. Rutenberg, *in* "Starch: Chemistry and Technology," R. L. Whistler and E. F. Paschall, eds., Academic Press, New York, 1967, Vol. II, p. 369.
(*21*) C. H. Hullinger, *in* "Starch: Chemistry and Technology," R. L. Whistler and E. F. Paschall, eds., Academic Press, New York, 1967, Vol. II, p. 445.
(*22*) J. W. Evans, U.S. Patent 2,806,026 (1957); *Chem. Abstr.,* **52,** 612 (1958).
(*23*) T. J. Schoch, *Wallerstein Lab. Commun.,* **32,** 149 (1969).
(*24*) E. R. Morris, "Polysaccharide Conformation as a Basis of Food Structure," Unilever Research, Sharnbrook, Beds, U.K.
(*25*) P. Shildneck and C. E. Smith, *in* "Starch: Chemistry and Technology," R. L. Whistler and E. F. Paschall, eds., Academic Press, New York, 1967, Vol. II, p. 217.
(*26*) T. J. Schoch, *Glass Packer Process.,* **45,** 18 (1966).
(*27*) L. J. Filer, Jr., *Nutr. Rev.,* **29,** 55 (1971).
(*28*) C. Mercier and P. Feillet, *Cereal Chem.,* **52,** 283 (1975).

CHAPTER XX

STARCH AND DEXTRINS IN PREPARED ADHESIVES

By H. M. Kennedy*

Research Department, Acme Resin Company, Division of CPC International, Forest Park, Illinois

AND

A. C. Fischer, Jr.

Technical Sales Service, Corn Products, Summit-Argo, Illinois

I. Introduction

The earliest known use of starch as an adhesive was for bonding papyrus strips by the Egyptians nearly 6000 years ago (*1*). Starch still finds use today in prepared adhesives, but to a much lesser extent than do dextrins. It was the

*Present address: Grain Processing Corporation, Muscatine, Iowa

STARCH, 2nd ed.

ISBN 0-12-746270-8

Table I

Starches for Paper Converting

Product	*Paste characteristics*	*Applications*
Unmodified starch	Low solids (5–15%) Slow tack and drying Sets to a stiff gel Strong final bond	Corrugated board Bag-bottom paste
Thin-boiling starches	Low viscosity High solids Good film strength	High-solids bag-bottom pastes Bag seam pastes One-tank corrugating carrier portion
Starch ethers	Stable viscosity Good film clarity Flexible film	Bag-seam pastes
Waxy starches	Little or no setback High gloss Good film clarity Reduced blocking Excellent remoistening characteristics	Gummed tape Bottle labeling Viscosity stabilizer in general adhesive formulas
Pregelatinized starches		
a. Thick-boiling	Low solids Slow tack and drying Sets to a soft gel	One-tank corrugating carrier portion
b. Thin-boiling	Improved adhesion Controlled setback Fast tack Thixotropic	Envelope back seam High-solids bag-bottom pastes Library pastes
c. Chemically modified (borated)	Higher use solids Fast tack Good adhesion	General cold-water adhesive

partial burning of starch stored in a textile factory in Dublin, Ireland that led to brown, sticky starch powder and ultimately to commercial pyrodextrin production (*2*). Subsequently, it was discovered that dextrins having a wide range of solution properties could be made by dry-roasting starch alone or in the presence of an acidic or alkaline catalyst.

In 1977, 158 million pounds (71.7 million kg) of prepared dextrin pastes and 38 million pounds (17.2 million kg) of adhesives based on starch were sold in the United States, according to the Bureau of Census (*3*). These numbers do not include dextrins and starches sold directly to the paper converters. The capacity to formulate adhesives of higher solids level generally accounts for the greater use of dextrins over starches. Dextrins give faster tack development, more rapid drying, and increased production rates.

Starches and dextrins are used in numerous adhesive applications, such as tube winding, laminating, case and carton sealing, bottle labeling, flat gumming, envelope sealing, and gummed tape manufacture. These products are also used in cigarette seam adhesives, library pastes, wallpaper adhesives, and single and multi-wall bag adhesives, as well as numerous other applications. Tables I and II list the types of starches and dextrins used in a variety of adhesive applications in the paper-converting industry (*4*). Paste and adhesive characteristics often associated with specific starches or dextrins are also described.

Table II

Dextrins for Paper Converting

Product	*Paste characteristics*	*Applications*
White dextrins	Light-colored pastes Excellent adhesion Wide range of solubilities	Bag-seam pastes Tube winding Case and carton sealing Laminating Gummed sheets and labels Envelope back seams
Canary dextrins	Stable viscosity High solids Excellent remoistening properties	Gummed tape and sheets Envelope front seals Stamps Case and carton sealing Laminating Tube winding
British gums	Dark color Stable viscosity Fast tack Wide range of solubilities Higher viscosity than white dextrin of comparable solubility	Solid fiber laminating Bag-seam pastes Convolute tube winding
Waxy starch dextrins	Excellent viscosity stability High solids Excellent remoistening properties	Envelope front seal Gummed sheets U.S. Government and trading stamps
Dextrin/silicate blends	Fast tack Rigid films Maximum adhesion Low viscosity	High-speed case sealing Spiral tube winding Fiber foil cans
Borated dextrins	Fast tack Maximum adhesion Stable viscosity	Case and carton sealing Non-water-resistant bag-seam pastes Spiral tube winding Fiber foil cans Laminating

Table III

Manufacturing Conditions for Dextrins

	Dextrin type		
	White	*Canary*	*British gum*
Roasting temperature	110°–130°	135°–160°	150°–180°
Roasting time (h)	3–7	8–14	10–24
Amount of catalyst	high	medium	low
Solubility	low to high	high	low to high
Viscosity	low to high	low	low to high
Color	white to cream	buff to dark tan	light to dark tan

II. Dextrin Manufacture

Dextrins can be prepared which, depending on manufacturing conditions, range in color from white to tan, have low to high solubilities in cold water, and give pastes that vary widely in viscosity. Manufacturing variables include type of starch and moisture content, roasting time and temperature, and the type and amount of catalyst used. Dry-roasting causes both depolymerization of starch and condensation of starch fragments. The result is more highly branched molecules with greater solution stability than the parent starch has.

Dextrins are commonly placed in three categories: white, canary, and British gums. Relative conditions for manufacture of these dextrins and their physical properties are given in Table III.

1. Type of Starch

Dextrins are produced from all commercial grain and tuber starches. The conversion process is essentially the same for all starches for manufacture of a given dextrin, but ease of conversion varies with starch type and quality.

Potato starch is generally regarded as the easiest to convert, followed closely by tapioca and sago starches. Corn starch and other cereal starches require longer converting times and higher temperatures to reach a given level of dextrin conversion than do potato or tapioca starch. Corn starch, however, is the major source for dextrins in the United States because of its low cost and ready availability.

2. Redrying

Depending upon the type of dextrin desired, the moisture content of the starting starch should range from 1–5%. This moisture range is obtained either in a dextrin converter or in a converter operated under vacuum so that moisture is removed rapidly with minimal heating. High moisture in the starch is undesirable because it promotes starch hydrolysis and suppresses condensation reactions. Condensation reactions are usually negligible unless starch moisture content is 3% or less.

3. Acidification

Acidification is accomplished by spraying powdered starch with a dilute solution of acid, usually hydrochloric acid, or acid salts. The primary concern in acidification is to avoid non-uniform distribution of the catalyst in the starch. To obtain uniform distribution, a mixer, either horizontal or vertical, is used. After the proper amount of catalyst is added, sufficient mixing time is employed to ensure that the catalyst is evenly distributed throughout the starch. Black speck formation is minimized when the starch moisture is below 5%.

Hydrochloric acid (or hydrogen chloride gas) is the most common catalyst used in dextrin conversions because of its high activity, low cost, volatility, and availability. As dextrinization proceeds, the acid will volatize to some extent, and neutralization of the finished product sometimes can be omitted.

Oxidative catalysts, such as chlorine, have been used to obtain dextrins that are more stable to retrogradation or setback. Oxidation during dextrinization results in the formation of a small number of carboxyl groups which retard retrogradation and increase paste clarity. Use of chlorine gas is claimed by Berquist (*5*), Fuller (*6, 7*), and Kerr (*8*) to produce superior dextrin products. Treatment of starch with monochloroacetic acid followed by chlorine gas was found by Bulfer and Gapen (*9*) to be especially effective in producing white dextrins of improved quality. Nitric acid, which functions both as an acid and oxidizing agent, has been used in dextrinization (*10*).

Kerr (*11*) claimed that addition of 0.05–0.50% aluminum chloride to starch at pH of 2.8–3.4 resulted in improved products with shorter converting times. Use of calcium chloride was suggested by Bloede (*12*), who claimed that it acts as both a catalyst and modifying agent.

Dextrin manufacture under alkaline conditions is patented (*13*). Solutions of sodium carbonate, ammonium hydroxide, and urea (*14*) are blended with the starch in quantities sufficient to maintain the pH at the desired level throughout the heating cycle.

4. Converting

Conversion is usually made in vertically or horizontally mounted mixers holding anywhere from 100 lb (45 kg) to 10,000 lb (4500 kg) of starch and heated either by steam or oil jackets or by direct heating. The rate of temperature increase and the residual moisture content during the heating cycle are critical for proper conversion. High moisture content, coupled with a slow rise in temperature, results in low-viscosity products with high reducing sugar levels. A stream of air is sometimes used to reduce the moisture content more rapidly. Heating rate is limited by the maximum temperature differential that can be tolerated without charring the starch next to the walls of the heating elements. A low-pressure atmosphere during the converting phase of the process is suggested by Staerkle and Meier (*15*).

Heating and agitation are continued to develop specified levels of viscosity, soluble material, and color. The dextrin is then transferred to another mixer for immediate cooling to prevent over-conversion. To facilitate cooling, the mixer not only is water-jacketed, but cold water may be circulated through the ribbon-type agitator. Once cooled, the dextrin may be neutralized by the addition of ammonia or other alkaline material, or it may be packed and shipped without neutralization.

5. Equipment

Bulk-type converters in which several thousand pounds of starch can be converted at one time are common for the manufacture of dextrins. During the roasting period, the mass of acidified starch is slowly stirred by a gear-driven ribbon agitator. Heating is effected by steam, hot oil, or gas combustion products being circulated through the side and bottom walls of the converter. With bulk-type converters, heatup is slow and it is difficult to produce a uniform temperature throughout the entire batch of starch. Good agitation is important to facilitate the rate of heat transfer through the converter. Rowe and Hagen (*16*) patented an improved steam-jacketed cooker containing a vertical agitator. By dividing the jacket into zones having both a steam inlet and outlet, a more uniform transfer of heat to the starch is achieved. Phillips (*17*) developed a continuous conversion process in which a moving belt carries a thin layer of acidified starch through a heated oven.

In the patented process of Staerkle and Meier (*18*), starch is dextrinized in a nonoxidizing environment to avoid discoloration. The process uses reduced pressure or an atmosphere of nitrogen, sulfur dioxide or carbon dioxide, and application of acid to the starch as a vapor after partial drying. It is stated that other chemicals may be added as desired and that product derivatization may be concurrent with dextrinization.

Ziegler and co-workers (*19*) devised a method for roasting dextrin in which the starch is passed through a heated spiral-tube conveyor. Starch is moved through the tube by means of vibrations having a frequency of 1250/min and an amplitude of 2–4 mm, with the dextrin being discharged to a conveyor.

Frederickson (*20*) recommends converting starch to dextrins in a fluid bed. Subsequently, 1% calcium phosphate is added to promote fluidization of the starch (*21*). Improvements in time, quality, and product uniformity are claimed for the fluid bed process. Idaszak (*22, 23*) added mechanical agitators at both the upper and lower ends of the fluidizing chamber of a fluid bed. These agitators provide vigorous mixing that eliminate dead zones where the starch could overheat causing fires and explosions. Also, the added agitators eliminate channeling of materials through the bed.

III. Adhesive Preparation

To be effective adhesives, granular starches must be cooked (gelatinized) in water to form a paste. Only in this state can the starch granule fragments and molecules of amylose and amylopectin be expected to adhere to the substrates and subsequently rejoin one another for effective bonding. In essence, a bond is a film between the two substrates which may vary in thickness depending on the type of adhesive used and its applications.

Surprisingly, even completely cold-water-soluble dextrins require cooking to maximize their hydration. The differences in viscosities shown in Table IV demonstrate the effect of cooking on a nominally 100% soluble canary dextrin. In the comparison, two slurries of dextrin were prepared at 33% solids. One was pasted by heating to 95° and allowed to cool to room temperature, while the other remained at ambient temperature. The pasted sample showed gel structure development on standing, whereas the unpasted sample was without structure, showing only an increase in viscosity with time. The higher viscosities of the pasted sample were measured after the initial gel structure was destroyed by stirring.

Starch and dextrin adhesives are usually cooked in upright cylindrical or horizontal trough-like tanks. The tanks are equipped with a propeller, anchor, or ribbon-type agitator and are fitted for either direct steam injection or with a jacket for heating and cooling. Baffles are considered to be beneficial if the agitator rotates slowly. High-speed agitators should be avoided since severe shearing

Table IV

Effect of Cooking on Viscosity[a] of a Canary Dextrin

Time	*Cold slurry, cP*	*Cooked paste, cP*
2 hours	90	800
1 day	160	1000
1 week	230	1000
3 weeks	400	1000

[a] Brookfield viscosity in centipoise (cP); Model RVF at 20 rpm, 25°.

action can damage the paste by breaking the starch molecules. After pasting, the adhesive is either packaged into containers for sale or is used immediately.

IV. Modifiers

Starch and dextrin-based adhesives may be used alone or in combination with modifiers that impart additional desired characteristics to the adhesive. These modifiers are classified according to their function. The following discussion explains the function of commonly used modifiers.

Tackifiers are added to an adhesive to increase tack or adhesivity. Sodium tetraborate (borax), sodium hydroxide (caustic soda), and sodium metaborate are the most common tackifiers. They not only increase tack, but also increase the viscosity and paste color. Common practice is to add borates before cooking to achieve maximum performance. Sodium hydroxide, on the other hand, is usually added after cooking to minimize loss of viscosity due to alkaline degradation and color buildup. Typical addition levels are 10% for borates and 0.5% for sodium hydroxide, both based on total solids. Excessive use of these chemicals can result in extreme cohesiveness that can cause machinability problems.

Plasticizers are added to impart finished film characteristics to dextrin-based adhesives. Three types are effective: chemicals that form solid solutions with the dextrins, humectants that control the amount of moisture in the film, and fatty compounds that lubricate the film. Chemicals such as urea, sodium nitrate, dicyandiamide, salicylic acid, thiocyanates, formaldehyde, iodides, and guanidinium salts belong to the first group because they decrease the viscosity of dextrin solutions. In addition, they impart plasticity by keeping the dextrin from separating from the film. Urea seems to be most generally used. Additive levels range from 1% to 10% depending upon the chemical used and the property sought.

Humectants, such as glycerol, ethylene glycol, invert sugar, D-glucose, and sorbitol, are added to adhesives so that the film dries slowly without becoming brittle. The quality of the humectant and its level of use must be such that it gives the desired bond flexibility without excessive stickiness or blocking. Traditionally, sugars rather than urea or sodium nitrate are added to retard the drying rate. Glycerol and ethylene glycol are better humectants than are sugars and do not cause darkening of the bond with age.

Lubricants, another form of plasticizer, include sulfonated castor oil, sulfated alcohols, and soluble soaps. These have the advantage of imparting permanent flexibility regardless of atmospheric conditions; but they must be used at low levels to prevent weakening of the adhesive bonds. Clays and bentonites are inert fillers normally added to the adhesive to reduce cost. Their addition increases adhesive solids, inhibiting penetration of the adhesive into the substrate. Selection of fillers must take into account their variation in particle size, color, and dispersibility.

Bleaches are added to decrease film color. Sodium bisulfite, hydrogen peroxide, sodium peroxide, and sodium perborate are used primarily with high-soluble white dextrins, canary dextrins, and British gums in applications where film color is critical. Titanium dioxide is often added to an adhesive to prevent discoloration of the glue line. To increase adhesion to wax-treated or wax-impregnated paper, solvents may be added to the adhesive. Because of safety and toxicity problems with certain solvents, special precautions may be required. To minimize evaporation, solvent usually is added to cooked pastes after cooling. Suitable solvents are toluene, carbon tetrachloride, and trichloroethylene.

Preservatives are sometimes needed to control microbiological growth in starch and dextrin-based adhesives. Formaldehyde is most commonly used; chlorinated hydrocarbons are also effective.

Insolubilizing agents are used in applications where water resistance is required. Thermosetting resins, such as urea–formaldehyde and resorcinol–formaldehyde, give the greatest degree of water resistance. Control of pH is essential to obtain maximum reaction and water resistance for the particular resin used. Thermoplastic resins, such as poly(vinyl acetate) and acrylics, blend easily with starch pastes and give, not only moderate water resistance to the adhesive, but also improved machinability. Poly(vinyl alcohol) is also used to impart water resistance to the adhesive bond.

Addition of defoamer to starch and dextrin-based adhesives reduces foam formation during cooking and use of the adhesive. Compatibility with both the adhesive and other adhesive components is a necessary criteria for defoamers. Colloid stabilizers retard the tendency of adhesives to retrograde or set back to high viscosity. Soaps and sodium chloride are commonly used stabilizers.

V. Adhesive Applications and Formulations

1. Case and Carton Sealing

Case sealing consists of sealing the top and bottom flaps of fiberboard and corrugated boxes while carton sealing involves the closure of the ends of smaller boxes, normally made of fiberboard. Both applications require fast tacking adhesives with good machinability for high-speed equipment. Borated dextrins, sometimes modified with caustic, are usually used for case and carton sealing. Adhesives at 30–40% solids and 1000–3000 cP (25°) viscosity are common. A typical formulation is shown in Table V.

Sealing of cartons using a high-fluidity starch rather than a high-soluble white or canary dextrin has been demonstrated using the following formulation (*24*), in parts by weight: acid-thinned starch (75 fluidity), 35; water, 70; 25% sodium hydroxide, 14. This is a cold, caustic conversion which only requires stirring for about an hour to paste the starch. Up to 200% of filler clay may be added.

For normal case sealing, borated dextrins are used at a solids content of

Table V

Formulation of Case-Sealing Adhesive[a]

	Parts by weight
High-soluble white dextrin	37.5
Sodium metaborate	5.0
Borax (10 mol)	1.3
Water	56.2

[a] Viscosity ~2000 cP at 20°–25°.

20–50% and viscosities of 500–3000 cP. If water resistance is required, poly(vinyl alcohol) may be added to the adhesive.

2. Laminating

Laminating is divided into two categories: general and solid fiber. General laminating involves the bonding of paper to paper or paperboard for making products such as poster boards and display mountings. Important requirements are low adhesive penetration, non-slipping, high tack, and noncurling or lay-flat characteristics. Noncurling during drying of chipboard laminated with paper is achieved using highly plasticized white dextrin adhesives (*25*). A typical formula in parts by weight is shown in Table VI. After pasting and cooling to room temperature, the viscosity of this system is ~500 cP. Adhesives as high as 5000 cP, made by incorporating modified starches into the adhesive, have been successfully employed (*26*).

Solid fiber laminating involves the bonding of several plies of paperboard to form multiple laminations, as found in shipping containers. This application usually requires water resistance and good lay-flat, as well as good dry-bond, characteristics. Different types of starch and dextrin meet these requirements. For example, cold-water-soluble starch adhesive to which 5–25% urea–formaldehyde resin (based on starch) is added will provide these properties with certain

Table VI

Formulation of Noncurling Adhesive

	Parts by weight
High-soluble dextrin	20.2
Clay	13.5
Urea	6.7
Borax (10 mol)	5.0
Water	54.6

substrates (*27*). More often, however, a blend of starch and poly(vinyl alcohol) is used along with clay to give a fast-tacking adhesive with good water resistance (*25*). When metal foil is laminated to paper or paperboard, adhesive pH should be alkaline (sometimes as high as pH 12) to overcome any trace of oil present on the metal (*27*). These highly alkaline systems should be used only where very fast drying can be accomplished.

3. Tube Winding

Tube winding involves the formation of paper tubes, cans, cores, and cones. Windings may be of the convolute or spiral type. Spiral tubes require fast tacking adhesives, usually at high solids. According to Roland and McGuire (*25*), adhesives for tube winding are often applied at 50°–60° so that adhesives at higher solids can be used. Fiber–foil cans are typical of compositions that require high solids adhesives. Convolute tubes do not require fast tack, but they require a longer open time to permit the formed tube to be removed from the mandrel after being wound.

4. Bag Adhesives

One of the largest consumers of starch adhesives is the paper bag industry with most of the starch used in the manufacture of grocery and multi-wall paper bags (*28*). Two basic, but completely different adhesives, are used to make bags (the side-seam adhesive and the bottom paste) and both are made with or without water resistance.

a. *Side-Seam Adhesives*

The first operation in the manufacture of a single-ply bag is the production of an open sack on the tubing machine. The side-seam adhesive closes the paper tube and should be a slow-drying, low solids, nonpenetrating adhesive that provides a high dry-bond strength (*29*). While white dextrins and thin-boiling (high-fluidity) starches were used at 10–15% starch solids as late as 1965, they are now being used at 20–22% solids at viscosities in the 2000–4000 cP range (*25*). Viscosity should not exceed 6000 cP because the seam adhesive is usually pumped to the applicator through a small diameter tubing. An example of a non-water-resistant adhesive used for side seams is shown in Table VII. If water resistance is required, a thermoplastic or thermosetting resin (*30*) may be employed (*34*). Examples of water-resistant resins that may be used are poly(vinyl acetate), acrylate polymers, poly(vinyl alcohol), and urea–, melamine–, or resorcinol–formaldehyde resins. A good adhesive formulation with water resistance (in parts by weight) is given in Table VIII. The salt improves viscosity stability and allows the adhesive to dewater faster in the drying process.

Table VII

Non-Water-Resistant Adhesive Formula

	Parts by weight
High-soluble white dextrin	43.0
Borax (10 mol)	3.2
Boric acid	2.8
Water	51.0

b. *Bottom-Paste Adhesive*

Bag-bottom pastes are much higher in viscosity than side-seam pastes. Without agitation, the paste thickens to the extent that it can be ladled into the glue pan with a flat board. This thixotropic behavior prevents the adhesive from running through the applicator nip onto the bags passing below. In spite of its heavy consistency, the paste becomes very fluid when subjected to shear and is very easily applied to the paper by applicator rolls. Bottom paste must have good viscosity stability and tack to prevent the bottoms from opening before they are bundled and wrapped (*31*). The adhesive must be short so it does not throw during application, and finally, it must have a very strong dry bond.

Starches used in bottom pastes range from unmodified to low viscosity, cold-water-soluble. An adhesive based on the latter has the following composition in parts by weight: pregelled, low-viscosity starch, 36; borax (10 mol), 7; glycerol, 3; water, 54. This non-water-resistant paste is first mixed at high speed to achieve a smooth consistency and then held overnight before using. Paste viscosity ranges from 100,000 to 140,000 cP.

A water-resistant-type based on unmodified corn starch has the following composition in parts by weight: unmodified corn starch, 12.6; poly(vinyl alcohol), 4.5; poly(vinyl acetate), 0.9; soap, 0.1; water, 81.9. Ingredients should be slurried, heated to 90° and held 15 min before cooling and addition of preservative. Viscosity is 60,000–70,000 cP at 25°.

c. *Cross Paste*

This paste is used only in the manufacture of multiwall bags because its sole purpose is to glue the various plies together before the tube is formed and the seam paste applied. The cross paste is very similar to seam pastes except that there must be no penetration of adhesive through the outer layers of paper to prevent the finished, stacked sacks from sticking together. Some manufacturers add poly(vinyl acetate) latex to a regular water-resistant seam adhesive to prevent penetration.

5. Library Pastes

This paste is often formulated with a mixture of starch and white dextrin, plasticizer, and deodorizer. It is high in viscosity, smooth in texture, and white in color. A typical formula in parts by weight is as follows: low-soluble white dextrin, 45; unmodified corn starch, 5; glycerol, 5; water, 45. After cooking and cooling, this paste sets back to a smooth, firm texture.

6. Bottle-Label Adhesives

This adhesive is usually a high-viscosity, heavy-bodied, cohesive paste having good adhesion to glass. The viscosity should be stable, and the paste should be nonstringy. These pastes usually contain 40–50% solids with viscosities in the 80,000–150,000 cP range. Waxy corn starch gelatinized with caustic or plasticized, acid-modified waxy starch are generally used.

Unmodified waxy corn starch is gelatinized with caustic in a cold-water slurry under agitation to give a clear, thick paste. By neutralizing the sodium hydroxide with nitric acid, sodium nitrate, which plasticizes the finished paste, is formed. Salicylic acid or sodium salicylate can be added to improve viscosity stability. A typical acid-modified waxy corn starch-based adhesive, in parts by weight, consists of waxy starch, 40-fluidity, 38; urea, 3; sodium nitrate, 3; water, 56.

Beer manufacturers require a label adhesive that will withstand several days of immersion in ice water. The adhesive, however, must be soluble in hot water so that labels can be removed by washing prior to bottle reuse. An adhesive based on unmodified potato starch and poly(vinyl alcohol) is reported to have properties suited to modern, high-speed bottling machines (up to 60,000 beer bottles per hour) and to be water resistant (*32*).

7. Envelope Adhesives

Adhesives for envelope sealing consist of a back gum adhesive that holds the envelope together and a front seal adhesive that the consumer moistens to close the envelope. The back gum adhesive must be at high solids to prevent adhesive

Table VIII

Water-Resistant Adhesive Formulation

	Parts by weight
High-fluidity starch	19
High-fluidity, high-soluble starch	7
Urea–formaldehyde resin	3
Salt	1
Water	70

penetration and to provide good lay-flat properties. A very light color which does not darken on aging obviously is necessary for white envelopes. Either a heavily plasticized white dextrin or a heavily plasticized pregelled starch is used. Adhesive solids are in the 60–70% range and viscosity is normally around 2500–5000 cP. A formula that includes a poly(vinyl acetate) latex, in parts by weight, is pregelled, low-viscosity starch, 30–35; corn sugar (dextrose), 10–35; sodium nitrate, 8–15; urea, 0–8; water, 27–30; poly(vinyl acetate), 5–15. All ingredients, except the latex, are heated with agitation to 90° and held 15 min prior to cooling. Latex and preservatives are added at 50°.

Remoistenability is a prime requirement for an effective envelope front seal adhesive. High-soluble dextrins made from waxy corn, tapioca, or potato starch are all satisfactory. The adhesive must have a stable viscosity and not dry or "cotton" too quickly during application to the envelope. It must provide good lay-flat, high gloss, and nonblocking characteristics without discoloring the envelope. Poly(vinyl acetate) may be added to improve specific adhesion, nonblocking, and noncottoning properties (*25*). An example of a formula in parts by weight is as follows: high-soluble dextrin from waxy corn, 63; sodium bisulfite, 1; carbowax 4000 (Union Carbide), 0.5; water, 35.5.

8. Flat Gumming

Adhesives for articles such as labels and trading stamps are products of the flat gumming industry. There are two methods for adhesive application to paper substrates. In one, the conventional process, the adhesive is applied from an aqueous solution and, in the other, from an organic solvent. The conventional process calls for a cooked dextrin paste, usually of high-soluble white or canary dextrins, that is applied at elevated temperatures to the paper stock. Solids can be as high as 50% and viscosity close to 1000 cP. Final adhesive film should be clear, light in color, and glossy.

In the solvent process, a high-soluble dextrin or an acid-thinned, pregelled starch is suspended in an organic solvent along with a synthetic binder. The slurry usually is ~40% solids, and viscosity is less than 1000 cP at room temperature. When the slurry is coated on the substrate and the solvent is evaporated, a discontinuous film remains. Because of film discontinuity, the adhesive film is free from the curl problems often observed when the conventional process is used.

9. Gummed Tape

Remoistenable adhesive tapes common to gummed tape manufacture are the regular sealing tape, reinforced sealing tape, and box tape. Regular sealing tape is made by depositing an adhesive on 35–90-lb kraft paper and is used to seal boxes and cartons. An adhesive preparation for this type of tape has the following composition in parts by weight: thin-boiling waxy starch, 44; urea, 6; water,

50. Adhesive viscosity at 65° is close to 500 cP. Sodium nitrate, instead of urea, is often used as the plasticizer.

Reinforced sealing tape is used for applications where stronger bond strengths are required. The adhesive must be tackier and give higher bond strength than in regular sealing tape. Reinforced paper stock is used, which means a layer of reinforcing fibers is present between two sheets of paper to give very strong finished tape. Adhesives based on animal glue extended with dextrin are commonly used for this tape.

A patented adhesive for reinforced sealing tape is one based on a combination of waxy starch, dextrin, and poly(acrylamide) as binders (*33*). Dextrin is used to regulate the open time and the poly(acrylamide) to improve final adhesion. The formulation in parts by weight is as follows: thin-boiling waxy starch, 39.5; canary dextrin, 17.0; poly(acrylamide), 2.0; dispersing agent, 0.4; water, 41.1.

Box tape is used in the formation of boxes from corrugated stock and is the strongest of the three tapes. Animal glue is the only suitable adhesive for this tape.

10. Wall Covering Adhesives

These packaged pastes used by contractors and home owners alike must have excellent viscosity stability, good slip for easy application, and good wet tack, and be strippable after drying. Plasticizers are added to achieve good slip; borates are added to increase tack. Level of addition of these two additives must be carefully balanced, for basically their activities oppose one another. Clay addition lends strippability to the dry bond, a requirement for easy removal of the wall covering. A typical formula in parts by weight is as follows: acid-modified pregelled starch, 25.0; Hydrite R clay, 20.0; urea, 3.75, sodium metaborate, 1.25; water, 50.0.

VI. Adhesive Terminology

Adhesion.—Adhesion is the condition by which two surfaces are held together in one of the following ways: (a) physically, that is, the components are held together by interlocking due to physical penetration; (b) chemically, that is, the components are bound together by valence forces of the same type which give rise to cohesion.

Adhesive.—An adhesive is a substance capable of holding materials together either by chemical or mechanical attraction. A general term including cement, glue, mucilage, paste, thermoplastics, etc., to which are applied descriptive adjectives indicating: (a) physical form, for example, liquid or tape adhesives; (b) chemical type, for example, silicate and resin adhesives; (c) materials bonded, for example, paper and can-label adhesives; (d) conditions of use, for example, hot setting, thermoplastic adhesive.

Animal glue.—Animal glue is proteinaceous material of animal origin (hides, bones, etc.) which is processed for use in adhesives.

Blocking.—Blocking is an undesired adhesion between touching layers of material, such as might occur under moderate pressure, temperature, or high relative humidity during storage or use (critical for gummed tape).

Bond time.—Bond time is the time it takes for a laminate to develop fiber tearing bond after initial preparation.

Consistency.—Consistency is that property of a liquid adhesive by virtue of which it tends to resist deformation or flow. It is not a fundamental property but is comprised of viscosity, plasticity, and other phenomena.

Color.—Color is a physical property of adhesive pastes which is critical for certain applications.

Cottoning.—Cottoning is a phenomenon observed during machine application of an adhesive characterized by the formation of weblike filaments of adhesive between machine parts themselves or between machine parts and the receiving surface during transfer of the liquid adhesive onto the receiving surface.

Delamination.—Delamination is the separation of layers in a laminate.

Drying time.—Drying time is the time between the application of an adhesive and the time it reaches equilibrium with the moisture in the stock or in the air.

Feel.—Feel refers to general adhesive characteristics (tack, drying time, stringiness, cottoning, rolling tendency, shortness, consistency) as measured on the "fingers."

Fiber tear.—Fiber tear is the tearing of fiber in a paper assembly rather than the separation of adhesive. If the adhesive film is weak, unit separation will occur at the paper interface rather than in the paper fiber.

Filler.—Filler is the substance, such as clay, used to increase solids and reduce penetration.

Humectant.—Humectant is a substance that promotes retention of moisture.

Laminate.—A laminate is a product made by bonding two or more layers of material or materials.

Lay flat.—Lay flat refers to non-warping or non-wrinkling charactersitics.

Machineability.—Machineability is the ability of an adhesive to be handled and applied to an adherend by a machine.

Open time.—Open time is the maximum time lapse between applying the adhesive and bringing the substrates together within which a satisfactory bond can be obtained.

Penetration.—Penetration is the passage of an adhesive into a substrate. It is measured on a comparative basis by forcing adhesive through a stack of filter papers under constant pressure with the use of a mounting press.

Phase separation.—Phase separation is a condition in which an adhesive forms two distinct layers.

Plasticizer.—A plasticizer is a material which adds film flexibility and/or wet tack to an adhesive.

Porous.—Porous is the ability of the substrate to absorb.

Pot life.—Pot life is the period of time during which an adhesive remains suitable for spreading before hardening in the receptacle.

Rate of set (speed of fiber tearing bond development).—In working with a paper substrate, this is usually the time required after application of a film of adhesive under prescribed conditions to give fiber tear.

Remoistenable.—Remoistenable refers to the addition of moisture to a dried adhesive film to give tack.

Resin.—Resin is a synthetic product used to give specific adhesive characteristics.

Setting speed.—Setting speed is the rate at which an adhesive increases in tack after application to the substrate.

Short-breaking.—Short-breaking is a lack of stringiness.

Shortness.—Shortness is the non-stringing property of an adhesive.

Slipping.—Slipping is the undesirable condition which exists when the plies of an adhesive laminate slide.

Stock.—Stock is another term for substrate.

Tack.—Tack is stickiness of an adhesive which may vary with time, temperature, film thickness, etc. Also tack may be defined as that property of an adhesive that enables it to form a bond of measurable strength immediately after adhesive and adherend are brought into contact under low pressure.

Texture.—Texture is consistency and body of an adhesive.

Thixotropic.—Thixotropic refers to a fluid of viscous suspension which shows a time-dependent decrease in viscosity with an increase in rate of shear and a time-dependent increase in viscosity with a decrease in the rate of shear.

Water resistance.—Water resistance is a property by which the adhesive has a degree of resistance to permeability or to any damage caused by water in liquid form.

Viscosity.—Viscosity is the resistance of an adhesive to flow.

Viscosity stability.—Viscosity stability describes the ability of an adhesive to maintain a workable consistency for an extended period of time for use in a given application.

VII. References

(*1*) R. L. Whistler, *in* "Starch, Chemistry and Technology," E. F. Paschall and R. L. Whistler, eds., Academic Press, New York, 1965, Vol. I, p. 1.

(*2*) V. G. Bloede, *in* "A Comprehensive Survey of Starch Chemistry" R. P. Walton, ed., The Chemical Catalog Company, New York, 1928, Vol. I, p. 159.

(*3*) Anon., *Adhesives Age* (Atlanta) **22,** No. 9, 36 (1979).

(*4*) L. Roland and E. P. McGuire, *Paper Trade J.*, **149,** No. 4, 37 (1965).

(*5*) C. Berquist, U.S. Patent 1,851,749 (1932); *Chem. Abstr.*, **26,** 3135 (1932).

(*6*) A. D. Fuller, U.S. Patent 1,937,752 (1933); *Chem. Abstr.*, **28,** 1214 (1932).

(*7*) A. D. Fuller, U.S. Patent 1,942,544 (1934); *Chem. Abstr.*, **28,** 1888 (1934).

(*8*) R. W. Kerr, U.S. Patent 2,108,862 (1938): *Chem. Abstr.*, **32,** 3188 (1938).

(*9*) A. J. Bulfer and C. C. Gapen, U.S. Patent 2,287,599 (1942); *Chem. Abstr.*, **37,** 280 (1943).
(*10*) C. O'Neill, British Patent 1,861 (1838).
(*11*) R. W. Kerr, U.S. Patent 2,503,053 (1950): *Chem. Abstr.*, **44,** 6180 (1950).
(*12*) V. G. Bloede, U.S. Patent 536,260 (1895).
(*13*) J. E. Clegg, U.S. Patent 2,127,205 (1938).
(*14*) G. V. Caesar, U.S. Patent 2,131,724 (1938); *Chem. Abstr.*, **32,** 9341 (1938).
(*15*) M. A. Staerkle and E. Meier, U.S. Patent 2,698,937 (1955); *Chem. Abstr.*, **49,** 4315 (1955).
(*16*) W. J. Rowe and C. Hagen, U.S. Patent 2,332,345 (1943); *Chem. Abstr.*, **38,** 1660 (1944).
(*17*) N. C. Phillips, U.S. Patent 1,894,570 (1933); *Chem. Abstr.*, **27,** 2599 (1933).
(*18*) M. A. Staerkle and E. Meier, U.S. Patent 2,698,818; *Chem. Abstr.*, **49,** 4314 (1955).
(*19*) C. Ziegler, R. Kohler, and H. Riiggeberg, U.S. Patent 2,818,357 (1957); *Chem. Abstr.*,**52,** 12438 (1958).
(*20*) R. E. C. Fredrickson, U.S. Patent 2,845,368 (1958); *Chem. Abstr.*, **53,** 2658 (1959).
(*21*) R. E. C. Fredrickson, U.S. Patent 3,003,894 (1961); *Chem. Abstr.*, **56,** 2624 (1962).
(*22*) L. R. Idaszak, U.S. Patent 3,967,975 (1976); Ger. Offen. 2,552,891 (1976); *Chem. Abstr.*, **86,** 6665t (1977).
(*23*) L. R. Idaszak, U.S. Patent 4,021,927 (1977); Ger. Offen. 2,552,891 (1976); *Chem. Abstr.*, **86,** 6665t (1977).
(*24*) National Starch and Chemical Corporation, Netherlands Appl. 66/03507 (1966); *Chem. Abstr.*, **66,** 30206b (1967).
(*25*) L. Roland and E. P. McGuire, *Paper Trade J.*, **149,** No. 9, 32 (1965).
(*26*) D. J. Iwinski, "Paper-to-Paper Laminating," brochure, Corn Products Company, Division of CPC International, p. 2.
(*27*) E. F. W. Dux, *in* "Starch: Chemistry and Technology," E. F. Paschall and R. L. Whistler, eds., Academic Press, New York, 1965, Vol. II, p. 550.
(*28*) G. V. Caesar, *in* "Handbook of Adhesives," I. Skeist, ed., Reinhold, New York, 1962, p. 175.
(*29*) E. F. W. Dux, reference *27*, p. 547.
(*30*) E. F. W. Dux, reference *27*, p. 543–4.
(*31*) G. V. Caesar, reference *28*, p. 176.
(*32*) H. G. Sebel, U.S. Patent 4,008,116 (1900); *Ger. Offen.* 2,364,438 (1975); *Chem. Abstr.*, **83,** 148668h (1975).
(*33*) R. W. Monte, U.S. Patent 4,105,824 (1978).

CHAPTER XXI

GLUCOSE- AND FRUCTOSE-CONTAINING SWEETNERS FROM STARCH

By Norman E. Lloyd* and William J. Nelson*

Clinton Corn Processing Company, A Division of Standard Brands Inc., Clinton, Iowa

I. Introduction

There is great interest in recently developed technology for the preparation of starch hydrolyzates by enzymic means, enzymic isomerization to prepare D-fructose-containing products, and methods for increasing the D-fructose content. The following definitions and abbreviations are used.

Dextrose Equivalent (DE) is an indication of total reducing sugars calculated as D-glucose on a dry-weight basis. The DE value is inversely related to the degree of polymerization (DP). Unhydrolyzed starch has a DE of virtually zero, whereas the DE of anhydrous D-glucose is defined as 100.

*Present address: Corporate Technology Group, Wilton Technology Center, Nabisco Brands, Inc., Wilton, Connecticut 06897.

STARCH, 2nd ed.

ISBN 0-12-746270-8

Corn syrup (glucose syrup) is the purified concentrated aqueous solution of nutritive saccharides of DE 20 or more obtained by hydrolysis of edible starch (see Section II.3).

Maltodextrin is a mixture of purified nutritive saccharides obtained by hydrolysis of starch having a DE of less than 20. As an article of commerce, it is usually dried, but may also occur as a concentrated solution (see Section II.2).

High-fructose corn syrup (HFCS) is corn syrup produced with the additional step of enzymic conversion of a portion of the D-glucose to D-fructose. Products containing more than about 50% D-fructose on a dry basis must be obtained by D-fructose enrichment methods, which involve the separation of D-glucose and D-fructose. Articles of commerce are characterized by D-fructose contents of 42, 55, and 90% (dry basis) (see Sections III and IV).

Dextrose is the trivial name given to crystalline D-glucose. The term is used here in reference to the commercially available crystalline forms of D-glucose, namely, anhydrous α-D-glucopyranose and α-D-glucopyranose monohydrate (see Section V.1).

Retrogradation is the formation of an insoluble precipitate through the association of poorly soluble linear dextrins present in starch solutions and low DE hydrolyzates (see Chap. VIII).

Reversion is the process of forming di- and higher oligosaccharides through the condensation of reducing sugars.

Craving for sweetness is an inborn trait, rather than a learned preference. A baby at birth prefers sweet substances to those with other tastes. Even the 4-month fetus increases its rate of swallowing when saccharin is injected into the amniotic fluid (*1*), and many other animals are known to prefer sweet substances. For millenia, man's ''sweet tooth'' could be satisfied only by naturally available fruits and berries. The first concentrated source of sweetener was honey, which for centuries was available only to a few privileged people. Highly concentrated sweeteners became generally available in the nineteenth and twentieth centuries when cane and beet sugar could be produced in large quantity.

Sucrose occurs in free form in cane and sugar beets and can be extracted and purified by relatively simple processes to yield high-quality products. Practical methods for producing nutritive sweeteners from starch required a longer period of development because of the more complex technology needed.

The production of syrups and sugars from starch in the United States has been fostered by a bountiful supply of corn. Over 8 billion bushels of corn were produced in the United States in 1982, and of this, only about 5–6% was used for production of starch and sweeteners. Corn is an ideal raw material because of its high starch content, plentiful supply, and year around availability. Though sucrose remains the most widely used sweetener, growth rate in the production of sweeteners from starch in the U.S. increased from 2.5 billion lb (1.1×10^9 kg) per year in 1960 to 11.2 billion lb (5.1×10^9 kg) per year in 1982 (*2*). Starch-derived sweeteners now constitute 39% of U.S. consumption of caloric sweeteners (*151*). Most of this increase owes to the development of commercial processes for the enzyme-catalyzed conversion of D-glucose to D-fructose leading to the production of high-fructose corn syrups (HFCS) with sweetness levels equivalent on a dry basis to that of sucrose.

A generalized pathway for the production of sweeteners from starch is shown in Figure 1. Two primary processes are used, depending on the initial degree of conversion of starch. Corn syrups, corn syrup solids, and maltodextrins in the DE 10–70 range are produced by partial hydrolysis of starch. These products are used in foods primarily for their functional properties rather than for sweetness. They are important as moisture conditioners, food plasticizers, crystallization inhibitors, stabilizers, carriers, and bulking agents. Products of low DE possess little sweetness. They are mixtures of nutritive saccharides with a clean, pleasant taste comprised almost entirely of D-glucose and D-glucose oligomers.

Crystalline dextrose and HFCS are derived from high-conversion starch hy-

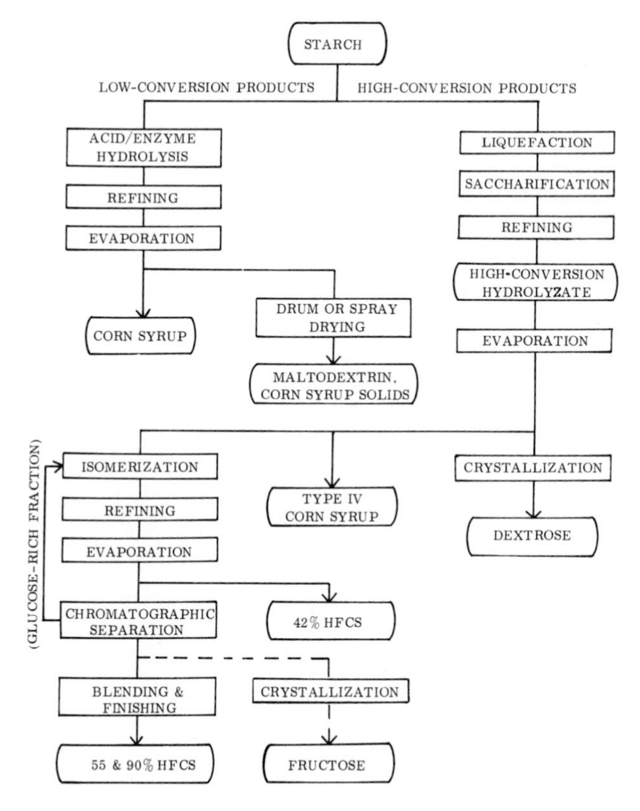

FIG. 1.—Generalized scheme for the production of sweeteners from starch.

drolyzates containing at least 93% D-glucose on a dry basis. The primary function of these products is to provide sweetness and a source of fermentable sugars for various foods and beverages.

II. Refined Starch Hydrolyzates

1. Hydrolyzate Classification

Commercial starch hydrolyzates are classified on the basis of DE. Hydrolyzates below 20 DE are identified as maltodextrins. Those above 20 DE are designated as corn syrups and are further classified into four types (*2*): Type I, 20–38 DE; Type II, 38–58 DE; Type III, 58–73 DE; and Type IV, ≥73 DE.

Type I corn syrups consist mainly of high-molecular-weight branched and linear dextrins. Type II syrups contain 50–75% of low-molecular-weight saccharides, including D-glucose, maltose, and maltotriose. Type III corn syrups, sometimes referred to as high-conversion or high-fermentable corn syrups, contain 75–85% of D-glucose, maltose and maltotriose. Type IV conventional corn syrups, containing no D-fructose, consist mainly of D-glucose.

Corn syrups are sometimes identified by the method of hydrolysis used in their preparation, that is, according to whether they are produced solely by acid conversion, a combination of acid and enzyme conversion, or solely by enzyme conversion.

2. Maltodextrins (see also Chapters IV and V)

a. *Hydrolysis*

Maltodextrins are manufactured by processes substantially identical to those used in the production of corn syrups, except that hydrolysis is controlled to keep the DE below 20. Hydrolysis may be catalyzed using an acid (usually HCl) or an enzyme (α-amylase). Some hydrolysis schemes employ both acid- and enzyme-catalyzed hydrolysis.

Maltodextrins made by acid-catalyzed hydrolysis have a strong tendency to retrograde, forming haze. Linear starch fragments in low-DE acid hydrolyzates large enough to reassociate form insoluble aggregates. These retrograded particles cause maltodextrin solutions to become hazy, a condition undesirable for some applications.

To overcome haze formation and to obtain maltodextrins with low hygroscopicity and high water solubility, it is desirable to use acid–enzyme and enzyme catalysis. The hydrolysis pattern of enzymes (bacterial α-amylase) are such that the linear dextrins responsible for haze formation are hydrolyzed in preference to branched dextrins. Thus, Armbruster and Harjes (*3*) recommend an initial acid-catalyzed hydrolysis of slurry to 5–15 DE. The hydrolyzate is then neutralized and subjected to further hydrolysis with a bacterial α-amylase such as that obtained from *Bacillus subtilis* or *Bacillus mesentericus*. Maltodextrins derived from such procedures are haze-free and exhibit no retrogradation upon storage.

Non-hazing maltodextrins are produced also by processes employing only enzyme-catalyzed hydrolysis (*4, 5*). The starch slurry is initially liquefied by heating to 70–90° at neutral pH in the presence of a bacterial α-amylase to a DE of 2–15. The liquefied starch hydrolyzate is then autoclaved at 110–115° to completely gelatinize remaining insoluble starch and, on cooling, subjected to further enzymic hydrolysis to reach the desired DE.

b. *Refining and Concentration*

Following hydrolysis, the pH of the crude maltodextrin solution is adjusted to about 4.5, and the solution filtered to remove small amounts of fiber, lipids, and proteins. Filtration is difficult because of the high viscosities of maltodextrin solutions, and substantial process losses can occur at this stage.

The clarified solution is then refined by one or more of the procedures used for corn syrups (see Section II.3c). The refined solution is concentrated in vacuum evaporators to finished syrups containing about 75% solids or is spray-dried to obtain a white powder containing 3–5% moisture.

c. *Composition and Properties*

Two types of maltodextrins are in commercial use: those ranging from about 10 to 14 DE and those ranging from about 15 to 19 DE (*6*). The compositions of these products depend not only on DE but, as noted previously, upon the method of hydrolysis employed in manufacture.

The saccharide compositions of maltodextrins obtained by acid-catalyzed hydrolysis are different from those obtained by acid–enzyme- or enzyme-catalyzed hydrolysis, as shown in Table I, even though both possess the same DE. Acid hydrolyzates contain a greater proportion of high-molecular-weight dextrins, material which retrogrades easily.

The solubility of maltodextrins varies with the DE and the method of hydrolysis. Water solubility data have been reported for products in the DE ranges (*7*) shown in Table II. Enzyme-hydrolyzed products contain lower concentrations of the high-molecular-weight saccharides and are more water soluble than acid-hydrolyzed maltodextrins of equivalent DE.

Maltodextrins are relatively non-hygroscopic compared to corn syrups, and those having the lowest DE exhibit the least tendency to absorb moisture.

The high viscosity of maltodextrins, an important property in many applications, is due to high levels of high-molecular-weight saccharides. Ueberbacker (*8*) has reported viscosity values for two types of products in general use, as shown in Table III. The viscosity of a low-DE corn syrup is given for compari-

Table I

Distribution According to Molecular Weight of Saccharides of Maltodextrins Prepared by Acid- vs. Enzyme-Catalyzed Hydrolysis (5)

	Hydrolyzate, % by weight	
Saccharides	*15 DE acid*	*15 DE enzyme*
Mono	3.7	0.7
Di	4.4	5.5
Tri	4.4	6.9
Tetra	4.5	5.2
Penta	4.3	5.5
Hexa	3.3	10.6
Higher	75.4	65.6

son. Maltodextrıns are quite bland, exhibiting virtually no sweetness, so then main contribution is a "bodying effect" resulting from their relatively high viscosity.

3. Corn Syrups

Conventional corn syrups (syrups other than high-fructose corn syrups) comprised about 37% of the caloric sweeteners derived from starch consumed in the U.S. in 1982 (*151*). Descriptions of conventional corn syrup manufacturing processes and the preparation of purified starch used in their manufacture have been reported by MacAllister (*10*). A general manufacturing procedure for corn syrups is outlined in Figure 2.

a. *Acid-Catalyzed Hydrolysis*

The first commercial corn syrup produced was made by acid-catalyzed hydrolysis, and a major portion of conventional corn syrups is still made this way. Although batch converters are used in the industry, more modern installations employ continuous converters because they are more efficient, allow better con-

Table II

Solubility of Maltodextrins

DE range	*Approximate solubility, % solids at room temp*
9–12	40
13–17	60
17–20	70

Table III

Viscosity of Maltodextrin Solutions (8)

	Viscosity at 37.8°, cP		
	Maltodextrins		*Corn syrup*
Solids (%)	*10–15 DE*	*15–20 DE*	*25–30 DE*
50	125	12.5	1.2
60	1,250	125	12
70	20,000	2000	200

trol of DE, and minimize formation of objectionable color in the conversion product (*11, 12*). In a typical continuous conversion, a 40% starch slurry is acidified to pH 1.8–2.0, usually with hydrochloric acid, and pumped into the converter simultaneously with steam injection to heat the slurry. Steam injection and pumping rates are controlled to attain about 140° for approximately 10 min.

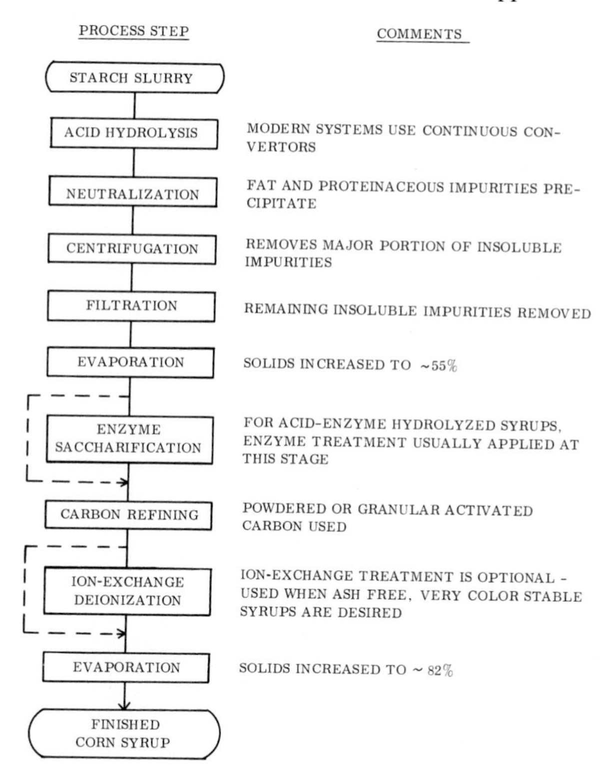

FIG. 2.—Typical process for manufacturing conventional corn syrups.

Table IV

Saccharide Composition of Syrups Made by the Acid-Catalyzed Hydrolysis of Corn Starch (14)

	Saccharides, % by weight							
DE	Mono	Di	Tri	Tetra	Penta	Hexa	Hepta	Higher
10	2.3	2.8	2.9	3.0	3.0	2.2	2.1	81.7
15	3.7	4.4	4.4	4.5	4.3	3.3	3.0	72.4
20	5.5	5.9	5.8	5.8	5.5	4.3	3.9	63.3
25	7.7	7.5	7.2	7.2	6.5	5.2	4.6	54.1
30	10.4	9.3	8.6	8.2	7.2	6.0	5.2	45.1
35	13.4	11.3	10.0	9.1	7.8	6.5	5.5	36.4
40	16.9	13.2	11.2	9.7	8.3	6.7	5.7	28.3
45	21.0	14.9	12.2	10.1	8.4	6.5	5.6	21.3
50	25.8	16.6	12.9	10.0	7.9	5.9	5.0	15.9
55	30.8	18.1	13.2	9.5	7.2	5.1	4.2	11.9
60	36.2	19.5	13.2	8.7	6.3	4.4	3.2	8.5
65	42.5	20.9	12.7	7.5	5.1	3.6	2.2	5.5
67	45.1	21.4	12.5	6.9	4.6	3.2	1.8	4.5

The α-D-(1→4) linkages undergo hydrolysis more easily than do the α-D-(1→6) linkages. Also, linkages nearer the nonreducing end of the starch polymer are hydrolyzed more rapidly than bonds located in the polymer interior (*13*). These differences in hydrolysis rates are small, however, and the component distributions obtained by acid catalysis are substantially that expected from a random hydrolysis. The saccharide compositions of acid-converted starch hydrolyzates at various DE levels are shown in Table IV.

b. *Acid–Enzyme-Catalyzed Hydrolysis*

Corn syrups of DE higher than about 55 are not commercially produced by acid conversion, because side reactions during hydrolysis generate excessive color and undesirable bitter flavored substances not removable by ordinary refining. Of particular note is the bitter disaccharide gentiobiose formed in high-DE hydrolyzates through reversion. Such components limited the commercial potential of corn sweeteners for many years. In 1940, a dual acid–enzyme conversion process was patented, opening a new era in conventional syrup production (*15*).

By employing enzyme catalysis, corn syrups containing a much wider variety of saccharide mixtures are attainable. The action patterns of enzymes used in commercial corn syrup manufacture have been discussed by MacAllister (*10*). Purified commercial enzyme preparations used for converting starch are commonly obtained from fermentations of *Aspergillus oryzae*. These enzyme preparations contain a mixture of β-amylases and glucoamylases. They are usually applied to acid hydrolyzates to increase the D-glucose and maltose contents and, consequently, the syrup sweetness level. This process tends to eliminate carbohydrate degradation products and β-linked reversion products such as gentibiose (*16*).

The major portion of acid–enzyme converted corn syrups is made in the 58–73 DE range (Type III classification) by the process described by Dale and Langlois (*15*). In the production of a typical 65 DE dual conversion syrup, a starch slurry containing about 40% solids is hydrolyzed with acid to ~ 44 DE. The hydrolyzate is then neutralized and filtered to remove the major portion of proteinaceous and fatty materials introduced as impurities present in the starch. The filtered hydrolyzate is next concentrated to about 55% solids. Concentration minimizes saccharification equipment needed and helps to inhibit microbial growth during the enzyme treatment step which is typically performed at ~ 55° and a pH of ~ 5.2. Because the commercial fungal amylases used for hydrolysis are potent, only small quantities are needed. The hydrolysis time is usually about 48 h, but may be varied according to production needs. Upon attaining the desired DE, the hydrolyzate pH is reduced to ~ 3.0 to inactivate the enzyme and prevent further hydrolysis.

Patented processes have been described (*17, 18*) for production of highly fermentable, non-crystallizing corn syrups with levels of D-glucose and maltose higher than those in 65 DE syrup. One process for making highly fermentable syrup uses a triple enzyme conversion process and claims a D-glucose plus maltose content of at least 85%, with the D-glucose level less than 45% (*19*). The initial conversion stage uses α-amylase for starch liquefaction rather than acid catalysis.

Commercial syrups containing high levels of maltose are obtained by acid-catalyzed hydrolysis of starch to ~20 DE and subsequent hydrolysis with β-amylase, usually from barley malt. A purified, precipitated malt amylase is commonly used to minimize colored impurities introduced when crude barley malt extracts are used.

High conversion (95 DE) corn syrup is obtained by a process equivalent to those used for the manufacture of high-fructose corn syrups and crystalline dextrose (see Section II.4). Such syrup contains a D-glucose content of ⩾93% (dry substance basis) and requires a lower finished syrup dry solids level than other corn syrups and a higher than normal storage and shipping temperature to prevent crystallization.

c. *Refining and Concentration*

The following description of syrup refining treatments generally applies to all corn syrup manufactured. Minor variations may be applied to meet the particular refining requirements of special hydrolyzates.

After hydrolysis to the desired DE, the starch hydrolyzate is neutralized to pH 4.6–4.8 with sodium carbonate. Close pH adjustment is necessary for maximum precipitation of lipids and proteinaceous material. In the proper pH range, flocculation of impurities is sharp and the solids are easily removed.

Precipitated impurities may be removed by centrifugation or by skimming operations. Solid impurities can be removed by passing the liquor through deep tanks with a weir that separates the fat and protein floc from the liquor. Subsequently, the hydrolyzate is further clarified by filtration using a filter aid such as diatomaceous earth or crushed perlite. In some systems, activated, powdered carbon that has had prior use in other refining stages is applied to effect some decolorization and make more efficient use of the decolorizing carbon used in the next step.

The clear, slightly yellow filtrate with ~ 40% solids is concentrated to ~ 55% dry substance using steam-heated, multi-effect evaporators. The concentrated syrup liquor is refined using activated carbon to remove color, color precursors, and undesirable off-flavored materials. Carbon treatment removes most of the soluble proteinaceous material present and substantially all the 5-(hydroxymethyl)-2-furaldehyde formed during the acid treatment. Also, many commercial activated carbons are effective in removal of heavy metals, such as iron and copper, that can act as catalysts for developing color. Carbon treatment may be a 2- or 3-stage countercurrent batch application of activated powdered carbon or a countercurrent application of activated granular carbon in cylindrical columns. Both methods are in use, but most new installations use the granular carbon treatment because it can be conveniently re-activated, yielding more favorable economics (*20*).

Although carbon refining is adequate for purification of most conventional corn syrups, some applications require syrups that are ash-free, have essentially no taste other than sweetness, and are more color stable than can be produced by carbon treatment alone. In such cases, the carbon refined liquor is ion-exchange deionized. Such treatment removes substantially all remaining soluble nitrogenous compounds, including amino acids and peptides that contribute color body formation via the Maillard reaction with reducing sugars. In addition, heavy metals and weakly acidic organic constituents that can affect syrup color development on storage are removed (*21*).

A typical ion-exchange deionization system consists of six fixed-bed columns (three pairs of cation and anion exchange resin columns, two pairs of which are in service and one which is out of service for regeneration). The cation-exchange resins used are strong acid exchangers (sulfonated resins in the hydrogen form), and the anion exchangers usually are weak base resins (tertiary amine in the free base form). The anion-exchange resins remove acids generated by reaction of the salts in the syrup liquor with the cation-exchange resins.

After refining, the corn syrup liquor is adjusted to ~ 82% dry solids in a single effect vacuum evaporator controlled so that temperature does not exceed 60°. Near the end of evaporation, a buffer solution may be added to adjust the pH to

~ 4.8 and to provide buffering capacity required in some applications. Usually an acetate buffer is introduced. Some corn syrups are not buffered and adjustment to the final pH is made by addition of alkali or acid.

The dry substance content is carefully controlled within ±0.2% to meet specifications for specific gravity. Following evaporation, the concentrated syrup is transferred to a cooling vessel where it is checked for required composition, and the syrup temperature is reduced to the level desired for storage and transport.

Some corn syrups are sold in a dry form. These are prepared by drying the refined corn syrup liquor in spray or vacuum drum driers to a moisture content of ~ 2.5%. The 36 and 42 DE acid converted syrups are typical of products offered in a dry form. Because of their hygroscopicity, dried corn syrups must be packaged in moisture proof containers for storage and distribution.

d. *Composition and Properties*

The saccharide compositions of the major corn syrups produced, other than high-fructose corn syrups, are shown in Table V. Sweetness of corn syrups is dependent on the amounts of low-molecular-weight saccharides (primarily D-glucose and maltose) present, the concentration of solids, the temperature at which the product is used, and on other substances, such as salts resulting from the manufacturing procedure, which may be present. Sweetness increases with increasing solids concentration. As DE is an indicator of the relative concentrations of mono- and disaccharides in corn syrup, the DE value also relates to sweetness. In Table VI, the approximate relative sweetness levels of major commercial corn syrups are compared (*22*). The effect of concentration on

Table V

Saccharide Compositions of Typical Corn Syrups (2)

Type of corn syrup[a]	*DE*	*% saccharides, carbohydrate basis*						
		Mono	*Di*	*Tri*	*Tetra*	*Penta*	*Hexa*	*Higher*
AC	27	9	9	8	7	7	6	54
AC	36	14	12	10	9	8	7	40
AC	42	20	14	12	9	8	7	30
AC	55	31	18	12	10	7	5	17
HM,DC	43	8	40	15	7	2	2	26
HM,DC	49	9	52	15	1	2	2	19
DC	65	39	31	7	5	4	3	11
DC	70	47	27	5	5	4	3	9
DC,E	95	93	3	1	1	—	—	2

[a] AC = acid conversion; DC = dual conversion (acid–enzyme); HM = high-maltose; E = enzyme conversion.

Table VI

Relative Sweetness of Corn Syrups

Sweetener	Sweetness relative to sucrose
36 DE corn syrup[a]	35–40
42 DE corn syrup[a]	45–50
54 DE corn syrup[a]	50–55
62 DE corn syrup[b]	60–70
Dextrose	70–80
Sucrose	100

[a] Acid converted.
[b] Acid-enzyme converted.

sweetness level as reported by Nieman (*23*) is shown in Table VII for 42 DE acid converted and 64 DE acid–enzyme converted syrups.

Corn syrups are readily soluble in water and can be concentrated to high dry substance levels without crystallization, if the D-glucose content is less than 45% dry basis. Normally, conventional corn syrups are stored and shipped at solids levels of from 77 to 85% to minimize transportation costs and to prevent microbial growth.

Viscosity is an important physical property of corn syrup, because it is the primary factor influencing the maximum concentration at which conventional corn syrups can be handled. Viscosity is dependent on concentration, saccharide composition, and temperature. Viscosities of some commercial corn syrups are shown over a temperature range of from 15° to 60° in Table VIII.

The average molecular weight of corn syrups affects the freezing point of

Table VII

Effect of Concentration on Sweetness of Corn Syrups

Dry substance, %	Sweetness relative to sucrose	
	42 DE	64 DE
5	30.5	42
10	33	49
15	36	56
20	39	63.5
30	44	75.5
40	50.5	85
50	58	91

Table VIII

Effect of DE and Temperature on Corn Syrup Viscosity (24)

			Viscosity, cP				
DE	Dry substance,[a] %	Type of conversion	15.6°	26.7°	37.4°	48.9°	60.0°
36	79.9	Acid	—	800	180	60	23
42	80.3	Acid	3000	590	124	42	17
42		Acid–enzyme[b]	3400	580	145	49	—
48		Acid–enzyme[b]	—	425	115	38	—
54	81.1	Acid	1770	365	70	25	10
64		Acid–enzyme	—	220	60	21	—
70		Acid–enzyme	—	170	48	17	—

[a] Dry substance levels are equivalent to 43° commercial Baumé.
[b] High-maltose syrups.

solutions. Effects of the concentration of 42 and 63 DE corn syrups and other sweeteners on the freezing point are shown in Figure 3. Corn syrup of 55 DE has about the same average molecular weight as sucrose and, hence, produces a similar freezing point depression (*25*).

The hygroscopic character of corn syrups is a useful property, but it also affects storage and handling of dried syrup. The equilibrium relative humidity of dried syrups (ERH), that is, the relative humidity at which moisture is neither gained nor lost, is influenced by composite average molecular weight and solids concentration. High-DE syrups are more hygroscopic and exhibit a lower ERH than do low-DE syrups. The ERH values of some corn syrups at various temperatures have been reported by Norrish (*26*).

Initial color and color stability need to be maintained for substantial periods of time, perhaps as much as 6 months at temperatures of ~ 25°. Initially, corn syrups are essentially colorless and, when adequately refined, will remain color stable.

4. High-Conversion Hydrolyzates

High-conversion hydrolyzates are prepared almost exclusively by the use of enzymes. Acid-catalyzed hydrolysis of starch is not capable of giving practical hydrolyzates with more than about 90% D-glucose (*10*), owing to acid-catalyzed reversion and dehydration reactions resulting in a sizable loss of D-glucose. Enzyme-catalyzed hydrolysis is conducted at relatively low temperatures and at pH 4–5 where reducing sugars have maximum stability. However, enzyme-catalyzed reversion reactions still remain.

There are two principal applications for high-conversion hydrolyzates. One is a source of D-glucose for the manufacture of high-fructose corn syrup (HFCS)

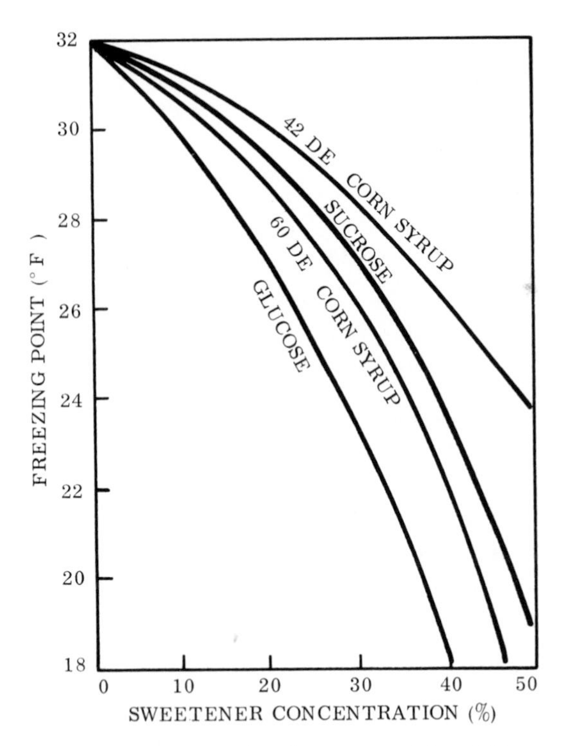

FIG. 3.—Freezing point of sweetener solutions (25).

and the second is a source of crystalline dextrose. In both applications, it is economically and functionally advantageous to provide the highest practical dry substance content and the highest degree of conversion to D-glucose without formation of impurities such as organic acids, ash, or colored products. A typical flow diagram showing the steps involved in production of a refined, high-conversion starch hydrolyzate is shown in Figure 4.

a. *Liquefaction*

The objective of the liquefaction process is to convert a concentrated suspension of starch granules into a solution of soluble dextrins of low viscosity for convenient handling in ordinary equipment and for easy conversion to glucose by glucoamylase.

Typically, a suspension of starch in water is treated with calcium hydroxide (slaked lime) to pH 6–7, optimal for α-amylase. Lime is used because it serves as a source of calcium ion needed by most α-amylases as an activator and a stabilizer. A solution of bacterial α-amylase is then added, and the suspension is pumped into a steam jet where the temperature is raised instantaneously to

80°–115°. The starch is immediately gelatinized and, in the presence of the α-amylase, is depolymerized rapidly to a fluid mass that is easily handled. In a variation of this process, the starch slurry containing a calcium compound is heated to 120°–180° to gelatinize the starch, and α-amylase is added following cooling to 90°–105° (*27*).

The maximum temperature to which the starch slurry containing the α-amylase can be heated safely in the steam jet is dependent on the source of α-amylase. Ordinary bacterial α-amylase such as that produced by *Bacillus subtilis* or *Bacillus mesentericus* can be used for a few minutes at temperatures up to about 94°. The α-amylase from *Bacillus licheniformis* is unusually heat resistant and can be used at temperatures up to about 115° for short periods (*28*). Moreover, this enzyme has an exceedingly low requirement for calcium, so that calcium ion beyond the amount normally present need not be added. Tamuri and co-workers (*152*) describe an α-amylase from *Bacillus stearothermophilus* that is both heat- and acid-stable and can be used at temperatures above 100° at pH 4.5. This pH is also optimal for glucoamylase action so that pH adjustment can be minimized for the overall starch hydrolysis process.

After heating in the steam jet, the starch slurry is held at the high temperature for a few minutes to facilitate liquefaction, after which it is reduced to < 90° and

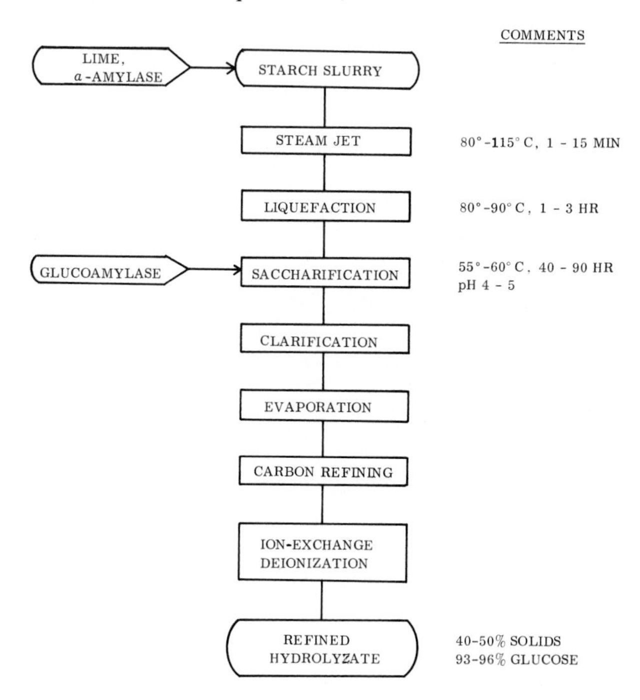

FIG. 4.—Typical process for the production of refined, high-conversion starch hydrolyzates.

maintained at that temperature for 1–3 h or until hydrolysis has progressed to 15–20 DE.

The hydrolyzate from cereal starches is not stable at less than DE 13; below this value, retrogradation occurs, forming insoluble starch particles that are difficult to remove by filtration. DE's up to ~ 45 may be attained by prolonged α-amylase treatment, but there is no economic advantage in doing so if the liquefied starch is to be converted to D-glucose. Komaki (*29*) showed that insoluble starch particles can form during liquefaction and persist through saccharification with glucoamylase. Such particles can be hydrolyzed by bacterial α-amylase if they are first solubilized by heating in water under high pressure.

Armbruster (*5*) describes a dual liquefaction process. The starch slurry is treated as described above to provide a solution of 2–15 DE which is heated at 110°–150°, cooled, and further liquefied with α-amylase at < 85°. The dual treatment results in a solution which is more easily filtered.

To a limited extent, two-stage liquefaction processes employing a first treatment with dilute acid to obtain a low-DE hydrolyzate (preferably < 3 DE) which is further treated with α-amylase to yield a stable liquefied starch have been used. When the preparation is to be further converted with glucoamylase, acid liquefaction of starch should be avoided, for reversion reactions wherein disaccharides with β-D-glucosidic linkages are formed are catalyzed by acid. The latter are not susceptible to hydrolysis by glucoamylase and constitute barriers which limit conversion to glucose (*30*).

Liquefaction with bacterial α-amylase according to the dual process described by Armbruster yields hydrolyzates containing small proportions of lipids and proteinaceous matter in suspension. The hydrolyzates typically have an oligosaccharide composition shown in Table IX. An appreciable proportion of the

Table IX

Distribution of Saccharides in Low-DE Starch Hydrolyzates Derived by Enzymic Liquefaction (5)

Degree of polymerization	*% by weight, carbohydrate dry basis*		
	10 DE	*15 DE*	*20 DE*
1	0.3	0.7	1.4
2	3.4	5.5	7.6
3	4.3	6.9	9.4
4	3.5	5.2	6.9
5	3.6	5.5	7.4
6	7.0	10.6	14.3
>6	77.9	65.6	53.0

Table X

Interaction of Saccharification Variables

Variable	*Feasible range*	*Comments*
Substrate concentration	25–35%	Range dictated by cost of removing excess water from hydrolyzates to prepare HFCS or crystalline dextrose balanced against higher conversion to glucose attainable at low substrate concentration.
Enzyme–substrate ratio	0.15–0.5 GU/g[a]	Cost of enzyme balanced against saccharification time and equipment required for long saccharifications.
Saccharification		
Time	40–90 h	Dependent on enzyme–substrate ratio.
Temperature	50–63°	Upper limit dependent on glucoamylase stability, lower limit on need to inhibit microbial contamination during long saccharification times.
pH	4.0–5.0	Dependent on source of glucoamylase. Reducing sugars most stable in this pH range.

[a] One GU (glucoamylase unit) catalyzes the conversion of starch to glucose at the rate of 1 g/h at 60°.

oligosaccharides of DP 4 and higher are branched when derived from starches containing amylopectin.

b. *Saccharification*

The objective of saccharification is to convert starch to D-glucose in yields as high as possible while observing restraints bearing on economics. Using glucoamylase it is possible to convert starch almost totally (> 99%) to D-glucose, but it is not economically feasible to do so. Several important technical and economic variables interact to limit conditions to those allowing a maximum conversion to about 93–96% D-glucose when corn starch is converted with glucoamylase. These variables and interactions are shown in Table X.

The kinetics of the saccharification of liquefied starch by glucoamylase is complicated because, at any given time in the hydrolysis, a wide array of linear and branched dextrins are present causing many simultaneous reactions, each with a different rate. The system as a whole has defied rigorous kinetic description. It is instructive nevertheless to consider the basic reactions involved in terms of their relative rates, as shown in Figure 5, to gain a qualitative understanding of the forces limiting complete conversion under commercially feasible conditions.

The amylose and amylopectin portions of cereal starch are converted by α-amylase during liquefaction to a collection of linear and branched dextrins. The linear dextrins are rapidly and almost totally converted to D-glucose by glucoamylase. The branched dextrins are much less susceptible to hydrolysis, no

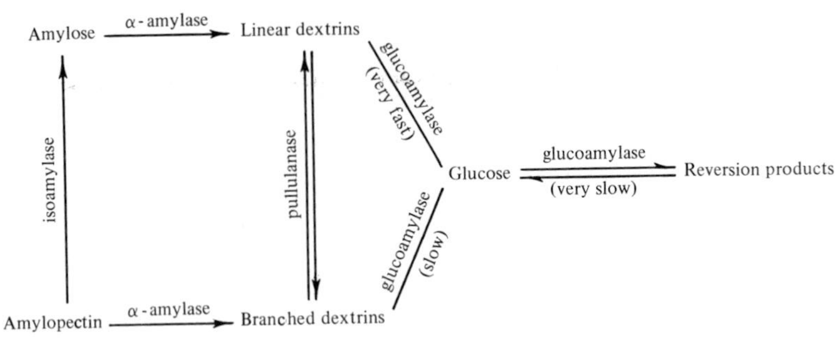

Fig. 5.—Hydrolysis/condensation reactions controlling the enzymic conversion of starch to D-glucose.

doubt owing to the lower rate at which glucoamylase cleaves the α-(1→6) D-glucosidic linkage, as compared to cleavage of the α-D-(1→4) linkage (*31*). This is illustrated in Figure 6 which compares the hydrolysis of linear dextrins with the hydrolysis of liquefied corn starch containing both linear and branched dextrins.

For practical purposes, the dextrin hydrolysis reactions are irreversible. [The primary exception is hydrolysis of the disaccharides maltose and isomaltose. These species can arise both from dextrin degradation and from the condensation of glucose (reversion).] However, hydrolysis to D-glucose is not complete be-

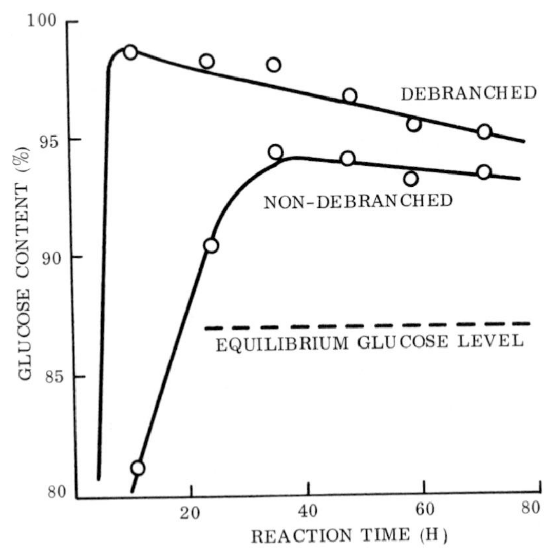

Fig. 6.—Enzymic hydrolysis of liquefied corn starch and of linear dextrins derived therefrom by debranching enzymes. Hydrolysis conditions: 60°, 0.3 glucoamylase units/g substrate, initial substrate concentration 31%.

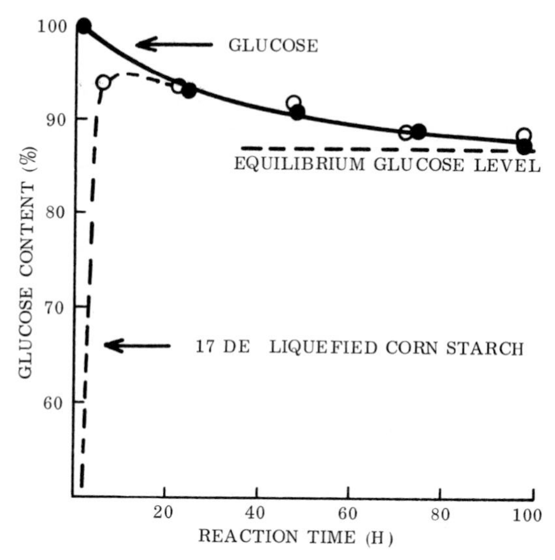

FIG. 7.—Action of glucoamylase on glucose and liquefied corn starch solutions. Hydrolysis conditions: 60°, 1.8 glucoamylase units/g substrate, initial substrate concentration 30%.

cause simultaneous condensation reactions occur wherein D-glucose is condensed to reversion products. Reversion is simply a manifestation of the reversability of hydrolysis. Both D-glucose and liquefied corn starch approach the same equilibrium D-glucose content when treated with a large amount of glucoamylase (Fig. 7). The composition of the final mixture is substantially the same whether D-glucose or starch is the starting material. Table XI shows the

Table XI

Equilibrium Composition of Saccharides Derived by Treatment of 2.78 M Glucose Solution with Glucoamylase at 60°

Species	*Dry basis, %*	*Equilibrium molar concentration*	*Molar equilibrium constant*[a]
Glucose	78.2	2.15	
Isomaltose	13.4	0.194	K_I = 1.6
Isomaltotriose	2.3	0.0221	K_3 = 2.0
Disaccharide A	1.9	0.0267	K_A = 0.22
Maltose	1.6	0.0225	K_M = 0.19
Disaccharide B	1.1	0.0156	K_B = 0.13
Isomaltotetraose	0.6	0.0043	K_4 = 3.4

[a] K_I, K_A, K_M, K_B = $[H_2O]$[Disaccharide]/$[\text{Glucose}]^2$, K_3 = $[H_2O]$[Trisaccharide]/[Glucose][Isomaltose], K_4 = $[H_2O]$[Tetrasaccharide]/[Glucose][Trisaccharide].

equilibrium composition of the reversion products derived from the action of glucoamylase on a 50% D-glucose solution. The unknown disaccharides A and B are tentatively identified as nigerose and kojibiose. Glucoamylase is able to hydrolyze nigerose (*32*), thus it is reasonable to assume that it could be formed through reversion. Glucoamylase forms α-linked saccharides with the (1→6)-linked compounds predominating.

The tri- and tetrasaccharides formed by reversion consist predominantly of isomaltotriose and isomaltotetraose, but no doubt include traces of other saccharides from the condensation of D-glucose with maltose and disaccharides A and B.

Both Figures 7 and 8 show that "overshoot" occurs when starch is hydrolyzed, that is, the equilibrium D-glucose concentration is exceeded in the first phases of the reaction where scission of α-D-(1→4) bonds predominate and thermodynamic equilibrium is approached in the later phase of the reaction as D-glucose condensation occurs.

The maximum quantity of D-glucose may be increased by treating starch with debranching enzymes such as isoamylase (glycogen 6-glucanohydrolase, EC 3.2.1.68) and pullulanase (pullulan 6-glucanohydrolase, EC 3.2.1.41) to reduce the number of (1→6) α-D-glucosidic linkages that impede rapid hydrolysis of starch by glucoamylase (*33, 34*). Hurst (*35*) developed a procedure for attaining high dextrose (98.5%) contents through hydrolysis of enzyme-liquefied starch at 30% solids with a mixture of glucoamylase and pullulanase (from *Klebsiella pneumoniae*). To achieve these results, it is necessary to conduct the hydrolysis at pH 5.9–6.3, although glucoamylase (from *A. niger*) action is optimal at pH 4.3. The higher pH is necessary because of the poor activity and stability of this type of pullulanase at the lower pH values (*36*). Recently, a pullulanase from a *Bacillus* source has been developed that is stable at pH 4.3 and is more efficiently utilized during concurrent action with glucoamylase (*153, 154*).

Glucoamylase—Glucoamylase preparations used commercially are usually mixtures of several amylases. Preparations from *Aspergillus niger* contain two glucoamylase isozymes, at least one α-amylase, and a transglucosylase (*37*). Early technology used glucoamylase preparations from the *Aspergillus* group containing significant amounts of transglucosylase activity, so it was necessary to remove the transglucosylase to achieve maximum conversion of starch to glucose (*38*). *Aspergillus* strains have since been found which give high yields of glucoamylase and little transglucosylase (*39*).

It has been reported that the glucoamylase isozymes differ in their ability to hydrolyze the α-D-(1→6) linkages in amylopectin (*33, 40*), but the implications of this for the commercial production of D-glucose from starch are not clear. Marshall and Whelan (*41*) have shown that amylopectin in dilute solution is only partially hydrolyzed by glucoamylase preparations that are free of all detectable α-amylase. In the presence of a small amount of α-amylase, however, hydrolysis is complete. This suggests that starch may contain bonds resistant to pure glucoamylase which are bypassed by α-amylase.

In the ensuing discussions, glucoamylase connotes commercial preparations that contain a mixture of glucoamylase isozymes and α-amylases, but which are substantially free of transglucosylase.

Immobilized Amylases—Numerous procedures have been revealed for the immobilization of enzymes (*42*). Significant advantages in the use of immobilized enzyme are its recovery and re-use and the convenience and low cost of a continuous reaction system. Because high effective concentrations of catalyst can be economically utilized, reaction times are short, and unwanted side reactions are minimized.

Though several methods have been described for the immobilization of α-amylases (*43*), none has found commercial application in the liquefaction of starch for the reasons given below.

1. Immobilized α-amylase systems are not applicable to the extremely viscous solutions produced when starch is gelatinized at commercial concentrations.
2. Adequate continuous liquefaction processes employing soluble α-amylase require only a few hours to produce stable solutions of hydrolyzed starch.
3. The cost of using soluble α-amylase to liquefy starch is so low that there is no economic incentive to develop a workable immobilized α-amylase system.

Though immobilized α-amylase does not seem suited to large-volume starch liquefaction, other applications have been developed. An example is the use of α-amylase on insoluble carriers to prepare retrogradation-resistant starch solutions through preferential hydrolysis of the amylose in gelatinized starch dispersions (*44*).

Many processes for the immobilization of glucoamylase and its use for starch hydrolysis have been described (*45*). Lee and co-workers (*46*) reported an extensive pilot plant study of the use of glucoamylase attached to alkylamine porous silica for the conversion of α-amylase-hydrolyzed dextrin. Best results are obtained when freshly prepared solutions of α-amylase-hydrolyzed dextrins of 25–30 DE are used as feed for the immobilized glucoamylase bed. Conversions of up to about 94% D-glucose are obtained at a feed concentration of 27%. The immobilized glucoamylase bed displays excellent stability when used at 40°, though occasional cleaning by backflushing with a sterilant solution is required to control microbiological contamination. Maximum dextrose attainable with the immobilized glucoamylase is usually about 1% less than that attained by treating the same feed with a soluble glucoamylase preparation. This phenomenon is typical of immobilized glucoamylase systems and is attributable to diffusion limitations.

The lower conversions of liquefied starch with immobilized glucoamylase compared to those obtained with soluble preparations was interpreted by Thompson and co-workers (*47*) as chiefly due to inadequate expression of α-amylase activity in commercial immobilized glucoamylase. Supplementation of immobilized glucoamylase with saccharifying α-amylases is more effective than supplementation with liquefying types.

Industry has been reluctant to adapt immobilized glucoamylase technology for the production of high-conversion hydrolyzates owing to poor economic incentive. Soluble glucoamylase is relatively inexpensive and constitutes only a small fraction of the cost of producing high-conversion hydrolyzates. Adequate capacity exists within the industry to produce hydrolyzates by batch reaction with soluble glucoamylase. A major impediment is the lack of a commercially available immobilized glucoamylase usable at temperatures high enough ($> 55°$) to ensure control of microbiological contamination.

c. *Refining*

Once the saccharification reaction has proceeded to the extent desired, the clarification and refining treatments shown in Fig. 4 are usually applied. The crude liquor obtained from enzymic saccharification typically contains less than 1% of lipids and proteinaceous materials. These impurities are removed through filtration (*10*), usually on continuous, rotary, vacuum filters coated with perlite filter aid or diatomaceous earth. In some installations, centrifugation preceeds filtration. The filtrate is a clear, slightly yellow liquor containing about 0.1% nitrogenous material and 0.2–0.4% ash.

The clarified solution may be concentrated under vacuum to 40–60% dry substance using multiple effect evaporators, and is then treated with carbon to remove the major part of the color and nitrogenous material. Typical procedures for these steps as well as for ion exchange deionization have been described in Section II.3c.

If the decolorized hydrolyzate is intended for the production of crystalline dextrose, further refining may not be necessary. For HFCS production, ion-exchange deionization is required to remove calcium ions and other as yet unidentified impurities which can result in the inhibition and destabilization of glucose isomerase (*48*). Deionization may be achieved by passing the decolorized liquor through a bed of strong acid, cation-exchange resin in the acid form and then through a bed of weak base anion-exchange resin in the free base form (*49*). This may be followed by a second treatment with cation- and anion-exchange resins either as single or mixed beds.

d. *Composition*

Deionization reduces the color to practically zero and lowers the ash and nitrogen contents to a few ppm. The refined hydrolyzate is a solution containing practically pure carbohydrate with a taste only of sweetness. A typical composition is shown in Table XII. In this composition, the disaccharides are derived almost entirely through reversion, whereas the tetra- and higher saccharides are

Table XII

Composition of High-Conversion Starch Hydrolyzate

Saccharide	*Carbohydrate dry basis, %*
Glucose	94.0
Maltose	1.0
Isomaltose	1.2
Other disaccharides	1.3
Trisaccharides	0.4
Tetrasaccharides	0.1
Higher saccharides	2.0

derived from branched dextrins which are only slowly hydrolyzed by glucoamylase and, therefore, survive the saccharification treatment.

III. Syrups Containing Fructose

The market for sweeteners made from starch can be substantially enlarged by an efficient method for converting D-glucose to D-fructose. Invert sugar, a composition of approximately equal parts of D-glucose and D-fructose derived by hydrolysis of sucrose, was formerly used as a sweetener in many food applications. An equivalent composition from isomerized D-glucose had been an industry goal for many years. Several attempts were made to exploit the Lobry de Bruyn–Alberda van Ekenstein reaction to convert D-glucose to D-fructose by base catalysis (*50–52*), but all attempts failed. Hydroxide ion-catalyzed isomerization of D-glucose is non-specific, resulting in the production of D-mannose as a major by-product. In addition, secondary isomerization of D-fructose gives D-psicose (D-allulose) as the most prevalent keto hexose impurity. Because of the low stability of reducing sugars at the pH and temperature required for alkaline isomerization, relatively large amounts of degradation products which contribute to ash and colored impurities are produced. The high level of impurities and relatively low yields of D-fructose obtained from alkaline isomerates require extensive refining and separation systems (*53*) to produce products of acceptable color and flavor. Even when highly refined, such products contain appreciable quantities of unwanted non-glucose, non-fructose saccharides.

1. Enzymic Isomerization

Events leading to the successful development of high-fructose corn syrup (HFCS) have been reviewed by MacAllister (*54*) and by Casey (*55*). Development in the 1950s of efficient enzymic means for converting starch to solutions of substantially pure carbohydrate containing 93% or more D-glucose was one of the key elements of this development in providing high quality substrate for the enzymic isomerization processes that were developed later. The discoveries which led to the eventual development of low-cost glucose isomerase (D-xylose ketol-isomerase, EC 5.3.1.5) were also key elements. Of particular importance were the discovery of an enzyme which catalyzes the direct interconversion of an aldose (D-erythrose) and a ketose (D-*glycero*-tetrulose) without need for a phosphate ester intermediate (*56*), the recognition by Marshall and Kooi that D-xylose ketol-isomerase catalyzes the interconversion of D-glucose and D-fructose as well as that of D-xylose and D-xylulose (*57*), and the identification of certain *Streptomyces* sp. as rich sources of heat-stable glucose isomerase (*58, 59*). A further factor in the commercialization of HFCS was the development of low-cost methods for the immobilization of glucose isomerase.

a. *Isomerase Production*

Glucose isomerase is produced intracellularly by certain species of microorganisms. Antrim (*60*) lists 52 species of organisms which produce glucose isomerase of the D-xylose ketol-isomerase type. [This enzyme is not to be confused with D-glucose ketol-isomerase, EC 5.3.1.18, which is an NAD-requiring intramolecular oxidoreductase that catalyzes glucose-fructose interconversion (*62*).] Of the large number listed, those of the genus *Streptomyces, Bacillus, Arthrobacter,* and *Actinoplanes* are of greatest commercial interest because of their ability to produce thermostable isomerase in high yield.

In the early phases of the development of isomerization technology, the organisms producing glucose isomerase required D-xylose in the growth medium as an inducer to obtain meaningful quantities of isomerase (*57, 62*). As D-xylose is a relatively expensive sugar, an important early development was Takasaki's discovery of *Streptomyces* strains that were capable of producing large yields of isomerase when grown on inexpensive, xylan-containing media such as wheat bran, corn cobs, or corn hulls (*59*). The isomerases commercially available today are produced from constitutive isolates or mutants which do not require xylose for isomerase production. Some examples are those described by Shieh (*63*), Ottrup (*64*), Armbruster (*65*), and Lee (*66*).

The procedure for culturing microorganisms for isomerase production varies according to the type of microorganism involved. All procedures use pure-culture, submerged, aerated fermentations conducted in several stages on media designed to promote maximum growth of microorganisms and maximum yield of isomerase. The process starts with the inoculation of a small amount of media in a shake flask from a slant containing a pure culture of the desired microorganism. The process may proceed with various substages as seed to inoculate succeeding stages culminating in a final production fermentation with a fermenting broth volume of 50,000 gallons (189,250 L). At all stages of the process, diligence is required to exclude contaminating microorganisms. Dworschack and Lamm (*67*) have described such a multiple stage process.

b. *Isomerase Immobilization*

Glucose isomerase is used almost entirely in immobilized form. The literature on isomerase immobilization and the use thereof has been reviewed by Antrim (*60*) and MacAllister (*54*). Though numerous methods for enzyme immobilization have been applied to isomerase, only two principal methods are so far economically feasible; (*a*) adsorption on an insoluble carrier, and (*b*) fixation of the isomerase within the cellular microorganism in which it occurs.

The adsorption technique requires extraction of the intracellular isomerase from cells by maceration or sonic treatment or through destruction of cell wall material by lytic enzymes (*68*). The solubilized isomerase may then be adsorbed on an insoluble carrier such as DEAE-cellulose without further treatment. The isomerase in adsorbed form can retain all the activity of the soluble isomerase (*69*). Typically, however, recovery of isomerase activity is substantially less than complete.

Immobilization of isomerase intracellularly may be done by simply heating the harvested mycelia at a temperature which destroys the lytic mechanism of the cells (*70*). This method can be used only if the isomerase is more heat stable than the lytic enzymes. Several methods employ binding of the heat-treated intact cells to various polymeric agents (gelatin, chitosan, flocculants, or cell contents obtained by partial lysis) followed by strengthening the binding through cross-linking with glutaraldehyde. Table XIII summarizes the industrially important immobilization techniques.

c. *Isomerase Reaction Kinetics*

Available evidence indicates that α-D-glucopyranose is the form that participates in enzymically catalyzed aldose-ketose isomerization (*80*). Information on the enzymically active form of D-fructose is more ambiguous, but mechanistic studies suggest that one of the two furanose forms is involved (*81*). Thus the reaction can involve three sets of equilibria.

$$\alpha\text{-D-glucopyranose} \rightleftharpoons \textit{aldehydo}\text{-D-glucose} \rightleftharpoons \beta\text{-D-glucopyranose} \tag{1}$$

$$\alpha\text{-D-glucopyranose} \rightleftharpoons \text{D-fructofuranose} \tag{2}$$

$$\begin{array}{ccc} \beta\text{-D-fructofuranose} & & \alpha\text{-D-fructofuranose} \\ \nwarrow\!\searrow & & \swarrow\!\nearrow \\ & \textit{keto}\text{-D-fructose} & \\ \swarrow\!\nearrow & & \nwarrow\!\searrow \\ \alpha\text{-D-fructopyranose} & & \beta\text{-D-fructopyranose} \end{array} \tag{3}$$

Aldehydo-D-glucose, *keto*-D-fructose, and α-D-fructopyranose are present in trace amounts. In practice, D-glucose and D-fructose mutarotations are faster than enzymic isomerization, so that treatment of the kinetics can be simplified to consideration of the D-glucose and D-fructose equilibrium. In the temperature range of interest, the equilibrium between α-D- and β-D-glucopyranose is un-

Table XIII

Industrially Important Techniques for the Immobilization of Glucose Isomerase

Immobilization technique	*Enzyme source*	*Form*[a]	*Ref.*
Adsorption on anion-exchange cellulose or anion-exchange resin	*Streptomyces* sp.	Powder or granular	(*71*) (*72*)
Adsorption on porous alumina	*Streptomyces* sp.	Granular	(*73*)
Heat treatment of whole cells at temperatures that destroy lytic enzymes without inactivating isomerase	*Streptomyces* sp.	Powder	(*70*)
Glutaraldehyde cross-linking of partially lysed cells	*Bacillus coagulans*	Granular	(*74*)
Entrapment of whole cells in glutaraldehyde cross-linked gelatin	*Actinoplanes missouriensis*	Granular	(*75*)
Binding of whole cells with flocculating agents	*Arthrobacter* sp.	Granular	(*76,77*)
Binding of whole cells with chitin and cross-linking with glutaraldehyde	*Streptomyces* sp.	Granular	(*78,79*)

[a] Immobilized isomerases in powder form are usually < 100 mesh; granular isomerse preparations are > 100 mesh.

affected. The equilibrium among the various D-fructose anomers is shifted toward a higher proportion of keto and furanose forms (*82*). Increasing temperature also shifts the glucose–fructose equilibrium toward a higher proportion of D-fructose as shown in Figure 8.

Unlike the kinetics of starch saccharification, the kinetics of D-glucose isomerization are well established owing to the simplicity of the reaction. Enzyme-catalyzed isomerization has been shown to follow reversible Michaelis–Menten kinetics over the entire course of the reaction (*83, 84*), and Lloyd (*69*) has shown that substantially the same kinetics apply to immobilized isomerase prepared by adsorption of the enzyme on DEAE-cellulose. The isomerization reaction can be depicted by the following equation:

$$\mathrm{Glc} + \mathrm{E} \underset{k_2}{\overset{k_1}{\rightleftharpoons}} \mathrm{ES} \underset{k_4}{\overset{k_3}{\rightleftharpoons}} \mathrm{E} + \mathrm{Fru} \qquad (1)$$

where Glc, E, ES, and Fru stand for D-glucose, enzyme, enzyme–substrate complex, and D-fructose, respectively. The rate of conversion of glucose to fructose is given by the following:

$$\frac{dI}{dt} = \frac{k_3(E)(1 - I/I_e)}{(C)[1 + K_G/C + I(K_G/K_F - 1)]} \qquad (2)$$

where I = mole fraction of fructose; I_e = mole fraction of fructose at equilibrium; E = total concentration of enzyme; K_G,K_F = Michaelis constants for glucose and fructose, respectively; and C = total concentration of D-glucose and D-fructose.

In practice, the isomerase has a finite lifetime (on the order of a few hundred hours). For most purposes, it is sufficient to assume that loss of enzymic activity follows first order decay as expressed by the following:

$$E = E_o \exp(-\ kt_s) \quad (3)$$

where t_s = time that the immobilized isomerase is in service and k = first-order decay constant for enzyme inactivation.

Isomerization is typically performed by pumping a solution of D-glucose through a bed of the immobilized isomerase. The reaction is heterogenous and reaction time, t, is the effective contact time between the solid-phase active enzyme in the reactor and the substrate solution and is governed by the following simple expression:

$$t = V/R \quad (4)$$

where V = volume of liquid phase in the void spaces of the packed bed and R = flow rate through the bed. Enzyme concentration may be treated successfully as the equivalent of the ratio of active solid-phase enzyme in the bed (E_s) to the volume of liquid phase therein.

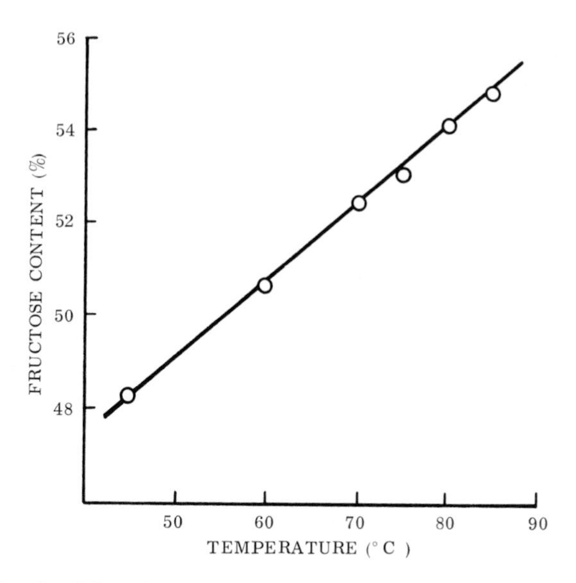

FIG. 8.—Effect of temperature on D-glucose–D-fructose equilibrium.

$$E = E_s/V \tag{5}$$

By setting $k_3/I_e(K_G/C+1) = k_f$ and by making the simplifying assumption that $K_G = K_F$, a working equation can be derived which shows the relationship between the most important elements controlling the dynamics of isomerization with fixed beds of immobilized isomerase.

$$\ln\left[\frac{I_e - I_o}{I_e - I}\right] = \frac{k_f E_{so} \exp(-kt_s)}{RC} \tag{6}$$

where I_o = mole fraction of fructose in solution (based on G + F) entering the reactor; I = mole fraction of D-fructose in solution (based on G + F) exiting the reactor; and E_{so} = total isomerase activity in the bed when first placed in service. This applies only for isothermal operation when the pH and composition of the substrate solution are held constant.

d. *Reactors and Composition Control*

The types of reactor systems which might be applied for the enzymic isomerization of D-glucose have been reviewed by Vieth (*85*). While a number of reactor system designs have been demonstrated experimentally, most commercial systems use packed-bed reactors which differ depending on whether powdered or granular immobilized isomerase is used.

Immobilized isomerase preparations in powdered form can be utilized in reactors employing leaf filters in which a relatively shallow layer [< 12 in. (30 cm)] of powder-type immobilized isomerase is built up on the filter leaves. D-Glucose solution is then pumped through the isomerase. Several reactors may be connected in series to correct for channeling which can diminish reaction efficiency when single shallow packed beds are used (*86*). By periodically replacing the oldest reactors in the series with fresh ones, a relatively constant level of activity is maintained while achieving maximum utilization of the immobilized isomerase. Individual reactors may be kept on stream through three or more half-lives, whereupon the isomerase activity has diminished to less than 10% of the initial value.

Granular-type isomerase preparations are generally used in deep-bed reactors with cylindrical columns 5–15 ft (1.5–4.6 m) high (*48, 75, 87*). Such reactors are of simpler design than are filter leaf reactors. The relatively large particle size of the granular type isomerases required for such reactors can result in loss of reaction efficiency owing to a limitation of diffusion into pores which diminishes the accessibility of the isomerase held within the inner regions of the granules (*88*). Such diffusion limitation is absent in powdered-type isomerase held on DEAE-cellulose (*69*) because of the high surface area and the proximity of the isomerase to the surface.

Table XIV

Typical Operating Conditions for Isomerization in Packed-Bed Immobilized Glucose Isomerase Reactors

Glucose concentration	40–60%
pH	7.0–8.5
Reaction temperature	55–65°
Isomerase activator	0.001–0.005 M Mg^{2+}
Reactor residence time	<4 h

Conditions common to most commercial isomerization reaction systems are given in Table XIV. D-Glucose solution used as feed to the reactor is typically a refined, high-conversion starch hydrolyzate prepared as described in Section II.4. Refining is required to reduce the calcium content of the hydrolyzate to levels that do not seriously inhibit isomerase activity (usually < 20 ppm) and to remove other unidentified substances which can seriously diminish the stability of the isomerase. Recently, a process has been described wherein starch hydrolyzates suitable for enzymic isomerization are produced under conditions which avoid the introduction and formation of isomerase inhibitors and destabilizers (*155*). Such hydrolyzates need not be refined prior to isomerization. Other activators and stabilizers may be added to the substrate solution in small amounts as, for example, ferrous salts (*89*), sulfite salts (*90*), or combinations of such compounds (*91*). The reaction temperature is maintained high enough to inhibit growth of microorganisms.

Once the reactor conditions are established, the degree of conversion of D-glucose to D-fructose is controlled by changing the flow rate according to the relationship given by Eq. (6). Because high potency immobilized isomerase is used, the reaction period need be no more than about 4 h and is usually less than 1 h. Under such mild conditions, there is practically no production of the sugar degradation products that typify alkali-catalyzed isomerization.

2. Refining, Concentration, Composition

The isomerized products contain little else than water, carbohydrate, and the various activator substances added prior to isomerization. They are, nonetheless, further refined to remove traces of salts, colored substances, color precursors, and off-flavor. Because the isomerate contains D-fructose, it is necessary to refine at a pH and temperature that avoids by-products such as D-psicose (*92*) and reversion products (*93*).

After refining, the isomerized syrup product is concentrated at low temperature in vacuum evaporators to solids contents high enough to ensure resistance to microbiological spoilage but low enough to prevent crystallization of D-glucose.

Table XV

Characteristics of Typical First Generation High-Fructose Corn Syrup (HFCS)

Solids	71%
pH	4.0
Carbohydrate content (dry solids basis)	
D-Glucose	52%
D-Fructose	42%
Di- and higher saccharides	5–6%
D-Psicose	<0.1%
Nitrogen	<10 ppm
Ash	<0.05%
Appearance	Clear, water white
Off-flavors	None
Relative sweetness[a] (15% aqueous solution)	100

[a] Based on sucrose as 100. Perceived relative sweetness depends on test conditions, in particular, temperature and concentration. For a discussion of this topic, see reference (*94*).

The resulting product is a clear, colorless syrup comprising substantially pure carbohydrate and water with a clean, sweet taste. Properties of the most commonly produced article of commerce are shown in Table XV.

IV. Syrups with Supraequilibrium Fructose Contents

1. Equilibrium Limitations

The maximum degree of conversion of D-glucose to D-fructose in a catalyzed reaction (wherein the rate of interconversion is substantially less than the rate of mutarotation) is fixed at about 50% by the thermodynamic equilibria between the various molecular forms of D-glucose and D-fructose in solution. This equilibrium is only slightly affected by temperature as shown in Figure 8.

Several methods for achieving compositions with D-fructose contents greater than equilibrium amounts have been devised and can be classified according to whether kinetic, chemical, or physical means are used. The kinetic and chemical methods can result in direct conversion to supraequilibrium D-fructose contents. The physical methods involve separation of a glucose–fructose mixture to obtain at least two fractions, a D-fructose-rich and a D-glucose-rich fraction. The D-glucose-rich fraction is usually reisomerized and reprocessed for economic reasons. In spite of this disadvantage, physical methods are favored over kinetic or chemical methods for production of supra-equilibrium fructose syrups because of the low cost of the high-efficiency, large-scale chromatographic methods that have been developed (see Section IV.5).

2. Equilibrium Overshoot

The composition of an equilibrium mixture of D-glucose and D-fructose at 60° is as follows:

$$\underset{(33.5\%)}{\beta\text{-D-Glucopyranose}} \rightleftharpoons \underset{(16.5\%)}{\alpha\text{-D-Glucopyranose}} \rightleftharpoons \underset{(50\%)}{\text{D-Fructose (mutarotated)}}$$

Thus, the equilibrium between the α-anomer of D-glucose and the fully mutarotated mixture of D-fructose anomers is in the ratio 16.5 to 50. Such a mixture contains aboot 75% D-fructose. It should be possible theoretically to convert pure α-D-glucose to a mixture with > 50% D-fructose under conditions where the rate of mutarotation to β-glucose is much less than the rate of conversion to D-fructose. The reaction could be stopped before full mutarotation of D-glucose occurred to produce an ''equilibrium overshoot'' analagous to that discussed in Section II.4b for the enzymic saccharification of starch with glucoamylase. Schray and Rose (*95*) obtained marked overshoot on treating freshly dissolved α-D-xylose with xylose isomerase or β-L-arabinose with arabinose isomerase and have discussed the necessary conditions.

The overshoot method can produce D-fructose contents greater than 50%, but substantially less than 75%. Though crystalline α-D-glucose is available in large quantities, its cost is greater than the cost of enriched D-fructose compositions obtainable by chromatographic fractionation; hence, the overshoot approach is not economically feasible.

3. D-Fructose Complexation

Various methods have been put forth for increasing the amount of D-fructose attainable in both enzyme- and nonenzyme-catalyzed isomerizations, wherein the reaction mixture contains a soluble salt of an oxyanion capable of reacting preferentially with the D-fructose as it is formed. Such practice does not alter the equilibrium between free D-glucose and D-fructose, but results in mixtures wherein the sum of the free and combined D-fructose is substantially greater than obtainable in the absence of complexant. Thus, isomerization in the presence of sodium stannate gives 91% fructose (*96*), sodium aluminate gives 67% fructose (*97*), salts of arylboronic acids give up to 68% fructose (*98*), and sodium borates give up to 90% (*99*) conversion. These procedures are effective because the oxyanion complexes more strongly with D-fructose than with D-glucose (*100*).

Very little is known of the structure of such complexes. However, the reaction of polyhydroxy compounds with oxyanions such as borate to form cyclic complexes is well known (*101*), and the greater complexing ability of D-fructose could owe to the ability of both D-fructofuranose and D-fructopyranose to assume conformations in which the dihedral angles of the hydroxyl groups on C-1 and C-2 are favorable for oxyanion complexation.

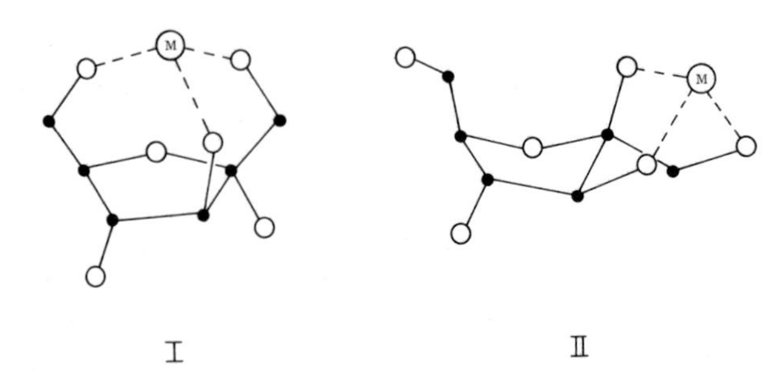

FIG. 9.—Conformers of D-fructose capable of complexing metal ions: I, α-D-fructofuranose metal complex; II, β-D-fructopyranose metal complex.

All methods that depend on complexation of D-fructose suffer the same disadvantage, namely the subsequent separation from the reaction mixture. The strongest complexants result in the largest yield of D-fructose, but require the most difficult conditions for removal of the complexant, often the addition of one equivalent of acid per mole of complexant.

Takasaki (*102*), in an attempt to circumvent the need for complexant removal, passed 50:50 D-glucose–D-fructose solutions through columns containing anion-exchange resin in the borate form mixed with an immobilized glucose isomerase to obtain an effluent solution containing 60–66% fructose (*102*). The increase in D-fructose above the normal equilibrium value is undoubtedly due to leaching of borate into solution, most likely by a site-sharing mechanism (*103*), to account for the apparent shift in equilibrium.

4. Non-Chromatographic Separations

Several processes for enriching D-glucose–D-fructose mixtures to obtain compositions with supraequilibrium D-fructose contents have been described. These are based on differences in solubility of the sugars in various solvents, and differences in complexation with salts or adsorbents. A straightforward procedure is the partial crystallization of D-glucose from aqueous solutions of invert sugar (equimolar concentrations of D-glucose and D-fructose) to obtain a mother liquor enriched in D-fructose (*104*). Only partial enrichment, to a practical maximum of about 65% D-fructose, is possible by this approach, owing to the high solubility of D-glucose.

Addition of monohydric alcohols (*105*) or mixtures of monohydric and polyhydric alcohols (*106*) to reduce D-glucose solubility have been proposed. Addition of sodium chloride to promote crystallization of D-glucose as the salt complex (*107*) or of calcium chloride to promote crystallization of the D-fructose–$CaCl_2$ complex (*108*) are also described. These methods result in contamination of sugar fractions with large amounts of salts or organic solvents.

Other methods for D-glucose–D-fructose separation are liquid-liquid extraction (*109*) and parametric pumping (*110*). The latter is a cyclic process which utilizes, as the driving force for the separation, changes in the ability of a cation-exchange resin in an alkaline earth metal form to adsorb D-fructose at different temperatures.

In another process, the D-glucose in invert sugar is oxidized to a mixture composed principally of a D-gluconate salt and D-fructose (*111*). The oxidation uses a mold, such as *Aspergillus niger,* rich in glucose oxidase (β-D-glucose:oxygen 1-oxidoreductase, EC 1.1.3.4) under aerobic conditions. The D-gluconate salt is isolated by crystallization from aqueous methanol. D-Fructose is crystallized from the mother liquor after various refining steps. D-Fructose prepared in this manner has been offered commercially, but is no longer available.

5. Chromatographic Methods

a. *Adsorbents*

An ideal process for D-glucose–D-fructose separation is one that requires no addition of chemicals, obviating the need for secondary purification. Such ideality is closely approached by present chromatographic separation processes based on ion-exchange materials as the adsorbent. Use of ion-exchange materials for the separation of non-ionizing, water-soluble compounds was generally set forth by Wheaton (*112*). Later, Serbia (*113*) discovered that the calcium form of sulfonated polystyrene resins preferentially adsorb D-fructose from aqueous solution in preference to D-glucose. Lefevre (*114*) showed that sulfonated resin salts of barium, strontium, or silver also had this property. Since then, zeolites, crystalline aluminosilicates, of the Y-type and Z-type in an alkaline earth metal form have been found to serve as preferential adsorbents for D-fructose (*115, 116*). All these adsorbents are ion-exchange materials capable of reversibly binding metal ions.

Angyal (*117*) discusses the requirements for the complexation between polyhydroxy compounds and metal ions. In general, divalent cations complex polyhydroxy compounds more strongly than do monovalent ions. The polyhydroxy compounds must be capable of assuming stable conformers, wherein the oxygen atoms of three hydroxyl groups can form an equilateral triangle with the oxygen atoms approximately 3 Å apart. Pyranose sugars having conformers with contiguous hydroxyl groups in an axial, equatorial, axial arrangement fulfill this requirement. The preferred conformers of D-glucose do not have this property, nor do the pyranose forms of D-fructose. Both furanose forms of D-fructose, however, are able to assume conformations wherein three hydroxyl group oxygen atoms can form an equilateral triangle with the requisite spacing as depicted in Figure 9.

The adsorption of D-fructose in preference to D-glucose by ion-exchange mate-

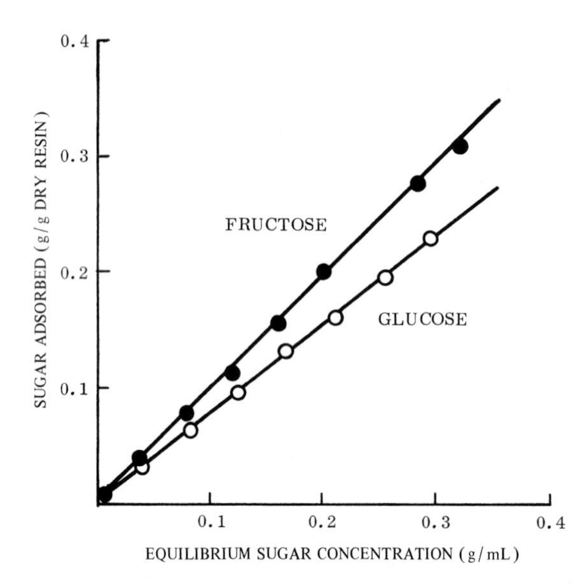

FIG. 10.—Isotherms for the adsorption of D-glucose and D-fructose on a cation-exchange resin (Dowex 50X-4) in the calcium form at 60°.

rials is due to the greater tendency of D-fructose to form complexes with the metal counterions contained therein. In practical applications, calcium forms of ion-exchange adsorbents are preferred because of their low cost, their non-toxic nature, and their strong complexing ability.

A particularly advantageous property of the ion-exchange resins as adsorbents is their linear adsorption equilibria, as demonstrated in Figure 10. This characteristic is desirable because it leads to the development of symmetrical chromatographic peaks over a broad range of concentrations which, in turn, allows efficient separations at high sugar concentrations.

Anion-exchange resins in the bisulfite form have been proposed as separations media (*118*), but have not proved suitable because of non-linear adsorption characteristics.

b. *Apparatus*

A number of chromatographic systems have been proposed for the large-scale separation of D-glucose and D-fructose employing chromatographic columns several feet in diameter capable of holding thousands of cubic feet of adsorbent. Large-scale systems must have a high degree of resolution and high reliability, high productivity, and low operating expense. A general classification of systems is given in Table XVI.

The highly competitive nature of the manufacture of D-fructose-containing products has virtually eliminated the use of simple chromatographic separations in favor of the more efficient recycle and simulated moving-bed systems.

Table XVI

Large-Scale Chromatographic Systems for Glucose–Fructose Separation

Type	*Identifying characteristics*	*Advantages*	*Disadvantages*	*Ref.*
Simple	Single-pass elution chromatography using a single-packed bed	Simplicity	Low productivity and high operating costs owing to large adsorbent inventory and high desorbent usage	(*113,119*)
Recycle	Uses single-packed bed. Effluent fractions are too dilute to be of value and poorly separated fractions are recycled to the inlet of the column	Reduced requirement for adsorbent and desorbent. Improved production rate	Slightly more complex than simple chromatography	(*120,121*)
Simulated moving bed	Several short chromatographic columns are connected in series to form a closed circuit with means for introducing feed or desorbent, or taking off glucose-rich or fructose-rich streams continuously at the junctures between columns. Entrance and exit points are moved periodically to coincide with progression of chromatographic gradient pattern around the circuit. Simulates movement of adsorbent countercurrent to the flow of liquid	Same as recycle chromatography. Capable of continuous operation	High degree of complexity requiring more sophisticated control measures than other forms of chromatography	(*122,123*) (*124,125*)

In all systems, pure water is used both as the solvent for the mixture of sugars introduced as feedstock for the systems and as the desorbent for elution of D-fructose-rich and D-glucose-rich fractions. Because the adsorbents in use are ion-exchange materials typically containing divalent calcium as the counterion, it is necessary to fully deionize feed streams to avoid stripping calcium ion from the adsorbent with consequent loss of separating power. Effluent streams emanating from the chromatographic separation systems are practically pure carbohydrate and water.

6. Final Processing and Composition

Final processing is simplified by the high quality of the chromatographic fractions and may consist of blending the D-fructose-rich stream with other streams, such as unfractionated feedstock, to obtain the desired D-fructose level. Final refining and evaporation of excess water may be needed. The D-glucose-rich fraction can be reprocessed through isomerization and refining to produce more feed stock.

Table XVII gives the composition of the two types of commercially available syrup products with supraequilibrium D-fructose contents, the so-called "second generation" high-fructose corn syrups.

V. Crystalline Sweeteners from Starch

1. Sweeteners Containing Predominantly D-Glucose

Crystalline sweeteners containing D-glucose may contain one or more of the three crystalline forms of D-glucose, α-D-glucopyranose monohydrate, anhydrous α-D-glucopyranose, and anhydrous β-D-glucopyranose. Of these, the primary product of commerce is α-D-glucopyranose monohydrate which is commonly referred to in the trade as dextrose monohydrate or just dextrose. About one billion pounds (4.5×10^8 kg) of dextrose monohydrate are sold each year in the United States (*9*). Anhydrous α-D-glucose is also marketed at a substantially lower volume. Anhydrous β-D-glucose is not commercially available from U.S. producers. In addition to the manufacture of pure crystalline products, a "total sugar" product which is a refined starch hydrolyzate containing a high D-glucose content, crystallized and dried without separation of non-D-glucose saccharides, is made and sold.

a. *Total Sugar*

Although substantially all crystalline dextrose sold is produced by processes in which the crystalline dextrose is separated from the mother liquor, there have been recent process developments in which improved "total sugar" type products are made. As with the crude solidified products of a century ago, these

Table XVII

Characteristics of Typical Second Generation High-Fructose Corn Syrups (HFCS)

Characteristics	*55% fructose*	*90% fructose*
Solids	77%	80%
pH	3.6	4.0
Carbohydrate content (dry solids basis)		
D-Glucose	41%	7%
D-Fructose	55%	90%
Di- and higher saccharides	<4%	<3%
D-Psicose	<0.1%	<0.1%
Nitrogen	<10 ppm	<10 ppm
Ash	<0.5%	<0.5%
Appearance	Clear, water white	Clear, water white
Off-flavors	None	None
Relative sweetness[a]	100	120–160

[a] Based on sucrose as 100. Perceived relative sweetness depends on test conditions, in particular, temperature and concentration. For a discussion of this topic see reference *94*.

"total sugar" products contain all the non-D-glucose sugars present in the starch hydrolyzate since no mother liquor is separated from the solidified carbohydrate mixture. These new total sugar products are substantially improved over past art in that the dextrose content is much higher because of improved starch hydrolysis methods. The total sugars that were made before enzyme-catalyzed hydrolysis contained less dextrose and, hence, exhibited less sweetness. Also, the acid-converted hydrolyzates used in earlier total sugar processes contained impurities which imparted a bitter taste. The process for manufacture of high conversion hydrolyzates currently used for production of crystalline "total sugar" is described in Section II.4.

A process patented by Black (*126*) describes the crystallization of a high-DE starch hydrolyzate to yield a total sugar containing predominantly α-D-glucose monohydrate. A refined enzyme-converted starch hydrolyzate of 96 DE is concentrated to 70% solids, cooled to 25°, seeded with crystals of dextrose monohydrate, and crystallized rapidly until an equilibrium is established between the dextrose monohydrate crystals and the saturated solution at which about 50% of the total-D-glucose is crystallized as microcrystals of the monohydrate. This massecuite is mixed with recycled dried total sugar of ~ 9% moisture (predominantly water of crystallization associated with dextrose crystals) and further crystallization is effected. Subsequently, the mixed product is dried below 50° to ~ 9% water content, yielding a product in which substantially all the dextrose is present as the α-monohydrate. A total sugar containing a mixture of the crystalline monohydrate and anhydrous α-D-forms of dextrose can be obtained by final drying at a temperature higher than 50°.

A process for making a total sugar containing a mixture of anhydrous α-D-glucopyranose and anhydrous β-D-glucopyranose is disclosed by Wilson and Frankel (*127*). A refined starch hydrolyzate containing 85–95% D-glucose is concentrated to 75% solids and spray dried using recycled dried particles as seed for crystallizing the concentrated liquor. Drier temperatures (> 50°) are controlled to crystallize a significant amount of dextrose in the anhydrous β-form. Crystalline β-dextrose does not revert to the anhydrous α-form or the α-monohydrate at moisture contents less than 4%, for the oligosaccharides present have a higher affinity for the water and prevent the β-dextrose from dissolving and crystallizing out in the α-anomeric form.

A more recent disclosure by Mise and co-workers (*128*) defines a total sugar process that produces high levels of anhydrous β-dextrose. Products containing > 85% anhydrous β-dextrose can be made from crystalline α-dextrose (monohydrate or anhydrous) or high-DE starch hydrolyzates. The dextrose-containing solution is concentrated under reduced pressure to 94–98%. Subsequently, the concentrated liquor is efficiently mixed with seed crystals of anhydrous β-dextrose in a kneader containing viscous D-glucose syrup at a temperature, preferably at about 90° or above, conducive for anhydrous β-dextrose crystallization. As crystallization proceeds during the kneading action, the product is dehydrated under reduced pressure until substantially all the D-glucose is crystallized. The stirring action of the kneader causes the dehydrated mass to become a flowable powder which is subsequently dried under reduced pressure.

b. *Crystalline Dextrose*

α-D-Glucose Monohydrate.—Prior to about 1960 all commercial crystalline dextrose produced in the United States was obtained by acid-catalyzed hydrolysis of corn starch. Processes for manufacturing crystalline dextrose from such hydrolyzates have been reviewed by Dean and Gottfried (*129*). Since then, improved processes using acid–enzyme- or enzyme-catalyzed hydrolysis have been developed to improve the quality and yields. Refining systems for processing D-glucose liquors prior to crystallization have also been modified to include ion-exchange deionization treatment to remove ionic impurities and improve color stability.

Figure 11 shows the process steps in the production of crystalline dextrose monohydrate. The conversion of starch to high-glucose liquors preferred for crystallization is described in Section II.4. Subsequent refining is essentially identical to corn syrup refining as described in Section II.3c. Most commercial systems include ion-exchange deionization treatment to obtain liquors for crystallization that are ash-free and low in color.

After refining and concentration to the desired dry substance content (70–78%), the hydrolyzate is cooled and pumped to a crystallizer, which commonly consists of a large horizontal cylindrical tank fitted with a cooling jacket and a

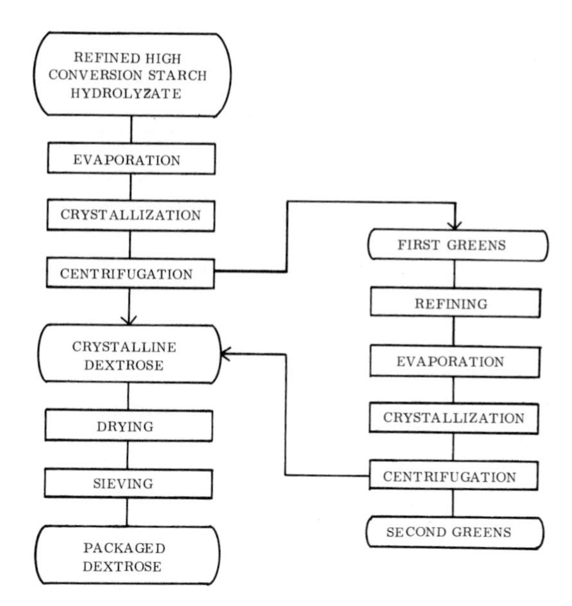

FIG. 11.—Typical process for dextrose monohydrate production.

slowly rotating cooling coil. Such crystallizers are operated batchwise. More modern installations use continuous crystallizers in which the massecuite is continuously transferred through a series of crystallizing vessels. The crystallizer feed liquor is mixed with massecuite from a previous crystallization which provides seed crystals. Normally, about 20% of the previous crystallizer batch is retained for seeding.

Conditions for crystallization described by Rentshler and co-workers (*130*) include control of D-glucose content, dry solids basis, and the dry substance content of the crystallizer liquor. Regulation of crystallizer dry substance feed is necessary to avoid undesirable "shock seeding" and concomitant production of small crystals that are difficult to separate. Also, good control is needed to obtain optimum production rate and yield.

Data in Table XVIII illustrate the dry substance–glucose content relationship when crystallization is begun at 40°. When starting at other crystallization temperatures in the range of 20–54°, the crystallizer feed liquor dry substance is determined by the expression:

$$S_T = S_{40} + 0.8\ (T - 40)$$

where T = temperature, S_{40} = percent dry substance (see Table XVIII) for a given range of D-glucose content, and S_T = percent dry substance in the crystallizer feed at an initial crystallization temperature of 20–54°. During crystallization, it is important to maintain control of supersaturation so that the D-

Table XVIII

Optimum Dry Substance Levels for Crystallization of α-D-Glucose Monohydrate[a]

D-Glucose content, % dry basis	Dry substance in crystallizer feed, %
82.0–85.9	76
86.0–89.9	75
90.0–92.9	74
93.0–94.9	73
95.0–95.9	72

[a] These glucose content and dry substance levels apply only when crystallization is begun at 40°.

glucose is constantly crystallizing at a rate producing the desired crystal size. A supersaturation ratio between 1.02 and 1.20 is maintained during crystallization. Supersaturation is defined as the ratio of the observed concentration of D-glucose in the crystallizer, neglecting non-D-glucose solids present, to the known equilibrium concentration of D-glucose in water at the same temperature. When crystallizer feed liquor is initially mixed with the seed massecuite, the supersaturation ratio is 1.1–1.2; but as crystallization proceeds, this ratio decreases to ~ 1.02, the final value at which the crystalline solids and mother liquor are separated. Supersaturation is conveniently maintained by controlling the massecuite temperature.

D-Glucose crystallization is an exothermic reaction and a significant amount of heat exchange is necessary to continuously maintain the desired supersaturation throughout the crystallization period. The massecuite is slowly cooled with continuous slow agitation during 3–4 days to ~ 20–30°. At the end of the crystallization period, about 55–60% of the crystallizer feed liquor solids are crystallized as α-D-glucose monohydrate. The weight of solid phase with respect to total massecuite weight approximates 50%.

At the end of crystallization, the massecuite is transported to large centrifugals lined with perforated screens in which the crystalline solids are separated from the mother liquor. During the latter part of the centrifugal operation, the wet crystalline cake is washed with water to remove residual mother liquor. The wet dextrose cake contains about 14% moisture including water of crystallization. Subsequently the cake is dried to a water content somewhat less than the theoretical 9.1% for dextrose monohydrate, usually to ~ 8.5%. The dried crystals are then classified according to particle size and packaged.

The mother liquor obtained from the first crystallization is commonly identified as "first greens." It may be reprocessed as shown in Figure 11 to recover a second crop of wet dextrose monohydrate crystals which have purity substantially equivalent to that obtained from the first crystallization. The yield of crystalline solids from the second crystallization is 55–60%, giving a total yield of 80–85% for the two-stage crystallization. The first greens may also be partially recycled to the initial crystallization stage to recover the total production as a single crop of crystals.

The mother liquor obtained from the second crystallization is known in the trade as "second greens" or "hydrol." This mother liquor may be recycled to recover a higher yield of dextrose crystals; however, it is usually sold as a raw material for food ingredients. Second greens are a primary raw material source for caramel color manufacture.

The yield and quality of crystalline dextrose is dependent on the dextrose content of the starch hydrolyzate and the refining treatment applied. In addition to a material balance accounting, the yield may be calculated from the dry substance composition of in-process materials according to the following equation (*10*).

$$Y = (G_c - G_g)(100)/(G_p - G_g)$$

where Y = percent yield of dry substance crystalline dextrose; G_c = percent dextrose in crystallizer liquor, G_g = percent dextrose in the mother liquor (first greens for estimate of yield in first crystallization or second greens for estimate of combined yield from both crystallization steps), and G_p = percent dextrose in crystalline dextrose.

Anhydrous α-D-Glucose.—The commercial production of anhydrous α-dextrose is substantially less than that of dextrose monohydrate. Anhydrous α-dextrose is generally made by redissolving and recrystallizing α-D-glucopyranose hydrate crystals in a batch vacuum evaporator/crystallizer (*129, 131, 132*) in a manner similar to crystallization of cane or beet sugar. Evaporation during crystallization maintains the required supersaturation as D-glucose solution is added. Seeding of the batch is accomplished by initially concentrating a limited quantity of the D-glucose solution at about 65° until crystal nuclei of anhydrous α-dextrose form. These nuclei are then made to grow by controlled addition of D-glucose solution to the evaporator/crystallizer. The crystallization time is short compared to crystallization of α-dextrose hydrate, usually ~ 5 h. The massecuite containing anhydrous α-dextrose crystals and mother liquor is separated by centrifugation. After substantially all mother liquor is purged, the crystals are washed to give a high purity product which is then dried to < 0.1% moisture. The yield of crystalline product is about 50%, and the mother liquor may be used for α-glucose hydrate manufacture.

A more recent process has been described, wherein pure anhydrous α-dextrose can be crystallized from purified starch hydrolyzates such as are currently used for α-dextrose hydrate manufacture (*133*). The process improves the economics

Table XIX

Physical Properties of D*-Glucose*

	Anhydrous α-D-glucose	*α-D-Glucose monohydrate*	*Anhydrous β-D-glucose*
Melting point	146°	83°	150°
Solubility at 25°, % by wt	62	30.2	72
Specific optical rotation,[a] $[\alpha]^{20}{}_{D}$	112°	112.2°	18.7°
Heat of solution at 25°, cal/g	−14.2	−25.2	−6.2

[a] Specific optical rotation of equilibrated solution is 52.7°.

of anhydrous α-dextrose production and simplifies equipment requirements. A refined starch hydrolyzate containing at least 93% D-glucose is concentrated to 80–90% solids in an ordinary single-effect vacuum pan, transferred to a batch crystallizer, and seeded. Usually, a large quantity of seed is employed, for example, a massecuite volume equal to ~ 60% of the crystallizer volume, to provide adequate surface for controlling crystal growth. Crystallization is isothermal or at programmed temperatures in the range of 55–75°. The crystalline anhydrous α-dextrose is separated from the mother liquor by centrifugation and further purified by washing with water at 55–60°. Mother liquor from this process may also be used to produce α-dextrose hydrate.

Anhydrous β-D-Glucose.—Compared to the two crystalline forms of α-dextrose, the higher dissolution rate and solubility of anhydrous β-dextrose make it attractive for certain applications. However, a commercial product has not developed primarily because of its higher cost of manufacture, difficulties with color formation at the high temperatures required for crystallization, and the tendency to revert to α-dextrose under slightly humid conditions.

c. *Properties*

Properties of concern in the commercial production and use of dextrose have been reported by others (*134–137*) and only a brief discussion of major physical and functional properties are given here. Table XIX lists the melting points, initial solubilities, optical rotations, and heats of solution for the three crystalline forms of dextrose (*137*). Solubilities are those observed upon initial dissolution. The solubility of an equilibrated mixture of the α and β anomers (about 38% α-D-glucose and 62% β-D-glucose) is 51% at 25°. The rate of dissolution of the three crystalline forms corresponds to their initial solubilities, that is, anhydrous β-D-glucose > anhydrous α-D-glucose > α-D-glucose monohydrate. The solubility of D-glucose in water versus temperature is shown in Figure 12.

Viscosity of D-glucose solutions at various concentrations and temperatures is given in Table XX. Viscosity values are substantially lower than for corn syrups

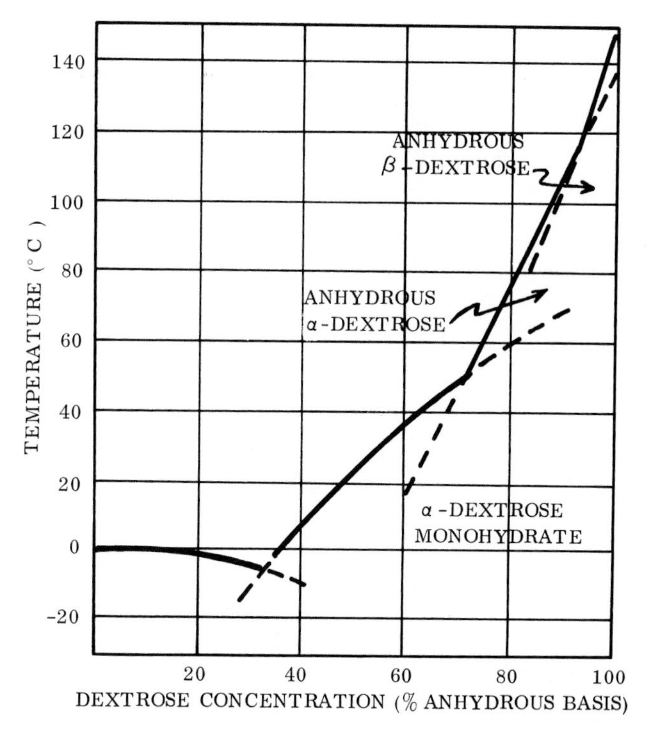

FIG. 12.—Solubility of dextrose in water (*136*).

shown in Table VIII as the molecular weight of D-glucose is much lower than the average molecular weight of the saccharide mixture in corn syrups.

Crystalline dextrose is hygroscopic and absorbs moisture from the atmosphere if not properly packaged and stored. The equilibrium relative humidity value at 25° for anhydrous dextrose is about 60% (*138*).

Table XX

Viscosity of Glucose Solutions as Affected by Concentration and Temperature (135)

Concentration, %	Viscosity, cP						
	15.6°	*26.7°*	*37.4°*	*48.9°*	*60.0°*	*71.1°*	*82.2°*
20	2.4	2.4	1.3	1.0	0.8	0.7	0.6
35	5.3	3.4	2.4	1.8	1.4	1.2	0.9
50	16.3	9.3	6.1	4.4	3.1	2.4	1.9
65	100.0	44.9	29.5	18.0	11.2	7.9	5.5
75	—	—	200.0	94.4	49.0	28.6	17.5

Table XXI

Effect of Concentration on Sweetness of Dextrose (139)

Concentration, g/100 mL	*Sweetness relative to sucrose*[a]
2.5	56
5.0	56
10.0	63
20.0	72
30.0	77
40.0	83

[a] Sucrose = 100.

The sweetness of dextrose relative to sucrose and conventional corn syrups is noted in Table VI. As with corn syrups, the relative sweetness increases with solids concentration as shown in Table XXI.

Dextrose is a readily fermentable sugar and is a source of food for yeasts in many food applications.

2. Sweeteners Containing Predominantly D-Fructose

a. *Total Sugar*

With the advent of isomerized D-glucose syrups, interest has increased in the commercial production of dry crystalline sweeteners containing D-fructose, and a number of patented processes (*140–143*) have issued relating to the production of dry total-sugar-type products which contain crystalline D-fructose admixed with other sugars, primarily dextrose. Currently, however, no total sugar products containing D-fructose are marketed.

b. *Crystalline D-Fructose*

The crystalline D-fructose of commerce is anhydrous β-D-fructopyranose which is the only known crystalline form suitable for commercial exploitation. Two other crystalline forms of D-fructose, the dihydrate and the hemihydrate of β-D-fructopyranose, exist but neither is stable at temperatures of normal storage and use (*144*). The raw material for crystalline D-fructose manufactured in the United States is refined, isomerized starch hydrolyzate.

Figure 13 is a simplified diagram for production of crystalline D-fructose based on recent patent literature. A solution containing $>$ 90% D-fructose is obtained by chromatographic separation (*120, 145*) of commercially available high D-fructose corn syrup or invert sugar. The fraction is concentrated to dry substance $\geq$ 90% and crystallized from either water (*146, 147*) or an alcohol-

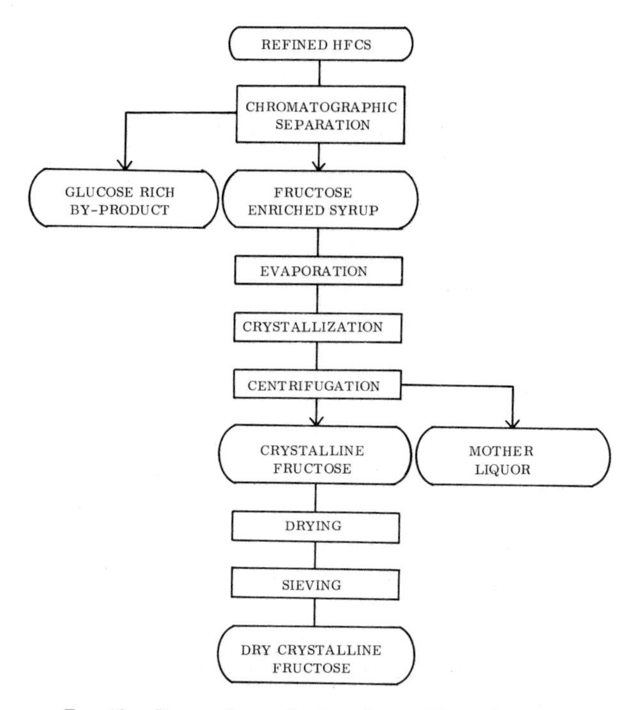

FIG. 13.—Process for production of crystalline D-fructose.

water system (*148*). Crystalline anhydrous D-fructose is separated from the mother liquor by centrifugation, washed with solvent during centrifugation to remove residual mother liquor, and subsequently dried. The mother liquor may be recycled to increase yields.

c. *Properties*

The properties of D-fructose have been summarized by MacAllister and coworkers (*149*) and Verstraeten (*150*). D-Fructose is substantially sweeter than

Table XXII

Sweetness of D-Fructose (139)

Concentration, g/100 mL	*Sweetness relative to sucrose*[a]
2.5	132
5.0	133
10.0	123
20.0	124

[a] Sucrose = 100.

sucrose as shown in Table XXII. Because of its greater sweetness, D-fructose offers the potential for a reduced calorie diet without sacrificing sweetness or the preference for a natural sugar as the sweetener. Crystalline D-fructose is more hygroscopic than dextrose and requires special packaging for storage and handling.

VI. References

(1) L. M. Beidler, *in* "Sweeteners and Dental Caries," J. H. Shaw and G. G. Roussos, eds., Special Supplement to Feeding, Weight & Obesity Abstracts, Information Retrieval Inc., Washington, D.C., 1978, pp. 33–39.

(2) Anon., *in* "Nutritive Sweeteners from Corn," Corn Refiners Association, Inc., Washington, D.C., 2nd Ed., 1979.

(3) F. C. Armbruster and C. F. Harjes, U.S. Patent 3,560,343 (1971); F. C. Armbruster, E. R. Kooi, and C. F. Harjes, S. African Patent 6,707,132 (1968); *Chem. Abstr.*, **70,** 48867k (1969).

(4) A. L. Morehouse, R. C. Malzahn, and J. T. Day, U.S. Patent 3,663,369 (1972); *Chem. Abstr.* **77,** 33147w (1972).

(5) F. C. Armbruster, U.S. Patent 3,853,706 (1974); F. C. Armbruster, E. R. Kooi, and C. F. Harjes, S. African Patent 6,707,132 (1968); *Chem. Abstr.*, **70,** 48867k (1969).

(6) C. J. Bates, *in* "Seminar Proceedings. Products of the Corn Refining Industry in Food," Corn Refiners Association, Inc., Washington, D.C., 1978, p. 34.

(7) W. R. Junk and H. M. Pancoast, "Handbook of Sugars," AVI Publishing Company, Westport, Connecticut, 1963, p. 168.

(8) R. L. Ueberbacher, *in* "Symposium Proceedings. Products of the Wet Milling Industry in Food," Section V, Corn Refiners Association, Washington, D.C., 1970, p. 2.

(9) J. Harness, *in* "Seminar Proceedings. Products of the Corn Refining Industry in Food," Corn Refiners Association, Washington, D.C., 1978, p. 7.

(10) R. V. MacAllister, *Adv. Carbohydr. Chem. Biochem.*, **36,** 15 (1979).

(11) G. G. Taylor, U.S. Patent 3,169,083 (1965); *Chem. Abstr.*, **62,** 10656a (1965).

(12) G. G. Taylor, U.S. Patent 3,348,972 (1967).

(13) J. N. BeMiller, *Adv. Carbohydr. Chem.*, **22,** 85 (1967).

(14) Anon., *in* "Critical Data Tables," Corn Refiners Association, Inc., Washington, D.C., 3rd Ed., 1969, p. 7.

(15) J. K. Dale and D. P. Langlois, U.S. Patent 2,201,609 (1940).

(16) L. D. Ough, *Anal. Chem.*, **34,** 660 (1962).

(17) T. L. Hurst and A. W. Turner, U.S. Patent 3,137,639 (1964); A. E. Staley Mfg. Co., Belgian Patent 631,666 (1963); *Chem. Abstr.*, **61,** 2041a (1964).

(18) I. Ehrenthal and G. J. Block, U.S. Patent 3,067,066 (1962); *Chem. Abstr.*, **58,** 3858g (1963).

(19) D. A. Bodnar, C. W. Hinman, and W. J. Nelson, U.S. Patent 3,644,126 (1972); *Chem. Abstr.* **76,** 125449d (1972).

(20) J. K. Conlee, *Staerke*, **23,** 366 (1971).

(21) G. J. de Jong, *Staerke*, **13,** 43 (1961).

(22) W. R. Junk and H. M. Pancoast, "Handbook of Sugars," AVI Publishing Company, Westport, Connecticut, 1973, p. 148.

(23) C. Nieman, *Manuf. Confect.*, **40,** 2 (1960).

(24) W. R. Junk and H. M. Pancoast, "Handbook of Sugars," AVI Publishing Company, Westport, Connecticut, 1973, p. 142.

(25) J. E. Long, *in* "Seminar Proceedings. Products of the Corn Refining Industry in Food," Corn Refiners Association, Washington, D.C., 1978, p. 39.
(26) R. S. Norrish, *Confect. Prod.*, **30**(10), 769 (1964).
(27) R. V. Vance, A. O. Rock, and P. W. Carr, U.S. Patent 3,654,081 (1972); Union Starch and Refining Co., Inc., British Patent 1,200,817 (1970); *Chem. Abstr.*, **74**, 4884s (1971).
(28) N. H. Aschengreen, *Process Biochem.*, **10**(4), 17 (1975).
(29) T. Komaki, *Agr. Biol. Chem.*, **32**, 123 (1968).
(30) L. J. Denault and L. A. Underkofler, *Cereal Chem.*, **40**, 618 (1963).
(31) M. Abdullah, I. D. Fleming, P. M. Taylor, and W. J. Whelan, *Biochem. J.*, **89**, 35P (1963).
(32) J. H. Pazur and K. Kleppe, *J. Biol. Chem.*, **237**, 1002 (1962).
(33) S. Ueda, R. Ohba, and S. Kano, *Staerke*, **26**, 374 (1974).
(34) M. Abdullah, B. J. Catley, E. Y. C. Lee, J. Robyt, K. Wallenfels, and W. J. Whelan, *Cereal Chem.*, **43**, 111 (1966).
(35) T. L. Hurst, U.S. Patent 3,897,305 (1975); Ger. Offen., 1,943,096 (1970); *Chem. Abstr.*, **73**, 119231y (1970).
(36) H. Bender and K. Wallenfels, *Biochem. Z.*, **334**, 79 (1961).
(37) J. H. Pazur and T. Ando, *J. Biol. Chem.*, **234**, 1966 (1959).
(38) E. R. Kooi, C. Harjes, and J. S. Gilkison, U.S. Patent 3,042,584 (1962); *Chem. Abstr.*, **57**, 8790a (1962).
(39) M. C. Cadmus, L. G. Jayko, D. E. Hensley, H. J. Gasdorf, and K. L. Smiley, *Cereal Chem.*, **43**, 658 (1966).
(40) K. L. Smiley, D. E. Hensley, M. J. Smiley, and H. J. Gasdorf, *Arch. Biochem. Biophys.*, **144**, 694 (1971).
(41) J. J. Marshall and W. J. Whelan, *FEBS Letters*, **9**, 85 (1970).
(42) O. R. Zaborsky, "Immobilized Enzymes," CRC Press, Ohio, 1973, 175 pp.
(43) T. Fukushi and T. Isemura, *J. Biochem.*, **64**, 283 (1968).
(44) B. T. Hofreiter, K. L. Smiley, and J. A. Boundy, U.S. Patent 4,121,974 (1978).
(45) S. A. Barker, P. J. Somers, R. Epton, and J. V. McLaren, *Carbohydr. Res.*, **14**, 287 (1970).
(46) D. D. Lee, Y. Y. Lee, P. J. Reilly, E. V. Collins, Jr., and G. T. Tsao, *Biotechnol. Bioeng.*, **18**, 253 (1976).
(47) K. N. Thompson, R. A. Johnson, and N. E. Lloyd, U.S. Patent 4,011,137 (1977); Ger. Offen. 2,538,322 (1967); *Chem. Abstr.* **84**, 178228c (1976).
(48) L. Zittan, P. B. Poulsen, and St. H. Hemmingsen, *Staerke*, **27**, 236 (1975).
(49) I. M. Abrams and L. Benezra, *in* "Encyclopedia of Polymer Science and Technology," Wiley, New York, 1967, Vol. 7, pp. 726–727.
(50) E. Katz, I. Ehrenthal, and B. L. Scallet, U.S. Patent 3,690,948 (1972); *Chem. Abstr.*, **77**, 154234f (1972).
(51) S. M. Cantor and K. C. Hobbs, U.S. Patent 2,354,664 (1944).
(52) K. Kainuma and S. Suzuki, *Staerke*, **19**, 60 (1967).
(53) B. L. Scallet and I. Ehrenthal, U.S. Patent 3,305,395 (1967); *Chem. Abstr.*, **66**, 77229b (1967).
(54) R. V. MacAllister, *in* "Immobilized Enzymes for Food Processing," W. H. Pitcher, Jr., ed., CRC Press, Boca Raton, Florida, 1980, pp. 81–111.
(55) J. P. Casey, *Res. Manag.*, **19**, 27 (1976).
(56) S. Akabori, K. Uehara, and I. Muramatsu, *Proc. Jpn. Acad., Ser. B*, **28**, 39 (1952).
(57) R. O. Marshall and E. R. Kooi, *Science*, **125**, 648 (1957).
(58) T. Sato and S. Tsumura, Japan. Patent 17,649/66 (1966); *Chem. Abstr.*, **68**, 86143n (1968).
(59) Y. Takasaki and O. Tanabe, U.S. Patent 3,616,221 (1971); Japan, Bureau of Industrial Techniques, Meth. Appl. 6,602,982 (1966); *Chem. Abstr.*, **76**, 84776f (1972).
(60) R. L. Antrim, W. Colilla, and B. J. Schnyder, *in* "Applied Biochemistry and Bioengineer-

ing,'' L. B. Wingard, E. Katchalski-Katzir, and L. Goldstein, eds., Academic Press, New York, 1979, Vol. 2, pp. 97–155.

(61) Y. Takasaki and O. Tanabe, *Hakko Kyokaishi,* **20,** 449 (1962); *Chem. Abstr.*, **60,** 806f (1964).

(62) N. Tsumura and T. Sato, *Agr. Biol. Chem.*, **29,** 1123 (1965).

(63) K. K. Shieh, H. A. Lee, and B. J. Donnelly, U.S. Patent 3,834,988 (1974); Ger. Offen., 2,351,443 (1974); *Chem. Abstr.*, **81,** 24191m (1974).

(64) H. Outtrup, U.S. Patent 3,979,261 (1976); Ger. Offen., 2,400,323 (1974); *Chem. Abstr.*, **81,** 118544f (1974).

(65) F. C. Armbruster, R. E. Heady, and R. P. Cory, U.S. Patent 3,813,318 (1974); Gen. Offen., 2,245,402 (1973); *Chem. Abstr.*, **79,** 3843u (1973).

(66) C. K. Lee, L. E. Hayes, and M. E. Long, U.S. Patent 3,645,848 (1972); Gen. Offen., 2,055,515 (1971); *Chem. Abstr.*, **75,** 33955s (1971).

(67) R. G. Dworschack and W. R. Lamm, U.S. Patent 3,666,628 (1972); *Chem. Abstr.*, **77,** 112513r (1972).

(68) Y. Takasaki, Y. Kosugi and A. Kamibayashi, *in* ''Fermentation Advances,'' D. Perlman, ed., Academic Press, New York, 1969, p. 561.

(69) N. E. Lloyd and K. Khaleeluddin, *Cereal Chem.*, **53,** 270 (1976).

(70) Y. Takasaki and A. Kamibayashi, U.S. Patent 3,753,858 (1973); Japan Patent 72 #19,032 (1972); *Chem. Abstr.*, **77,** 60041p (1972).

(71) K. N. Thompson, R. A. Johnson, and N. F. Lloyd, U.S. Patent 3,788,945 (1974); *Chem. Abstr.*, **81,** 24193p (1974).

(72) R. F. Sutthoff, R. V. MacAllister, and K. Khaleeluddin, U.S. Patent 4,110,164 (1978).

(73) R. A. Messing, U.S. Patent 3,850,751 (1974); *Chem. Abstr.*, **81,** 165517d (1974).

(74) S. Amotz, T. K. Nielsen, N. O. Thiesen, U.S. Patent 3,980,521 (1976); Ger. Offen. 2,405,353 (1974); *Chem. Abstr.*, **84,** 178251e (1976).

(75) J. V. Hupkes and R. van Tilburg, *Staerke,* **28,** 356 (1976).

(76) M. E. Long, U.S. Patent 3,935,069 (1976); *Chem. Abstr.*, **85,** 19099r (1976).

(77) C. W. Nystrom, U.S. Patent 3,935,068 (1976); *Chem. Abstr.*, **85,** 19098q (1976).

(78) T. Kasumi, M. Tsuji, K. Hayashi, and N. Tsumura, *Agr. Biol. Chem.*, **41,** 1865 (1977).

(79) N. Tsumura, T. Kasumi, and M. Ishikawa, *Starch/Staerke,* **30,** 420 (1978).

(80) M. S. Feather, V. Deshpande, and M. J. Lybyer, *Biochem. Biophys. Res. Commun.*, **38,** 859 (1970).

(81) A. S. Mildvan, *in* ''Bioinorganic Chemistry,'' Advances in Chemistry Series No. 100, R. F. Gould, ed., American Chemical Society, Washington, D.C., 1971, p. 404.

(82) S. J. Angyal and G. S. Bethell, *Aust. J. Chem.*, **29,** 1249 (1976).

(83) Y. Takasaki, *Agr. Biol. Chem.*, **31,** 309 (1967).

(84) N. B. Havewala and W. H. Pitcher, *in* ''Enzyme Fngineering,'' E. K. Pye and L. B. Wingard, eds., Plenum Press, New York, 1974, Vol. II, pp. 315–378.

(85) W. R. Vieth, K. Venkatasubramanian, A. Constantinides, and B. Davidson, *in* ''Applied Biochemistry and Bioengineering,'' L. B. Wingard, Jr., E. Katchalski-Katzir, and L. Goldstein, eds., Academic Press, New York, 1976, Vol. I, pp. 222–322.

(86) N. E. Lloyd, L. T. Lewis, R. M. Logan, and D. N. Patel, U.S. Patent 3,964,314 (1976).

(87) J. Oestergaard and S. L. Knudsen, *Staerke,* **28,** 350 (1976).

(88) J. A. Roels and R. van Tilburg, *Starch/Staerke,* **31,** 17 (1979).

(89) F. Yoshimasa, A. Matsumoto, H. Ishikawa, T. Hishida, H. Takamisawa, U.S. Patent 4,008,124 (1977).

(90) W. P. Cotter, N. E. Lloyd, and C. W. Hinman, U.S. Patent Reissue 28,885 (1976); *Chem. Abstr.*, **85,** 145151q (1976).

(*91*) T. L. Hurst, U.S. Patent 4,133,565 (1978).
(*92*) K. Khaleeluddin, R. F. Sutthoff, and W. J. Nelson, U.S. Patent 3,834,940 (1974); *Chem. Abstr.* **81,** 171854h (1974).
(*93*) V. K. Heyns, *Starch/Staerke,* **30,** 345 (1978).
(*94*) H. W. Spencer, *in* "Sweetness and Sweeteners," G. G. Birch, L. F. Green, and C. B. Coulson, eds., Applied Science Publishers, London, 1971, pp. 112–129.
(*95*) K. J. Schray and I. A. Rose, *Biochemistry,* **10,** 1058 (1971).
(*96*) Y. Nitta, Y. Nakajima, A. Momose, E. Tomita, and N. Sakamoto, Japanese Patent Spec. 3487/67 (1967); *Chem. Abstr.,* **67,** 117205 (1967).
(*97*) C. F. Boehringer and Soehne G.m.b.H., British Patent 949,293 (1964); Ger. Offen., 1,163,307 (1964); *Chem. Abstr.,* **60,** 14598a (1964).
(*98*) S. A. Barker, P. J. Somers, and B. W. Hatt, U.S. Patent 3,875,140 (1975); Ger. Offen., 2,229,064 (1973); *Chem. Abstr.,* **80,** 83524z (1974).
(*99*) Y. Takasaki, *Agr. Biol. Chem.,* **35,** 1371 (1971).
(*100*) S. Barker, *Carbohydr. Res.,* **26,** 33 (1973).
(*101*) J. Boeseken, *Adv. Carbohydr. Chem.,* **4,** 189 (1949).
(*102*) Y. Takasaki, U.S. Patent 3,689,362 (1972).
(*103*) F. Helfferich, "Ion Exchange," McGraw Hill, New York, 1973, p. 148.
(*104*) T. Kubo and F. Naito, Japanese Patent 48,146/75 (1975); *Chem. Abstr.,* **83,** 149478 (1975).
(*105*) J. C. Mahoney, U.S. Patent 2,357,838 (1944).
(*106*) K. Hara, M. Samoto, M. Sawai, and S. Nakamura, U.S. Patent 3,704,168 (1972); Ger. Offen., 2,031,252 (1971); *Chem. Abstr.,* **74,** 127944u (1971).
(*107*) R. Tatuki, U.S. Patent 3,671,316 (1972).
(*108*) T. Kubo and R. Tatuki, U.S. Patent 3,666,647 (1972); R. Taluki, and T. Kubo, Ger. Offen., 2,007,227 (1970); *Chem. Abstr.,* **73,** 89418v (1970).
(*109*) H. H. Hatt and A. C. K. Triffett, *J. Appl. Chem.,* **15,** 556 (1965).
(*110*) V. S. H. Liu, N. E. Lloyd, and K. Khaleeluddin, U.S. Patent 4,096,036 (1978).
(*111*) A. G. Holstein and G. C. Holsing, U.S. Patent 3,050,444 (1962).
(*112*) R. M. Wheaton, U.S. Patent 2,911,362 (1959).
(*113*) G. R. Serbia and P. R. Aguirre, U.S. Patent 3,044,904 (1962).
(*114*) L. J. Lefevre, U.S. Patent 3,044,905 (1962).
(*115*) R. W. Neuzil and J. W. Priegnitz, U.S. Patent 4,024,331 (1977).
(*116*) H. Odawara, Y. Noguchi, and M. Ohno, U.S. Patent 4,014,711 (1977).
(*117*) S. J. Angyal, *in* "Carbohydrates in Solution," H. S. Isbell, ed., American Chemical Society, Washington, D.C., 1973, pp. 106–120.
(*118*) Y. Takasaki, U.S. Patent 3,806,363 (1974).
(*119*) A. J. Melaja, L. Hamalainen, and L. Rantanen, U.S. Patent 3,928,193 (1975).
(*120*) R. F. Sutthoff and W. J. Nelson, U.S. Patent 4,022,637 (1977).
(*121*) H. W. Keller, A. C. Reents, and J. W. Laraway, reported at the International Biochemical Symposium, Toronto, Canada, 1977.
(*122*) H. Ishikawa, Japanese Patent 101,140/76 (1976).
(*123*) D. B. Broughton, *in* "Kirk-Othmer Encyclopedia of Chemical Technology," Wiley, New York, 1978, Vol. 1, pp. 569–572, 576.
(*124*) H. J. Bieser and A. J. de Rosset, *Staerke,* **29,** 392 (1977).
(*125*) H. Odawara, M. Ohno, T. Yamazak, and M. Kanaoka, U.S. Patent 4,157,267 (1979).
(*126*) W. C. Black, U.S. Patent 3,582,399 (1971); Ger. Offen., 1,935,890 (1970); *Chem. Abstr.,* **72,** 102079m (1970).
(*127*) A. L. Wilson and I. Frankel, U.S. Patent 2,854,359 (1958).
(*128*) Y. Mise and E. Tomimura, South African Patent 5,869/77 (1977).

(*129*) G. R. Dean and J. B. Gottfried, *Adv. Carbohydr. Chem.*, **5,** 127 (1950).
(*130*) D. F. Rentshler, D. P. Langlois, R. F. Larson, and L. H. Alverson, U.S. Patent 3,039,935 (1962); *Chem. Abstr.*, **57,** 10085i (1962).
(*131*) E. R. Kooi and F. C. Armbruster, *in* "Starch: Chemistry and Technology," R. L. Whistler and E. F. Paschall, eds., Academic Press, New York, 1967, Vol. II, pp. 553–568.
(*132*) W. B. Newkirk, U.S. Patent 1,976,361 (1934).
(*133*) K. Richter and H. Mueller, German Offen. 2,144,406 (1973); *Chem. Abstr.*, **78,** 160060h (1973).
(*134*) T. C. Garren, *in* "Seminar Proceedings. Products of the Corn Refining Industry in Food," Corn Refiners Association, Washington, D.C., 1978, pp. 41–46.
(*135*) E. R. Erickson, R. A. Bentsen, and M. A. Eliason, *J. Chem. Eng. Data*, **11,** 485 (1966).
(*136*) Anon., *in* "Critical Data Tables," Corn Refiners Association, Washington, D.C., 3rd Ed., 1969, pp. 141–209.
(*137*) E. R. Kooi, *in* "Kirk-Othmer Encyclopedia of Chemical Technology," Wiley, New York, 1965, Vol. 6, pp. 919–926.
(*138*) A. E. Sloan and T. P. Labuza, *Food. Prod. Dev.*, **9**(7), 75 (1975).
(*139*) S. Yamaguchi, T. Yoshikawa, S. Ikeda, and T. Ninomiya, *Agr. Biol. Chem.*, **34,** 181 (1970).
(*140*) J. Lundquist, P. Veltman, and E. Woodruff, U.S. Patent 3,956,009 (1976).
(*141*) A. J. Melaja, U.S. Patent 3,816,175 (1974); Ger. Offen., 2,333,513 (1974); *Chem. Abstr.*, **80,** 108816m (1974).
(*142*) T. Kusch, W. Gosewinkel, and G. Stoeck, U.S. Patent 3,513,023 (1970); C. F. Boehringer und Soehne G.m.b.H., British Patent 1,117,903 (1968); *Chem. Abstr.*, **69,** 44703k (1968).
(*143*) T. Yamauchi, U.S. Patent 3,929,503 (1975); Ger. Offen., 2,426,437 (1974); *Chem. Abstr.*, **82,** 126879e (1975).
(*144*) F. E. Young, F. T. Jones, and H. J. Lewis, *J. Phys. Chem.*, **56,** 1093 (1952).
(*145*) A. J. Melaja, U.S. Patent 3,692,582 (1972); Ger. Offen. 2,406,663 (1974); *Chem. Abstr.*, **74,** 127945v (1971).
(*146*) K. H. Forsberg, L. Hamalainen, A. J. Melaja, and J. J. Virtanen, U.S. Patent 3,883,365 (1975).
(*147*) T. Yamauchi, U.S. Patent 3,928,062 (1975); Ger. Offen., 2,406,663 (1974); *Chem. Abstr.*, **82,** 5544h (1975).
(*148*) K. Lauer and P. Stephan, U.S. Patent 3,607,392 (1971).
(*149*) R. V. MacAllister and E. K. Wardrip, *in* "Encyclopedia of Food Science," M. S. Peterson and A. H. Johnson, eds., AVI Publishing Company, Westport, Connecticut, 1978, Vol. 3, pp. 329–332.
(*150*) L. M. J. Verstraeten, *Adv. Carbohydr. Chem.*, **22,** 230 (1967).
(*151*) "Sugar and Sweetner Outlook and Situation," U.S.D.A. Economic Research Service, U.S. Department of Agriculture, Washington, D.C., December, 1982, pp. 10–12.
(*152*) M. Tamuri, M. Kanno, and Y. Ishii, U.S. Patent 4,284,722 (1981).
(*153*) B. E. Norman, *Starch/Staerke*, **34,** 340 (1982).
(*154*) S. Nielsen, I. Diers, H. Outtrup, and B. E. Norman. British Patent 2,097,405 (1982).
(*155*) L. S. Hurst and N. E. Lloyd, U.S. Patent 4,376,824 (1983).

CHAPTER XXII

INDUSTRIAL MICROSCOPY OF STARCHES

By Eileen Maywald Snyder

Moffett Technical Center, Corn Products, Summit-Argo, Illinois

I. Introduction

Currently, corn starch is the primary cereal starch refined by wet milling in the United States. Approximately three billion pounds of starch is produced annually from regular yellow dent corn, while waxy and high-amylose varieties of corn account for another half-billion pounds. The United States and Canada also refine some wheat and potato starch, while world markets provide a variety of other starches including tapioca, rice (both normal and waxy), arrowroot, sweet potato, sago and canna. From time to time new sources are uncovered. For example, babassu starch is a potential commercial starch. The babassu nut, fruit of the babassu palm (*Orbygnia martinana*), is native to and widely spread in Brazil (*1, 2*). Tons of nuts are harvested each year for the oil contained in the kernels, which represents only about 7% of the total weight of the nut, leaving over a half million tons of by-product starch currently not used.

Many starch varieties are chemically or physically modified to improve the properties for specific uses, and two or more unmodified varieties may be mixed to form a single product with properties different from either of the individual components. The light microscope provides a useful, rapid means of identifying the various native starches and detecting mixtures of different starches, as well as a means of monitoring processing conditions and modifications.

Commercialization of a practical scanning electron microscope (SEM) in the late 1960s has resulted in a proliferation of articles (see references in Chapter

STARCH, 2nd ed.

ISBN 0-12-746270-8

XXIII) and SEM photomicrographs commencing in about 1968 (*3*). However, the light microscope provides the advantages of color through selective staining and details on internal structure not available by SEM microscopy, and light microscopy remains a useful, if not necessary, adjunct to any electron microscopy study. It must be emphasized that the extent of information obtained is directly related to the microscopist's experience and knowledge of starch. Pertinent information may be otherwise misinterpreted or overlooked. On the other hand, an experienced microscopist can identify from memory literally hundreds of particles on sight.

II. Identification of Starch Species

Most of the common starches are readily and unequivocally identifiable under a polarizing microscope, using the criteria of granule size and shape, form and position (centric or eccentric) of the hilum (the botanical center of the granule), and brilliance of the interference cross under polarized light. Kerr (*4*), Radley (*5*), the classic works of Reichert (*6*), and most recently, McCrone (*7*) and Seidemann (*8*), all provide excellent photomicrographs of various species which are helpful. However, it is preferable to maintain a collection of authentic samples of the various starches and to make side-by-side comparisons with unknown samples. Most of the starch species are easily distinguishable; corn and sorghum are an exception, and, of course, the waxy varieties of the cereal starches, such as barley, corn, rice, and sorghum, can be separated visually from the normal varieties by iodine staining.

Granule size is frequently useful information, and measurements can be made either manually or with great rapidity and accuracy by such means as particle size counters or image analyzers. The accuracy of particle size data is dependent upon both size and number of particles counted, and automated particle counters have the marked advantage of sizing and counting hundreds of particles in minutes. Excellent summaries of the subject and comparison of the various methods, including manual microscopy, is given by Stockham and Fochtman (*9*) and by Allen (*21*).

Table I shows the difference in granule size range and diameters of a granule of average weight for many of the common commercial unmodified starches using a Coulter particle size counter. Currently, wheat starch may show a narrower range and larger average diameter due to removal of smaller granules by centrifugal separation. Unclassified wheat shows a relatively broad particle size range of 5–35 μm.

Particle size data can also be useful in monitoring the effect of various processing changes or modifications of granular starch products. For example, Table II shows that neither regular nor waxy varieties of unmodified corn starch undergo any substantial change in granular size due to normal commercial drying opera-

Table I

Granule Size Analysis of Commercial Unmodified Starches

Starch species	Granule size range, μm — Light microscope	Granule size range, μm — Coulter counter	Diameter of granule of av. wt. μm (Coulter counter)
Waxy rice (Italian)	2–10	2–13	5.4, 5.6
High-amylose corn			
70% amylose	—	4–22	9.6, 10.0
55% amylose	5–35	3–22	10.3, 10.7
Corn	5–30	5–25	13.8, 14.3, 14.5
Waxy corn	5–25	4–28	13.9, 14.2
Tapioca (Brazilian)	—	3–28	13.8, 14.2
Grain sorghum	—	3–27	16.0, 16.2
Waxy sorghum	—	4–27	15.0, 16.5, 16.9
Wheat	5–35	3–34	16.4, 16.6
Barley	2–3, 35–40[a]	6–35	16.5, 16.7
Sweet potato (Japanese)	—	4–40	18.0, 18.6
Garbanzo (chick pea)	—	7–54	20.3, 21.1
Shoti	10–60[b]	9–34	21.0, 22.3
Arrowroot	10–50	9–40	22.9, 23.9
Sago flour	20–65	15–50	33.1
Potato	10–100	10–70	34.0, 36.0
Canna (Australian arrowroot)	30–130	22–85	53.0, 53.0

[a] Sharply bimodal.
[b] Non-spherical, flat granules.

tions. However, ionic derivatives may show an increase in both particle size range and average granule diameter; this is particularly true of derivatized potato starch.

III. Granule Aggregation and Gelatinized Granules

Most commercial starches are in the form of single granules. For economic reasons, most domestic starches are currently flash-dried, and occasionally aggregated products are obtained by accidental overheating of the wet starch. Surface gelatinization causes individual granules to stick together and form clumps ranging in size from a few granules to hundreds. Normally, the clumps are only slightly bonded and dissociate readily into discrete granules when suspended in water. Large aggregates that do not readily disperse indicate excessive heat damage, and usually at least a portion of the granules in such products will be either gelatinized or cold-water swelling which is detrimental in many applications. On the other hand, some products are deliberately aggregated or compact-

Table II

Effect of Processing and Chemical Modification on Granule Size Distribution

Sample identity	Source	Coulter counter data: Granule size range, μm	Diameter of granule of av. wt., μm
Unmodified corn starch			
Prime starch control		5–25	14.0
Process samples	Centrifuge supply, slurry	4–25	13.3
	Centrifuge underflow, slurry	4–25	13.1, 13.6
	Same, laboratory dried	4–25	13.8, 14.0
Anionic corn starch			
Process samples	Neutralized reaction product, laboratory dried	6–55	17.5, 18.2
	Spray-dried product	7–55	20.4, 21.2
Unmodified waxy corn starch			
Prime starch control		4–28	13.9, 14.2
Process samples	Centrifuge supply, slurry	5–27	14.2
	Same, laboratory dried	5–25	14.1, 14.5
	Centrifuge underflow, slurry	5–25	14.0
	Same, laboratory dried	5–25	13.9, 13.9
Cationic potato starch			
Unmodified, control		10–70	33.0, 34.0, 36.0
Process samples	Centrifuge supply, laboratory dried	13–100	51.0
	Centrifuge overflow, laboratory dried	16–90	46.8
	Centrifuge underflow, laboratory dried	12–90	46.1, 49.3
	Spray-dried product	13–85	45.0, 47.5

ed for applications requiring dustless starches, such as brewing adjuncts and sugar grinding aids.

To check for aggregation and gelatinized granules, a dilute starch suspension (0.2–0.3%) should be examined in both water and a nonaqueous medium (glycerol or light mineral oil are preferred). Starch that is highly aggregated in nonaqueous media and does not disperse readily in water has been heat-damaged during drying. The extent of damage can be estimated by noting the condition of the granules in the hilum area under crossed polarizers. Initial gelatinization appears as a darkened or enlarged hilum in nonaqueous media, and if damaged excessively, these granules will hydrate either partially or totally in water, and the interference cross will fade and eventually disappear as the granule swells. With experience, the amount of gelatinized granules can be estimated with an accuracy of ± 1%. If more accurate information is needed, the proportion of gelatinized granules may be actually counted using a hemacytometer. It is necessary to count a minimum of 200–300 granules to obtain reproducible data. When the hilum appears enlarged in oil or glycerol, and the granules remain birefringent in water under polarized light, the starch has probably undergone modification such as derivatization or dextrinization. In any case, when the granules remain insoluble and do not swell in water at room temperature, the properties of the product are not in any way impaired.

IV. Kofler Gelatinization Temperature

When starch is heated in an excess of water to progressively higher temperatures, the granules hydrate with almost simultaneous loss of their polarization crosses. The microscopist refers to this transition as gelatinization. While each individual granule gelatinizes quite sharply, not all of the granules in any one starch species gelatinize at the same temperature, but rather over a range of about 8°–10°. This behavior reflects differences in internal bonding forces within individual granules, and the resulting gelatinization range is a specific characteristic of each starch species. The term gelatinization is used loosely throughout the starch literature. It is often used interchangably with what is more correctly termed the pasting temperature of starch, or the point at which a starch slurry begins to thicken and show an increase in viscosity when heated. The pasting temperature is normally several degrees higher than the endpoint of the gelatinization range (see this Volume, Chap. VII and IX).

A Kofler hot-stage may be used to measure the gelatinization range accurately and to observe the changes that occur in an aqueous starch slurry with increasing temperature (*10*). In practice, some modifications of the microscope and its normal optical system may be necessary to accommodate the hot-stage. For instance, the stage of the microscope itself may need to be drilled to anchor the positioning "feet" of the hot-stage, and a special 10× objective with a long focal distance must be used to compensate for the thickness of the added apparatus; the resulting low magnification is counterbalanced by the use of a 20× ocular. The experimental procedure has been detailed previously by Watson (*11*).

After the gelatinization range of an unknown starch product has been determined, it is often helpful to plot the percentage of gelatinized granules against increasing temperature, particularly if the range is distorted from a normal spread of 8°–10°. A normal unmodified starch should give a fairly symmetrical sigmoid curve. However, if the sample is a blend of two starches of markedly different gelatinization ranges, it will give a double sigmoid curve, and the proportion of components can be approximated from the transition point. If the starch has been derivatized with a chemical group, such as the hydroxyethyl which lowers the gelatinization temperature, the entire sigmoid curve is displaced uniformly to the left, or to a lower temperature range than that shown by the parent starch. Sometimes the gelatinization curve of derivatized products is attenuated, with a greatly reduced initiation point, but with the termination temperature substantially the same as that of the parent unmodified starch. Such a curve is definite

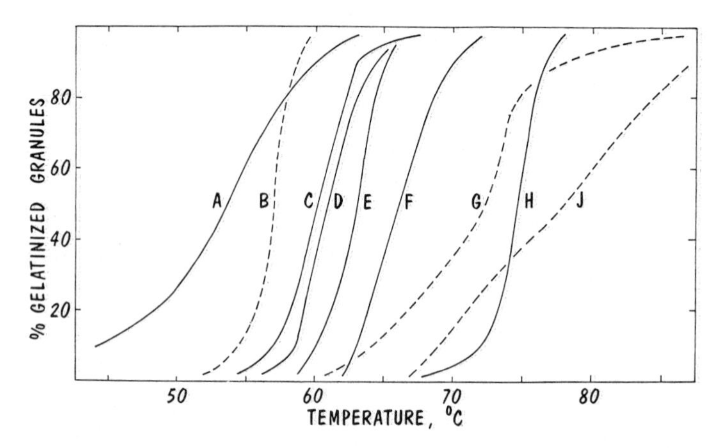

FIG. 1.—Kofler gelatinization temperature ranges of various starches. A, sulfodicarboxylic ester of corn starch. B, barley starch. C, cationic corn starch, 0.05 DS. D, potato starch. E, hydroxyethyl sorghum starch, 0.1 DS. F, corn starch. G, cationic high-amylose corn starch. H, sorghum starch. J, high-amylose corn starch, 55% amylose.

evidence of non-uniform derivatization, indicating that some of the granules are over-derivatized while other granules are virtually untouched. The gelatinization curve is the only known method whereby this situation can be diagnosed. Examples of each type of curve are shown in Figure 1.

When the last granules are gelatinized, useful information on the swelling behavior of the starch can be obtained by continuing to raise the temperature toward 100°. Chemically cross-bonded starches have essentially the same gelatinization range as the parent starch, but granule swelling is very inhibited or restricted at higher temperatures. Quantitative conformation of the extent of cross-bonding can be obtained by determining the swelling power and solubility at 95°, using the method of Leach and co-workers (*12*). Starches which have been modified to give low-viscosity, high-solubility pastes, such as acid-modified and oxidized starches, or some of the white dextrins, show extensive granule dissolution as the temperature is raised. In fact, products that have been highly acid-modified (over 70-fluidity) dissolve and separate into quadrants without swelling while the gelatinization temperature is being determined.

Occasionally, a food-grade potato or tapicoa starch may have a gelatinization temperature substantially higher than normal. When heated above the gelatinization range, the granules do not swell freely but show an inhibited lobe-type of swelling. The granules in a restricted potato starch product open up by a radial split and then contort into peculiar shrimp-like shapes. If the swelling power is determined at 95°, it will be restricted far below normal. Such products are usually heat-moisture treated, physically modified as opposed to being chemically modified, probably by maintaining high humidity during the drying operation. This causes a rearrangement in the crystalline pattern within the granule, similar to the heat-moisture treatment of Sair (*13*), or to hot digestion in 70% diacetone alcohol described by Leach (*12*). The purpose of heat-moisture treatment and chemical cross-bonding is to restrict granule swelling and thus stabilize the paste viscosity, particularly in acid systems or at retort temperatures used in some food applications. While most starches will undergo some increase in gelatinization temperature on heat-moisture treatment (Table III), it is particularly effective in altering and modifying the paste properties of the high-swelling root and tuber starches such as tapioca and potato (see also Chap. VII). Table III also shows the gelatinization temperatures of various chemically modified starches. The following significant features should be noted:

a. Chemical cross-bonding does not significantly alter the gelatinization temperature of starch.

b. Introduction of ionizable groups, such as cationic or anionic groups into the starch granule, decreases the gelatinization temperature in direct proportion to the degree of substitution. Mutual repulsion between similarly charged ionic sites within the granule may even induce gelatinization in water at room temperature.

c. Oxidized starches show a progressive decrease in gelatinization temperature with increasing oxidation. Similarly, the introduction of derivative groups, such as hydroxyethyl groups, lowers the gelatinization temperature by decreasing the intermolecular bonding within the granule.

d. Lower degrees of acid modification, such as the 40- and 60-fluidity thin-boiling starches, do not significantly alter the gelatinization temperature. Higher levels of acid modification, over 80-fluidity, surprisingly increase the gelatinization temperature.

A Kofler hot-stage can likewise be used to determine the effect of various additives on the gelatinization temperature of starch. Table III includes gelatinization temperatures of corn starch in aqueous solutions of sucrose and various salts. Depending on their position in the Hofmeister lyotropic series, salts, such as thiocyanate and iodide, may decrease the gelatinization temperature, while others, such as sulfate and carbonate, may increase it. Concentrations of sucrose of up to 10% do not hinder the gelatinization of corn starch, while high concentrations greatly increase the temperature of gelatinization. Hence, the starch in foods that are high in sugar and low in moisture is not completely swollen. Seidemann's Staerke-Atlas (*8*) contains additional useful information on the effect of numerous agents, such as acids and organic and inorganic salts, on the gelatinization temperature of starch.

V. Iodine Staining

Iodine staining of granular cereal starch is routine practice to distinguish between the red-staining waxy varieties and their normal blue-staining counterparts. Most of the commercial waxy starches contain up to 10% of blue-staining granules due to field contamination. The percentage of each variety in a mixture

Table III

Kofler Gelatinization Temperatures of Various Starches (Initiation, Midpoint, and Completion in °)

Unmodified Starches			
Corn	62–67–72	Waxy maize	63–68–72
Sorghum	68–73.5–78	Waxy sorghum	67.5–70.5–74
Wheat	58–61–64	Barley	51.5–57–59.5
Tapioca:		Rye	57–61–70
Brazilian	49–57–64.5	Pea (green garden)	57–65–70
Dominican	58.5–64.5–70	Rice	68–74.5–78
Siamese	62–68–73	High–amylose corn	67–80–?[a]
Potato	50–63–68		
Modified Starches			
Acid–modified corn starch:		Anionic corn starch, high viscosity:	
40-fluidity	62–67–72	0.036 DS	52–57–63.5
60-fluidity	63.5–69–73.5	Hydroxyethyl corn starch, thick–boiling:	
80-fluidity	68–72–77	0.05 DS	58–63–68
Oxidized corn starch:		0.09 DS	55–60–64
Low-converted	55–64–73	Heat–moisture treated starches:	
Medium-converted	54–60–69.5	Potato:	
High-converted	52–59–68	Untreated	56–61–67
Cross–bonded starches:		Treated[c]	65–71–77
Waxy maize	63–70–76	Tapioca:	
Corn	62–69–74	Untreated	51–58–66
Cationic corn starch, high viscosity:		Treated[c]	62–65–69
0.046 DS[b]	52–58–65	Corn:	
0.11 DS	room temperature	Untreated	62–65–69
		Treated[c]	68.5–72–76
Corn Starch in Various Aqueous Media[d]			
Water, control	62–66–72	NaOH, 0.2%	55.5–64–69.5
		0.3%	49–59–65
Sucrose, 5%	60.5–67–72.5		
10%	60–67–74	NaCl, 1.5%	67.5–72–77
20%	65.5–72–78	3.0%	69.5–74–78.5
30%	69.5–74–81	6.0%	75–79.5–82.5
40%	72–79.5–85	Na_2CO_3, 5%	64–70–75
50%	76–85–90.5	10%	67–72–76
60%	84–90.5–96.5	20%	77.5–82–87
		30%	92–98–103

[a] Complete gelatinization of high-amylose corn starch is not effected in boiling water.

[b] DS = degree of substitution.

[c] Heat-moisture treated by refluxing in 70% diacetone alcohol, according to Leach and coworkers (*12*).

[d] The sample of corn starch was used in these tests in various aqueous media.

can be readily determined by lightly staining the sample with iodine, using a 0.3% iodine-potassium iodide solution, and counting the proportion of red and blue granules in a hemacytometer. With experience, it is possible to judge the proportion of contamination with reasonable accuracy. Over-staining with iodine should be avoided since it obliterates certain pertinent features of the granules, and both red- and blue-staining granules may appear black. In such cases, the method recommended by M. M. MacMasters (*14*) is employed, whereby a sample is mounted in water, and a drop of dilute iodine solution is placed at the outer edge of the cover glass and drawn through the sample by touching the opposite edge with a piece of dry filter paper to set up a concentration gradient. As the advancing iodine boundry is examined microscopically, all gradations of staining will be seen. Light staining does not obscure the interference cross under polarized light, and a mixture of waxy and normal starches then appears as red and blue granules on a dark field. This method also enhances the detection of any incipient or extensive gelatinization of either or both of the components.

Only the cereal starches have naturally occurring waxy counterparts that stain red with iodine due to a genetically low amylose content. Commercial root or tuber starches, such as potato and tapioca, may stain red with iodine because they have been modified by the addition of fatty acids, monoglycerides, or similar compounds to decrease hydration rate and to stabilize paste viscosity. If the fatty adjunct has not been deliberately complexed with the amylose component (*15*), it can be removed by cold extraction with a suitable solvent such as hexane. Extraction with hot alcohol, the standard starch defatting procedure (*16*), will remove deliberately complexed fatty materials. After the fatty adjunct is removed, the starch will then stain blue with iodine.

Extreme pH conditions are also known to affect iodine absorption adversely and should be avoided.

VI. Dye Staining

Anionic starches stain with positively charged organic dyes, and conversely, cationic starches stain with negative dyes (*10*). The degree of staining is a qualitative indication of the ionic charge on the starch. For instance, potato starch contains natural ionic phosphate ester groups, and therefore stains a uniform moderately-dark blue with methylene blue. Among the more common anionic starches are the oxidized products, various phosphate ester derivatives, and in Europe, carboxymethylstarch. Preferred practice is to add 25–50 mg of the starch to a 0.1% aqueous methylene blue solution in a small test-tube, mix by shaking, and centrifuge. The supernatant dye solution is decanted, the sedimented starch washed twice by shaking with distilled water followed by centrifuging, and the stained sample then examined under the microscope. An important point is the uniformity of staining. If some granules are strongly stained and the remainder unstained, either a deliberate blend or accidental contamination is indicated. If all degrees of staining are observed in a single sample, the method of modification has not been uniform and homogeneous. For example, oxidized starch produced by a semi-dry process shows a wide range of staining due to non-uniformity of the reaction between the individual granules and the oxidizing agent.

While cationic starches are never encountered in food use, they are widely used for internal and surface sizing of paper and for retention of filler in paper manufacturing. The preferred stain for these products is "Light Green SF Yellowish," obtained as a certified biological stain; acid fuschin is a suitable alternate. The same conditions apply with respect to depth and uniformity of staining as with methylene blue.

Both light green SF yellowish and methylene blue dye can be stripped from starch with a 1% saline solution. Reactive dyes such as the Levafix E type which chemically bond and therefore stain starch permanently can also be used with similar results (*17, 18*). The more complex dye staining procedures carried over from botanical and biological microscopy and summarized by Moss (*19*) are not necessary for routine starch analysis in an industrial laboratory or plant where rapid identification is usually desirable, if not mandatory.

VII. Pregelatinized Starches

Pregelatinized starches are "convenience products," precooked and dried by the manufacturer and reconstituted in water to give viscous pastes. They have practical importance as constituents of packaged premixes and as viscosity agents for users who do not have the facilities for cooking starch. Typical applications are as a bodying agent for oil-well drilling mud, as an adhesive in foundry cores for metal casting, and as a thickener for puddings, sauces, and pie fillings. Food premixes are generally complex mixtures, and a microscope is frequently all that is needed to identify the components (*20*). If conformation of the microscopic analysis is necessary, the individual components can be isolated and identified by more sophisticated and time consuming means, such as x-ray diffraction for the crystalline portion, nuclear magnetic resonance and infrared for the organic components. Such methods require very small samples for positive identification. In any event, the microscope is frequently the key to the best means of isolating the individual components.

The mode of drying can usually be determined by mounting the sample in glycerol and examining under low power magnification (100–200×).

a. *Spray-dried:* These products consist of distorted hollow spheres, usually with an air cell enclosed at the center. They are made by first cooking the starch in water and then by spraying the hot paste into a drying chamber or tower.

b. *Roll-dried:* Particles appear as transparent, flat, irregular platelets, quite similar to broken shards of window glass. In general, such products are simultaneously cooked and dried on heated rolls, using either a closely set pair of squeeze rolls or a single roll with a closely set doctor blade. In either case, a paper-thin flake, which is then ground to the desired screen size, is obtained.

c. *Extruded or drum-dried:* Individual particles from either process are much thicker and more irregular in dimensions than roll-dried products. Drum-drying is similar to roll-drying except that a thicker coating of starch paste is applied to the heated rolls, and the dried product is then ground to the desired particle size. In the extruded process, moistened starch is forced through a superheated chamber under very high shear, then "exploded" and simultaneously dried by venting to atmospheric pressure. In both of these processes, particularly in extrusion where very limited amounts of water are used, the starch is not completely gelatinized, and the starch species can be easily identified by looking for intact residual granules. In cases where the starch has been put through a cooking step prior to drying, no residual granules will remain intact.

One of the most important characteristics of a pregelatinized starch is particle size. Finely ground products generally give the highest viscosity when reconstituted with water, and properly prepared pastes show a good surface gloss. However, such products are difficult to disperse in water since they tend to hydrate too rapidly, thus giving lumps and clots that contain unwetted starch in the center. In practice, they are premixed with other ingredients, such as sucrose, to slow the hydration rate. Coarsely ground products reconstitute much more readily in cold water, but the pastes are of lower viscosity and have a grainy texture.

To be effective as a thickening agent, substantially all the starch granules in a pregelatinized product should be well gelatinized. This can be qualitatively determined by slurrying the product in cold water and examining it under the polarizing microscope. There should be an insignificant number of ungelatinized granules that show polarization crosses. Commercial products have actually been encountered containing 25% or more of raw ungelatinized granules which retained their polarization crosses. Obviously, such materials have poor efficiency as cold-pasting products and require additional heating to cook out the ungelatinized portion.

The species of a pregelatinized starch can frequently be identified by the fact that a few granules usually escape gelatinization during manufacture. However, these cannot be detected in the mass of pregelatinized flakes, but if the sample is treated with a strong solution of liquefying α-amylase for 30–60 min at 30°–40°, the pregelatinized material is digested first, leaving the gelatinized granules relatively intact. The mixture can then be centrifuged, and the trace of insoluble sediment examined under the polarizing microscope. This procedure will readily detect one part per thousand of ungelatinized granules, and thus identify the starch origin. Any of the enzyme manufacturers, such as Rohm and Haas,

Takamine, and Novo, can supply a suitable enzyme, but it is mandatory to remove any insoluble material from the enzyme solution by careful filtration prior to use. In addition, the method must be used with considerable discrimination. For example, if a manufacturer processes several species of pregelatinized and granular starches, trace contamination by the latter may occur, either airborne or from the use of common bagging equipment. If the sediment shows honeycomb structures, these are empty cell wall residues from which the starch granules have been removed. These structures do not survive a wet-milling process, and their presence is indicative of dry-milled cereal.

A physical mixture of several pregelatinized starches can sometimes be detected by iodine staining. A few milligrams of the dry material is placed on a microscope slide and covered with a cover slip. A drop of very dilute iodine solution is allowed to seep in from the edge of the cover slip and the color of the particles at the advancing wet boundary is observed. Particles of waxy starch will stain red or brownish-red, corn or wheat particles a purplish-blue, and potato or tapioca a strong bright blue. This method is only applicable to dry-blended mixtures of several pregelatinized starches or to a single species of pregelatinized starch. If a mixture of two or more granular starches, such as corn and potato, is blended and then roll-dried, the identity of the composite particle cannot be determined. Amylose-complexing agents, such as monoglyceride, are sometimes added to pregelatinized starches to slow hydration and thus improve the dispersibility of the product in water. Since such adjuncts block iodine uptake by the starch as previously noted, a small portion of the sample should be extracted with hot alcohol several times in a test tube, dried, and then tested with iodine under the microscope.

Occasionally, useful information can be obtained by staining a pregelatinized starch with methylene blue or with light green SF dye. The technique is the same as for iodine staining, except that the excess dye should be subsequently removed by drawing a droplet of distilled water through the sample with a piece of filter paper. For example, this technique can be used to distinguish between pregelatinized potato and tapioca starches, since the latter does not bind the stain.

VIII. References

(*1*) F. R. T. Rosenthal and A. M. C. Espindola, *Revista Brasileira Technologia,* **6,** 307 (1975).

(*2*) Y. K. Park, E. Marancenbaum, and A. A. Morga, *J. Agr. Food Chem.,* **26,** 1198 (1978).

(*3*) C. Aranyi and E. J. Hawrylewicz, *Cereal Chem.,* **45,** 500 (1968).

(*4*) R. W. Kerr, *in* "Chemistry and Industry of Starch," R. W. Kerr, ed., Academic Press, Inc., New York, 2nd Ed., 1950, pp. 18–25.

(*5*) E. Young, *in* "Starch and Its Derivatives," J. A. Radley, ed., Chapman and Hall, Ltd., London, 3rd Ed., 1953, Vol. 2, p. 443ff.

(*6*) E. T. Reichert, "The Differentiation and Specificity of Starches in Relation to Genera, Species, etc.," Carnegie Institute of Washington, Washington, D.C., Publication 173, Parts 1 and 2, 1913.

(*7*) W. C. McCrone and J. G. Delly, "The Particle Atlas," Ann Arbor Science Publishers, Ann Arbor, Michigan, 2nd Ed., 1973, Vol. 2, pp. 457–462.
(*8*) J. Seidemann, "Staerke-Atlas," Paul Parey, Berlin, 1966, pp. 173–326.
(*9*) J. D. Stockham and E. G. Fochtman, eds., "Particle Size Analysis," Ann Arbor Science Publishers, Ann Arbor, Michigan, 1977.
(*10*) T. J. Schoch and Eileen C. Maywald, *Anal. Chem.*, **28,** 382 (1956).
(*11*) S. A. Watson, *Meth. Carbohydr. Chem.*, **4,** 240 (1964).
(*12*) H. W. Leach, L. D. McCowen, and T. J. Schoch, *Cereal Chem.*, **36,** 534 (1959).
(*13*) L. Sair and W. R. Fetzer, *Ind. Eng. Chem.*, **36,** 205 (1944).
(*14*) M. M. MacMasters, *Meth. Carbohydr. Chem.*, **4,** 237 (1964).
(*15*) V. M. Gray and T. J. Schoch, *Staerke,* **7,** 239 (1962).
(*16*) T. J. Schoch, *Meth. Carbohydr. Chem.*, **4,** 56 (1964).
(*17*) R. Stute, *Staerke,* **25,** 409 (1973).
(*18*) R. Stute and H. U. Woelk, *Staerke,* **26,** 2 (1974).
(*19*) G. E. Moss, *in* "Examination and Analysis of Starch and Starch Products," J. A. Radley, ed., Applied Science Publishers, London, 1976, pp. 6–8.
(*20*) "Food Microscopy," J. G. Vaughan, ed., Academic Press, New York, 1979.
(*21*) T. Allen, "Particle Size Measurement," Chapman and Hall, London, 1981.

CHAPTER XXIII

PHOTOMICROGRAPHS OF STARCHES

BY LARRY E. FITT AND EILEEN MAYWALD SNYDER

Moffett Technical Center, Corn Products, Summit-Argo, Illinois

I. INTRODUCTION

Different varieties of unmodified starch exhibit wide variations in granule appearance. Sizes and shapes range from the small, angular granules of rice to the large, smooth, oval granules of potato. Ordinary light microscopy, with a resolution of 2000–2500 Å (0.2–0.25 μm) (*1, 2*) can be used to examine the hilum of granules as well as granule shape, size, and size distribution, as illustrated in the publication by Reichert (*3*, see also Chap. VII, Section III.1). However, as shown by Hall and Sayre (*2*), observations under the light microscope are sometimes difficult to interpret owing to problems in distinguishing surface and internal structures.

Use of a scanning electron microscope (SEM) for examining granule morphology has two major advantages over use of a light microscope (see also Chap. VII, Section III.2). First, it has hundreds of times greater depth of focus than the optical microscope, and second, resolution is on the order of 70 Å or less depending on the instrument used. This instrument permits study of fine surface structure of starch and cereal products in general. SEM has been described by Gallant and Sterling (*4*) with special regard to operation and application to starch.

Figures 1, 2, 5, 7, 9, 11, 13, and 15 are optical photomicrographs of various starches at a magnification of 350 diameters (350×). Corn starch is shown in both normal and polarized light to illustrate the internal order within native granular starches. Figures 3, 4, 6, 8, 10, 12, 14, and 16 are electron photomicrographs of the same starches. The magnifications selected vary because of granule shape, size, or surface feature being emphasized.

STARCH, 2nd ed.

ISBN 0-12-746270-8

II. Photographic Methods

1. Optical (Visual Light) Microscopy

Each of the starches was dispersed in a 1:1 v/v glycerol–water mixture to minimize evaporation and granule movement in the field (*5, 6*). Brownian movement, sometimes observed when very small granules are being studied, can be minimized by adding a trace of sodium chloride to the medium. Enough pressure was applied to the cover slip to disperse the starch uniformly, but without fracturing (*7, 8*), maintaining granule integrity. In the case of corn starch (Fig. 2), the same field was photographed with polarized light by merely inserting a polarizing lens into the body tube of the microscope.

A Zeiss polarizing microscope (Model WL) fitted with a Zeiss 25× planapochromat objective and a Zeiss 10× Komplan ocular was used for this work. This optical system, or its equivalent, is essential for a flat "in-focus" field and good definition (*6*). A Kodak No. 58-B green filter was employed to provide monochromatic light for improved detail. By means of a stage micrometer slide (1 mm graduated in 0.01 mm divisions), the bellows of the camera were adjusted to give exactly 400× magnification on the ground-glass viewing plate. Prints were then enlarged to a final magnification of exactly 700×. To ensure against error, a photomicrograph of the micrometer slide was processed in the same fashion as were the starch samples. Kodak 5 × 7 inch panchromatic Tri X film having an ASA rating of 320 was used. Illumination was determined with a Gossen Lunasix exposure meter. Exposure time was approximately 0.2 sec with normal lighting, and 0.5–2.0 sec with polarized light. To improve contrast, exposure was decreased to one-quarter the calculated time, and the film was then overdeveloped in Kodak Polydol for 13.5 min at 20° instead of the normal 8 min.

2. Scanning Electron Microscopy

A general description of the electron microscopy and its application to starch and starch products has been given previously (*4*).

Each unmodified starch specimen was prepared by permitting the granules to fall onto double-backed Scotch tape, No. 665, previously attached to a specimen stub, as suggested by Hall and Sayre (*5*). Aranyi and Hawyrlewicz (*1*) found that cereal samples should be coated with a thin layer of metal under high vacuum to prevent the build-up of a primary electron negative charge on otherwise nonconductive starch; our samples were coated using a Technics Model Hummer V Sputter Coater. This coater was used because it is rapid, omnidirectional, and eliminates the need for rotational motion. An added advantage is the negligible temperature rise during the coating step, which is important with temperature-sensitive samples. An International Scientific Instruments (ISI) Model 100 scanning electron microscope was used for the analysis of unmodified starch granules. The following instrumental conditions were used: working distance, 8 mm; tilt angle, 35°; accelerating voltage, 15 kV. The condensing lens was in position 3, and the spot size was in the 1:00 o'clock position.

The micron marker displayed on all SEM photomicrographs provides a permanent record of the magnification on the micrograph and is coupled to all operating conditions. This feature produced accurate magnification at all times. The micron bar, located in the upper left corner, is coded to represent 100 μm when preceded by 3 dots, 10 μm when preceded by 2 dots, 1 μm when preceded by 1 dot, and 0.1 μm when preceded by 0 dot. Micron bars found on the last two plates do not follow the code, but are self-explanatory. The magnifications selected for the unmodified starches are 1500×, 3000×, and 5000× on the original photographs. Changes in magnification because of reproduction and publishing are noted as follows. If the micron bar is preceded by 2 dots and is 6 mm long, the magnification is 600× (length in mm × 100 = magnification). If the micron bar is preceded by 1 dot and is 6 mm long, the magnification is 6000× (length in mm × 1000 = magnification).

The film used for the SEM was Polaroid 4″ × 5″ Land Film Type 55 (positive/negative). Type 55 film has a 50 ASA rating and a development time of

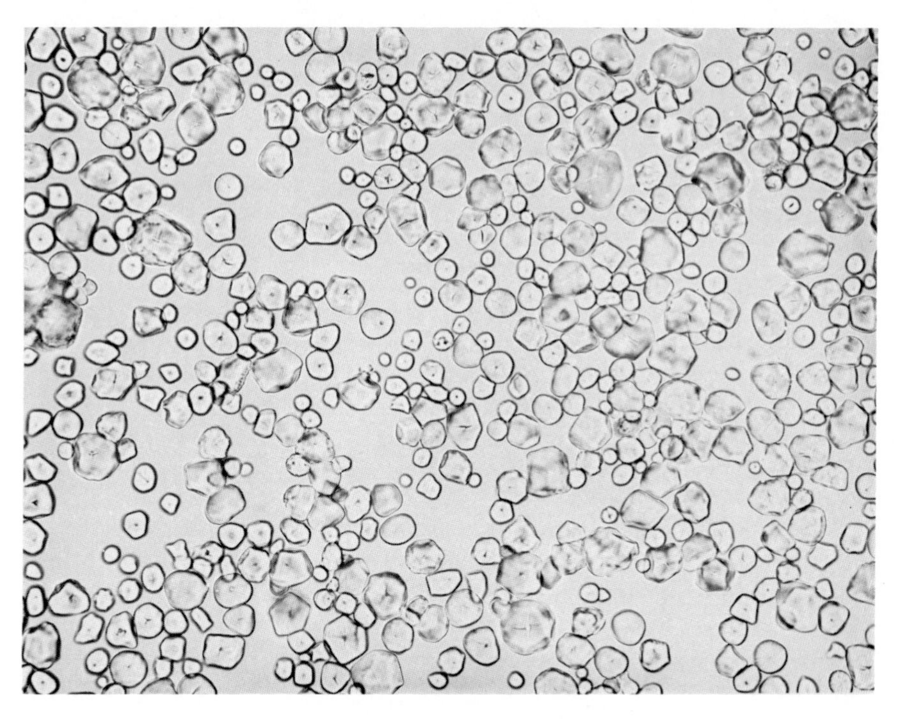

FIG. 1.—Corn starch, 350×.

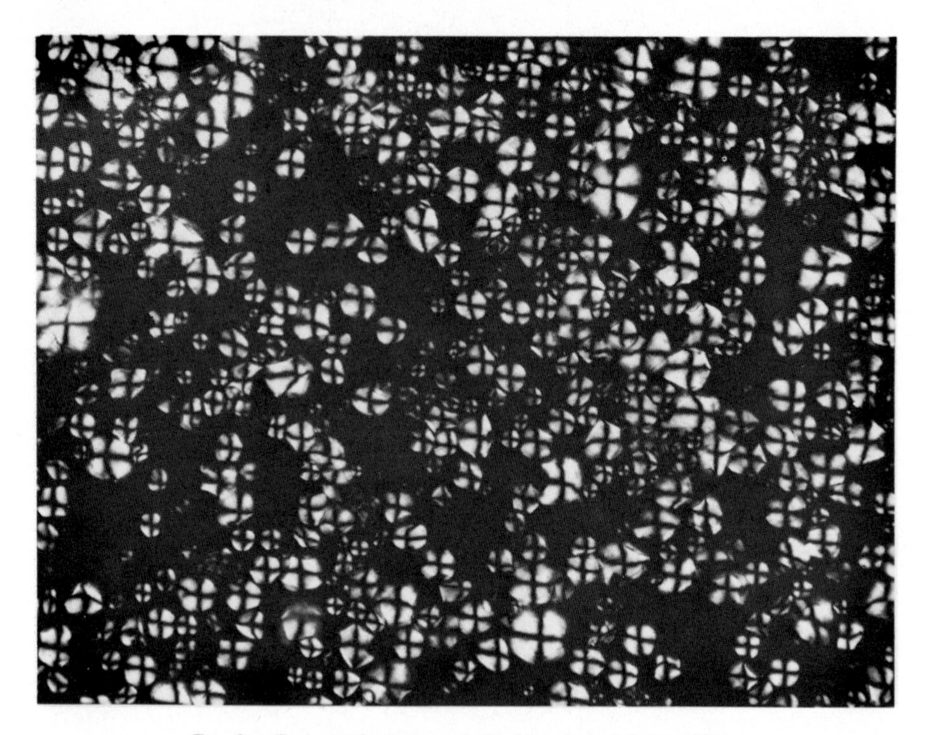

FIG. 2.—Corn starch photographed with polarized light, 700×.

20–25 sec at 70°. High-contrast photographic enlargements were made from each negative and reduced to comply with the printing format.

III. DESCRIPTION OF STARCHES

Corn (maize) starch, obtained from the seed of *Zea mays* var. *indentata,* is a mixture of rounded granules from the floury endosperm and angular granules, usually four- or five-sided, from the horny endosperm. Angular granules show pronounced pressure facets from field-drying. The light microscope shows a centric hilum and polarization crosses that are moderate to strong in brightness (Figs. 1, 2).

SEM also shows the mixture of rounded and angular granules. Surface features of corn starch are particularly interesting, especially the crater-shaped impressions that are caused by zein bodies pressing into the soft endosperm of the growing corn kernel (*9, 10*). The more spherical granules usually have smooth or more regular surfaces compared to those of the angular granules, which are often grooved or dimpled (Figs. 3, 4).

Babassu starch, derived from *Orbignya martinana,* which is plentiful in tropical climates such as Northern Brazil, has a variety of particle shapes.Compound

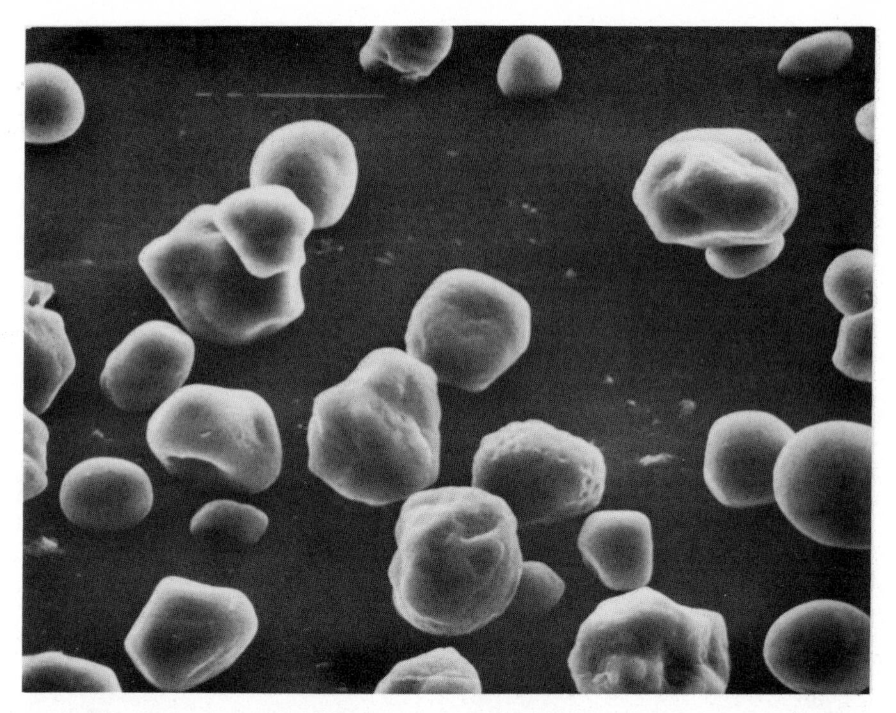

Fig. 3.—Corn starch, 1500×.

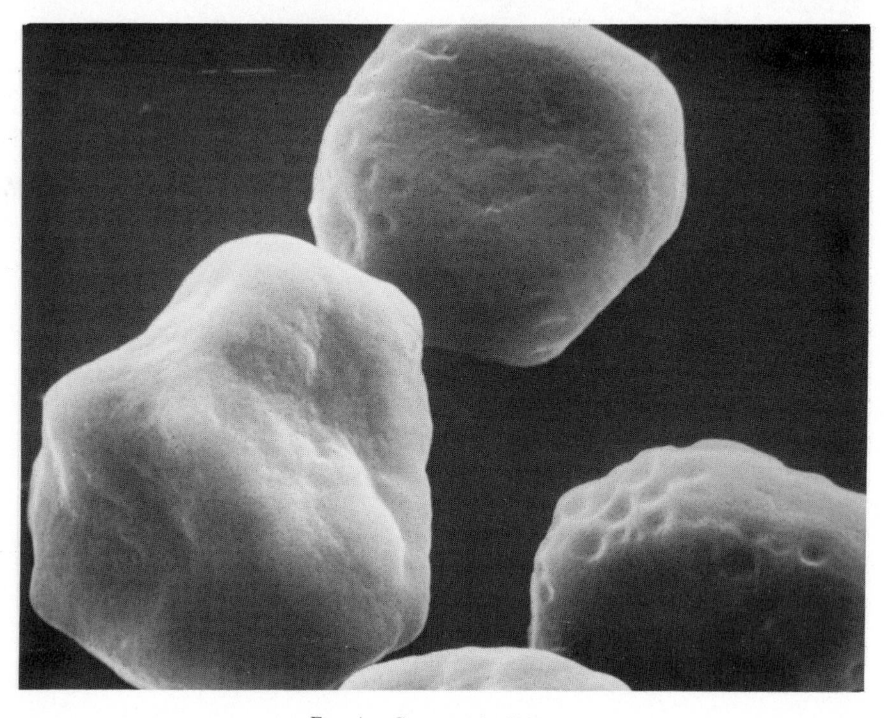

Fig. 4.—Corn starch, 5000×.

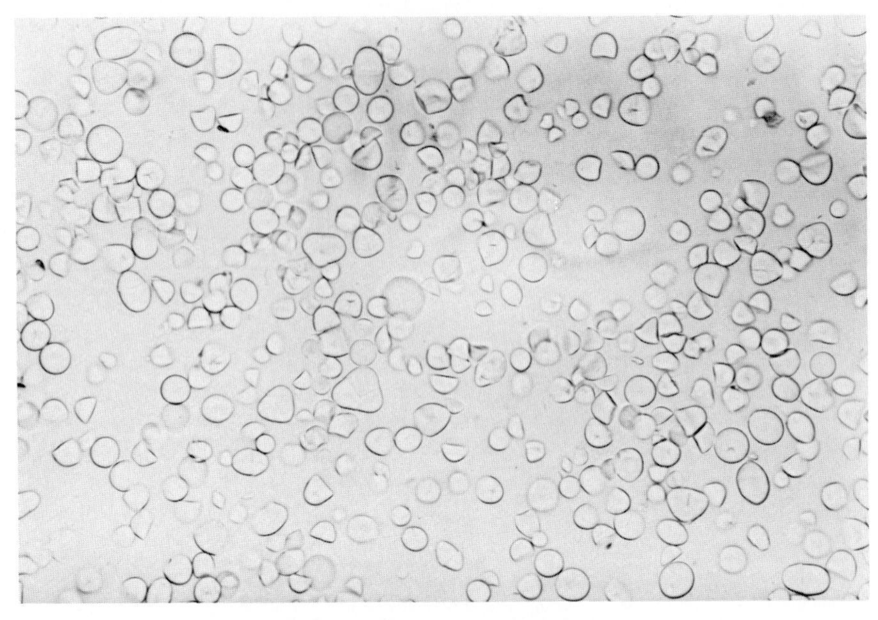

FIG. 5.—Babassu starch, 350×.

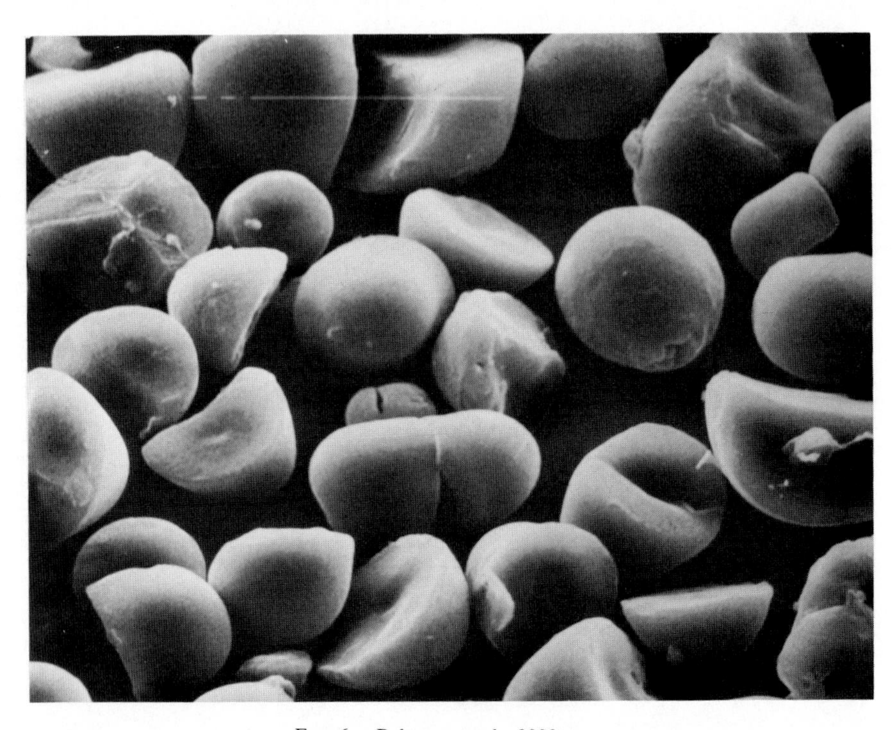

FIG. 6.—Babassu starch, 3000×.

granules are common, both oval (twins) or truncated triplets. Smaller irregularly shaped pieces that appear to be fragments under polarized light are actually whole granules. The SEM photographs show that the surfaces of the granules are quite smooth except for the dimpled or hollowed-out areas (Figs. 5, 6).

Granule shapes of high-amylose corn starches obtained from *Zea mays* seed varieties containing 55–85% of the linear starch fraction (amylose) range from nearly smooth and spherical, through the lobed or budded variety, to the ''snake-like'' or elongated type. ''Deformed'' granules have been found to be more abundant as the amylose content increases. It is not clear why granules are deformed (Figs. 7, 8) (see also this Volume, Chap. III and VII).

Potato starch, obtained from tubers of *Solanum tuberosum,* consists mostly of large, oval granules. Optical microscopy shows a highly eccentric hila and pronounced lamellations similar to the striations that can be observed in oyster-shell. SEM also shows the large, oval granules with surfaces that are among the smoothest of the granules examined. The bright horizontal crease on the granule in the foreground is occasionally observed and not a photographic artifact (Figs. 9, 10).

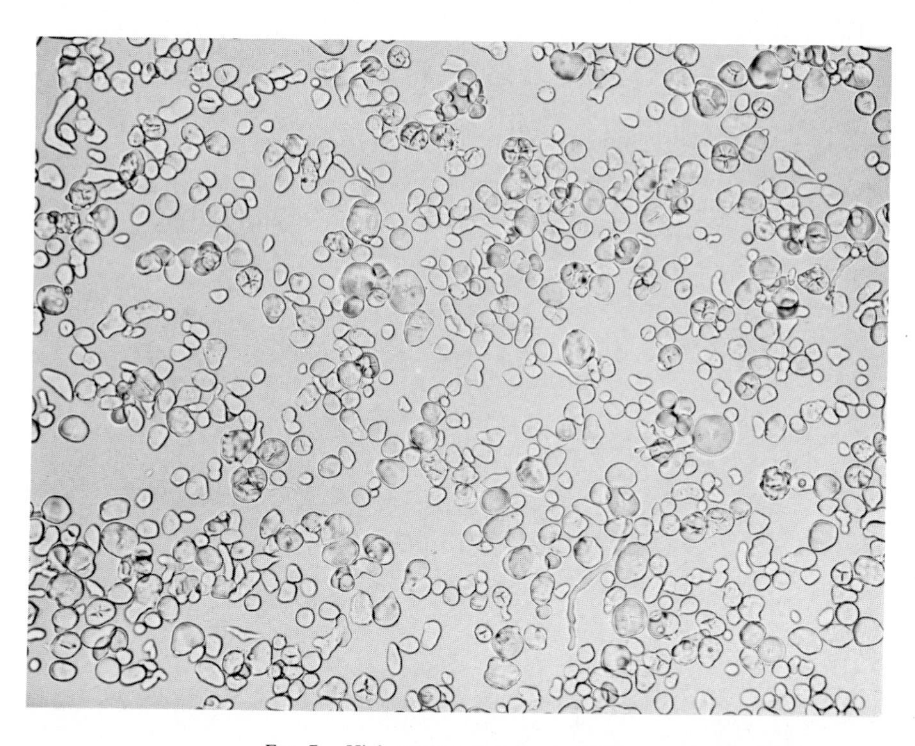

FIG. 7.—High-amylose corn starch, 350×.

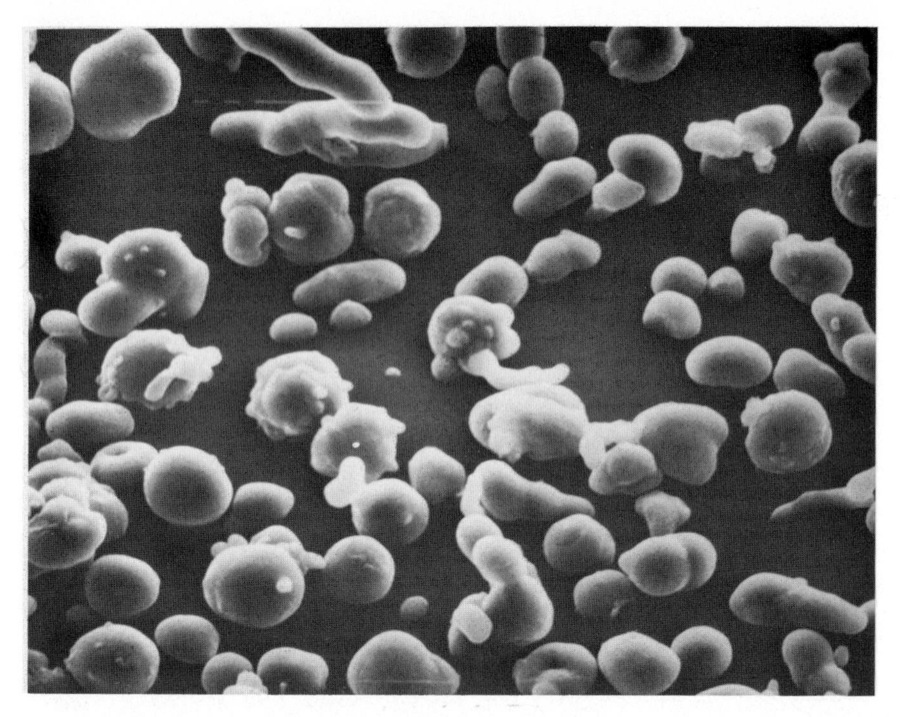

FIG. 8.—High-amylose corn starch, 1500×.

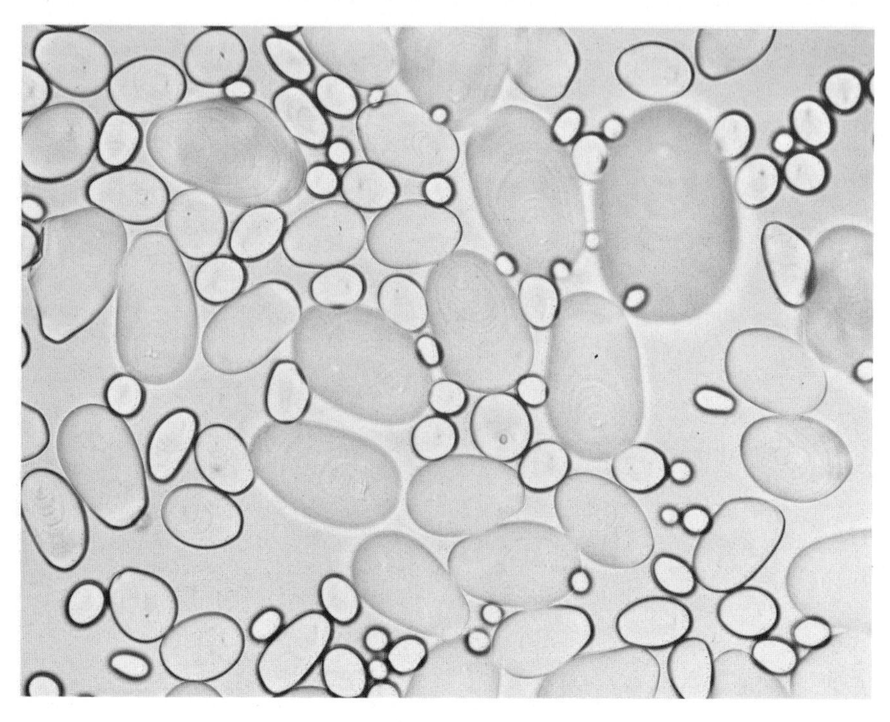

FIG. 9.—Potato starch, 350×.

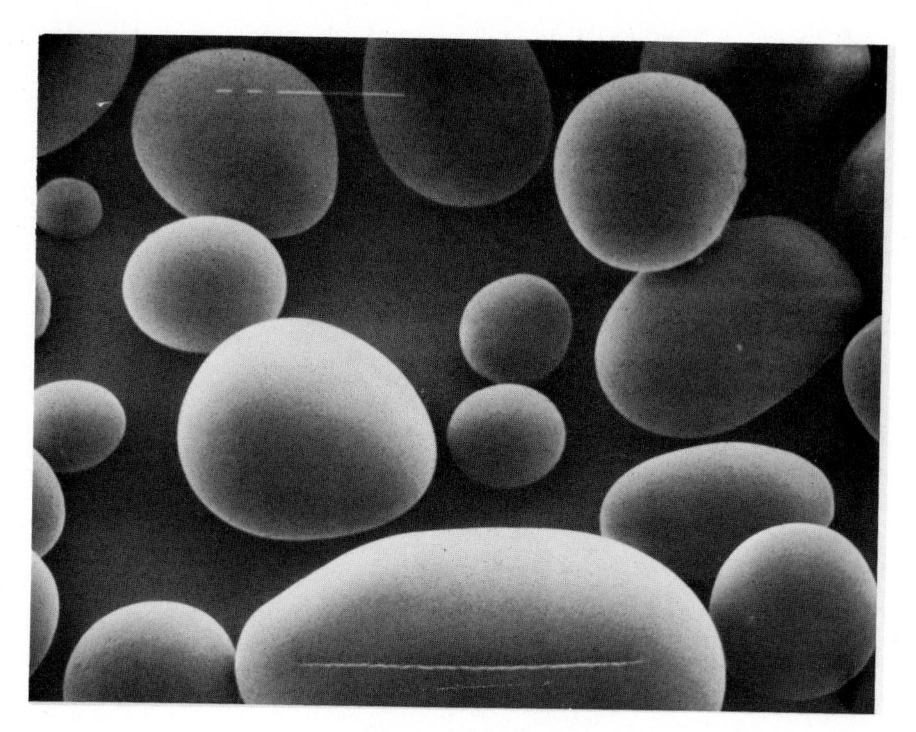

FIG. 10.—Potato starch, 1500×.

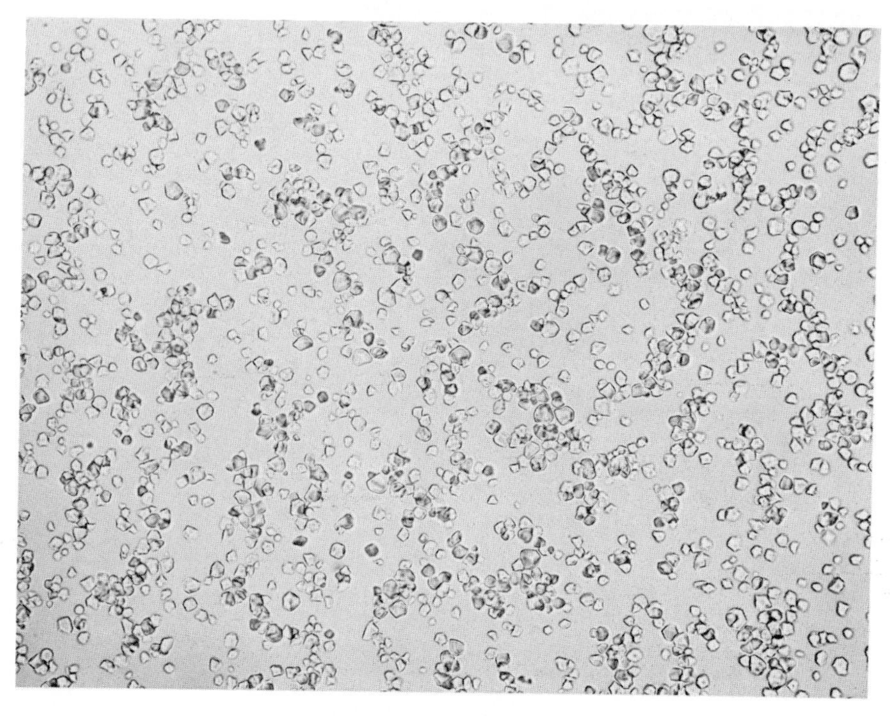

FIG. 11.—Rice starch, 350×.

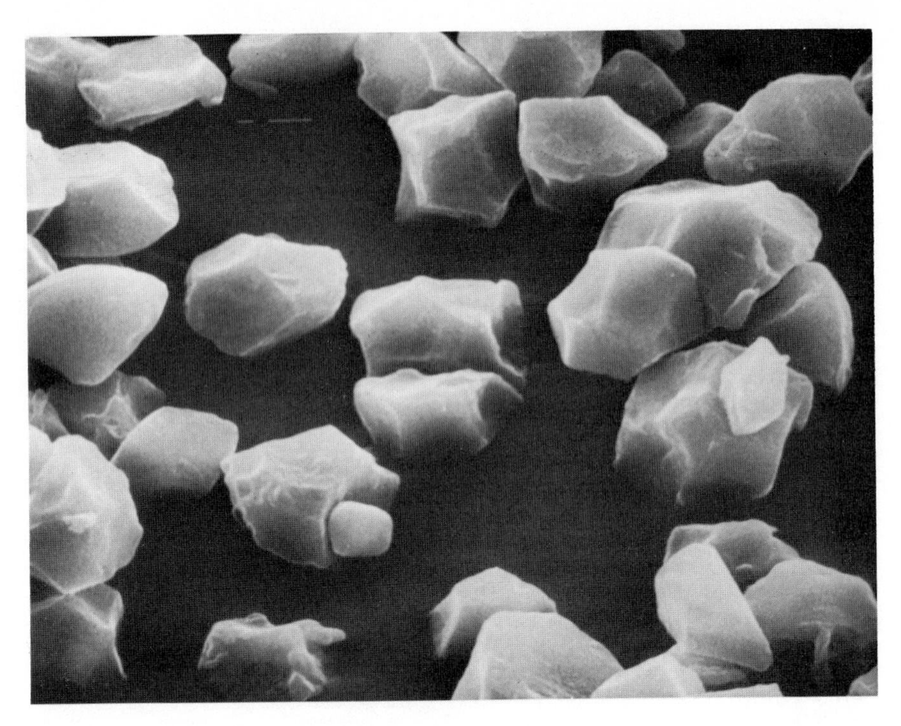

Fig. 12.—Rice starch, 5000×.

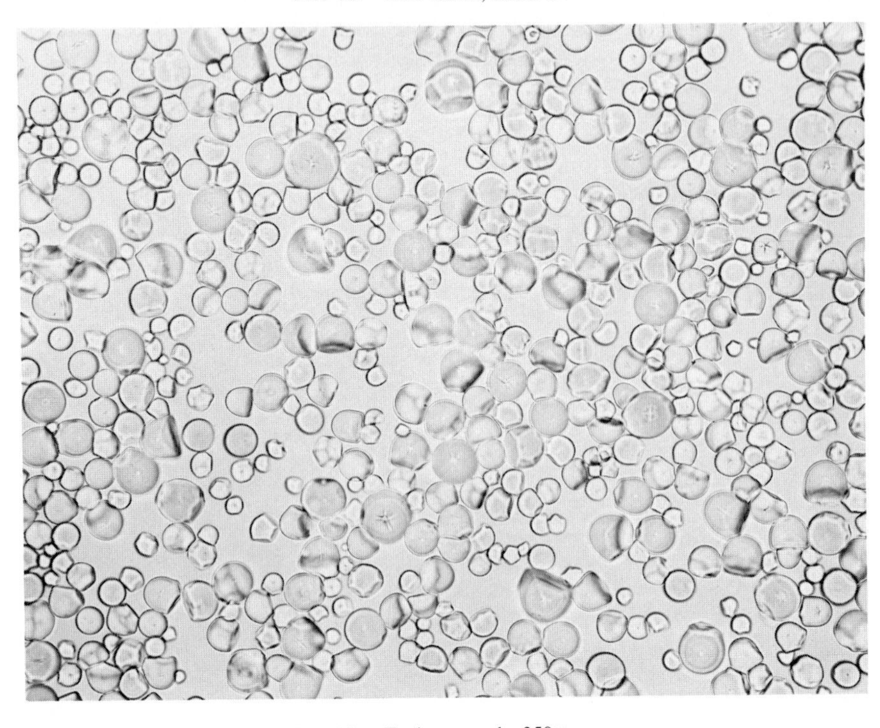

Fig. 13.—Tapioca starch, 350×.

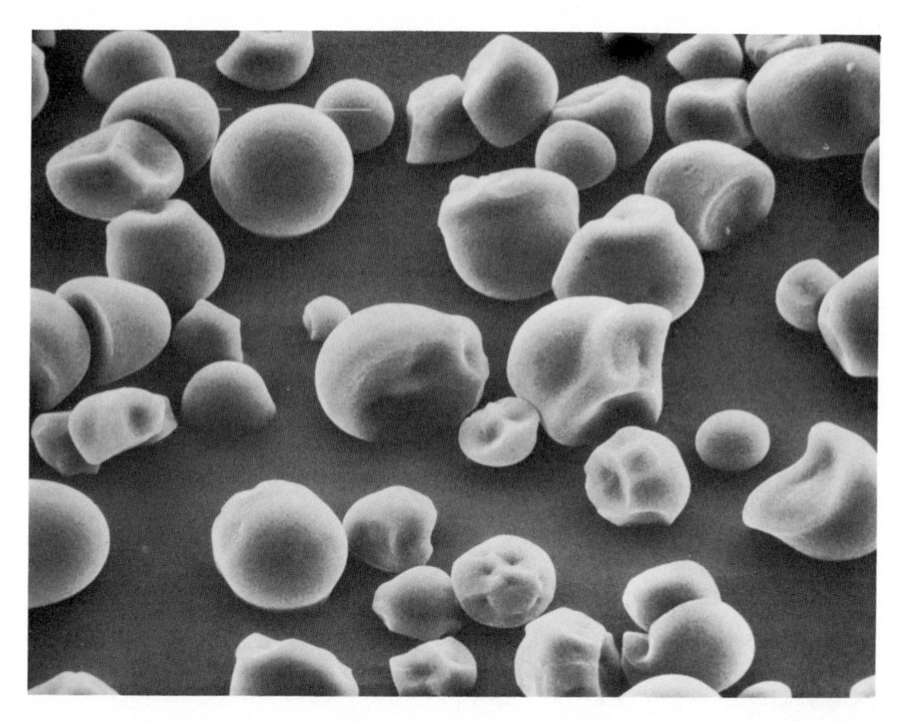

FIG. 14.—Tapioca starch, 1500×.

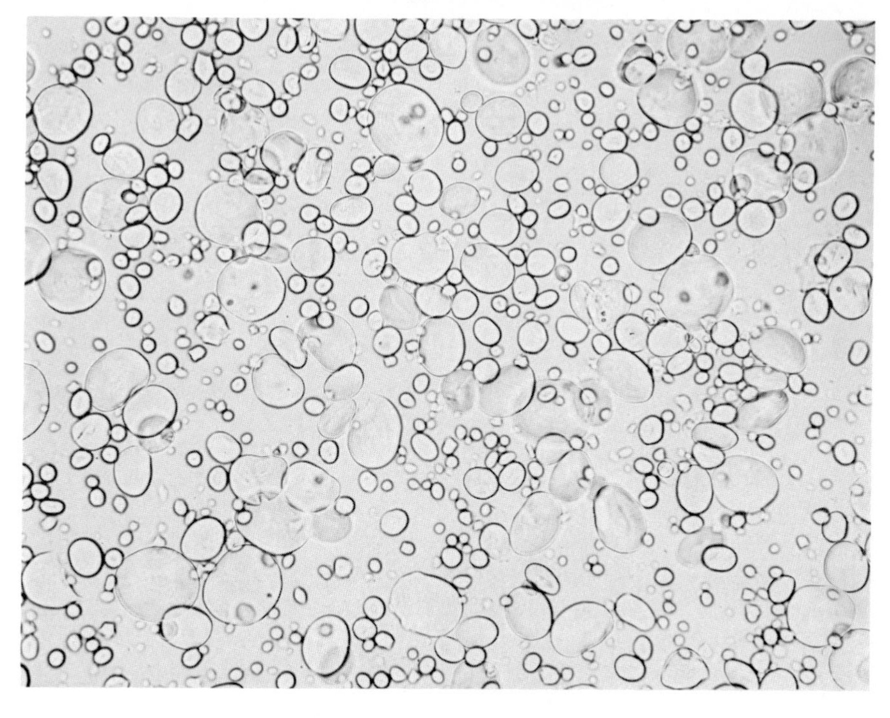

FIG. 15.—Wheat starch, 350×.

Rice starch, obtained from the seed of *Oryza sativa,* has the smallest granules of the common commercial starches. They are very angular (usually five-sided), frequently aggregated into large clusters owing to the steeping or drying conditions used during manufacture, and have a centric hilum and low birefringence. SEM photographs emphasize the sharp, angular nature of the granules and the irregular, layered appearance of the surfaces (Figs. 11, 12).

Tapioca starch, obtained from the root of *Manihot utilissima,* is also known as cassava or manioc starch and (improperly) "Brazilian arrowroot." Granules vary in shape and may be round, truncated, egg-shaped, or cap-shaped. Hila are centric, sometimes slightly fissured. In addition, SEM shows that fractured sides are often dimpled and have a more textured appearance than the outer surface (Figs. 13, 14).

Wheat starch, obtained from the seed of *Triticum aestivum* (*T. sativum*), contains both large, round, lens-shaped granules and small, spherical granules. The hilum is not visible. The SEM view shows a wide variety of features as in corn. Some particles are smooth while others are striated or dimpled (Figs. 15, 16).

For the most part, starches are gelatinized by cooking and used in a swollen or

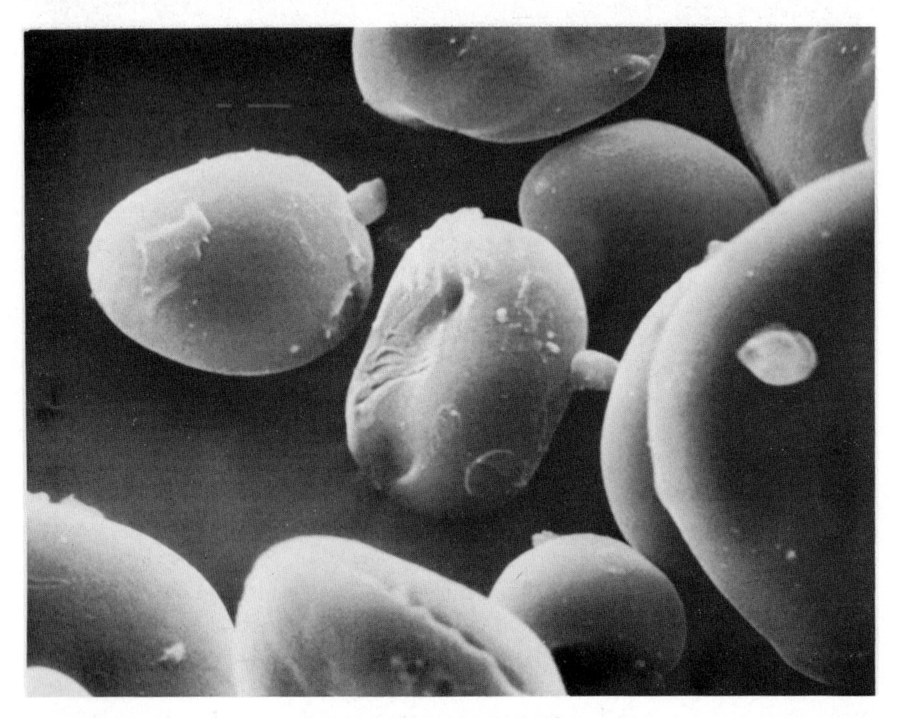

FIG. 16.—Wheat starch, 5000×.

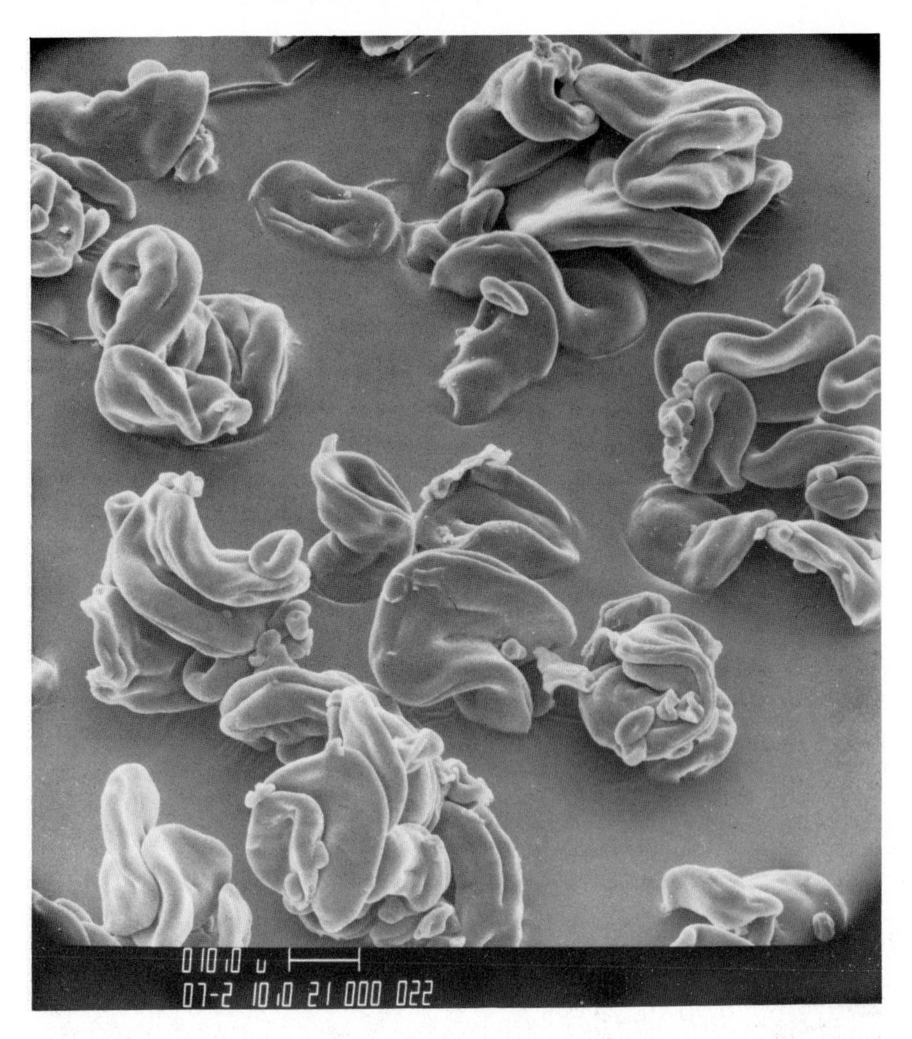

FIG. 17.—Wheat starch, heat swollen (*11*).

dispersed state rather than in the native forms depicted in Figures 1–16. Depending on the gelatinization conditions and the intended end-use of a starch, the degree of granular disruption and swelling can vary widely. Figure 17 illustrates wheat starch granules that were swollen at 90° by heating a 10% suspension of starch in water and isolating the swollen granules by centrifugation. The SEM picture shows a variety of granules, nearly all of which have collapsed following gelatinization. The dimples and pits often present on ungelatinized granules are no longer visible at 90°. The original article (*11*) gives the details of specimen preparation on this as well as on several other examples of swollen wheat starch.

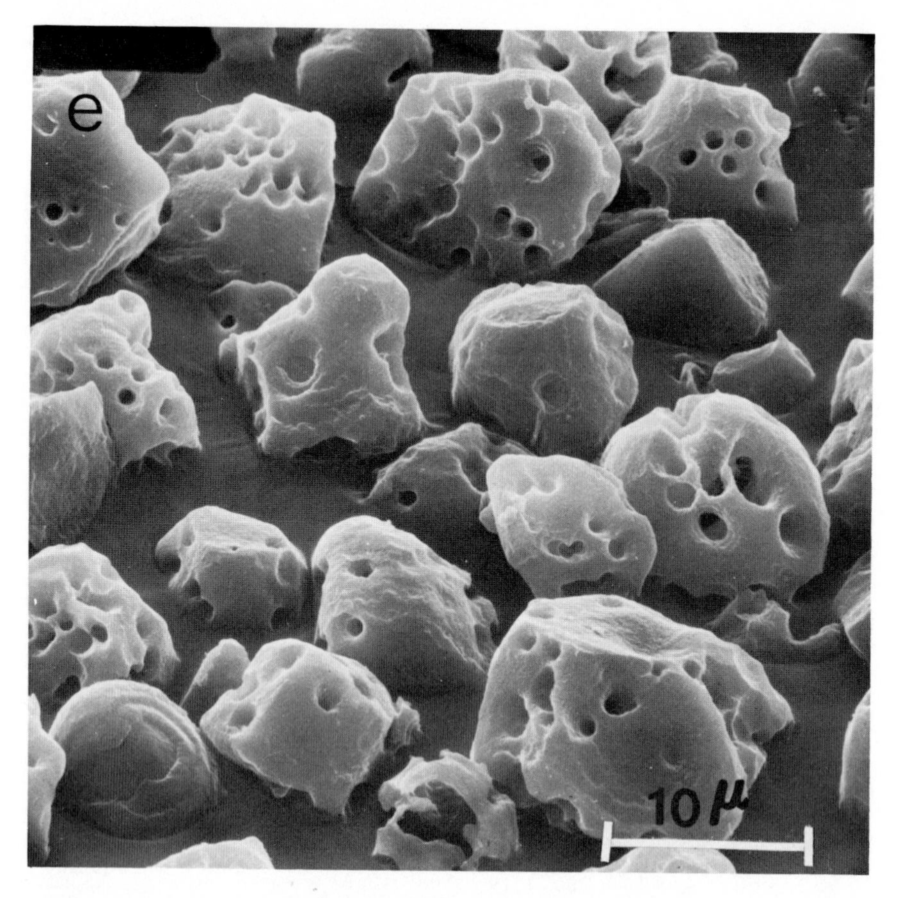

FIG. 18.—Corn starch exposed to glucoamylase attack (*12*).

The native granular structure can also be disrupted by enzymic attack. Such action occurs in the digestion of animal feed and during the germination of cereal grains. Figure 18 shows corn starch granules after being subjected to attack by glucoamylase (from *Aspergillus niger*) for 16 h (*12*). The granules show two apparent forms of attack. In most granules, deep holes that often penetrate to the center of the granule are formed. Also, surface erosion occurs that exposes a layered structure due to some areas being more resistant to enzymic attack than others. Studies along this line provide insight into granule organization and its formation.

ACKNOWLEDGMENTS

THANKS TO G. P. WIVINIS AND T. K. ATKINSON FOR THEIR ADVICE AND WORK IN THE TECHNICAL PHOTOGRAPHY.

IV. References

(*1*) C. Aranyi and E. J. Hawrylewicz, *Cereal Sci. Today,* **14,** 230 (1969).
(*2*) D. M. Hall and J. G. Sayre, *Textile Res. J.,* **41,** 880 (1971).
(*3*) C. T. Reichert, ''The Differentiation and Specificity of Starches in Relation to Genera, Species, etc.,'' Carnegie Institution of Washington, D.C., Publication 173, Part 1, 1913.
(*4*) D. J. Gallant and C. Sterling, *in* ''Examination and Analysis of Starch and Starch Products,'' J. A. Radley, ed., Applied Science Publication Ltd., London, 4th Ed., 1976, p. 33.
(*5*) D. M. Hall and J. G. Sayre, *Textile Res. J.,* **40,** 147 (1970).
(*6*) G. P. Wivinis and E. C. Maywald, *in* ''Starch: Chemistry and Technology,'' R. L. Whistler and E. F. Paschall, eds., Academic Press, New York, Vol. II, 1967, p. 650.
(*7*) R. M. Sandstedt and H. Schroeder, *Food Technol.,* **14,** 257 (1960).
(*8*) O. A. Sjostrom, *Ind. Eng. Chem.,* **28,** 63 (1963).
(*9*) J. L. Robatti, R. C. Hoseney, and C. E. Wassom, *Cereal Chem.,* **51,** 173 (1974).
(*10*) D. D. Christianson, H. C. Nielson, V. Khoo, M. J. Wolf, and J. S. Wall, *Cereal Chem.,* **46,** 372 (1969).
(*11*) R. C. Hoseney, W. A. Atwell, and D. R. Lineback, *Cereal Foods World,* **22,** 56 (1977).
(*12*) Unpublished photomicrograph from the M.S. thesis of J. S. Smith, Pennsylvania State University, University Park, Pennsylvania (supplied by Professor D. R. Lineback).

INDEX

B

C

D

E

H

H

I

K

L

M

S

X

Z